# CELESTIAL MECHANICS
## Volume I
## Dynamical Principles and Transformation Theory

# CELESTIAL MECHANICS

Yusuke Hagihara

VOLUME I

# Dynamical Principles and Transformation Theory

The MIT Press

CAMBRIDGE, MASSACHUSETTS, AND LONDON, ENGLAND

*Designed by Dwight E. Agner. Set in Monotype Baskerville. Printed and bound in the United States of America by The Maple Press Company.*



*ISBN 0 262 08037 0* (*hardcover*)

*Library of Congress catalog card number: 74-95280*

# Preface

The rather ambitious aim of the present series is to recapitulate the results of the whole field of celestial mechanics and the associated branches of science during the last hundred years. During this period the classical method of formal expansion into series, due to Laplace and Le Verrier, has been severely criticized; a new class of orbits called periodic solutions has been introduced, with applications to Hill's lunar theory. The trigonometric series employed in the ordinary perturbation theory are expanded in powers of the small quantities characteristic of actual natural bodies, and with arguments proportional to time. Bruns and Poincaré have proved that such series are not uniformly convergent, although the existence of the solution has been established on function-theoretic grounds by Sundman. The newly discovered class of functions called "quasi-periodic" and "almost periodic" can be expanded in similar but uniformly convergent form. The question naturally arises: What conditions should be satisfied by the equations of motion of the $n$-body problem? There has been some endeavor to solve these equations by resorting to integral equations and infinite nonlinear analysis. The question of stability of motion, especially of the solar system, has been discussed by means of the linearized system of equations according to the ideas of Poincaré and Liapounov. The problem is to study stability behavior by means of nonlinear equations, even by dealing with non-convergent expansions in the vicinity of the equilibrium point. Birkhoff has extended the idea of periodic solutions of Hill and Poincaré to recurrent motions, central motions, and wandering motions, on the basis of measure theory. Another important aspect of the modern trend in celestial mechanics is the topological theory initiated by Poincaré and

Birkhoff, for example the invariant-point theorem and the ergodic theorem. The present series of volumes undertakes the program of treating all such questions, extending over the whole of celestial mechanics and associated fields, and of giving suggestions for future development of the subject, stimulated by the launching of artificial celestial bodies. This first volume presents introductory dynamical principles as Part I, and transformation theory as Part II.

Part I begins with the principles of analytical dynamics. From the variational principle of Hamilton, and using the calculus of variations, we deduce Hamilton's canonical form of the differential equations. As an introduction to the topological study of dynamical trajectories, Riemannian geometry is described as it pertains to the quadratic form of the kinetic energy. The notions of invariance, covariance, and contravariance are introduced for use in dealing with the canonical transformations. The Hamilton-Jacobi partial differential equation is derived both from the canonical equations and from a hydrodynamical analogy. Liouville's and Stäckel's theorems are proved, and the condition for separability of the variables in the Hamilton-Jacobi equation is obtained by using Riemannian geometry; the theorems obtained are also applied to the Schrödinger equation in wave mechanics.

Chapter 2 deals with quasi-periodic motions. We define a libration, a revolution, and an asymptotic motion for a simple integrable dynamical system; these definitions serve as keys to much more complicated types of motion in celestial mechanics. The multiply periodic motions and conditionally periodic motions are discussed on the basis of Weierstrass's preparatory theorem in the theory of functions of many variables and Kronecker's theorem in number theory. The motion of molecules in a cubic vessel is solved as a concrete example of ergodicity. Bohl's quasi-periodic functions and Bohr's almost periodic functions are defined, and certain of their properties which will be useful later are stated. Schwarzschild's relativistic one-body problem in the gravitational field and the problem of the motion of an earth satellite under the action of the distorted earth are solved by using the Hamilton-Jacobi equation and quasi-periodic motions. Various other methods of treating such problems are subjects of the subsequent volumes. The theory of algebraic integrals is briefly sketched as an introduction to the ideas used in the text. Chapter 3 is dedicated to particular solutions of the three-body and many-body problems. Euler's and Lagrange's types of particular solutions are obtained for the $n$-body problem in a general manner, and the nature of the motion is fully analyzed. The isosceles-triangular solution is discussed on the basis of the theory of analytic functions.

Part II of this volume begins in Chapter 4 with Lie's theory of continuous groups of transformations, with application to the $n$-body problem. The theorems due to Poisson, Liouville, and Lie are proved on the basis of Lie's ideas of complete systems, involution, Lie's function groups and distinguished functions; then these theorems are applied to the integrals of the $n$-body problem. Integral invariants and adiabatic invariants are introduced for the theoretical reduction of the order of the differential equations. The existence of adiabatic invariants is proved in a general abstract space for use in later arguments. A summary of the modern theory of Lie groups is given with a view to suggesting a new trend of development.

In Chapter 5, the differential equations of the $n$-body problem are reduced by using the known integrals. By Lagrange's, Levi-Civita's, and Poincaré's transformations, the differential equations are reduced to the lowest order. The two-body problem, especially, is discussed in full; and various expansions are obtained, using the theory of functions as applied to Keplerian motion to establish the convergence of the expansions. Then the equations for the variation of elements, Delaunay's, Poincaré's, and Kepler's, are derived as the fundamental principle in perturbation theory. The role of the angular-momentum integral is stressed in view of the elimination of the node by Jacobi. Various kinds of canonical variables are defined, along with their dynamical meanings.

Chapter 6 is devoted to discussions of Bruns's and Poincaré's theorems. The generality of Bruns's proof of the theorem of the nonexistence of new algebraic integrals is extended from three bodies to $n$ bodies. The proof of Painlevé's extension of Bruns's theorem is given. The integrals of the restricted three-body problem are discussed with special reference to the ergodic theorem. Poincaré's theorem concerning the nonexistence of new uniform integrals is proved and discussed in full detail. Exceptions to Poincaré's theorem are also dealt with. The top motions of Euler, Lagrange, and Kowalewski are discussed from the point of view of Bruns's and Poincaré's theorems, and the solution, given with integrals, is based on the theory of algebraic functions, especially on the Riemann $\vartheta$-functions and the hyperelliptic functions.

Volume II, which will be published shortly after Volume I, will contain perturbation theories in classical celestial mechanics for the motions of planets, asteroids, satellites, comets, and the moon, that is, theories of Laplace, Lagrange, Delaunay, von Zeipel, Hansen, Gyldén, Hill, Brown, and others, and their applications to the motions of artificial celestial bodies. Questions will be raised on the gaps in the asteroidal distribution and on the capture of comets.

In Volume III, the form of the solutions of these perturbation theories and the convergence of the series expansions for the solutions will be discussed on mathematical grounds. The question of the third integral of motion will be touched upon. Differential equations with periodic coefficients and with quasi-periodic coefficients will be used in connection with perturbation theory. Krylov-Bogoliubov's averaging method and Diliberto's periodic surface theory will be applied to the motion of celestial bodies.

In Volume IV, Poincaré's theory of periodic solutions will be discussed in full with various modern versions of the theory and with practical applications to the three- and many-body problems. Periodic solutions will be dealt with on the basis of Wintner's nonlinear infinite analysis and Lichtenstein's nonlinear integral equations. Motions in the neighborhood of an equilibrium point will be discussed according to the theories of Birkhoff and Siegel. The conditional existence of quasi-periodic solutions in the manner of Kolmogorov, Arnold, and Moser will be shown. The stability theory of Liapounov and its modern development will be described.

Volume V will contain proofs of the existence of the solution of the three-body problem by Sundman, Levi-Civita, Chazy, and Merman, as well as the proof of the rigorous capture theory of Merman; finally, there will be discussed the topological theory of Poincaré and Birkhoff, including the theory of characteristics, manifolds of motion, surface transformations, the invariant-point theorem, the ergodic theorem, and almost periodic motions.

The author wishes to express his appreciation for a unique editorial collaboration: Dr. Peter Musen of the National Aeronautics and Space Administration's *Goddard Space Flight Center* (GSFC) kindly read the manuscript and offered aid and encouragement as the work progressed. Subsequently, Dr. Musen gave generously of his time and wisdom to Mr. R. L. Tanner, chief editor of GSFC, and his associate Mr. P. Barr, who with NASA support jointly prepared the work for the publishers.

Yusuke Hagihara
Tokyo, July 1968

# Contents

**Introduction** 1

**PART I DYNAMICAL PRINCIPLES** 13

**Chapter 1 Principles of Analytical Dynamics** 15

VARIATIONAL PRINCIPLES 15

1.1 Hamilton's Principle 15
1.2 Relation to the Calculus of Variations 19
1.3 Canonical Equations 22
1.4 Covariance 29
1.5 Geometry of Motion 36
1.6 Equivalence of Dynamical Systems 40

CANONICAL TRANSFORMATIONS 42

1.7 Bilinear Covariants 42
1.8 Contact Transformations 47
1.9 Lagrange's and Poisson's Brackets 48
1.10 Canonical Transformations 51
1.11 Nonholonomic Constraint 59

THE HAMILTON-JACOBI PARTIAL DIFFERENTIAL EQUATION 61

1.12 The Hamilton-Jacobi Theorem 61
1.13 Hydrodynamical Proof of Jacobi's Theorem 65
1.14 Variation of Constants 67
1.15 Liouville's Theorem 70
1.16 Stäckel's Theorem 73
1.17 Condition for Separability 77

| | | |
|---|---|---|
| | HAMILTONIAN MATRIX | 82 |
| 1.18 | Hamiltonian Matrix | 82 |
| 1.19 | Canonical Transformation | 85 |
| 1.20 | Pfaff's Criterion | 88 |
| 1.21 | Polar Factorization | 90 |
| 1.22 | Pfaffian Dynamical System | 94 |
| **Chapter 2** | **Quasi-Periodic Motion** | 104 |
| | SIMPLY PERIODIC MOTION | 104 |
| 2.1 | Lebesgue's Integrals | 104 |
| 2.2 | Fourier Series | 113 |
| 2.3 | Lagrange's Theorem | 117 |
| 2.4 | Asymptotic Motion and Libration | 119 |
| | CONDITIONALLY PERIODIC MOTION | 123 |
| 2.5 | Multiply Periodic Functions | 123 |
| 2.6 | Weierstrass's Auxiliary Theorem | 128 |
| 2.7 | Conditionally Periodic Motion | 132 |
| 2.8 | Kronecker's Theorem | 137 |
| 2.9 | Motion of Molecules in a Cube | 142 |
| 2.10 | Quasi-Periodic Function of Bohl | 144 |
| | ALMOST PERIODIC FUNCTIONS | 149 |
| 2.11 | Definition | 149 |
| 2.12 | Properties of Almost Periodic Functions | 151 |
| 2.13 | Base Field | 159 |
| 2.14 | Quasi-Periodic Functions and Almost Periodic Functions | 162 |
| | RELATIVISTIC ONE-BODY PROBLEM | 166 |
| 2.15 | Transformation of the Problem | 166 |
| 2.16 | Relativistic Field of Schwarzschild | 168 |
| 2.17 | The Complete Integral | 171 |
| 2.18 | Solution by Elliptic Functions | 174 |
| 2.19 | Discussion of the Motion | 178 |
| 2.20 | Algebraic Integrals | 185 |
| | APPLICATION TO AN EARTH SATELLITE | 196 |
| 2.21 | Earth Potential | 196 |
| 2.22 | Oblate Spheroidal Coordinates | 199 |
| 2.23 | Integration by Elliptic Integrals | 205 |
| 2.24 | Reduction to the Two-Fixed-Center Problem | 215 |
| **Chapter 3** | **Particular Solutions of the Many-Body Problem** | 228 |
| | HOMOGRAPHIC MOTIONS | 228 |
| 3.1 | Fundamental Equations | 228 |

STRAIGHT-LINE HOMOGRAPHIC MOTIONS 232
3.2 Straight-Line Configuration 232
PLANE HOMOGRAPHIC MOTIONS 237
3.3 Condition for Plane Configurations 237
3.4 Central Figures 242
3.5 Properties of Central Figures 246
SPACE HOMOGRAPHIC MOTIONS 256
3.6 Condition for Space Configurations 256
ISOSCELES-TRIANGULAR SOLUTIONS 259
3.7 Regular and Irregular Singularities 259
3.8 Existence of the Solutions 268
3.9 Nature of the Isosceles-Triangular Solutions 276

**PART II TRANSFORMATION THEORY**

**Chapter 4 Continuous Groups of Transformations** 291
INFINITESIMAL TRANSFORMATIONS 291
4.1 The First Integrals 291
4.2 Infinitesimal Transformations 296
4.3 Poisson's Theorem 301
LIE'S THEOREM 303
4.4 Elimination of Time 303
4.5 Involution 304
4.6 Invariantive Relations 305
4.7 Liouville's Theorem 310
4.8 Lie's Theorem 312
FUNCTION GROUPS 316
4.9 Complete Systems 316
4.10 Function Groups 320
4.11 Canonical Form of a Function Group 323
INTEGRALS OF THE $n$-BODY PROBLEM 328
4.12 Eulerian Integrals 328
4.13 Distinguished Functions 333
4.14 Parametrization 336
4.15 Separating Parametrization 338
4.16 First Integrals 342
INTEGRAL INVARIANTS 343
4.17 Integrals and Integral Invariants 343
4.18 Condition for an Integral Invariant 346
4.19 Integral Invariants along Trajectories 350

| | | |
|---|---|---|
| 4.20 | Variational Equations | 352 |
| 4.21 | Integral Invariants of the $n$-Body Problem | 359 |
| 4.22 | Cartan's Exterior Multiplication | 363 |
| | ADIABATIC INVARIANTS | 381 |
| 4.23 | Adiabatic Invariants | 381 |
| 4.24 | Adiabatic Approximation | 391 |
| 4.25 | Existence of Adiabatic Invariants | 401 |
| | LIE GROUPS | 406 |
| 4.26 | Lie's Fundamental Theorems | 406 |
| 4.27 | Adjoint Groups | 419 |
| 4.28 | Simple and Semisimple Groups | 428 |
| 4.29 | Representation of a Lie Group | 436 |
| **Chapter 5** | **Reduction of the *n*-Body Problem** | 447 |
| | TRANSFORMATION OF LAGRANGE | 447 |
| 5.1 | Transformation of Lagrange | 447 |
| 5.2 | Levi-Civita's Transformation | 453 |
| | TRANSFORMATION OF POINCARÉ | 466 |
| 5.3 | Transformation of Poincaré | 466 |
| 5.4 | Elimination of the Nodes | 468 |
| 5.5 | Transformation $\alpha$ | 470 |
| 5.6 | Transformation $\beta$ | 472 |
| 5.7 | Transformation $\gamma$ | 475 |
| 5.8 | Reduction from $6n - 6$ to $6n - 10$ | 476 |
| | ELLIPTIC MOTIONS | 482 |
| 5.9 | Motion in Conic Sections | 482 |
| 5.10 | Elliptic Motions | 487 |
| 5.11 | Cauchy's First Theorem | 490 |
| 5.12 | Cauchy's Second Theorem | 496 |
| 5.13 | Expansion in Several Variables | 503 |
| 5.14 | Radius of Convergence | 510 |
| 5.15 | Andoyer's Expansion | 521 |
| | VARIATION OF CONSTANTS | 526 |
| 5.16 | Delaunay's Elements | 526 |
| 5.17 | Poincaré's Elements | 531 |
| 5.18 | Keplerian Elements | 537 |
| 5.19 | Angular Momentum Integrals | 540 |
| 5.20 | Other Kinds of Canonical Elements | 545 |
| **Chapter 6** | **Algebraic Integrals and Uniform Integrals** | 556 |
| | THEOREM OF BRUNS | 556 |
| 6.1 | Integrals Independent of Time | 556 |

6.2 Lemma I 559
6.3 Lemma II 563
6.4 Integrals of Momentum and Angular Momentum 568
6.5 Energy Integral 577
6.6 Integrals Containing Time 581

PAINLEVÉ'S EXTENSION 584

6.7 Preparatory Lemmas 584
6.8 Properties of (S) 588
6.9 Properties of $(S_0)$ 590
6.10 Properties of the Integrals of (S) 596
6.11 Application to the *n*-Body Problem 600

INTEGRALS OF THE RESTRICTED THREE-BODY PROBLEM 604

6.12 Poincaré's Recurrence Theorem 604
6.13 Lemma 607
6.14 Proof of Siegel's Theorem 611

THEOREM OF POINCARÉ 613

6.15 Case of Nonvanishing Hessian 613
6.16 Case of Vanishing Hessian 616
6.17 Application to the *n*-Body Problem 624
6.18 Discussion of Poincaré's Theorem 629
6.19 Integrals without Parameters 632

KOWALEWSKI'S TOP MOTION 639

6.20 Rigid-Body Rotation 639
6.21 Solution in Algebraic Integrals 642
6.22 Abel's Lemmas 648
6.23 Kowalewski's Solution 653
6.24 Riemann $\vartheta$-Functions 661
6.25 Hyperelliptic Functions 671

**Retrospect of Volume 1** 683

**Index** 685

# Introduction

Celestial mechanics, founded by Laplace, has been the most elegant of the mathematical sciences. It is a typically deductive science, treating the dynamical characteristics—that is, the motion and the figure of celestial bodies—by pure mathematical deduction based on the laws of dynamics. According to Poincaré, celestial mechanics convinces us of the existence of the laws of nature and teaches us how infinitely accurate those laws are. Until the present century, mathematical analysis and celestial mechanics progressed side by side, stimulating and helping each other, ever since the discovery of differential and integral calculus by Newton. Recent developments in technology now facilitate computation of the ephemerides of celestial bodies by means of electronic computers. The whole program of planning the orbits of artificial satellites, and the tracking of such satellites, relies at present on such machine computation. The three-body problem and also the $n$-body problem may be solved numerically; for example, we may solve numerically dynamical systems consisting of 100 or 200 stars in stellar clusters, and verify their global characteristics derived from statistical mechanics.

At this stage of the science we are compelled to discuss the foundations of the methods of celestial mechanics. The value of fundamental theory lies in its potential not only for enabling us to reflect on the processes and methods we have pursued and to attach physical meaning to the labyrinth of mere numbers, but also helping us to comprehend their merits and demerits, for steering us toward a completely harnessed logical system of theories. It is the aim of the present series of volumes to examine on rigorous mathematical grounds the question of whether the deductive process adopted in celestial mechanics can be logically justified.

Before a theory is fully justified on logical grounds, its results cannot safely be compared with observations in order to verify the laws of nature. For example, there is the great question whether the Newtonian laws, with their Galilean notion of time and space, or the Einsteinian relativistic theories of time, space, and gravitation are the right ones; the question, that is, whether and how we should amend our notion of time, space, and gravitation and finally our ultimate world-picture.

The motion of two bodies attracting each other by the Newtonian law of gravitation has been completely solved. There are three degrees of freedom and the differential equations of motion are integrated with six constants of integration. The orbit is known to be a conic section on a single plane and the motion is Keplerian. The usual method is to take as the integration constants: (1), the inclination of the orbital plane; (2), the longitude of the node; (3), the semimajor axis; (4), the eccentricity; (5), the longitude of the pericenter; and (6), the epoch of, say, the pericenter passage. These six constants are called the *Keplerian elements.*

The problem of computing the motion of *three* bodies attracting each other by the Newtonian law of gravitation, with arbitrary masses and arbitrary initial conditions, is called the *three-body problem.* Its fame is due to the difficulty of solving it: the labors of many eminent mathematicians in the past have borne hardly any fruit in this respect. Two kinds of particular solution have been obtained through the discoveries of Euler and Lagrange in the case of arbitrary masses: the straight-line equilibrium solution and the equilateral-triangular equilibrium solution. In the case of two equal masses we have the isosceles-triangular solution. These are the only exact solutions ever discovered.

An elementary method for solving a system of differential equations is to find out as many independent integrals as the order of the differential equations. The twelve classical integrals are: one integral for the energy, three for the angular momentum, six for the center of mass — these ten are called the classical Eulerian integrals — and finally one integral (discovered by Jacobi) for the elimination of the node and one for the elimination of time. Thus, by using these twelve integrals, the order of the differential equations for the $n$-body problem is reduced from $6n$ to $6n - 12$. Lie's theory of continuous groups of transformations, in particular the ideas of contact transformations, Pfaffians, involution, function groups, and distinguished functions, has been applied to reduce the order of the given differential equations, and hence to transform them to those of a lower order. The theory has revealed no means for the further reduction of the order. On the other hand, Bruns proved (and the proof has been extended since by Painlevé) the nonexistence of further integrals which are algebraic in the momenta; and Poincaré proved the nonexistence of further integrals which are uniform in the Keplerian elements. An exception to Poincaré's theorem is Kowalewski's top motion, which can be solved explicitly by means of hyperelliptic functions. Thus we must abandon our previous attempts to solve the $n$-body problem by the elementary method of finding further independent integrals and have recourse to some transcendental procedure. Recently the existence of the third integral has been under discussion with regard to the stellar dynamics of clusters and galaxies. The integral as well as the adelphic integral of Whittaker is nonuniform and its existence does not invalidate Poincaré's theorem.

The question whether there exists a solution of the differential equations for the three-body problem has been settled completely by Sundman. The differential equations for the motion become singular at a collision of two of the bodies or at a simultaneous collision of the three bodies. Painlevé has shown the condition for a collision to be transcendental. The condition was explicitly deduced by Bisconcini and Levi-Civita before Sundman. A simultaneous collision of three bodies occurs when and only when the total angular momentum of the three-body system is zero, as was proved by Sundman. Sundman regularized the differential equations in the sense of Klein and Koebe and proved that the coordinates and the momenta of the three bodies are expanded in ascending powers of the variable $\tau = (t - t_0)^{2/3}$, where $t_0$ denotes the time of the collision. Recently several methods for the regularization have been proposed, particularly for numerical solutions of the problem. Thus the existence of the solution has been proved in the sense of the Weierstrassian theory of analytic functions and, indeed, both before and after the collision; that is, the analytic continuation of the solution has been proved possible beyond the instant of the collision. We cannot lay too much stress on the importance of this theorem by Sundman. It is the basis of the topological study of the trajectories for an infinite time interval. This result has been extended to the $n$-body problem. The behavior of the motion at infinity has also been studied by Chazy and recently by Merman. The theorem by Poincaré, Schwarzschild, von Zeipel, and Chazy regarding the impossibility of a capture has been challenged by Merman as to its rigor for justifying some of the cosmogonical hypotheses on the origin of the solar system.

As has been stated, we must at the present stage rely on some transcendental process for the solution of the three-body problem. The process presently employed is the iteration of successive approximations, the so-called theory of *perturbations*. A primitive procedure for computing the perturbations is to take the Keplerian motion as a first approximation by supposing the third body to be of small mass or to be at a large distance (referring to the actual configuration of the solar system). The result is expanded in powers of the disturbing mass. Then the Keplerian elements are no longer constants but vary slowly, the rate of variation depending on the disturbing mass. We determine the variation of these elements by solving the differential equations that define their variations with time. This method is called the *variation of elements* of Lagrange and was first applied to the motion of the major planets by Le Verrier. Theories of the motion of the planets in the solar system are founded not only on such expansions in powers of the disturbing masses but also on expansions in powers of the eccentricities and orbital inclinations, which are in fact small owing to the present configuration of the solar system. Naturally, the convergent nature of these expansions is essential for the practical success of the theory of the motion of the particular planet in question. However, the new development of artificial celestial bodies compels us to make a fresh start, to investigate methods of completely new character, since the eccentricities and the orbital inclinations are not so very small.

Denote one of the Keplerian elements by $a$; then the differential equation for determining the variation of the element is represented, if we consider only one disturbing planet, in the form

$$\frac{da}{dt} = \mu \sum_{k_1, k_2} a_{k_1 k_2} \cos (k_1 l_1 + k_2 l_2 + g),$$

where

$$l_1 = n_1 t + \epsilon_1, \qquad l_2 = n_2 t + \epsilon_2.$$

The right-hand member represents the perturbation, which is, as has been just described, expanded in a formal series proceeding in powers of small quantities such as eccentricities and inclinations. The small constant $\mu$ is of the order of the disturbing mass; $k_1$, $k_2$ are positive or negative integers including zero, and the sum is extended over all combinations of $k_1$, $k_2$ from $-\infty$ to $+\infty$. The quantities $a_{k_1 k_2}$, $g$, $n_1$, $n_2$, $\epsilon_1$, $\epsilon_2$ are functions of the elements; and $n_1$, $n_2$ are called the mean motions. Obviously the right-hand member is expanded in a trigonometric series with linear functions of time as arguments. Here the elements whose variation we seek are contained in the right-hand member. But, since the variation is small owing to the factor $\mu$, we consider the elements in the right-hand member to be constant and integrate the equation with regard to $t$, which appears explicitly:

$$\delta a = \sum_{k_1, k_2} \frac{a_{k_1 k_2}}{k_1 n_1 + k_2 n_2} \sin (k_1 l_1 + k_2 l_2 + g) + \text{constant}.$$

This is called the perturbation of the first order. After substituting this solution for the elements appearing in the right-hand member of the differential equations for the variation of elements, we integrate the equations again by considering $t$ explicitly appearing in the right-hand member as the only variable, and we obtain the perturbation of the second order; and so on.

Such a method of formal integration fails for $k_1n_1 + k_2n_2 = 0$. The terms for which $k_1 = k_2 = 0$ are called *secular*. The terms for which $k_1n_1 + k_2n_2 = 0$ without $k_1 = k_2 = 0$ are called *critical*. In such cases terms multiplied by $t$ in front of the trigonometric function appear. The sum $k_1n_1 + k_2n_2$ can be made as small as we please according to Kronecker's theorem in number theory by making the integers $k_1$, $k_2$ sufficiently large. A small divisor $k_1n_1 + k_2n_2$ appears then in the denominator in the above method of integration. The period of such a term is very long—$2\pi/(k_1n_1 + k_2n_2)$—and the term is said to be of long period. If such a circumstance exists in a term of comparatively lower degree in the eccentricities and the inclinations — especially when the mean motions of the two planets are nearly in commensurable ratio of a simple fraction — then the magnitude of the coefficient of the term becomes large and the series obtained by such formal integration looks as though it were divergent. This is due to the inadequacy of the formal integration procedure.

It was by such a method that Poisson proved the invariability of the semimajor axes as regards the secular perturbation of the second order; and on such a method Laplace and Lagrange based the theory of secular perturbations by showing that they are formally represented by trigonometric series of very long periods. This idea has been extended by Newcomb, and the solution of the differential equations of the planetary theory is proved to be represented by formal trigonometric series having linear functions of $t$ as arguments and coefficients proceeding in powers of the eccentricities, of the inclinations, and (above all) of the disturbing masses. Expansions of the solution in series of similar forms have been obtained by Lindstedt, Bohlin, Gyldén, and others. Unfortunately, as Poincaré proved, such series are not uniformly convergent. Hence they cannot by any means represent the real solution of the problem, although they may represent the solution in a certain degree of approximation within present-day accuracy of the observations during a limited time interval. Such series were called *semiconvergent* by Poincaré and may be useful for practical purposes. It is essential, at the present stage, to be informed of the accuracy of the approximation within a certain finite time interval by means of such semiconvergent series as compared with the accuracy of observing techniques. It may happen that the series, when truncated at a certain term, represents at present the solution within an error negligible compared with observational errors; in that event, it might deviate more and more when truncated at a higher-order term in order to obtain more accurate values, similarly to Stirling's asymptotic formula for a factorial. Indeed, the accuracy of the observing technique is rapidly progressing. Thus, we are confronted with urgent need for new approaches to a more accurate solution of the problem. The averaging method due to Krylov and Bogoliubov, originally invented for use in electrical engineering and now in use for the motion of artificial celestial bodies, may enable us to estimate the error of approximations. The case in which the mean motions are commensurable has been worked out. The method of Birkhoff, since discussed by Siegel and Moser, is applicable to the estimation of the motions in the neighborhood of an equilibrium point in a dynamical system. Ritz-Galerkin's approximation method might be considered in this respect. Two-variable asymptotic expansion technique and polynomial expansion procedure are being used in computer programming of the motion of space vehicles. It is remarked here that the existence of mean motions in the Laplace-Lagrange theory of secular perturbation has been a topic of interest in mathematics recently. Weyl, following Bohl, has completed the proof in a mathematical sense on the basis of probability distributions, and Bohr and Jessen on the theory of almost periodic functions.

Trigonometric series with linear functions of time as the arguments, when they are uniformly convergent, represent *quasi-periodic* functions in the sense of Bohl and Esclangon. They belong to a class of functions named *almost periodic* by Bohr. If the solution of the planetary motion were known to satisfy the condition for almost periodicity, then the series formally obtained in the ordinary perturbation theory would be uniformly convergent. Gyldén has discussed the probability for an arbitrary number to be rational and argued that the series obtained were convergent except for a set of measure zero. But Poincaré, after Bruns, proved that they are not uniformly convergent. The Borel series and the Liouville numbers play an important role in this problem. It is shown that the series is divergent if $n_1 : n_2$ in the foregoing formulas is a Liouville's number or if it is a $U$-number of Mahler. Petersson and Siegel recently tried to refine the proof of the nonconvergence of the series. Indeed the planetary motion does not always satisfy the condition for almost periodicity. Hence we can say nothing about the physical characteristics of the motion in a very long time interval, both in the past and in the future, unless the idea of the measure theory is introduced and the cases of Lebesgue measure zero are ignored, as has been done by Arnold. The idea of Kolmogorov and Arnold is that if the ratio $n_1:n_2$ satisfies certain inequalities, then the perturbed motion can be represented by quasi-periodic functions and is everywhere dense on the distorted torus obtained (by the Newtonian approximation process to avoid the appearance of small divisors) from the torus for the nonperturbed system.

In the actual motion in the solar system the mean motions of Jupiter and Saturn are nearly in the commensurable ratio 5:3 and those of Uranus and Neptune nearly 2:1. There are corresponding terms with long periods, which are collectively called the *great inequality*, of comparatively large amplitude. We find gaps in the distribution of the mean motions of asteroids in commensurable ratios to that of Jupiter, for example, 2:1, and a clustering tendency a little off from such exact commensurability. These are called Kirkwood's gaps. For certain commensurable ratios, such as 3:2 and 1:1, there are asteroids at nearly exactly commensurable mean motions. This state of affairs cannot be explained by the presence of small divisors in the formal series solution; indeed, it is one of the unsolved enigmas in celestial mechanics. It is here that we urgently need mathematical discussion, based on purely logical foundations, of the nature of our formal method of integration. There are also several cases of commensurable ratios of the mean motions in satellite systems. Several gaps appear in Saturn's rings. The inner three of the four Galilean satellites of Jupiter are most remarkable. Their motions are very complicated and interesting. Using the latest developments of radar technique, work on the resonance problem in the rotation of planets, besides the new determination of astronomical constants, is now progressing.

The usual method for computing the perturbations fails in such cases. We have several methods for computing the ephemerides of such planets and satellites. Hansen, Delaunay, Newcomb, Gyldén, Brown, and von Zeipel each invented his own method. Von Zeipel's method is currently applied to the motion of artificial and even natural celestial bodies. Yet the series obtained by these methods are not uniformly convergent. The methods have been applied to the motion of Jupiter and Saturn, of Uranus and Neptune, and of some of the asteroids having mean motions nearly commensurable with that of Jupiter. Gyldén's heuristic method is worth noting for its originality. Instead of taking the Keplerian motion as the first approximation, he took as his intermediary orbit the absolute orbit with moving perihelion obtained by dropping all short-period terms from the perturbations. He integrated the equations by means of elliptic functions and thought that the solution written in series in the theory of elliptic functions was convergent. But Poincaré proved that these series, too, are not uniformly convergent. Bohlin expanded the solution of the general three-body problem in a different kind of series and called the method *autologous*.

A revolutionary trend in celestial mechanics was inaugurated toward the end of last century by Hill in his theory of the motion of the moon. Hill took as his intermediary orbit a periodic orbit of a fictitious, though very nearly true, system of the sun, moon, and earth. Hill's theory has been completed by Brown for the actual numerical calculation of the lunar ephemerides. A linear differential equation with periodic coefficients and a determinant with an infinite number of rows and columns play important roles in Hill's theory. The series Hill obtained for this periodic orbit are uniformly convergent (the radius of convergence has been derived on the basis of nonlinear infinite analysis), but the series representing the deviation of the actual orbit from this intermediary orbit are not uniformly convergent. This ingenious idea of Hill's was generalized by Poincaré in his theory of periodic solutions and has been applied to the motion of those asteroids with mean motions nearly commensurable with that of the main disturbing body, Jupiter, and also to the motion of the Galilean satellites of Jupiter by de Sitter. Poincaré's theory of periodic solutions has been cultivated mathematically by Perron, Lefschetz, Siegel, and others in conjunction with the theory of differential equations. Numerical computations for periodic solutions of idealized systems have been worked out by Darwin, Hill, and Strömgren, among others, and are now in progress with the help of electronic computers. Indeed, following Strömgren, Hénon, among others, has made with computers a complete survey of orbits in the restricted three-body problem (in which the second finite mass rotates around the first finite mass in a circular orbit and the third mass is negligible) and studied the genealogy of the periodic solutions. The existence of horseshoe periodic orbits in the equilateral-triangular case of this problem, as conjectured by Brown, is now cleared up by means of computers. It is remarkable that even the analytic manipulation of multiple Fourier series can now be achieved by the proper use of computers. Poincaré's theory of periodic and asymptotic solutions can be discussed on the basis of the theory of nonlinear integral equations developed by Schmidt, Lichtenstein, and Wintner. The theory of bifurcation can be founded on this consolidated ground.

What is then the right form of the solution of the three-body problem, in particular of the planetary theory? The solutions so far obtained are all in trigonometric series with linear functions of time as arguments. As has been remarked, such series are not uniformly convergent and hence they are not the mathematical solution of the problem. If they were uniformly convergent, then the motion represented by such forms of the solution would be the superposition of periodic motions; hence the present configuration of the solar system would remain essentially unaltered for an unlimited time — that is, the system would be stable. At this stage we might ask: Is it not possible to try the solution in series proceeding in some sort of eigenfunctions, with due care for the nature of the singularities of the solution, on a function-theoretical basis? Owing to the lack of such work, we are not in a position to say anything about the stability of the motion in a rigorous sense.

The stability of a periodic solution has been discussed by Poincaré on the basis of his characteristic exponents, that is, by studying the behavior of nearby motions on the basis of the linearized equations for the variations of the generating periodic solutions. Solutions close to the particular solutions of the three-body problem and to the motion of particles forming the Saturnian rings have been discussed in this manner. Poisson's theorem on the invariability of the major axes in planetary motion, and the Laplace-Lagrange theory of secular perturbation establish in a certain sense the stability of the solar system. Liapounov has defined stability in a more rigorous mathematical sense. For the complete discussion of the nonlinear equations we must refer to topological studies. The criterion for stability is that of Lagrange, and the condition derived from the discussion of small deviations is one of the necessary conditions but not a sufficient condition. In the theory of the figures of a rotating liquid mass, we must refer to the energy criterion. Liapounov, Poincaré, and Schwarzschild discussed the condition for stability by developing the idea of Dirichlet. Liapounov's direct method is now greatly cultivated in theories and in application to control and optimization problems as well as topological dynamics. Search for the converse of Liapounov's theorems has led to various refined ideas on stability, even in abstract topological spaces.

Since the formal series obtained for the solution of our problem are not uniformly convergent, we can draw no conclusion as to the physical behavior of the motion merely on the basis of such formal series expansions. They may well be valid in a limited time interval within observational accuracy. Sundman's proof of the existence of the solution is based on the Weierstrassian theory of analytic functions, and the radius of convergence of his series is so small that nothing can be said about the physical characteristics of the motion, although the radius of convergence of Sundman's series as evaluated by Belorizky has been enlarged recently by reference to the polynomial expansion theory of arbitrary functions.

Thus the topological study of the trajectories in the manifold of motion was initiated by Poincaré. He discussed in the large the geometrical behavior of the solution curves of a nonlinear differential equation in the manifold of motion of any connectivity and with any Betti and torsion numbers, without reference to these series expansions, and hence built up his whole theory of topology. Hadamard, Poincaré, and Morse discussed the topological characteristics of geodesics on a surface, since a dynamical system devoid of any external force field can be mapped on a curved surface of a suitable number of dimensions. It is remarkable that Denjoy found for the characteristics on the surface of a torus the dependence of the ultimate behavior of a dynamical system on the degree of regularity of the functions that characterize the motion. Poincaré's theory on the existence of periodic solutions by means of the minimizing principle has been generalized by Birkhoff to the minimax principle and by Morse to functional topology and the calculus of variations in the large. Birkhoff applied the idea of Poincaré to the topological study of the manifold of motion in the restricted three-body problem. Poincaré enunciated his famous invariant-point theorem, proved later by Birkhoff and extended by Brouwer, Birkhoff, and Nielsen. Poincaré and Birkhoff both applied the theorem to the proof of the existence of periodic orbits in the three-body problem. Birkhoff further studied the stability characters of the trajectories in the neighborhood of an invariant point of the surface transformations by using formal expansions.

Poincaré's idea of periodic motions has been extended by Birkhoff to recurrent motions, nonwandering motions, and central motions based on the point-set theory of Cantor. Poincaré proved his recurrence theorem which states that any trajectory can pass as near as we please to its initial position an infinite number of times if the domain of the trajectories is always bounded and if there exists a positive integral invariant such as the volume of the domain in the phase space. Consider a point set $E$ which is invariant for a transformation $T$ in a manifold of motion $\Sigma$. If there exists no measurable invariant set $E$ whose measure in the sense of Lebesgue is smaller than the measure of $\Sigma$ and different from zero, then the transformation is said to be *metrically transitive*. Such a motion has been studied by using modular groups, Fuchsian groups, and general topological groups. The recurrence theorem of Poincaré has been refined with the idea of point-set theory by Carathéodory, Khintchine, von Neumann, Birkhoff, Hopf, Wiener, and others to the modern ergodic theorem, the idea of mixture, homogeneous chaos, and statistical regularity. The ergodic theorem is closely related to the Markoff chain in the theory of probability, as Poincaré's recurrence theorem first appeared in his lectures on probability. It is the foundation of the kinetic theory of gases.

The trend of celestial mechanics at this stage of machine computers should be directed toward studying the general behavior of trajectories on rigorous mathematical foundations, so as to check the results of numerical computation, and at the same time toward estimating the errors involved in using nonconvergent series expansions within the accuracy to be expected from the limited observation time. Thus the science of celestial mechanics is now confronting a new revolutionary era by abruptly moving from reliance on the old weapon of series expansions toward rigorous mathematical discussion of the physical behavior of the trajectories for an unlimited time. The artificial satellites and planets are offering us the more difficult and more general problems of studying motions widely different from those of natural objects hitherto discussed in celestial mechanics. This new aspect completely surpasses the apparent elegance of the old-fashioned celestial mechanics, and its importance cannot be overemphasized.

# Part I
# Dynamical Principles

# Principles of Analytical Dynamics

## VARIATIONAL PRINCIPLES

### 1.1 HAMILTON'S PRINCIPLE

If an arbitrary infinitesimal change of each coordinate in a dynamical system is a permissible displacement of the dynamical system, then the dynamical system is said to be nonconstrained. If there is a *constraint*, an arbitrary displacement is not always permissible. When the constraint is expressed by a finite relation among the coordinates that define the position of the dynamical system, the dynamical system is said to be *holonomic*. In other words, when the relation that must exist among the infinitesimal displacements because of the constraint is integrable as a function of the coordinates and time, then the system is said to be holonomic according to Hertz (1910). Otherwise, it is nonholonomic. Hertz proved that the motion of a holonomic dynamical system is uniquely determined by the Lagrangian equations. When the constraint contains time $t$ only implicitly, the constraint is said to be *scleronomic* and otherwise *rheonomic*, according to Boltzmann (1922).

If the work done by an external force on the dynamical system depends only on the initial and final configuration of the system, the force is called *conservative*. In such cases there exists the *potential*, $V$. Let one of the coordinates be $q$. Then the force $\psi$ is represented by

$$\psi = -\frac{\partial V}{\partial q} = \frac{\partial U}{\partial q}.$$

$U$ is called the *force function*. Denote by $T$ the kinetic energy. Helmholtz called $L = T - V$ the *kinetic potential*, but it is also called the *Lagrangian function*. Take a conservative holonomic system. Put $\dot{q}_i = dq_i/dt$ $(i = 1, 2, \ldots, n)$, and consider $L(q_1, q_2, \ldots, q_n; \dot{q}_1, \dot{q}_2, \ldots, \dot{q}_n; t)$. Assume that $L$ is at least twice differentiable with regard to any one of the variables. Suppose these differential coefficients to be continuous. Write

$$W = \int_{t_0}^{t_1} L\, dt. \tag{1}$$

This is called the *principal function*. According to *Hamilton's Principle*, $W$ should be stationary for the permissible motions of the dynamical system. The word "stationary" means that $W$ is a maximum, a minimum, or a minimax, that is, an extremum. In order that this extremum be a minimum, the Weierstrassian $E$-function should be zero or positive between conjugate points, as will be shown later.

By varying $t_0$, $t_1$, $q_i$ $(i = 1, 2, \ldots, n)$ independently in (1) we obtain

$$\delta W = L_1\, \delta t_1 - L_0\, \delta t_0 + \int_{t_0}^{t_1} \sum_{i=1}^{n} \left(\frac{\partial L}{\partial q_i}\, \delta q_i + \frac{\partial L}{\partial \dot{q}_i}\, \delta \dot{q}_i\right) dt,$$

where $L_1$, $L_0$ and $\delta t_1$, $\delta t_0$ are the values of $L$ and $\delta t$ for $t = t_1$ and $t = t_0$, respectively. Since

$$\delta \dot{q}_i\, dt = \delta\left(\frac{dq_i}{dt}\right) dt = \frac{d}{dt}\,(\delta q_i)\, dt = d(\delta q_i),$$

we have

$$\delta W = L_1\, \delta t_1 - L_0\, \delta t_0 + \left[\sum_i \frac{\partial L}{\partial \dot{q}_i}\, \delta q_i\right]_{t_0}^{t_1} + \int_{t_0}^{t_1} \sum_i \left[\frac{\partial L}{\partial q_i} - \frac{d}{dt}\left(\frac{\partial L}{\partial \dot{q}_i}\right)\right] \delta q_i\, dt.$$

Consider a varying parameter $\alpha$ such that any displacement except $\delta t_0$ and $\delta t_1$ is due to a variation of $\alpha$, that is, such that the variation of $q_i$ for a fixed value of $t$ is due to $\delta\alpha$. Then, since

$$[\delta q_i]_{t_0} = \frac{\partial}{\partial \alpha}\, q_i(t_0, \alpha)\, \delta\alpha, \qquad \delta(q_i^0) = \frac{\partial q_i(t_0, \alpha)}{\partial t}\, \delta t_0 + \frac{\partial q_i(t_0, \alpha)}{\partial \alpha}\, \delta\alpha,$$

we have

$$[\delta q_i]_{t_0} = \delta(q_i^0) - \dot{q}_i^0\, \delta t_0, \qquad [\delta q_i]_{t_1} = \delta(q_i^1) - \dot{q}_i^1\, \delta t_1,$$

where $q_i^0$, $\dot{q}_i^0$ are the values of $q_i$, $\dot{q}_i$ for $t = t_0$ and $q_i^1$, $\dot{q}_i^1$ are the values for $t = t_1$. Thus, the variation $\delta W$ of (1) takes the form

$$\delta W = \left\{ \sum_i \left( \frac{\partial L}{\partial \dot{q}_i} \right)_1 \delta(q_i^1) - \left[ \sum_i \dot{q}_i \frac{\partial L}{\partial \dot{q}_i} - L \right]_1 \delta t_1 \right\}$$

$$- \left\{ \sum_i \left( \frac{\partial L}{\partial \dot{q}_i} \right)_0 \delta(q_i^0) - \left[ \sum_i \dot{q}_i \frac{\partial L}{\partial \dot{q}_i} - L \right]_0 \delta t_0 \right\}$$

$$+ \int_{t_0}^{t_1} \sum_i \left[ \frac{\partial L}{\partial q_i} - \frac{d}{dt} \left( \frac{\partial L}{\partial \dot{q}_i} \right) \right] dt \, \delta q_i. \tag{2}$$

Hamilton's principle states that $W$ for an actual motion should be an extremum for fixed $t_0$ and $t_1$. Hence the variation (2), as the Eulerian equations in the calculus of variations (Bolza, 1909; Carathéodory, 1935; Kneser, 1925), leads to

$$\frac{d}{dt} \left( \frac{\partial L}{\partial \dot{q}_i} \right) - \frac{\partial L}{\partial q_i} = 0 \qquad (i = 1, 2, \ldots, n). \tag{3}$$

These are the *Lagrangian equations*. The so-called Legendre transformation gives the Hamiltonian function

$$H = \sum_i \dot{q}_i \frac{\partial L}{\partial \dot{q}_i} - L. \tag{4}$$

The above variational principle requires that

$$W = \int_{t_0}^{t_1} \left( \sum_i \dot{q}_i \frac{\partial L}{\partial \dot{q}_i} - H \right) dt \tag{1a}$$

be an extremum. When $L$ contains $t$ only implicitly, $H$ is constant and is, in fact, the energy integral. As the end points $t_0$ and $t_1$ are fixed, the variation of $H$ is zero and the variational principle requires that the integral

$$\int_{t_0}^{t_1} \left( \sum_i \dot{q}_i \frac{\partial L}{\partial \dot{q}_i} \right) dt \tag{1b}$$

have an extremum, that is, the variation of the integral must vanish.

In order to prove this theorem, consider a motion with

$$H = h = \text{constant}$$

and a displaced motion with $H = h + \delta h$:

$$\begin{aligned}\delta \int_{t_0}^{t_1} \sum_i \dot{q}_i \frac{\partial L}{\partial \dot{q}_i}\, dt &= \delta \int_{t_0}^{t_1} (L + h)\, dt \\ &= (h + \delta h)(t_1 + \delta t_1 - t_0 - \delta t_0) \\ &\quad - h(t_1 - t_0) + \left[\sum_i \frac{\partial L}{\partial \dot{q}_i} \delta q_i - h\, \delta t\right]_{t_0}^{t_1} \\ &= \left[\sum_i \frac{\partial L}{\partial \dot{q}_i} \delta q_i + t\, \delta h\right]_{t_0}^{t_1}.\end{aligned}$$

Put $\delta t_0 = \delta t_1 = 0$ and $\delta h = 0$; then the $\delta q_i$ at $t_1$ and $t_0$ are zero and the variation of (1b) is zero. The integral (1b) is called the *action* and this variational principle is called the *principle of least action* of Maupertuis (Jacobi, 1884; Darboux, 1890; Wintner, 1930; Murnaghan, 1931).

Write $L$ as a sum of homogeneous terms of the same degree with respect to $\dot{q}_i$ in the form $L = L_2 + L_1 + L_0$. Then $H = L_2 - L_0 = h$. If $h = 0$, then $L_2 = L_0$ along the path of the motion. In such a case, consider

$$W^* = W - \int_{t_0}^{t_1} (\sqrt{L_2} - \sqrt{L_0})^2\, dt = \int_{t_0}^{t_1} (2\sqrt{L_0 L_2} + L_1)\, dt.$$

Then the variation of $W^*$ is

$$\delta W^* = \delta W - 2\int_{t_0}^{t_1} (\sqrt{L_2} - \sqrt{L_0})(\delta\sqrt{L_2} - \delta\sqrt{L_0})\, dt.$$

For a motion with $h = 0$, $\delta W^* = \delta W$. Since $\delta W = 0$, then $\delta W^* = 0$. But $W^*$ being homogeneous of degree one, does not depend on the value of $t$. Hence, if we take

$$\bar{t} = \int_0^t (\sqrt{L_2}/\sqrt{L_0})\, dt$$

in place of $t$, we obtain the energy $H = h$ in the form

$$H = L_2 - L_0 = h = 0.$$

Thus, if we take $\bar{t}$ in place of $t$, our variational principle requires the extremum of

$$W^* = \int_{t_0}^{t_1} (2\sqrt{L_0 L_2} + L_1)\, dt. \tag{1c}$$

This variational principle is due to Birkhoff (1917). When the Hamiltonian $H$ contains linear terms with respect to $\dot{q}_i$, the dynamical system is called *irreversible*; when $H$ contains $\dot{q}_i$ only in quadratic form, the system is called *reversible*. Equation (1c) gives a method for discussing an irreversible system, such as that of the restricted three-body problem, by the principle of least action. That is to say, if the term $L_1\,dt$ is a perfect differential with respect to $t$, it can be omitted in the discussion.

## 1.2 RELATION TO THE CALCULUS OF VARIATIONS

Put

$$\omega_\delta = \sum_i \frac{\partial L}{\partial \dot{q}_i}\,\delta q_i - H\,\delta t. \tag{5}$$

Cartan (1922) called this the *momentum-energy tensor*. Equation (2) can be written

$$\delta W = [\omega_\delta]_1 - [\omega_\delta]_0 + \int_{t_0}^{t_1} \sum_i \left[\frac{\partial L}{\partial q_i} - \frac{d}{dt}\left(\frac{\partial L}{\partial \dot{q}_i}\right)\right] \delta q_i\,dt,$$

in which the last integral vanishes along the path taken by the dynamical system. We consider $\int \omega_\delta$ to be the integral along a closed curve in a space formed by the points representing the states of displaced motions of the dynamical system which are obtained by varying $q_i$ and $t$. This integral $\int \omega_\delta$ has the same value at every instant along any motion; Cartan (1922) called it an *integral invariant*. In Figure 1, let $A_0$ and $A_1$ be two positions

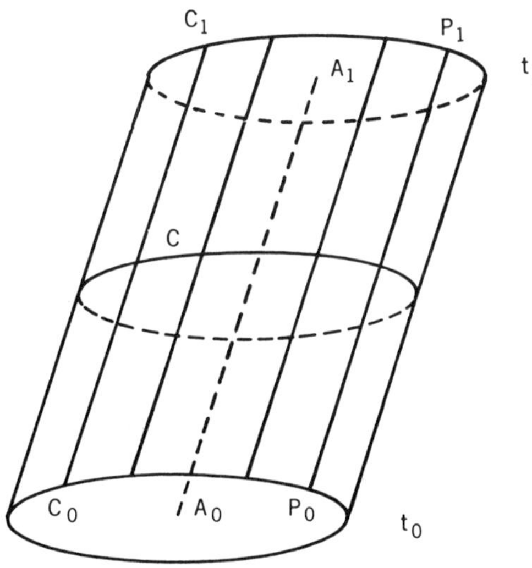


FIGURE I

When $L$ is quadratic in $\dot{q}$, which is the case for most dynamical systems, $\Delta \equiv 0$ means that the quadratic form of $L$ is *degenerate*—that is, it can be expressed as a sum of squares of less than $n$ variables, the rest of the variables being reduced to constants; $L$ in this case describes a constrained system.

Thus, if the Lagrangian function is normal, the Lagrangian system is equivalent to the canonical system described by equations (7). Let us consider the inverse of this theorem. If $\Delta \neq 0$, we solve (6) for $\dot{q}$ and obtain $\dot{q}_i = \partial H/\partial p_i$. In order that this system of equations can be solved for $p_i$, it is necessary that $\Delta_1 = \|\partial^2 H/\partial p_i\, \partial p_j\|$ not be identically zero. But

$$\sum_k \frac{\partial p_i}{\partial \dot{q}_k} \frac{\partial \dot{q}_k}{\partial p_j} = \delta_{ij} \qquad (i, j = 1, 2, \ldots, n),$$

with Kronecker's symbol $\delta_{ij}$, such that $\delta_{ii} = 1$, $\delta_{ij} = 0$ $(i \neq j)$. Substitute (6) for $p_i$ and the second equation of (7) for $\dot{q}_k$; then

$$\sum_k \frac{\partial^2 L}{\partial \dot{q}_i\, \partial \dot{q}_k} \cdot \frac{\partial^2 H}{\partial p_k\, \partial p_j} = \delta_{ij},$$

or

$$\Delta \cdot \Delta_1 = 1.$$

If $\Delta \neq 0$, then $\Delta_1 \neq 0$; and if $\Delta_1 \neq 0$, then $\Delta \neq 0$. Hence, if (3) is normal, then (3) is equivalent to (7), and (7) is equivalent to (3). Actually, $\dot{q}_i = \partial H/\partial p_i$ can be solved for $p$ because $\Delta_1 \neq 0$; also, $p_i = v_i(q, \dot{q}, t)$. If we put $v$ for $p$ in $L = \sum_i p_i \dot{q}_i - H$, then $L$ can be expressed in terms of $q$, $\dot{q}$, $t$.

The $2n$-dimensional space with Cartesian coordinates $p_i$, $q_i$ is called the *phase space* according to Gibbs. The solution $p_i = p_i(t)$, $q_i = q_i(t)$ of (7) is an *integral curve*. This is represented by a curve, that is, a *trajectory*, in the phase space as a function of a parameter $t$. In general, there is one and only one trajectory with the initial condition $p_i(t_0)$, $q_i(t_0)$ at $t = t_0$, by Cauchy's existence theorem of the solution of a differential equation. When a relation of the form $f(p, q, t) =$ constant is identically satisfied by (7), such a relation is called an *integral*, or $f$ itself is called an integral. When $H$ depends on $t$ only implicitly, there is an integral $H =$ constant, and this is called the *energy integral*. In fact,

$$\frac{dH}{dt} = \sum_i \left( \frac{\partial H}{\partial p_i} \dot{p}_i + \frac{\partial H}{\partial q_i} \dot{q}_i \right) + \frac{\partial H}{\partial t}.$$

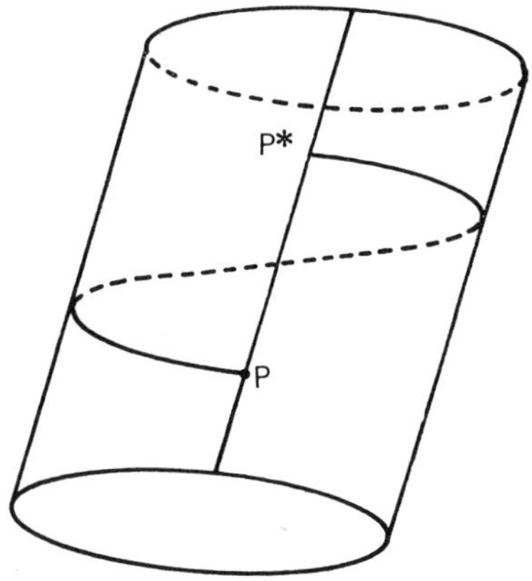


FIGURE 2

of manifolds $M_n$ of $n$ dimensions with one parameter in the space of $n + 1$ dimensions. This manifold $M_n$ has a transversal relation with the family of trajectories corresponding to $[\omega_\delta] = 0$ with $n$ parameters. Any two manifolds of $M_n$ cut from the trajectory a part with the same value of $\int[\omega_\delta]$. A bundle of trajectories that form such $M_n$ is called a *field* in the sense of the calculus of variations. In general, the word "field" is used when any two trajectories of the bundle intersect once and only once. The transversal manifold in our case is given by

$$W(q_1, q_2, \ldots, q_n, t) = \int_{P_0}^{P} \left(\sum_i \frac{\partial L}{\partial \dot{q}_i}\, \delta q_i - H\, \delta t\right) = \int_{P_0}^{P} [\omega_\delta] = \text{constant}.$$

This integral, extended from a point $P_0(q_i^0, t_0)$ to a point $P(q_i, t)$ in the field, is a function of the field, and does not depend on the path of integration. It can be written

$$\int_{P_0}^{P} \left[L + \sum_i \frac{\partial L}{\partial \dot{q}_i}\left(\frac{\delta q_i}{\delta t} - \dot{q}_i\right)\right] \delta t.$$

This is the *independency integral* of Hilbert (1906).

Consider a transversal manifold $M_n$ in the field that passes through $P_0$ (Figure 3). The trajectory through any point $P$ crosses the transversal manifold at a point $P_1$. The curve $P_0P_1$ lies on the transversal manifold $M_n$ and $[\omega_\delta]$ is zero on this manifold. On a trajectory of the field,

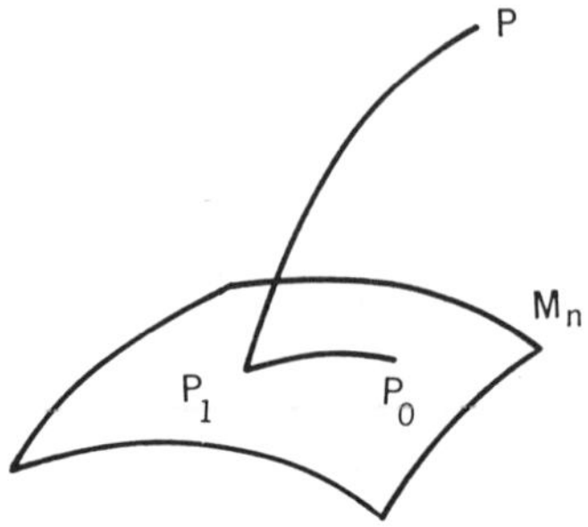


FIGURE 3

$\delta q_i/\delta t = \dot{q}_i$ $(i = 1, 2, \ldots, n)$. Hence, $[\omega_\delta] = L$. Accordingly, we can write

$$\int_{P_0}^{P} L\, dt = W(q_1, q_2, \ldots, q_n, t).$$

The coordinate $(q_i^0, t_0)$ of $P_0$ is a function of the coordinates $(q_i, t)$ of $P$ of the field. The differential of this function is $[\omega_\delta]$. Hence, as $P$ varies, $P_1$ moves but always on $M_n$. Thus we must have $[\omega_\delta] = 0$. But as $\delta W = [\omega_\delta] - [\omega_\delta]_0$, we should have $\delta W = [\omega_\delta]$. By referring to (5) we write this in the form

$$\left.\begin{aligned} \frac{\partial W}{\partial q_i} &= \frac{\partial L}{\partial \dot{q}_i} = p_i, \\ \frac{\partial W}{\partial t} &= -H, \end{aligned}\right\} \qquad (i = 1, 2, \ldots, n). \tag{5a}$$

These are the Hamilton-Jacobi relations, as we shall see immediately (Bolza, 1909; Carathéodory, 1935; Kneser, 1925; Koschmieder, 1933; Forsyth, 1927; Hadamard, 1910; Tonelli, 1921, 1923; de Donder, 1930).

## 1.3 CANONICAL EQUATIONS

If the determinant

$$\Delta = \left\| \frac{\partial^2 L}{\partial \dot{q}_i\, \partial \dot{q}_j} \right\|,$$

called the Hessian of $L$, is not identically zero, Equation (3) is said to be normal, according to Levi-Civita and Amaldi (1927). This is what is called *Legendre's condition* in the calculus of variations. Put

$$p_i = \frac{\partial L}{\partial \dot{q}_i} \qquad (i = 1, 2, \ldots, n). \tag{6}$$

$p_i$ is called the *conjugate variable* to $q_i$. As $\Delta \neq 0$, equations (6) are seen to be $n$ independent equations for $\dot{q}_i$ and can be solved by the theorem of implicit functions. Let the result be

$$\dot{q}_j = u_j(p_i, q_i, t) \qquad (j = 1, 2, \ldots, n); \tag{6a}$$

then (3) may be written

$$\dot{p}_i = \left(\frac{\partial L}{\partial q_i}\right)_{\dot{q}=u} \qquad (i = 1, 2, \ldots, n). \tag{3a}$$

Equations (3a) and (6a) give a system of $2n$ differential equations of order one, which is equivalent to (3). If we substitute (6) in place of $\dot{q}_i$ in the function defined by (4), then we have Legendre's transformation

$$H(p, q, t) = \sum_i p_i \dot{q}_i - L(q, \dot{q}, t), \tag{4a}$$

in which both members are expressed as functions of $p$, $q$, $t$.

If we fix $t$ and vary $p$ and $q$ by the amount $\delta p$ and $\delta q$, respectively, then

$$\delta H = \sum_i \left(\frac{\partial H}{\partial p_i}\,\delta p_i + \frac{\partial H}{\partial q_i}\,\delta q_i\right),$$

or by (4a)

$$\delta H = \sum_i \left[\dot{q}_i\,\delta p_i - \frac{\partial L}{\partial q_i}\,\delta q_i + \left(p_i - \frac{\partial L}{\partial \dot{q}_i}\right)\delta \dot{q}_i\right],$$

or by (6) and (6a)

$$\delta H = \sum_i \left[u_i\,\delta p_i - \left(\frac{\partial L}{\partial q_i}\right)_{\dot{q}=u}\delta q_i\right].$$

Hence,

$$u_i(p, q, t) = \frac{\partial H}{\partial p_i}, \qquad -\left(\frac{\partial L}{\partial q_i}\right)_{\dot{q}=u} = \frac{\partial H}{\partial q_i},$$

or

$$\frac{dp_i}{dt} = -\frac{\partial H}{\partial q_i}, \qquad \frac{dq_i}{dt} = \frac{\partial H}{\partial p_i} \qquad (i = 1, 2, \ldots, n). \tag{7}$$

This system of differential equations is called *canonical* or *Hamiltonian* and such a pair of variables $p$, $q$ *canonical variables*. The variables $p_i$ and $q_i$ are *canonically conjugate*.

When $L$ is quadratic in $\dot{q}$, which is the case for most dynamical systems, $\Delta \equiv 0$ means that the quadratic form of $L$ is *degenerate*—that is, it can be expressed as a sum of squares of less than $n$ variables, the rest of the variables being reduced to constants; $L$ in this case describes a constrained system.

Thus, if the Lagrangian function is normal, the Lagrangian system is equivalent to the canonical system described by equations (7). Let us consider the inverse of this theorem. If $\Delta \neq 0$, we solve (6) for $\dot{q}$ and obtain $\dot{q}_i = \partial H/\partial p_i$. In order that this system of equations can be solved for $p_i$, it is necessary that $\Delta_1 = \|\partial^2 H/\partial p_i\, \partial p_j\|$ not be identically zero. But

$$\sum_k \frac{\partial p_i}{\partial \dot{q}_k} \frac{\partial \dot{q}_k}{\partial p_j} = \delta_{ij} \qquad (i, j = 1, 2, \ldots, n),$$

with Kronecker's symbol $\delta_{ij}$, such that $\delta_{ii} = 1$, $\delta_{ij} = 0$ $(i \neq j)$. Substitute (6) for $p_i$ and the second equation of (7) for $\dot{q}_k$; then

$$\sum_k \frac{\partial^2 L}{\partial \dot{q}_i\, \partial \dot{q}_k} \cdot \frac{\partial^2 H}{\partial p_k\, \partial p_j} = \delta_{ij},$$

or

$$\Delta \cdot \Delta_1 = 1.$$

If $\Delta \neq 0$, then $\Delta_1 \neq 0$; and if $\Delta_1 \neq 0$, then $\Delta \neq 0$. Hence, if (3) is normal, then (3) is equivalent to (7), and (7) is equivalent to (3). Actually, $\dot{q}_i = \partial H/\partial p_i$ can be solved for $p$ because $\Delta_1 \neq 0$; also, $p_i = v_i(q, \dot{q}, t)$. If we put $v$ for $p$ in $L = \sum_i p_i \dot{q}_i - H$, then $L$ can be expressed in terms of $q$, $\dot{q}$, $t$.

The $2n$-dimensional space with Cartesian coordinates $p_i$, $q_i$ is called the *phase space* according to Gibbs. The solution $p_i = p_i(t)$, $q_i = q_i(t)$ of (7) is an *integral curve*. This is represented by a curve, that is, a *trajectory*, in the phase space as a function of a parameter $t$. In general, there is one and only one trajectory with the initial condition $p_i(t_0)$, $q_i(t_0)$ at $t = t_0$, by Cauchy's existence theorem of the solution of a differential equation. When a relation of the form $f(p, q, t) =$ constant is identically satisfied by (7), such a relation is called an *integral*, or $f$ itself is called an integral. When $H$ depends on $t$ only implicitly, there is an integral $H =$ constant, and this is called the *energy integral*. In fact,

$$\frac{dH}{dt} = \sum_i \left( \frac{\partial H}{\partial p_i} \dot{p}_i + \frac{\partial H}{\partial q_i} \dot{q}_i \right) + \frac{\partial H}{\partial t}.$$

By $\partial H/\partial t = 0$, and because of (7), $dH/dt = 0$. When $H$ contains neither $t$ nor $q_{n-r+1}, \ldots, q_n$, then, from (7),

$$p_{n-r+j} = \frac{\partial T}{\partial \dot{q}_{n-r+j}} = c_j(\text{constant}) \qquad (j = 1, 2, \ldots, r).$$

Hertz (1910) called such coordinates $q_{n-r+j}$ *ignorable coordinates*. Let

$$H^*(\dot{q}_1, \ldots, \dot{q}_{n-r}, q_1, \ldots, q_{n-r}, c_1, \ldots, c_r) = H - \sum_{j=1}^{r} \dot{q}_{n-r+j} c_j$$

and let $q_1, \ldots, q_{n-r}$ take, respectively, constant values $\gamma_1, \ldots, \gamma_{n-r}$; then, from the second part of (7), the ignorable coordinates are expressed by $q_{n-r+j} = t + b_j$ $(j = 1, \ldots, r)$. Levi-Civita and Amaldi (1927) further generalized these relations.

From (2),

$$\delta W = \left\{\sum_i \left(\frac{\partial L}{\partial \dot{q}_i}\right)_1 \delta(q_i^1) - \left[\sum_i \dot{q}_i \frac{\partial L}{\partial \dot{q}_i} - L\right]_1 \delta t_1\right\} - \left\{\sum_i \left(\frac{\partial L}{\partial \dot{q}_i}\right)_0 \delta(q_i^0) - \left[\sum_i \dot{q}_i \frac{\partial L}{\partial \dot{q}_i} - L\right]_0 \delta t_0\right\}.$$

Hence, because of (6) and (4),

$$\frac{\partial W}{\partial q_i^0} = -\left(\frac{\partial L}{\partial \dot{q}_i}\right)_0 = -p_i^0, \qquad \frac{\partial W}{\partial t_0} = H_0,$$
$$\frac{\partial W}{\partial q_i^1} = \left(\frac{\partial L}{\partial \dot{q}_i}\right)_1 = p_i^1, \qquad \frac{\partial W}{\partial t_1} = -H_1, \tag{8}$$

where $q_i^0$, $p_i^0$, $H_0$ denote the values of $q_i$, $p_i$, $H$, respectively, at $t_0$; and $q_i^1$, $p_i^1$, $H_1$ denote the corresponding values at $t_1$. The effect of the displacement at $t_1$ on the values of $p_i$ at $t_1$ defined by (6) is derived by differentiating these relations (8) again:

$$\frac{\partial p_i^1}{\partial q_j^1} = \frac{\partial p_j^1}{\partial q_i^1} = \frac{\partial^2 W}{\partial q_i^1\, \partial q_j^1},$$
$$\frac{\partial p_i^1}{\partial t_1} = -\frac{\partial H_1}{\partial q_i^1} = \frac{\partial^2 W}{\partial q_i^1\, \partial t_1}, \tag{9}$$
$$\frac{\partial H_1}{\partial q_i^1} = \sum_j \left(\frac{\partial H}{\partial p_j}\right)_1 \frac{\partial p_j^1}{\partial q_i^1} + \left(\frac{\partial H}{\partial q_i}\right)_1.$$

Analogous relations can be obtained for $t_0$. Similarly, the effect of the displacement at $t_0$ on the values of $p_i$ at $t_1$ and the reverse effect are given by

$$\frac{\partial p_i^1}{\partial q_j^0} = -\frac{\partial p_j^0}{\partial q_i^1} = \frac{\partial^2 W}{\partial q_i^1\, \partial q_j^0},$$

$$\frac{\partial p_i^1}{\partial t_0} = \frac{\partial H_0}{\partial q_i^1} = \frac{\partial^2 W}{\partial q_i^1\, \partial t_0}, \qquad \frac{\partial p_i^0}{\partial t_1} = \frac{\partial H_1}{\partial q_i^0} = -\frac{\partial^2 W}{\partial q_i^0\, \partial t_1}, \tag{9a}$$

$$\frac{\partial H_0}{\partial q_i^1} = \sum_j \left(\frac{\partial H}{\partial p_j}\right)_0 \frac{\partial p_j^0}{\partial q_i^1}, \qquad \frac{\partial H_1}{\partial q_i^0} = -\sum_j \left(\frac{\partial H}{\partial p_j}\right)_1 \frac{\partial p_j^1}{\partial q_i^0}.$$

Equations (9) and (9a) are the reciprocity relation of Helmholtz and were derived by Hamilton in his theory of geometrical optics (Carathéodory, 1937; Herzberger, 1931; Synge, 1960).

Let $n$ be the refractive index of the medium and $s$ be the length in the direction of light wave propagation. Fermat's principle in optics states that the path of the light ray is given by

$$\delta \int n(\alpha, \beta, \gamma, x, y, z)\, ds = 0,$$

$$\alpha = \frac{dx}{ds}, \qquad \beta = \frac{dy}{ds}, \qquad \gamma = \frac{dz}{ds},$$

where $x$, $y$, $z$ are the rectangular coordinates, and $\alpha$, $\beta$, $\gamma$ denote the vector of the light ray. As in Euler's equation in the calculus of variations, we obtain

$$\frac{d}{ds}\left(\frac{\partial n}{\partial \alpha}\right) - \frac{\partial n}{\partial x} = 0,$$

$$\frac{d}{ds}\left(\frac{\partial n}{\partial \beta}\right) - \frac{\partial n}{\partial y} = 0,$$

$$\frac{d}{ds}\left(\frac{\partial n}{\partial \gamma}\right) - \frac{\partial n}{\partial z} = 0.$$

Consider a function of the field called the *characteristic function* of Hamilton

$$W(x, y, z) = \int_{P_0}^{P} n(\alpha, \beta, \gamma, x, y, z)\, ds.$$

The surface $W$ = constant is the locus of points of equal elapsed time for the light rays and is called the *wave surface.* It can be shown by Fermat's principle that the normal of the wave surface is given by

$$\frac{\partial W}{\partial x} = \frac{\partial n}{\partial \alpha}, \qquad \frac{\partial W}{\partial y} = \frac{\partial n}{\partial \beta}, \qquad \frac{\partial W}{\partial z} = \frac{\partial n}{\partial \gamma}.$$

The direction of light propagation does not coincide with the normal of the wave surface. But they are in transversal relation. Hamilton obtained, by eliminating $\alpha$, $\beta$, $\gamma$ from these equations, a partial differential equation for the characteristic function

$$H\left(\frac{\partial W}{\partial x}, \frac{\partial W}{\partial y}, \frac{\partial W}{\partial z}, x, y, z\right) = 0.$$

The light ray vector along each individual ray is $(\alpha/n, \beta/n, \gamma/n)$, and its reciprocals represent the components of light velocity. The quantities

$$\pi = \frac{\partial n}{\partial \alpha}, \qquad \kappa = \frac{\partial n}{\partial \beta}, \qquad \rho = \frac{\partial n}{\partial \gamma}$$

are the components of the wave vector, which is normal to the wave surface. It can be shown that these equations may be written

$$\frac{\alpha}{n} = \frac{\partial H}{\partial \pi}, \qquad \frac{\beta}{n} = \frac{\partial H}{\partial \kappa}, \qquad \frac{\gamma}{n} = \frac{\partial H}{\partial \rho},$$

where

$$\pi \frac{\partial H}{\partial \pi} + \kappa \frac{\partial H}{\partial \kappa} + \rho \frac{\partial H}{\partial \rho} = 1.$$

By referring to the relation

$$dW = n(\alpha, \beta, \gamma, x, y, z)\, ds,$$

Hamilton obtained

$$\frac{dx}{dW} = \frac{\partial H}{\partial \pi}, \qquad \frac{dy}{dW} = \frac{\partial H}{\partial \kappa}, \qquad \frac{dz}{dW} = \frac{\partial H}{\partial \rho},$$

$$\frac{d\pi}{dW} = -\frac{\partial H}{\partial x}, \qquad \frac{d\kappa}{dW} = -\frac{\partial H}{\partial y}, \qquad \frac{d\rho}{dW} = -\frac{\partial H}{\partial z}$$

(Carathéodory, 1930, 1937; Herzberger, 1931).

Take a point $P_0(x_0, y_0, z_0)$ of the object space and a point $P_1(x_1, y_1, z_1)$ of the image space. Let $W(x_0, y_0, z_0, x_1, y_1, z_1)$ be Hamilton's characteristic function. Take a point $Q_0(\xi_0, \eta_0, \zeta_0)$ on the light ray in the object space and a point $Q_1(\xi_1, \eta_1, \zeta_1)$ in the image space. Let $\rho_0$ be the distance $P_0Q_0$ and $\rho_1$ be the distance $P_1Q_1$. Then,

$$x_0 = \xi_0 + \rho_0\alpha_0, \qquad y_0 = \eta_0 + \rho_0\beta_0, \qquad z_0 = \zeta_0 + \rho_0\gamma_0,$$

$$x_1 = \xi_1 + \rho_1\alpha_1, \qquad y_1 = \eta_1 + \rho_1\beta_1, \qquad z_1 = \zeta_1 + \rho_1\gamma_1;$$

and

$$\begin{aligned} dW = & -n_0(\alpha_0\,dx_0 + \beta_0\,dy_0 + \gamma_0\,dz_0) \\ & + n_1(\alpha_1\,dx_1 + \beta_1\,dy_1 + \gamma_1\,dz_1), \end{aligned}$$

or

$$\begin{aligned} dW = & -n_0(\alpha_0\,d\xi_0 + \beta_0\,d\eta_0 + \gamma_0\,d\zeta_0 + d\rho_0) \\ & + n_1(\alpha_1\,d\xi_1 + \beta_1\,d\eta_1 + \gamma_1\,d\zeta_1 + d\rho_1). \end{aligned}$$

From these equations the reciprocity relations corresponding to (9) can be derived.

Suppose that $\xi_0 = \xi_1 = 0$. Replacing $x_0, y_0, z_0$ by $\eta_0, \zeta_0, \rho_0$ and $x_1, y_1, z_1$ by $\eta_1, \zeta_1, \rho_1$ and writing

$$dE = -n_0(\beta_0\,d\eta_0 + \gamma_0\,d\zeta_0) + n_1(\beta_1\,d\eta_1 + \gamma_1\,d\zeta_1),$$

we obtain

$$dW = dE - n_0\,d\rho_0 + n_1\,d\rho_1,$$

$$\frac{\partial W}{\partial \rho_0} = -n_0, \qquad \frac{\partial W}{\partial \eta_0} = \frac{\partial E}{\partial \eta_0} = -n_0\beta_0, \qquad \frac{\partial W}{\partial \zeta_0} = \frac{\partial E}{\partial \zeta_0} = -n_0\gamma_0,$$

$$\frac{\partial W}{\partial \rho_1} = n_1, \qquad \frac{\partial W}{\partial \eta_1} = \frac{\partial E}{\partial \eta_1} = n_1\beta_1, \qquad \frac{\partial W}{\partial \zeta_1} = \frac{\partial E}{\partial \zeta_1} = n_1\gamma_1.$$

The function $E(\eta_0, \zeta_0, \eta_1, \zeta_1)$ is called the *eikonal* by Bruns.

Put

$$\begin{aligned} V = W & + n_0[(x_0 - \xi_0)\alpha_0 + (y_0 - \eta_0)\beta_0 + (z_0 - \zeta_0)\gamma_0] \\ & - n_1[(x_1 - \xi_1)\alpha_1 + (y_1 - \eta_1)\beta_1 + (z_1 - \zeta_1)\gamma_1]; \end{aligned}$$

then

$$\begin{aligned} dV = & \; n_0[(x_0 - \xi_0)\,d\alpha_0 + (y_0 - \eta_0)\,d\beta_0 + (z_0 - \zeta_0)\,d\gamma_0] \\ & - n_1[(x_1 - \xi_1)\,d\alpha_1 + (y_1 - \eta_1)\,d\beta_1 + (z_1 - \zeta_1)\,d\gamma_1]. \end{aligned}$$

The function $V(\sqrt{1 - \beta_0^2 - \gamma_0^2}, \beta_0, \gamma_0, \sqrt{1 - \beta_1^2 - \gamma_1^2}, \beta_1, \gamma_1)$, which is a function of $\beta_0, \gamma_0, \beta_1, \gamma_1$, is called the *angular eikonal* by Schwarzschild (Carathéodory, 1937).

The propagation of electromagnetic waves in the earth's ionosphere can be discussed in a similar manner.

## 1.4 COVARIANCE

In the relation

$$x_k = \phi_k(y_1, \ldots, y_n) \qquad (k = 1, 2, \ldots, n),$$

between $x_1, x_2, \ldots, x_n$ and $y_1, y_2, \ldots, y_n$, differentiation gives

$$dx_k = \sum_i \frac{\partial \phi_k}{\partial y_i}\, dy_i \qquad (k = 1, 2, \ldots, n).$$

Assume that the Jacobian $\partial(x_1, \ldots, x_n)/\partial(y_1, \ldots, y_n) \neq 0$. Then,

$$y_j = \psi_j(x_1, \ldots, x_n) \qquad (j = 1, 2, \ldots, n).$$

For an arbitrary function $F(x_1, x_2, \ldots, x_n)$, we have

$$\frac{\partial F}{\partial x_l} = \sum_j \frac{\partial \psi_j}{\partial x_l}\frac{\partial F}{\partial y_j} \qquad (l = 1, 2, \ldots, n).$$

In this case we say that $\partial F/\partial x_l$ and $dx_k$ vary *contragradiently*. The relation

$$\sum_k \frac{\partial F}{\partial x_k}\, dx_k = \sum_k \left\{ \sum_i \sum_j \frac{\partial \psi_j}{\partial x_k}\frac{\partial \phi_k}{\partial y_i}\, dy_i\, \frac{\partial F}{\partial y_j} \right\}$$

shows that

$$\sum_k \frac{\partial F}{\partial x_k}\, dx_k = \sum_i \frac{\partial F}{\partial y_i}\, dy_i,$$

since

$$\sum_k \frac{\partial \psi_j}{\partial x_k}\frac{\partial \phi_k}{\partial y_i} = \delta_{ij}.$$

The expression for $dF$ does not change its form under the transformation. Hence $dF$ is said to be an *invariant*.

Let an arbitrary function $A(x)$ be transformed to $A'(y)$ and suppose that

$$A_k(x) = \sum_i \frac{\partial \phi_k}{\partial y_i} A'_i(y), \quad \text{or} \quad A'_i(y) = \sum_k \frac{\partial \psi_i}{\partial x_k} A_k(x); \tag{10}$$

then we call $A_k(x)$ a *cogradient*. If for functions $B(x)$, $B'(y)$ we have the relation

$$B_l(x) = \sum_j \frac{\partial \psi_j}{\partial x_l} B'_j(y), \quad \text{or} \quad B'_j(y) = \sum_l \frac{\partial \phi_l}{\partial y_j} B_l(x), \tag{11}$$

then we call $B_i(x)$ a *contragradient*. In this case

$$\sum_k A_k(x)B_k(x) = \sum_i A'_i(y)B'_i(y); \tag{12}$$

this last expression is seen to be an invariant.

Conversely, when $\sum_k A_k(x)C_k(x)$ is invariant, $C_k(x)$ is a contragradient if $A_k(x)$ is a cogradient, and $C_k(x)$ is a cogradient if $A_k(x)$ is a contragradient. Eddington (1923) called this the *quotient theorem*. The proof is easy. From the relation

$$\sum_k A_k(x)C_k(x) = \sum_i A'_i(y)C'_i(y) = \sum_k \sum_j \frac{\partial \phi_k}{\partial y_j} A'_j(y)C_k(x)$$

we obtain

$$\sum_i A'_i(y)\left[C'_i(y) - \sum_k \frac{\partial \phi_k}{\partial y_i} C_k(x)\right] = 0.$$

Hence, $C_i(x)$ is a contragradient.

Next consider a quadratic differential form

$$ds^2 = \sum_{i,j}^{1,\ldots,n} g_{ij}\, dx_i\, dx_j, \qquad g_{ij} = g_{ji}. \tag{13}$$

Let the result of substitution

$$y_i = f_i(x_1, x_2, \ldots, x_n) \qquad (i = 1, 2, \ldots, n)$$

be

$$ds^2 = \sum_{i,j}^{1,\ldots,n} g'_{i,j}\, dy_i\, dy_j.$$

Then

$$ds^2 = \sum_{i,j} \sum_{\alpha,\beta} g'_{i,j} \frac{\partial f_i}{\partial x_\alpha} \frac{\partial f_j}{\partial x_\beta}\, dx_\alpha\, dx_\beta = \sum_{\alpha,\beta} \sum_{i,j} g'_{\alpha,\beta} \frac{\partial f_\alpha}{\partial x_i} \frac{\partial f_\beta}{\partial x_j}\, dx_i\, dx_j,$$

and hence

$$g_{ij} = \sum_{\alpha,\beta} g'_{\alpha,\beta} \frac{\partial f_\alpha}{\partial x_i} \frac{\partial f_\beta}{\partial x_j}.$$

Thus, $g_{ij}$ is a contragradient. Denote by $g^{ji}$ the minor of $g_{ij}$ in the determinant $\|g_{ij}\|$; then

$$\sum_j g^{ij} g_{jk} = \delta_{ik}, \qquad g^{ij} = g^{ji}.$$

By the quotient theorem, $g^{ij}$ is a cogradient.

Let

$$T = \frac{1}{2} \sum_{i,j} g_{ij}\dot{q}_i\dot{q}_j$$

in the Lagrangian function and let $Q_i$ be a component of the external force. Then

$$\frac{d}{dt} \sum_{j} g_{ij}\dot{q}_j - \frac{1}{2} \sum_{j,k} \frac{\partial g_{jk}}{\partial q_i} \dot{q}_j\dot{q}_k = Q_i.$$

If we write

$$[{}^{j\,k}_{\ i}] = \frac{1}{2} \left[\frac{\partial g_{ij}}{\partial q_k} + \frac{\partial g_{ik}}{\partial q_j} - \frac{\partial g_{jk}}{\partial q_i}\right],$$

which is Christoffel's symbol of the first kind, then

$$\sum_{j} g_{ij}\ddot{q}_j + \sum_{j,k} [{}^{j\,k}_{\ i}]\dot{q}_j\dot{q}_k = Q_i.$$

We may also write

$$\{{}^{j\,k}_{\ i}\} = \sum_{k} g^{hi}[{}^{j\,k}_{\ h}],$$

which is Christoffel's symbol of the second kind. Then

$$\ddot{q}_i = -\sum_{j,k} \{{}^{j\,k}_{\ i}\}\dot{q}_j\dot{q}_k + Q^i, \tag{14}$$

where $Q^i = \sum_j g^{ij}Q_j$. If the external force $Q_i$ is zero, then (14) represents the geodesic (Ricci and Levi-Civita, 1900; Levi-Civita, 1923, 1925). We have

$$\frac{d^2q_i}{ds^2} + \sum_{j,k} \{{}^{j\,k}_{\ i}\} \frac{dq_j}{ds}\frac{dq_k}{ds} = 0 \tag{14a}$$

in the Riemannian space described by equations (13). If there exists an integral $F(\dot{q}_1, \dot{q}_2, \ldots, \dot{q}_n; q_1, q_2, \ldots, q_n; t) =$ constant for (14), then

$$\frac{\partial F}{\partial t} + \sum_{i} \left[\frac{\partial F}{\partial q_i}\dot{q}_i - \frac{\partial F}{\partial \dot{q}_i}\left(Q^i - \sum_{j,k} \{{}^{j\,k}_{\ i}\}\dot{q}_j\dot{q}_k\right)\right] = 0.$$

The problem of determining the $Q^i$ which satisfy this equation identically when $T$ is given, is called *Bertrand's problem*.

Take

$$L = \frac{1}{2}\sum_{i,j} g_{ij}\dot{q}_i\dot{q}_j - V(q_1, q_2, \ldots, q_n), \qquad \|g_{ij}\| \neq 0.$$

Then, by our definition,

$$p_i = \frac{\partial L}{\partial \dot{q}_i} = \sum_j g_{ij}\dot{q}_j,$$

or

$$\dot{q}_j = \sum_i g^{ji}p_i.$$

But

$$\sum_{i,j} g_{ij}\dot{q}_i\dot{q}_j = \sum_{i,j}\sum_{\alpha,\beta} g_{ij}g^{i\alpha}g^{j\beta}p_\alpha p_\beta = \sum_{\alpha,\beta} g^{\alpha\beta}p_\alpha p_\beta.$$

Hence

$$H = \sum_i p_i\dot{q}_i - L = \sum_{i,j} g_{ij}\dot{q}_i\dot{q}_j - L$$

$$= \frac{1}{2}\sum_{i,j} g_{ij}\dot{q}_i\dot{q}_j + V = T + V = \frac{1}{2}\sum_{\alpha,\beta} g^{\alpha\beta}p_\alpha p_\beta + V.$$

The principle of least action of Maupertuis (Poincaré, 1893), when $H = h$, is

$$\delta\int_{t_0}^{t_1}\left(\sum_i \dot{q}_i\frac{\partial L}{\partial \dot{q}_i}\right)dt = 0,$$

or

$$\delta\int_{t_0}^{t_1} 2T\,dt = 0.$$

We may eliminate $t$ from this by noting

$$\left(\frac{ds}{dt}\right)^2 = \sum_{i,j} g_{ij}\dot{q}_i\dot{q}_j = 2T, \tag{15}$$

that is, $dt = ds/\sqrt{2T}$; then

$$\delta\int_{s_0}^{s_1}\sqrt{2T}\,ds = 0,$$

or, with $U = -V$,

$$\delta\int_{s_0}^{s_1}\sqrt{2(U+h)}\,ds = 0. \tag{16}$$

The variable $s$ is the arc length defined by

$$ds^2 = \sum_{i,j} g_{ij}\,dq_i\,dq_j$$

in the $n$-dimensional Riemannian space of equations (13). If we fix $t_0$ and $t_1$ in this space, the extremal of the integral of $2\sqrt{(U + h)}$ — which is a function of $q_1, q_2, \ldots, q_n$ over the arc length — represents the actual motion. This expression appears in Jacobi's lectures (Jacobi, 1884) on dynamics and is the foundation of the geometry of motion (Synge, 1926).

As a special case, consider the motion of a particle not subject to any external force. Then $U = 0$ and the variational principle is $\delta \int ds = 0$, that is, the geodesic line of the shortest arc length on the Riemannian space defined by the above $ds^2$. However, this theorem holds only in limited intervals; for example, the arc length larger than $\pi$ of a great circle on a spherical surface is not the shortest distance. The use of a minimum is limited to the comparison among all nearby curves. Let $B$ be the pole of a point $A$ (Figure 4). The great circles between $A$ and $B$ are all of the same arc length. That is,

$$A\alpha BC = A\beta B + BC = A\beta + \beta B + BC.$$

Let $\beta$ be a nearby point of $B$ and draw a great circle $\beta C$. Now

$$\beta C < \beta B + BC.$$

Hence, $A\beta + \beta C <$ the great circle $A\alpha BC$. Thus, there occur two cases when we attempt to draw the shortest paths in each direction from a point on the surface.

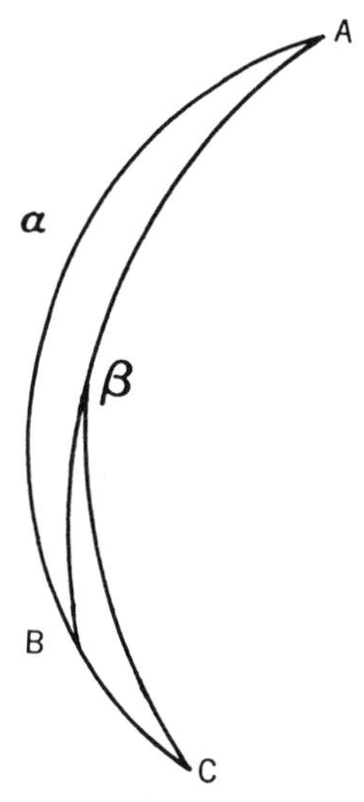


FIGURE 4

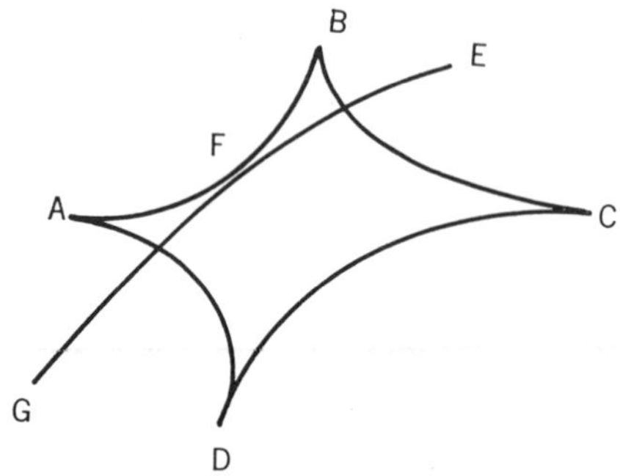


FIGURE 5

Two nearby shortest curves either remain close to each other but never intersect, or they do intersect; in the latter case the shortest curves have an envelope. The first case occurs on a developable surface or on a concave-convex surface—for example, on a hyperboloid of one sheet or on a hyperbolic paraboloid. The second case occurs, for instance, on an ellipsoid of revolution. In cases of the latter kind, the shortest curves starting from the same point on a surface do not intersect in the same way as do those on the surface of a sphere. They form an envelope around the pole of the point. It looks as though there were no shortest curve inside the envelope. But there is at least a curve something like *EFG* of Figure 5. The curve *EFG* does not represent the shortest curve beyond the point *F* (Jacobi, 1884). The works of Hadamard, Poincaré, and Birkhoff in the topological study of a dynamical system discuss the principle of least action geometrically by geodesic lines.

Now put

$$d\tau = U\frac{ds}{\sqrt{2T}}, \quad \text{or} \quad ds = \frac{d\tau\sqrt{2T}}{U}, \quad \text{or} \quad d\tau = U\,dt.$$

Then the principle of the least action takes the form

$$\delta\int_{t_0}^{t_1}\sqrt{2(U+h)T}\,dt = 0, \tag{17}$$

or

$$\delta\int_{\tau_0}^{\tau_1}\sqrt{2\left(1+\frac{h}{U}\right)\frac{T}{U}}\,d\tau = 0.$$

Write

$$\mathfrak{H} = 1, \qquad \mathfrak{U} = \frac{h}{U}, \qquad \mathfrak{T} = \frac{T}{U} = \frac{1}{2}\sum_{i,j} Ug_{ij}\frac{dq_i}{d\tau}\frac{dq_j}{d\tau}.$$

This is the case in which we take $\tau$ in place of $t$. The variable $\tau$ is that employed by Sundman for the regularization of the three-body problem. The Lagrangian function is now $\mathfrak{L} = \mathfrak{T} + \mathfrak{U}$. If we put

$$r_i = \frac{\partial \mathfrak{L}}{\partial(dq_i/d\tau)},$$

Then we may write

$$\mathfrak{H} = \frac{1}{2}\sum_{i,j} Ug_{ij}\frac{dq_i}{d\tau}\frac{dq_j}{d\tau} - \frac{h}{U} = \frac{1}{2}\sum_{i,j}\frac{g^{ij}}{U}r_ir_j - \frac{h}{U},$$

and the Hamiltonian equations become

$$\frac{dq_i}{d\tau} = \frac{\partial\mathfrak{H}}{\partial r_i}, \qquad \frac{dr_i}{d\tau} = -\frac{\partial\mathfrak{H}}{\partial q_i}.$$

Here, $\mathfrak{H} = 1$. Suppose that $V$ is a homogeneous form of degree $-1$ in $q$, and that $g^{ij}$ is homogeneous of degree zero in $q$. This is the case of the three-body problem. Then,

$$\sum_i q_i\frac{\partial\mathfrak{H}}{\partial q_i} = \mathfrak{H} = 1, \qquad \sum_i r_i\frac{\partial\mathfrak{H}}{\partial r_i} = 2\mathfrak{H} = 2.$$

From the Hamiltonian equations we obtain

$$\sum_i q_i\frac{dr_i}{d\tau} = -1, \qquad \sum_i r_i\frac{dq_i}{d\tau} = 2.$$

Hence, by integrating $d(\sum_i q_ir_i)/d\tau = 1$, we obtain

$$U\cdot\sum_{i,j} g_{ij}q_i\frac{dq_j}{d\tau} = \tau - \tau_0,$$

or

$$\sum_{i,j} g_{ij}q_i\frac{dq_j}{dt} = \int_{t_0}^{t} U\,dt.$$

This formula has been deduced by Dziobek (1900). Rewriting this, we put

$$\sum_i p_iq_i = \int_{t_0}^{t} U\,dt.$$

This expression has been obtained by Wintner (1936).

When $L$ contains linear terms in $\dot{q}_i$, that is, in the case of an irreversible dynamical system, we write

$$L = \frac{1}{2} \sum_{i,j} g_{ij}\dot{q}_i\dot{q}_j + \sum_i a_i\dot{q}_i - V(q_1, q_2, \ldots, q_n),$$

and assume that $g_{ij}$ and $a_i$ are functions of $q_1, q_2, \ldots, q_n$. Then the energy integral becomes

$$H = \sum_i p_i\dot{q}_i - L = \sum_i \frac{\partial L}{\partial \dot{q}_i}\dot{q}_i - L$$

$$= \frac{1}{2} \sum_{i,j} g_{ij}\dot{q}_i q_j + V(q_1, q_2, \ldots, q_n) = h,$$

and we see that the linear terms in $\dot{q}_i$ do not contribute to the energy; that is, the linear terms represent only the Coriolis force. In such a case,

$$\frac{\partial L}{\partial \dot{q}_i} = p_i = \sum_j g_{ij}\dot{q}_j + a_i, \qquad \dot{q}_j = \sum_k (p_k - a_k)g^{kj},$$

and

$$L = \frac{1}{2} \sum_{\alpha,\beta} g^{\alpha\beta}p_\alpha p_\beta - \frac{1}{2} \sum_{\alpha,\beta} g^{\alpha\beta}a_\alpha a_\beta - V,$$

$$H = \frac{1}{2} \sum_{\alpha,\beta} g^{\alpha\beta}p_\alpha p_\beta - \sum_i a^i p_i + V + \frac{1}{2} \sum_{\alpha,\beta} g^{\alpha\beta}a_\alpha a_\beta,$$

$$T = \frac{1}{2} \sum_{i,j} g_{ij}\dot{q}_i\dot{q}_j = \frac{1}{2} \sum_{\alpha,\beta} g^{\alpha\beta}(p_\alpha - a_\alpha)(p_\beta - a_\beta),$$

where

$$\sum_i a_i g^{ik} = a^k.$$

## 1.5 GEOMETRY OF MOTION

The principle of least action in $n$-dimensional Riemannian space (15) has been expressed in the form (16). A problem of dynamics is thus reduced to the study of the trajectories in the space (15) by the variational principle (16). Synge (1926) called the line element $ds$ given by (15) the *kinematical line element.* The line element $ds$ determined by the variational principle (17)

$$ds^2 = 2(h - V)T, \qquad dt^2 = (h - V) \sum_{i,j} g_{ij}\, dq_i\, dq_j$$

is called the *action line element*. We can also consider the $(n + 1)$-dimensional Riemannian space

$$ds^2 = 2L\,dt^2 = \sum_{i,j} g_{ij}\,dq_i\,dq_j - 2V\,dt^2$$

and the $(n + 1)$-dimensional Finsler space (Cartan, 1934). Any problem of dynamics can be related to suitably chosen curves.

Consider a kinematical space defined by (15). Let $C$ and $C^*$ be two curves touching mutually at some point $O$, and $P$ and $P^*$ be two points on each of these curves such that $OP = OP^* = s$. The expression

$$\kappa(C, C^*) = \lim_{s \to 0} 2\,\frac{\overline{PP^*}}{s^2}$$

is called the relative curvature of $C$ and $C^*$, where $\overline{PP^*}$ is the shortest distance between the two points. If $C^*$ is a geodesic line, then $\kappa$ gives the curvature of $C$. Put

$$\kappa_i = \frac{d^2 q_i}{ds^2} + \sum_{j,k} \left\{ {j\,k \atop i} \right\} \frac{dq_j}{ds}\frac{dq_k}{ds};$$

then $\kappa_i$ indicates the direction of the principal normal of the curve, and

$$\kappa^2 = \sum_{i,j} g_{ij}\kappa_i\kappa_j.$$

The relative curvature, if we define $\kappa^*$ similarly with respect to $C^*$, is

$$[\kappa(C, C^*)]^2 = \sum_{i,j} g_{ij}(\kappa_i - \kappa_i^*)(\kappa_j - \kappa_j^*).$$

The velocity vector $v_i = dq_i/dt$ is represented by

$$v = \left(\sum_{i,j} g_{ij}\,\frac{dq_i}{dt}\frac{dq_j}{dt}\right)^{1/2} = \frac{ds}{dt} = \sqrt{2T}\,.$$

Define

$$X_{/i} = \frac{dX_i}{dt} + \sum_{j,k} \left\{ {j\,k \atop i} \right\} X_j \frac{dq_k}{dt}$$

as the covariant derivative of $X_i$; then,

$$f_i = v_{/i} = \frac{dv_i}{dt} + \sum_{j,k} \left\{ {j\,k \atop i} \right\} v_j v_k = \frac{d^2 q_i}{dt^2} + \sum_{j,k} \left\{ {j\,k \atop i} \right\} \frac{dq_j}{dt}\frac{dq_k}{dt},$$

or

$$f_i = \frac{d^2 s}{dt^2}\frac{dq_i}{ds} + \left(\frac{ds}{dt}\right)^2 \kappa_i,$$

that is, the acceleration vector is on the same plane as the tangent $dq_i/ds$ and the principal normal $\kappa_i$, and is expressed by

$$f = \left(\sum_{i,j} g_{ij} f_i f_j\right)^{1/2},$$

or by

$$f_i = \frac{dT}{ds}\frac{dq_i}{ds} + 2\kappa_i T,$$

since $ds = \sqrt{2T}\cdot dt$. The equations of motion $f_i = Q_i$ are linear with respect to $T$ and $Q_i$. Hence, if several external forces act on a holonomic dynamical system simultaneously, the kinetic energy of the system equals the sum of kinetic energies for the dynamical system acted on by each single external force separately. This is called *Bonnet's theorem* (Bianchi, 1899, 1922).

If the trajectories in question are geodesics, then $\kappa_i = 0$, and the vector $f_i$ lies in the tangential direction. Hence, if $Q_i$ has a component only in the direction tangential to the trajectory, then and only then does the motion take place on a geodesic.

Let $C^*$ be the actual trajectory. Let $C$ and $C^*$ be in contact at a point $P$. Then

$$4T^2(\kappa_i - \kappa_i^*) = 2T(f_i - f_i^*) - \left(\frac{dT}{dt} - \frac{dT^*}{dt}\right)\frac{dq_i}{dt}.$$

By substituting in the expression for $\kappa(C, C^*)$ we obtain

$$8T^3[\kappa(C, C^*)]^2 = 2T\sum_{j,k} g_{jk}(f_j - Q_j)(f_k - Q_k),$$

since

$$f_i^* = Q_i, \qquad \frac{dT}{dt} = \sum_j \frac{dq_j}{dt} Q_j.$$

Consider three curves, all with the same velocity vector at a point $R$ and all touching at that point. Let the first curve be an arbitrary trajectory $C$ satisfying the condition of constraint, let the second be the actual trajectory $C^*$ satisfying the condition of constraint, and let the third be the actual trajectory $C^{**}$ in the case of no constraint. The condition for conservation of energy,

$$\frac{dT}{dt} = \sum_j Q_j \frac{dq_j}{dt},$$

is satisfied for $C^*$. Let the equations of motion be

$$f_i^* = Q_i + P_i$$

and the condition of constraint be

$$\sum_i P_i \, \delta q_i = 0.$$

Then

$$\sum_{j,k} g_{jk} f_j^* \, \delta q_k = \sum_{j,k} g_{jk}(Q_j + P_j) \, \delta q_k$$
$$= \sum_{j,k} g_{jk} Q_j \, \delta q_k.$$

Denote an arbitrary infinitesimal displacement by $\delta\tau$, and let $\delta q_i = (f_i - f_i^*) \, \delta\tau$ satisfy the condition of constraint; then

$$\sum_{j,k} g_{jk} f_j^*(f_k - f_k^*) = \sum_{j,k} g_{jk} Q_j (f_k - f_k^*).$$

Thus

$$8T^3\{[\kappa(C, C^{**})]^2 - [\kappa(C^*, C^{**})]^2\}$$
$$= 2T \sum_{j,k} g_{jk}[f_j f_k - f_j^* f_k^* - 2(f_j - f_j^*) Q_k]$$
$$- \left(\frac{dT}{dt} - \sum_{j,k} g_{jk} Q_j \frac{dq_k}{dt}\right)^2$$
$$= 2T \sum_{j,k} g_{jk}(f_j - f_j^*)(f_k - f_k^*)$$
$$- \left(\frac{dT}{dt} - \sum_{j,k} q_{jk} Q_j \frac{dq_k}{dt}\right)^2.$$

If the conservation of energy is satisfied also for $C$, the last term vanishes and the right-hand member is positive. Hence

$$\kappa(C, C^*) > \kappa(C^*, C^{**}).$$

Thus, if the constraint is imposed on a holonomic dynamical system, the actual trajectory has the smallest curvature of all curves with the same tangent that satisfy the condition of constraint. This is called the *principle of smallest curvature.*

### 1.6 EQUIVALENCE OF DYNAMICAL SYSTEMS

Consider two dynamical systems represented, respectively, by the following Lagrangian equations

$$\frac{d}{dt}\left(\frac{\partial T}{\partial \dot{q}_i}\right) - \frac{\partial T}{\partial q_i} = Q_i \qquad (i = 1, 2, \ldots, n) \tag{18}$$

$$\frac{d}{dt}\left(\frac{\partial \mathfrak{T}^*}{\partial \dot{\mathfrak{q}}_i}\right) - \frac{\partial \mathfrak{T}}{\partial \mathfrak{q}_i} = \mathfrak{Q}_i^* \qquad (i = 1, 2, \ldots, n). \tag{19}$$

If these two equations can be brought to the same form by the transformation

$$\mathfrak{q}_i = \phi_i(q_1, q_2, \ldots, q_n) \qquad (i = 1, 2, \ldots, n),$$

then the two dynamical systems are said to be *equivalent* according to Stäckel (1891).

Suppose that the tensors $g_{ij}$ and $\mathfrak{g}_{ij}^*$ in

$$T = \frac{1}{2}\sum_{i,j} g_{ij}\dot{q}_i\dot{q}_j, \qquad \mathfrak{T}^* = \frac{1}{2}\sum_{i,j} \mathfrak{g}_{ij}^*\dot{\mathfrak{q}}_i\dot{\mathfrak{q}}_j$$

do not depend on $t$ but only on $q$ and $\mathfrak{q}$, respectively. Let

$$ds^2 = \sum_{i,j} g_{ij}\, dq_i\, dq_j, \qquad d\mathfrak{s}^{*2} = \sum_{i,j} \mathfrak{g}_{ij}^*\, d\mathfrak{q}_i\, d\mathfrak{q}_j,$$

and the result of substitution of $\mathfrak{q}_i = \phi_i(q_1, q_2, \ldots, q_n)$ in $\mathfrak{T}^*$ and $\mathfrak{Q}^*$ be $\mathfrak{T}$, $\mathfrak{Q}$, respectively, and denote them by

$$\mathfrak{T} = \frac{1}{2}\sum_{j,k} \mathfrak{g}_{jk}\dot{q}_j\dot{q}_k,$$

$$\mathfrak{Q}_i = \sum_j \mathfrak{Q}_j^* \frac{\partial \phi_j}{\partial q_i}, \qquad d\mathfrak{s}^2 = \sum_{j,k} \mathfrak{g}_{jk}\, dq_j\, dq_k;$$

then the Lagrangian equations take the form

$$\ddot{q}_\rho + \sum_{\lambda,\mu} \left\{{}^{\lambda}{}_{\rho}{}^{\mu}\right\}\dot{q}_\lambda\dot{q}_\mu = Q^\rho, \qquad Q^\rho = \sum_\sigma g^{\rho\sigma}Q_\sigma,$$

$$\ddot{q}_\rho + \sum_{\lambda,\mu} \left\{{}^{\lambda}{}_{\rho}{}^{\mu}\right\}^*\dot{q}_\lambda\dot{q}_\mu = \mathfrak{Q}^\rho, \qquad \mathfrak{Q}^\rho = \sum_\sigma \mathfrak{g}^{\rho\sigma}\mathfrak{Q}_\sigma,$$

Here $\left\{{}^{\lambda}{}_{\rho}{}^{\mu}\right\}^*$ is the Christoffel symbol for $d\mathfrak{s}^2$. In order that the two dynamical systems be equivalent we should have

$$\left\{{}^{\lambda}{}_{\rho}{}^{\mu}\right\} = \left\{{}^{\lambda}{}_{\rho}{}^{\mu}\right\}^*, \qquad Q^\rho = \mathfrak{Q}^\rho \qquad (\lambda, \mu, \rho = 1, 2, \ldots, n).$$

These equations are satisfied by $\mathfrak{g}_{ik} = cg_{ik}$ ($c$ = constant) and

$$\mathfrak{Q}_\rho = \sum_\lambda \mathfrak{g}_{\rho\lambda} Q^\lambda = \sum_\lambda \mathfrak{g}_{\rho\lambda} \sum_\mu g^{\lambda\mu} Q_\mu = cQ_\rho.$$

These are the conditions for the identity of the two dynamical systems in space and in time. However, if we consider only the orbital curves, the condition for the equivalence of two dynamical systems is different.

Put $\dot{\mathfrak{q}}_i = d\mathfrak{q}_i/d\tau$ in (18) and (19). Then (19) becomes

$$\frac{d}{d\tau}\left(\frac{\partial \mathfrak{T}}{\partial \dot{q}_i^*}\right) - \frac{\partial \mathfrak{T}}{\partial q_i} = \mathfrak{Q}_i, \qquad \dot{q}_i^* = \frac{dq_i}{d\tau},$$

$$\mathfrak{T} = \frac{1}{2}\left(\frac{d\mathfrak{s}}{d\tau}\right)^2 = \frac{1}{2}\sum_{\lambda,\mu} \mathfrak{g}_{\lambda\mu}\dot{q}_\lambda^*\dot{q}_\mu^*, \qquad \mathfrak{Q}_i = \mathfrak{Q}_i(q_1, \ldots, q_n).$$

Let

$$dt = \psi(s)\, ds, \qquad d\tau = \chi(\mathfrak{s})\, d\mathfrak{s}.$$

Let the arc length and the time correspond in the two equivalent orbital curves (Appell, 1892). Then

$$d\tau = \frac{dt}{f(q_1, q_2, \ldots, q_n, \dot{q}_1, \dot{q}_2, \ldots, \dot{q}_n)}.$$

This is a transformation of time. The simplest such transformation is $d\tau = dt/\sqrt{c}$. In particular, if $c = -1$, then $d\tau = -i\, dt$; such motions are called *conjugate*. Put $d\mathfrak{s}^2 = C\, ds^2$, $\mathfrak{Q}_i = cQ_i$; then $d\tau = \sqrt{C/c}\, dt$ and is what Painlevé (1894) called the *ordinary correspondence*. When

$$Q_i = -\frac{V(q_1, q_2, \ldots, q_n)}{q_i}, \qquad T + V = h,$$

then the orbit is a geodesic line with arc length $ds^* = \sqrt{(h - V)}\, ds$. Denote the corresponding quantities of the equivalent systems $(d\mathfrak{s}, \mathfrak{V})$ and $(ds, V)$ by

$$d\mathfrak{s} = \sqrt{\alpha V + \beta}\, ds, \qquad \mathfrak{V} = \frac{\gamma V + \delta}{\alpha V + \beta},$$

or

$$h = -\frac{\beta \mathfrak{h} - \delta}{\alpha \mathfrak{h} - \gamma};$$

then

$$d\mathfrak{s}^* = 2(\mathfrak{h} - \mathfrak{V})\, d\mathfrak{s} = \sqrt{\gamma - \alpha\mathfrak{h}}\, ds^*.$$

This Darboux's transformation expresses the ordinary correspondence between $(ds, V)$ and $(d\mathfrak{s}, \mathfrak{V})$ in those cases where the potential exists (Prange, 1935).

When there is no external force, the function $f$ in the transformation of time is

$$f = C\left(\frac{G}{\mathfrak{G}}\right)^{1/(n+1)},$$

where $G$ and $\mathfrak{G}$ are the determinants of $g$ and $\mathfrak{g}$, respectively. Then the condition for the correspondence between $ds^2$ and $d\mathfrak{s}^2$ is according to Levi-Civita (1896, 1936), and Datzeff (1936)

$$\left\{{r\,s \atop j}\right\}^* = \left\{{r\,s \atop j}\right\}, \qquad \left\{{r\,s \atop r}\right\}^* = \left\{{r\,s \atop r}\right\} - \frac{1}{2}\frac{\partial \log f}{\partial q_s},$$

$$\left\{{r\,r \atop r}\right\}^* = \left\{{r\,r \atop r}\right\} = \frac{\partial \log f}{\partial q_r}.$$

We have defined the covariant derivative of $X_i$ to be

$$X_{/i} = \frac{dX_i}{dt} + \sum_{i,k} \left\{{j\,k \atop i}\right\} X_j \frac{dq_k}{dt}.$$

Substituting in $ds^2$ the covariant derivative of $f^2\mathfrak{g}_{\lambda\mu}$, we obtain

$$(f^2\mathfrak{g}_{rs})_{/t} + (f^2\mathfrak{g}_{st})_{/r} + (f^2\mathfrak{g}_{tr})_{/s} = 0.$$

The integral for the geodesic line in the space with line element $ds$ is expressed by

$$\left(\frac{G}{\mathfrak{G}}\right)^{2/(n+1)} \sum_{\lambda,\mu} \mathfrak{g}_{\lambda\mu}\dot{q}_\lambda\dot{q}_\mu = \text{constant}.$$

We can discuss the case when an extremal force is acting on the system by writing

$$d\tau^2 = \frac{dt^2}{f^2(q_1, q_2, \ldots, q_n)}\left\{1 - \sum_{r,s} c_{rs}\dot{q}_r\dot{q}_s\right\}.$$

Thus, the existence of a quadratic integral in momentum shows the existence of a corresponding line element.

## CANONICAL TRANSFORMATIONS

### 1.7 BILINEAR COVARIANTS

A linear homogeneous expression of differentials

$$\pi_\alpha = \sum_{i=1}^{n} X_i(x_1, x_2, \ldots, x_n)\, dx_i$$

is called a *Pfaffian* (Weber, 1900; Goursat, 1922; Cartan, 1922, 1937; Schouten and v. d. Kulk, 1949). Consider a transformation

$$x_k = \phi_k(y_1, y_2, \ldots, y_n) \qquad (k = 1, 2, \ldots, n),$$

with the Jacobian

$$\partial(x_1, x_2, \ldots, x_n)/\partial(y_1, y_2, \ldots, y_n) \neq 0.$$

A Pfaffian is transformed to a Pfaffian. As

$$dx_k = \sum_i \frac{\partial \phi_k}{\partial y_i} dy_i,$$

we have

$$\sum_k X_k\, dx_k = \sum_k \sum_i X_k \frac{\partial \phi_k}{\partial y_i} dy_i.$$

Consider another differential $\delta x_j$; then,

$$\delta x_k = \sum_k \frac{\partial \phi_k}{\partial y_i} \delta y_i.$$

Furthermore, since

$$d\delta x_k = \sum_i \frac{\partial \phi_k}{\partial y_i} d\delta y_i + \sum_{i,j} \frac{\partial^2 \phi_k}{\partial y_i\, \partial y_j} dy_j\, \delta y_i,$$

$$\delta d x_k = \sum_i \frac{\partial \phi_k}{\partial y_i} \delta d y_i + \sum_{i,j} \frac{\partial^2 \phi_k}{\partial y_i\, \partial y_j} \delta y_j\, dy_i,$$

we obtain

$$d\delta x_k - \delta d x_k = \sum_i \frac{\partial \phi_k}{\partial y_i} (d\delta y_i - \delta d y_i).$$

Thus, $d\delta x_k - \delta dx_k$ varies cogradiently to $dx_k$ or $\delta x_k$. The terms $dx_k$ and $(d\delta x_k - \delta dx_k)$ are said to be *covariants*. The expression $d\delta x_k - \delta dx_k$ is linear in $d$ and $\delta$. Hence we call it a *bilinear covariant*. Let $f$ be a function of $x_1, x_2, \ldots, x_n$; then

$$\sum_k (d\delta x_k - \delta dx_k) \frac{\partial f}{\partial x_k} = \sum_i (d\delta y_i - \delta d y_i) \frac{\partial f}{\partial y_i}$$

is an *invariant*. Denote the Pfaffian by

$$\pi_d = \sum_i X_i\, dx_i = \sum_k Y_k\, dy_k; \tag{20}$$

then, by the above theorem,

$$\sum_k X_k(d\delta x_k - \delta dx_k) = \sum_i Y_i(d\delta y_i - \delta dy_i). \tag{21}$$

Applying the operator $\delta$ on (20), we obtain

$$\delta\pi_d = \sum_k \delta X_k\, dx_k + \sum_k X_k\, \delta dx_k = \sum_i \delta Y_i\, dy_i + \sum_i Y_i\, \delta dy_i.$$

Similarly,

$$d\pi_\delta = \sum_k dX_k\, \delta x_k + \sum_k X_k\, d\delta x_k = \sum_i dY_i\, \delta y_i + \sum_i Y_i\, d\delta y_i.$$

The difference of these two expressions is

$$d\pi_\delta - \delta\pi_d = \sum_k (dx_k\, \delta x_k - \delta X_k\, dx_k) = \sum_i (dY_i\, \delta y_i - \delta Y_i\, dy_i). \tag{21a}$$

If (20) exists between two Pfaffians, then we have (21a). In other words

$$\sum_{j,k} \left(\frac{\partial X_k}{\partial x_j} - \frac{\partial X_j}{\partial x_k}\right) dx_j\, \delta x_k = \sum_{i,l} \left(\frac{\partial Y_i}{\partial y_l} - \frac{\partial Y_l}{\partial y_i}\right) dy_l\, \delta y_j. \tag{22}$$

Either side of (22) is called a bilinear covariant of $\pi_d$. This has been studied by Lipschitz and Frobenius (see Weyl, 1939). Lie did not employ bilinear covariants, but Poincaré (1893) and Cartan (1922) discussed them with regard to integral invariants.

Suppose that $x_1, x_2, \ldots, x_n$ are independent and not connected by any functional relation. Then $dx_1, dx_2, \ldots, dx_n, \delta x_1, \delta x_2, \ldots, \delta x_n$ are mutually independent differentials and so are $dy_1, dy_2, \ldots, dy_n, \delta y_1, \delta y_2, \ldots, \delta y_n$. If the differentials $\delta$ are all arbitrary, then the necessary and sufficient condition for

$$\delta\pi_d - d\pi_\delta = \sum_{i,j} \left(\frac{\partial X_i}{\partial x_j} - \frac{\partial X_j}{\partial x_i}\right) dx_i\, \delta x_j = 0$$

is that

$$\psi_j \equiv \sum_i \left(\frac{\partial X_i}{\partial x_j} - \frac{\partial X_j}{\partial x_i}\right) dx_i = 0. \tag{23}$$

This expression is called the *associated Pfaffian* to the Pfaffian $\sum_i X_i\, dx_i$. Similarly, if the Jacobian $\partial(x_1, x_2, \ldots, x_n)/\partial(y_1, y_2, \ldots, y_n) \neq 0$, then the differentials $\delta$ are arbitrary, and the necessary and sufficient condition for

$$\delta\pi_d - d\pi_\delta = \sum_{i,j} \left( \frac{\partial Y_i}{\partial y_j} - \frac{\partial Y_j}{\partial y_i} \right) dy_i \, \delta y_j = 0$$

is that

$$\psi_j \equiv \sum_i \left( \frac{\partial Y_i}{\partial y_j} - \frac{\partial Y_j}{\partial y_i} \right) dy_i = 0. \tag{23a}$$

The necessary and sufficient condition for the Pfaffian $\sum_i X_i \, dx_i$ to be a perfect differential is, as we shall immediately see,

$$\frac{\partial X_i}{\partial x_j} - \frac{\partial X_j}{\partial x_i} = 0 \qquad (i, j = 1, 2, \ldots, n).$$

Hence, the necessary and sufficient condition for the Pfaffian $\sum_i X_i \, dx_i$ to be a perfect differential is $\psi_j = 0$, or $\delta\pi_d - d\pi_\delta = 0$.

*Theorem:* The necessary and sufficient condition for a Pfaffian to be a perfect differential is that the bilinear covariants are zero.

*Corollary:* The bilinear covariant of a Pfaffian which is a perfect differential is zero.

We now prove the theorem. The necessary and sufficient condition for a Pfaffian $\sum_i X_i \, dx_i$ to be a perfect differential is that

$$\frac{\partial X_i}{\partial x_j} - \frac{\partial X_j}{\partial x_i} = 0 \qquad (i, j = 1, 2, \ldots, n). \tag{24}$$

If $\sum_i X_i \, dx_i$ is a perfect differential $dZ$, then $X_i = \partial Z / \partial x_i$ and hence

$$\frac{\partial X_i}{\partial x_j} = \frac{\partial^2 Z}{\partial x_i \, \partial x_j} = \frac{\partial X_j}{\partial x_i}.$$

Thus, the condition is necessary. In order to prove that the condition is sufficient, that is, that $\sum_i X_i \, dx_i$ is a perfect differential if (24) is true, we carry out Mayer's transformation

$$x_1 = x_1^0 + \lambda_1, \qquad x_i = x_i^0 + \lambda_1 \lambda_i \qquad (i = 1, 2, \ldots, n).$$

Then $dx_1 = d\lambda_1$, $dx_i = \lambda_i \, d\lambda_1 + \lambda_1 \, d\lambda_i$. Denote the result of this transformation in $X_i$ by $[X_i]$; then,

$$\sum_{i=1}^{n} X_i \, dx_i = \lambda_1 \sum_{j=2}^{n} [X_j] \, d\lambda_j + \sum_{j=2} \lambda_j [X_j] \, d\lambda_1 + [X_1] \, d\lambda_1.$$

As the associated Pfaffian is zero, by our assumption, so is the right-hand side of this last equation. Hence

$$-\frac{\partial}{\partial\lambda_i}\left(\sum_{j=2}^{n}\lambda_j[X_j] + [X_1]\right) + \frac{\partial}{\partial\lambda_1}\left(\lambda_1\sum_{i=2}^{n}[X_i]\right) = 0 \quad (i, j = 2, 3, \ldots, n),$$

that is,

$$[X_i] + \lambda_1\frac{\partial[X_i]}{\partial\lambda_1} - [X_i] - \sum_{j=2}^{n}\lambda_j\frac{\partial[X_j]}{\partial\lambda_i} - \frac{\partial[X_1]}{\partial\lambda_i} = 0 \quad (i = 2, 3, \ldots, n).$$

Thus

$$\lambda_1\frac{\partial[X_i]}{\partial\lambda_1} = \sum_{j=2}^{n}\lambda_j\frac{\partial[X_j]}{\partial\lambda_i} + \frac{\partial[X_1]}{\partial\lambda_i} \qquad (i = 2, 3, \ldots, n). \tag{25}$$

To prove that $\sum_i X_i\, dx_i$ is a perfect differential, it is sufficient to show that there exists a function $U$ such that $\partial U/\partial\lambda_1 = \sum_{j=2}^{n}\lambda_j[X_j] + [X_1]$, that is,

$$U = \int_0^{\lambda_1}\left(\sum_{j=2}^{n}\lambda_j[X_j] + [X_1]\right)d\lambda_1,$$

or

$$\frac{\partial U}{\partial\lambda_i} = \int_0^{\lambda_1}\left\{\sum_{j=2}^{n}\lambda_j\frac{\partial}{\partial\lambda_i}[X_j] + [X_i] + \frac{\partial[X_1]}{\partial\lambda_i}\right\}d\lambda_1 \qquad (i = 2, 3, \ldots, n),$$

or, by (25),

$$\frac{\partial U}{\partial\lambda_i} = \int_0^{\lambda_1}\left\{[X_i] + \lambda_1\frac{\partial}{\partial\lambda_1}[X_i]\right\}d\lambda_1 = \lambda_1[X_i] \qquad (i = 2, 3, \ldots, n).$$

We now have

$$\begin{aligned} dU &= \lambda_1\sum_{i=2}^{n}[X_i]\,d\lambda_i + \sum_{j=2}^{n}\lambda_j[X_j]\,d\lambda_1 + [X_1]\,d\lambda_1 \\ &= \sum_{i=2}^{n}[X_i](\lambda_1\,d\lambda_i + \lambda_i\,d\lambda_1) + [X_1]\,d\lambda_1 \\ &= \sum_{i=1} X_i\,d(x_i - x_i^0). \end{aligned}$$

Hence

$$\sum_{i=1}^{n} X_i\,dx_i = dU(x_1 - x_1^0, x_2 - x_2^0, \ldots, x_n - x_n^0),$$

a perfect differential.

## 1.8 CONTACT TRANSFORMATIONS

A transformation from $(q_1, q_2, \ldots, q_n; p_1, p_2, \ldots, p_n)$ to

$$(Q_1, Q_2, \ldots, Q_n; P_1, P_2, \ldots, P_n)$$

is called a *contact transformation* if it makes $\sum_{i=1}^{n} P_i\, dQ_i - \sum_{i=1}^{n} p_i\, dq_i$, expressed as a function of $(q_1, q_2, \ldots, q_n; p_1, p_2, \ldots, p_n)$, a perfect differential of a function of $(q_1, q_2, \ldots, q_n; p_1, p_2, \ldots, p_n)$.

For a contact transformation there exists a perfect differential such that

$$\sum_i (P_i\, dQ_i - p_i\, dq_i) = dW. \tag{26}$$

In other words, the difference of two differentials $\sum_i P_i\, dQ_i$ and $\sum_j p_j\, dq_j$ is a perfect differential. The necessary and sufficient condition for a Pfaffian to be a perfect differential is that the bilinear covariant is zero. The condition is expressed by taking the bilinear covariant of (26) as

$$\sum_i (\delta P_i\, dQ_i - dP_i\, \delta Q_i) = \sum_i (\delta p_i\, dq_i - dp_i\, \delta q_i). \tag{27}$$

Hence we obtain the following

*Theorem:* $\sum_i (\delta p_i\, dq_i - dp_i\, \delta q_i)$ is invariant by a contact transformation.

Conversely, if the relation (27) is given, the bilinear covariant $\sum_i(\delta P_i\, dQ_i - dP_i\, \delta Q_i) - \sum_i (\delta p_i\, dq_i - dp_i\, \delta q_i)$, when transformed as a function of $(q_i, p_i)$ by the transformation $(Q_i, P_i) \to (q_i, p_i)$, is zero. Hence the difference $\sum_i P_i\, dQ_i - \sum_i p_i\, dq_i$ of the Pfaffians corresponding to each member of (27) must be a perfect differential. Therefore, this transformation is a contact transformation.

*Theorem:* The necessary and sufficient condition for a contact transformation is that $\sum_i (\delta p_i\, dq_i - dp_i\, \delta q_i)$ be an invariant.

The most general form of a contact transformation is

$$dZ - \sum_i P_i\, dQ_i = \rho(z, p_i, q_i) \cdot \left(dz - \sum_i p_i\, dq_i\right), \qquad \rho \neq 0,$$

as discussed by Lie. Goursat (1923) called this an *inhomogeneous* contact transformation, and the form just discussed a *homogeneous* contact transformation. An inhomogeneous contact transformation, if we put

$$z = q_{n+1}, \qquad P_{n+1} = \frac{p_{n+1}}{\rho}, \qquad -p_i p_{n+1} = \bar{p}_i \qquad (i = 1, 2, \ldots, n),$$

$$Z = Q_{n+1}, \qquad -P_i P_{n+1} = \bar{P}_i \qquad (i = 1, 2, \ldots, n),$$

can be reduced to a homogeneous contact transformation

$$\sum_{i=1}^{n} \bar{P}_i\, dQ_i + P_{n+1}\, dQ_{n+1} = \sum_{i=1}^{n} \bar{p}_i\, dq_i + p_{n+1}\, dq_{n+1}.$$

Hence, an inhomogeneous contact transformation can be treated as a special case of a homogeneous contact transformation.

## 1.9 LAGRANGE'S AND POISSON'S BRACKETS

Supposing that $(q_1, q_2, \ldots, q_n; p_1, p_2, \ldots, p_n)$ are functions of $u$ and $v$; we call the expression

$$[u, v] = \sum_i \left\{ \frac{\partial q_i}{\partial u}\frac{\partial p_i}{\partial v} - \frac{\partial q_i}{\partial v}\frac{\partial p_i}{\partial u} \right\} \tag{28}$$

the *Lagrangian bracket*. If $u = v$, then $[u, v] \equiv 0$. It is also obvious that $[u, v] = -[v, u]$. Carry out the transformation $(q_i, p_i) \to (Q_i, P_i)$ in the bilinear covariant $\sum_i (\delta p_i\, dq_i - dp_i\, \delta q_i)$. Then

$$\begin{aligned}
\sum_i (\delta p_i\, dq_i - dp_i\, \delta q_i) &= \frac{1}{2} \sum_{j,k} [Q_j, Q_k](\delta Q_k\, dQ_j - dQ_k\, \delta Q_j) \\
&\quad + \frac{1}{2} \sum_{j,k} [Q_j, P_k](\delta P_k\, dQ_j - dP_k\, \delta Q_j) \\
&\quad + \frac{1}{2} \sum_{j,k} [P_j, Q_k](\delta Q_k\, dP_j - dQ_k\, \delta P_j) \\
&\quad + \frac{1}{2} \sum_{j,k} [P_j, P_k](\delta P_k\, dP_j - dP_k\, \delta P_j).
\end{aligned}$$

The necessary and sufficient condition for $(q_i, p_i) \to (Q_i, P_i)$ to be a contact transformation is equality of the bilinear covariants:

$$\sum_i (\delta p_i\, dq_i - dp_i\, \delta q_i) = \sum_j (\delta P_j\, dQ_j - dP_j\, \delta Q_j).$$

Comparing these two equations with arbitrary $d$ and $\delta$ leads to the following

*Theorem:* The necessary and sufficient condition for a contact transformation is

$$[Q_i, P_j] = \delta_{ij}, \qquad [P_i, P_j] = 0, \qquad [Q_i, Q_j] = 0,$$
$$[P_i, Q_j] = -\delta_{ij} \qquad (i, j = 1, 2, \ldots, n). \tag{29}$$

The expression

$$(u, v) = \sum_i \left( \frac{\partial u}{\partial q_i} \frac{\partial v}{\partial p_i} - \frac{\partial u}{\partial p_i} \frac{\partial v}{\partial q_i} \right) \tag{30}$$

is called the *Poisson bracket*. We see that $(u, u) = 0$ and $(u, v) = -(v, u)$. Let $(u_1, \ldots, u_n, u_{n+1}, \ldots, u_{2n})$ be mutually independent functions of $(q_i, p_i)$, such that

$$\partial(u_1, \ldots, u_n, u_{n+1}, \ldots, u_{2n}) / \partial(q_1, \ldots, q_n, p_1, \ldots, p_n) \neq 0.$$

Then, conversely, $(q_i, p_i)$ are mutually independent functions of $(u_1, \ldots, u_n, u_{n+1}, \ldots, u_{2n})$. Since

$$\sum_{\alpha=1}^{2n} \frac{\partial q_j}{\partial u_\alpha} \frac{\partial u_\alpha}{\partial q_i} = \delta_{ij}, \qquad \sum_{\alpha=1}^{2n} \frac{\partial p_j}{\partial u_\alpha} \frac{\partial u_\alpha}{\partial p_i} = \delta_{ij},$$
$$\sum_{\alpha=1}^{2n} \frac{\partial p_j}{\partial u_\alpha} \frac{\partial u_\alpha}{\partial q_i} = 0, \qquad \sum_{\alpha=1}^{2n} \frac{\partial q_j}{\partial u_\alpha} \frac{\partial u_\alpha}{\partial p_i} = 0 \qquad (i, j = 1, 2, \ldots, n),$$

we can prove, by adding the partial products of (28) and (30), that

$$\sum_{\alpha=1}^{2n} (u_\alpha, u_r)[u_\alpha, u_s] = \sum_i \left( \frac{\partial u_r}{\partial p_i} \frac{\partial p_i}{\partial u_s} + \frac{\partial u_r}{\partial q_i} \frac{\partial q_i}{\partial u_s} \right) = \delta_{rs}.$$

Hence, the product of the determinant formed of $(u_i, u_j)$ and the determinant formed of $[u_i, u_j]$ is unity. A matrix in which the rows and columns are exchanged is called a *transposed matrix* of the original matrix. The matrix formed of the Lagrangian brackets, and the transposed matrix of the matrix formed of the Poisson brackets, are *reciprocal*. Since the determinant formed of $[u_i, u_j]$ is nonzero, reciprocity leads to the

*Theorem:* The necessary and sufficient condition for a contact transformation is

$$(Q_i, P_j) = \delta_{ij}, \qquad (P_i, P_j) = 0, \qquad (Q_i, Q_j) = 0,$$
$$(P_i, Q_j) = -\delta_{ij} \qquad (i, j = 1, 2, \ldots, n). \tag{31}$$

To prove this theorem directly, write $\sum_j (\delta P_j\, dQ_j - dP_j\, \delta Q_j)$ as a function of $\delta p_i$, $dp_i$, $\delta q_i$, $dq_i$ with the help of the Poisson brackets and show that it is equal to $\sum_j (\delta p_j\, dq_j - dp_j\, \delta q_j)$. In the course of this proof we see that the value of each determinant is equal to

$$\|(u_i, u_j)\| = (-1)^{n^2+n}, \qquad \|[u_i, u_j]\| = (-1)^{n^2+n}.$$

The condition for an inhomogeneous transformation to be a contact transformation is expressed by

$$((Q_i, Q_j)) = 0, \qquad ((P_i, P_j)) = 0, \qquad ((Z, Q_i)) = 0,$$
$$((P_i, Z)) = \rho P_i, \qquad ((P_i, Q_j)) = \delta_{ij}\cdot\rho,$$

where

$$\frac{\delta}{\delta q_i} = \frac{\partial}{\partial q_i} + p_i \frac{\partial}{\partial z}, \qquad ((U, V)) = \sum_i \left(\frac{\partial U}{\partial q_i}\frac{\partial V}{\delta q_j} - \frac{\partial V}{\partial q_i}\frac{\partial U}{\delta q_j}\right).$$

The Poisson brackets have an important bearing on Dirac's quantum mechanics (1925). By Dirac's $q$-numbers, which do not obey the commutative rule,

$$\frac{ih}{2\pi}(q_r, q_s), \qquad \frac{ih}{2\pi}(p_r, p_s), \qquad \frac{ih}{2\pi}(q_r, p_s)$$

are supposed to correspond to Heisenberg's uncertainty relations

$$q_r q_s - q_s q_r = 0, \qquad p_r p_s - p_s p_r = 0, \qquad q_r p_s - p_s q_r = \frac{ih}{2\pi}\delta_{rs}$$

and are obtained by contact transformations. The Poisson brackets are useful in the application of special relativity theory, to the theory of electrons, radiation fields, and meson fields (Synge, 1960).

A linear canonical transformation is called exceptional if it does not satisfy (29). Ura (1951) proved the following necessary and sufficient conditions for the existence of a linear canonical exceptional transformation: that constants $\lambda_{ij}$ and $\mu_{ij}$ $(i = 1, \ldots, n;\ j = 1, \ldots, l \leqq 2n - 2)$ can be found such that if

$$X_j = \sum_{i=1}^{n} \lambda_{ij} p_i - \sum_{i=1}^{n} \mu_{ij} q_i \qquad (j = 1, \ldots, l \leqq 2n - 2),$$

and, inversely, if the $2n$ variables $q_i$, $p_i$ are substituted in the Hamiltonian function $H$ as functions of $l$ variables $X_j$, then $H$ is expressed as functions of these $l$ variables

$$H(q_1, \ldots, q_n, p_1, \ldots, p_n) = K(X_1, \ldots, X_l).$$

He also showed that if a linear canonical exceptional transformation is known, then two linearly independent integrals can be found and, inversely, if two linear independent integrals are known, then a linear canonical exceptional transformation can be derived from these two integrals.

### 1.10 CANONICAL TRANSFORMATIONS

Consider a Pfaffian

$$\omega_d = \sum_i p_i \, dq_i - H \, dt \tag{32}$$

and form the corresponding bilinear covariant

$$\sum_i (\delta p_i \, dq_i - dp_i \, \delta q_i) - \delta H \, dt - dH \, \delta t.$$

Substituting

$$\delta H = \sum_i \frac{\partial H}{\partial p_i} \delta p_i + \sum_i \frac{\partial H}{\partial q_i} \delta q_i + \frac{\partial H}{\partial t} \delta t$$

in this expression, we obtain

$$\delta\omega_d - d\omega_\delta = \sum_i \left\{ -\left(dp_i + \frac{\partial H}{\partial q_i} dt\right) \delta q_i + \left(dq_i - \frac{\partial H}{\partial p_i} dt\right) \delta p_i \right\} + \left(dH - \frac{\partial H}{\partial t} dt\right) \delta t.$$

Equate the coefficients of $\delta q_i$, $\delta p_i$, $\delta t$ in this expression to zero—that is, equate the Pfaffian associated with (23) to zero; then, corresponding to (23), we obtain

$$\frac{dp_i}{dt} = -\frac{\partial H}{\partial q_i}, \qquad \frac{dq_i}{dt} = \frac{\partial H}{\partial p_i}, \qquad \frac{dH}{dt} = \frac{\partial H}{\partial t}. \tag{33}$$

The last equation has been proved to be a result of the former pair of $2n$ equations. Hence, the Hamiltonian equations (33) are the associated Pfaffians corresponding to (23) and can be replaced by the bilinear covariant (32).

Consider a contact transformation

$$\sum_i p_i \, dq_i - H \, dt = \sum_i p_i \, dQ_i - K \, d\tau + dW, \tag{34}$$

where $P_i$, $Q_i$, $K$, $\tau$, $W$ are all functions of $q_i$, $p_i$, $H$, $t$. The bilinear covariant on the right-hand side is, corresponding to (26),

$$\sum_i (\delta P_i\, dQ_i - dP_i\, \delta Q_i) - \delta K\, d\tau + dK\, \delta\tau.$$

As the transformation is a contact transformation, this bilinear covariant should equal the bilinear covariant of (32). Since $\delta P_i$, $\delta Q_i$, $\delta K$, $\delta\tau$, and $\delta p_i$, $\delta q_i$, $\delta H$, $\delta t$ are arbitrary and mutually independent, the associated Pfaffian

$$\frac{dP_i}{d\tau} = -\frac{\partial K}{\partial Q_i}, \qquad \frac{dQ_i}{d\tau} = \frac{\partial K}{\partial P_i}, \qquad \frac{dK}{d\tau} = \frac{\partial K}{\partial \tau}$$

is the consequence of (33). Hence the canonical form of the differential equations is preserved by a contact transformation. In particular, put $\tau = t$; then

$$\sum_i p_i\, dq_i - H\, dt = \sum_i P_i\, dQ_i - K\, dt + dW \tag{35}$$

and

$$p_i = \frac{\partial W}{\partial q_i}, \qquad P_i = -\frac{\partial W}{\partial Q_i}, \qquad H = K - \frac{\partial W}{\partial t}. \tag{36}$$

This is a condition for a contact transformation, provided that

$$\left\| \frac{\partial^2 W}{\partial q_i \partial Q_j} \right\| \neq 0.$$

Schering calls $W$ the *substitution function.*

Rewriting (35) in the form

$$\sum_i p_i\, dq_i - H\, dt = d\Big(\sum_i P_i Q_i\Big) - \sum_i Q_i\, dP_i - K\, dt + dW \tag{37}$$

gives us another form of the condition for canonical transformation:

$$p_i = \frac{\partial S}{\partial q_i}, \qquad Q_i = \frac{\partial S}{\partial P_i}, \qquad H = K - \frac{\partial S}{\partial t}. \tag{38}$$

provided that

$$\left\| \frac{\partial^2 S}{\partial q_i\, \partial P_j} \right\| \neq 0.$$

Here $\sum_i P_i Q_i + W = S$, and $S$ is called the *principal function.*

Thus, a contact transformation defined by (36) or (38) preserves the canonical form of the differential equations (33). In general, a transformation that preserves the canonical form of differential equations is called a *canonical transformation.* A contact transformation is a canonical transformation, but a canonical transformation is not necessarily a contact transformation. Either (36) or (38) is a sufficient but not a necessary condition for canonicity.

Suppose that the differential equations

$$\frac{dp_i}{dt} = -\frac{\partial H}{\partial q_i}, \qquad \frac{dq_i}{dt} = \frac{\partial H}{\partial p_i} \qquad (i = 1, 2, \ldots, n)$$

are transformed to

$$\frac{dP_i}{dt} = -\frac{\partial K}{\partial Q_i}, \qquad \frac{dQ_i}{dt} = \frac{\partial K}{\partial P_i} \qquad (i = 1, 2, \ldots, n),$$

and that the transformation is not a contact transformation. Lie (1922) tried to obtain the transformation in the form

$$\sum_i p_i \, dq_i = \sum_i G_i \, dH_i + dV.$$

He considered $H_1$ among $H_i$ to be a given function and $\lambda_i$, $\mu_i$, $U$, to be arbitrary functions of $G_1, \ldots, G_n, H_1, \ldots, H_n$. His result is that if $L_i = \lambda_i + \partial U/\partial H_i$ and $M_i = \mu_i + \partial U/\partial G_i$, then $\sum_i L_i \, dH_i + \sum_i M_i \, dG_i$ can be most generally written in the form $\sum_i P_i \, dQ_i + dW$. According to Carathéodory (1930) this problem has nothing to do with canonicity. Consider two integral curves

$$q_i = q_i(t, c_1, \ldots, c_{2n}), \qquad p_i = p_i(t, c_1, \ldots, c_{2n}) \qquad (i = 1, 2, \ldots, n),$$

and

$$Q_j = Q_j(t, C_1, \ldots, C_{2n}), \qquad P_j = P_j(t, C_1, \ldots, C_{2n}) \qquad (j = 1, 2, \ldots, n),$$

and the relations

$$c_k = \phi_k(C_1, \ldots, C_{2n}) \qquad (k = 1, 2, \ldots, 2n),$$

where the $\phi_k$ are arbitrary functions. The two integral curves are made to correspond by these relations. If we eliminate $4n$ constants $C_l$, $c_l$ from these $6n$ equations, then we obtain a transformation (not necessarily canonical). Carathéodory (1935) then defined a canonical transformation by

$$\sum_i P_i \, dQ_i = \sum_i p_i \, dq_i + d\Omega$$

and called $\Omega$ the *generating function.* If

$$\sum_j p_j \frac{\partial Q_i}{\partial p_j} = 0, \qquad P_i = \sum_j p_j \frac{\partial P_i}{\partial p_j} \qquad (i, j = 1, 2, \ldots, n),$$

then $Q_i$ and $P_i$ are homogeneous of order zero and one respectively in $p_i$, and the canonical transformation is called homogeneous. Carathéodory proved that a homogeneous canonical transformation is also a homogeneous contact transformation. Next, he took $r$ pairs from $p_i$, $q_i$ and put

$$\xi_\alpha = q_\alpha, \qquad \eta_\alpha = p_\alpha, \qquad \xi_\kappa = p_\kappa, \qquad \eta_\kappa = -q_\kappa$$
$$(\alpha = 1, 2, \ldots, r; \kappa = r + 1, \ldots, n)$$

and

$$Q_i(q_j, p_j) = \Xi_i(\xi_j, \eta_j)$$

and then proved that we can arrange matters so that

$$\frac{\partial(\Xi_1, \ldots, \Xi_n)}{\partial(p_1, \ldots, q_n)} \neq 0.$$

Solving

$$Q_i = Q_i(q_j, p_j), \qquad P_i = P_i(q_j, p_j)$$

for such $p_1, \ldots, p_r, q_{r+1}, \ldots, q_n$, we obtain

$$\frac{\partial \chi_\alpha}{\partial q_\beta} = \frac{\partial \chi_\beta}{\partial q_\alpha}, \qquad \frac{\partial \chi_\alpha}{\partial p_\kappa} = \frac{\partial \chi_\kappa}{\partial q_\alpha}, \qquad \frac{\partial \chi_\kappa}{\partial p_\lambda} = \frac{\partial \chi_\lambda}{\partial p_\kappa} \qquad (\alpha, \beta = 1, \ldots, r),$$

where

$$p_\alpha = \chi_\alpha(q_\beta, p_\lambda, Q_i), \qquad q_\kappa = -\chi_\kappa(q_\beta, p_\lambda, Q_i) \qquad (\kappa, \lambda = r + 1, \ldots, n).$$

It can be shown that the generating function $\Omega(q_\beta, p_\lambda, Q_i)$ is determined by

$$-\chi_\alpha = \frac{\partial \Omega}{\partial q_\alpha}, \qquad -\chi_\kappa = \frac{\partial \Omega}{\partial p_\kappa}.$$

Let $\Omega_0$ be a function satisfying these equations and $W(Q_i)$ be an arbitrary function of $Q_i$. Then the most general generating function $\Omega$ is written

$$\Omega(q_\alpha, p_\kappa, Q_i) = \Omega_0(q_\beta, p_\lambda, Q_i) + W(Q_i).$$

Hence, the most general canonical transformation, according to Carathéodory, is

$$p_\alpha = -\frac{\partial \Omega}{\partial q_\alpha}, \qquad q_\kappa = \frac{\partial \Omega}{\partial p_\kappa}, \qquad P_i = \frac{\partial \Omega}{\partial Q_i} \qquad (i = 1, 2, \ldots, n).$$

Ura (1946) proved that a canonical transformation is identical with a contact transformation if the Hamiltonian function is kept invariant for $n = 1$; he also deduced for a case of general $n$ the necessary and sufficient condition for a canonical transformation, by using a null divisor in the abstract algebra. Miyahara (1954) considered noncontact transformations in general that keep the transformed equations canonical.

Now we return to our main topic and suppose that $W$ does not depend on $t$ in (35) or (36). In this case we obtain

$$\sum_i p_i\,dq_i = \sum_i P_i\,dQ_i + dV,$$

and $V$ does not depend on $t$. Levi-Civita and Amaldi (1927) called such a transformation a *completely canonical transformation.* From (36) we obtain

$$\sum_i p_i\,dq_i - \sum_i P_i\,dQ_i = (H - K)\,dt + dW.$$

As the left-hand side contains neither $t$ nor $dt$, we obtain

$$H - K + \frac{\partial W}{\partial t} = 0.$$

Put $W = V + T$, where $V$ does not depend on $t$, and $T$ is a function of $t$ and is zero, by our assumption. Hence, $H = K$; also, $(q_i, p_i) \to (Q_i, P_i)$ is a contact transformation.

A canonical transformation does not necessarily contain differentials. Write

$$a = \sum_i q_i^2, \qquad b = \sum_i q_i p_i, \qquad c = \sum_i p_i^2,$$

$$A = \sum_j Q_j^2, \qquad B = \sum_j Q_j P_j, \qquad C = \sum_j P_j^2.$$

Then a regularizing transformation of Levi-Civita

$$A = ac^2, \qquad B = -b, \qquad C = \frac{1}{c},$$

is a canonical transformation.

Consider a contact transformation $(q_i, p_i) \to (Q_i, P_i)$

$$p_i = \frac{\partial U}{\partial q_i}, \qquad -P_i = \frac{\partial U}{\partial Q_i} \qquad (i = 1, 2, \ldots, n),$$

with an arbitrary function $U$ under the condition that there exist $r$ relations

$$\phi_j(q_1, \ldots, q_n, Q_1, \ldots, Q_n) = 0 \qquad (j = 1, \ldots, r)$$

among $q_1, \ldots, q_n; Q_1, \ldots, Q_n$. We use in place of $U$ an expression $U + \lambda_1\phi_1 + \cdots + \lambda_r\phi_r$ with $r$ parameters $\lambda_1, \ldots, \lambda_r$, which we call Lagrange's *indeterminate multipliers*. Then we have a contact transformation

$$p_i = \frac{\partial U}{\partial q_i} + \lambda_1 \frac{\partial \phi_1}{\partial q_i} + \cdots + \lambda_r \frac{\partial \phi_r}{\partial q_i},$$

$$P_i = -\frac{\partial U}{\partial Q_i} - \lambda_1 \frac{\partial \phi_1}{\partial Q_i} - \cdots - \lambda_r \frac{\partial \phi_r}{\partial Q_i} \qquad (i = 1, 2, \ldots, n).$$

If we eliminate $\lambda_1, \ldots, \lambda_r$ from these $2n$ equations and the $r$ relations $\phi_j = 0$, then we obtain $q_i$ and $p_i$ as functions of $Q_i$ and $P_i$, and the Pfaffian can be written

$$\sum_i P_i\, dQ_i - H\, dt + dU + \lambda_1\, d\phi_1 + \cdots + \lambda_r\, d\phi_r.$$

We can easily derive equations (36) by using the integral invariants discussed in Section 1.2. Consider a trajectory as an integral curve in the $(2n + 1)$-dimensional space for the Hamiltonian equations

$$\frac{dq_i}{dt} = \frac{\partial H}{\partial p_i}, \qquad \frac{dp_i}{dt} = -\frac{\partial H}{\partial q_i}, \qquad \frac{dH}{dt} = \frac{\partial H}{\partial t} \qquad (i = 1, 2, \ldots, n),$$

and the variational curves from this trajectory as functions of two parameters $\alpha$ and $\beta$. Such variational curves form a cylinder around the original trajectory. Let $\beta$ = constant be a closed curve $C$ surrounding this cylinder. From Cartan's integral invariant

$$I = \int_C \omega_\delta = \int_C \sum_i p_i\, \delta q_i - H\, \delta t$$

we obtain

$$dI = \int_C \sum_i (dp_i\, \delta q_i - dq_i\, \delta p_i) - dH\, \delta t + dt\, \delta H.$$

As $\delta$ is the symbol of variation due to the variation of the parameter $\alpha$, $dI$ should equal zero in order that $I$ be an integral invariant for any arbitrary value of $\alpha$, that is, for any values of $\delta q_i$, $\delta p_i$, $\delta t$. Hence, by putting $2n + 1$ coefficients of $\delta$ in the expression for $dI$ equal to zero, we obtain

$$dp_i + \frac{\partial H}{\partial q_i} dt = 0, \qquad -dq_i + \frac{\partial H}{\partial p_i} dt = 0, \qquad -dH + \frac{\partial H}{\partial t} dt = 0$$

$$(i = 1, 2, \ldots, n),$$

in other words, the Hamiltonian canonical equations (33).

Next consider an integral invariant

$$I = \int \varpi_\delta, \qquad \varpi_\delta = \sum_i P_i \, dQ_i - K \, dt,$$

associated with

$$\frac{dQ_i}{dt} = \frac{\partial K}{\partial P_i}, \qquad \frac{dP_i}{dt} = -\frac{\partial K}{\partial Q_i}, \qquad \frac{dK}{dt} = \frac{\partial K}{\partial t} \qquad (i = 1, 2, \ldots, n),$$

and suppose that the transformation from $(q_i, p_i, H, t)$ to $(Q_i, P_i, K, t)$ makes the difference of $\omega_\delta$ and $\varpi_\delta$ a perfect differential. Since an integral invariant along a closed curve is unaltered by the addition of a perfect differential, the canonical form of the differential equations is preserved after this transformation. The condition for form preservation is expressed by

$$\sum_i p_i \, \delta q_i - \sum_i P_i \, \delta Q_i - (H - K) \, \delta t = \delta W,$$

and we obtain a condition for this that is equivalent to (36) for contact transformations.

Again we vary both parameters $\alpha$ and $\beta$. Let the points $A$, $B$, $C$, and $D$ correspond to $(\alpha, \beta)$, $(\alpha + \delta\alpha, \beta)$, $(\alpha, \beta + \delta\beta)$, and $(\alpha + \delta\alpha, \beta + \delta\beta)$, respectively, and take $\int \omega$ along the closed curve $ABCD$. Then we have $\omega_\delta$ along $AB$, $\omega'_\delta$ along $AC$, $\omega_\delta + \delta'\omega_\delta$ along $CD$, and $\omega'_\delta + \delta\omega'_\delta$ along $BD$. Hence

$$\int_{ABCD} = \delta\omega'_\delta - \delta'\omega_\delta = \omega'(\delta, \delta').$$

The integral along the closed curve of this bilinear expression remains invariant when we move the closed curve along the corresponding trajectory. It is a double integral

$$\iint \omega' = \iint \sum_i \delta p_i \, \delta q_i - \delta H \, \delta t$$

over a two-dimensional area. If we apply an extended form of Stokes' theorem, then this expression equals the integral

$$\int \omega = \int \Big(\sum_i p_i\,\delta q_i - H\,\delta t\Big)$$

integrated along the closed curve forming the boundary of the two-dimensional area. In the sense of Poincaré's integral invariants (1889, 1893), which are integrals over the values at the same $t$, $\iint \sum_i \delta p_i\,\delta q_i$ is an integral invariant. Poincaré calls $\iint \omega'$ an *absolute integral invariant* and $\int \omega$ a *relative integral invariant*. The former is not restricted to either a closed or an open domain of integration; the latter is restricted to a closed domain. The fact that an integral invariant $\iint \omega'$ exists is equivalent to the fact that the given differential equations have the momentum-energy tensor $\omega_\delta$.

Taylor (1935, 1937) and Lewis (1937) considered integral invariants that are intermediate between those of Poincaré and Cartan. Denote by $x_j$ a set of $q_i$ and $p_i$. Integrals of $x_j = f_j[t; x_k(t)]$ over the points $t = t(u) + t_0$, $x_j = f_j[t(u) + t_0; x_k(u)]$ are called by Taylor *associated integral invariants*. When $t$ is fixed, the integral invariant reduces to Poincaré's. When $t_0$ is not a constant and is an arbitrary function of $u_1, u_2, \ldots, u_p$, it reduces to Cartan's (see Weitzenböck, 1923).

Birkhoff (1927, 1928) considered a variational principle of Pfaff:

$$\delta \int_{t_0}^{t_1} \left\{ \sum_{i=1}^{2m} X_i \dot{x}_i + Z \right\} dt = 0.$$

The integral is linear with respect to all the derivatives of first order; also their coefficients $X_1, \ldots, X_{2m}, Z$ are arbitrary functions of $x_1, x_2, \ldots, x_{2m}$, provided that the determinant

$$\left\| \frac{\partial X_i}{\partial x_j} - \frac{\partial X_j}{\partial x_i} \right\| \neq 0 \qquad (i, j = 1, 2, \ldots, 2m).$$

If we develop the variational expression, we obtain

$$\sum_{j=1}^{2m} \left( \frac{\partial X_i}{\partial x_j} - \frac{\partial X_j}{\partial x_i} \right) \frac{dx_j}{dt} - \frac{\partial Z}{\partial x_i} = 0 \qquad (i = 1, 2, \ldots, 2m).$$

This integral is unchanged by transformations under the integral sign from variables $x_j$ to $y_j$. If we multiply each of these equations by $dx_i/dt$ and add, then we obtain $dZ/dt = 0$ or $Z =$ constant. It corresponds to the energy integral of the Hamiltonian differential equations. Féraud (1929) discussed the relation between this principle of Birkhoff's and the integral invariants of Cartan.

## I.II NONHOLONOMIC CONSTRAINT

Consider a nonholonomic constraint (Section 1.1)

$$d'\kappa_i = a_{i1}\,dq_1 + a_{i2}\,dq_2 + \cdots + a_{in}\,dq_n + a_{i0}\,dt = 0 \qquad (i = 1, 2, \ldots, k), \tag{39}$$

where $q_i$, $\dot{q}_i$ are functions of $t$. This is not a perfect differential according to the definition. The variables $\kappa_i$ are called quasi-coordinates. Any possible displacement of the dynamical system must always satisfy (39). Any increment of velocity must satisfy $\delta\dot{q}_i = d(\delta q_i)/dt$. That is, the incremental operators $d$ and $\delta$ must satisfy a commutative law $\delta dq_i = d\delta q_i$. Since any neighboring motion must also satisfy the constraint, we have

$$d(a_{i1}\,\delta q_1 + a_{i2}\,\delta q_2 + \cdots + a_{in}\,\delta q_n + a_{i0}\,\delta t) = 0.$$

By (39),

$$\delta(a_{i1}\,dq_1 + a_{i2}\,dq_2 + \cdots + a_{in}\,dq_n + a_{i0}\,dt) = 0.$$

Owing to the commutative law for $d$ and $\delta$, we obtain

$$d'(\delta'\kappa_i) - \delta'(d'\kappa_i) = \sum_{j,k}\left(\frac{\partial a_{ik}}{\partial q_j} - \frac{\partial a_{ij}}{\partial q_k}\right)(dq_j\,\delta q_k - \delta q_j\,dq_k) + dt\sum_j\left(\frac{\partial a_{ij}}{\partial t} - \frac{\partial a_{i0}}{\partial q_j}\right)\delta q_j.$$

The right-hand member becomes zero when and only when $d'\kappa_i$ is a perfect differential. Denote the result of solving (39) by

$$dq_\lambda = A_{1\lambda}\,d'\kappa_1 + A_{2\lambda}\,d'\kappa_2 + \cdots + A_{k\lambda}\,d'\kappa_k + A_{k+1,\lambda}\,d'\kappa_{k+1} + \cdots + A_{n\lambda}\,d'\kappa_n + A_{0\lambda}\,d'\kappa_0 \qquad (\lambda = 1, 2, \ldots, n).$$

Substituting this in the preceding equation, we obtain

$$\begin{aligned} d'(\delta'\kappa_i) - \delta'(d'\kappa_i) &= \sum_{\sigma,\tau}\left[\sum_{\lambda,\mu}\left(\frac{\partial a_{i\lambda}}{\partial q_\mu} - \frac{\partial a_{i\mu}}{\partial q_\lambda}\right)A_{\sigma\mu}A_{\tau\lambda}\right] \\ &\quad\times (d'\kappa_\sigma\,\delta'\kappa_\tau - \delta'\kappa_\sigma\,d'\kappa_\tau) \\ &\quad + dt\sum_\sigma\left[\sum_{\lambda,\mu}\left(\frac{\partial a_{i\lambda}}{\partial q_\mu} - \frac{\partial a_{i\mu}}{\partial q_\lambda}\right)(A_{0\mu}A_{\sigma\lambda} - A_{0\lambda}A_{\sigma\mu})\right. \\ &\qquad\qquad \left. + \sum_\lambda\left(\frac{\partial a_{i\lambda}}{\partial t} - \frac{\partial a_{i0}}{\partial q_\lambda}\right)A_{\sigma\lambda}\right]\delta'\kappa_\sigma \\ &= \sum_{\sigma,\tau} c^i_{\sigma\tau}(d'\kappa_\sigma\,\delta'\kappa_\tau - \delta'\kappa_\sigma\,d'\kappa_\tau) \\ &\quad + dt\sum_\sigma c^i_{0\sigma}\,\delta'\kappa_\sigma, \end{aligned}$$

where

$$c^i_{\sigma\tau} = -c^i_{\tau\sigma}, \qquad c^i_{\sigma\sigma} = 0.$$

This equation has been called by Heun (1901) the transitivity equation. The constants $c^i_{\sigma\tau}$ correspond to the structure constants in Lie's theory of transformation groups (see Section 4.2).

If (39) does not contain $t$ explicitly, then Lagrange's equations are

$$\frac{d}{dt}\left(\frac{\partial T}{\partial \dot{q}_i}\right) - \frac{\partial T}{\partial q_i} = Q_i + \sum_{j=1}^{k}\lambda_j a_{ji} \qquad (i = 1, 2, \ldots, n),$$

$$\sum_{i=1}^{n} a_{1i}\,dq_i = 0, \ldots, \sum_{i=1}^{n} a_{ki}\,dq_i = 0.$$

If we use the quasi-coordinates

$$\delta'\kappa_{n-k+1} = 0, \ldots, \delta'\kappa_n = 0,$$

$$\delta q_i = A_{1i}\,\delta'\kappa_1 + A_{2i}\,\delta'\kappa_2 + \cdots + A_{n-k,i}\,\delta'\kappa_{n-k},$$

we can cast the Lagrangian equations in the form (Hamel, 1924)

$$\sum_{i=1}^{n-k}\left\{\sum_{\lambda=i}^{n} A_{i\lambda}\left[\frac{d}{dt}\left(\frac{\partial T}{\partial \dot{q}_\lambda}\right) - \frac{\partial T}{\partial q_\lambda} - Q_\lambda\right]\right\}\delta'\kappa_i = 0.$$

Thus,

$$\sum_{\lambda=1}^{n} A_{i\lambda}\left[\frac{d}{dt}\left(\frac{\partial T}{\partial \dot{q}_\lambda}\right) - \frac{\partial T}{\partial q_\lambda}\right] = \sum_{\lambda=1}^{n} A_{i\lambda} Q_\lambda \qquad (i = 1, 2, \ldots, n - k),$$

$$\sum_{\lambda=1}^{n} a_{j\lambda}\, dq_\lambda = 0 \qquad (j = n - k + 1, \ldots, n).$$

It has been seen that a holonomic dynamical system can be studied by referring to a Riemannian space (Section 1.5). Similarly, a nonholonomic dynamical system can be discussed in a nonholonomic space. Suppose that there exist $n - m$ Pfaffians

$$dx_i = \sum_{\alpha=m+1}^{n} \lambda_i^\alpha\, d's_\alpha \qquad (\alpha, \beta = m + 1, \ldots, n),$$

in a space $R_n$ such that

$$d's_\alpha = \sum_{i=1}^{n} \lambda_\alpha^i\, dx_i = 0,$$

where

$$\sum_{\alpha=m+1}^{n} \lambda_i^\alpha \lambda_\alpha^j = \delta_{ij}, \qquad \sum_{i=1}^{n} \lambda_\alpha^i \lambda_i^\beta = \delta_{\alpha\beta} \qquad (i, j = 1, 2, \ldots, n),$$

and $d's_\alpha$ is not a perfect differential. If $d's_\alpha$ were a perfect differential, then it would be possible to choose $x_i$ so that $ds_\alpha = dx_\alpha = 0$. Thus $R_n$ would reduce to $R_m$ and the line element of the space would be such that

$$ds^2 = ds_1^2 + ds_2^2 + \cdots + ds_m^2.$$

The space $R_n$ with $n$ quasi-coordinates $x_i$ constitutes a nonholonomic space $V_n^m$. The geometry of nonholonomic spaces has been studied by Vranceanu (1929, 1935, 1961), Schouten (1929, 1954), and Yano (1945, 1949).

## THE HAMILTON-JACOBI PARTIAL DIFFERENTIAL EQUATION

### 1.12 THE HAMILTON-JACOBI THEOREM

If we apply a contact transformation

$$p_i = \frac{\partial}{\partial q_i} W(q_i, Q_i, t), \qquad P_i = -\frac{\partial}{\partial Q_t} W(q_i, Q_i, t), \tag{40}$$

$$H = K - \frac{\partial}{\partial t} W(q_i, Q_i, t),$$

to Hamilton's canonical equations

$$\frac{dq_i}{dt} = \frac{\partial}{\partial p_i} H(q_i, p_i, t), \qquad \frac{dp_i}{dt} = -\frac{\partial}{\partial q_i} H(q_i, p_i, t) \qquad (i = 1, 2, \ldots, n), \tag{41}$$

then these equations (41) become Hamilton's canonical equations

$$\frac{dQ_i}{dt} = \frac{\partial K}{\partial p_i}, \qquad \frac{dP_i}{dt} = -\frac{\partial K}{\partial Q_i} \qquad (i = 1, 2, \ldots, n). \tag{42}$$

Suppose that $K \equiv 0$, then the solution is

$$Q_i = \text{constants} = \beta_i, \qquad P_i = \text{constants} = \alpha_i \qquad (i = 1, 2, \ldots, n). \tag{43}$$

This solution represents an equilibrium position; that is, the state of motion in the variables $Q_i$, $P_i$ is an equilibrium state. Hence, if we can find such a function $W(q_i, Q_i, t)$, then we can obtain the solution of the original Hamilton equations (41) from the equilibrium solution (43) by the transformation (40), provided that

$$\left\| \frac{\partial^2 W}{\partial q_i \, \partial Q_j} \right\| \neq 0, \tag{44}$$

that is, provided that the second equation of (40) can be solved for $q_i$. Thus the solution of (41) reduces to obtaining $W(q_i, Q_i, t)$.

Put $K \equiv 0$ in (40); then

$$\frac{\partial W}{\partial t} + H = 0.$$

Since $H$ is a function of $q_i, p_i, t$, we consider $q_i$ and $t$ as the independent variables and substitute the first equation of (40) in the place of $p_i$. Then we obtain a partial differential equation for determining the unknown function $W$ as a function of $n + 1$ variables $q_i$ and $t$:

$$\frac{\partial W}{\partial t} + H\left(q_i, \frac{\partial W}{\partial q_i}, t\right) = 0. \tag{45}$$

This is called the *Hamilton-Jacobi partial differential equation.* A function $W(q_i, p_i, t)$ satisfying equation (44) with $n$ arbitrary independent constants $\alpha_1, \alpha_2, \ldots, \alpha_n$, besides $q_i$, $t$ (any additive constant to $t$ is not counted) is called a *complete integral.* By "independent constants" is meant that we can choose the $\alpha_i$ such that the $(\partial W/\partial q_i)_{t=0}$ can have arbitrary values, and in this case (44) is satisfied.

Hence, when we know a complete integral $W(q_i, Q_i, t)$ and when

$$\Delta \equiv \left\| \frac{\partial^2 W}{\partial q_i \, \partial \alpha_j} \right\| \neq 0, \tag{44a}$$

then we can obtain the *general integral* of (41) in the form $p_i(t, \alpha_i, \beta_i)$, $q_i(t, \alpha_i, \beta_i)$, by solving the equations

$$p_i = \frac{\partial}{\partial q_i} W(q_i, \alpha_i, t), \qquad \beta_i = -\frac{\partial}{\partial \alpha_i} W(q_i, \alpha_i, t) \qquad (i = 1, 2, \ldots, n) \tag{41a}$$

derived from (41). Transforming back to the original variables $q_i$, $p_i$, we obtain the solution of (41); and, indeed, this is the general integral, as this solution contains $2n$ integration constants $\alpha_i$, $\beta_i$. This is called *Jacobi's theorem.*

As a corollary of Jacobi's theorem we see that if $H$ does not contain $t$ explicitly, then the complete integral of (45) is

$$W = -ht + V(q_1, \ldots, q_n, \alpha_1, \ldots, \alpha_{n-1}, h), \tag{46}$$

where $h$ is one of the integration constants.

If we put (46) in (45), we obtain

$$H\left(q_1, \ldots, q_n, \frac{\partial V}{\partial q_1}, \ldots, \frac{\partial V}{\partial q_n}\right) = h. \tag{47}$$

If $V(q_1, \ldots, q_n, \alpha_1, \ldots, \alpha_{n-1}, h)$ is an integral of (47) for an arbitrary value of $h$, then $K \equiv 0$ in (40); and substituting (46) in (40) gives

$$\begin{aligned} p_i &= \frac{\partial V}{\partial q_i} \qquad (i = 1, 2, \ldots, n), \\ \beta_j &= -\frac{\partial V}{\partial \alpha_j} \qquad (j = 1, 2, \ldots, n-1), \\ t - t_0 &= \frac{\partial V}{\partial h}. \end{aligned} \tag{48}$$

The integration constants are $\alpha_1, \ldots, \alpha_{n-1}, h, \beta_1, \ldots, \beta_{n-1}, t_0$. Equations (48) can be solved for $p_i$, $q_i$ as functions of

$$\alpha_1, \ldots, \alpha_{n-1}, h, \beta_1, \ldots, \beta_{n-1}, t - t_0,$$

provided that (44a) does not equal zero when $h$ is substituted for $\alpha_1$.

That the integral $V$ is the complete integral of (47) follows from $\Delta \neq 0$. It may be seen in the following manner. Put

$$\Delta_1 = \left\| \frac{\partial^2 V}{\partial q_i \, \partial \alpha_j} \right\| \qquad (i, j = 1, 2, \ldots, n - 1). \tag{49}$$

Then (44a) can be written

$$\Delta = \left\| \begin{matrix} \Delta_1 & \dfrac{\partial^2 V}{\partial q_i \, \partial h} \\ \dfrac{\partial^2 V}{\partial q_n \, \partial \alpha_j} & \dfrac{\partial^2 V}{\partial q_n \, \partial h} \end{matrix} \right\|.$$

Multiply each row by $\partial H/\partial p_i$ and suppose that $H$ contains at least one of the $p_i$, say $p_n$, so that $\partial H/\partial p_n \neq 0$. Then

$$\sum_r \frac{\partial^2 V}{\partial q_r \, \partial \alpha_j} \frac{\partial H}{\partial p_r} = \sum_r \frac{\partial H}{\partial p_r} \frac{\partial p_r}{\partial \alpha_j} \qquad (j = 1, 2, \ldots, n - 1).$$

Here it is supposed that $H$ in $H = h$ is a function of $q_i$ and $\partial V/\partial q_i$, and that $V$ is a function of $q_i$ and $\alpha_i$. As $H$ contains $\alpha_i$ only through $\partial V/\partial q_i$, that is, through $p_i$, this expression is the derivative of $H$ with respect to $\alpha_j$. Since the right-hand side of $H = h$ gives $\partial h/\partial \alpha_j = 0$, this expression should equal zero. From the last column we obtain

$$\sum_r \frac{\partial^2 V}{\partial q_r \, \partial h} \frac{\partial H}{\partial p_r} = \sum_r \frac{\partial H}{\partial p_r} \frac{\partial p_r}{\partial h} = 1.$$

Hence,

$$\Delta = \frac{1}{\partial H/\partial p_n} \left\| \begin{matrix} \Delta_1 & \dfrac{\partial^2 V}{\partial q_i \, \partial h} \\ \dfrac{\partial H}{\partial p_n} \dfrac{\partial^2 V}{\partial q_n \, \partial \alpha_j} & \dfrac{\partial H}{\partial p_n} \dfrac{\partial^2 V}{\partial q_n \, \partial h} \end{matrix} \right\|$$

$$= \frac{1}{\partial H/\partial p_n} \left\| \begin{matrix} \Delta_1 & \dfrac{\partial^2 H}{\partial q_i \, \partial h} \\ 0 & 1 \end{matrix} \right\| = \frac{1}{\partial H/\partial p_n} \Delta_1.$$

Thus, the condition that $V$ should be the complete integral is $\Delta_1 \neq 0$. Here $H = h$ expresses that the total energy of the system is constant.

Instead of supposing that $h$ is a function of only one independent variable $\alpha_i$ as in the foregoing treatment, we now suppose that $h$ is a function of $n$ independent variables $\alpha_1, \alpha_2, \ldots, \alpha_n$, and that $h$ does not contain $q_i$ and $t$. Then (46) may be written

$$W = -h(\alpha_1, \ldots, \alpha_n)t + V(q_1, \ldots, q_n, \alpha_1, \ldots, \alpha_n) \tag{46a}$$

and (41a) takes the form

$$p_i = \frac{\partial V}{\partial q_i}, \qquad \frac{\partial V}{\partial \alpha_i} = n_i t - \beta_i, \qquad n_i = \frac{\partial h}{\partial \alpha_i} \qquad (i = 1, 2, \ldots, n). \tag{41b}$$

In this form, which Poincaré employed, all $\alpha_i$ appear symmetrically in the formulas.

Another corollary of Jacobi's theorem is that if $H$ does not contain $q_1, \ldots, q_k$ $(k < n)$, and hence can be written

$$H = H(t, q_{k+1}, \ldots, q_n, p_1, \ldots, p_n),$$

then, putting

$$W = \alpha_1 q_1 + \cdots + \alpha_k q_k + V(t, q_{k+1}, \ldots, q_n, \alpha_1, \ldots, \alpha_n)$$

in the Hamilton-Jacobi partial differential equation (45), we obtain

$$\frac{\partial V}{\partial t} + H\left(t, q_{k+1}, \ldots, q_n, \alpha_1, \ldots, \alpha_k, \frac{\partial V}{\partial q_{k+1}}, \ldots, \frac{\partial V}{\partial q_n}\right) = 0.$$

If we can obtain $V(t, q_{k+1}, \ldots, q_n, \alpha_1, \ldots, \alpha_k, \alpha_{k+1}, \ldots, \alpha_n)$ with arbitrary constants $\alpha_1, \alpha_2, \ldots, \alpha_n$, then $W = \alpha_1 q_1 + \cdots + \alpha_k q_k + V$ is the complete integral of (45), and the motion is described by

$$\beta_i = -q_i - \frac{\partial V}{\partial \alpha_i}, \qquad p_i = \alpha_i \qquad (i = 1, \ldots, k),$$

$$\beta_j = -\frac{\partial V}{\partial \alpha_j}, \qquad p_j = \frac{\partial V}{\partial q_j} \qquad (j = k + 1, \ldots, n).$$

Finally, if $H$ contains neither $t$ nor $q_1, \ldots, q_k$, then we can proceed in a similar manner by putting

$$W = -ht + \alpha_1 q_1 + \cdots + \alpha_k q_k + V(q_{k+1}, \ldots, q_n, \alpha_1, \ldots, \alpha_n).$$

Lehmann-Filhés (1904) discussed the case of incomplete integrals, that is, integrals of the Hamilton-Jacobi partial differential equation that do not contain the necessary number of integration constants.

## 1.13 HYDRODYNAMICAL PROOF OF JACOBI'S THEOREM

We consider Hamilton's canonical equations (41) as describing the motion of the molecules of a fluid. The molecules are at $q_i = q_i^{(0)}$,

$p_i = p_i^{(0)}$ at $t = t_0$, and we seek the values of $q_i$ and $p_i$ at time $t$. This is the Lagrangian problem in the hydrodynamic equations of motion. If, instead, we seek the values of $p_i$ for the molecules passing through the points $q_i$ at time $t$, then the problem is Euler's.

According to Euler's point of view, we consider the $q_i$ and $t$ as the independent variables and suppose $p_i$ to be functions of these $n + 1$ variables. Then the second equation of (41) can be written

$$\frac{\partial p_i}{\partial t} + \frac{\partial H}{\partial q_i} + \sum_r \frac{\partial H}{\partial p_r}\frac{\partial p_r}{\partial q_i} = 0.$$

As $H$ is a known function of $q_i$, $p_i$, and $t$, so $\partial H/\partial q_i$ and $\partial H/\partial p_i$ are known functions of $q_i$, $p_i$, and $t$. These equations are now considered a system of partial differential equations for $n$ unknown functions $p_i$, that is, for $n$ unknown functions of $q_i$ and $t$. Suppose that the $p_i$ can be derived from a potential, that is, they satisfy

$$\frac{\partial p_i}{\partial q_k} - \frac{\partial p_k}{\partial q_i} = 0 \qquad (i, k = 1, 2, \ldots, n);$$

then there exists potential $W$ such that $p_i = \partial W/\partial q_i$.

Thus, the above partial differential equations for the $p_i$ are reduced to a partial differential equation for $W$, that is, the second term in the parentheses of the equation

$$\frac{\partial p_i}{\partial t} + \left(\frac{\partial H}{\partial q_i} + \sum_r \frac{\partial H}{\partial p_r}\frac{\partial p_r}{\partial q_i}\right) = 0$$

is the partial derivative $\partial \overline{H}/\partial q_i$ in which the $p_i$ are supposed to be functions of the $n + 1$ independent variables $q_i$ and $t$. By substituting $p_i = \partial W/\partial q_i$ in this equation we obtain

$$\frac{\partial}{\partial q_i}\left(\frac{\partial W}{\partial t} + \overline{H}\right) = 0.$$

Hence we determine $W$ from $\partial W/\partial t + \overline{H} = 0$ and let $W$ be the complete integral with constants of integration $\alpha_1, \ldots, \alpha_n$. Even if we replace the $\alpha_i$ by $\alpha_i + d\alpha_i$ in $W$, the partial differential equation is satisfied; therefore

$$\frac{\partial^2 W}{\partial t\, \partial \alpha_i} + \sum_r \frac{\partial H}{\partial p_r}\frac{\partial^2 W}{\partial q_r\, \partial \alpha_i} = 0,$$

because the $\alpha_i$ are contained in $H(q_i\ p_i, t)$ only through $p_i$, that is, only through $\partial W/\partial q_i$ $(i = 1, 2, \ldots, n)$. Furthermore, from the first set of

Hamilton's canonical equations (41), we have

$$\left(\frac{\partial}{\partial t} + \sum_r \frac{dq_r}{dt}\frac{\partial}{\partial q_r}\right)\frac{\partial W}{\partial \alpha_i} = 0,$$

or

$$\frac{d}{dt}\frac{\partial W}{\partial \alpha_i} = 0.$$

Hence,

$$\frac{\partial W}{\partial \alpha_i} = \beta_i = \text{constant}.$$

Thus Jacobi's theorem has been proved on hydrodynamical foundations. This proof is due to Einstein (1917).

## 1.14 VARIATION OF CONSTANTS

In order to solve the canonical equations

$$\frac{dq_i}{dt} = \frac{\partial H}{\partial p_i}, \qquad \frac{dp_i}{dt} = -\frac{\partial H}{\partial q_i} \qquad (i = 1, 2, \ldots, n), \tag{50}$$

put

$$H = H_0 + H_1, \tag{51}$$

where the solution of

$$\frac{dq_i}{dt} = \frac{\partial H_0}{\partial p_i}, \qquad \frac{dp_i}{dt} = -\frac{\partial H_0}{\partial q_i} \qquad (i = 1, \ldots, n) \tag{52}$$

is assumed to be known. This solution is obtained by solving the Hamilton-Jacobi partial differential equation

$$\frac{\partial W}{\partial t} + H_0\left(q_i, \frac{\partial W}{\partial q_i}, t\right) = 0. \tag{53}$$

Let the solution be

$$\begin{aligned} q_i &= q_i(t, \alpha_1, \ldots, \alpha_n, \beta_1, \ldots, \beta_n), \\ p_i &= p_i(t, \alpha_1, \ldots, \alpha_n, \beta_1, \ldots, \beta_n), \end{aligned} \tag{54}$$

with $2n$ integration constants $\alpha_1, \ldots, \alpha_n, \beta_1, \ldots, \beta_n$.

Next, take the Hamiltonian function $H = H_0 + H_1$ and consider $\alpha_i$ and $\beta_i$ to be no longer constant but to vary according to the differential equations that we have already derived. We employ our usual procedure to solve the differential equations in $\alpha_i$ and $\beta_i$; that is, we drop terms of small magnitude and integrate; then, inserting the neglected terms, we consider the integration constants so obtained as variable, and thus form a new system of differential equations for the variation of these constants. This principle is called the *variation of constants*, or sometimes the *Lagrangian method*.

Suppose that we have obtained the complete integral $W(q_i, \alpha_i, t)$ of (53). If $\|\partial^2 W/\partial q_i\, \partial\alpha_j\| \neq 0$ in

$$p_i = \frac{\partial W}{\partial q_i}, \qquad \beta_i = -\frac{\partial W}{\partial \alpha_i} \qquad (i = 1, 2, \ldots, n), \tag{55}$$

then we can solve for $p_i$ and $q_i$ as functions of $\alpha_i$, $\beta_i$, and $t$ from (55). By the result in Section 1.12, the original differential equations (50) are transformed to

$$\frac{dQ_i}{dt} = \frac{\partial K}{\partial P_i}, \qquad \frac{dP_i}{dt} = -\frac{\partial K}{\partial Q_i} \qquad (i = 1, 2, \ldots, n)$$

by the transformation

$$p_i = \frac{\partial W}{\partial q_i}, \qquad P_i = -\frac{\partial W}{\partial Q_i}, \qquad K = H + \frac{\partial W}{\partial t} \qquad (i = 1, 2, \ldots, n).$$

If $H = H_0$, then the solution is

$$P_i = \beta_i, \qquad Q_i = \alpha_i \qquad (i = 1, 2, \ldots, n).$$

If we substitute

$$K = H + \frac{\partial W}{\partial t} = H - H_0 = H_1$$

for $K$, then we obtain

$$\frac{d\alpha_i}{dt} = \frac{\partial H_1}{\partial \beta_i}, \qquad \frac{d\beta_i}{dt} = -\frac{\partial H_1}{\partial \alpha_i} \qquad (i = 1, 2, \ldots, n). \tag{56}$$

If $H = H_0$, then $\alpha_i$ and $\beta_i$ are constants. The equations (56) give the variation of these constants. $H_1$ is taken in general to be small compared with $H_0$; it is called the *perturbation*.

In the foregoing we have considered the variation of canonical constants. Let $A_1, A_2, \ldots, A_{2n}$, be any set of independent integration constants that are functions of the canonical constants

$$\alpha_1, \ldots, \alpha_n, \beta_1, \ldots, \beta_n.$$

Then

$$\begin{aligned}\frac{dA_r}{dt} &= \sum_i \frac{\partial A_r}{\partial \alpha_i}\frac{d\alpha_i}{dt} + \sum_i \frac{\partial A_r}{\partial \beta_i}\frac{d\beta_i}{dt} \\ &= \sum_i \left(\frac{\partial A_r}{\partial \alpha_i}\frac{\partial K}{\partial \beta_i} - \frac{\partial A_r}{\partial \beta_i}\frac{\partial K}{\partial \alpha_i}\right) \\ &= \sum_{i=1}^{n}\sum_{s=1}^{2n} \left(\frac{\partial A_r}{\partial \alpha_i}\frac{\partial K}{\partial A_s}\frac{\partial A_s}{\partial \beta_i} - \frac{\partial A_r}{\partial \beta_i}\frac{\partial K}{\partial A_s}\frac{\partial A_s}{\partial \alpha_i}\right) \\ &= \sum_{s=1}^{2n} \frac{\partial K}{\partial A_s}(A_r, A_s),\end{aligned}$$

where $(A_r, A_s)$ is the Poisson bracket. Hence

$$\frac{dA_r}{dt} = \sum_{s=1}^{2n} (A_r, A_s)\frac{\partial H_1}{\partial A_s} \qquad (r = 1, 2, \ldots, 2n). \tag{57}$$

In the usual perturbation calculation in celestial mechanics we take, for example, the six integration constants in the elliptic motion of a planet around the sun — neglecting the perturbations of the orbit due to a third body. The six integration constants $A_r$ which we take are the semimajor axis, eccentricity, longitude of the perihelion, orbital inclination to the fundamental plane, longitude of the ascending node, and epoch of the perihelion passage; these are called the *Keplerian elements*. These Keplerian elements are functions of the canonical constants in the Keplerian elliptic motion. The perturbing Hamiltonian $H_1$ is taken to represent perturbations due to the action of other bodies or due to the derivation of the figure of the attracting body from a spherical mass distribution. In general, $H_1$ is multiplied by the perturbing mass and hence is small compared with the solar mass, in the order of one in several thousands. $H_1$ is also small when the perturbing body is much farther away than the attracting body.

A similar consideration is important also in quantum mechanics. When we superpose the effect of an electromagnetic field or a light quantum or some other particle on the motion of an electron around an atomic nucleus, the superposed effect is considered a perturbation added to the attraction between the electron and the nucleus. The equation of motion in the unperturbed case, that is, the Schrödinger equation or the Dirac equation, is integrated. When the perturbation is taken into account, the integration constants are supposed to vary; we obtain the variation by solving the differential equations for the perturbation (Corson, 1951).

In celestial mechanics and quantum mechanics we can reach a better approximation by including a part of the perturbation in the first integration and applying the principle of the variation of constants to the remaining part. The equations of motion in the perturbation calculations of celestial mechanics can be solved to a high degree of approximation because of the smallness of the perturbing masses. But in quantum mechanics there already appears in the second approximation a trace of the nonconvergence peculiar to this method. Even in celestial mechanics there is a serious difficulty in that the method does not lead to uniform convergence, as will be shown in later chapters.

## 1.15 LIOUVILLE'S THEOREM

The Lagrangian function $L$ in the Hamiltonian $H = \sum_i (\partial L/\partial \dot{q}_i)\dot{q}_i - L$ is divided into terms of homogeneous degree with regard to $\dot{q}_i$ and is written $L = L_2 + L_1 + L_0$, so that

$$\sum_i (\partial L_2/\partial \dot{q}_i)\dot{q}_i = 2L_2, \qquad \sum_i (\partial L_1/\partial \dot{q}_i)\dot{q}_i = L_1, \qquad \sum_i (\partial L_0/\partial \dot{q}_i)\dot{q}_i = 0.$$

Then, $H = L_2 - L_0$. If $T$ is homogeneous of degree 2 with regard to $\dot{q}_i$ and if the external force function $U$ does not contain $\dot{q}_i$, then we can take $L_2 = T$, $L_0 = U$ in $L = T + U$. Hence, $H = T - U$.

Scholars of the Paris school—Darboux, Bertrand, and others—studied dynamical systems that admit integrals homogeneous of degree 1 or 2 in the velocity. As an example Liouville considered a dynamical system with $T$ and $U$ of the form

$$2T = [\phi_1(q_1) + \cdots + \phi_n(q_n)][A_1(q_1)\dot{q}_1^2 + \cdots + A_n(q_n)\dot{q}_n^2],$$

$$U = \frac{\psi}{\phi} = \frac{\psi_1(q_1) + \cdots + \psi_n(q_n)}{\phi_1(q_1) + \cdots + \phi_n(q_n)}.$$

In this case we can solve the problem without integrating the differential equations but with quadrature alone. This is called *Liouville's theorem.* We have

$$H = T - U = \tfrac{1}{2}\phi[A_1(q_1)\dot{q}_1^2 + \cdots + A_n(q_n)\dot{q}_n^2] - \frac{\psi}{\phi},$$

$$p_i = \frac{\partial T}{\partial \dot{q}_i} = \phi A_i(q_i)\dot{q}_i.$$

This theorem can be extended to an irreversible dynamical system. Put $U = (\psi/\phi) + \sum_i a_i(q_i)\dot{q}_i$; then $L_1 = \sum_i a_i(q_i)\dot{q}_i$. But $L_1\,dt$ is a perfect differential; hence we can omit this term, according to the theorem by Birkhoff described in Section 1.1.

Take the Hamilton-Jacobi partial differential equation for the Liouville case

$$\frac{1}{2\phi}\left[\frac{(\partial V/\partial q_1)^2}{A_1(q_1)} + \cdots + \frac{(\partial V/\partial q_n)^2}{A_n(q_n)}\right] - \frac{\psi_1(q_1) + \cdots + \psi_n(q_n)}{\phi_1(q_1) + \cdots + \phi_n(q_n)} = h,$$

that is,

$$\sum_i \frac{1}{A_i(q_i)}\left(\frac{\partial V}{\partial q_i}\right)^2 = 2\sum_i [\psi_i(q_i) + h\phi_i(q_i)].$$

Let the complete integral of the partial differential equation be $V = \sum_i V_i(q_i)$, that is, the sum of functions each containing a single variable $q_i$. Then,

$$V_i(q_i) = \int dq_i \sqrt{2A_i(q_i)[\psi_i(q_i) + h\phi_i(q_i) + \alpha_i]},$$

with the condition that $\alpha_1 + \alpha_2 + \cdots + \alpha_n = 0$ among the integration constants $\alpha_i$. As $\alpha_n = -(\alpha_1 + \cdots + \alpha_{n-1})$, the motion, according to (48), is given by

$$\frac{\partial V}{\partial \alpha_j} - \frac{\partial V}{\partial \alpha_n} = -\beta_j \qquad (j = 1, 2, \ldots, n-1),$$

$$t - t_0 = \frac{\partial V}{\partial h}, \qquad \frac{\partial V}{\partial q_i} = p_i \qquad (i = 1, 2, \ldots, n).$$

Write

$$F_j(q_j) = \sqrt{\psi_j(q_j) + h\phi_j(q_j) + \alpha_j} \qquad (j = 1, 2, \ldots, n-1);$$

then,

$$\int \sqrt{\frac{A_1(q_1)}{2F_1(q_1)}}\, dq_1 + \beta_1 = \int \sqrt{\frac{A_2(q_2)}{2F_2(q_2)}}\, dq_2 + \beta_2 = \cdots$$

$$= \int \sqrt{\frac{A_{n-1}(q_{n-1})}{2F_{n-1}(q_{n-1})}}\, dq_{n-1} + \beta_{n-1}$$

$$= \int \sqrt{\frac{A_n(q_n)}{2F_n(q_n)}}\, dq_n,$$

$$t - t_0 = \int \phi_1(q_1)\sqrt{\frac{A_1(q_1)}{2F_1(q_1)}}\, dq_1 + \cdots + \int \phi_n(q_n)\sqrt{\frac{A_n(q_n)}{2F_n(q_n)}}\, dq_n.$$

Thus we obtain the solution by taking $V$ as the sum of functions each containing a single variable $q_i$. Such a procedure is called the *separation of variables*; and most practical problems in dynamics can be solved by this means. Examples are Keplerian elliptic motion, the one-body problem in general relativity theory, the motion of a particle attracted by two fixed mass centers, the Stark effect in the classical quantum theory, the Sommerfeld theory on the fine structure of spectral lines due to the relativity shift, and the nonperturbed quantum-mechanical Kepler problem. Vinti's treatment of the motion of an earth satellite in the gravitational field of the distorted earth is another example (see Section 2.21). Robertson (1928) and Eisenhardt (1934) applied the separation of variables to the Schrödinger equation. In the following we enumerate a few examples. Such motions have an interesting bearing on quasi-periodic functions and almost periodic functions, as will be seen in the next chapter (Wintner, 1938).

For the motion of a third body around two fixed centers with masses $m$ and $m'$ we employ elliptic coordinates (Charlier, 1906, 1906a; Hiltebeitel, 1911; Badaljan, 1934; see also Section 2.24):

$$\lambda = \tfrac{1}{2}(r + r'), \qquad \mu = \tfrac{1}{2}(r - r'),$$

where $r$ and $r'$ denote the distances of the third body from the two fixed centers, which are in turn separated by a distance $2c$. Then

$$2T = (\lambda^2 - \mu^2)\left(\frac{\dot{\lambda}^2}{\lambda^2 - c^2} + \frac{\dot{\mu}^2}{c^2 - \mu^2}\right),$$

$$U = \frac{1}{\lambda^2 - \mu^2}\,[(m + m')\lambda - (m - m')\mu].$$

The problem studied by Weierstrass (1894) of the geodesics on the surface of an ellipsoid is a special case of this problem.

The motion of an electron with mass $m$ and electric charge $-e$ in the Coulomb field of an atomic nucleus of charge $Ze$ superposed on the uniform electrostatic field of intensity $E$ can be described, if we suppose that the atomic nucleus is at rest and denote the distance to the electron by $r$, by the Hamiltonian function (Schwarzschild, 1916; Epstein, 1916, 1919, 1922)

$$H = \frac{m}{2}\,(\dot{x}^2 + \dot{y}^2 + \dot{z}^2) - \frac{e^2 Z}{r} + eE.$$

Transform to parabolic coordinates by

$$z + ir = \tfrac{1}{2}(\xi + i\eta)^2, \qquad x^2 + y^2 = r^2,$$

or by

$$x = \xi\eta \cos \phi, \qquad y = \xi\eta \sin \phi, \qquad z = \tfrac{1}{2}(\xi^2 - \eta^2), \qquad r = \xi\eta;$$

then,

$$T = \frac{m}{2}\{(\xi^2 + \eta^2)(\dot{\xi}^2 + \dot{\eta}^2) + \xi^2\eta^2\dot{\phi}^2\},$$

$$U = \frac{2e^2Z}{\xi^2 + \eta^2} - \tfrac{1}{2}eE(\xi^2 - \eta^2).$$

The motion of an electron with charge $-e$ around an atomic nucleus of charge $Ze$ was treated in accordance with the special relativity theory by Sommerfeld (1916, 1924), and by Gleich (1926):

$$T = c^2(m - m_0) = c^2m_0\left(\frac{1}{\sqrt{1 - \beta^2}} - 1\right),$$

and

$$H = m_0c^2\left(\frac{1}{\sqrt{1 - \beta^2}} - 1\right) - \frac{e^2Z}{r},$$

where

$$m = \frac{m_0}{\sqrt{1 - \beta^2}}, \qquad \beta = \frac{v}{c},$$

and $m_0$ denotes the rest mass of the electron and $c$ the velocity of light.

## 1.16 STÄCKEL'S THEOREM

Suppose that a Hamilton-Jacobi equation

$$H\left(q_1, \ldots, q_n, \frac{\partial V}{\partial q_1}, \ldots, \frac{\partial V}{\partial q_n}\right) = h \tag{58}$$

can be integrated by the separation of variables, so that

$$V = \sum_{i=1}^{n} V^{(i)}(q_i). \tag{59}$$

Then (58) takes the form

$$\frac{1}{2}\sum_k A_k\left(\frac{\partial V}{\partial q_k}\right)^2 - (U + h) = 0. \tag{60}$$

How shall we write $A_k$ and $U$ as functions of $q_i$ in order to solve the problem by the separation of variables in the form (59)?

Denote, for convenience, $\partial V/\partial q_k = V_k$, where $V_k$ is a function of $q_k$ only. Then (60) becomes

$$\sum_k A_k V_k^2 = 2(U + h). \tag{61}$$

The complete integral $V$ of this equation is a function of $q_1, \ldots, q_n$ and $\alpha_1, \ldots, \alpha_n$. Differentiate (61) with respect to $\alpha_k$:

$$A_1 \frac{\partial(V_1^2)}{\partial\alpha_k} + A_2 \frac{\partial(V_2^2)}{\partial\alpha_k} + \cdots + A_n \frac{\partial(V_n^2)}{\partial\alpha_k} = 2\frac{\partial h}{\partial\alpha_k} \quad (k = 1.\, 2, \ldots, n). \tag{62}$$

As $V$ is the complete integral, we have $\Delta = \|\partial^2 V/\partial q_i\, \partial\alpha_j\| \neq 0$. But $\Delta_2 = \|\partial(V_j^2)/\partial\alpha_i\| = 2^n V_1 V_2 \cdots V_n \Delta$. Because $\Delta \neq 0$, we have $\Delta_2 \neq 0$. Hence we can solve the linear simultaneous equations (62) for $A_1, \ldots, A_n$:

$$A_i = \frac{2}{\Delta_2} \sum_{j=2}^{n} \frac{\partial h}{\partial\alpha_j} \frac{\partial\Delta_2}{\partial[\partial(V_i^2)/\partial\alpha_j]}. \tag{63}$$

Then (61) takes the form

$$U = -h + \sum_{i=1}^{n} \sum_{j=1}^{n} \frac{V_j^2}{\Delta_2} \frac{\partial\Delta_2}{\partial[\partial(V_i^2)/\partial\alpha_j]} \frac{\partial h}{\partial\alpha_j}.$$

Now

$$\frac{1}{\Delta_2} \frac{\partial\Delta_2}{\partial[\partial(V_i^2)/\partial\alpha_j]}$$

is a minor of $\Delta_2$ with regard to $\partial(V_i^2)/\partial\alpha_j$, and

$$\Delta_2 = \sum_j \frac{\partial(V_i^2)}{\partial\alpha_j} \frac{\partial\Delta_2}{\partial[\partial(V_i^2)/\partial\alpha_j]}.$$

Hence

$$U = \sum_j \sum_i \left[V_i^2 \frac{\partial h}{\partial\alpha_j} - h\frac{\partial(V_i^2)}{\partial\alpha_j}\right] \frac{1}{\Delta_2} \frac{\partial\Delta_2}{\partial[\partial(V_i^2)/\partial\alpha_j]}. \tag{64}$$

We have made no assumption in (63) and (64) other than that $V$ is the complete integral. Assume now that the problem can be solved by the separation of variables in the form (59). Then, because of (61), $V_i$ is a function of $q_i$ only. As $\Delta_2 \neq 0$, we can solve (62), and the result is expressed in the form (63). The derivative $\partial(V_i^2)/\partial\alpha_j$ is a function of $q_i$ only. Thus,

$$\Delta_2 = \|\phi_{ij}(q_i)\|, \tag{65}$$

where $\phi_{ij}(q_i) = \partial(V_i^2)/\partial\alpha_j$. Put

$$\psi_{ij}(q_i) = V_i^2 \frac{\partial h}{\partial \alpha_j} - h \frac{\partial(V_i^2)}{\partial \alpha_j},$$

which is a function of $q_i$ only; then we obtain the following

*Theorem:* In order that

$$H = \frac{1}{2} \sum_i A_i \left(\frac{\partial V}{\partial q_i}\right)^2 - U = h$$

can be solved by the separation of variables, there must exist two sets of single-valued finite continuous functions $\phi_{ij}(q_i)$ and $\psi_{ij}(q_i)$, such that $\Delta_2$, in (65), formed with these functions does not vanish; that any $\phi_{ij}(q_i)$ does not change its sign; and that $A_i$ and $U$ are of the forms

$$A_i = \frac{2}{\Delta_2} \sum_{j=1}^{n} \frac{\partial h}{\partial \alpha_j} \frac{\partial \Delta_2}{\partial \phi_{ij}}, \qquad U = \sum_{j=1}^{n} \sum_{i=1}^{n} \psi_{ij}(q_i) \frac{1}{\Delta_2} \frac{\partial \Delta_2}{\partial \phi_{ij}}.$$

This is called *Stäckel's theorem* (Stäckel, 1905; Kneser, 1894; Ahrens, 1895).

Substitute (63) and (64) in (60). Referring to the relation

$$\sum_i \frac{1}{\Delta_2} \frac{\partial \Delta_2}{\partial \phi_{ij}} \phi_{ij} = 1,$$

we have

$$\frac{1}{\Delta_2} \sum_i \sum_j \frac{\partial \Delta_2}{\partial \phi_{ij}} \left[\left(\frac{\partial V}{\partial q_i}\right)^2 \frac{\partial h}{\partial \alpha_j} - \psi_{ij}(q_i) - \frac{h}{n} \phi_{ij}\right] = 0.$$

Hence,

$$\left(\frac{\partial V}{\partial q_i}\right)^2 = \sum_j^n \frac{1}{\partial h/\partial \alpha_j} \left[\psi_{ij}(q_i) + \frac{h}{n} \phi_{ij}(q_i)\right], \tag{66}$$

or

$$V = \sum_{i=1}^{n} \int \sqrt{\sum_{j=1}^{n} \left\{\left[\psi_{ij}(q_i) + \frac{h}{n} \phi_{ij}(q_i)\right] \bigg/ \frac{\partial h}{\partial \alpha_j}\right\}}\, dq_i. \tag{67}$$

This $V$ satisfies (61), and contains the independent variables $\alpha_1, \alpha_2, \ldots, \alpha_n$ (independent because $\Delta_2 \neq 0$). Hence, the solution of our problem is

$$\frac{\partial V}{\partial \alpha_k} = \nu_k t - \beta_k = \sum_{i=1}^{n} \int \frac{(1/2n)(\partial h/\partial \alpha_k)\phi_{ik}(q_i)\, dq_i}{\sqrt{\sum_{j=i}^{n}[\psi_{ij}(q_i) + (h/n)\phi_{ij}(q_i)]}}, \tag{68}$$

where $\nu_k = \partial h/\partial \alpha_k$ and $\beta_k$ are the necessary $n$ integration constants. Putting

$$\Phi_i(q_i) \equiv 4n^2\left[\sum_{j=1}^{n} \psi_{ij}(q_i) + \frac{h}{n}\phi_{ij}(q_i)\right],$$

we obtain

$$dt = \sum_{i=1}^{n} \frac{\phi_{ik}(q_i)\, dq_i}{\sqrt{\Phi_i(q_i)}} \qquad (k = 1, 2, \ldots, n). \tag{69}$$

If we write $x_k = \nu_k t - \beta_k = (\partial h/\partial \alpha_k)t - \beta_k$, then $q_1, q_2, \ldots, q_n$ are expressed as functions of $x_1, x_2, \ldots, x_n$, which, in turn, are linear in $t$. In particular, we can write

$$\Phi_i(q_i) = (q_i - a_i)(b_i - q_i)\Psi_i(q_i), \tag{70}$$

where $\phi_{ij}(q_i)$ and $\Psi_i(q_i)$ are single-valued, finite, and continuous in $a_i \leqq q_i \leqq b_i$, and $\phi_{ij}(q_i)$ does not change its sign. The function $\Psi_i(q_i)$ is always positive and never becomes zero. The determinant $\|\phi_{ij}(\dot{q}_i)\|$ is always positive or always negative and never becomes zero. In such a case

$$x_j = \sum_{i=1}^{n} \int_{a_i}^{b_i} \frac{\phi_{ij}(q_i)\, dq_i}{\sqrt{(q_i - a_i)(b_i - q_i)\Psi_i(q_i)}}, \tag{71}$$

and $q_i$ is single-valued, finite, and continuous, and has $n$-fold period

$$2\omega_{i1}, 2\omega_{i2}, \ldots, 2\omega_{in},$$

where

$$\omega_{ij} = \int_{a_i}^{b_i} \frac{\phi_{ij}(q_i)\, dq_i}{\sqrt{(q_i - a_i)(b_i - q_i)\Psi_i(q_i)}}. \tag{72}$$

Note that a dynamical system that can be integrated by Liouville's theorem can be integrated by the separation of variables according to Stäckel's theorem; however, the converse is only true for dynamical systems with two degrees of freedom (Rao, 1935, 1935a; Vrkljan, 1929, 1930).

In quantum mechanics the separation of variables has an important bearing on the quantization. From $p_i = \partial V/\partial q_i$ the *action*

$$V_i = \int p_i \, dq_i = \int p_i \dot{q}_i \, dt$$

is computed by integrating over a closed curve in the variable $q_i$ and quantization is applied to the action itself. Write $V_i$ as $J_i$ and put $J_i = n_i h$ with Planck's constant $h$. The constant $n_i$ is a positive integer. We assume that there is a limit function of $J = \sum_{i=1}^{n} J_i(q_i)$ as $n$ approaches infinity. The integration constant $\alpha_i$ measures the energy of the quantum state, and $\beta_i$ measures the instantaneous electron configuration in the atomic system defined by the values of $\alpha_i$. The $n$ variables $\alpha_i$ are not independent, as the total energy of the system is fixed. Let $\nu_i$ be the frequency and $E_i = \nu_i J_i$ be the energy of the $i$th quantum state according to Bohr, then the total energy is $E = \sum_{i=1}^{n} E_i$. Transform to the conjugate variables $w_i$ and $J_i$ by

$$w_i = \frac{\partial V}{\partial J_i}, \qquad p_i = \frac{\partial V}{\partial q_i};$$

then $q_i$ is a multiply periodic function with period $2\pi$ for each $w_i$. As we obtain these variables in the form

$$w_i = \nu_i t - \beta_i, \qquad \nu_i = \frac{\partial H}{\partial J_i},$$

$q_i$ is expressed in multiple Fourier series

$$q_i = \sum_{j_1, \ldots, j_n} C^{(i)}_{j_1 j_2 \ldots j_n} \exp\,[2\pi\sqrt{-1}(j_1 w_1 + \cdots + j_n w_n)],$$

where $C^{(i)}_{j_1 j_2 \ldots j_n}$ is a function of $J_1, J_2, \ldots, J_n$ only. Poincaré calls the variable $w_i$ the *angular variable* and $J_i$ the *linear variable*. In quantum mechanics $J_i$ is called the *action variable*. Write $E_j - E_i = h\nu_{ij}$, according to Einstein; this is the energy of the light quantum absorbed or emitted by the transition from the $i$th quantum state to the $j$th. This is the quantum theory of radiation of Bohr and Einstein (Bohr, 1923, 1924; Kramers, 1920). Kneser (1921) has discussed the separation of variables as regards the quantization according to Sommerfeld (1916), Schwarzschild (1916) and Epstein (1916, 1919).

## 1.17 CONDITION FOR SEPARABILITY

The condition for the separability of variables in the Hamiltonian equations has been obtained by Morera (1881) for $n = 2$ and by

Dall'Aqua (1908, 1912) for $n = 3$. Liouville's and Stäckel's cases are concrete examples. Both theorems are concerned with the existence of integrals that are quadratic in the momenta (cf. Liouville, 1895; Levi-Civita, 1896; Painlevé, 1897; Fues, 1925; Weinacht, 1928).

In order that the solution of the Hamilton-Jacobi equation

$$H(p_1, \ldots, p_n, q_1, \ldots, q_n) = h, \qquad p_i = \frac{\partial V}{\partial q_i},$$

can be written in the form $V = \sum_{i=1}^{n} V_i$, it is necessary and sufficient that

$$\frac{dp_i}{dq_j} = 0 \qquad (i \neq j) \tag{73}$$

for all combinations of $i, j = 1, 2, \ldots, n$, in which $p_i$ is considered a function of $q_1, \ldots, q_n$. Under this condition

$$\frac{\partial H}{\partial q_i} + \frac{\partial H}{\partial p_i}\frac{\partial p_i}{\partial q_i} = 0.$$

If none of the $\partial H/\partial p_i$ vanish, then

$$\frac{\partial p_i}{\partial q_i} = -\frac{\partial H}{\partial q_i}\bigg/\frac{\partial H}{\partial p_i}. \tag{74}$$

Hence the condition for the separability of variables is

$$\frac{d}{dq_j}\left(\frac{\partial H}{\partial q_i}\bigg/\frac{\partial H}{\partial p_i}\right) = 0 \qquad (i \neq j),$$

where

$$\frac{d}{dq_j} = \frac{\partial}{\partial q_j} + \frac{\partial p_j}{\partial q_j}\frac{\partial}{\partial p_j}.$$

By using (74) for $\partial p_i/\partial q_i$ we have as the necessary and sufficient condition for the separability of variables that $H$ should satisfy the $n(n-1)/2$ equations

$$\frac{\partial H}{\partial p_i}\frac{\partial H}{\partial p_j}\frac{\partial^2 H}{\partial q_i\,\partial q_j} - \frac{\partial H}{\partial p_i}\frac{\partial H}{\partial q_j}\frac{\partial^2 H}{\partial q_i\,\partial p_j} - \frac{\partial H}{\partial q_i}\frac{\partial H}{\partial p_j}\frac{\partial^2 H}{\partial p_i\,\partial q_j} + \frac{\partial H}{\partial q_i}\frac{\partial H}{\partial q_j}\frac{\partial^2 H}{\partial p_i\,\partial p_j} = 0. \tag{75}$$

Let

$$T = \frac{1}{2}\sum_{r,s} a_{rs}\dot{q}_r\dot{q}_s$$

and let the contravariant quantity to $a_{rs}$ be $a^{sr}$; then,

$$\mathfrak{T} = \frac{1}{2}\sum_{r,s} a^{rs}p_r p_s, \qquad H = \mathfrak{T} - U.$$

Now insert

$$\dot{q}_i = \frac{\partial \mathfrak{T}}{\partial p_i}, \qquad \left(\frac{\partial \mathfrak{T}}{\partial q_i}\right)_p = -\left(\frac{\partial T}{\partial q_i}\right)_{\dot{q}}$$

in (75) and put the coefficients of the fourth, second, and zero-th degree terms equal to zero; then

$$\frac{\partial \mathfrak{T}}{\partial p_i}\frac{\partial \mathfrak{T}}{\partial p_j}\frac{\partial^2 \mathfrak{T}}{\partial q_i\,\partial q_j} - \frac{\partial \mathfrak{T}}{\partial p_i}\frac{\partial \mathfrak{T}}{\partial q_j}\frac{\partial^2 \mathfrak{T}}{\partial q_i\,\partial p_j} - \frac{\partial \mathfrak{T}}{\partial q_i}\frac{\partial \mathfrak{T}}{\partial p_j}\frac{\partial^2 \mathfrak{T}}{\partial p_i\,\partial q_j} + \frac{\partial \mathfrak{T}}{\partial q_i}\frac{\partial \mathfrak{T}}{\partial q_j}\frac{\partial^2 \mathfrak{T}}{\partial p_i\,\partial p_j} = 0, \tag{76}$$

$$\frac{\partial \mathfrak{T}}{\partial p_i}\frac{\partial \mathfrak{T}}{\partial p_j}\frac{\partial^2 U}{\partial q_i\,\partial q_j} - \frac{\partial \mathfrak{T}}{\partial p_i}\frac{\partial^2 \mathfrak{T}}{\partial q_i\,\partial p_j}\frac{\partial U}{\partial q_j} - \frac{\partial \mathfrak{T}}{\partial p_j}\frac{\partial^2 \mathfrak{T}}{\partial q_j\,\partial p_i}\frac{\partial U}{\partial q_i} + \frac{\partial^2 \mathfrak{T}}{\partial p_i\,\partial p_j}\left(\frac{\partial \mathfrak{T}}{\partial q_i}\frac{\partial U}{\partial q_j} + \frac{\partial \mathfrak{T}}{\partial q_j}\frac{\partial U}{\partial q_i}\right) = 0, \tag{77}$$

$$\frac{\partial^2 \mathfrak{T}}{\partial p_i\,\partial p_j}\frac{\partial U}{\partial q_i}\frac{\partial U}{\partial q_j} = a^{ij}\frac{\partial U}{\partial q_i}\frac{\partial U}{\partial q_j} = 0 \qquad (i \neq j). \tag{78}$$

Equation (76) determines the form of $\mathfrak{T}$, and (77) and (78) determine $U$.

Let us transform (76) first. Write

$$\sigma_{ij} = -\frac{\partial \mathfrak{T}}{\partial p_j}\frac{\partial^2 \mathfrak{T}}{\partial p_i\,\partial q_j} + a^{ij}\frac{\partial \mathfrak{T}}{\partial q_j}.$$

As $\partial^2\mathfrak{T}/\partial p_i\,\partial p_j = a^{ij}$, equation (76) becomes, when $i$ is fixed,

$$\frac{\partial \mathfrak{T}}{\partial p_i}\left(\frac{\partial \mathfrak{T}}{\partial p_j}\frac{\partial^2 \mathfrak{T}}{\partial q_i\,\partial q_j} - \frac{\partial \mathfrak{T}}{\partial q_j}\frac{\partial^2 \mathfrak{T}}{\partial q_i\,\partial p_j}\right) + \frac{\partial \mathfrak{T}}{\partial q_i}\sigma_{ij} = 0.$$

Here $\partial\mathfrak{T}/\partial p_i$ is linear with respect to $p$, and $\partial\mathfrak{T}/\partial q_i$ and $\sigma_{ij}$ are quadratic with respect to $p$. In order that this formula shall hold identically, one of the latter two expressions must be divisible by $\partial\mathfrak{T}/\partial p_i$. But

$$\frac{\partial \mathfrak{T}}{\partial q_i} = -\frac{\partial T}{\partial q_i} = -\sum_{r,s}\frac{\partial a_{rs}}{\partial q_i}\dot{q}_r\dot{q}_s.$$

Hence, in order that $\partial\mathfrak{T}/\partial q_i$ be divisible by $\partial\mathfrak{T}/\partial p_i$, we should have

$$\frac{\partial a_{rs}}{\partial q_i} = 0 \qquad (r, s \neq i). \tag{79}$$

Furthermore

$$\sigma_{ij} = \dot{q}_j \sum_l a^{il} \frac{\partial^2 T}{\partial \dot{q}_l \, \partial q_j} - a^{ij} \frac{\partial T}{\partial q_j}.$$

Hence, in order that this expression be divisible by $\partial \mathfrak{T}/\partial p_i = \dot{q}_i$, the following must be true:

$$a^{ij} \frac{\partial a_{rs}}{\partial q_j} = 0, \qquad \sum_{l=1}^{n} a^{il} \frac{\partial a_{rl}}{\partial q_j} - a^{ij} \frac{\partial a_{rj}}{\partial q_j} = 0,$$
$$\sum_l a^{il} \frac{\partial a_{jl}}{\partial q_j} - \tfrac{1}{2} a^{ij} \frac{\partial a_{jj}}{\partial q_j} = 0 \qquad (j, r, s \neq i; r, s \neq j). \tag{80}$$

Thus either (79) or (80) must be valid.

In order that all expressions $\partial \mathfrak{T}/\partial q_i$ shall be divisible by $\partial \mathfrak{T}/\partial p_i$, the Christoffel symbol of the first kind must be zero, i.e.,

$$\left[{i\,j \atop r}\right] = 0, \tag{81}$$

or the Riemannian curvature tensor (Bianchi, 1899, 1922, 1924)

$$F_{rs,ij} = \frac{\partial}{\partial q_j} \left\{{r\,i \atop s}\right\} - \frac{\partial}{\partial q_i} \left\{{r\,j \atop s}\right\} + \sum_l \left( \left\{{r\,i \atop l}\right\}\left\{{l\,j \atop s}\right\} - \left\{{r\,j \atop l}\right\}\left\{{l\,i \atop s}\right\} \right)$$

should be zero, that is,

$$F_{rs,ij} = 0. \tag{82}$$

In other words, the Riemannian curvature of the Riemann space $ds^2 = \sum_{r,s} a_{rs} \, dx_s \, dx_r$ should be zero, i.e., the space should be differentially transformed to a Euclidean $n$-dimensional flat space. Several examples with $ds^2$ for the expressions of $T$ which satisfy either of these conditions were given by Levi-Civita (1904) and Dall'Aqua (1908, 1912) (cf. Kneser, 1921; Weinacht, 1928; Fues, 1925). Weinacht (1928) has given several separable examples and solved the problem by referring to elliptic coordinates.

In order that the wave equation in Schrödinger's wave mechanics,

$$\sum_{i=1}^{n} \sqrt{H} \frac{\partial}{\partial x_i} \left( \frac{H_i}{\sqrt{H}} \frac{\partial \Psi}{\partial x_i} \right) + k^2 (E - V) \Psi = 0,$$

$$H = \prod_{i=1}^{n} H_i(x_i),$$

be solvable by the separation of variables in the form $\Psi = \prod_{i=1}^{n} \Psi_i(x_i)$, it is necessary that the coefficients of $\partial^2\Psi/\partial x_i^2$ and $\partial\Psi/\partial x_i$ differ by a factor which is a function of only a single variable $x_i$. Hence

$$\frac{\sqrt{H}}{H_i}\frac{\partial}{\partial x_i}\left(\frac{H_i}{\sqrt{H}}\right)$$

is a function of only a single variable $x_i$, and

$$H_i = \sqrt{H}\, a_i(x_1, x_2, \ldots, x_{i-1}, x_{i+1}, \ldots, x_n) f_i(x_i),$$

and the wave equation is of the form

$$\sum_i \sqrt{H}\frac{a_i}{\Psi}\frac{\partial}{\partial x_i}\left(f_i\frac{\partial\Psi}{\partial x_i}\right) + k^2(E - V) = 0.$$

This equation has a solution containing $n$ constants $\alpha_1, \alpha_2, \ldots, \alpha_n$ beside $E$. We construct

$$\phi_{ij}(x_i) = \frac{\partial}{\partial\alpha_j}\left[\frac{1}{\Psi_i}\frac{\partial}{\partial x_i}\left(f_i\frac{\partial\Psi_i}{\partial x_i}\right)\right]$$

by choosing $\alpha_1, \alpha_2, \ldots, \alpha_n$ so that $\|\phi_{ij}(x_i)\| \neq 0$. Then the wave equation becomes

$$\sum_i \sqrt{H} a_i \phi_{ij} = -k^2\,\delta_{1j}$$

and $\sqrt{H}\, a_i = -k^2\phi^{i1}$. Hence, $H_i = -k^2 f_i(x_i)\phi^{i1}$, where $\phi\phi^{ij}$ is the minor of $\|\phi_{ij}\|$ and does not depend on $x_i$ in the same way as $a_i$. Hence $\sqrt{H}\,\phi = c =$ constant. Thus we should have

$$V = \sum_i \phi^{i1}\chi_i(x_i),$$

$$\chi_i = \alpha_1\phi_{i1}(x_i) - \frac{1}{\Psi_i}\frac{d}{dx_i}\left(f_i\frac{d\Psi_i}{dx_i}\right),$$

or, by putting

$$\Phi_{ij}(x_i) = -k^2 f_i(x_i)^{-1}\phi_{ij}(x_i),$$

$$F_i(x_i) = -k^2 C^{-1/n} f_i(x_i),$$

$$X_i(x_i) = -k^2 f_i(x_i)\chi_i(x_i),$$

it is necessary that

$$H_i = \Phi^{i1}, \qquad V = \sum_{i=1}^{n} \Phi^{i1} X_i(x_i), \qquad \sqrt{H}\,\Phi = \prod_{i=1}^{n} (F_i(x_i))^{-1}.$$

To prove that this condition is sufficient, first note that

$$\sum_{i=1}^{n} \Phi^{i1}\Omega_i(x_i) = 0,$$

where

$$\Omega_i(x_i) = \frac{1}{F_i \Psi_i} \frac{d}{dx_i}\left(F_i \frac{d\Psi_i}{dx_i}\right) + k^2(\alpha_1 \Phi_{i1} - X_1).$$

This relation represents the circumstance that a vector $(\Psi_1, \Psi_2, \ldots, \Psi_n)$ is orthogonal to a vector $(\Phi^{11}, \ldots, \Phi^{n1})$. It is known that the vector $(\Phi_{1j}, \ldots, \Phi_{nj})$ $(j = 2, 3, \ldots, n)$ has such a property. Hence, the vector $(\Psi_1, \ldots, \Psi_n)$ should be represented by a linear combination of the vectors $\Phi$. We write the constant coefficients of this linear expression as $-k^2\alpha_2, \ldots, -k^2\alpha_n$; then

$$\Omega_i(x_i) = -k^2 \sum_{j=1}^{n} \alpha_j \Phi_{ij}(x_i).$$

Hence $\Psi_i$ is determined by the following equation (Robertson, 1928; Brillouin, 1926, 1926a, 1935; Eisenhardt, 1934, 1948, 1949; Rao, 1935, 1935a; Kemble, 1937):

$$\frac{1}{F_i(x_i)} \frac{d}{dx_i}\left[F_i(x_i) \frac{d\Psi_i}{dx_i}\right] + k^2\left[\sum_{j=1}^{n} \alpha_j \Phi_{ij}(x_i) - X_i(x_i)\right]\Psi_i = 0.$$

The convergence of the solution in eigenfunction expansions in the separable cases of the Schrödinger equation has been proved by Titchmarsh (1955, 1958).

## HAMILTONIAN MATRIX

### 1.18 HAMILTONIAN MATRIX

Operations on a matrix obey the rules for a determinant. A matrix obtained by an interchange of the rows and the columns of a matrix $M$ is called a *transposed matrix* and denoted by $M'$. When the number of rows equals the number of columns, the array is called a square matrix and the number of rows or columns is called the order. A matrix whose diagonal elements are all 1 while the remaining elements are all 0, is a unit matrix $E$ or $\epsilon$. A matrix $M^{-1}$ such that $MM^{-1} = E$ is the *inverse matrix* of $M$. The determinant formed of a square matrix $M$ is denoted by det $M$. Obviously det $E = 1$. A matrix such that det $M \neq 0$ is said to be *nonsingular*; otherwise, *singular*. A matrix whose elements are all zero is a *null matrix* $\omega$.

Two fixed nonsingular matrices $A$ and $B$ satisfy the symmetry relation $(AHB)' = AHB$ for any symmetric matrix $H = H'$ if and only if there exists a number $s \neq 0$ such that $A' = sB$. Put $M' = A'B^{-1}$; then the condition $(AHB)' = AHB$ is equivalent to $MH = HM'$, where $H = H'$ is arbitrary and may be chosen to be $E$. Hence $M = M'$. Now $MH = HM$ holds for any $H = H'$ if and only if $M$ is of the form $sE$. This is evident if $M$ is a diagonal matrix. If it is not diagonal, then $M = M'$ is transformed by means of an orthogonal matrix $R$ into a diagonal matrix $RMR^{-1}$, whereas $R$ transforms the set of all matrices $H = H'$ into itself. Thus $RMR^{-1}$ is a diagonal matrix and is equal to $sE$. Finally, $s \neq 0$, because $sE = M = M' = A'B^{-1}$ is the product of two nonsingular matrices. Hence $A' = sB$, $s \neq 0$ (Wintner, 1934; Williamson, 1936, 1937, 1939, 1940).

Consider a matrix composed of a null matrix $\omega$ of order $n$ and a unit matrix $\epsilon$ of order $n$, such as

$$G = \begin{pmatrix} \omega & -\epsilon \\ \epsilon & \omega \end{pmatrix},$$

in contrast to

$$E = \begin{pmatrix} \epsilon & \omega \\ \omega & \epsilon \end{pmatrix};$$

then,

$$G = -G^{-1} = -G'.$$

This is the matrix formed by the coefficients of $\delta$ and $d$ in a bilinear covariant expression (Section 1.7) and is naturally nonsingular.

A system of $2n$ ordinary differential equations of the first order that are linear and homogeneous and do not contain $t$ explicitly is *canonical* if and only if there exists a symmetric matrix $H$ such that the differential equations can be written in the form

$$G\dot{x} = Hx.$$

In fact, this is (up to the factor 2) the matrix of the $2n$-ary quadratic form representing the Hamiltonian function. Since $G^{-1} = -G$, we can write

$$\dot{x} = -GHx, \qquad H = H'.$$

A nonsingular matrix $C$ independent of $t$ is called *Hamiltonian* if the transformation $x = Cy$ transforms any differential system of this form to a differential system of the same form (Section 1.3).

If we substitute $x = Cy$ in the above equation, we obtain $\dot{y} = -GKy$, where $K = -GC^{-1}GHC$, since $G = -G^{-1}$. Hence, $C$ is a Hamiltonian matrix if and only if $K$ is the matrix of a $2n$-ary quadratic form, that is, if and only if $K = K'$ whenever $H = H'$. We can write $K = K'$ in the form $(AHB)' = AHB$ by putting $A = C^{-1}G$ and $B = CG$, so that $\det A \neq 0$, $\det B \neq 0$, i.e., $A' = sB$. Hence $C'GC = sG$,

$$|\det C| = |s|^n > 0,$$

if we write $s$ instead of $-s^{-1}$.

If a matrix $C$ of order $2n$ with real elements satisfies

$$C'GC = sG, \tag{83}$$

then it is said to be Hamiltonian (Wintner, 1937, 1947; van Kampen and Wintner, 1936). When $s$ is a constant different from zero, $s$ is called the *multiplier* of $C$ (van Kampen and Wintner, 1936, 1938; Wintner, 1937). If $\det G \neq 0$, then $|\det C| = |s|^n$.

The matrix

$$K = -GC^{-1}GHC$$

belonging to the transformed Hamiltonian function takes the form

$$K = s^{-1}C'HC.$$

Thus the necessary and sufficient condition (83) for a Hamiltonian matrix $C$ is that the bilinear form in cogradient variables belonging to the matrix $G$ be a relative invariant of a nonsingular linear substitution if and only if the matrix of this substitution is a Hamiltonian matrix. The group of the Hamiltonian matrices, in contrast to the rotation and reflection groups of Euclidean space, cannot be characterized by the absolute invariance of a bilinear form (Section 4.2). For example, if

$$C = \begin{pmatrix} \epsilon & \omega \\ \omega & -\epsilon \end{pmatrix},$$

then $C'GC = -G \neq G$ although $|\det C| = |(-1)^n| = 1$.

Since $s \neq 0$, a Hamiltonian matrix is nonsingular. If $C$ is Hamiltonian, then $C'$ is also Hamiltonian and has the same multiplier. When $s = 1$, then $C$ and $C^{-1}$ have the same eigenvalues; that is, the roots of

$$\det (C - \lambda E) = 0$$

and the roots of $\det (C^{-1} - \lambda E) = 0$ are the same. For, by the definition of a Hamiltonian matrix, $C' = GC^{-1}G^{-1}$ or $G^{-1}C'G = C^{-1}$. Hence

$$G^{-1}(C' - \lambda E)G = C^{-1} - \lambda E.$$

The totality of eigenvalues is called the *spectrum*; it may be discrete or continuous (Section 20.1).

Take a set of $m$ functions $Y(y_1, \ldots, y_m)$ and consider the square matrix $A = A(Y)$, of order $m$, and a vector $B = B(Y)$. A vector is a matrix with a single column consisting of the vector components. Suppose that there exists a scalar function $g = g(Y)$ satisfying

$$B + A \operatorname{grad} f = \operatorname{grad} g, \tag{84}$$

in the whole domain of the vector space $Y$ for any scalar function $f = f(Y)$. Then for all functions $f$ and for a given pair $A$ and $B$, there exists a function $g$ satisfying (84) if and only if $B$ has the form of the gradient of a scalar function and $A$ has the form $sE$, where $s$ is a scalar independent of $Y$ and $E$ is a unit matrix of order $m$. It is assumed that $A$, $B$ and $f$, $g$ all have continuous partial derivatives of first and second order.

The sufficiency of these conditions is obvious. To prove their necessity: suppose that there exists a function $g$ for any function $f$. Choose $f = f(Y)$ independent of $Y$; then, from (84), $B$ is a gradient. Hence, if we choose $f = f(y_1, \ldots, y_m)$ as an arbitrary polynomial $q(y_i)$ of the $i$th vector-component of $Y$, and denote by $A_i = A_i(Y)$ the vector whose components are the elements of the $i$th column of $A = A(Y)$, then the vector $q(y_i)A_i(Y)$ is a gradient for every polynomial $q(y_i)$. Hence $A(Y)$ is a diagonal matrix, since the integrability condition is satisfied by a gradient, and the $i$th diagonal element of $A(Y)$ is a function of $y_i$ alone. Furthermore, since $A \operatorname{grad} f$ is a gradient for any $f$, we see — by choosing $i$ and $j$ arbitrarily and putting $f(Y) = y_i y_j$ — that the diagonal elements of $A(Y)$ are all equal and hence $A = sE$ for a constant $s$.

## 1.19 CANONICAL TRANSFORMATION

Put

$$x_i = q_i, \qquad x_{n+i} = p_i \qquad (i = 1, 2, \ldots, n)$$

and take a vector $X = (x_1, \ldots, x_{2n})$; then Hamilton's equations can be written

$$G\dot{X} = \operatorname{grad}_X H, \qquad H = H(X, t), \tag{85}$$

where a dot denotes total differentiation with respect to the time $t$, and the subscript of "grad" refers to the space in which the partial differentiations are performed.

Now transform the phase space $X$ for any fixed $t$ into a phase space $Y$ by means of

$$Y = Y(X; t), \qquad Y = (y_1, \ldots, y_{2n}), \qquad X = (x_1, \ldots, x_{2n}). \tag{86}$$

Assume that these $2n$ functions have continuous partial derivatives of the second order considered as functions of $2n$ variables and $t$, and that the transformation builds a locally one-to-one correspondence between the two phase spaces for any fixed $t$, the Jacobian of order $2n$ being everywhere different from zero. A transformation satisfying these properties is called *canonical* if it transforms any canonical system (85) into a canonical system such that

$$G\dot{Y} = \text{grad}_Y K, \qquad K = K(Y, t). \tag{87}$$

Let $C$ be the Jacobian matrix of the transformation (86) for a fixed $t$, so that the $j$th element in the $k$th column of the nonsingular $2n$ order square matrix $C$ is the partial derivative of $y_j$ with respect to $x_k$. The matrix $C$ can be considered a function $C(Y, t)$ of $Y$ and $t$ rather than a function of $X$ and $t$. By direct differentiation we obtain

$$\dot{Y} = C\dot{X} + Y_t, \tag{88}$$

where the subscript $t$ refers to a partial differentiation with respect to $t$; furthermore

$$X = X(Y, t) \tag{86a}$$

is the inverse transformation of (86). Also, for any function $f$,

$$\text{grad}_X f = C' \text{grad}_Y f, \tag{89}$$

where $C'$ is the transposed matrix of $C$; finally,

$$\text{grad}_Y Y_t = C_t C^{-1}. \tag{90}$$

From (88) and (89) we see that (86), which is not assumed to be a constant transformation, transforms (85) into

$$GC^{-1}(\dot{Y} - Y_t) = C' \text{grad}_Y H.$$

Referring to the definition of $G$, we can put this equation in the form

$$G\dot{Y} = GY_t - GCGC' \text{grad}_Y H.$$

Comparing this equation with (87), we see that (86) is a canonical transformation if and only if there exists a $K = K(Y, t)$ for any $H$, such that

$$GY_t - GCGC' \text{grad}_Y H = \text{grad}_Y K, \tag{91}$$

where $H = H(X, t)$ is expressed as a function of $Y$ and $t$ by (86a). Compare (91) with (84) by keeping $t$ fixed; then we see that (86) is a canonical transformation if and only if the vector $B = GY_t$ is the gradient of a scalar function with respect to $Y$ for any fixed $t$, and that the matrix $A = -GCGC'$ is of the form $sE$ for any fixed $t$, where $s$ is a scalar function independent of $Y$, and $E$ is the unit matrix of order $2n$. Hence the transformation (86) is canonical if and only if there exist two scalars

$$R = R(Y, t), \qquad s = s(t),$$

such that

$$GY_t = \operatorname{grad}_Y R, \qquad C'GC = sG. \tag{92}$$

If this condition is satisfied, then the Hamiltonian function for (87), because of (91), is

$$K = sH + R. \tag{93}$$

The first condition of (92) cannot be applied directly to a given transformation (86), since $R$ is unknown. But it can be shown that there exists for a given transformation (86) a function $R$ satisfying this condition if and only if the matrix $C'GC$, considered as a function of $Y$ and $t$, is independent of $t$. In fact, the function $R$ exists for a given transformation (86) satisfying this condition if and only if $\operatorname{grad}_Y GY_t$ is a symmetric matrix at every point of the $(2n + 1)$-dimensional domain in question. Since we can verify that $\operatorname{grad}_Y GY_t = G \operatorname{grad}_Y Y_t$, we have $\operatorname{grad}_Y GY_t = GC_tC^{-1}$ by (90). Hence there exists a function $R$ if and only if $GC_tC^{-1}$ is a symmetric matrix, that is, is equal to the matrix $(GC_tC^{-1})'$; in other words, if and only if

$$C_t'GC + C'GC_t \equiv (C'GC)_t$$

is the null-matrix for any $Y$ and $t$, that is, if and only if $C'GC$ is a function of $Y$ alone.

The scalar $s$ in the necessary and sufficient condition (92) is independent of the vector $Y$ in the phase space for any fixed $t$. This fact is due to (84). The pair of conditions (92) implies that the scalar $s$ should be independent both of $Y$ and $t$. A transformation (86) can be shown to be canonical if and only if there exists an $s \neq 0$ independent of both $Y$ and $t$, such that the Jacobian matrix $C = C(Y, t)$ of (86) is a Hamiltonian matrix for any $Y$ and $t$ with the constant $s$ as the multiplier. This is because (86) is canonical if and only if (92) can be satisfied by a suitable choice of the functions $R$ and $s$. The first condition of (92) is satisfied if and only if $C'GC$ is independent of $t$, that is, $s$ is a constant by the second condition of (92), since $s \neq 0$. Hence, a transformation (86) is canonical if and only if its Jacobian matrix $C = C(Y, t)$ determines a linear substitution under which the bilinear form in cogradient variables belonging to the skew-symmetric matrix $G$ is a relative invariant with a multiplier $s \neq 0$ and is independent both of $Y$ and $t$.

It is clear from the definition of a canonical transformation that the set of canonical transformations defined in a common $(2n + 1)$-dimensional domain forms a group. If $R_1 = R_1(Y, t)$ belongs to a canonical transformation which is substituted into another canonical transformation characterized by $R_2 = R_2(Y, t)$ and $s_2$, then the function $R$ characterizing the composite transformation is $R = s_2R_1 + R_2$. If $C$, $R$, $s$ characterize a canonical transformation, then $C^{-1}$, $-R/s$, $1/s$ characterize the inverse transformation. The multiplier $s$ is uniquely determined by the canonical transformation, while the function $R$ remains indeterminate up to an additive term which is a function of $t$ alone.

If (86) satisfies the first condition of (92) for any $t$ and $Y$, and the second condition for a fixed $t$ and any $Y$, then (86) is a canonical transformation. For, if the first condition of (92) is satisfied for any $t$ and $Y$, then $C'GC$ is independent of $t$, so that the second condition of (92) holds for any $t$ and $Y$ if it holds for a single $t$ and any $Y$. Call a transformation (86) *conservative* if it does not contain $t$ explicitly. If (86) is a contact transformation and $t_0$ is a fixed value of $t$, then the conservative transformation defined by $Y = Y(X, t_0)$ is canonical. In fact, any conservative transformation satisfies the first condition of (92) if we put $R \equiv 0$; it also satisfies the second condition for any $t$ or no $t$. Thus (86) need not be canonical even if the conservative transformation $Y = Y(X, t_0)$ is canonical for any fixed $t = t_0$ (van Kampen and Wintner, 1936).

### 1.20 PFAFF'S CRITERION

Put

$$y_i = v_i, \qquad y_{n+i} = u_i \qquad (i = 1, 2, \ldots, n)$$

in (86); then (86) is a canonical transformation if and only if there exists a function $R$ for which the Pfaffian (Section 1.7)

$$s \sum_{i=1}^{n} p_i \, dq_i - \sum_{i=1}^{n} u_i \, dv_i + R \, dt \tag{94}$$

is a perfect differential by (86). Consider a Pfaffian

$$\tfrac{1}{2} s X'G \, dX - \tfrac{1}{2} Y'G \, dY + R \, dt, \tag{95}$$

where an expression such as $Z'AW$ denotes the bilinear form belonging to the square matrix $A$ and the vectors $Z$ and $W$, so that $Z'AW$ is considered as the product of three matrices. Whether the transformation (86) be canonical or not, the difference of the two Pfaffians (94) and (95) is a perfect differential of the function

$$s \sum_{i=1}^{n} p_i q_i - \sum_{i=1}^{n} u_i v_i.$$

Now (95) is a perfect differential in virtue of (86) if and only if (86) is canonical. In fact, (95) becomes a Pfaffian

$$\tfrac{1}{2}(sX'G - Y'GC)\,dX + (R - \tfrac{1}{2}Y'GY_t)\,dt$$

by (86), where $Y$ and $Y_t$ are considered as functions of $X$ and $t$. This last Pfaffian is a perfect differential of a function of $2n + 1$ variables $X$ and $t$ if and only if (i) the partial derivatives of the covariant vector

$$\tfrac{1}{2}(sX'G - Y'GC)$$

with respect to $t$ is the covariant gradient of the scalar $R - \frac{1}{2}Y'GY_t$ with respect to the vector $X$ and (ii) the Stokes alternating derivative of this covariant vector with respect to the vector $X$ is identically zero. It can be shown that conditions (i) and (ii) are equivalent to the two conditions of (92).

Actually, condition (i) is satisfied if and only if

$$-\tfrac{1}{2}(Y'GC)_t = (\operatorname{grad}_X R)' - \tfrac{1}{2}[\operatorname{grad}_X (Y'GY_t)]',$$

or, when transposed,

$$-\tfrac{1}{2}(C'G'Y)_t = \operatorname{grad}_X R - \tfrac{1}{2}C'GY_t + \tfrac{1}{2}C_t'GY,$$

that is, if and only if

$$C'GY_t = C' \operatorname{grad}_Y R.$$

Since this can be written in the form of the first condition of (92), condition (i) is equivalent to this first condition of (92). Furthermore, since an alternating derivative is transformed by any transformation (86) as a tensor, and since the alternating derivative of $X'G$ with respect to $X$ is $G$ by definition, then the alternating derivative of $Y'GC$ with respect to $X$ is $C'GC$ and condition (ii) is satisfied if and only if the matrix $\frac{1}{2}(sG - C'GC)$ is identically zero. Hence, condition (ii) is equivalent to the second condition of (92) (van Kampen and Wintner 1936; Wintner 1947).

### 1.21 POLAR FACTORIZATION

A matrix with complex elements is called *symplectic* according to Weyl (1939). If two symplectic matrices $A = (a_{jk})$ and $B = (b_{jk})$ are such that $a_{jk} = b_{jk}$ for all $j$ and $k$, then $A = B$. Let $\bar{A} = (\bar{a}_{jk})$ be a conjugate matrix of $A(a_{jk})$ such that $a_{kj} = \bar{a}_{jk}$, where a bar means the complex conjugate, so that $A' = \bar{A}$ and $\overline{A'} = A$. The matrix $A$ determines a bilinear form

$$A = A(x, y) = \sum_{k,j} a_{jk} y_j x_k$$

and

$$A(x_j) = \frac{\partial A(x, y)}{\partial y_j} = \sum_k a_{jk} x_k,$$

that is,

$$A(x) = (a)(x).$$

If $a_{kj} = \bar{a}_{jk}$, then $A(x, y)$ is called a *Hermitian form*. The Hermitian form $A(x, \bar{x}) = \overline{A(x, \bar{x})'} = \overline{A'}(x, \bar{x})$ is real. If $A(x, \bar{x})$ equals zero when all $x$ are equal to zero and never changes its sign, then it is called *definite*. If the sign is positive, i.e., $A(x, \bar{x}) > 0$, then it is called a Hermitian form. The product of two bilinear forms $A(x, y)$ and $B(x, y)$ is defined by

$$C = \sum_l \frac{\partial A}{\partial x_l} \frac{\partial B}{\partial y_l} = \sum_l \sum_j a_{jl} y_j \sum_k b_{lk} x_k$$

$$= \sum_{j,k} y_j x_k \sum_l a_{jl} b_{lk} = (c_{jk}),$$

where

$$c_{jk} = \sum_l a_{jl} b_{lk}.$$

Operate with

$$B(x_l) = \sum_k b_{lk} x_k$$

on

$$A(x_j) = \sum_l a_{jl} x_l.$$

Then we obtain

$$AB(x_j) = \sum_{l,k} a_{jl} b_{lk} x_k = \sum_k x_k \sum_l a_{jl} b_{lk} = \sum_k c_{jk} x_k,$$

that is, a bilinear form $C$; also $(AB)' = B'A'$, and $\overline{(AB)} = \bar{A}\bar{B}$.

The simplest Hermitian form is the unit-form $E = E(x, y) = \sum \bar{x}y$. We have $E(x, \bar{x}) = \sum (x)^2$. A form $U$ such that $\overline{U}'U = E$, $\overline{U}' = U^{-1}$, $U \neq 0$, is called *unitary*.

Take

$$R(z, t) = \sum_{j,k} t_{jk}t_j z_k, \qquad x_j = \frac{\partial R(z, t)}{\partial t_j},$$

and transform a Hermitian form $A$ to $R^{-1}AR$. Then $X = A(x, \bar{x})$ is transformed to $Z = \overline{R}'AR(z, \bar{z}) = X$, where $X$ is a Hermitian form. For, $X$ equals zero if all $x$ are zero and $Z$ equals zero if all $z$ are zero. $X$ is positive for all values of $x$, and $Z$ is positive for all $z$. Hence,

$$\overline{Z'} = \overline{(\overline{R}'AR)'} = Z.$$

Thus, a transformation $R$ transforms a Hermitian form $A$ to a Hermitian form $\overline{R}'AR$. Let a transformation of a Hermitian form $A$ to a Hermitian form $R^{-1}AR$ be $R = U$, $\overline{R}' = R^{-1}$; then the new Hermitian form gives a new Hermitian transformation.

Let the $N$ transformations of a group $G_n$ of a finite order $n$ be $A_l$ $(l = 1, 2, \ldots, N)$, and $H$ be any Hermitian form. The Hermitian form

$$\text{H} = \sum_l \overline{A}_l'HA_l(x, \bar{x})$$

is an absolute invariant by all transformations of $G_n$. If the invariant of $G_n$ is $E$, then $G_n$ is formed exclusively by unitary transformations and is itself unitary. Consider a transformation $T$ such that $\overline{T}' = T$. Let $\sigma$ be a real parameter and $i = \sqrt{-1}$. Then it can be proved that a transformation

$$\begin{aligned} U &= (\sigma E + iT)^{-1}(\sigma E - iT) \\ &= (\sigma E - iT)(\sigma E + iT)^{-1} \end{aligned}$$

is unitary. A unitary transformation forms a unitary group. It can be shown that this group is transitive (Section 4.27; Weyl, 1923, 1931, 1939; Wintner, 1929).

A Hermitian transformation $H$ and its transform $U^{-1}HU$ generated by a unitary transformation are considered to be equivalent. Take $H = (h_{jk})$, $h_{jk} = \bar{h}_{kj}$, $\overline{H}' = H$; then $\det H \neq 0$. Let $r$ be a root of the characteristic equation $\Delta = \Delta(r) = \det(H - rE) = 0$. There exists at least one point $\xi$ such that $\sum_k h_{jk}\xi_k = r\xi_j$ or $H(\xi) = r\xi$. Hence, a Hermitian transformation keeps at least one point $\xi$ invariant in the space of $\xi$.

By a suitable unitary transformation a Hermitian form of order $n$ can be decomposed into a Hermitian form $H_{n-1}$ of order $n - 1$ and a Hermitian form $h_{nn}x_ny_n$ of order $n$, as can be proved by considering a unitary transformation $W = [w_{jk}]$ and its subtransformation

$$W_{jk} = \partial|W|/\partial w_{jk},$$

such that $\overline{W'}W = E$, $W'^{-1} = \overline{W}$, that is, $(\det W)^{-1}W_{jk} = \bar{w}_{jk}$.

A matrix $C = (c_{jk})$ is called *normal* if $c_{jk} = 0$ for $j \neq k$, or $C$ is of the form $C = \sum_j c_{jj}x_jy_j$. If $R^{-1}AR$ formed of two matrices $R$ and $A$ is normal, then $R$ is called a *normalizing matrix* of $A$. When the characteristic equation for the transformation has only simple roots, then there always exists a normalizing matrix for $A$. When the characteristic equation has multiple roots, then $A$ cannot generally be normalized. Jordan (Turnbull and Aitken, 1932; Wedderburn, 1934) has proved that any transformation $A$ of finite order $n$ can be normalized. If the characteristic equation $\Delta = \Delta(r) = \det(A - rE) = 0$ has a root $r$ of multipicity $p$, then the minor determinants of $\Delta$ of orders $n - p + 1, n - p + 2, \ldots, n - 1, n$ vanish for this value of $r$. This property holds also for any Hermitian matrix, as well as for any unitary matrix.

It can be proved that the matrix obtained by multiplying a Hermitian matrix by itself $m$ times is also Hermitian, and that for a given Hermitian matrix $H$ there exists one and only one Hermitian matrix defined by $A^m = H$.

Let $P$ be an arbitrary matrix with $\det P \neq 0$; then $H = \bar{P}'P$ is Hermitian. Conversely, any Hermitian matrix can be put in the form $H = \bar{P}'P$. The matrix $P$ is of the form $P = U\sqrt{H}$ with a unitary matrix $U$ (Autonne, 1902).

A Hermitian form does not in general form a group. If a group $\mathrm{G}$ has a Hermitian form $H$ as an absolute invariant, then the group $H^{1/2}\mathrm{G}H^{-1/2}$ is unitary and has the Hermitian $E$ as its absolute invariant. It can be proved that a group that has two distinct Hermitian forms as its invariants is decomposable.

When we limit ourselves to the real domain, then a Hermitian form $H = (h_{jk})$ — such that $h_{jk} = h_{kj} = \bar{h}_{jk}$ — is real and symmetric at the same time; and a unitary transformation $U$ — such that $\overline{U} = U$ — is orthogonal, $U'U = E$. An orthogonal group is transitive in the real domain. Thus a Hermitian form in the real domain has at least an orthogonal normal form, that is, a real unitary form. A multiple root of a real Hermitian form is also a real Hermitian form (Autonne, 1902; Williamson, 1935).

In virtue of this result of *polar factorization* it can be seen that there exists for any nonsingular matrix $A$ one and only one positive-definite and symmetric matrix $P = P'$ and one and only one orthogonal matrix $R = R'^{-1}$ such that $A = PR$. It can be proved that both these polar factors $P$ and $R$ of any Hamiltonian matrix $C$ are Hamiltonian matrices (Wintner, 1934).

In fact, if we substitute $C = PR$ (where $P = P'$, $R' = R^{-1}$) in the formula (83), with $|\det C| = |s|^n > 0$, then we obtain

$$R^{-1}PGPR = sG.$$

Hence

$$P_1R_1 = P_2R_2,$$

where

$$R_1 = G, \qquad R_2 = \pm RGR^{-1},$$

$$P_1 = P, \qquad P_2 = \pm sR_2P^{-1}R_2^{-1} \qquad (\pm s = |s|).$$

Orthogonal matrices form a group containing $R$ and $\pm G$, since

$$G' = -G = G^{-1},$$

so that $R_1$ and $R_2$ are orthogonal matrices. On the other hand, $P_1$ and $P_2$ are positive-definite matrices, because the reciprocal matrix of a positive-definite matrix $P$ is positive-definite and remains so after an orthogonal transformation and multiplication by a positive number $\pm s = |s|$. Accordingly, $R_1 = R_2$ and $P_1 = P_2$, owing to the uniqueness of the polar factorization of the nonsingular matrix $P_1R_1 = P_2R_2$. Hence

$$G = \pm RGR^{-1}, \qquad P = \pm sGP^{-1}G^{-1},$$

where

$$R^{-1} = R', \qquad P' = P.$$

Consequently

$$C'GC = sG, \qquad |\det C| = |s|^n > 0$$

is satisfied by $C = R$ and by $C = P$, if we choose the multiplier $s$ equal to $\pm 1$, $\pm s$, respectively. Thus $R$ and $P$ are Hamiltonian matrices.

Since the product of two Hamiltonian matrices is Hamiltonian, it follows that it is sufficient for normalization to know those positive-definite or orthogonal Hamiltonian matrices, that is, those matrices that represent dilatations and rotations or reflections in a Euclidean space of $2n$ dimensions. The projection of an orthogonal transformation of the $2n$-dimensional phase space on the $n$-dimensional coordinate space need not be an orthogonal transformation of this $n$-dimensional space. Write a Hamiltonian matrix $C$ in the form

$$C = \begin{pmatrix} \alpha & \beta \\ \gamma & \delta \end{pmatrix},$$

so that $\alpha$, $\beta$, $\gamma$, $\delta$ are matrices of order $n$, and substitute the definition of $G$ of Section 1.18 and of this matrix $C$ in condition (83) with $C' = C^{-1}$ of $2n$-ary orthogonality; then, $\gamma = -\beta$, $\delta = \alpha$ if $s > 0$, and $\gamma = \beta$, $\delta = -\alpha$ if $s < 0$. Hence, if $C$ is a Hamiltonian matrix, it is a $2n$-ary orthogonal matrix if and only if it has either of the particular forms

$$C = \begin{pmatrix} \alpha & \beta \\ -\beta & \alpha \end{pmatrix} \qquad (s = 1),$$

or

$$C = \begin{pmatrix} \alpha & \beta \\ \beta & -\alpha \end{pmatrix} \qquad (s = -1).$$

Williamson (1936a) extended such polar representation to singular matrices. He also (1935) discussed the nonsingular pencils of Hermitian matrices.

### 1.22 PFAFFIAN DYNAMICAL SYSTEM

Birkhoff (1928) noticed that Hamiltonian equations in dynamics can be regarded as a special type arising from the more general Pfaffian variational principle

$$\delta \int_{t_0}^{t_1} \left( \sum_{j=1}^{n} P_j \dot{p}_j + Q \right) dt = 0,$$

in which the integral is linear in all of the first derivatives with arbitrary functions $P_1, \ldots, P_n$, $Q$ of the variables $p_1, \ldots, p_n$ as coefficients, and $n$ is even. In fact, the Hamiltonian equations are derived from the variational problem

$$\delta \int_{t_0}^{t_1} \left( \sum_{j=1}^{n} p_j \dot{q}_j - H \right) dt = 0,$$

or

$$\delta \int_{t_0}^{t_1} L\, dt = 0, \qquad L = L_0 + L_1 + L_2,$$

$$H = -L + \sum_{j=1}^{n} p_j \dot{q}_j,$$

where

$$\frac{dq_i}{dt} = \frac{\partial H}{\partial p_i}, \qquad \frac{dp_i}{dt} = -\frac{\partial H}{\partial q_i} \qquad (i = 1, 2, \ldots, n).$$

If we develop this general *Pfaffian variational principle* explicitly, then we obtain

$$\sum_{j=1}^{n} \left(\frac{\partial P_i}{\partial p_j} - \frac{\partial P_j}{\partial p_i}\right) \frac{dp_j}{dt} - \frac{\partial Q}{\partial p_i} = 0 \qquad (i = 1, 2, \ldots, n).$$

These equations are those of a degenerate Lagrangian problem with $L_2 = 0$, $L_1 = \sum_j P_j \dot{p}_j$, $L_0 = Q$, so that there exists a particular integral $Q =$ constant, which reduces to the energy integral in the Hamiltonian case.

The conservative Hamiltonian principle

$$\delta \int \left(\sum_{i=1}^{n} p_i \frac{dq_i}{dt} - H(p_1, \ldots, p_n, q_1, \ldots, q_n)\right) dt = 0$$

has been generalized by Birkhoff (1928) to the Pfaffian principle

$$\delta \int \left(\sum_{i=1}^{2n} R_i \, dx_i - H \, dt\right) = 0, \tag{96}$$

or

$$\sum_{j=1}^{2n} \left(\frac{\partial R_i}{\partial x_j} - \frac{\partial R_j}{\partial x_i}\right) \frac{dx_j}{dt} = \frac{\partial H}{\partial x_i} \qquad (i = 1, 2, \ldots, 2n), \tag{97}$$

where

$$H = H(x_1, \ldots, x_{2n}), \qquad R_i = R_i(x_1, \ldots, x_{2n})$$

are $2n + 1$ given functions of $2n$ coordinates, and

$$\det \left| \frac{\partial R_i}{\partial x_j} - \frac{\partial R_j}{\partial x_i} \right| \neq 0.$$

Equations (97), which admit the integral $H =$ constant, are invariant for any coordinate transformation. Most of the essential properties of the Hamiltonian system hold also for the Pfaffian cases.

The variational equations (Section 4.20) belonging to the given periodic solution $x_i = x_i(t)$ $(i = 1, 2, \ldots, 2n)$,

$$\sum_{j=1}^{2n} \left(\frac{\partial R_i}{\partial x_j} - \frac{\partial R_j}{\partial x_i}\right) \frac{d\xi_i}{dt} + \sum_{j=1}^{2n} \sum_{l=1}^{2n} \left(\frac{\partial^2 R_i}{\partial x_l \, \partial x_j} - \frac{\partial^2 R_j}{\partial x_l \, \partial x_i}\right) \xi_l \frac{dx_j}{dt}$$
$$= \sum_{j=1}^{2n} \frac{\partial^2 H}{\partial x_j \, \partial x_i} \xi_j \qquad (i = 1, 2, \ldots, 2n) \tag{98}$$

can be written

$$\frac{d\xi_i}{dt} = \sum_{j=1}^{2n} a_{ij}(t) \xi_j \qquad (j = 1, 2, \ldots, 2n). \tag{99}$$

Interchange $i$ and $j$, and write $\eta$ in place of $\xi$ in (98); then,

$$\sum_{i=1}^{2n}\left(\frac{\partial R_i}{\partial x_j}-\frac{\partial R_j}{\partial x_i}\right)\frac{d\eta_i}{dt}+\sum_{i=1}^{2n}\sum_{l=1}^{2n}\left(\frac{\partial^2 R_i}{\partial x_l\,\partial x_j}-\frac{\partial^2 R_j}{\partial x_l\,\partial x_i}\right)\eta_l\frac{dx_i}{dt}$$

$$=-\sum_{i=1}^{2n}\frac{\partial^2 H}{\partial x_i\,\partial x_j}\eta_i \qquad (j=1,2,\ldots,2n). \quad (100)$$

Furthermore, the expression

$$\sum_{i=1}^{2n}\sum_{j=1}^{2n}\sum_{l=1}^{2n}\left(\frac{\partial^2 R_i}{\partial x_l\,\partial x_j}-\frac{\partial^2 R_j}{\partial x_l\,\partial x_i}\right)\left(\xi_j\eta_i\frac{dx_l}{dt}-\xi_l\eta_i\frac{dx_j}{dt}-\xi_j\eta_l\frac{dx_i}{dt}\right) \quad (101)$$

is the sum of six trilinear forms, three of which are the negatives of the three others. Multiplying (98) by $\eta_i$ and (100) by $\xi_j$, we obtain $4n$ relations. Adding these, we obtain

$$\sum_{i=1}^{2n}\sum_{j=1}^{2n}\left(\frac{\partial R_i}{\partial x_j}-\frac{\partial R_j}{\partial x_i}\right)\left(\frac{d\xi_j}{dt}\eta_i+\frac{d\eta_i}{dt}\xi_j\right)$$

$$+\sum_{i=1}^{2n}\sum_{j=1}^{2n}\sum_{l=1}^{2n}\left(\frac{\partial^2 R_i}{\partial x_l\,\partial x_j}-\frac{\partial^2 R_j}{\partial x_l\,\partial x_i}\right)\left(\xi_l\eta_i\frac{dx_j}{dt}+\xi_j\eta_l\frac{dx_i}{dt}\right)=0. \quad (102)$$

The sum of the two vanishing expressions (101) and (102) is the time derivative of the bilinear covariant

$$\sum_{i=1}^{2n}\sum_{j=1}^{2n}s_{ij}(t)\xi_j(t)\eta_i(t), \quad (103)$$

where

$$s_{ij}(t)=\frac{\partial R_i}{\partial x_j}-\frac{\partial R_j}{\partial x_i},$$

which is associated with the Pfaffian

$$\sum_{i=1}^{2n}R_i\,dx_i-H\,dt.$$

The value of this bilinear covariant (Section 1.7) is independent of $t$ for two arbitrary solutions $\xi$, $\eta$ of the variational equations belonging to the solution $x_i=x_i(t)$ $(i=1,2,\ldots,2n)$.

It follows that the characteristic equation associated with (98) or (99) is always reciprocal. This is Wintner's (1931) generalized form of Birkhoff's theorem. The skew-symmetric nonsingular bilinear form (103) is the Pfaffian generalization of Poincaré's bilinear differential invariant corresponding to the Hamiltonian-type integral invariant

$$\iint \sum_{i=1}^{n} \delta p_i \, \delta q_i.$$

The integral invariants of the Pfaffian systems have been studied by Féraud (1929).

The property of polar factorization can be extended to such Pfaffian dynamical systems. In the linear conservative case this has the form

$$\dot{y} = S^{-1}Fy, \tag{104}$$

where $F = F'$ and $S$ is nonsingular and equal to $-S'$ but not necessarily equal to $G$, so that (104) is more general than the equations

$$\dot{x} = -GHx. \tag{104a}$$

It can be proved that these Pfaffian systems may be reduced to the Hamiltonian. If $y = Tx$, where $T$ is a nonsingular matrix, then (104) is equivalent to

$$\dot{x} = (T'ST)^{-1}(T'FT)x.$$

Now $H = T'FT = H'$, since $F = F'$. If $S = -S'$ and $\det S \neq 0$, then there exists a nonsingular matrix $T$ such that $T'ST = G = -G^{-1}$ (Weyl, 1931). Using this matrix $T$ in $y = Tx$, we obtain (104a).

The definition of $G$ gives, for any $H = H'$ and any integer $m \geqq 0$,

$$G(-GH)^m G^{-1} = -GG(-HG)^{m-1}HG^{-1} = (-1)^m(HG)^m$$

and

$$[(GH)^m]' = (H'G')^m = (-1)^m(HG)^m.$$

Hence,

$$G(-GH)^m G^{-1} = [(GH)^m]'.$$

Consequently

$$G[\exp(-GH)]G^{-1} = [\exp(GH)]',$$

where $\exp A$ is defined for any $A$ by means of the exponential series. Since $\exp(-A) = (\exp A)^{-1}$, it is clear that $C'GC = sG$,

$$|\det C| = |s|^m > 0,$$

is satisfied by $C = \exp(GH)$, $s = 1$. Hence, $\exp(GH)$ is for any $H = H'$ a Hamiltonian matrix of determinant $+1$ (Williamson, 1939, 1940).

On the other hand, any solution $x = x(t)$ of $\dot{x} = Bx$, where $B$ is independent of $t$, is related to the vector $x(0)$ of the initial constant by the linear transformation

$$x(t) = L(t)x(0),$$

where $L(t)$ denotes the matrix exp $(Bt)$. If we apply this to (104a), with $H = H'$ and $B = -GH$, then — since $-tH$ is, like $H$, real and symmetric for $-\infty < t < +\infty$ — we obtain this result: the cyclic transformation group $L(t)$ generated by the infinitesimal transformation (104a) consists of Hamiltonian matrices and $\det L(t) = +1$ for any finite $t$ (see Section 4.26 et seq. on the Lie groups).

Consider such interesting subgroups of $C$ as

$$C_1 = \begin{pmatrix} \alpha & \omega \\ \omega & \delta \end{pmatrix}, \qquad C_2 = \begin{pmatrix} \omega & \beta \\ \gamma & \omega \end{pmatrix}.$$

We obtain the usual condition of contragradiency for $C_1$

$$\alpha' = s\delta^{-1}, \qquad s \neq 0;$$

and for $C_2$

$$\beta' = -s\gamma^{-1}, \qquad s \neq 0,$$

as the necessary and sufficient condition for the Hamiltonian character of the matrices (Wintner, 1934). (See Section 4.14).

## REFERENCES


Ahrens, W.

1895 *Ueber eine Gattung n-fach periodischer Functionen von n reellen Veränderlichen*, Dissertation, Rostock.

Appell, P.

1892 *J. reine angew. Math.*, **110**, 37.

Autonne, L.

1902 *Rendiconti Cir. Mat. Palermo*, **16**, 104.

Badaljan, G.

1934 *Russian Astron. J.*, **11**, 376.

Bianchi, L.

1899 *Vorlesungen über Differentialgeometrie*, Teubner, Leipzig.

1922 *Lezioni di Geometria Differenziale*, I, Spoerri, Pisa.

1923 *Ibid.*, II, Part 1.

1924 *Ibid.*, II, Part 2.

Birkhoff, G. D.

1917 *Trans. Am. Math. Soc.*, **18**, 199.

1927 *Am. J. Math.*, **49**, 1.

1928 *Dynamical Systems*, Am. Math. Soc., New York.

Bohr, N.
1923 *Z. Physik*, **13**, 117.
1924 *Proc. Cambridge Phil. Soc.*, **22**, Supplement.

Boltzmann, L.
1922 *Vorlesungen über die Prinzipien der Mechanik*, I, II, 2nd ed., Barth, Leipzig.

Bolza, O.
1909 *Vorlesungen über Variationsrechnung*, Teubner, Leipzig.

Brillouin, L.
1926 *Compt. Rend.*, **183**, 24.
1926a *J. de Phys.*, **7**, 353.
1935 *Quantenstatistik*. Springer, Berlin.

Carathéodory, C.
1930 *Rendiconti Accad. Lincei*, ser. 6, **12**, 353.
1935 *Variationsrechnung*, Teubner, Leipzig.
1937 *Geometrische Optik*, Springer, Berlin.

Cartan, E.
1922 *Leçons sur les Invariants Intégraux*, Hermann, Paris.
1934 *Les Espaces de Finsler*, Gauthier-Villars, Paris.
1937 *Théorie des Groupes Finis et Continus*, Gauthier-Villars, Paris.

Charlier, C. V. L.
1906 *Mechanik des Himmels*, I, II, W. de Gruyter, Leipzig. Second rev. ed. published as *Die Mechanik des Himmels für Vorlesung*, I, II, W. de Gruyter, Leipzig, 1927.
1906a *Arkiv Mat. Astron. Fys.*, **2**, no. 29.

Corson, E. M.
1951 *Perturbation Methods in the Quantum Mechanics of n-Electron Systems*, Blackie, London.

Dall'Aqua, F. A.
1908 *Math. Annalen*, **66**, 394.
1912 *Rendiconti Cir. Mat. Palermo*, **33**, 341.

Darboux, G.
1890 *Théorie des Surfaces*, III, Book 6, Chap. V, Gauthier-Villars, Paris.

Datzeff, A.
1936 *Compt. Rend.* **203**, 300.

de Donder, T.
1901 *Rendiconti Cir. Mat. Palermo*, **15**, 65.
1902 *Ibid.*, **16**, 155.
1930 *Théorie Invariantive du Calcul des Variations*, Gauthier-Villars, Paris.

Dirac, P. A. M.
1925 *Proc. Roy. Soc. London*, ser. A, **109**, 642.

Dziobek, O.
1900 *Astron. Nachr.*, **152**, 33.

Eddington, A. S.
1923 *The Mathematical Theory of Relativity*, Cambridge University Press, London.

Einstein, A.
1917 *Sitzber. preuss. Akad. Wiss. Berlin* **1917**, 606.

Eisenhardt, L. P.
1934 *Annals of Math.*, **35**, 284; *Phys. Rev.*, **45**, 427.
1948 *Phys. Rev.*, **74**, 87.
1949 *Proc. Natl. Acad. Sci. U.S.*, **35**, 412, 490.
Epstein, P.
1916 *Ann. Physik*, **50**, 489; **51**, 168.
1919 *Ibid.*, **58**, 553.
1922 *Z. Physik.* **8**, 211, 305; **9**, 92.
Féraud, L.
1929 *Trans. Am. Math. Soc.*, **32**, 817.
Forsyth, A. R.
1927 *Calculus of Variations*, Cambridge University Press, London.
Fues, E.
1925 *Z. Physik*, **34**, 788.
Gleich, G. v.
1926 *Z. Physik*, **35**, 7.
Goursat, E.
1908 *J. Math. pures appl.*, ser. 6, **4**, 333.
1915 *Ibid.*, ser. 6, **11**, 241.
4922 *Compt. Rend.*, **174**, 1089.
1923 *Problème de Pfaff*, Gauthier-Villars, Paris.
Hadamard, J.
1910 *Calcul des Variations*, Hermann, Paris.
Hamel, G.
1924 *Math. Annalen*, **92**, 33.
Hertz, H. von,
1910 *Prinzipien der Mechanik in neuen Zusammenhang dargestellt*, Barth, Leipzig.
Herzberger, M.
1931 *Strahlenoptik*, Springer, Berlin.
Heun, K.
1901 *Arch. Math. Phys.*, ser. 3, **2**, 57, 298.
Hilbert, D.
1906 *Math. Annalen*, **62**, 351.
Hiltebeitel, A. M.
1911 *Am. J. Math.*, **33**, 337.
Jacobi, C. G. J.
1884 *Vorlesungen über Dynamik*, Reimer, Berlin.
Kemble, E. C.
1937 *The Fundamental Principles of Quantum Mechanics*, McGraw-Hill, New York.
Kneser, A.
1894 *Math. Annalen*, **45**, 446.
1925 *Lehrbuch der Variationsrechnung*, Vieweg, Brunswick.
Kneser, H.
1921 *Math. Annalen*, **84**, 277.
Koschmieder, G.
1933 *Variationsrechnung*, Gruyter, Berlin.

Kramers, H. A.
1920 *Z. Physik*, **3**, 199.
Lehmann-Filhés, R.
1904 *Astron. Nachr.*, **165**, 209.
Levi-Civita, T.
1896 *Annali Mat. pura appl.*, ser. 2, **24**, 255.
1904 *Math. Annalen*, **59**, 383.
1923 *Der absolute Differentialkalkül*, Springer, Berlin.
1925 *Lezioni di Calcolo Differenziale Assoluto*, Zanichelli, Bologna.
1936 *Recueil Math.*, **1**, 123.
Levi-Civita, T., and Amaldi, U.
1927 *Lezioni di Meccanica Razionale*, II, Part 2, 318. Zanichelli, Bologna.
Lewis, D. C.
1937 *Bull. Am. Math. Soc.*, **43**, 277.
Lie, S.
1922 *Gesammelte Werke*, III, 295, Teubner, Leipzig.
Liouville, R.
1895 *Acta Math.*, **19**, 251
Miyahara, S.
1954 *Pub. Astron. Soc. Japan*, **6**, 9.
Morera, G.
1881 *Atti Accad. Torino*, **16**, 276.
Murnaghan, F. D.
1931 *Proc. Natl. Acad. Sci. U.S.*, **17**, 128.
Painlevé, P.
1894 *J. Math. pures appl.*, ser. 4, **10**, 5.
1897 *Compt. Rend.*, **124**, 221.
Poincaré, H.
1899 *Acta Math.*, **13**, 1. Reprinted in *Oeuvres de Henri Poincaré*, VII, 286. Gauthier-Villars, Paris, 1952.
1893 *Méthodes Nouvelles de la Mécanique Céleste*, III, Gauthier-Villars, Paris.
Prange, G.
1935 *Die allgemeine Integrationsmethode der analytischen Mechanik*, Teubner, Leipzig.
Rao, B. S. Medhava.
1935 *Math. Annalen*, **111**, 459.
1935a *Z. Physik*, **93**, 320.
Reed, F. W.
1918 *Am. J. Math.*, **40**, 97.
Ricci, A., and Levi-Civita, T.
1900 *Math. Annalen*, **54**, 125.
Robertson, H. P.
1928 *Math. Annalen*, **98**, 749.
Schouten, J. A., and v. d. Kulk, W.
1949 *Pfaff's Problem and its Generalizations*, Clarendon, Oxford.
Schouten, J. A.
1929 *Math. Z.*, **30**, 149.

1954 *Ricci-Calculus*, Springer, Berlin.

Schwarzschild, K.
1916 *Sitzber. preuss. Akad. Wiss. Berlin*, **1916**, 548.

Sommerfeld, A.
1916 *Ann. Physik*, **51**, 1, 125.
1924 *Atombau und Spektrallinien*, Vieweg, Brunswick.

Stäckel, P.
1891 *J. reine angew. Math.*, **107**, 319.
1905 *Ibid.*, **128**, 222.

Synge, J. L.
1926 *Phil. Trans. Roy. Soc. London*, ser. A, **226**, 31.
1960 *Encyclopedia of Physics*, III-1, *Classical Dynamics*, Springer, Berlin.

Taylor, A. E.
1935 *Bull. Am. Math. Soc.*, **41**, 735.
1937 *J. Math. pures appl.*, ser. 9, **16**, 15.

Titchmarsh, E. C.
1955 *Proc. London Math. Soc.*, ser. 3, **5**, 1.
1958 *Eigenfunction Expansions Associated with Second Order Differential Equations*, Part II, Clarendon, Oxford.

Tonelli, L.
1921 *Fondamenti di Calcolo di Variazioni*, I, Zanichelli, Bologna.
1923 *Ibid.*, II.

Turnbull, W. H., and Aitken, A. C.
1932 *An Introduction to the theory of Canonical Matrices*, Blackie, London.

Ura, T.
1946 *Proc. Imp. Acad. Japan*, **22**, 1.
1951 *Bull. Astron.*, ser. 2, **16**, 333.

Van Kampen, E. R. and Wintner, A.
1936 *Am. J. Math.*, **58**, 851.
1938 *Trans. Am. Math. Soc.*, **44**, 168.

Vranceanu, G.
1929 *Annali Mat. pura appl.*, ser. 4, **6**, 9.
1935 *Les Espaces Non-Holonomes*, Gauthier-Villars, Paris.
1961 *Vorlesungen über Differentialgeometrie*, II, Akad. Verlag, Berlin.

Vrkljan, V. S.
1929 *Z. Physik*, **59**, 718.
1930 *Ibid.*, **65**, 280.

Weber, H. v.
1900 *Vorlesungen über das Pfaff'sche Problem*, Teubner, Leipzig.

Wedderburn, J. H. M.
1934 *Lectures on Matrices*, Amer. Math. Soc. Colloq. Publ., New York.

Weierstrass, K.
1894 *Gesammelte Werke*, I, 257, Mayer u. Müller, Berlin.

Weinacht, J.
1928 *Math. Annalen*, **91**, 279.

Weitzenböck, R.
1923 *Invariantentheorie*, Noordhoff, Groningen.

Weyl, H.
1923 *Mathematische Analyse des Raumproblems*, Springer, Berlin.
1931 *Gruppentheorie und Quantenmechanik*, 2nd edition, Hirzel, Leipzig.
1939 *The Classical Groups*, Princeton University Press, Princeton, N.J.
Williamson, J.
1935 *Am. J. Math.*, **57**, 475.
1936 *Ibid.*, **58**, 141.
1936a *Bull. Am. Math. Soc.*, **41**, 118.
1937 *Am. J. Math.*, **59**, 599.
1939 *Bull. Am. Math. Soc.*, **45**, 920.
1940 *Am. J. Math.*, **62**, 881.
Wintner, A.
1929 *Spektraltheorie der unendlichen Matrizen*, S. Hirzel, Leipzig.
1930 *Berichte Math.-Phys. Kl. sächs. Akad. Wiss.*, **82**, 3.
1931 *Am. J. Math.*, **53**, 623.
1934 *Annali Mat. pura appl.*, ser. 4, **13**, 104.
1936 *Quart. J. Math.*, **7**, 214.
1937 *Proc. Edinburgh Math. Soc.*, ser. 2, **5**, 90.
1938 *Am. J. Math.*, **60**, 463.
1947 *The Analytical Foundation of Celestial Mechanics*, Princeton University Press, Princeton, N.J.
Yano, K.
1945 *Proc. Imp. Acad. Japan*, **21**, 171, 248, 261.
1949 *J. Faculty Sci. Univ. Tokyo*, **6**, 1.

# Quasi-Periodic Motion

## SIMPLY PERIODIC MOTION

### 2.1 LEBESGUE'S INTEGRALS

A collection of definite elements related to their totality by a certain law is called a *set*. Any two sets that can be made to correspond to each other are *equivalent*. A set of natural numbers is said to have the power of *denumerable* numbers. A set of rational numbers and a set of algebraic numbers determined as the roots of an algebraic equation with rational coefficients have the power of denumerable numbers. By representing such a number as the coordinate of a point on the straight line interval between 0 and 1, we can find a number belonging to this set in any small but finite segment in this interval. Such being the case, the rational numbers and the algebraic numbers are said to be *everywhere dense* in the interval. But the set of all points in the interval has the power of the *continuum*, because such a set cannot be made to correspond to any set of rational or algebraic numbers.

Describe a circle with a small but finite radius $r$ around a point $P$. The region inside this circle is called the *r-neighborhood*. If there is no point belonging to the set $M$ in the neighborhood of $P$, then $P$ is said to be *isolated*. If there are infinitely many points of the set $M$ other than $P$ in the $r$-neighborhood of $P$, then $P$ is called an *accumulation* point or a *limit point* of the set $M$. A set consisting of accumulation points is called a *derived* set. A set containing its derived set is said to be *closed*. A set coinciding with its derived set is called a *perfect* set. A perfect set cannot have the power of denumerable numbers but has the power of the *continuum* according to Cantor.

A *sum* set (*union, Vereinigungsmenge*) of the sets $M_1, M_2, \ldots, M_k$ is a set $M$ containing all points that are in *any* of these $k$ sets. A *product* set of the sets $M_1, M_2, \ldots, M_k$ (*intersection, Durchschnitt*) is a set of points that are common to *all* these $k$ sets. A set that is a part of a set is called a *subset*. A subset $M_1$ of $M$ (written $M_1 \subset M$) consisting of all points not contained in a subset $M_2$ of $M$ is called the *complement of $M_2$ in $M$* when $M_1 + M_2 = M$. A complementary set of a closed set is an open set.

If there is a number larger or smaller than any member of the set $M$, then the number is called the *upper bound* or the *lower bound* of $M$. If the upper and the lower bound both exist, then the set $M$ is said to be *bounded*. There should be at least one accumulation point for any bounded set consisting of an infinite number of elements (*Bolzano-Weierstrass theorem*). If there is only one accumulation point in a bounded set, then the set is said to be *convergent* and the accumulation point is called the *limit*. The largest and smallest accumulation points are denoted respectively by "$\overline{\lim}\, M$" or "lim sup $M$" and "$\underline{\lim}\, M$" or "lim inf $M$," and are called the *superior* and *inferior* limits, respectively. A set consisting of only interior points of a set $M$ is said to be *open*. A bounded closed set can be covered by a finite number of open sets. This is *Heine-Borel's covering theorem*.

If all subsets consisting of an infinite number of elements selected from a set $M$ possess accumulation points, then the set $M$ is called *compact*. A number $\|x, y\|$ can be made to correspond to a pair of two points $x$ and $y$, such that $\|x, x\| = 0$, $\|x, y\| = \|y, x\| > 0$ for $x \neq y$, and

$$\|x, y\| + \|y, z\| \geqq \|x, z\|.$$

This number is called the *distance* of the two points, and such a set is said to be *metrical*. If the set of points $(x_1, x_2, \ldots, x_n)$ is metrical, then the set constitutes a metrical space. When

$$\|x, y\| = [(x_1 - y_1)^2 + (x_2 - y_2)^2 + \cdots + (x_n - y_n)^2]^{1/2},$$

the space is called *Euclidean*. The necessary and sufficient condition for a set $M$ of a Euclidean space $R_n$ to be compact is that a finite number $A$ can be found such that $|x_i| < A$ $(i = 1, 2, \ldots, n)$, for any point $(x_1, x_2, \ldots, x_n)$ of $M$.

If every point of a set is itself an accumulation point, then the set is called *dense in itself*. A perfect set is dense in itself. A set $M$ is nowhere dense in a set $N$ if $M$ is not dense in any subset of $N$. If a set $M$ cannot be represented by a sum of its closed subsets, then $M$ is called *connected*. A sum of connected sets that are nonoverlapping — i.e., any two of which have no common point — is connected. Any two points $a'$ and $a''$

of a connected set $M$ can be connected by a finite number of sequences $a' = a_0, a_1, \ldots, a_{l-1}, a_l = a''$, belonging to the set $M$ such that

$$\|a_{i-1}, a_i\| \leqq \rho \qquad (i = 1, 2, \ldots, l).$$

A closed connected set is called a *continuum*. A connected set is dense in itself. A continuum is a perfect set.

If there is in $M$ a denumerably infinite number of subsets that are dense in themselves, then $M$ is said to be *separable*. A compact set is separable. Any set of a Euclidean space is separable. If there exists an $n$ for a given $\epsilon$ such that $\|a_{n'}, a_{n''}\| < \epsilon$ for $n' \geqq n$, $n'' \geqq n$, then the sequence $\{a_n\}$ is said to be a *Cauchy sequence*. If a Cauchy sequence from a set $M$ possesses an accumulation point belonging to $M$, then it is called *complete*. A complete set is closed and a compact closed set is complete. A closed subset of a Euclidean space is complete.

Let $A$ be a system of sets $M$ in an $n$-dimensional metrical space $R_n$. Let a function $\phi(M)$ be such that: (i) $\phi(M) \geqq 0$ for all $M$ belonging to $A$; (ii) $\phi(M) = 0$ for any empty set, i.e., for any set containing no element; (iii) $\phi(M) \geqq \phi(M_1)$ for a subset $M_1$ of $M$, that is, for $M_1 \subset M$; and (iv) $\phi(M) \leqq \phi(M_1) + \phi(M_2) + \cdots$ for $M_1 + M_2 + \cdots = M$, where $M_1, M_2, \ldots$ are a denumerably infinite number of sets in the system $A$. Then the function $\phi(M)$ is called a *measure function*. Let $m(J) = \prod_{k=1}^{n} (b_k - a_k)$ be the measure function of an open domain $J$ of an $n$-dimensional space $a_k < x_k < b_k$ $(k = 1, 2, \ldots, n)$. Let an arbitrary set $M$ in the metrical space $R_n$ be covered by at most a denumerable number of open sets, by various methods. The sum set of such open sets contains $M$. The sum of the measure functions of such open sets may differ according to the method of covering. Call the lower bound of these sums the *outer measure*; that is, the lower bound of the outer measures of all open sets containing $M$ is called the outer measure of $M$ and denoted by $m_a(M)$. Let a set $M$ be covered by an open set $O$. The difference between the outer measure of $O$ and the outer measure of the complementary set of $M$, $O - M$, is called the *inner measure* $m_i(M)$; that is, $m_a(O) - m_a(O - M) = m_i(M)$. In general: $m_i(M) \leqq m_a(M)$. If $M_1 \subset M$, then $m_i(M_1) \leqq m_i(M)$. A set $M$ such that

$$m_i(M) = m_a(M) = m(M)$$

is called *measurable* according to Lebesgue.

Any set constructed by a finite or a denumerable number of operations such as addition, subtraction, multiplication — finite and infinite — from measurable sets is measurable. Such sets are called *Borel sets*. The operations just mentioned constitute a field called the *Borel field*.

Let $M_1 + M_2$ and $M_1M_2$ be respectively the sum set and the product set of two sets $M_1$ and $M_2$. We have

$$m_a(M_1 + M_2) + m_a(M_1M_2) \leqq m_a(M_1) + m_a(M_2),$$

$$m_i(M_1 + M_2) + m_i(M_1M_2) \geqq m_i(M_1) + m_i(M_2).$$

If $M_1$ and $M_2$ are both measurable, that is, if

$$m_i(M_1) = m_a(M_1) = m(M_1), \qquad m_i(M_2) = m_a(M_2) = m(M_2),$$

then

$$m_i(M_1 + M_2) = m_a(M_1 + M_2), \qquad m_i(M_1M_2) = m_a(M_1M_2),$$

and the sum set and the product set are measurable. We have

$$m(M_1 + M_2) + m(M_1M_2) = m(M_1) + m(M_2).$$

In particular, if $M_1$ and $M_2$ are nonoverlapping, i.e., possess no common point, then

$$m(M_1 + M_2) = m(M_1) + m(M_2), \qquad m(M_1M_2) = 0.$$

The sum set and the product set of a denumerably infinite number of measurable sets are both measurable. A measure cannot be negative. The measure of an empty set is zero. The measure of the set containing all points between 0 and 1 is one. The measure of the set containing a denumerably infinite number of points, such as rational numbers or algebraic numbers, is zero.

For a point $a$ of $M$, i.e., $a \in M$, the set of values of $f(a)$ is called a *function* $f$. For various sequences $\{a_n\}$ converging to a point $a$ of $M$, the point function $\limsup_{n\to\infty} f(a_n)$ is denoted by $G(a;f, M)$, and

$$\liminf_{n\to\infty} f(a_n)$$

by $g(a;f, M)$. If $G(a;f, M) = g(a;f, M)$, then $f$ is said to be continuous at $a \in M$. A function that is continuous at all points of a set $M$ — that is, a function for which $G(f, M) = g(f, M)$ — is said to be *continuous* in $M$. If there exists a $\rho > 0$ for a given $\epsilon > 0$ such that $|f(a') - f(a'')| < \epsilon$ for any two points $a'$ and $a''$ of $M$ whose distance apart is less than $\rho$, then the bounded function $f$ is said to be *uniformly continuous* in $M$.

If $M$ is compact and closed, then any function continuous in $M$ is uniformly continuous. The function

$$\omega(a;f, M) = G(a;f, M) - g(a;f, M)$$

is called the *fluctuation* of $f$ in $M$. In order that $\omega(a;f, M) = 0$, it is necessary and sufficient that $f$ should be continuous in $M$. Otherwise, $f$ is said to be *discontinuous.*

If the discontinuity point of $f$ is compact in $M$, then the finite but discontinuous function $f$ is said to be *totally discontinuous*; if it is not compact, then it is said to be *pointwise discontinuous.* A function continuous on a perfect bounded set $M$ is called a function of *Baire's class 0.* A continuous function represented as the limit function of a sequence $\{f_n\}$ of continuous functions $f_n$ is a function of *Baire's class 1.* If $f_n$ belongs to Baire's class $\alpha$ but $f = \lim_{n\to\infty} f_n$ does not belong to any Baire's class lower than $\alpha$, then $f$ is said to be of Baire's class $\alpha$. A discontinuous function which is represented as the limit of functions of Baire's class $\alpha$ is said to be of Baire's class $\alpha + 1$.

Let $f(x)$ be a nonnegative bounded function of $x$. Draw a graph with coordinates $x$ and $y$ such that $y = f(x)$. The set of values of $y$ such that $0 \leqq y \leqq f(x)$ for a measurable $n$-dimensional set $M$ of $x$ is called an *ordinate set* and denoted by $\Omega\{M; f(x)\}$. Denote the measure of this measurable space of $x$, $y$ by $\mu$; then

$$\mu_a[\Omega\{M; f(x)\}] = \overline{\int_M f(x)\, dx}, \qquad \mu_i[\Omega\{M; f(x)\}] = \underline{\int_M f(x)\, dx}.$$

Call these integrals the upper and lower Lebesgue integrals of $f$ over the set $M$. If $\Omega\{M; f(x)\}$ is measurable, then these two integrals coincide. If the measure is finite, then $f(x)$ is called *integrable* or summable in Lebesgue's sense. $\mu[\Omega\{M; f(x)\}] = \int_M f(x)\, dx$ is called a *Lebesgue integral* of $f$ on $M$.

When $f(x)$ is a constant, that is, $f = h > 0$, then $\Omega(M; h)$ is an open ordinate set represented by a cylinder of height $h$ with base $M$. The $(n + 1)$-dimensional measure of this cylinder is $\mu[\Omega\{M; h\}] = h\cdot m(M)$. It is equal to zero for $m(M) = 0$. For $h = \infty$ and for a measurable $M$ with $m(M) > 0$, the measure of $\Omega\{M; \infty\}$ is measurable with infinite measure. If $m(M) = 0$, then

$$\mu[\Omega\{M; \infty\}] = \lim_{k\to\infty} \mu[\Omega\{M; k\}] = \lim_{k\to\infty} [k\cdot m(M)] = 0;$$

that is, an ordinate set erected over a set of measure zero is itself of measure zero.

Denote a subset of $M$, such that $f < \alpha$, by $M(f < \alpha) = M_\alpha$. It can be shown that $M_\alpha$ is measurable for all $\alpha \geqq 0$. Since $M(f > 1), M(f > 2), \ldots$ are measurable, $M(x = \infty)$ is measurable. If $M(f = 0)$ is *measurable* then $M(f \geqq \alpha)$ is measurable for all nonnegative $\alpha$. The set $M(f < \alpha)$ is also measurable. Functions such as $f$ are said to be measurable according to Lebesgue. Thus the measurability of a function can be deduced from the measurability of an ordinate set. Conversely, the existence of the integral $\int_M f(x)\,dx$ can be deduced, if the measure is finite, from the measurability of the function $f(x)$. Since $m(M_\infty) = 0$ (as is now evident), this function $f$ becomes infinite only on sets of measure zero, if $f$ is integrable on $M$.

Let $f(x)$ be a measurable nonpositive function defined on a set $M$ of $x$ with finite measure. Construct the mirror image of this function with respect to $y = 0$. Denote the ordinate set of a nonnegative function $-f$ by $\Omega^+\{M; -f\}$, and let its mirror image be $\Omega^-\{M;f\}$. If $\Omega^-\{M;f\}$ is measurable, then take

$$\int_M f\,dx = -\mu[\Omega^-\{M;f\}].$$

Since

$$\mu[\Omega^-\{M;f\}] = \mu[\Omega^+\{M; -f\}],$$

we obtain

$$\int_M f\,dx = -\int_M (-f)\,dx.$$

Hence the properties for a nonnegative function can be extended to those for a nonpositive function, and consequently to those for any function with arbitrary sign. Thus it can be shown that $M(f \leqq -\alpha)$, $M(f < -\alpha)$, $M(f \geqq -\alpha)$, $M(f > -\alpha)$ are measurable if $f \leqq 0$ is integrable on $M$.

Consider a function which can have either sign on a measurable set $M$ with finite measure. Denote $f^+ = \max(f, 0)$, $f^- = \min(0, f)$; then, $f = f^+ + f^-$. Put

$$\Omega^+\{f^+\} = \Omega^+\{f\}, \qquad \Omega^-\{f^-\} = \Omega^-\{f\}.$$

Then,

$$\mu_a[\Omega^+\{f\}] - \mu_i[\Omega^-\{f\}] = \overline{\int_M} f\,dx,$$

$$\mu_i[\Omega^+\{f\}] - \mu_a[\Omega^-\{f\}] = \underline{\int_M} f\,dx$$

are called the upper and lower integrals of $f$, respectively. If these two integrals are equal, then

$$\int_M f\,dx = \mu[\Omega^+\{f\}] - \mu[\Omega^-\{f\}] = \int_M f^+\,dx + \int_M f^-\,dx$$

is called the *integral* of $f$. For $f$ to be integrable it is necessary that $f^+$ and $f^-$ are both integrable. We say that $f$ is measurable on $M$ if $\Omega^+\{f\}$ and $\Omega^-\{f\}$ are both measurable.

The sum of a finite number of measurable functions $f_1, f_2, \ldots, f_n$ on $M$, for which the operation $\sum_{i=1}^{\infty} f_i$ is applicable, is measurable on a set $M^* \subset M$. There exists $\lim_{n\to\infty} f_n$, which is on a measurable set and is measurable on the set. Suppose that an infinite sequence of finite and integrable functions $f_1, f_2, \ldots$ on $M$ converges to a finite function $f$ on $M$ (except sets of measure zero). Let $f_1 + f_2 + \cdots + f_n = S_n$; then the $S_n$ are integrable. If there exists an integrable function $\Psi$ that is the upper bound of the absolute values of all $S_n$, then $f$ is integrable and

$$\int_M f\,dx = \int_M \sum_{n=1}^{\infty} f_n\,dx = \sum_{n=1}^{\infty} \int_M f_n\,dx.$$

Thus a convergent infinite series of finite integrable functions can be integrated term by term. If, in particular, there is a constant $c$ independent of $n$ such that $|S_n| < c$, that is, if the partial sums $S_n$ are uniformly bounded, then the series can be integrated term by term.

Lebesgue defined an auxiliary function $f_n$ for an integer $n$ corresponding to a nonnegative measurable function $f$ on $M$ by the condition $f_n = f$ for any point $P$ where $f \leqq n$, and $f_n = n$ for any point $P$ where $f > n$. The function $f_n$ is bounded and measurable. We have

$$\int_M f\,dP = \lim_{n\to\infty} \int_M f_n\,dP.$$

If this limit is finite, then $f$, according to Lebesgue, is *summable* on $M$. If $f$ is nonpositive, a definition of summability can still be made. If $M$ is the sum of a finite or a denumerably infinite number of nonoverlapping sets $M_1, M_2, \ldots$, then the integral on $M$ is equal to the sum of integrals on $M_1, M_2, \ldots$. Thus the sum of summable functions is summable, and the integral of the sum equals the sum of the integrals of each function. If $f$ is summable on a set $M$ and $M^* \subset M$, then the integral of $f$ on $M^*$ tends to zero with the measure of $M^*$. If, further, the functions of a sequence $f_1, f_2, \ldots, f_n, \ldots$ are summable on a set $M$, if their absolute values are smaller than a positive summable function $\Phi$ independent of $n$, and if the sequence admits a limit function $F$, then $F$ is summable and

$$\lim_{n\to\infty} \int_M f_n\,dP = \int_M F\,dP.$$

This theorem is due to Lebesgue.

Let a function $f$ be defined on a measurable set $M$ with finite measure and be integrable. Consider a function $\bar{f}$ which coincides with $f$ on $M$ but is equal to zero outside $M$, i.e., on a complementary set of $M$. A set function on a set such that $F(U) = \int_U \bar{f}\, dx$ with respect to an arbitrary measurable set $U$ in $R_n$ is measurable and is called an *indefinite integral* of $\bar{f}$. Let a function $F$ of finite values on a set $U$ with finite measure be such that $F(U) = F(U_1) + F(U_2) + \cdots$ for $U = U_1 + U_2 + \cdots$, where $U$ is the sum set of denumerably infinite nonoverlapping measurable sets $U_1, U_2, \ldots$. This property is called the *complete additivity* of the set function $F(U)$. Let $F$ be such that with a given positive $\epsilon$ there exists a constant $\delta$ so that $|F(U)| < \epsilon$ for $m(U) < \delta$. Then $F$ is said to be *totally continuous*. An indefinite integral has three properties: (*i*), it has a finite value for any finite set of finite measure; (*ii*), it is completely additive; and (*iii*), it is totally continuous. If the inequality

$$g \cdot m(U) \leqq \int_U f\, dx = F(U) \leqq G \cdot m(U)$$

(*the mean value theorem*), holds in a set $U$ of an integrable function $f$, then

$$g \leqq \frac{F(U)}{m(U)} \leqq G.$$

Let the set function $F(U)$ possess the above three properties. In a sequence of sets $U_1, U_2, \ldots$ which converges to $X$ for $n \to \infty$, we consider a set $K_n$ containing $X$ and one of $U_n$. If $m(U_n)/m(K_n) \geqq \alpha$ for a positive $\alpha$, and if $\lim_{n\to\infty} [F(U_n)/m(U_n)]$ exists, then the limit is called the *derivative* of $F(U)$ at $X$, and $F$ is *differentiable* at $X$. Consider at each point $X$ of a set $M$ of finite outer measure an admissible sequence of closed sets $V$ convergent to $X$. Then finite or denumerably infinite sets $V_1, V_2, \ldots$ can be chosen from the sequence so that: (i), $V_1, V_2, \ldots$ are nonoverlapping; (ii), $M$ can be covered (except sets of measure zero) by the sets $V_1, V_2, \ldots$ so that $m_a[M - M(\sum_n V_n)] = 0$; and (iii), for a given $\epsilon > 0$ we have $\sum_n m(V_n) < m_a(M) + \epsilon$. This is called *Vitali's covering theorem*. On the basis of this theorem Lebesgue has shown that a finite, totally continuous, and completely additive set function $F$ is almost everywhere differentiable (i.e., except for sets of measure zero) and is equal to an indefinite integral of its derivative.

If a function $\Phi$ can be found such that

$$\Phi(J) = \Phi(J_1) + \cdots + \Phi(J_n) \quad \text{for} \quad J = J_1 + J_2 + \cdots + J_n,$$

with nonoverlapping sets $J_1, J_2, \ldots$ and that with a given $\epsilon > 0$ a constant $\delta$ can be chosen such that

$$\left|\sum_{i=1}^{n} \Phi(J_i)\right| < \epsilon \quad \text{if} \quad \sum_{i=1}^{n} m(J_i) < \delta,$$

then $\Phi$ is called an *interval function.* The displayed properties are called complete additivity and total continuity, respectively. Let $f$ be an integrable function in $a \leqq x \leqq b$, and $a \leqq \alpha < \beta \leqq b$ be a closed interval $J = [\alpha, \beta]$. Then

$$\int_J f\,dx = \int_\alpha^\beta f\,dx,$$

and

$$\int_\alpha^\beta f\,dx = \int_\alpha^\gamma f\,dx + \int_\gamma^\beta f\,dx$$

for $\alpha < \gamma < \beta$. The indefinite integral, owing to the relation

$$\Phi(J) = \Phi[\alpha, \beta] = \int_\alpha^\beta f\,dx,$$

is represented by an interval function and can be written

$$\Phi(J) = F(\beta) - F(\alpha), \qquad F = \int_\gamma^x f\,dx.$$

Consider $a \leqq x_1 < x_1' < x_2 < \cdots < x_n < x_n' \leqq b$ contained in the interval $J^o = [a, b]$. If $|\sum_{i=1}^{n} \Phi(x_i, x_i')|$ is bounded, then

$$\left|\sum_{i=1}^{n} \{F(x_i') - F(x_i)\}\right|$$

is bounded. The function is then said to be of *bounded variation* on $J^o$.

The integral of a bounded function $f(x)$ in the interval $[a, b]$ depends on the interval. This interval function obeys the mean value theorem relative to $f(x)$ and has the property of complete additivity. Such an integral is called a *Riemann integral,* which is an ordinary integral. If $f(x)$ is integrable in the Riemann sense then $f(x)$ is measurable and the Lebesgue integral of $f$ coincides with the Riemann integral (Vallée Poussin, 1916; Hahn, 1921; Baire, 1925; Hobson, 1926; Sierpinski, 1928, 1934; Schlesinger and Plessner, 1926; Kestelman, 1937; Saks, 1937; Lebesgue, 1930; Halmos, 1950; McShane, 1947; Thielman, 1953; Natanson, 1955, 1960; Whitney, 1957; Bourbaki, 1963).

## 2.2 FOURIER SERIES

If we can determine $N = N(\epsilon)$ independent of $x$ for a given $\epsilon$ such that

$$|f_{n+1}(x) + \cdots + f_{n+k}(x)| < \epsilon,$$

with an integer $k$ and an integer $n > N$ for all values of $x$ in the given interval, then the sum $\sum_n f_n(x)$ is said to be *uniformly convergent* in that interval. If $\sum_n f_n(x)$ is uniformly convergent and $f_n(x)$ is continuous at $x = x_0$, then $F(x) = \sum_n f_n(x)$ is also continuous at $x = x_0$ and is summable term by term, that is,

$$\lim_{x=x_0} \sum_n f_n(x) = \sum_{n=0}^{\infty} \lim_{x=x_0} f_n(x).$$

If a series is uniformly convergent and each of its terms is integrable, then $F(x)$ is also integrable and summable term by term. If each term of $\sum_n f_n(x)$ is differentiable and $\sum_n f_n'(x)$ is uniformly convergent, (where $f'(x)$ is the derivative of $f(x)$) and if $\sum_n f_n(x)$ is uniformly convergent at least at one point, then $\sum_n f_n(x)$ is uniformly convergent everywhere, and the derived function of $F(x)$ is $F'(x) = \sum_n f_n'(x)$; that is, the series is differentiable term by term (Knopp, 1931).

A series of the form

$$f(x) = \frac{a_0}{2} + \sum_{k=1}^{\infty} (a_k \cos kx + b_k \sin kx) \tag{1}$$

is called a *trigonometric series*. If the series is uniformly convergent it represents a periodic function of period $2\pi$. Then a sum of finite number of terms of (1) is also called a trigonometric series and represents a single-valued continuous function. If the absolute value of such a function $f(x)$ is integrable, then

$$a_k = \frac{1}{\pi}\int_0^{2\pi} f(t) \cos kt\, dt, \qquad b_k = \frac{1}{\pi}\int_0^{2\pi} f(t) \sin kt\, dt; \tag{2}$$

these are called the *Fourier coefficients*. When $k = 0$, the function equals $a_0/2$. In such cases the expression of the form (1) is called a *Fourier series*.

Let $f(x)$ be continuous in a closed interval $[a, b]$ and let $x_1, x_2$ be any two points in this interval. Write the maximum of $|f(x_2) - f(x_1)|$, with $|x_2 - x_1| \leqq \delta$, as $\omega(\delta)$. The quantity $\omega(\delta)$ is called the *modulus of continuity* of $f(x)$ in $[a, b]$. If $\omega(\delta) \leqq M\delta$ (where $M$ is a constant), it is said to satisfy the *Lipschitz condition*. If the modulus of continuity $\omega(\delta)$ of a function $f(x)$ satisfies the Dini-Lipschitz condition

$$\lim_{\delta\to 0} \omega(\delta) \log (1/\delta) = 0,$$

then the Fourier series of $f(x)$ converges uniformly to $f(x)$.

If $|f(x)|$ is integrable in $[a, b]$, then

$$\lim_{k\to\infty}\int_a^b f(x)\sin kx\,dx = 0, \qquad \lim_{k=\infty}\int_a^b f(x)\cos kx\,dx = 0.$$

If $f(x)$ is continuous and stepwise differentiable in $[0, 2\pi]$ — that is, if $f(x)$ has at most a finite number of discontinuity points in any finite interval — and if $|f'(x)|$ is integrable, then

$$\lim_{k\to\infty} ka_k = 0, \qquad \lim_{k\to\infty} kb_k = \frac{f(+0) - f(-0)}{\pi}.$$

Thus, a Fourier series is integrable term by term.

If $f(x)$ is continuous in $[0, 2\pi]$ and is such that $f(0) = f(2\pi)$, then the finite sum of trigonometric functions

$$T_n(x) = \frac{a_0}{2} + \sum_{k=1}^{n} (\alpha_k \cos kx + \beta_k \sin kx) \tag{3}$$

exists for a given $\epsilon > 0$, such that the relation $|f(x) - T_n(x)| \leqq \epsilon$ would hold uniformly in $[0, 2\pi]$. This is called the *Weierstrassian approximation theorem.* If $f(x)$ is continuous in $[a, b]$, then for a given $\epsilon$ there exists a polynomial $P(x)$ such that $|f(x) - P(x)| \leqq \epsilon$ uniformly in $[a, b]$. This is another form of the Weierstrassian approximation theorem, using not a trigonometric series but by a series of polynomials.

The Fourier coefficients $a_k$ and $b_k$ can be obtained by minimizing $\int_0^{2\pi} (f(x) - T_n(x))^2\,dx$ among the sum of trigonometrical functions of the form (3), or by

$$\lim_{n\to\infty}\int_0^{2\pi} (f(x) - T_n(x))^2\,dx = 0.$$

This is sometimes called the *minimum property* of the Fourier coefficients. If $f(x)$ is a quadratically integrable function, then

$$\frac{a_0}{2} + \sum_{k=1}^{\infty} (a_k^2 + b_k^2)$$

converges to an integral

$$\frac{1}{\pi}\int_0^{2\pi} (f(x))^2\,dx.$$

This is called the *completeness theorem* of Fourier series. Accordingly, if the Fourier coefficients of $f(x)$ are all zero, then $f(x)$ is identically zero, and two such series with the same corresponding Fourier coefficients are identical.

The sum

$$S_n = \frac{a_0}{2} + \sum_{k=1}^{n} a_k$$

is called the *Fourier partial sum*. Then

$$S_n(\xi) = \frac{1}{2\pi}\int_{\xi-\pi}^{\xi+\pi} f(x)\,\frac{\sin\,(2n+1)\,\dfrac{x-\xi}{2}}{\sin\dfrac{x-\xi}{2}}\,dx$$

$$= \frac{1}{\pi}\int_0^{\pi/2} [f(\xi+2t)+f(\xi-2t)]\,\frac{\sin\,(2n+1)t}{\sin t}\,dt.$$

Riemann proved that this $S_n(\xi)$ depends only on the property of $f(x)$ at $x = \xi$ and that the integral on the right-hand side of this equation (if the lower limit *0* of the integral is replaced by a fixed $\delta$) converges uniformly to zero as $n$ approaches infinity, for all values of $\xi$.

In general, when $\int_a^b f(x)\,dx$ exists, $f(x)$ is said to belong to class $L_1$; when $\int_a^b |f(x)|^2\,dx$ exists, $f(x)$ is said to belong to class $L_2$. The integrals are taken in the Lebesgue sense. If $[a, b]$ is finite, then $L_2$ contains $L_1$; because, if $|f(x)|$ is large, then so is $|f(x)|^2$. But $L_2$ and $L_1$ are not correlated in the infinite interval $[-\infty, +\infty]$. If $f(x)$ and $g(x)$ both belong to class $L_2$, then $f(x)g(x)$ belongs to $L_1$ and

$$\left(\int_a^b |f(x)g(x)|\,dx\right)^2 \leqq \int_a^b |f(x)|^2\,dx \int_a^b |g(x)|^2\,dx.$$

This is called the *Schwarz inequality*. In fact, since $(|f(x)| + \lambda|g(x)|)^2$ is integrable, the inequality $2|f(x)g(x)| \leqq |f(x)|^2 + |g(x)|^2$ gives

$$0 \leqq \int_a^b (|f(x)| + \lambda|g(x)|)^2\,dx$$

$$= \int_a^b |f(x)|^2\,dx + 2\lambda\int_a^b |f(x)g(x)|\,dx + \lambda^2\int_a^b |g(x)|^2\,dx,$$

a quadratic expression in $\lambda$, in which the discriminant cannot be positive.

We now introduce the *Riesz-Fischer theorem:*

If $\sum_{n=-\infty}^{+\infty} |a_n|^2$ is convergent, then there exists a function $f(x)$ in $[-\pi, +\pi]$, belonging to class $L_2$, such that

$$\lim_{n\to\infty}\int_{-\pi}^{\pi}\left|f(x) - \sum_{k=-n}^{n} a_k e^{ikx}\right|^2 dx = 0.$$

Suppose that we are given a sequence of functions $\{f_n(x)\}$ of the class $L_2$ in $[a, b]$ or in $[-\infty, +\infty]$ and that

$$\lim_{m,n\to\infty} \int_a^b |f_m(x) - f_n(x)|^2 \, dx = 0.$$

Then there exists a function $f(x)$ of class $L_2$ that satisfies

$$\lim_{n\to\infty} \int_a^b |f_n(x) - f(x)|^2 \, dx = 0.$$

In such a case $f_n(x)$ is said to *converge* to $f(x)$ *in the mean*, and is written $f(x) = \text{l.i.m.}_{n\to\infty} f_n(x)$. This means that we exclude a set of points of measure zero. It does not mean that $f(x)$ converges at all points $x$.

By stating that a sequence of functions $\{\phi_n(x)\}$ is *orthogonal* we mean that the functions are of class $L_2$ in $[a, b]$ or $[-\infty, +\infty]$ and satisfy $\int \phi_m(x)\overline{\phi_n(x)} \, dx = 0$ $(m \neq n)$ in that interval, where $\overline{\phi(x)}$ denotes the conjugate complex function of $\phi(x)$. If $\int |\phi_n(x)|^2 \, dx = 1$, then the sequence is said to be *normal*. The minimum value of

$$\int \left| f(x) - \sum_{k=-n}^{n} a_k\phi_k(x) \right|^2 dx$$

is attained at $a_k = \int f\phi_k \, dx$. There exists a function $f(x)$ of class $L_2$ such that $\lim_{n\to\infty} \int |f_n - f|^2 \, dx = 0$. If $f(x)$ is of class $L_2$ and $\{\phi_n\}$ is a sequence of normal functions, then there exists a function $g(x)$ of class $L_2$ such that

$$g(x) \sim f(x) - \sum_{n=1}^{\infty} \phi_n(x) \int f(x)\overline{\phi_n(x)} \, dx.$$

Furthermore, $\int g(x)\overline{\phi_n(x)} \, dx = 0$ for all values of $n$. If $f(x)$ belongs to class $L_2$ and

$$f(x) \sim \sum_{n=1}^{\infty} \phi_n(x) \int f(x)\overline{\phi_n(x)} \, dx,$$

then the sequence of functions $\{\phi_n\}$ is said to be closed or complete. In order that $\{\phi_n\}$ be complete, it is necessary and sufficient that for a given $\epsilon > 0$ there should exist a function $f(x)$ of class $L_2$, an integer $N$, and a sum $\sum_{k=1}^{N} a_k\phi_k(x)$ such that

$$\int \left| f(x) - \sum_{k=1}^{N} a_k\phi_k(x) \right|^2 dx < \epsilon.$$

Also, in such a case,

$$\int |f(x)|^2 \, dx = \sum_{n=1}^{\infty} \left| \int f\bar{\phi}_n \, dx \right|^2.$$

This is called the *Parseval theorem*.

In the theories of perturbation in celestial mechanics we usually employ trigonometric series at present. Those theorems and properties are indispensable for criticism of the theories discussed in the following chapters (Schlesinger and Plessner, 1926; Zygmund, 1935; Hobson, 1926; Tonelli, 1928; Kaczmarz and Steinhaus, 1935; Lebesgue, 1906; Vallée Poussin, 1919; Bernstein, 1926; Hilb and Riesz, 1924; Rogosinski, 1930; Denjoy, 1941, 1949; Hardy and Rogosinski, 1950; Natanson, 1955, 1960; Kolmogorov and Fomin, 1961).

## 2.3 LAGRANGE'S THEOREM

Let $f(x)$ be a regular function inside a domain $A$ bounded by a simple closed curve $C$ in a complex plane of $x$, and $\theta(z)$ be a regular function inside and on the boundary of $C$. Let $a$ be a zero point of $\theta(z)$ inside $C$, such that $a$ is a simple root of $\theta(z) = 0$. By a "simple curve" we mean a continuous curve without any double or any multiple point; that is, a curve that does not intersect itself. Let $x$ be a given point inside $A$ and $|\theta(x)| < |\theta(z)|$ for all points $z$ inside $C$. Then, by Cauchy's theorem in the theory of analytic functions, the complex integral

$$\frac{1}{2\pi i}\int_C f(z)\,\frac{\theta'(z)}{\theta(z)}\,dz$$

represents the sum of the residues inside $C$. In fact, if we write the development of $\theta(z)$ around $a$

$$\theta(z) = A_1(z-a)^r + A_2(z-a)^{r+1} + \cdots,$$

then

$$\theta'(z) = A_1 r(z-a)^{r-1} + \cdots, \qquad f(z) = f(a) + (z-a)f'(a) + \cdots,$$

and, in addition,

$$\frac{f(z)\theta'(z)}{\theta(z)} - \frac{rf(a)}{z-a}$$

is regular at $a$.

Hence

$$\frac{1}{2\pi i}\int_C \frac{dz}{z-a} = 1, \qquad \frac{1}{2\pi i}\int_C \frac{dz}{(z-a)^r} = 0 \qquad (r \neq 1),$$

and, hence, the residue of $f(z)\theta'(z)/\theta(z)$ is $rf(a)$. From this follows the theorem. In particular, if we put $f(z) = 1$, then the integral

$$\frac{1}{2\pi i}\int_C \frac{\theta'(z)}{\theta(z)}\,dx$$

gives the total number of zero points of $\theta(z)$ inside $C$.

In the statement of this theorem we put $\theta(z) - \theta(x)$ in place of $\theta(z)$. Then, since $|\theta(x)| < |\theta(z)|$ on $C$, we can easily show that the total number of zero points of $\theta(z) - \theta(x)$ inside $C$ is

$$\frac{1}{2\pi i}\int_C \frac{\theta'(z)\,dz}{\theta(z) - \theta(x)} = \frac{1}{2\pi i}\left[\int_C \frac{\theta'(z)}{\theta(z)}\,dz + \theta(x)\int_C \frac{\theta'(z)}{(\theta(z))^2}\,dz + \cdots\right]$$

$$= \frac{1}{2\pi i}\int_C \frac{\theta'(z)\,dz}{\theta(z)}.$$

This integral represents the number of zero points of $\theta(z) - \theta(x)$ inside $C$. By our assumption, $\theta(z)$ has only one root — indeed, one simple root — $a$ inside $C$. Thus $\theta(z) - \theta(x)$ has only a simple zero point $z = x$ inside $C$. Hence, by Cauchy's theorem,

$$f(x) = \frac{1}{2\pi i}\int_C \frac{f(z)\theta'(z)\,dz}{\theta(z) - \theta(x)}.$$

We expand the integrand, since $|\theta(x)| < \theta(z)$, in the form

$$f(x) = \frac{1}{2\pi i}\left[\sum_{n=0}^{\infty} (\theta(x))^n \int_C \frac{f(z)\theta'(z)}{(\theta(z))^{n+1}}\,dz\right]$$

$$= \sum_{n=0}^{\infty} A_n(\theta(x))^n, \qquad A_n = \frac{1}{2\pi i}\int_C \frac{f(z)\theta'(z)\,dz}{(\theta(z))^{n+1}}.$$

Or, from integration by parts,

$$A_n = \frac{1}{2\pi i n}\int_C \frac{f'(z)}{(\theta(z))^n}\,dz.$$

Now we write $\theta(z) = (z - a)\theta_1(z)$. The function $\theta_1(z)$ is regular on and inside $C$ and does not vanish. Hence,

$$A_n = \frac{1}{2\pi i n}\int_C \frac{f'(z)\,dz}{(z - a)^n(\theta_1(z))^n} = \frac{1}{n!}\frac{d^{n-1}}{da^{n-1}}\left[\frac{f'(a)}{(\theta_1(a))^n}\right],$$

because

$$\frac{1}{2\pi i}\int_C \frac{f(z)\,dz}{(z - a)^n} = \frac{1}{(n-1)!}f^{(n-1)}(a).$$

Especially for $n = 0$, using Cauchy's residue theorem again gives

$$A_0 = \frac{1}{2\pi i}\int_C \frac{f(z)\theta'(z)}{\theta_1(z)}\frac{dz}{z - a} = f(a).$$

Thus, if we put

$$\theta_1(z) = \frac{1}{\phi(z)}, \qquad \theta(x) = t, \qquad \frac{z-a}{\phi(z)} = \theta(z),$$

then

$$|\theta(x)| < \left|\frac{z-a}{\phi(z)}\right| \qquad \text{on } C.$$

This leads to the following *theorem of Lagrange* (Whittaker and Watson, 1924; Goursat, 1918):

Let $f(z)$, $\phi(z)$ be regular on and inside a closed curve $C$ around $a$, and let $t$ be such that $|t\phi(z)| < |z - a|$ holds for all points $z$ on $C$. Then the equation for $x$, $x = a + t\phi(x)$, admits one root inside $C$. Any function regular on and inside $C$ can be expanded in an integral power series of $t$ in the form

$$f(x) = f(a) + \sum_{n=1}^{\infty} \frac{t^n}{n!} \frac{d^{n-1}}{da^{n-1}} [f'(a)(\phi(a))^n]. \tag{4}$$

## 2.4 ASYMPTOTIC MOTION AND LIBRATION

Consider a dynamical system with one degree of freedom

$$\frac{dq}{dt} = \frac{\partial H}{\partial p}, \qquad \frac{dp}{dt} = -\frac{\partial H}{\partial q}, \qquad H = \tfrac{1}{2}Ap^2 - U(q).$$

The corresponding Hamilton-Jacobi partial differential equation is

$$A(q)\left(\frac{\partial V}{\partial q}\right)^2 = 2(U + \alpha),$$

and the solution is

$$V = \int \sqrt{\frac{2(U+\alpha)}{A(q)}}\, dq, \qquad t - t_0 = \frac{\partial V}{\partial a} = \int \frac{dq}{\sqrt{2(U+\alpha)A(q)}}.$$

Take as the independent variable in place of $t$ a new variable called a *mechanical variable*, which increases monotonically and continuously with $t$ and becomes infinite together with $t$. In our case the mechanical variable is $q$; writing $2(U + \alpha)A(q) = F(q)$ gives

$$\left(\frac{dq}{dt}\right)^2 = F(q).$$

By our assumption, $F(q)$ is continuous; it either does or does not vanish for some values of $q$.

First suppose that $F(q)$ does not vanish and is always positive. If $F(q)$ were negative, then the motion would be imaginary. Further, $F(q)$ is supposed to be an even function of $q$, and to remain finite for any finite value of $q$, and, in particular, to be periodic with period $2\pi$ with regard to $q$. As $F(q) \neq 0$, it can be expanded by the Weierstrassian approximation theorem of Section 2.2 in a Fourier series

$$(F(q))^{-1/2} = A_0 + A_1 \cos q + A_2 \cos 2q + \cdots.$$

By our assumption this does not vanish, so that

$$|A_0| > |A_1| + |A_2| + \cdots,$$

which is a sufficient condition. Hence

$$t - t_0 = \int (F(q))^{-1/2}\, dq = A_0 q + A_1 \sin q + \tfrac{1}{2}A_2 \sin 2q + \cdots,$$

and

$$|A_0| > |A_1| + \tfrac{1}{2}|A_2| + \cdots.$$

Since the first term is the largest, we integrate this expression with respect to $q$ from 0 to $2\pi$, obtaining $2T = 2\pi A_0$. By Lagrange's theorem of Section 2.3 we can expand $q$ in the form

$$q = \frac{t - t_0}{A_0} + \sum_{j=1} A_j \sin\left(j\frac{t - t_0}{A_0}\right),$$

if $|\phi(q)| < |t - t_0 - A_0 q|$, where $t - t_0 = A_0 q + \phi(q)$. Hence, when $t$ increases by $2T$, the variable $q$ increases by $2\pi$. Thus $q$, regarded as an angular variable, rotates through the whole circumference of a circle from $q = 0$ to $q = 2\pi$. This is called the case of *revolution.*

Next, suppose that $F(q)$ can become zero. $q$ regarded as an angular variable cannot take all values but it *oscillates* between two values of $q$ smaller than $2\pi$; that is, it cannot pass through the value $a$ for which $F(a) = 0$. In fact, if we put $F(q) = (q - a)^m \phi(a)$, $\phi(a) \neq 0$, then $\phi(a)$ can be expanded in a Taylor series and

$$\frac{dq}{(q - a)^{m/2}} [C_0 + C_1(q - a) + C_2(q - a)^2 + \cdots] = \pm dt.$$

In particular, if $m$ is odd, then

$$\frac{2}{(q - a)^{m/2-1}} \left[\frac{C_0}{m - 2} + \frac{C_1}{m - 4}(q - a) + \cdots\right] = \mp(t - t_0);$$

if $m$ is even, then

$$C_{m/2-1} \log (q - a) + \frac{2}{(q - a)^{m/2-1}} \left[\frac{C_0}{m - 2} + \frac{C_1}{m - 4}(q - a) + \cdots\right] = \mp(t - t_0).$$

When $q$ approaches the value $a$, the absolute value of the left-hand side increases; hence, $t$ increases, provided that $m/2 \geqq 1$. The variable $q$ does not approach $a$ for any finite value of $t$. Suppose that $q$ does not pass through $a$. This is called a case of *asymptotic* motion. If $m = 1$ and $a$ is a simple root of $F(q) = 0$, then

$$q - a = A_1(t - t_0)^2 + A_2(t - t_0)^4 + \cdots.$$

For $q = a$, $t = t_0$; thus $q$ reaches $a$ for a finite value of $t$ but never *through* $a$; it turns back and follows the same route in the opposite sense. Such a point is called an *ordinary* point.

Let two consecutive roots of $F(q) = 0$ be $a$ and $b$. Then

$$\left(\frac{dq}{dt}\right)^2 = (q - a)^m(b - q)^n\psi(q), \qquad \psi(a) \neq 0, \qquad \psi(b) \neq 0;$$

or

$$\left(\frac{dq}{dw}\right)^2 = \epsilon^2(q - a)^m(b - q)^n, \qquad \left(\frac{dw}{dt}\right)^2 = \frac{1}{\epsilon^2}\psi(q).$$

Take $w$ such that $w = 0$ for $t = 0$, and suppose that $dt$ and $dw$ are of the same sign. The variable $w$ is the mechanical variable. If $q$ is initially between $a$ and $b$, then $q$ always lies between $a$ and $b$.

Let $m = n = 1$, then

$$\frac{dq}{dw} = \epsilon\sqrt{(q - a)(b - q)} \quad \text{and} \quad q = a\cos^2\frac{\epsilon w}{2} + b\sin^2\frac{\epsilon w}{2}.$$

Hence, $q$ is a periodic function of $w$ with period $2\pi/\epsilon$ with respect to $w$. The variable $q$ is in this case a periodic function of $t$, also. Actually,

$$\int_0^w \frac{\epsilon\, dw}{\sqrt{\psi(q)}} = t.$$

The variable $q$ does not change when $w$ increases by $2\pi/\epsilon$, while $t$ increases by $2T$. Since $q$ is a periodic function of $\epsilon w$,

$$T = \int_0^\pi \frac{\epsilon\, dw}{\sqrt{\psi(a\cos^2(\epsilon w/2) + b\sin^2(\epsilon w/2))}}$$

$$= \int_a^b \frac{dq}{\sqrt{(q - a)(b - q)\psi(q)}}.$$

In order that the periods with respect to $w$ and to $t$ be equal, $\epsilon$ should equal $\pi/T$. Thus $q$ is periodic with respect to $t$; from the Weierstrassian approximation theorem,

$$q = \tfrac{1}{2}B_0 + \sum_{j=1}^{\infty} B_j \cos j\frac{\pi t}{T}.$$

In this case $q$ does not make the whole revolution on the circumference of a circle; it oscillates between two limits smaller than $2\pi$ around $q = \frac{1}{2}B_0$. This is called a *libration*. In a region where $F(q)$ is positive between two consecutive simple roots of $F(q) = 0$, there is a libration. It is necessary to state in what variable the libration takes place. The motion with regard to $q$ is a libration, but in our case it is also a revolution with regard to $w$ (Hagihara, 1944; Brown, 1932). The phenomenon of libration is very important in celestial mechanics. The moon librates around its axis of rotation with respect to the line connecting the moon and the earth; this is called the *physical* libration and causes more than half the surface area of the moon to be visible from the earth. The perisaturnia of some satellites of Saturn librate around the perisaturnia or the nodes of other satellite orbits. The librations of the satellites of Jupiter are most remarkable. Librations also occur in the motion of some asteroids.

If $m$ or $n$ or both equal or exceed 2, an asymptotic motion takes place. The moving point approaches the limit point but never passes through it in the limit as $t$ approaches infinity. When two simple zero points $q = a$ and $q = b$ coincide and become a double zero point, then the moving point $C$ comes to rest at that double point if the domain of real motion degenerates to this single point; but the real motion is asymptotic from both sides of the double zero point if the domain of imaginary motion degenerates to this single point. When the root is multiple in an odd order, then the moving point is either at rest at a point in the domain of real motion or approaches it asymptotically. Furthermore, in the domain between the largest real root and $+\infty$ and in the domain between the smallest real root and $-\infty$, the motion is asymptotic for $q = +\infty$ and $q = -\infty$, respectively. The motion, beginning with $q = +\infty$ or $q = -\infty$ at $t = -\infty$, reaches the largest or the smallest real simple zero point in a finite time and returns to the initial point for $t = +\infty$. When the real zero point is multiple, the motion is asymptotic to that point.

In order to study the motion at $q = \pm\infty$, we write $q = 1/u$, as in the theory of analytic functions, and study the motion at $u = 0$. Put

$$F(q) = C_n q^n + C_{n-1} q^{n-1} + \cdots + C_0;$$

then,

$$dt = -\frac{du}{u^{2-n/2}\sqrt{C_n + C_{n-1}u + \cdots + C_0 u^n}}.$$

If $n \neq 2$, the motion is asymptotic at $u = 0$ or $q = \infty$. If $n = 2$, the motion is also asymptotic. But, if the domain of real motion extends from the largest real simple zero point to $+\infty$, or from $-\infty$ to the smallest real simple zero point, then we can transform the real domain to a real finite domain by

$$q = \frac{\alpha v + \beta}{\gamma v + \delta}$$

as in the theory of elliptic functions, modular functions, and automorphic functions, the motion is a libration with regard to the variable $v$ in that domain, that is, the singular character at $q = \infty$ with regard to $v$ is only an apparent one. For example, if we take $F(q) = (q - a)(b - q)$, then if we write $q = c + 1/v$ $(a < c < b)$, the motion is a libration between $-1/(c - a)$ and $+1/(b - c)$ with regard to $v$.

When $F(q)$ is a third-degree polynomial with respect to $q$, $\psi(q)$ may become zero and the above condition is not satisfied. When $F(q)$ is a fourth degree polynomial and has four real zero points $a < b < c < d$, then the real motion of libration takes place in the real domain in which the initial position is situated. But, if the domain extends to values smaller than $a$ or larger than $d$, the motion is asymptotic at $q = -\infty$ or $q = +\infty$. The polynomial $F(q)$ can then be transformed to its canonical form

$$F(q) = 4q^3 - g_2 q - g_3,$$

in the theory of elliptic functions, where $g_2$ and $g_3$ are the invariants of quartic forms. One zero point is driven to infinity (Hagihara, 1931; Appell and Lacour, 1922; Halphen, 1886).

More generally, when $F(q)$ is an algebraic polynomial, the behavior of the motion can be discussed as it is in the theory of hyperelliptic functions (Appell and Goursat, 1895; Baker, 1897, 1907; Bliss, 1933; Osgood, 1932; Roberts and Hartford, 1938; see also Sections 6.23, 6.25).

## CONDITIONALLY PERIODIC MOTIONS

### 2.5 MULTIPLY PERIODIC FUNCTIONS

When a real function $F(x)$ satisfies the condition $F(x) = F(x + \omega)$ with a real constant $\omega$ for all real values of $x$, the function $F(x)$ is said to have a *period* $\omega$. With any positive or negative integer $\tau$ we have

$$F(x) = F(x + \tau\omega);$$

that is, if $\omega$ is a period, then its positive integral multiple or its negative integral multiple is also a period. Suppose that $F(x)$ has another real period $\omega'$. If $F(x) = F(x + \omega')$, then $\tau'\omega'$ with any positive or negative integer $\tau'$ is also a period. Hence, $\tau\omega + \tau'\omega'$ is also a period.

Assume that $F(x)$ has no infinitely small period. If $F(x)$ has two real periods $\omega$ and $\omega'$, then we can choose $\tau\omega + \tau'\omega'$ as small as we please by taking two integers $\tau$ and $\tau'$ suitably, as will be described in Kronecker's theorem in Section 2.8. But, since by our assumption $F(x)$ has no infinitely small period, the two periods $\omega$ and $\omega'$ must be commensurable; that is, they must be in the ratio of two integers. Consider now the ratio $p/q$ of two mutually prime integers. If $\omega$ is a period, then any other period must be in the form $p\omega/q$. If we choose $\tau$ and $\tau'$ such that

$$\tau q + \tau' p = 1,$$

then $\tau\omega + \tau' p\omega/q = \omega/q$, and $\omega/q$ is also a period. By assumption there should be the smallest number $\omega/q_1$ among all periods of the form $\omega/q$. Hence, there should be necessarily a smallest period if there is no infinitely small period.

Next we consider a function $F(x_1, x_2, \ldots, x_n)$ of $n$ variables $x_1, x_2, \ldots, x_n$, and suppose that when $x_1, x_2, \ldots, x_n$ increase by $\omega_{i1}, \omega_{i2}, \ldots, \omega_{in}$, respectively, the function $F(x)$ has the same value. Then we say that $F(x)$ is *multiply periodic* with the period $(\omega_i) = (\omega_{i1}, \omega_{i2}, \ldots, \omega_{in})$. The result of multiplying every element $\omega_{ik}$ — $k = 1, 2, \ldots, n$ — of $(\omega_i)$ by a fixed integer is also a multiple period. If $(\omega_i)$ and $(\omega_j)$ are both multiple periods of $F(x)$, then $(\omega_i + \omega_j) = (\omega_{i1} + \omega_{j1}, \ldots, \omega_{in} + \omega_{jn})$ is also a multiple period of $F(x)$. $(\tau\omega_i + \tau'\omega_j) = (\tau\omega_{i1} + \tau'\omega_{j1}, \ldots, \tau\omega_{in} + \tau'\omega_{jn})$ is also a multiple period. Generally,

$$\left(\sum_{k=1}^{m} \tau_k \omega_{k1}, \sum_{k=1}^{m} \tau_k \omega_{k2}, \ldots, \sum_{k=1}^{m} \tau_k \omega_{kn}\right) \qquad (m \leqq n)$$

is also a multiple period. When there is no relation of the form

$$\frac{\tau_1}{\omega_{1j}} + \frac{\tau_2}{\omega_{2j}} + \cdots + \frac{\tau_m}{\omega_{mj}} = 0 \qquad (j = 1, 2, \ldots, m),$$

with positive or negative integers $\tau_1, \tau_2, \ldots, \tau_m$ (all of which do not vanish at the same time), then the multiple periods $(\omega_i)$ (all of which do not vanish at the same time for $i = 1, 2, \ldots, m$), are said to be *independent*. When a multiple period $(\omega_l)$ is connected with other multiple periods $(\omega_i)$ by a linear homogeneous relation, then the multiple period $(\omega_l)$ is said to belong to a *field of multiple periods* $(\omega_i)$, and the $m$ multiple periods $(\omega_i)$ are called the *primitive periods of order m* (Esclangon, 1902, 1903, 1904, 1916).

In general, a system admitting the operations of addition and multiplication, excluding the commutative law in multiplication, is called a *ring*. If a ring has at least one nonzero element, if $ax = b$ $(a \neq 0)$ can be solved for $x$, and if the ring obeys the commutative law, then it is called a *field*. By this definition there must be a unit element in a field and also a reciprocal element (Van der Waerden, 1937). When $F(x_1, x_2, \ldots, x_n)$ cannot be reduced to a function of a smaller number of variables $y_1, y_2, \ldots, y_p$, $p < n$, by a linear transformation from $x_1, x_2, \ldots, x_n$, the function is $F$ called *irreducible*.

The multiple period $(\omega_{i1}, \omega_{i2}, \ldots, \omega_{in})$ can be represented as the components of a vector $(\omega_i)$ in an $n$-dimensional space. The function $F(x_1, x_2, \ldots, x_n)$ can be considered to belong to a point $(x_1, x_2, \ldots, x_n)$ of that space. Suppose that there is at least one point at which the function $F(x_1, x_2, \ldots, x_n)$ is continuous for all variables $x_1, x_2, \ldots, x_n$ on the straight line $\Delta$:

$$x_j = a_j + \omega_{ij}t \qquad (j = 1, 2, \ldots, n)$$

in this space. Any such function with this property — in particular, a function that is continuous and irreducible — must be a constant; otherwise, it cannot have an infinitely small period.

For the proof of this theorem of Esclangon we first suppose that $F$ *does* have an infinitely small period. (Eventually we will show that this is impossible.) As it is irreducible, the function

$$\Phi(t) = F(a_1 + \omega_{i1}t, a_2 + \omega_{i2}t, \ldots, a_n + \omega_{in}t)$$

would not be a constant on the corresponding straight line $\Delta$ for any set of values of $\omega_{i1}, \omega_{i2}, \ldots, \omega_{in}$. In fact, if it were a constant on a certain straight line, it would be expressible as a function of a smaller number of variables and become reducible by a transformation that makes the straight line one of the coordinate axes. Now we draw a sphere with radius 1 and the origin $O(a_1, a_2, \ldots, a_n)$ as center, and let the points $\omega_{ij}/|\omega_i|$, where $|\omega_i| = |\sqrt{\omega_{i1}^2 + \omega_{i2}^2 + \cdots + \omega_{in}^2}|$ $(j = 1, 2, \ldots, n)$ correspond to the points on the surface of the sphere. Let $\{\epsilon_1\}, \{\epsilon_2\}, \ldots, \{\epsilon_n\}$ be a set of sequences, each tending to zero. Then there will be a multiple period $(\omega_i)$ with its absolute values $|\omega_i|$ each smaller than the corresponding $\epsilon$. Denote them by $(\omega')$, $(\omega'')$, $\ldots$, $(\omega^{(k)})$, $\ldots$, and represent them by the respective points $P'$, $P''$, $\ldots$, $P^{(k)}$, $\ldots$ on the surface of the sphere. The sequence of points $P'$, $P''$, $\ldots$, $P^{(k)}$, $\ldots$ will have its limit point $P^{(o)}(\omega^{(o)})$ on the sphere. There should be at least one multiple period with its absolute value smaller than $\epsilon$ inside a small circle on the sphere with $P^{(o)}$ as its center. Take a point $Q$ in the space, and a point $Q'$

on the straight line through $Q$ drawn parallel to $OP^{(o)}$. It is required to prove that there is a point $Q_1'$ in the neighborhood of $Q$ such that

$$F(Q_1') = F(Q).$$

Draw straight lines $\Delta_k$, $k = 1, 2, \ldots$, which pass, respectively, through points $(\omega_k), 2(\omega_k), \ldots, p(\omega_k)$, $k = 1, 2, \ldots$, with $Q$ as origin. These points form a point set $E$. For example, all points on $QQ'$ are the limit points of this point set $E$. We draw, with $Q'$ as center, a hemisphere $S$ with radius $\rho$ and a hemisphere $S'$ with radius $\rho/2$. Since the straight lines $\Delta_k$ have their limit $QQ'$, there are an infinite number of points belonging to the hemisphere $S'$. Among these points corresponding to the multiple periods $(\omega_{kj})$ there should certainly be at least one of absolute value less than $\rho/2$. Let $\Delta_k$ be the straight line corresponding to one such point. Then the hemisphere $S$ should contain in its interior one point $Q_1'$ at the ends of the lines through $(\omega_{kj}), 2(\omega_{kj}), \ldots$. As we can make the radius of $S$ as small as we please, $Q'$ should be a limit point of $E$. Hence, all points on $QQ'$ are the limit points of $E$. Now at the points belonging to the point set $E$, the function $F$ takes the value $F(Q)$. If $F$ is continuous at $Q'$, then its value should be $F(Q)$. Since it is irreducible, $F$ cannot be a constant on a straight line parallel to $QQ'$. Let $QQ'$ be a straight line on which $F$ is not a constant. On our assumption concerning $F$, there must certainly be a point $Q'$ at which $F$ is continuous on the line $QQ'$. Let $Q$ be another arbitrary point on the straight line; then

$$F(Q) = F(Q')$$

must hold. Thus $F$ must be a constant on $QQ'$. This is contradictory to our assumption. Hence, $F$ cannot have an infinitely small period.

An irreducible function with a finite multiple period does not necessarily satisfy the above assumption. To do this (so Esclangon proved) the function should be a continuous function, or a *pointwise discontinuous* function (as defined by Baire) or a *totally discontinuous* function (Section 2.1; Baire, 1925; de la Vallée Poussin, 1916; Hahn, 1921). When the points of discontinuity of a function $f$ are *compact* — that is, when any partial set with an infinite number of elements taken out from a point set composed of such discontinuity points has always its limit point — then the discontinuous function $f$ is said to be *totally discontinuous* and, when the discontinuity points of $f$ are not compact, $f$ is said to be *pointwise discontinuous*, according to Baire. However, if an irreducible function has an infinitely small period, then the function must be a totally discontinuous function, in Baire's sense, at least on a certain straight line. We do not consider such a function with an infinitely small period. Esclangon calls an irreducible continuous function with a finite multiple period a *uniformly continuous quasi-periodic function.* From now on we consider only uniformly continuous quasi-periodic functions and call them merely quasi-periodic functions. Then

$$\Phi(t) = F(a_1 + \omega_{i1}t, \ldots, a_n + \omega_{in}t)$$

should be periodic for $t$ in the same direction, that is, along $(\omega_i)$, whatever the values of $a_1, a_2, \ldots, a_n$ are. The smallest common period of $\Phi(t)$ as we vary $(a_1, a_2, \ldots, a_n)$ should be in the form $(\omega_i)/q$, where $q$ is an integer. Hence, $F$ has a multiple period $(\omega_{i1}/q, \omega_{i2}/q, \ldots, \omega_{in}/q)$. Of course, the case when $\Phi(t)$ is continuous in $t$ is a particular case of our assumption; therefore $F$ should have the smallest multiple periods.

Consider only the minimum periods among the periods of the same direction in the $n$-dimensional space. We construct a parallelogram with $(\omega_i)$ and $(\omega_j)$ as the two sides and a lattice system obtained by translating $(\omega_i)$ and $(\omega_j)$ by positive and negative integral multiples of the remaining $(\omega)$'s. The periods $\tau_i(\omega_i) + \tau_j(\omega_j)$ are all represented by this lattice system. In general, we take $(\omega_k)$, $k = 1, 2, \ldots, m$, as the basis period and obtain a lattice system consisting of an $n$-dimensional parallelopiped. We call the parallelopiped formed by the $m$ vectors $(\omega_k)$ the *basis parallelopiped.* When $n = m$, this lattice system fills up the $n$-dimensional space. We transform this by a linear homogeneous transformation with positive or negative integers $\tau_{jk}$ as the coefficients:

$$(\omega_j') = \sum_{k=1}^{n} \tau_{jk} \cdot (\omega_k) \qquad (j = 1, 2, \ldots, n).$$

The transformed period $(\omega_j')$ also belongs to the basis parallelopiped. Moreover, if $\|\tau_{ij}\| = \pm 1$, then the volume of the basis parallelopiped is invariant. This is a *unimodular* transformation. Here the vector $(\omega_j')$ is also independent and is a basis period of this parallelopiped. Let $(\omega_j')$ be transformed into $(0, 0, \ldots, 0, \omega_{jj}', 0, \ldots, 0)$ by such a transformation. Put $\omega_{jj}' = \omega_j$. Then the transformed basis periods are in the canonical form:

$$\begin{matrix} \omega_1 & 0 & \cdots & 0 \\ 0 & \omega_2 & \cdots & 0 \\ \vdots & & & \\ 0 & 0 & \cdots & \omega_n. \end{matrix} \tag{5}$$

Apply the same unimodular transformation to $x_1, x_2, \ldots, x_n$ and denote the result by $w_1, w_2, \ldots, w_n$. Then

$$w_j = \sum_{k=1}^{n} \tau_{jk} x_k \qquad (j = 1, 2, \ldots, n).$$

This means that the vectors representing the original periods $(\omega_j)$ become the new coordinate axes $w_k$. These $w$-axes are not necessarily orthogonal. If $m < n$, then only $n - m$ coordinate axes are transformed by this operation. We further change the scale of length of these coordinates such that the transformed parallelopiped has the sides

$$\begin{matrix} 2\pi & 0 & \cdots & 0 \\ 0 & 2\pi & \cdots & 0 \\ \vdots & & & \\ 0 & 0 & \cdots & 2\pi. \end{matrix} \tag{6}$$

Denote the result of this transformation of $F(x_1, x_2, \ldots, x_n)$ by

$$F(x_1, x_2, \ldots, x_n) = \Psi(w_1, w_2, \ldots, w_n).$$

This can be written by Weierstrass's auxiliary theorem, which is proved in Section 2.6, immediately following, in the form

$$\Psi(w_1, w_2, \ldots, w_n) = \sum_{\nu_1, \nu_2, \ldots, \nu_n} C_{\nu_1 \nu_2 \cdots \nu_n} \exp\left[\sqrt{-1}\,(\nu_1 w_1 + \cdots + \nu_n w_n)\right]. \tag{7}$$

This result is of great importance in the quantum theory of Bohr and Kramers.

### 2.6 WEIERSTRASS'S AUXILIARY THEOREM

Suppose that an integral function $g(v, x_1, \ldots, x_n)$ of $n + 1$ variables $v, x_1, \ldots, x_n$ is periodic with period 1 with respect to $v$. We have then $g(v + m, x_1, \ldots, x_n) = g(v, x_1, \ldots, x_n)$, for any arbitrary integer $m$. An integral function is a function that can be developed in a Taylor series with respect to all the independent variables. Now put

$$g_0(v, x_1, \ldots, x_n) = \tfrac{1}{2} g(v, x_1, \ldots, x_n) + \tfrac{1}{2} g(-v, x_1, \ldots, x_n),$$

$$g_1(v, x_1, \ldots, x_n) = \frac{1}{\sin 2\pi v}\left(\tfrac{1}{2} g(v, x_1, \ldots, x_n) - \tfrac{1}{2} g(-v, x_1, \ldots, x_n)\right).$$

Then $g_0$ and $g_1$ are both integral functions of $v, x_1, \ldots, x_n$. Actually the denominator of $g_1$ vanishes only when $2v$ is equal to an arbitrary integer $m$. Writing $v = (m/2) + h$ gives

$$\sin 2\left(\frac{m}{2} + h\right)\pi = \pm \sin 2h\pi,$$

$$g\left(-\frac{m}{2} - h, x_1, \ldots, x_n\right) = g\left(\frac{m}{2} - h, x_1, \ldots, x_n\right).$$

Hence

$$g\left(\frac{m}{2} + h, x_1, \ldots, x_n\right) - g\left(-\frac{m}{2} - h, x_1, \ldots, x_n\right) = hP(h, x_1, \ldots, x_n),$$

where $P$ is itself an integral function of its variables. Thus, $g_1(v, x_1, \ldots, x_n)$ is finite at $v = m/2$.

Now define $G_0(s, x_1, \ldots, x_n)$ by varying $s$ arbitrarily in the following manner. If

$$G_0(s, x_1, \ldots, x_n) = g_0(v, x_1, \ldots, x_n),$$

with the variable $v$ satisfying $s = \cos 2\pi v$ for any finite value of $s$, then $G_0(s, x_1, \ldots, x_n)$ has a definite finite value for any finite values of $s, x_1, \ldots, x_n$. Let $v'$ be one of the values of $v$ satisfying $s' = \cos 2\pi v'$ for a given $s'$; then any other value of $v$ satisfying this equation is represented by $\pm v' + m$. For all such values of $v$ the function $g_0(v, x_1, \ldots, x_n)$ has the same value. Take $s$ and $s'$ nearly equal to each other and put $h = 2\pi(v - v')$; then, since

$$s - s' = -\sin 2\pi v'\left(h - \frac{h^3}{3!} + \cdots\right) - \cos 2\pi v'\left(\frac{h^2}{2!} - \cdots\right),$$

we can determine the value of $v - v'$ satisfying this equation if

$$\sin 2\pi v' \neq 0.$$

When $\sin 2\pi v' = 0$, then $(v - v')^2$ is expressed as an integral power series of $s - s'$. In the latter case, $v' = m/2$ and

$$\begin{aligned} g_0(v, x_1, \ldots, x_n) &= \tfrac{1}{2}g\left(-\frac{m}{2} + (v - v'), x_1, \ldots, x_n\right) \\ &\quad - \tfrac{1}{2}g\left(-\frac{m}{2} - (v - v'), x_1, \ldots, x_n\right) \\ &= \tfrac{1}{2}g\left(\frac{m}{2} + (v - v'), x_1, \ldots, x_n\right) \\ &\quad - \tfrac{1}{2}g\left(\frac{m}{2} - (v - v'), x_1, \ldots, x_n\right). \end{aligned}$$

Hence, the expansion of $G_0(v, x_1, \ldots, x_n)$ has only even powers of $v - v'$. In either case $G_0(s, x_1, \ldots, x_n)$ can be expanded in an integral power series of $s - s', x_1, \ldots, x_n$. Hence, $G_0(s, x_1, \ldots, x_n)$ is an integral function of $s, x_1, \ldots, x_n$ and is equal to $g_0(v, x_1, \ldots, x_n)$, where

$$s = \cos 2\pi v.$$

Similarly, it can be proved that there exists an integral function $G_1(s, x_1, \ldots, x_n)$ that becomes $g_1(v, x_1, \ldots, x_n)$ if $s = \cos 2\pi v$. Hence,

$$\begin{aligned} g(v, x_1, \ldots, x_n) &= G_0(\cos 2\pi v, x_1, \ldots, x_n) \\ &\quad + G_1(\cos 2\pi v, x_1, \ldots, x_n) \sin 2\pi v. \end{aligned}$$

We now apply this theorem to the function $\phi(v_1, v_2, \ldots, v_n)$ satisfying

$$\begin{aligned} \phi(v_1 + 1, v_2, \ldots, v_n) &= \phi(v_1, v_2, \ldots, v_n), \\ \phi(v_1, v_2 + 1, \ldots, v_n) &= \phi(v_1, v_2, \ldots, v_n), \\ &\vdots \\ \phi(v_1, v_2, \ldots, v_n + 1) &= \phi(v_1, v_2, \ldots, v_n), \end{aligned} \tag{8}$$

and, in general,

$$\phi(v_1 + m_1, v_2 + m_2, \ldots, v_n + m_n) = \phi(v_1, v_2, \ldots, v_n)$$

with positive or negative integers $m_1, m_2, \ldots, m_n$. This function can be written in the form

$$G_0(\cos 2\pi v_1, v_2, \ldots, v_n) + G_1(\cos 2\pi v_1, v_2, \ldots, v_n) \sin 2\pi v_1,$$

and $G_0$ and $G_1$ each have period 1 with respect to $v_2, \ldots, v_n$. Hence

$$\begin{aligned} G_{\epsilon,1}(\cos 2\pi v_1, v_2, \ldots, v_n) &= G_{\epsilon,0}(\cos 2\pi v_1, \cos 2\pi v_2, v_3, \ldots, v_n) \\ &\quad + G_{\epsilon,1}(\cos 2\pi v_1, \cos 2\pi v_2, v_3, \ldots, v_n) \cos 2\pi v_2, \end{aligned}$$

where $\epsilon = 0, 1$. Both $G_{\epsilon,0}$ and $G_{\epsilon,1}$ have period 1 with respect to $v$. By repeating this procedure we show that $\phi(v_1, v_2, \ldots, v_n)$ can be written in the form

$$\begin{aligned} \sum_{\epsilon_1=0}^{1} \sum_{\epsilon_2=0}^{1} \cdots \sum_{\epsilon_n=0}^{1} & G_{\epsilon_1,\epsilon_2,\ldots,\epsilon_n}(\cos 2\pi v_1, \cos 2\pi v_2, \ldots, \cos 2\pi v_n) \\ & \times (\sin 2\pi v_1)^{\epsilon_1} (\sin 2\pi v_2)^{\epsilon_2} \cdots (\sin 2\pi v_n)^{\epsilon_n}. \end{aligned}$$

The coefficient $G_{\epsilon_1,\epsilon_2,\ldots,\epsilon_n}$ is an integral function of

$$\cos 2\pi v_1, \cos 2\pi v_2, \ldots, \cos 2\pi v_n.$$

Writing

$$\cos 2\pi v_\alpha = \tfrac{1}{2}\exp(2\pi\sqrt{-1}\,v_\alpha) + \tfrac{1}{2}\exp(-2\pi\sqrt{-1}\,v_\alpha)$$
$$\sqrt{-1}\sin 2\pi v_\alpha = \tfrac{1}{2}\exp(2\pi\sqrt{-1}\,v_\alpha) - \tfrac{1}{2}\exp(-2\pi\sqrt{-1}\,v_\alpha)$$

permits the preceding expression to be written in a convergent power series of $n$ exponential functions

$$\exp(2\pi\sqrt{-1}\,v_1),\ \exp(-2\pi\sqrt{-1}\,v_1), \ldots, \qquad \exp(2\pi\sqrt{-1}\,v_n),\ \exp(-2\pi\sqrt{-1}\,v_n).$$

Hence

$$\phi(v_1, v_2, \ldots, v_n) = \sum_{\lambda_1,\mu_1,\ldots,\lambda_n,\mu_n} A_{\lambda_1\mu_1\cdots\lambda_n\mu_n} \times \exp\{2\pi\sqrt{-1}\,[(\lambda_1 - \mu_1)v_1 + \cdots + (\lambda_n - \mu_n)v_n]\}$$
$$(\lambda_1, \mu_1, \ldots, \lambda_n, \mu_n = 0, 1, \ldots, \infty).$$

Since this series is unconditionally convergent, we replace

$$\lambda_1 - \mu_1, \ldots, \lambda_n - \mu_n$$

by $\nu_1, \nu_2, \ldots, \nu_n$, respectively. This gives a convergent series

$$\phi(v_1, v_2, \ldots, v_n) = \sum_{\nu_1,\nu_2,\ldots,\nu_n} C_{\nu_1\nu_2\cdots\nu_n} \exp[2\pi\sqrt{-1}\,(\nu_1 v_1 + \cdots + \nu_n v_n)]$$
$$(\nu_1, \nu_2, \ldots, \nu_n = -\infty, \ldots, +\infty). \quad (9)$$

Since this is, moreover, uniformly convergent,

$$\int_0^1\cdots\int_0^1 \phi(v_1, v_2, \ldots, v_n) \exp[-2\pi\sqrt{-1}\,(\nu_1 v_1 + \cdots + \nu_n v_n)]\,dv_1\cdots dv_n$$
$$= \sum_{\mu_1,\ldots,\mu_n} C_{\mu_1\cdots\mu_n} \int_0^1\cdots\int_0^1 \exp\{2\pi\sqrt{-1}\,[(\mu_1 - \nu_1)v_1 + \cdots + (\mu_n - \nu_n)v_n]\}\,dv_1\cdots dv_n$$
$$= C_{\nu_1\cdots\nu_n} \qquad (\mu_1, \mu_2, \ldots, \mu_n = -\infty, \ldots, +\infty).$$

Hence the coefficients of the series (9) are expressed, similarly to the

Fourier coefficients, in the form (Weierstrass, 1866, 1895; Osgood, 1924; Hobson, 1926)

$$C_{\nu_1\cdots\nu_n} = \int_0^1\cdots\int_0^1 \phi(v_1, \ldots, v_n)$$
$$\times \exp\left[-2\pi\sqrt{-1}\,(\nu_1 v_1 + \cdots + \nu_n v_n)\right] dv_1\cdots dv_n. \quad (10)$$

## 2.7 CONDITIONALLY PERIODIC MOTION

A conditionally periodic motion has been studied by Staude (1887, 1889) for a dynamical system with two degrees of freedom and by Stäckel (1891, 1893) for a general case.

From Stäckel's assumption (Section 1.16) concerning the integral functions $\phi_{ik}(q_i)$ and $\Phi_i(q_i)$, there follows

$$dt = \sum_{i=1}^{n} \frac{\phi_{ik}(q_i)\,dq_i}{\sqrt{\Phi_i(q_i)}} \qquad (k = 1, 2, \ldots, n). \qquad (11)$$

Denote the consecutive zero points of $\Phi_i(q_i) = 0$ by

$$a_i < b_i \qquad (i = 1, 2, \ldots, n),$$

and let the initial value of $q_i$ be in this interval.

Writing

$$\frac{dq_i}{(q_i - a_i)^{m_i/2}(b_i - q_i)^{n_i/2}} = \epsilon_i\,dw_i,$$

$$\Phi_i(q_i) = (q_i - a_i)^{m_i}(b_i - q_i)^{n_i}\Psi_i(q_i), \qquad F_{ik}(q_i) = \frac{\phi_{ik}(q_i)}{\sqrt{\Psi_i(q_i)}},$$

we may rewrite equation (11) to read

$$dt = \sum_{i=1}^{n} F_{ik}(q_i)\epsilon_i\,dw_i \qquad (k = 1, 2, \ldots, n). \qquad (12)$$

If $\Delta = \|\phi_{ik}\| \neq 0$, then we get

$$\frac{1}{E}\frac{\partial E}{\partial \phi_{ik}} = \frac{1}{\Delta}\frac{\partial \Delta}{\partial \phi_{ik}}$$

by formal logarithmic differentiation of $E = \Delta/\sqrt{\Psi_1\Psi_2\cdots\Psi_n}$. However,

$$\frac{1}{E}\frac{\partial E}{\partial F_{ik}} = \sqrt{\Psi_i(q_i)}\,\frac{1}{\Delta}\frac{\partial \Delta}{\partial \phi_{ik}}$$

from the relation

$$\frac{\partial E}{\partial \phi_{ik}} = \frac{\partial E}{\partial F_{ik}}\frac{1}{\sqrt{\Psi_i(q_i)}}.$$

Since the expressions $A_i(q_i)$ of (60) and (61) of Section 1.16 cannot all be zero, at least one $\partial\Delta/\partial\phi_{ik}$ is not zero. Hence, at least one $\partial E/\partial F_{ik}$ $(i = 1, 2, \ldots, n)$ is not zero for any value of $k$. As, $\phi_{ik}$ in the interval $(a_i, b_i)$ becomes neither zero nor infinity, the determinant $\Delta$ — and, accordingly, $E$ — becomes neither zero nor infinity. Hence, solving (12) gives

$$\frac{dt}{E} = \frac{\epsilon_1\, dw_1}{E_1} = \cdots = \frac{\epsilon_n\, dw_n}{E_n},$$

where $E_j$ is the determinant in which all elements of the $j$th column of the determinant $E \equiv \|F_{ik}(q_i)\|$ are replaced by 1. If we choose suitably the signs of $\epsilon_1, \epsilon_2, \ldots, \epsilon_n$, then we can make $w_1, w_2, \ldots, w_n$ increase monotonically while $t$ increases from $-\infty$ to $+\infty$, because none of $E_i/E$ becomes zero or infinity in this interval. Thus $w_i$ is the mechanical variable and there will be an asymptotic motion or a libration according to the discussion in Section 2.4.

Assume that $m_i = n_i = 1$ $(i = 1, 2, \ldots, n)$; then

$$q_i = a_i \cos^2\frac{\epsilon_i w_i}{2} + b_i \sin^2\frac{\epsilon_i w_i}{2}.$$

Integrate (12) by putting $G_{ik}(q_i) = \int F_{ik}(q_i)\epsilon_i\, dw_i$, and denote the integration constants by $\beta_1, \beta_2, \ldots, \beta_n$. Then

$$t + \beta_k = \sum_{i=1}^{n} G_{ik}(w_i) \qquad (k = 1, 2, \ldots, n). \tag{13}$$

Take $\epsilon_i = \pm 1$ and let $w_i$ increase by $2\pi$. Then

$$t + \beta_k + 2\omega_{jk} = G_{jk}(w_j + 2\pi) + \sideset{}{'}\sum_{\substack{i=1\\ i\neq j}}^{n} G_{ik}(w_i) \qquad (j, k = 1, 2, \ldots, n),$$

where we exclude $i = j$ in the summation $\sum'$. Thus

$$2\omega_{jk} = G_{jk}(w_j + 2\pi) - G_{jk}(w_j) = \int_{w_j}^{w_j+2\pi} F_{jk}(q_j)\cdot\epsilon_j\, dw_j,$$

where $q_j$ is a periodic function of $w_j$. Hence,

$$\omega_{jk} = \int_0^{\pi} F_{jk}(q_j)\, dw_j \qquad (j, k = 1, 2, \ldots, n),$$

or

$$\omega_{jk} = \int_{a_j}^{b_j} \frac{\phi_{jk}(q_j)\, dq_j}{\sqrt{(q_j - a_j)(b_j - q_j)\Psi_j(q_j)}}. \tag{14}$$

Thus $t + \beta_k$ increases by $\omega_{jk}$ as $w_j$ increases by $2\pi$. Conversely, by considering $w_1, w_2, \ldots, w_n$ as functions of $t + \beta_1, t + \beta_2, \ldots, t + \beta_n$, we see that $w_j$ increases by $2\pi$ while $w_1, w_2, \ldots, w_{j-1}, w_{j+1}, \ldots, w_n$ remain unaltered, as $t + \beta_1, t + \beta_2, \ldots, t + \beta_n$ increase, respectively, by

$$2\omega_{j1}, 2\omega_{j2}, \ldots, 2\omega_{jn}.$$

From (13) and from reference to (12),

$$d\beta_k = \sum_{i=1}^{n} F_{ik}(q_i)\epsilon_i \, dw_i \qquad (k = 1, 2, \ldots, n).$$

By solving these equations we obtain

$$dw_j = \sum_{i=1}^{n} E_{ji} \, d\beta_i \qquad (j = 1, 2, \ldots, n),$$

as $E \neq 0$, where $E_{ji}$ is the minor of the determinant $E$. Thus

$$\beta_1, \beta_2, \ldots, \beta_n$$

vary continuously as $w_1, w_2, \ldots, w_n$ vary continuously. Hence

$$w_1, w_2, \ldots, w_n$$

are single-valued functions of $t + \beta_1, t + \beta_2, \ldots, t + \beta_n$. If we set

$$q_k = f_k(t + \beta_1, t + \beta_2, \ldots, t + \beta_n) \qquad (k = 1, 2, \ldots, n),$$

then

$$\begin{aligned} &f_k(t + \beta_1 + 2\omega_{j1}, t + \beta_2 + 2\omega_{j2}, \ldots, t + \beta_n + 2\omega_{jn}) \\ &\qquad = f_k(t + \beta_1, t + \beta_2, \ldots, t + \beta_n) \qquad (j, k = 1, 2, \ldots, n). \end{aligned}$$

Such functions are multiply periodic functions (Section 2.5); the $\omega_{jk}$ are given by (14). By a transformation

$$\pi(t + \beta_j) = \sum_{i=1}^{n} \omega_{ij} u_i \qquad (j = 1, 2, \ldots, n), \tag{15}$$

we change to the variables $u_1, u_2, \ldots, u_n$, by assuming $\Omega \equiv \|\omega_{jk}\| \neq 0$, and suppose that the functions $f_k(t + \beta_1, t + \beta_2, \ldots, t + \beta_n)$ are changed to $g_k(u_1, u_2, \ldots, u_n)$. Then the above property is expressed by

$$\begin{aligned} g_k(u_1 + 2\pi, u_2, \ldots, u_n) &= g_k(u_1, u_2 + 2\pi, \ldots, u_n) = \cdots \\ &= g_k(u_1, u_2, \ldots, u_n + 2\pi) \\ &= g_k(u_1, u_2, \ldots, u_n) \qquad (k = 1, 2, \ldots, n). \end{aligned}$$

Generally, we have the relations with positive or negative integers $m_1, m_2, \ldots, m_n$:

$$g_k(u_1 + 2m_1\pi, u_2 + 2m_2\pi, \ldots, u_n + 2m_n\pi) = g_k(u_1, u_2, \ldots, u_n).$$

Such functions satisfy the condition (8) of Weierstrass and hence can be expanded in form (9) by Weierstrass's auxiliary theorem. We have

$$g_k = \sum_{\nu_1, \ldots, \nu_n} C^{(k)}_{\nu_1 \cdots \nu_n} \exp\left[\sqrt{-1}\,(\nu_1 u_1 + \cdots + \nu_n u_n)\right]$$
$$(\nu_1, \ldots, \nu_n = -\infty, \ldots, +\infty) \tag{16}$$

where

$$\pi^2 C^{(k)}_{\nu_1 \cdots \nu_n} = \int_0^\pi \cdots \int_0^\pi g_k(u_1, \ldots, u_n)$$
$$\times \exp\left[-\sqrt{-1}\,(\nu_1 u_1 + \cdots + \nu_n u_n)\right] du_1 \cdots du_n. \tag{17}$$

From (15)

$$\Omega u_j = \sum_{i=1}^{n} \frac{\partial \Omega}{\partial \omega_{ji}} \pi(t + \beta_i) \qquad (j = 1, 2, \ldots, n). \tag{18}$$

Thus $u_1, u_2, \ldots, u_n$ can be expressed by linear functions of $t$; actually, $u_1, u_2, \ldots, u_n$ increase by the amounts

$$\sum_{i=1}^{n} \frac{1}{\Omega} \frac{\partial \Omega}{\partial \omega_{ji}} 2\pi T,$$

respectively, as $t$ increases by $2T$. Hence, in order that the motion be periodic with period $2T$ with respect to $t$, these increments to $u_i$ should be integral multiples of $2\pi$, that is, either $n$ equations

$$\sum_{i=1}^{n} \frac{1}{\Omega} \frac{\partial \Omega}{\partial \omega_{ji}} = \frac{m_j}{T} \qquad (j = 1, 2, \ldots, n) \tag{19}$$

or $n$ equations

$$T = \sum_{i=1}^{n} m_i \omega_{ij} \qquad (j = 1, 2, \ldots, n) \tag{20}$$

should hold simultaneously. Staude (1887, 1889), Stäckel (1912), and Ahrens (1895) called such motions *conditionally periodic motions*. Our mode of presentation is different from Charlier (1906). It is a modified form of Weierstrass (1895).

The problem of solving the set of differential equations (11) and expressing $q_i$ as functions of $t + \beta_j$ is called the problem of *inversion*. It has been discussed since the end of the last century in the inversion problem of elliptic integrals and general algebraic integrals; the addition theorem of these integrals has been developed by Abel, not only in the theory of multiply periodic functions but also in its application to algebraic geometry. The formula (69) of Section 1.16 resembles the expression in the addition theorem of algebraic functions (Section 6.25), but there are important differences: the functions $\Phi_i$ in the same expression are all different, and the periods in our case are not complex, as in the theory of algebraic functions, but real. Moreover, the expressions $\Phi_i(q_i)$ and $\phi_{ik}(q_i)$ in our case are not polynomials. We can simplify polynomials by a projective transformation or a Cremona transformation. In the one-body problem of relativity theory, as will be shown in Section 2.17, the functions $\Phi_i(q_i)$ are cubic equations and the theory of elliptic functions can be applied directly. In the three-body problem of Newtonian mechanics a similar situation appears in the theory of libration.

Suppose that $n$ uniform (single-valued and continuous) integrals

$$C_i(q_i, \ldots, q_n, p_1, \ldots, p_n) = C_i \qquad (i = 1, 2, \ldots, n),$$

are mutually in involution (as will be defined in Chapter 4) and let $\Phi_{2n}$ be a phase space of $2n$ dimensions of $q_1, \ldots, q_n, p_1, \ldots, p_n$. Suppose that every point $P$ of $\Phi_{2n}$ is at a finite distance from any other, and that two arbitrary points can be connected by a path of finite length described on $\Phi_{2n}$, and, moreover, that at least one of the determinants of order $n$ formed from the matrix

$$\left(\frac{\partial C_i}{\partial q_j}, \frac{\partial C_i}{\partial p_k}\right) \qquad (i, j, k = 1, 2, \ldots, n)$$

is not zero. In such a case, write $V(P) = \int_{P_0}^{P_1} \sum_i p_i \, dq_i$, where the path of integration is chosen so as to be on $\Phi_{2n}$, and replace the assumption of Stäckel. $\int p_i \, dq_i$ has the dimensionality of action. In classical quantum theory this action is put equal to an integral multiple of Planck's constant in the case when the variables can be separated in Stäckel's manner (Born, 1925).

Put

$$\frac{\partial V}{\partial C_i} = w_i, \qquad \frac{\partial V}{\partial q_i} = p_i,$$

and transform from the canonical variables $p_i$, $q_i$ to $C_i$, $w_i$. Replace $C_1, C_2, \ldots, C_n$ by $n$ integrals in involution (Section 4.5) and suppose that the variables $u_1, u_2, \ldots, u_n$ determined by

$$\frac{\partial V}{\partial v_i} = u_i, \qquad \frac{\partial V}{\partial q_i} = p_i$$

have periods

$$\begin{matrix} 1 & 0 & \cdots & 0 \\ 0 & 1 & \cdots & 0 \\ \vdots & & & \\ 0 & 0 & \cdots & 1. \end{matrix}$$

Then $(q_i, p_i)$ can be expanded in trigonometric series with the sum of integral multiples of $2\pi u_i$ as arguments, as in (16). Then $v_i = \gamma_i$, $u_i = \nu_i t + \epsilon_i$, and we obtain a result similar to Stäckel's (Mineur, 1936). The assumption as to the form of $V(P)$ is fairly strong, like Stäckel's.

We have transformed the canonically conjugate variables from $p_i$, $q_i$ to $\alpha_i$, $\beta_i$. The variables $\beta_i$ are contained only in $u$, as can be seen from (18); hence, only in $w$. Their conjugate variables $\alpha_i$ appear in $F_{jk}$ in the expressions for $\omega_{jk}$, and in the coefficients of the transformations from $\beta$ to $u$ and from $u$ to $w$. We write the canonically conjugate variable of $u$ as $v$. The variable $q$ is periodic with respect to $u$, as may be seen from (16); and $u$ is contained only in the arguments of exponentials. Hence, the coefficients $C_{\nu_1 \cdots \nu_n}$ should be functions of $v$; actually, $u$ is linear in $t$ and periodic with period $2\pi$. Such a variable is called an *angular variable.* On the other hand, $v$ does not have this property and is called a *linear variable.* Thus the coordinates and momenta are expressed in the form

$$q_i = q_i(v_1, \ldots, v_n; u_1, \ldots, u_n), \qquad u_j = \nu_j t + \epsilon_j,$$

$$p_i = p_i(v_1, \ldots, v_n; u_1, \ldots, u_n),$$

where $\epsilon_j$ are the integration constants and $\nu_j$ are functions of $v$. These equations, $2n$ in number, give the relations between $p$, $q$, $u$, $v$. Eliminating $u$ gives $n$ relations among $p$, $q$, and $v$. These severe conditions are necessary for a conditionally periodic motion. This has been proved by Levi-Civita (1933) by the use of Weierstrass's auxiliary theorem (Section 2.6). Fues (1925) proved that the necessary and sufficient condition for a conditionally periodic motion is that the surface of action $V =$ constant can be divided into a finite number of distinct sheets of surfaces. This circumstance has been important in Bohr's quantum theory based on his correspondence principle (see Born, 1925).

## 2.8 KRONECKER'S THEOREM

Suppose that we are given real numbers $\mu_1, \ldots, \mu_m$, $\tau$ and a positive number $\epsilon$. Then there exists a constant $C = C(\epsilon, m)$, independent of $\tau$, and $m$ integers $g_1, \ldots, g_m$, such that $m$ inequalities

$$|t\mu_i - g_i| < \epsilon \qquad (i = 1, 2, \ldots, m) \tag{21}$$

hold for all values of $t$ in the interval $\tau < t < \tau + C$.

This is known as *Kronecker's theorem* (1899). It has been proved by Lettenmeyer (1923), Bohr (1923), and Thomas (1924) (see also Hardy and Littlewood, 1914; Weyl, 1916). The most elegant proof is by Bohr, later extended by Bacon (1934); this proceeds as follows:

Consider $m$ complex numbers $\exp(2\pi\sqrt{-1}\,t\mu_j)$. It is sufficient to prove that the differences of these numbers each from $\exp(0) = 1$ are at most $\epsilon$, that is, the upper bound $L_F$ of $|F(t)|$, where

$$F(t) = 1 + \exp(2\pi\sqrt{-1}\,t\mu_1) + \exp(2\pi\sqrt{-1}\,t\mu_2) + \cdots + \exp(2\pi\sqrt{-1}\,t\mu_m)$$

is, for all real values of $t$, equal to the upper bound $L_G = m + 1$ of $G(x_1, x_2, \ldots, x_m)$, where

$$G(x_1, x_2, \ldots, x_m) = 1 + \exp(2\pi\sqrt{-1}\,x_1) + \exp(2\pi\sqrt{-1}\,x_2) + \cdots + \exp(2\pi\sqrt{-1}\,x_m).$$

Here we suppose that the $x_i$ vary independently of each other in the interval $0 \leqq x_i \leqq 1$. Thus it is sufficient to prove that $L_F \geqq L_G = m + 1$.

We consider

$$[F(t)]^p = [1 + \exp(2\pi\sqrt{-1}\,t\mu_1) + \cdots + \exp(2\pi\sqrt{-1}\,t\mu_m)]^p$$

with a positive integer $p$. Any two terms of the expansion of this expression do not coincide, because $\mu_i$ are mutually linearly independent, that is, the number of terms in the expansion equals the number of terms in the expansion of $[G(x_1, x_2, \ldots, x_m)]^p$; indeed, corresponding terms have the same coefficients. Hence,

$$F_p = \lim_{t\to\infty} \frac{1}{2t} \int_{-t}^{t} |[F(t)]^p|^2\, dt$$

is equal to

$$G_p = \int_0^1 \cdots \int_0^1 |\{G(x_1, \ldots, x_m)\}^p|^2\, dx_1 \cdots dx_m,$$

and both are equal to the sum of squares of the coefficients in these two expansions. It is easy to see that

$$(G_p)^{1/2p} \to L_G = m + 1$$

for $p \to \infty$. Hence, $|G(x_1, \ldots, x_m)|$ attains its upper bound $L_G$ at the point $(0, 0, \ldots, 0)$ of the $m$-dimensional parallelopiped or the $m$-dimensional torus $0 \leqq x_1 \leqq 1, \ldots, 0 \leqq x_m \leqq 1$. Because $F_p = G_p$, therefore $F_p^{1/2p} \to L_G$. But, because $F_p^{1/2p} \leqq L_F$ for all values of $p$, we obtain $L_F \geqq L_G$. Thus the theorem is proved.

We can prove the following theorem as a corollary to Kronecker's theorem: Suppose that there is no linear homogeneous relation with integer coefficients among $\mu_1, \mu_2, \ldots, \mu_m$. Given arbitrary numbers

$k_1, \ldots, k_m$ and $\epsilon > 0$, we can obtain $t$ and the integers $g_1, \ldots, g_m$, such that

$$|\mu_i t - k_i - g_i| < \epsilon \qquad (i = 1, 2, \ldots, m).$$

In order to prove this theorem we take, following Bohl (1893), $\delta < |\mu_i|$ $(i = 1, 2, \ldots, m)$. According to Kronecker's Theorem we can choose $m^2$ expressions

$$s_i^{(j)} = \mu_i t^{(j)} - g_i^{(j)} \qquad (i, j = 1, 2, \ldots, m),$$

such that their absolute values are smaller than $\epsilon > 0$. If the determinant $\|s_i^{(j)}\| \neq 0$, then we take $\epsilon < 4\delta/m$ and determine $p_j$ by

$$p_1 s_i^{(1)} + p_2 s_i^{(2)} + \cdots + p_m s_i^{(m)} = k_i \qquad (i = 1, 2, \ldots, m).$$

If we replace each $p_j$ by an integer that differs from $p_j$ at most by $\frac{1}{2}$, then the difference between the left-hand side of this equation and $k_i$ becomes smaller than $\delta/2$ and we can take $s_i^{(j)}$ smaller than $\frac{1}{2}$. Hence the problem is solved.

But in order to see that $\|s_i^{(j)}\| \neq 0$, we tentatively assume that $\|s_i^{(j)}\| = 0$. This should give

$$\begin{vmatrix} 1 & 0 & \cdots & 0 \\ t^{(1)} & s_1^{(1)} & \cdots & s_m^{(1)} \\ t^{(2)} & s_1^{(2)} & \cdots & s_m^{(2)} \\ \vdots & & & \\ t^{(m)} & s_1^{(m)} & \cdots & s_m^{(m)} \end{vmatrix} = \begin{vmatrix} 1 & -\mu_1 & \cdots & -\mu_m \\ t^{(1)} & -g_1^{(1)} & \cdots & -g_m^{(1)} \\ t^{(2)} & -g_1^{(2)} & \cdots & -g_m^{(2)} \\ \vdots & & & \\ t^{(m)} & -g_1^{(m)} & \cdots & -g_m^{(m)} \end{vmatrix} = 0.$$

The second determinant is obtained by multiplying each element in the first column of the first determinant by the appropriate $\mu_i$ and subtracting it from each column. $g_i^{(j)}$ are integers, and there is no linear homogeneous relation with integer coefficients among $\mu_1, \mu_2, \ldots, \mu_m$. Hence, the coefficients of $t^{(j)}$ in the expansion of the determinant with respect to the first column are all different from zero. As the coefficient of 1 in this expansion is an integer, such a relation cannot exist, with $t^{(j)}$ either rational or irrational. Hence, the corollary is proved.

The estimate of $\epsilon$ and $C$ in (21) has been recently made by Siegel (Landau 1927, 1929, 1935), van der Corput (1921, 1931, 1932) and Schönberg (1928).

We have obtained in Section 2.7

$$f\left(t + \beta_1 + \sum_{i=1}^{n} 2m_i\omega_{i1}, t + \beta_2 + \sum_{i=1}^{n} 2m_i\omega_{i2}, \ldots, t + \beta_n + \sum_{i=1}^{n} 2m_i\omega_{in}\right)$$
$$= f(t + \beta_1, t + \beta_2, \ldots, t + \beta_n). \quad (22)$$

However, we can make

$$\sum_{i=1}^{n} 2m_i\omega_{ij} - \sum_{j=1}^{n} 2m_i\omega_{i1} \qquad (j = 2, 3, \ldots, n)$$

smaller than any arbitrarily given number $\eta$.

In fact, it is sufficient to prove

$$\sum_{i=1}^{n} 2m_i\omega_{ij} - \sum_{i=1}^{n} 2m_i\omega_{i1} = \eta_j, \qquad |\eta_j| < \eta \qquad (j = 2, 3, \ldots, n). \quad (23)$$

By solving these equations we see that at least one of the determinants of order $n - 1$ formed from the matrix

$$\begin{pmatrix} \omega_{12} - \omega_{11} & \omega_{22} - \omega_{21} & \cdots & \omega_{n2} - \omega_{n1} \\ \omega_{13} - \omega_{11} & \omega_{23} - \omega_{21} & \cdots & \omega_{n3} - \omega_{n1} \\ \vdots & & & \\ \omega_{1n} - \omega_{11} & \omega_{2n} - \omega_{21} & \cdots & \omega_{nn} - \omega_{n1} \end{pmatrix}$$

is not zero. Because, if all of these determinants were zero, then the determinant of order $n$ formed from this matrix by writing

$$\omega_{11}, \omega_{21}, \ldots, \omega_{n1}$$

in the first row would be $\Omega$. We have assumed that $\Omega \equiv \|\omega_{ij}\| \neq 0$. Hence, (23) is proved.

We now solve the equations (23) by writing

$$\Omega_1 = \|\omega_{ij} - \omega_{i1}\| \qquad (i, j = 2, 3, \ldots, n):$$

$$2m_i + 2m_n\left[\frac{1}{\Omega_1}\sum_{j=2}^{n}\frac{\partial\Omega_1}{\partial\omega_{ij}}(\omega_{nj} - \omega_{n1})\right] = \frac{1}{\Omega_1}\sum_{j=2}^{n}\frac{\partial\Omega_1}{\partial\omega_{ij}}\eta_j$$

$$(j = 2, 3, \ldots, n).$$

Hence, if we take

$$\mu_i = \frac{1}{\Omega_1}\sum_{j=2}^{n}\frac{\partial\Omega_1}{\partial\omega_{ij}}(\omega_{nj} - \omega_{n1}), \qquad t = -m_n, \quad g_i = m_i, \quad m = n - 1,$$

then we can solve $\epsilon_j$ from the given constants $\eta_j$ by

$$\frac{1}{\Omega_1}\sum_{j=2}^{n}\frac{\partial\Omega_1}{\partial\omega_{ij}}\eta_j = \epsilon_i \qquad (i = 2, 3, \ldots, m).$$

If we take $\epsilon$ such that $\epsilon_i < \epsilon$, then a direct application of Kronecker's theorem (21) shows that we can find $t, g_1, \ldots, g_m$ such that

$$|\mu_i t - g_i| < \epsilon_i \qquad (i = 2, 3, \ldots, m);$$

that is, we can find $m_1, m_2, \ldots, m_{n-1}$. Hence, there exists a set of integers $m_1, m_2, \ldots, m_n$ satisfying (23), however small we may take $\eta$.

With such a set of $m_1, m_2, \ldots, m_{n-1}, m_n$ we form $\sum_{j=1}^{n} 2m_j\omega_{j1} = T$; then we can write (22) in the form

$$f(t + T + \beta_1, t + T + \beta_2 + \eta_2, \ldots, t + T + \beta_n + \eta_n)$$
$$= f(t + \beta_1, t + \beta_2, \ldots, t + \beta_n).$$

But, as we have proved that the function $f$ is a continuous function of its variables, we obtain

$$|f(t + T + \beta_1, t + T + \beta_2 + \eta_2, \ldots, t + T + \beta_n + \eta_n)$$
$$- f(t + T + \beta_1, t + T + \beta_2, \ldots, t + T + \beta_n)|$$
$$< M \sum_{i=2}^{n} \eta_i < M(n-1)\eta,$$

where

$$M = \max \left| \frac{\partial f}{\partial \beta_i} \right| \qquad (i = 1, 2, \ldots, m).$$

Hence,

$$|f(t + T + \beta_1, t + T + \beta_2, \ldots, t + T + \beta_n)$$
$$- f(t + \beta_1, t + \beta_2, \ldots, t + \beta_n)| < M(n-1)\eta.$$

Thus, by taking $T$ large enough, we can make the difference of the values of $f$ for $t$ and $t + T$ smaller than the given number $\eta$, however small $\eta$ may be. In other words, there exists a constant $T$ such that the differences at $t$ and $t + T$ of the coordinates $q_i$ and the momenta $p_i$ $(i = 1, 2, \ldots, n)$ of the dynamical system in question can be made as small as we please. In other words, the dynamical system in question can return to a state of motion as near as we please to the initial state of motion, as $t$ increases indefinitely; that is, if any trajectory in the $2n$-dimensional phase space of Gibbs for $p_i$ and $q_i$ once recedes from its initial position, it returns again indefinitely close to this initial position.

By using the terminology of point-set theory we see that this phase space of Gibbs is filled up *everywhere densely* by the trajectory of motion (Section 2.1). This result is only for the case of simple motions, such as those discussed here (the general case will be discussed in a later chapter). One particular example is that of a two-dimensional simple pendulum describing a Lissajous figure when the ratio of the periods of the motion in the two perpendicular directions is irrational. In the next section, we will discuss the motion of molecules in a cube as the simplest example.

## 2.9 MOTION OF MOLECULES IN A CUBE

Take a square with its sides $x = 0, \frac{1}{2}$; $y = 0, \frac{1}{2}$. Suppose that any molecule moving in a straight line inside this square makes a perfect reflection when it strikes any of the sides of the square. The mirror image of the trajectory with respect to the reflecting side is also a straight line. Let $P(a, b)$ be a point inside the square. The point, of which the images are $(a + m, b + n)$, $(a + m, -b + n)$, $(-a + m, b + n)$, $(-a + m, -b + n)$ on the $xy$ plane, where $m$ and $n$ are integers, moves in a straight line $x = a + \alpha t$, $y = b + \beta t$ on that plane. In order that the point shall return to its initial position we should have $x = a + m$, $y = b + n$. Hence, $m = \alpha t$, $n = \beta t$; that is, the ratio $\alpha:\beta$ should be rational. If it is irrational, the moving point comes indefinitely near, but not exactly back to, its initial position. Denoting the point by $\xi$, $\eta$ we obtain

$$|a + \alpha t - (\xi + m)| < \epsilon, \qquad |b + \beta t - (\eta + n)| < \epsilon,$$

and we can determine two integers $m$ and $n$ such that these inequalities are satisfied with an arbitrarily given number $\epsilon$ by the corollary of Kronecker's theorem described in Section 2.8.

Next, consider a three-dimensional cube, with the sides $x = 0, \frac{1}{2}$; $y = 0, \frac{1}{2}$; $z = 0, \frac{1}{2}$. In order that the trajectory represented by the straight line $L$,

$$x = a + \alpha t, \qquad y = b + \beta t, \qquad z = c + \gamma t,$$

return to its initial position, that is, in order that the trajectory be periodic, it is necessary and sufficient that the double ratio $\alpha:\beta:\gamma$ should be rational. That is, there should exist the relations

$$A\alpha + B\beta + C\gamma = 0, \qquad A'\alpha + B'\beta + C'\gamma = 0,$$

with integers $A, B, C, A', B', C'$. It is supposed here that the coefficients are such that $A:B:C$ and $A':B':C'$ are both irreducible ratios. We specify two classes of nonperiodic trajectory: the first, in which there is only one such linear relation among $\alpha$, $\beta$, $\gamma$, and the second, in which there is no such relation.

First suppose that there is only one relation, $A\alpha + B\beta + C\gamma = 0$, with integer coefficients. The path of a moving point proceeding in the direction with direction cosines $A$, $B$, $C$ is perpendicular to the straight line $L$. The plane $S$, given by

$$A(x - a) + B(y - b) + C(z - c) = 0,$$

contains the straight line $L$ and is perpendicular to the straight line with direction cosines $A, B, C$. We carry out a symmetric transformation on $S$ with respect to the cube surface that is cut by $S$. This gives a plane $S'$. Carry out again a symmetric transformation on $S'$ with respect to the cube surface that is cut by the plane $S'$. Repeat these operations. The parts of the surfaces $S, S', \ldots$ inside the given cube form a closed polyhedron $\Pi$ of a finite number of faces. The edges are all on the surface of the cube. Let $Aa + Bb + Cc = D$. Then the equations for the planes containing the faces of $\Pi$ are

$$\Sigma\colon \pm Ax \pm By \pm Cz - (D \pm Am \pm Bn \pm Cp) = 0,$$

with integers $m, n, p$. If a point on some plane $\Sigma$ is inside the cube, then its distance from the origin $(0, 0, 0)$ is smaller than the length of the diagonal $\sqrt{3}/2$ of the cube; this gives

$$\left|\frac{D \pm Am \pm Bn \pm Cp}{\sqrt{A^2 + B^2 + C^2}}\right| < \sqrt{3}/2.$$

This moving point, the molecule, takes no value other than

$$D, D \pm 1, D \pm 2, \ldots.$$

There are only a finite number of absolute values less than

$$\sqrt{3}\,\sqrt{A^2 + B^2 + C^2}/2.$$

Hence, there are only a finite number of planes $\Sigma$ satisfying this inequality. Moreover, the parts of planes inside the cube are polygons whose sides always also belong to contiguous polygons. Thus, the polyhedron is closed.

Since the image of our trajectory $L$ is on the plane $S$, the part of $L$ inside the cube is a straight-line segment on the polyhedron $\Pi$; hence, the trajectory is always on the surface of the polyhedron $\Pi$. Thus, it is not everywhere dense inside the cube. It can be shown that the trajectory approaches indefinitely any point on the surface of $\Pi$ if it is not periodic. Take a point $(\xi, \eta, \zeta)$ on $S$ and consider

$$A(\xi - a) + B(\eta - b) + C(\zeta - c) = 0.$$

We can determine $t$ and integers $m, n, p$ such that

$$|a + \alpha t - (\xi + m)| < \epsilon, \qquad |b + \beta t - (\eta + n)| < \epsilon,$$
$$|c + \gamma t - (\zeta + p)| < \epsilon,$$

by the corollary of Kronecker's theorem. Hence, one of the ratios $\alpha/\beta$, $\beta/\gamma$, $\gamma/\alpha$ is irrational.

Next, suppose that there is no linear relation of the form

$$A\alpha + B\beta + C\gamma = 0.$$

In this case the trajectory returns infinitely near any given position. We can choose $t$ and integers $m, n, p$, such that

$$|a + \alpha t - (\xi + m)| < \epsilon, \qquad |b + \beta t - (\eta + n)| < \epsilon,$$
$$|c + \gamma t - (\zeta + p)| < \epsilon$$

by the corollary of Kronecker's theorem (König and Szücs, 1913).

In this example the trajectory is everywhere dense in a space or in a subspace of lower dimensionality contained in that space. But the case of rational ratios $\alpha:\beta:\gamma$ is very rare; that is, the probability of occurrence of rational ratios is infinitely small compared with the probability of occurrence of irrational ratios. The appearance of irrational ratios defines the *regional recurrence*. It will be discussed later, together with its Lebesgue measure. In the terminology of point-set theory the measure of the point set of rational numbers is zero, but the case in which the trajectory is everywhere dense except a point set of measure zero is called *quasi-ergodic* and will be discussed in a later chapter.

### 2.10 QUASI-PERIODIC FUNCTION OF BOHL

Let $\mu_1, \mu_2, \ldots, \mu_m$ be $m$ nonzero and linearly independent numbers; that is, there is no linear homogeneous relation with integer coefficients among these numbers. Let $t$ be a variable, and $U_j$ a rational function of $\sin 2\pi p_j t$ and $\cos 2\pi p_j t$ $(j = 1, 2, \ldots, m)$. Assume that a trigonometric series $U_1 + U_2 + \cdots + U_j + \cdots$ is uniformly convergent in the whole real domain $T$ of $t$; assume, that is, that with a positive integer $n$ for each $\epsilon > 0$, the expression $|U_{l+1} + U_{l+2} + \cdots + U_{l+s-1}|$ can be made smaller than $\epsilon$ for an integer $l > n$ and an integer $s > 0$ for all values of $t$ belonging to $T$. In other words, assume that this series represents a function $\psi(t)$ in $T$. Assume, furthermore, that we can choose an $\eta > 0$ so that $|\psi(t + \tau) - \psi(t)| < \epsilon$ for all numbers $\epsilon > 0$, with $t$ and $\tau$ both belonging to $T$, and that $\mu_i\tau$ $(i = 1, 2, \ldots, m)$ differ at most by $\eta$ from the corresponding integers. Then there exists a uniformly continuous function $\Psi(x_1, x_2, \ldots, x_m)$ of period 1 for each variable $x_1, x_2, \ldots, x_m$, such that

$$\Psi(\mu_1 t, \mu_2 t, \ldots, \mu_m t) = \psi(t)$$

for all values of $t$.

We now prove the first theorem of Bohl (1893, 1906; see also Esclangon, 1902, 1903, 1904, 1916).

When

$$a_i \equiv b_i \pmod 1 \qquad (i = 1, 2, \ldots, m),$$

we say that $(a_1, a_2, \ldots, a_m)$ and $(b_1, b_2, \ldots, b_m)$ belong to the same *class*. When $(\eta_1 t, \eta_2 t, \ldots, \eta_m t)$ belongs to a certain class, we say that $t$ belongs to that class. We now present the three following properties. First: there is one and only one class for each value of $t$. Second: we can choose $\mu_1, \mu_2, \ldots, \mu_m$ so that $\mu_1 t - k_1, \mu_2 t - k_2, \ldots, \mu_m t - k_m$ differ from integers $g_1, g_2, \ldots, g_m$, respectively, at most by $\epsilon$, according to Kronecker's theorem. Hence there exists a set $(\mu_1 t, \mu_2 t, \ldots, \mu_m t)$ such that

$$\mu_1 t, \mu_2 t, \ldots, \mu_m t$$

differ from $k_1, k_2, \ldots, k_m$, respectively, at most by $\epsilon$, however

$$k_1, k_2, \ldots, k_m$$

may be given. Third: when $(a_1, a_2, \ldots, a_m)$ and $(b_1, b_2, \ldots, b_m)$ are both smaller than $\eta > 0$, we denote the values of $t$ belonging to these two sets of numbers respectively by $t_a$ and $t_b$. Then $|\psi(t_a) - \psi(t_b)| < \epsilon$. For, since

$$|\mu_i(t_a - t_b)| = |b_i - a_i| < \eta \qquad (i = 1, 2, \ldots, m),$$

it should follow according to our assumption that

$$|\psi(t_a + t_b - t_a) - \psi(t_a)| < \epsilon.$$

Now construct the function $\Psi$. If there is a value of $t$ belonging to the class $(x_1, x_2, \ldots, x_m)$, then we can form $\psi(t)$. Otherwise proceed as follows. Let $\epsilon_1, \epsilon_2, \ldots$ be a sequence of positive numbers tending to zero; let $\eta_1, \eta_2, \ldots$ correspond respectively to $\epsilon_1, \epsilon_2, \ldots$. Let $(x_1', x_2', \ldots, x_m')$ be a class of numbers that differ from the corresponding $(x_1, x_2, \ldots, x_m)$ at most by $\eta_1/2$; let there be a value of $t$ belonging to that class. (This is possible because of the second property.) Denote the corresponding value of $\psi$ by $\psi_1$. Let $(x_1'', x_2'', \ldots, x_m'')$ be a class of numbers that differ from the corresponding $(x_1, x_2, \ldots, x_m)$ at most by $\eta_2/2$ and let there be a value of $t$ belonging to that class. Denote the corresponding $\psi$ by $\psi_2$. The sequence $\psi_1, \psi_2, \ldots$ obtained by repeating this process indefinitely converges to a definite limit. In fact, each number in

$$(x_1^{(\mu)}, x_2^{(\mu)}, \ldots, x_m^{(\mu)})$$

differs from the corresponding number in $(x_1^{(\nu)}, x_2^{(\nu)}, \ldots, x_m^{(\nu)})$ at most by $(\eta_\mu + \eta_\nu)/2$, which cannot be larger than either $\eta_\mu$ or $\eta_\nu$. Hence,

$$|\psi_\mu - \psi_\nu| < \epsilon_\mu, \epsilon_\nu.$$

As the sequence $\epsilon_1, \epsilon_2, \ldots$ converges to zero, we can choose an integer $N$ such that $\epsilon_n < \epsilon$ for $n > N$ for every $\epsilon$. Hence, $|\psi_\mu - \psi_\nu| < \epsilon$ for $\mu, \nu > N$, and the sequence $\psi_1, \psi_2, \ldots$ converges. Furthermore, $\psi_i$ converges to the same value, however the sequence $\epsilon_1, \epsilon_2, \ldots,$ and the corresponding sequence $\eta_1, \eta_2, \ldots$ may be chosen, because, if we choose another set of values $\bar{\epsilon}_1, \bar{\epsilon}_2, \ldots$ and $\bar{\eta}_1, \bar{\eta}_2, \ldots,$ we can choose $\epsilon_\nu$ and $\bar{\epsilon}_\nu$ smaller than $\epsilon$ for $\nu > n$. A number in $(\bar{x}_1^{(\nu)}, \bar{x}_2^{(\nu)}, \ldots, \bar{x}_m^{(\nu)})$ differs from the corresponding number in $(x_1^{(\nu)}, x_2^{(\nu)}, \ldots, x_m^{(\nu)})$ at most by $(\eta_\nu + \bar{\eta}_\nu)/2$. Hence, $|\bar{\psi}_\nu - \psi_\nu| < \epsilon$ for $\nu > n$, and the sequence

$$|\bar{\psi}_1 - \psi_1|, |\bar{\psi}_2 - \psi_2|, \ldots$$

converges to zero. Let the limit to which this sequence converges be the value of $\Psi$ at $(x_1, x_2, \ldots, x_m)$.

Next we must prove that the function $\Psi$ thus obtained is uniformly continuous. By assigning the corresponding $\eta > 0$ for a given $\epsilon > 0$ we consider $(x_1, x_2, \ldots, x_m)$, respectively, in the open domain

$$(a_1, a_1 + \eta), (a_2, a_2 + \eta), \ldots, (a_m, a_m + \eta).$$

The values of $\Psi$ in this open domain at two points — for which a $t$ exists belonging to the same class — differ from each other at most by $\epsilon$. If the two points in the open domain do not belong to the same class, or if one of the two points does not belong to the same class, then we choose two points of the same class by the second property above stated; we can make the values at these two points differ from each other at most by the given $\epsilon > 0$. Thus we can make the difference of the corresponding values of $\Psi$ arbitrarily close to the difference of the values of $\psi$ at these latter two points. Hence, the two values of $\Psi$ differ at most by $\epsilon$, and the function $\Psi$ is uniformly continuous.

Thus the function $\Psi$ has one determinate finite value at each point and has period 1 for each of the variables $x_1, x_2, \ldots, x_m$, and, furthermore, is uniformly continuous everywhere. Hence

$$\Psi(\mu_1 t, \mu_2 t, \ldots, \mu_m t) = \psi(t).$$

This proves the first theorem of Bohl.

We now come to the second theorem of Bohl.

Let $f(x_1, x_2, \ldots, x_m)$ be an everywhere single-valued continuous and periodic function, having period 1 with respect to each variable. Then we can choose a single-valued continuous and periodic function

$$\phi(x_1, x_2, \ldots, x_m)$$

such that the functions $\phi$ and $f$ differ at most by a given $\epsilon > 0$ everywhere; and such that

$$\frac{\partial^{\nu_1 + \cdots + \nu_m}}{\partial x_1^{\nu_1} \cdots \partial x_m^{\nu_m}} \phi$$

exists, if $\nu_1, \nu_2, \ldots, \nu_m \leqq 2$, and is everywhere finite, single-valued, and continuous.

In the case of one variable this theorem reduces to Fourier's and is obviously true. Supposing that this theorem is valid for $m - 1$ variables, we propose to prove by mathematical induction that it holds true for $m$ variables.

Let $n_1, n_2, \ldots, n_m$ be integers. Suppose that we have chosen the domain of $x_i$ to be $n_i\delta_i \leqq x_i \leqq (n_i + 1)\delta_i$ $(i = 1, 2, \ldots, m)$, such that the fluctuation (Section 2.1) of the values of $f$ in this domain is smaller than $\epsilon/m$. If $x_i/\delta_i$ $(i = 1, 2, \ldots, m)$ are integers, then by our assumption we can find a function $\phi(x_1, x_2, \ldots, x_{m-1}, n\delta_m)$ that satisfies the conditions of the theorem and differs from $f(x_1, x_2, \ldots, x_{m-1}, n\delta_m)$ at most by $(m - 1)\epsilon/m$ everywhere for all values of the integer $n$. Furthermore

$$\phi(x_1, x_2, \ldots, x_{m-1}, n\delta_m + 1) = \phi(x_1, x_2, \ldots, x_{m-1}, n\delta_m).$$

If $n\delta_m < x_m < (n + 1)\delta_m$, then $\phi$ is determinate for $x_m = n\delta_m$ if we put

$$\begin{aligned}
\phi(x_1, x_2, \ldots, x_m) & \\
= \tfrac{1}{2}[\phi(x_1, x_2, \ldots, x_{m-1}, (n+1)\delta_m) &+ \phi(x_1, x_2, \ldots, x_{m-1}, n\delta_m)] \\
+ \tfrac{1}{2}[\phi(x_1, x_2, \ldots, x_{m-1}, (n+1)\delta_m) &- \phi(x_1, x_2, \ldots, x_{m-1}, n\delta_m)] \\
\times \left[\frac{9}{8} \sin \frac{\pi}{\delta_m}\left(x_m - \frac{2n+1}{2}\delta_m\right) \right. &+ \left. \frac{1}{8} \sin \frac{3\pi}{\delta_m}\left(x_m - \frac{2n+1}{2}\delta_m\right)\right].
\end{aligned}$$

The expression in sines lies between $-1$ and $+1$ or equals either of them. Hence, $|f(x_1, \ldots, x_m) - \phi(x_1, \ldots, x_m)|$ is less than or equal to either $|f(x_1, \ldots, x_{m-1}, x_m) - \phi(x_1, \ldots, x_{m-1}, n\delta_m)|$ or

$$|f(x_1, \ldots, x_{m-1}, x_m) - \phi(x_1, \ldots, x_{m-1}, (n + 1)\delta_m)|.$$

Hence, $|f - \phi|$ must be smaller than $\epsilon$. We can prove the second property of the theorem in a similar manner.

Finally, we state the third theorem of Bohl.

Let $\phi(x_1, x_2, \ldots, x_m)$ be a single-valued continuous and periodic function for all variables $x_1, x_2, \ldots, x_m$; and let

$$\frac{\partial^{n_1 + \cdots + n_m}}{\partial x_1^{n_1} \cdots \partial x_m^{n_m}} \phi,$$

with $n_i \leqq M_i$, $M_i > 2$ $(i = 1, 2, \ldots, m)$, be single-valued continuous and finite everywhere. Then

$$\frac{\partial^{\nu_1 + \cdots + \nu_m}}{\partial x_1^{\nu_1} \cdots \partial x_m^{\nu_m}} \phi, \qquad \nu_i \leqq M_i - 2 \qquad (i = 1, 2, \ldots, m)$$

can be developed in a multiple Fourier series of $m$ variables.

This is Weierstrass's auxiliary theorem, proved in Section 2.6.

Summarizing these three theorems leads to the following

*Theorem:* Consider a positive number $\epsilon > 0$ and a function

$$f(x_1, x_2, \ldots, x_m)$$

that is single-valued and continuous everywhere and periodic with period 1 for all variables. Then we can find a uniformly and absolutely convergent trigonometric series

$$\sum A_{n_1 \cdots n_m}^{\epsilon_1 \cdots \epsilon_m} \cos(2\pi n_1 x_1 - \epsilon_1 \pi) \cos(2\pi n_2 x_2 - \epsilon_2 \pi) \cdots \cos(2\pi n_m x_m - \epsilon_m \pi)$$
$$(\epsilon_1, \epsilon_2, \ldots, \epsilon_m = 0, \tfrac{1}{2};$$
$$n_1, n_2, \ldots, n_m = 0, 1, 2, \ldots),$$

that differs from $f$ at most by $\epsilon$ everywhere.

This trigonometric series can be replaced by a trigonometric series with a finite number of terms. Take a sequence of positive numbers $\sigma_1, \sigma_2, \ldots$ converging to zero, and a trigonometric series $S_\nu$ of a finite number of terms with arguments

$$2\pi n_1 x_1 - \epsilon_1 \pi, 2\pi n_2 x_2 - \epsilon_2 \pi, \ldots, 2\pi n_m x_m - \epsilon_m \pi,$$

such that $S_\nu$ differs from our function $\Psi(x_1, x_2, \ldots, x_m)$ at most by $\sigma_\nu$ everywhere. Let $U_0 = S_1$, $U_1 = S_2 - S_1$, $U_2 = S_3 - S_2, \ldots$; then $U_\nu$ is a rational integral function of $\cos 2\pi x_\mu$ and $\sin 2\pi x_\mu$ $(\mu = 1, 2, \ldots, m)$, and $U_0 + U_1 + U_2 + \cdots$ is a uniformly convergent series that can be made to differ from $\Psi(x_1, x_2, \ldots, x_m)$ at most by a given $\epsilon$. Accordingly $\psi(t)$, also, can be approximated by a series of this form. Hence, the theorem of Bohl enunciated in the beginning has been proved.

Such a class of functions $\psi(t)$ is called a *quasi-periodic function of Bohl.* Bohl (1906) called it a periodic function in a wider sense. Esclangon (1902, 1904, 1916) called it a quasi-periodic function, and the function $\Psi(x_1, x_2, \ldots, x_m)$ the *associated function.* In the case where the number of the variables $x_1, x_2, \ldots, x_m$ is infinite, with the power of denumerable numbers, Bohr (1923, 1925) called this function $\psi(t)$ a *limit-periodic function*. When $x_n = n\xi$ $(n = 1, 2, \ldots)$, it is the ordinary multiply periodic function discussed in Section 2.5 and can be expanded in a multiple Fourier series. It is a special case of Bohr's *almost periodic functions* (1925, 1926). The important role of such a class of functions in celestial mechanics will be described in later chapters. It can easily be seen that the conditionally periodic motions discussed in Section 2.7 can be represented by quasi-periodic functions.

The trajectories of an $n$-dimensional motion represented by a quasi-periodic function fill an $n$-dimensional torus everywhere densely; that is, the motion is stable in Poisson's sense (Poincaré, 1899).

It can be shown that for a quasi-periodic function $f(x_1, x_2, \ldots, x_n)$ the time mean and the space mean both exist and are equal; that is,

$$\lim_{T\to\infty} \frac{1}{T}\int_0^T f(x_1, x_2, \ldots, x_n)\, dt$$

$$= \int_{x_1} \cdots \int_{x_n} f(x_1, x_2, \ldots, x_n)\, dx_1\, dx_2 \cdots dx_n.$$

This is the ergodic property of a quasi-periodic function. We shall discuss ergodicity in detail in a later chapter.

## ALMOST PERIODIC FUNCTIONS

### 2.11 DEFINITION

The necessary and sufficient condition for a function $f(x)$, defined in the interval $-\infty < x < +\infty$, to be an everywhere continuous simply periodic function with period $2\pi/k$, is that it should be uniformly approximated by a trigonometric series of the form $\sum_{n=-N}^{N} b_n e^{inkx}$ according to Weierstrass's theorem of Section 2.6. Let $k_1, k_2, \ldots, k_m$ be positive integers. The necessary and sufficient condition for a continuous function $f(x)$ to be uniformly approximated by a trigonometric series of the form

$$\sum_n b_n \exp\,[i(n_1k_1 + \cdots + n_mk_m)x]$$

is that, given $\epsilon > 0$, we should be able to find $\delta > 0$, for which

$$|f(x + \tau) - f(x)| \leqq \epsilon \qquad (-\infty < x < +\infty),$$

holds true for the values of $\tau$, such that $k_1\tau, k_2\tau, \ldots, k_m\tau$ differ from the corresponding integral multiples of $2\pi$ at most by $\delta$, according to Bohl, as discussed in Section 2.10. In such a case, $\tau = \tau(f, \epsilon)$ is called the *translation number*, belonging to $\epsilon$, of $f(x)$, and the exponents in this series are integral multiples of $k_1, k_2, \ldots, k_m$ by $n_1, n_2, \ldots, n_m$, respectively.

Now we consider an arbitrary sequence of numbers $\lambda_n$ as the exponents. A class of functions represented by a series of such a form has been considered by Bohr (1923a, 1925, 1925a, 1925b, 1926).

Construct a series of finite terms

$$s(x) = \sum_{n=1}^{N} a_n e^{i\lambda_n x}, \qquad (-\infty < x < +\infty), \tag{24}$$

with complex numbers $a_n$ and real numbers $\lambda_n$. A class of functions that can be uniformly approximated by such a sum $s(x)$ — that is, a class of continuous functions $f(x) = u(x) + iv(x)$, such that for a given $\epsilon > 0$ we can find an $s(x)$ that satisfies $|f(x) - s(x)| \leqq \epsilon$, $(-\infty < x < +\infty)$ — is called a *closed set* (*abgeschlossen Hülle*) of functions $\{s(x)\}$ and is written $H\{s(x)\}$. The sequence of numbers $\lambda_n$ is of power higher than the denumerable numbers and corresponds to $\lambda = nk$ $(n = 0, \pm 1, \pm 2, \ldots)$ for the case of simply periodic functions. A real number $\tau(f, \epsilon)$ that satisfies

$$|f(x + \tau) - f(x)| \leqq \epsilon \qquad (-\infty < x < +\infty) \tag{25}$$

is called the *translation number* (*Verschiebungszahl*), belonging to $\epsilon$, of $f(x)$. If there exists a length $L$ that contains at least one value of the set $E$ of real numbers $\tau$ in any interval $\alpha < x < \alpha + L$, then $E$ is said to be *relatively dense.* If there exists for any $\epsilon > 0$ a relatively dense set of translation numbers, belonging to $\epsilon$, of a function $f(x)$ that is continuous in $-\infty < x < +\infty$, then $f(x)$ is called an *almost periodic function.* To be precise, such a function $f(x)$ should be called uniformly almost periodic, but the adverb "uniformly" is omitted in the following (Section 2.5; Besicovitch, 1932). A quasi-periodic function is a special case of an almost periodic function. A simply periodic function is a special case of a quasi-periodic function.

If $f(x)$ is an almost periodic function, then $f(x + c)$ and $f(cx)$ with an arbitrary constant $c$ are almost periodic; also $u(x)$ and $v(x)$ are almost periodic. The sum and the product of two almost periodic functions $f(x)$ and $g(x)$, and $1/f(x)$ (if the lower bound of $|f(x)|$ is not zero) are almost periodic. More generally, the sum and the product of continuous almost periodic functions $f_1(x), f_2(x), \ldots, f_N(x)$ are almost periodic. In particular, the sum $s(x) = \sum_{n=1}^{N} a_n e^{i\lambda_n x}$ is almost periodic. The limit function $f(x)$ of almost periodic functions $f_1(x), f_2(x), \ldots$ to which they converge uniformly in $-\infty < x < +\infty$, and hence any function uniformly approximated by the sum $s(x)$, is also almost periodic. Hence, the class of functions belonging to the set $H\{s(x)\}$ is almost periodic. Bohr's theory of almost periodic functions is based on the proof for the identification of such almost periodic functions with the set $H\{s(x)\}$.

Substitute $x'$ for $x - \tau$ in (25), then, as $\tau$ is in $0 < x' < L$, the inequality (25) becomes $|f(x') - f(x)| \leqq \epsilon$. Let $B'$ be the upper bound of $f(x)$ in this interval; then it is evident that $|f(x)| \leqq B' + \epsilon$ and that $f(x)$ is bounded. Hence, an almost periodic function is bounded.

Choose $x_1$ and $x_2$ in the interval between $-L$ and $+L$. If $|x_2 - x_1| \leqq \delta$, then $|f(x_2) - f(x_1)| \leqq \epsilon$. Let $|f(x_1') - f(x_1)| \leqq \epsilon$, $|f(x_2') - f(x_2)| \leqq \epsilon$, $x_1' = x_1 - \tau$, $x_2' = x_2 - \tau$, and choose $x_2'$ and $x_1'$ in the interval between $-L$ and $+L$ such that $|x_2' - x_1'| < \delta$, then, since $|f(x_2') - f(x_1')| \leqq 3\epsilon$, it is evident that an almost periodic function $f(x)$ is uniformly continuous (Section 2.1) in the interval $-\infty < x < +\infty$.

Take a sequence of real numbers $h_1, h_2, \ldots$, that converge to zero, and let $f(x)$ be an almost periodic function. If $f'(x)$ is uniformly continuous in the sequence of functions

$$\phi_n(x) = \frac{1}{h_n}\{f(x + h_n) - f(x)\} = f'(x + \theta_n h_n), \qquad 0 < \theta_n < 1$$

$$(n = 1, 2, \ldots),$$

then we can find $\delta$ such that $|f'(x') - f'(x)| \leqq \epsilon$ if $|x' - x| \leqq \delta$ for a given $\epsilon$. As the sequence $\{h_n\}$ converges to zero, there exists an $N$ such that $|\theta_n h_n| \leqq \delta$ for $n \leqq N$. Since $|\phi(x) - f'(x)| \leqq \epsilon$ for such an $n$, the function $\phi_n(x)$ converges uniformly to $f'(x)$. Thus $\phi_n(x)$, accordingly its limit function, is almost periodic. Hence, the derived function $f'(x)$ of an almost periodic function $f(x)$ is almost periodic.

As an example, consider a simply periodic function $e^{i\lambda x}$ and a translation number $\tau$ such that $|e^{i\lambda\tau} - 1| \leqq \epsilon$. For $\epsilon < 1$ there exists $\delta$ such that $|e^{i\delta} - 1| < \epsilon$. Hence, $\delta$ satisfies $|\lambda\tau| \leqq \delta \pmod{2\pi}$. For $N$ almost periodic functions $e^{i\lambda_1 x}, \ldots, e^{i\lambda_N x}$ and $\delta < \pi$ there should exist $\epsilon(\delta)$ and at least one almost periodic function in an interval of length $L$. Hence, if $N$ inequalities

$$|\lambda_n \tau| \leqq \delta \pmod{2\pi} \qquad (n = 1, 2, \ldots, N),$$

and $\delta > 0$ are given, then there should exist at least one $\tau$ satisfying these inequalities in any interval of length $L$. This is Kronecker's theorem, proved in Section 2.8.

## 2.12 PROPERTIES OF ALMOST PERIODIC FUNCTIONS

We propose the following important theorems:

*Theorem:* The necessary and sufficient condition for the indefinite integral $F(x) = \int_0^x f(x)\,dx$ of an almost periodic function $f(x)$ to be almost periodic is that $F(x)$ should be bounded.

In fact, let the upper and the lower bounds of $F(x)$ be, respectively, $k_2$ and $k_1$, $k_2 > k_1$. If $k_1 = k_2$, then $F(x)$ is a constant. It is sufficient to prove that for a given $\epsilon > 0$ there exists an $\epsilon_1$ such that the translation number, belonging to $\epsilon$, of $F(x)$ is the translation number of $f(x)$, belonging to $\epsilon_1 = \epsilon_1(\epsilon, f(x))$.

There are two values $x_1$ and $x_2$ satisfying $F(x_1) < k_1 + \epsilon/6$, $F(x_2) > k_2 - \epsilon/6$. Denote the smaller of the two by $\xi$ and the distance $\|x_2 - x_1\|$ by $d$. Choose $l_0 = L(\xi/6d)$ so that in any interval of length $l_0$ there is at least a translation number $\tau = \tau(\epsilon/6d)$ of $f(x)$, belonging to $\epsilon_1 = \epsilon/6d$. We can choose $\tau$ so that $\xi + \tau$ is in the interval $(\alpha, \alpha + l_0)$ and accordingly $x_1 + \tau = z_1$, $x_2 + \tau = z_2$ are in the interval $(\alpha, \alpha + L_0)$ with $L_0 = l_0 + d$. Thus

$$\begin{aligned} F(z_2) - F(z_1) &= F(x_2) - F(x_1) + \int_{z_1}^{z_2} f(y)\,dy - \int_{x_1}^{x_2} f(y)\,dy \\ &= F(x_2) - F(x_1) + \int_{x_1}^{x_2} [f(y + \tau) - f(y)]\,dy \\ &\geqq F(x_2) - F(x_1) - d\frac{\epsilon}{6d} \\ &> k_2 - k_1 - \frac{2\epsilon}{6} - \frac{\epsilon}{6} = k_2 - k_1 - \frac{\epsilon}{2}. \end{aligned}$$

Since this inequality is satisfied for $F(z_1) < k_1 + \epsilon/2$, $F(z_2) > k_2 - \epsilon/2$ therefore the two values $z_1$ and $z_2$ satisfying these inequalities are in any interval $(\alpha, \alpha + L_0)$.

For any arbitrarily given $x$, take $z$ such that $F(z_1) < k_1 + \epsilon/2$ in the interval $(x, x + L_0)$; then

$$\begin{aligned} F(x + \tau) - F(x) &= F(z_1 + \tau) - F(z_1) + \int_{x}^{x+\tau} f(y)\,dy - \int_{z_1}^{z_1+\tau} f(y)\,dy \\ &= F(z_1 + \tau) - F(z_1) + \int_{x}^{z_1} f(y)\,dy - \int_{x+\tau}^{z_1+\tau} f(y)\,dy \\ &> k_1 - \left(k_1 + \frac{\epsilon}{2}\right) - \left|\int_{x}^{z_1} [f(y + \tau) - f(y)]\,dy\right| \\ &> -\frac{\epsilon}{2} - L_0\frac{\epsilon}{2L_0} = -\epsilon. \end{aligned}$$

Next, take $z_2$ such that $F(z_2) > k_2 - \epsilon/2$ in the interval $(x, x + L_0)$; then it can similarly be proved that

$$|F(x + \tau) - F(x)| < \epsilon.$$

From these two inequalities we obtain

$$|F(x + \tau) - F(x)| \leqq \epsilon.$$

*The Mean Value Theorem:* There exists the following mean value for an almost periodic function $f(x)$:

$$\lim_{T\to\infty} \frac{1}{T}\int_0^T f(x)\,dx = \mathfrak{M}\{f(x)\}.$$

In fact, let $L(\epsilon)$ be the translation number of $f(x)$, belonging to $\epsilon$, and $n$ be a positive integer; then there exists a translation number $\tau_\nu$ belonging to $\epsilon$ in the interval $(\nu T, \nu T + L)$ for

$$\frac{1}{nT}\int_0^{nT} f(x)\,dx = \sum_{\nu=0}^{n-1} \frac{1}{nT}\int_{\nu T}^{(\nu+1)T} f(x)\,dx.$$

Hence

$$\int_{\nu T}^{(\nu+1)T} f(x)\,dx = \int_{\nu T-\tau_\nu}^{(\nu+1)T-\tau_\nu} f(x + \tau_\nu)\,dx$$

$$= \left(\int_0^T + \int_{\nu T-\tau_\nu}^{0} + \int_T^{(\nu+1)T-\tau_\nu}\right) f(x + \tau_\nu)\,dx.$$

The lengths of the intervals in the last two integrals do not exceed $L$; that is, each of these integrals has a smaller absolute value than $\Gamma L$, where $\Gamma$ is the upper bound of $|f(x)|$. Owing to the inequality

$$\left|\int_0^T [f(x + \tau_\nu) - f(x)]\,dx\right| \leqq \epsilon T,$$

the first integral satisfies

$$\left|\frac{1}{nT}\int_{\nu T}^{(\nu+1)T} f(x)\,dx - \frac{1}{nT}\int_0^T f(x)\,dx\right| \leqq \frac{1}{n}\left(\epsilon + \frac{2\Gamma L}{T}\right).$$

Hence

$$\left|\frac{1}{nT}\int_0^{nT} f(x)\,dx - \frac{1}{T}\int_0^T f(x)\,dx\right|$$

$$\leqq \sum_{\nu=0}^{n-1} \left|\frac{1}{nT}\int_{\nu T}^{(\nu+1)T} f(x)\,dx - \frac{1}{nT}\int_0^T f(x)\,dx\right|$$

$$\leqq \epsilon + \frac{2\Gamma L}{T}. \tag{26}$$

Since

$$\left|\int_0^{T'} f(x)\,dx - \int_0^{nT} f(x)\,dx\right| \leqq \Gamma T$$

for $nT < T' \leqq (n + 1)T$, therefore

$$\left|\frac{1}{T'}\int_0^{T'} f(x)\,dx - \frac{1}{nT}\int_0^{nT} f(x)\,dx\right|$$

$$\leqq \left|\left(\frac{1}{T'}\int_0^{T'} - \frac{1}{T'}\int_0^{nT}\right) f(x)\,dx\right| + \left|\left(\frac{1}{T'}\int_0^{nT} - \frac{1}{nT}\int_0^{nT}\right) f(x)\,dx\right|$$

$$\leqq \frac{\Gamma}{n} + \frac{1}{n}\left|\frac{1}{nT}\int_0^{nT} f(x)\,dx\right|$$

$$\leqq \frac{\Gamma}{n} + \frac{1}{n}\left|\frac{1}{T}\int_0^{T} f(x)\,dx\right| + \frac{\epsilon}{n} + \frac{2\Gamma L}{nT}$$

$$\leqq \frac{2\Gamma}{n} + \frac{\epsilon}{n} + \frac{2\Gamma L}{nT}. \tag{27}$$

From (26) and (27) we obtain

$$\left|\frac{1}{T'}\int_0^{T'} f(x)\,dx - \frac{1}{T}\int_0^{T} f(x)\,dx\right|$$

$$\leqq \left|\left(\frac{1}{T'}\int_0^{T'} - \frac{1}{nT}\int_0^{nT}\right) f(x)\,dx\right| + \left|\left(\frac{1}{nT}\int_0^{nT} - \frac{1}{T}\int_0^{T}\right) f(x)\,dx\right|$$

$$\leqq \frac{2\Gamma}{n} + \left(\epsilon + \frac{2\Gamma L}{T}\right)\left(1 + \frac{1}{n}\right) \leqq \frac{2\Gamma}{n} + 2\left(\epsilon + \frac{2\Gamma L}{T}\right).$$

If we choose $T_1$ and $T_2$, respectively, in the intervals $[n_1 T, (n_1 + 1)T]$ and $[n_2 T, (n_2 + 1)T]$, where $n_1, n_2 > n$, then

$$\left|\frac{1}{T_1}\int_0^{T_1} f(x)\,dx - \frac{1}{T_2}\int_0^{T_2} f(x)\,dx\right|$$

$$\leqq \left|\left(\frac{1}{T_1}\int_0^{T_1} - \frac{1}{T}\int_0^{T}\right) f(x)\,dx\right| + \left|\left(\frac{1}{T_2}\int_0^{T_2} - \frac{L}{T}\int_0^{T}\right) f(x)\,dx\right|$$

$$\leqq \frac{4\Gamma}{n} + 4\left(\epsilon + \frac{2\Gamma L}{T}\right).$$

For a given $\epsilon$ we choose $T$ so that $2\Gamma L/T \leqq \epsilon$ and $n$ so that $4\Gamma/n \leqq \epsilon$; then, for $T_1, T_2 > nT$,

$$\left|\frac{1}{T_1}\int_0^{T_1} f(x)\,dx - \frac{1}{T_2}\int_0^{T_2} f(x)\,dx\right| \leqq 9\epsilon.$$

If $\epsilon$ is made sufficiently small, then $(1/T)\int_0^T f(x)\,dx$ converges uniformly to its limit for $T \to \infty$, according to Cauchy's theorem. Hence, there exists the limit function $\mathfrak{M}\{f(x)\}$ in the statement of the mean value theorem.

Since $f(x)$ and $f(x+a)$ are both almost periodic, and, since

$$\frac{1}{T}\int_0^T f(x+a)\,dx = \frac{1}{T}\int_a^{a+T} f(x)\,dx,$$

therefore

$$\mathfrak{M}_x\{f(x+a)\} = \mathfrak{M}_x\{f(x)\}. \tag{28}$$

Since $f(x+a)$ and $f(x)$ have the same translation number, we have

$$\left|\mathfrak{M}_x\{f(x)\} - \frac{1}{T}\int_a^{a+T} f(x)\,dx\right| \leqq \epsilon + \frac{2\Gamma L}{T};$$

that is, $(1/T)\int_a^{a+T} f(x)\,dx$ tends uniformly to $\mathfrak{M}\{f(x)\}$ as $T$ increases indefinitely. Hence, the mean value theorem is proved.

*Theorem:* If $f(x)$ is almost periodic, then $f(x+t)\bar{f}(t)$ is also almost periodic, where $\bar{f}(t)$ is the conjugate complex function of $f(t)$. The *convolution function* (*Faltung*)

$$g(x) = \mathfrak{M}_x\{f(x+t)\bar{f}(t)\} \tag{29}$$

is also almost periodic.

In fact, $f(x)$ being uniformly continuous, there exists a $\delta(\epsilon)$ for any $\epsilon$, such that $|f(x_1) - f(x_2)| \leqq \epsilon$ for $|x_1 - x_2| \leqq \delta = \delta(\epsilon)$. Hence, it should follow that

$$|g(x+h) - g(x)| = \left|\mathfrak{M}_t\{[f(x+h+t) - f(x+t)]\bar{f}(t)\}\right| \leqq \epsilon\Gamma$$

for $|h| \leqq \delta$. But, since

$$|g(x+\tau) - g(x)| = \left|\mathfrak{M}_t\{[f(x+\tau+t) - f(x+t)]\bar{f}(t)\}\right| \leqq \epsilon,$$

therefore the translation number $\tau$ of $f(x)$, belonging to $\epsilon/\Gamma$, is also the translation number of $g(x)$, belonging to $\epsilon$.

As a special case of this property, the function

$$g_T(x) = \frac{1}{T}\int_0^T f(x+t)\bar{f}(t)\,dt$$

is also almost periodic. Hence $\mathfrak{M}\{g(x)\} = \lim_{T\to\infty} \mathfrak{M}\{g_T(x)\}$. But, since

$$\begin{aligned}\mathfrak{M}\{g_T(x)\} &= \lim_{X\to\infty} \frac{1}{X}\int_0^X \left(\frac{1}{T}\int_0^T f(x+t)\bar{f}(t)\,dt\right) dx \\ &= \lim_{X\to\infty} \frac{1}{T}\int_0^T \left(\frac{1}{X}\int_0^X f(x+t)\,dx\right)\bar{f}(t)\,dt \\ &= \frac{1}{T}\int_0^T \underset{x}{\mathfrak{M}}\{f(x+t)\}\bar{f}(t)\,dt \\ &= \frac{1}{T}\int_0^T \underset{x}{\mathfrak{M}}\{f(x)\}\bar{f}(t)\,dt = \underset{x}{\mathfrak{M}}\{f(x)\}\cdot\frac{1}{T}\int_0^T \bar{f}(t)\,dt,\end{aligned}$$

therefore

$$\mathfrak{M}\{g(x)\} = \lim_{T\to\infty} \mathfrak{M}\{g_T(x)\} = \underset{x}{\mathfrak{M}}\{f(x)\}\cdot\underset{t}{\mathfrak{M}}\{\bar{f}(t)\}.$$

Hence

$$\mathfrak{M}\{g(x)\} = |\mathfrak{M}\{f(x)\}|^2. \tag{30}$$

*Theorem on the Fourier Coefficients:*

$$a(\lambda) = \mathfrak{M}\{f(x)e^{-i\lambda x}\}$$

is zero for any value of $\lambda$, except, at most, for a denumerable set of $\lambda$.

In fact, as the product $f(x)e^{-i\lambda x}$ of two almost periodic functions $f(x)$ and $e^{-i\lambda x}$ is almost periodic, the mean value

$$a(\lambda) = \mathfrak{M}\{f(x)e^{-i\lambda x}\}$$

exists with the real number $\lambda$. Consider $\sum_{n=1}^N b_n e^{i\lambda_n x}$, where $\lambda_n$ are all different. The function $|f(x) - \sum_{n=1}^N b_n e^{i\lambda_n x}|^2$ is almost periodic; thus

$$\begin{aligned}&\mathfrak{M}\left\{\left|f(x) - \sum_{n=1}^N b_n e^{i\lambda_n x}\right|^2\right\} \\ &\qquad = \mathfrak{M}\left\{\left[f(x) - \sum_{n=1}^N b_n e^{i\lambda_n x}\right]\left[\bar{f}(x) - \sum_{n=1}^N \bar{b}_n e^{-i\lambda_n x}\right]\right\}.\end{aligned} \tag{31}$$

However, the mean value of a sum is equal to the sum of the mean values. Hence

$$\mathfrak{M}\left\{\left|f(x) - \sum_{n=1}^{N} b_n e^{i\lambda_n x}\right|^2\right\} = \mathfrak{M}\{f(x)\bar{f}(x)\} - \sum_{n=1}^{N} \bar{b}_n \mathfrak{M}\{f(x)e^{-i\lambda_n x}\}$$
$$- \sum_{n=1}^{N} b_n \mathfrak{M}\{\bar{f}(x)e^{i\lambda_n x}\}$$
$$+ \sum_{m=1}^{N}\sum_{n=1}^{N} b_n \bar{b}_n \mathfrak{M}\{e^{i(\lambda_m - \lambda_n)x}\}.$$

Here $\mathfrak{M}\{e^{i\lambda_n x}\} = 0$ for $\lambda \neq 0$ and $= 1$ for $\lambda = 0$. Thus, this formula is written

$$= \mathfrak{M}\{|f(x)|^2\} - \sum_{n=1}^{N} \bar{b}_n a(\lambda_n) - \sum_{n=1}^{N} b_n \bar{a}(\lambda_n) + \sum_{n=1}^{N} b_n \bar{b}_n$$
$$= \mathfrak{M}\{|f(x)|^2\} - \sum_{n=1}^{N} |a(\lambda_n)|^2 + \sum_{n=1}^{N} |b_n - a(\lambda_n)|^2.$$

The expression (31) has a minimum value at $b_n = a(\lambda_n)$ with a given $\lambda_n$; thus

$$\mathfrak{M}\left\{\left|f(x) - \sum_{n=1}^{N} a(\lambda_n)e^{i\lambda_n x}\right|^2\right\} = \mathfrak{M}\{|f(x)|^2\} - \sum_{n=1}^{N} |a(\lambda_n)|^2.$$

But this right-hand member cannot be negative, and

$$\sum_{n=1}^{N} |a(\lambda_n)|^2 \leqq \mathfrak{M}\{|f(x)|^2\}. \tag{32}$$

Now denote the sets of values of $\lambda$ as follows:

$$|a(\lambda)| > 1 \quad \text{as} \quad \lambda_1, \lambda_2, \ldots, \lambda_{n_1}$$
$$1 \geqq |a(\lambda)| > \tfrac{1}{2} \quad \text{as} \quad \lambda_{n_1+1}, \lambda_{n_1+2}, \ldots, \lambda_{n_2}$$
$$\tfrac{1}{2} \geqq |a(\lambda)| > \tfrac{1}{3} \quad \text{as} \quad \lambda_{n_2+1}, \lambda_{n_2+2}, \ldots, \lambda_{n_3}$$

and so on.

Evidently the set of all values of $\lambda$ such that $|a(\lambda)| > 0$ is at most denumerable. Hence, $a(\lambda) = 0$ except, at most, for a denumerable set.

Let the values of $\lambda$ corresponding to these nonzero values of $a(\lambda)$ be $\Lambda_1, \Lambda_2, \ldots$. We call such $\Lambda_1, \Lambda_2, \ldots$ the *Fourier exponents* of $f(x)$, and the corresponding mean values $a(\Lambda_1) = A_1,\ a(\Lambda_2) = A_2, \ldots$ the *Fourier coefficients*. We write

$$f(x) \sim \sum_n A_n e^{i\Lambda_n x} \tag{33}$$

and call it the *Fourier series* of $f(x)$. If

$$\lim_{N\to\infty} \mathfrak{M}\left\{\left|f(x) - \sum_{n=1}^{N} A_n e^{i\Lambda_n x}\right|^2\right\} = 0,$$

then we say that the Fourier series $\sum_{n=1}^{N} A_n e^{i\Lambda_n x}$ *converges in the mean* to $f(x)$. If $\sum_{n=1}^{N} a_n e^{i\lambda_n x} e^{-i\lambda x}$ is uniformly convergent in $-\infty < x < +\infty$, then we can construct the mean values term by term, and

$$a(\lambda) = \mathfrak{M}\{f(x)e^{-i\lambda x}\} = \mathfrak{M}\left\{\sum_{n=1}^{N} a_n e^{i\lambda_n x}\cdot e^{-i\lambda x}\right\} \tag{34}$$

is equal to $a_n$ for $\lambda = \lambda_n$ and equal to zero for $\lambda \neq \lambda_n$. If (33) holds and $\sum_n |A_n|^2$ is convergent in (32), then

$$\sum_{n=1}^{N} |A_n|^2 \leqq \mathfrak{M}\{|f(x)|^2\}.$$

As we shall see later,

$$\sum_{n=1}^{N} |A_n|^2 = \mathfrak{M}\{|f(x)|^2\}.$$

This is called *Parseval's equation*. The theorem on a simply periodic function given in Section 2.2 can be applied also for an almost periodic function.

To prove that an almost periodic function can be uniformly approximated by a Fourier series we must first show that an almost periodic function can be uniquely determined by its Fourier series. This is called the *uniqueness theorem*. To prove the uniqueness theorem it is enough to show that all the coefficients of the Fourier series belonging to the difference $f(x) = f_1(x) - f_2(x)$ of two different almost periodic functions $f_1(x)$ and $f_2(x)$ are zero; that is,

$$a(\lambda) = \mathfrak{M}\{f(x)e^{-i\lambda_n x}\} = 0.$$

This is equivalent to Parseval's equation. In fact, if the equation holds, then $\mathfrak{M}\{|f(x)|^2\} = 0$ and $f(x) \equiv 0$, and conversely if

$$f(x) \sim \sum_n A_n e^{i\Lambda_n x},$$

then by writing $f(x)e^{-i\lambda x}$ instead of $f(x)$ in (30), we obtain, for the function $g(x)$ of (29),

$$\mathfrak{M}\{f(x+t)e^{-i\lambda(x+t)}\bar{f}(t)e^{i\lambda t}\} = g(x)e^{-i\lambda x}$$

and

$$\mathfrak{M}\{g(x)e^{-i\lambda x}\} = |\mathfrak{M}\{f(x)e^{-i\lambda x}\}|^2;$$

hence

$$g(x) = \sum_n |A_n|^2 e^{i\Lambda_n x};$$

that is

$$g(0) = \mathfrak{M}_t\{f(t)\bar{f}(t)\} = \mathfrak{M}_t\{|f(t)|^2\} = \sum_n |A_n|^2, \tag{35}$$

which is Parseval's equation.

In order to prove $\mathfrak{M}\{|f(x)|^2\} = 0$, Vallée Poussin (1916; see also Bohr, 1932; Besicovitch, 1932; Favard, 1933), constructed

$$G_T(x) = \frac{1}{T}\int_0^T F_T(x+t)\bar{F}_T(t)\,dt$$

with an almost periodic function

$$F_T(x) = \sum_{n=-\infty}^{\infty} \alpha_n e^{2\pi i n x/T}$$

and let $G_T(x)$ be approximated by $g(x)$ in $0 < x < T$. For the assumption $a(\lambda) = 0$ he substituted the assumption that the upper bound of $|\alpha_n|$ for $T \to \infty$ is zero in $-\infty < n < +\infty$ and proved from $\mathfrak{M}\{|g(x)|^2\} = 0$ that $g(x) \equiv 0$, and — see (35) — from $g(0) \equiv 0$ that $\mathfrak{M}_t\{|f(t)|^2\} = 0$, that is, $f(x) \equiv 0$. This leads to the following

*The Uniqueness Theorem:* An almost periodic function is determined uniquely by its Fourier series.

A simply periodic function is a special case of an almost periodic function. Hence, the uniqueness theorem is true for a simply periodic function, and also for the quasi-periodic function of Bohl, as it is also a special case of an almost periodic function. A similar situation arises for the approximation theorem we are going to prove. An almost periodic function thus belongs to a very wide class of functions.

## 2.13 BASE FIELD

Consider a denumerable sequence of real numbers $\Lambda_1, \Lambda_2, \ldots$. We say that $B = \{\beta_1, \beta_2, \ldots\}$ is the *base* if

(i) $\beta_1, \beta_2, \ldots$ are linearly independent, that is, there is no linear homogeneous relation

$$R_1\beta_1 + R_2\beta_2 + \cdots + R_m\beta_m = 0$$

with rational coefficients $R_1, R_2, \ldots, R_m$ for any integer $m$, and

(ii) any $\Lambda_n$ can be written uniquely in the form

$$\Lambda_n = r_1\beta_1 + r_2\beta_2 + \cdots + r_m\beta_m \qquad (n = 1, 2, \ldots) \tag{36}$$

with rational numbers $r_1, r_2, \ldots, r_m$ for a sufficiently large $m$. The numbers $\Lambda_1, \Lambda_2, \ldots$ belong to this *base field*, because they form a field as shown by (36). For constructing the base of an arbitrary sequence $\Lambda_1, \Lambda_2, \ldots,$ we take at first a nonzero $\Lambda_{n_1}$ and let it be $\beta_1$. Let $\Lambda_n$ such that $r_1\beta_1 + r_2\Lambda_n = 0$ be $\Lambda_{n_1+1}, \Lambda_{n_1+2}, \ldots$. We take as $\beta_2$ such $\Lambda_{n_2}$ as do not satisfy this relation. Let $\Lambda_n$ such that $r_1\beta_1 + r_2\beta_2 + r_3\Lambda_n = 0$ be $\Lambda_{n_2+1}, \Lambda_{n_2+2}, \ldots$. We repeat this process. If it comes to an end after a finite number of steps, $p$, then

$$\{\beta_1, \beta_2, \ldots, \beta_p\} = \{\Lambda_{n_1}, \Lambda_{n_2}, \ldots, \Lambda_{n_p}\}$$

is the base. It may be necessary to repeat this process a denumerable number of times. Then the sequence $\mathrm{B} = \{\beta_1, \beta_2, \ldots\}$ is the base. Hence from any sequence $\Lambda_1, \Lambda_2, \ldots$ we can construct the base $\mathrm{B} = \{\beta_1, \beta_2, \ldots\}$ in general.

Suppose that the Fourier exponents $\Lambda_n$ are expressed by the base B in the form (36) with rational coefficients. Assuming that $0 \leqq k_n^{(q)} \leqq 1$ for all $n$ and $q$, and $k_n^{(q)} \to 1$ for $q \to \infty$ with a fixed $n$, we form the finite sum

$$s_q(x) = \sum_n k_n^{(q)} A_n e^{i\Lambda_n x} \qquad (q = 1, 2, \ldots). \tag{37}$$

Then, by Parseval's equation,

$$\mathfrak{M}\{|f(x) - s_q(x)|^2\} = \sum_{n=1}^{\infty} (1 - k_n^{(q)})^2 |A_n|^2.$$

Hence we can prove that the finite sum (37) converges to zero for $q \to \infty$, and $s_q(x)$ converges to $f(x)$ in the mean.

With Fejér's nucleus

$$K_n(t) = \sum_{\nu=-n}^{n} \left(1 - \frac{|\nu|}{n}\right) e^{-i\nu t},$$

Bochner (1927) constructed

$$K^{(q)}(t) = \prod_{\nu=L}^{q} K_P\left(\frac{\beta_\nu}{Q} t\right)$$

$$= \sum_{\nu_1=-P}^{P} \cdots \sum_{\nu_q=-P}^{P} \left(1 - \frac{|\nu_1|}{P}\right) \cdots \left(1 - \frac{|\nu_q|}{P}\right)$$

$$\times \exp\left[-it\left(\frac{\nu_1}{Q}\beta_1 + \cdots + \frac{\nu_q}{Q}\beta_q\right)\right],$$

where $q$ is an integer and $Q = q!$, $P = qQ$, and considered

$$K^{(q)}(t) = \sum_n k_n^{(q)} e^{-i\Lambda_n t}.$$

Such a $k_n^{(q)}$ satisfies the above assumption that $0 \leqq k_n^{(q)} \leqq 1$ for all $n$ and $q$, and $k_n^{(q)} \to 1$ for $q \to \infty$ for a fixed $n$. Bochner proved that $f(x)$ is a *majorant sequence* of the sequence of functions $s_q(x)$ of (37) with such $k_n^{(q)}$; in other words, the translation number $\tau$ of $f(x)$, belonging to any $\epsilon > 0$, equals the translation number of $s_q(x)$, belonging to $\epsilon$. In general, if a sequence of functions $f_n(x)$ $(n = 1, 2, \ldots)$ converges to its majorant function $f(x)$ in the mean for $n \to \infty$, that is, if

$$\mathfrak{M}\{|f_n(x) - f(x)|^2\} \to 0,$$

then the upper bound of $|f_n(x) - f(x)|$ in $-\infty < x < +\infty$ converges to zero for $n \to \infty$. Hence $f_n(x)$ converges uniformly to $f(x)$. Thus it can be proved that the sequence of function $s_q(x)$ converges uniformly to $f(x)$ in the mean for $q \to \infty$. The proof is based on the formula

$$s_q(x) = \mathfrak{M}_t\{f(x + t)K^{(q)}(t)\},$$

and on the property that $K^{(q)}(t) \geqq 0$ and $\mathfrak{M}_t\{K^{(q)}(t)\} = 1$. This leads to the following

*Approximation Theorem:* An almost periodic function can be uniformly approximated in $-\infty < x < +\infty$ by the finite sum

$$s(x) = \sum_{n=-N}^{N} a_n e^{i\lambda_n x};$$

that is, there exists $s(x)$ such that for a given $\epsilon > 0$ the inequality $|f(x) - s(x)| \leqq \epsilon$ holds for all values of $x$.

As $s(x)$ itself is an almost periodic function, its limit function is also almost periodic. Hence, the closed set $H\{s(x)\}$ of the class of all functions $s(x)$ coincides with almost periodic functions.

Thus we have summarized Bohr's theory of almost periodic functions in the revised form of Vallée Poussin and Bochner; we have omitted the proof in order not to waste space on problems not directly concerned with celestial mechanics.

### 2.14 QUASI-PERIODIC FUNCTIONS AND ALMOST PERIODIC FUNCTIONS

We now consider the connection between almost periodic functions and Bohl's quasi-periodic functions (Bohr, 1925). Let $\mu_1, \mu_2, \ldots, \mu_N$ be nonzero real numbers. There is a segment of length $L$ containing in it the solution $t$ of $N$ *Diophantine inequalities* (Minkowski, 1907; Koksma, 1936)

$$|\mu_i t| < \delta \pmod{2\pi} \qquad (i = 1, 2, \ldots, N)$$

in the interval $\alpha < t < \beta$ for a given $\delta > 0$. Suppose that $\lambda$ is a real number that cannot be written in the form $n_1\mu_1 + \cdots + n_N\mu_N$ with integer coefficients $n_i$. Then, by Kronecker's theorem of Section 2.8, there is a value of $t$ in the segment of length $L$ that satisfies

$$|\mu_i t| < \delta \pmod{2\pi}, \qquad |\lambda t - \pi| < \frac{\pi}{2} \pmod{2\pi}$$

for a given $\delta > 0$. If there exists a $\delta$ such that the value of $\tau$ satisfying

$$|\tau\mu_i| < \delta \pmod{2\pi} \qquad (i = 1, 2, \ldots, N)$$

for a given $\epsilon > 0$ is the translation number of $f(x)$, belonging to $\epsilon$, then $f(x)$ is an almost periodic function with the exponents $\mu_1, \mu_2, \ldots, \mu_N$. Such a function is included in the class of almost periodic functions

$$f(x) \sim \sum_n A_n e^{i\Lambda_n x},$$

with all such exponents $\Lambda_i$ that can be written in the form

$$\Lambda_i = g_{i1}\mu_1 + g_{i2}\mu_2 + \cdots + g_{iN}\mu_N \tag{38}$$

with integer coefficients $g_{ij}$. Conversely, if the exponents $\Lambda_i$ of an almost periodic function are written in the form (38) with integer coefficients, then the function is multiply periodic with periods $\mu_1, \mu_2, \ldots, \mu_N$. In particular, if there is only one term in the base, then the function is simply periodic.

Take a base $\mathrm{B} = \{\beta_1, \beta_2, \ldots\}$ and express $\sum_n A_n e^{i\Lambda_n x}$ in the form

$$\sum_n A_n \exp\left[i(r_{n1}\beta_1 x + \cdots + r_n q_n \beta_{q_n} x)\right].$$

Write $e^{i\beta_1 x_1}, e^{i\beta_2 x_2}, \ldots$ in place of $e^{i\beta_1 x}, e^{i\beta_2 x}, \ldots$, respectively. Let $x_1, x_2, \ldots$ be denumerably infinite linearly independent real numbers, and construct

$$\sum_n A_n \exp\,[i(r_{n1}\beta_1 x_1 + \cdots)]e^{i\Lambda_n x} = \sum_n \mathrm{B}_n(x_1, x_2, \ldots)e^{i\Lambda_n x}. \quad (39)$$

If we put $x_1 = x_2 = \cdots = k$, then this expression becomes $\sum_n A_n e^{i\Lambda_n(x+k)}$, a Fourier series of an almost periodic function $f(x + k)$.

Write

$$r_{n1}\beta_1 x_1 + \cdots + r_{nq_n}\beta_{q_n} x_{q_n} = \theta_n.$$

According to Kronecker's theorem there exists a value of $k$ that satisfies $|e^{i\theta_n} - e^{i\Lambda_n x}| < \delta$ $(n = 1, 2, \ldots, N)$ for a given $N$ and $\delta > 0$. Hence, if we choose the sequence $k_1, k_2, \ldots$ so that it satisfies

$$\lim e^{i\Lambda_n k_n} = e^{i\theta_n},$$

then

$$\mathfrak{M}\{|f(x + k_p) - f(x + k_q)|^2\}$$

converges to zero for $p, q \to \infty$. Hence, $f(x + k_1), f(x + k_2), \ldots$ converge to the limit function $\phi(x)$ uniformly for all values of $x$. Since

$$\mathfrak{M}\{\phi(x)e^{-i\lambda x}\} = \lim_{m\to\infty} \mathfrak{M}\{f(x + k_m)e^{-i\lambda x}\},$$

$$= \lim_{m\to\infty} A_n e^{i\Lambda_n k_m} = \begin{cases} A_n e^{i\theta_n} & (\lambda = \Lambda_n) \\ 0 & (\lambda \neq \Lambda_n) \end{cases},$$

therefore, the function $\phi(x)$ has (39) as its Fourier series. Hence, (39) is the Fourier series of an almost periodic function $\phi(x) = F(x_1, x_2, \ldots)$, whatever we may choose for $x_1, x_2, \ldots$. It can be shown that this function $F(x_1, x_2, \ldots)$ is uniformly continuous in the space of denumerable dimensions with coordinates $x_1, x_2, \ldots$.

If a function $G(x_1, x_2, \ldots)$ of a denumerable number of variables is uniformly approximated by a continuous multiply periodic function of as many number of variables, that is, if there exists for a given $\epsilon > 0$ a continuous multiply periodic function $P(x_1, x_2, \ldots)$ that satisfies

$$|G(x_1, x_2, \ldots) - P(x_1, x_2, \ldots)| < \epsilon$$

in the whole space, then $G(x_1, x_2, \ldots)$ is called a *limit periodic function.*

A limit periodic function is uniformly continuous in the whole space, similarly to the function $P(x_1, x_2, \ldots)$. Let

$$x_1 = c_1 x + d_1, \qquad x_2 = c_2 x + d_2, \ldots$$

in $G(x_1, x_2, \ldots)$; then on the straight line the function

$$h(x) = G(c_1x + d_1, c_2x + d_2, \ldots) \tag{40}$$

is almost periodic. Indeed, as the function (40) is uniformly continuous, we have, for

$$|x'_m - x''_m| < \delta \quad (\text{mod } r_{q_m}) \qquad (m = 1, 2, \ldots, N)$$

with a rational number $r$, the inequality

$$|G(x'_1, x'_2, \ldots) - G(x''_1, x''_2, \ldots)| < \epsilon.$$

But we know from Kronecker's theorem of Section 2.8 that

$$|[c_m(x + \tau) - d_m] - [c_mx + d_m]| = |c_m\tau| < \delta \quad (\text{mod } r_{q_m})$$
$$(m = 1, 2, \ldots, N),$$

where $\tau$ is the translation number. We can also prove the existence of the mean value of $G(x_1, x_2, \ldots)$:

$$\underset{(x_1, x_2, \cdots)}{\mathfrak{M}} \{G(x_1, x_2, \ldots)\};$$

also the approximation theorem that the limit periodic function $G(x_1, x_2, \ldots)$ can be as nearly as possible approximated by the finite sum of Fourier series terms by taking $m$ sufficiently large, that is,

$$G(x_1, x_2, \ldots) \sim \sum_{n_1, n_2, \cdots, n_N} c_{n_1n_2\ldots n_N} \exp\,[i(n_1\beta_1x_1 + \cdots + n_N\beta_Nx_N)],$$

$$c_{n_1n_2\ldots n_N} = \underset{(x_1, x_2, \cdots)}{\mathfrak{M}} \{G(x_1, x_2, \ldots) \exp\,[-i(n_1\beta_1x_1 + \cdots + n_N\beta_Nx_N)]\},$$

$$\beta_i = \frac{2\pi}{q_i}. \tag{41}$$

This is an extension of Weierstrass's theorem of Section 2.6. In particular, in the case of multiply periodic functions the mean value is replaced by

$$\frac{1}{p_i}\int_0^{p_i} P(x_1, x_2, \ldots)\, dx_i;$$

this forms the foundation of Schuster's *periodogram analysis* (Brunt, 1917; Stumpff, 1937), which has been often employed in astronomy for discovering the periodicities of the sunspot cycles, of the latitude variation, and of solar-terrestrial phenomena.

Thus the relation between a multiply periodic function and an almost periodic function has been generalized to the relation between a limit periodic function and an almost periodic function.

If we can extract a sequence that uniformly converges to a limit periodic function in $-\infty < x < +\infty$ from a sequence

$$f(x + h_1), f(x + h_2), \ldots, f(x + h_n), \ldots \tag{42}$$

constructed by a continuous function $f(x)$ in $-\infty < x < +\infty$ and a sequence of arbitrary real numbers $h_1, h_2, \ldots,$ then $f(x)$ is called a *normal function* according to Bochner (1925, 1927, 1933; Favard, 1928, 1933). Bochner called the limit function a *super-function* (*Überfunktion*). An almost periodic function is a normal function and conversely a normal function is almost periodic. It resembles the normal family of functions of Montel (1927) in the theory of analytic functions. The idea of normal functions is useful in the theory of linear differential equations with almost periodic coefficients (see Chapter 12). Such types of equations are generalizations of the differential equations appearing in celestial mechanics.

When a sequence of almost periodic functions

$$f(x + k) \qquad (-\infty < k < +\infty),$$

contains its limit function, it is called a *closed set of functions* (*abgeschlossene Hülle*) according to Bohr (1926), and written $H\{f(x + k)\}$. If $g(x)$ belongs to the closed set $H\{f(x + k)\}$, then it can be shown that $H\{f(x + k)\}$ and $H\{g(x + k)\}$ are identical. If

$$f(x) \sim \sum_n A_n e^{i\Lambda_n x},$$

then any function $\phi(x)$ in $H$ is written

$$\phi(x) \sim \sum_n A_n e^{i\Theta_n} e^{i\Lambda_n x}.$$

In order that this Fourier series shall determine a function belonging to the closed set of functions $H\{f(x + k)\}$, it is necessary and sufficient that the points $(\Theta_1, \Theta_2, \ldots)$ (mod $2\pi$) in a space of a denumerable number of dimensions should be the accumulation points of the point set formed of the points $(\Lambda_1 k, \Lambda_2 k, \ldots)$, that is, that we should be able to determine $k_1, k_2, \ldots$ such that the Diophantine equations

$$\lim_{n\to\infty} \Lambda_n k_m = \Theta_n \pmod{2\pi} \qquad (m = 1, 2, \ldots)$$

are satisfied.

An almost periodic function, which can be written in the form

$$\sum_n A_n \exp\,[i(r_{n1}\beta_1 k_1 + \cdots + r_{nq_n}\beta_{q_n} k_{q_n})] e^{i\Lambda_n x}$$

with any sequence of variables $k_1, k_2, \ldots$ and of rational numbers $r_{ij}$, is said to belong to the closed set of functions $K\{f(x)\}$. The closed set of functions $K$ depends only on $f(x)$ but does not depend on any particular value of the base $\{\beta_1, \beta_2, \ldots\}$. The corresponding limit periodic functions satisfy in the whole space of a denumerable number of dimensions the equations

$$G(x_1, x_2, \ldots) = F(x_1 + k_1, x_2 + k_2, \ldots);$$

and the closed set of functions $K$ consists of all functions written in the form $g(x) = F(x + k_1, x + k_2, \ldots)$. If these two closed sets of functions $K$ and $H$ have a function $f(x)$ in common, then $K$ is contained in $H$, and $H$ is a closed set of functions of $K$. In order that $K\{f(x)\}$ and $H\{f(x + k)\}$ be coincident, it is necessary and sufficient that $\Lambda_n$ should be the base of integer coefficients, that is, all $\Lambda_n$ should be expressed by linear homogeneous equations with integer coefficients with respect to the base $\{\beta_1, \beta_2, \ldots\}$. Thus $K$ is the closed set of almost periodic functions and $H$ is the closed set of more general almost periodic functions. In this way the relation between a quasi-periodic function and an almost periodic function becomes clearer.

Bochner founded the theory of almost periodic functions on his normal functions. The idea of almost periodic functions has been generalized by Stepanoff—to square summable functions—and amended by Besicovitch. Weyl discussed almost periodic functions on the basis of group theory. We are concerned with uniformly almost periodic functions, as has been studied by Bohr and Bochner (Bohr, 1925 and 1932; Bochner, 1927; Besicovitch, 1932; Maak, 1950).

## RELATIVISTIC ONE-BODY PROBLEM

### 2.15 TRANSFORMATION OF THE PROBLEM

The relativistic one-body problem, that is, the motion of a particle with an infinitesimal mass in the relativistic gravitational field of Schwarzschild, can be solved as an example of quasi-periodic motions. The relativistic many-body problem, especially the two-body problem, can be approximately treated as a one-body problem by taking the center of mass as origin and replacing the mass in the one-body problem by the sum of the two masses, as has been shown by Eddington and Clark (1938), Einstein, Infeld, and Hoffmann (1938), Einstein and Infeld (1940), and Robertson (1938). Hence, the behavior of the motion can be outlined by solving the one-body problem. The following treatment is due to the author (Hagihara, 1931).

In order to apply the theory of Hamilton-Jacobi to this problem we consider the following auxiliary

*Theorem:* The variational principle

$$\delta \int ds = 0, \tag{43}$$

where

$$ds^2 = \sum_{\alpha,\beta} g_{\alpha\beta}\, dx_\alpha\, dx_\beta \qquad (\alpha, \beta = 1, 2, \ldots, n), \tag{44}$$

can be replaced by

$$\delta \int T\, d\sigma = 0, \tag{45}$$

where

$$T = \frac{1}{2} \sum_{\alpha,\beta} g_{\alpha\beta} \frac{dx_\alpha}{d\sigma} \frac{dx_\beta}{d\sigma} \qquad (\alpha, \beta = 1, 2, \ldots, n). \tag{46}$$

Here $x_i$ are at first supposed to be all independent; then the constant $C^2$ is determined by

$$\sum_{\alpha,\beta} g_{\alpha\beta} \frac{dx_\alpha}{d\sigma} \frac{dx_\beta}{d\sigma} = C^2 \qquad (\alpha, \beta = 1, 2, \ldots, n) \tag{47}$$

as a function of the integration constants in the second variational statement (45); finally $\sigma$ is eliminated by

$$s = C\sigma. \tag{48}$$

Since (46) is the expression for twice the kinetic energy if $\sigma$ is replaced by $t$, the motion of the dynamical system according to the second variational statement (45) is given by a Hamilton-Jacobi equation of the form

$$\frac{\partial W}{\partial \sigma} + \Theta\left(x_i, \frac{\partial W}{\partial x_i}\right) = 0, \tag{49}$$

where the momenta are defined by

$$\frac{\partial T(x_i, dx_i/d\sigma)}{\partial(dx_j/d\sigma)} = p_j \qquad (j = 1, 2, \ldots, n).$$

The *characteristic* of Cauchy for this equation, which will be defined in Chapter 4, is determined by

$$\frac{dx_i}{\partial\Theta/\partial p_i} = \frac{-dp_j}{\partial\Theta/\partial x_j + (\partial\Theta/\partial p_j)p_j} = \frac{dW}{\sum_k(\partial\Theta/\partial p_k)p_k} = d\sigma$$

$$(i, j, k = 1, 2, \ldots, n),$$

or by the Hamiltonian canonical equations

$$\frac{dx_i}{d\sigma} = \frac{\partial\Theta}{\partial p_i}, \quad \frac{dp_i}{d\sigma} = -\frac{\partial\Theta}{\partial x_i} \qquad (i = 1, 2, \ldots, n).$$

In particular, if $T$ is in the form (46), then the Hamilton-Jacobi equation (49) becomes

$$\frac{\partial W}{\partial \sigma} + \frac{1}{2} \sum_{\alpha,\beta} g^{\alpha\beta} \frac{\partial W}{\partial x_\alpha} \frac{\partial W}{\partial x_\beta} = 0 \qquad (\alpha, \beta = 1, 2, \ldots, n) \tag{49a}$$

and the Cauchy characteristic is

$$\frac{d^2 x_i}{d\sigma^2} + \sum_{\alpha,\beta} \left\{ {\alpha \, \beta \atop i} \right\} \frac{dx_\alpha}{d\sigma} \frac{dx_\beta}{d\sigma} = 0 \qquad (i = 1, 2, \ldots, n), \tag{50}$$

where $g^{\alpha\beta}$ is contragradient to $g_{\alpha\beta}$ and the determinant $\| g_{\alpha\beta} \|$ is assumed to be nonzero. The Christoffel symbol $\left\{ {\alpha \, \beta \atop i} \right\}$ is defined in Section 1.4.

Equation (48) is now written

$$2T - C^2 = 0. \tag{51}$$

When there is a constraint expressed by (51), the variational problem is further replaced (Bolza, 1909) by

$$\delta \int F \, d\sigma = 0,$$

with

$$F = T + \lambda(2T - C^2) = (1 + 2\lambda) T - \lambda C^2,$$

where $\lambda$ is the Lagrangian indeterminate multiplier. The corresponding Euler-Lagrange equation is

$$\frac{\partial T}{\partial x_i} - \frac{d}{d\sigma} \left[ \frac{\partial T}{\partial (dx_i/d\sigma)} \right] = 0 \qquad (i = 1, 2, \ldots, n), \tag{52}$$

with the restriction (51). Equation (52) admits an integral:

$$T - \sum_i \frac{\partial T}{\partial (dx_i/d\sigma)} \frac{dx_i}{d\sigma} = \text{constant}.$$

Since $T$ is a homogeneous quadratic with respect to $dx_i/d\sigma$, $T$ is constant. If we put this constant equal to $C^2/2$, then (47) is satisfied. On the other hand, the solution of (43) is known to be (50). Hence, the auxiliary theorem has been proved. Thus we need only consider the second variational principle (45), instead of the first one (43), and solve the Hamilton-Jacobi partial differential equation (49) (Synge, 1928; Infeld, 1931).

For the relativistic trajectory of a light ray we take $ds = 0$; that is, $C^2 = 0$ in the preceding formulas.

## 2.16 RELATIVISTIC FIELD OF SCHWARZSCHILD

We take a particle or the center of a spherically symmetrical celestial body as origin and adopt the time and space as measured by an observer located at the origin. Then the relativistic gravitational field for the one-body problem obtained by Schwarzschild (Schwarzschild, 1916; Weyl, 1919, 1919a; Einstein, 1914, 1915, 1916, 1916a; de Donder, 1930; Haag, 1931) is

$$ds^2 = c_0^2(1 - \alpha/r)\,dt^2 - \frac{1}{1 - \alpha/r}\,dr^2 - r^2\,d\phi^2 - r^2 \sin^2\phi\,d\theta^2, \quad (53)$$

where $r$, $\phi$, $\theta$ are the polar coordinates for the observer, $c_0$ is the light velocity in vacuum at infinity, and $m = \alpha/2$ is the mass of the body measured at an infinite distance in the coordinate system in which the particle is at rest.

Combining equations (49) and (53), we obtain

$$\frac{\partial W}{\partial \sigma} + \frac{1}{2c_0^2(1 - \alpha/r)}\left(\frac{\partial W}{\partial t}\right)^2 - \frac{1}{2}\left(1 - \frac{\alpha}{r}\right)\left(\frac{\partial W}{\partial r}\right)^2 - \frac{1}{2r^2 \sin^2\phi}\left(\frac{\partial W}{\partial \theta}\right)^2 - \frac{1}{2r^2}\left(\frac{\partial W}{\partial \phi}\right)^2 = 0, \quad (54a)$$

and the Cauchy characteristic of this equation is expressed by

$$\frac{dt}{d\sigma} = \frac{1}{c_0^2(1 - \alpha/r)}\frac{\partial W}{\partial t},$$

$$\frac{dr}{d\sigma} = -(1 - \alpha/r)\frac{\partial W}{\partial r},$$

$$\frac{d\phi}{d\sigma} = -\frac{1}{r^2}\frac{\partial W}{\partial \phi},$$

$$\frac{d\theta}{d\sigma} = -\frac{1}{r^2 \sin^2\phi}\frac{\partial W}{\partial \theta},$$

$$\frac{d}{d\sigma}\left(\frac{\partial W}{\partial t}\right) = 0, \quad (54b)$$

$$\frac{d}{d\sigma}\left(\frac{\partial W}{\partial r}\right) = \frac{\alpha/r^2}{2c_0^2(1 - \alpha/r)^2}\left(\frac{\partial W}{\partial t}\right)^2 + \frac{\alpha}{2r^2}\left(\frac{\partial W}{\partial r}\right)^2 - \frac{1}{r^3}\left(\frac{\partial W}{\partial \phi}\right)^2 - \frac{1}{r^3 \sin^2\phi}\left(\frac{\partial W}{\partial \theta}\right)^2,$$

$$\frac{d}{d\sigma}\left(\frac{\partial W}{\partial \phi}\right) = -\frac{\cos\phi}{r^2 \sin^3\phi}\left(\frac{\partial W}{\partial \theta}\right)^2,$$

$$\frac{d}{d\sigma}\left(\frac{\partial W}{\partial \theta}\right) = 0.$$

The variables $\theta$ and $t$ are *ignorable* coordinates and we can write

$$W = -K\sigma + ft + g\theta + V(r, \phi)$$

with the integration constants $K, f, g$. The Hamilton-Jacobi equation for $V(r, \phi)$ is

$$\frac{\partial V}{\partial \sigma} + \frac{f^2}{2c_0^2(1 - \alpha/r)} - \frac{g^2}{2r^2 \sin^2 \phi} - \frac{1}{2}\left(1 - \frac{\alpha}{r}\right)\left(\frac{\partial V}{\partial r}\right)^2 - \frac{1}{2r^2}\left(\frac{\partial V}{\partial \phi}\right)^2 = 0$$

and the Cauchy characteristic is

$$\begin{aligned}
\frac{dr}{d\sigma} &= -\left(1 - \frac{\alpha}{r}\right)\frac{\partial V}{\partial r}, \\
\frac{d\phi}{d\sigma} &= -\frac{1}{r^2}\frac{\partial V}{\partial \phi}, \\
\frac{d}{d\sigma}\left(\frac{\partial V}{\partial r}\right) &= \frac{(\alpha/r^2)f^2}{2c_0^2(1 - \alpha/r)^2} - \frac{g^2}{r^3 \sin^2 \phi} + \frac{\alpha}{2r^2}\left(\frac{\partial V}{\partial r}\right)^2 - \frac{1}{r^3}\left(\frac{\partial V}{\partial \phi}\right)^2, \\
\frac{d}{d\sigma}\left(\frac{\partial V}{\partial \phi}\right) &= -\frac{\cos \phi}{r^2 \sin^3 \phi}\, g^2.
\end{aligned} \tag{55}$$

To obtain $V$ we can put $g = 0$, because we know that there exists an integral $\theta =$ constant and that the trajectory is always on the plane $\theta =$ constant, which we take as the plane $\theta = 0$. Then, from (55),

$$r^2 \frac{d\phi}{d\sigma} = AC,$$

$$\frac{d}{d\sigma}\left(\frac{dr/d\sigma}{1 - \alpha/r}\right) = \frac{-(\alpha/r^2)f^2}{2c_0^2(1 - \alpha/r)^2} - \frac{\alpha}{2r^2}\frac{(dr/d\sigma)^2}{(1 - \alpha/r)^2} + \frac{1}{r^3}A^2C^2, \tag{56}$$

and, from (54b),

$$\frac{dt}{dr} = \frac{f}{c_0^2(1 - \alpha/r)}, \tag{57}$$

where $A$ is an integration constant. From (48), (53), (56), and (57),

$$\left(\frac{dr}{ds}\right)^2 = -\left(1 - \frac{\alpha}{r}\right)\left(1 + \frac{A}{r^2}\right) + \frac{f^2}{c_0^2 C^2}, \tag{58}$$

$$r^2 \frac{d\phi}{ds} = A, \tag{59}$$

$$\left(\frac{A}{r^2}\frac{dr}{d\phi}\right)^2 + \frac{A^2}{r^2} = \frac{f^2}{c_0^2 C^2} - 1 + \frac{\alpha}{r} + \frac{\alpha}{r}\frac{A^2}{r^2}. \tag{60}$$

For a light ray we put $A = \infty$ in these formulas. Equations (56) and (57) are of the same form as those usually obtained in relativistic mechanics. If we put $f/(c_0 C) = c$, then the formulas reduce to Eddington's (Eddington, 1923; cf. Einstein, 1915, 1916; de Sitter, 1916, 1917).

The above deduction fails for $\| g_{\alpha\beta} \| = 0$. The value of the determinant is $-c_0^2 r^4 \sin^2 \phi$. It vanishes when $\phi = 0$ or $\pi$. This difficulty can easily be avoided by taking suitable coordinate axes. Except in the rare case when the two bodies collide in a finite time, $r$ never becomes zero. Hence, we can suppose that the determinant never vanishes.

## 2.17 THE COMPLETE INTEGRAL

We start with (54a) and, by putting

$$W = -K\sigma + V_1(t) + V_2(r) + V_3(\phi) + V_4(\theta),$$

follow the method of the separation of variables treated in Section 1.16. Then, with the new integration constants $h_1$, $h_2$, $h_3$,

$$\begin{aligned} V_1 &= c_0 h_3 t, \\ V_2 &= \int \sqrt{\frac{2}{1 - \alpha/r}\left[1 - \frac{h_1^2}{2r^2} + \frac{h_3^2}{2(1 - \alpha/r)} - K\right]}\, dr, \\ V_3 &= \int \sqrt{h_1^2 - \frac{h_3^2}{\sin^2 \phi}}\, d\phi, \\ V_4 &= -h_2 \theta. \end{aligned} \tag{61}$$

If the canonically conjugate constants $\beta_0$, $\beta_1$, $\beta_2$, $\beta_3$ are put in place of $K$, $h_1$, $h_2$, $h_3$, then the solution, by the method of Jacobi (Section 1.10), is

$$\begin{aligned} \frac{\partial W}{\partial K} &= \beta_0, \qquad y_0 = \frac{\partial W}{\partial x_0}, \\ \frac{\partial W}{\partial h_i} &= \beta_i, \qquad y_i = \frac{\partial W}{\partial x_i}, \qquad (i = 1, 2, 3), \end{aligned} \tag{62}$$

or, by the auxiliary theorem of Section 2.15,

$$1 = c_0^2\left(1 - \frac{\alpha}{r}\right)\left(\frac{dt}{d\sigma}\right) - \frac{1}{1 - \alpha/r}\left(\frac{dr}{d\sigma}\right)^2 - r^2\left(\frac{d\phi}{d\sigma}\right)^2 - r^2 \sin^2 \phi \left(\frac{d\theta}{d\sigma}\right)^2. \tag{63}$$

Hence,

$$\begin{aligned}
\beta_0 &= -\sigma - \int \frac{dr}{\sqrt{R(r)}}, \\
\beta_1 &= -\int \frac{h_1\,dr}{r^2\sqrt{R(r)}} + \int \frac{h_1}{\sqrt{\Phi(\phi)}}\,d\phi, \\
\beta_2 &= -\int \frac{h_2\,d\phi}{\sin^2\phi\sqrt{\Phi(\phi)}} - \theta, \\
\beta_3 &= \int \frac{h_1 r\,dr}{(r-\alpha)\sqrt{R(r)}} + c_0 t,
\end{aligned} \tag{64}$$

where

$$R(r) = (h_3^2 - 2K) + \frac{2\alpha K}{r} - \frac{h_1^2}{r^2} + \frac{\alpha h_1^2}{r^3},$$

and

$$\Phi(\phi) = h_1^2 - \frac{h_2^2}{\sin^2\phi}.$$

Let

$$u = \frac{\alpha}{r}, \qquad U(u) = \frac{\alpha^2(h_3^2 - 2K)}{h_1^2} + \frac{2K\alpha^2}{h_1^2}u - u^2 + u^3;$$

then

$$\begin{aligned}
\frac{h_1\,d\sigma}{\alpha^2} &= \frac{du}{u^2\sqrt{U(u)}}, \\
\frac{h_1\,d\phi}{\sqrt{\Phi(\phi)}} &= -\frac{du}{\sqrt{U(u)}}, \\
d\theta &= -\frac{h_2}{\sin^2\phi}\frac{d\phi}{\sqrt{\Phi(\phi)}}, \\
c_0\,dt &= \frac{\alpha^2 h_3}{h_1}\frac{du}{(1-u)u^2\sqrt{U(u)}},
\end{aligned} \tag{65}$$

and our condition takes the form

$$d\sigma^2 = c_0^2(1-u)\,dt^2 + \frac{\alpha^2}{u^2}\,d\phi^2 + \frac{\alpha^2}{u^2}\sin^2\phi\,d\theta^2 + \frac{1}{1-u}\frac{\alpha^2}{u^4}\,du^2 = 0,$$

or, $K = \frac{1}{2}$. Hence, $\sigma$ is determined by

$$d\sigma^2 = c_0^2(1-\alpha/r)\,dt^2 - \frac{1}{1-\alpha/r}\,dr^2 - r^2\,d\phi^2 - r^2\sin^2\phi\,d\theta^2.$$

Thus we write $s$ in place of $\sigma$ and take $s = 0$ for $\sigma = 0$. Now

$$U(u) = u^3 - u^2 + \frac{\alpha^2}{h_1^2} u + \frac{\alpha^2(h_3^2 - 1)}{h_1^2}.$$

For a light ray the trajectory is determined by $ds = 0$, or by $K = 0$. In the above treatment, $K = \frac{1}{2}$. Thus the two cases are contradictory. Hence the trajectory of a light ray nowhere coincides with the trajectory of a particle of infinitesimal mass.

Now let $\cos I = h_2/h_1$, where $I$ denotes the inclination of the orbital plane with reference to the coordinate plane. Let $\cos \phi = \sin I \sin \psi$, where $\phi$ denotes the colatitude and $\psi$ the argument of latitude. Since

$$\frac{d\phi}{\sqrt{\Phi(\phi)}} = -\frac{d\psi}{h_1},$$

therefore,

$$\frac{du}{\sqrt{U(u)}} = d\psi.$$

We transform the third equation of (65) into the form

$$\begin{aligned} d\theta &= \frac{\cos I}{\cos^2 \phi} d\psi = \frac{\cos I}{2} \left( \frac{d\psi}{1 - \sin I \sin \psi} + \frac{d\psi}{1 + \sin I \sin \psi} \right) \\ &= d \cdot \tan^{-1} (\tan \psi \cos I). \end{aligned}$$

That is,

$$\tan (\theta + \beta_2) = \tan \psi \cos I.$$

This is a fundamental formula in spherical trigonometry; $\theta + \beta_2$ denotes the longitude counted from the ascending node of the orbit. We obtain finally

$$\begin{aligned} \frac{h_1}{\alpha^2} ds &= \frac{du}{u^2\sqrt{U(u)}}, \\ 0 &= \frac{du}{\sqrt{U(u)}} - d\psi, \qquad (66) \\ 0 &= \frac{du}{(1 - u)u^2\sqrt{U(u)}} - \frac{h_1 c_0}{\alpha^2 h_3} dt. \end{aligned}$$

This is a special case of the inversion problem in the theory of algebraic functions. The second formula of (66) is the Abelian integral of the first kind, and the first and the third formulas are the Abelian integrals of the third kind; they have logarithmic singularities (see Section 2.20).

## 2.18 SOLUTION BY ELLIPTIC FUNCTIONS

Denote the three roots of $U(u) = 0$ by $u_1$, $u_2$, $u_3$.

$$u_1 + u_2 + u_3 = 1,$$

$$u_2u_3 + u_3u_1 + u_1u_2 = \frac{\alpha^2}{h_1^2},$$

$$u_1u_2u_3 = -\frac{\alpha^2(h_3^2 - 1)}{h_1^2}.$$

Let

$$x = u - \tfrac{1}{3}, \qquad e_i = u_i - \tfrac{1}{3} \qquad (i = 1, 2, 3);$$

then

$$d\psi = \frac{dx}{\sqrt{(x - e_1)(x - e_2)(x - e_3)}}.$$

Using Weierstrass's $\wp$-function gives (Appell and Lacour, 1922)

$$x = \wp\left(\frac{\psi + \beta_1}{2}\right), \qquad u = \frac{\alpha}{r} = \wp\left(\frac{\psi + \beta_1}{2}\right) + \frac{1}{3}.$$

Hence, (66) becomes

$$\frac{h_1}{\alpha^2}\,ds = \frac{d\psi}{\left[\wp\left(\frac{\psi + \beta_1}{2}\right) + \frac{1}{3}\right]^2}$$

$$= \frac{dx}{(x + \frac{1}{3})^2\sqrt{(x - e_1)(x - e_2)(x - e_3)}},$$

$$\frac{h_1c_0}{\alpha^2h_3}\,dt = \frac{d\psi}{\left[\frac{2}{3} - \wp\left(\frac{\psi + \beta_1}{2}\right)\right]\left[\wp\left(\frac{\psi + \beta_1}{2}\right) + \frac{1}{3}\right]^2} \tag{67}$$

$$= \frac{dx}{(\frac{2}{3} - x)(x + \frac{1}{3})^2\sqrt{(x - e_1)(x - e_2)(x - e_3)}}.$$

From the formula of elliptic functions we have

$$\zeta(x + y) - \zeta(x - y) - 2\zeta(y) = -\frac{\wp'(y)}{\wp(x) - \wp(y)}. \tag{68}$$

Differentiation gives

$$\wp(x + y) + \wp(x - y) - 2\wp(y) = \frac{[\wp'(y)]^2}{[\wp(x) - \wp(y)]^2} + \frac{\wp''(y)}{\wp(x) - \wp(y)}. \tag{69}$$

Write $\wp(y) = -\frac{1}{3}$ and denote the invariants of elliptic functions by $g_2$, $g_3$, $\Delta$; i.e.,

$$g_2 = -4(e_1e_2 + e_2e_3 + e_3e_1) = 4\left(\frac{1}{3} - \frac{\alpha^2}{h_1^2}\right),$$

$$g_3 = 4e_1e_2e_3 = 4\left(\frac{2}{27} + \frac{2}{3}\frac{\alpha^2}{h_1^2} - \frac{\alpha^2 h_3^2}{h_1^2}\right),$$

$$\Delta = g_2^3 - 27g_3^2.$$

Then

$$[\wp'(y)]^2 = 4[\wp(y) - e_1][\wp(y) - e_2][\wp(y) - e_3] = \frac{4\alpha^2(h_3^2 - 1)}{h_1^2},$$

$$\wp''(y) = 6[\wp(y)]^2 - \tfrac{1}{2}g_2 = \frac{2\alpha^2}{h_1^2}.$$

Furthermore, by the formulas

$$\frac{d}{dx}\zeta(x) = -\wp(x), \qquad \frac{d}{dx}\log\sigma(x) = \zeta(x),$$

(68) and (69) are combined into

$$\int \frac{dx}{[\wp(x) - \wp(y)]^2} = -\zeta(x + y) - \zeta(x - y) - 2x\wp(y) + \frac{\wp''(y)}{\wp'(y)}\left[\log\frac{\sigma(x + y)}{\sigma(x - y)} - 2x\zeta(y)\right].$$

Hence the first formula of (67) becomes

$$\frac{h_1}{2\alpha^2}(s + \beta_0) = \frac{1}{[\wp'(y)]^2} \times \left\{-\zeta\left(\frac{\psi + \beta_1}{2} + y\right) - \zeta\left(\frac{\psi + \beta_1}{2} - y\right) - (\psi + \beta_1)\wp(y) + \frac{\wp''(y)}{\wp'(y)}\left[\log\frac{\sigma\left(\frac{\psi + \beta_1}{2} + y\right)}{\sigma\left(\frac{\psi + \beta_1}{2} - y\right)} - (\psi + \beta_1)\zeta(y)\right]\right\}. \tag{70}$$

Write $\wp(z) = \frac{2}{3}$; this gives $[\wp'(z)]^2 = 4\alpha^2 h_3^2/h_1^2$. Integrating the second equation of (67) gives

$$\frac{h_1}{\alpha^2 h_3}(c_0 t + \beta_3) = \frac{1}{\wp'(z)}\left\{(\psi + \beta_1)\zeta(z) - \log\frac{\sigma\left(\frac{\psi+\beta_1}{2} + z\right)}{\sigma\left(\frac{\psi+\beta_1}{2} - z\right)}\right\}$$

$$- \frac{1}{\wp'(y)}\left\{(\psi + \beta_1)\zeta(y) - \log\frac{\sigma\left(\frac{\psi+\beta_1}{2} + y\right)}{\sigma\left(\frac{\psi+\beta_1}{2} - y\right)}\right\}$$

$$\times\left\{1 - \frac{\wp''(y)}{[\wp'(y)]^2}\right\}$$

$$+ \frac{1}{[\wp'(y)]^2}\left\{\zeta\left(\frac{\psi+\beta_1}{2} + y\right) + \zeta\left(\frac{\psi+\beta_1}{2} - y\right)\right.$$

$$\left. + (\psi + \beta_1)\wp(y)\right\}.$$

Equations (70) and (71) show that the first and the third equations of (66) are Abelian integrals of the third kind.

By the formulas of elliptic functions (Halphen, 1886),

$$\wp(\omega) = e_1, \qquad \wp(\omega + \omega') = e_2, \qquad \wp(\omega') = e_3,$$

$$\zeta(\omega) = \eta, \qquad \zeta(\omega') = \eta',$$

$$\wp(w) = -\frac{\eta}{\omega} + \left(\frac{\pi}{2\omega}\right)^2 \frac{1}{\sin^2\frac{\pi w}{2\omega}} - 2\left(\frac{\pi}{\omega}\right)^2 \sum_{n=1}^{\infty} \frac{nq^{2n}}{1 - q^{2n}} \cos\frac{n\pi w}{\omega},$$

$$\zeta(w) = \frac{\eta w}{\omega} + \frac{\pi}{2\omega}\cot\frac{\pi w}{2\omega} + \frac{2\pi}{\omega}\sum_{n=1}^{\infty}\frac{q^{2n}}{1 - q^{2n}}\sin\frac{n\pi w}{\omega},$$

$$\log\sigma w = \frac{\eta w^2}{2\omega} + \log\left(\frac{2\omega}{\pi}\sin\frac{\pi w}{2\omega}\right) + 2\sum_{n=1}^{\infty}\frac{q^{2n}}{1 - q^{2n}}\frac{1 - \cos\frac{n\pi w}{\omega}}{n},$$

$$\eta\omega = \frac{1}{12}\pi^2 - 2\pi^2\sum_{n=1}^{\infty}\frac{q^{2n}}{(1 - q^{2n})^2},$$

$$q = \exp\left(-\frac{\pi\omega'}{\omega i}\right),$$

we transform (67) and (68) into the following series form:

$$\frac{h_1}{2\alpha^2}[\wp'(y)]^2(s+\beta_0)$$

$$=\left\{-\wp(y)-\frac{\eta}{\omega}+\frac{\wp''(y)}{\wp'(y)}\left[\frac{\eta y}{\omega}-\zeta(y)\right]\right\}(\psi+\beta_1)$$

$$-\frac{\pi}{\omega}\frac{\sin\frac{\pi}{2\omega}(\psi+\beta_1)}{\cos\frac{\pi y}{\omega}-\cos\frac{\pi}{2\omega}(\psi+\beta_1)}+\frac{\wp''(y)}{\wp'(y)}\log\frac{\sin\frac{\pi}{2\omega}\left(\frac{\psi+\beta_1}{2}+y\right)}{\sin\frac{\pi}{2\omega}\left(\frac{\psi+\beta_1}{2}-y\right)}$$

$$+\sum_{n=1}^{\infty}\left\{-\frac{\pi}{\omega}\cos\frac{n\pi}{\omega}y+\frac{\wp''(y)}{\wp'(y)}\frac{\sin\frac{n\pi}{\omega}y}{n}\right\}\frac{4q^{2n}}{1-q^{2n}}\sin\frac{n\pi}{2\omega}(\psi+\beta_1),$$

$$\frac{h_1}{\alpha^2h_3}(c_0t+\beta_3)$$

$$=\left\{\frac{1}{\wp'(z)}\left[\zeta(z)-\frac{\eta z}{\omega}\right]-\frac{1}{\wp'(y)}\left[1-\frac{\wp''(y)}{[\wp'(y)]^2}\right]\left[\zeta(y)-\frac{\eta y}{\omega}\right]\right.$$

$$\left.+\frac{1}{[\wp'(y)]^2}\left[\frac{\eta}{\omega}+\wp(y)\right]\right\}(\psi+\beta_1)$$

$$+\frac{\pi}{[\wp'(y)]^2\omega}\frac{\sin\frac{\pi}{2\omega}(\psi+\beta_1)}{\cos\frac{\pi y}{\omega}-\cos\frac{\pi}{2\omega}(\psi+\beta_1)}$$

$$-\frac{1}{\wp'(z)}\log\frac{\sin\frac{\pi}{2\omega}\left(\frac{\psi+\beta_1}{2}+z\right)}{\sin\frac{\pi}{2\omega}\left(\frac{\psi+\beta_1}{2}-z\right)}$$

$$+\frac{1}{\wp'(y)}\left[1-\frac{\wp''(y)}{[\wp'(y)]^2}\right]\log\frac{\sin\frac{\pi}{2\omega}\left(\frac{\psi+\beta_1}{2}+y\right)}{\sin\frac{\pi}{2\omega}\left(\frac{\psi+\beta_1}{2}-y\right)}$$

$$+\sum_{n=1}^{\infty}\left\{-\frac{1}{\wp'(z)}\frac{\sin\frac{n\pi z}{\omega}}{n}+\frac{1}{\wp'(y)}\left(1-\frac{\wp''(y)}{[\wp'(y)]^2}\right)\frac{\sin\frac{n\pi y}{\omega}}{n}\right.$$

$$\left.+\frac{\pi}{[\wp'(y)]^2\omega}\cos\frac{n\pi y}{\omega}\right\}\frac{4q^{2n}}{1-q^{2n}}\sin\frac{n\pi}{2\omega}(\psi+\beta_1),$$

$$u - \frac{1}{3} = -\frac{\eta}{\omega} + \left(\frac{\eta}{2\omega}\right) \frac{2}{1 - \cos \frac{\pi}{2\omega}(\psi + \beta_1)}$$

$$- 2\left(\frac{\pi}{\omega}\right)^2 \sum_{n=1}^{\infty} \frac{nq^{2n}}{1 - q^{2n}} \cos \frac{n\pi}{2\omega}(\psi + \beta_1).$$

The radius vector, longitude, and latitude are computed from these formulas and the relations

$$r = \frac{\alpha}{u},$$

$$\tan(\theta + \beta_2) = \tan \psi \cos I,$$

$$\cos \phi = \sin I \sin \psi.$$

## 2.19 DISCUSSION OF THE MOTION

From the properties of Weierstrass's $\sigma$-functions it is evident that $s$ and $t$ have singularities at

$$\begin{aligned} &\text{(A)} \quad \frac{\psi + \beta_1}{2} \pm y \equiv 0 \pmod{2\omega, 2\omega'}, \\ &\text{(B)} \quad \frac{\psi + \beta_1}{2} \pm z \equiv 0 \pmod{2\omega, 2\omega'}. \end{aligned} \tag{72}$$

Thus our integrals are of the third kind in Riemann's theory of algebraic functions (Section 2.20). The radius vector $r$ has singularities at $u = 0$, i.e.,

$$\wp\left(\frac{\psi + \beta_1}{2}\right) + \frac{1}{3} = 0, \quad \text{or} \quad \wp\left(\frac{\psi + \beta_1}{2}\right) = \wp(y);$$

and these singularities coincide with (A). Hence, when the three roots $e_1, e_2, e_3$ of the cubic equation $U(u) = 0$ are all different, these are the only singularities.

For $\Delta > 0$ the three roots are all real and $e_3 < e_2 < e_1$. We classify two cases:

(I) $h_3^2 > 1$, then $e_3 < -\frac{1}{3} < e_2 < e_1 < \frac{2}{3}$,

(II) $h_3^2 < 1$, then $-\frac{1}{3} < e_3 < e_2 < e_1 < \frac{2}{3}$.

For $\Delta < 0$, only one root is real and we have two cases:

(III) $h_3^2 > 1$, then $e_3 < -\frac{1}{3} < \frac{2}{3}$,

(IV) $h_3^2 < 1$, then $-\frac{1}{3} < e_k < \frac{1}{3}$ $(k = 1 \text{ or } 3)$.

We suppose that $x$ can take any real value extending to infinity.

Case I:

$$\wp'(y) = \frac{2\alpha}{h_1}\sqrt{h_3^2 - 1}, \qquad \wp'(z) = -\frac{2\alpha h_3}{h_1}, \qquad \wp''(y) = \frac{2\alpha^2}{h_1^2}.$$

Separate the real and imaginary parts:

$$y = y' + \omega', \qquad \beta_0 = \beta_0' + i\beta_0'', \qquad \beta_3 = \beta_3' + i\beta_3''.$$

Then

$$\frac{2(h_3^2 - 1)}{h_1}(s + \beta_0')$$
$$= \left\{\frac{1}{3} - \frac{\eta}{\omega} - \frac{2\alpha\pi}{\omega h_1\sqrt{h_3^2 - 1}}\sum_{n=1}^{\infty}\frac{q^n}{1 - q^{2n}}\sin\frac{n\pi y'}{\omega}\right\}(\psi + \beta_1)$$
$$- \sum_{n=1}^{\infty}\left\{\frac{4\pi}{\omega}\cos\frac{n\pi y'}{\omega} - \frac{4\alpha}{h_1\sqrt{h_3^2 - 1}}\frac{\sin\dfrac{2\pi y'}{\omega}}{n}\right\}$$
$$\times\frac{q^n}{1 - q^{2n}}\sin\frac{n\pi}{2\omega}(\psi + \beta_1),$$

$$\frac{2}{\alpha h_3}(c_0 t + \beta_3')$$
$$= -\left\{\frac{1}{h_3}\left[\zeta(z) - \frac{\eta z}{\omega}\right] + \frac{2h_3^2 - 3}{2(h_3^2 - 1)^{3/2}}\frac{2\pi}{\omega}\left(\sum_{n=1}^{\infty}\frac{q^n}{1 - q^{2n}}\sin\frac{n\pi y'}{\omega}\right)\right.$$
$$\left. - \frac{h_1}{2\alpha(h_3^2 - 1)}\left(\frac{\eta}{\omega} - \frac{1}{3}\right)\right\}(\psi + \beta_1)$$
$$+ \frac{1}{h_3}\log\frac{\sin\dfrac{\pi}{2\omega}\left(\dfrac{\psi + \beta_1}{2} + z\right)}{\sin\dfrac{\pi}{2\omega}\left(\dfrac{\psi + \beta_1}{2} - z\right)}$$
$$+ \sum_{n=1}^{\infty}\left\{\frac{4q^n}{h_3}\frac{\sin\dfrac{n\pi z}{\omega}}{n} + \frac{2(h_3^2 - 3)}{(h_3^2 - 1)^{3/2}}\frac{\sin\dfrac{n\pi y'}{\omega}}{n}\right.$$
$$\left. + \frac{2\pi h_1}{\alpha(h_3^2 - 1)\omega}\cos\frac{n\pi y'}{\omega}\right\}\frac{q^n}{1 - q^{2n}}\sin\frac{n\pi}{2\omega}(\psi + \beta_1),$$

$$u - \frac{1}{3} = -\frac{\eta}{\omega} + \left(\frac{\pi}{2\omega}\right)^2 \frac{2}{1 - \cos\frac{\pi}{2\omega}(\psi + \beta_1)}$$

$$- 2\left(\frac{\pi}{\omega}\right)^2 \sum_{n=1}^{\infty} \frac{nq^{2n}}{1 - q^{2n}} \cos\frac{n\pi}{\omega}(\psi + \beta_1),$$

$$\frac{2(h_3^2 - 1)}{h_1} \beta_0'' i = \frac{\alpha\pi}{h_1\sqrt{h_3^2 - 1}} i,$$

$$\frac{2}{\alpha h_3} \beta_3'' i = -\frac{2h_3^2 - 3}{2(h_3^2 - 1)^{3/2}} \pi i.$$

The series for $s$ has no singularity, but the series for $u$ does have singularities corresponding to collisions at the origin. The series for $t$ has singularities at

$$\psi_1 \equiv 2z - \beta_1 \pmod{4\omega}, \qquad \text{for } t \to \infty,$$

and

$$\psi_2 \equiv -2z - \beta_1 \pmod{4\omega}, \quad \text{for } t \to -\infty.$$

There are two kinds of trajectories in this case: (i) the trajectories in $\frac{2}{3} < x \leqq \infty$, i.e., $0 \leqq r < \alpha$; and (ii) those in $e_1 < x < \frac{2}{3}$, i.e., $\alpha < r \leqq \dfrac{\alpha}{e_1 + \frac{1}{3}}$. The first kind is an ejection orbit from the origin; either it asymptotically approaches $r = \alpha$, or it starts asymptotically from $r = \alpha$ in the infinite past and collides at the origin. The second kind starts asymptotically from $r = \alpha$ and reaches a maximum at $r = \dfrac{\alpha}{e_1 + \frac{1}{3}}$ and tends to $r = \alpha$ asymptotically again. However, there is at $r = \alpha$ a spherical surface through which any light ray cannot pass even in an infinite time, because the coefficient of $dt^2$ in the expression $ds^2$ of Schwarzschild becomes infinite at $r = \alpha$. This spherical surface $r = \alpha$ cannot lie outside the space occupied by the material, even if this has the maximum density known to present-day physics. For the surface $r = \alpha$ to lie on the surface of a celestial body, the mean density of the body must be of the order $10^{17}$ cgs. Stoner (1929, 1930) and Pilowski (1928) have deduced the density of a sphere composed of relativistically degenerate gas to be $10^{13}$. Andersen (1929, 1930) showed that density above $10^{17}$ cannot be attained at the center of a celestial body on the basis of relativistic cosmology. Oppenheimer and Volkoff (1939) estimated the density of a neutron star as $10^{15}$. Hence, we need not consider the sphere $r = \alpha$ for any physically realizable matter. We call the motion outside this sphere $r = \alpha$ *pseudo-elliptic*.

Case II: The situation is similar to Case I and the motion is still pseudo-elliptic.

Case III: The motion, which approaches the two branches of asymptotes at infinity, includes $r = \infty$. We call such motion *pseudo-hyperbolic*. In particular, if $e_3 = \frac{1}{3}$, the motion is *pseudo-parabolic*.

Case IV: Permissible motions are pseudo-elliptic.

Next, suppose that the values of $x$ do not extend to infinity, but that they are bounded between two finite limits.

Case I:

$$e_3 < -\tfrac{1}{3} < e_2 < e_1 < \tfrac{2}{3}, \qquad e_3 < x < e_2.$$

The trajectory has two asymptotes and lies between $r = \alpha/(e_2 + \frac{1}{3})$ and $\infty$. We call it *quasi-hyperbolic*. In particular, if $e_3 = -\frac{1}{3}$, it is *quasi-parabolic*. The point for the minimum $r$, that is, the pericenter, rotates in this relativistic trajectory, but in the corresponding Newtonian trajectories the pericenter is always at rest.

Case II:

$$-\tfrac{1}{3} < e_3 < e_2 < e_1 < \tfrac{2}{3}, \qquad e_3 < x < e_2.$$

Let

$$\psi = \psi' + 2\omega', \qquad y = iy', \qquad \beta_0 = \beta_0' + i\beta_0'', \qquad \beta_3 = \beta_3' + \beta_3'',$$

then

$$\wp'(y) = -i\frac{2\alpha}{h_1}\sqrt{1 - h_3^2}, \qquad \wp'(z) = -\frac{2\alpha h_3}{h_1}, \qquad \wp''(y) = \frac{2\alpha^2}{h_1^2},$$

and

$$\begin{aligned}\frac{2(h_3^2 - 1)}{h_1}(s + \beta_0') = {} & \left\{\frac{1}{3} - \frac{\eta}{\omega} - \frac{\alpha}{h_1\sqrt{1 - h_3^2}}\left[\frac{\eta y'}{\omega} + \zeta(y')\right]\right\}(\psi' + \beta_1) \\ & - \sum_{n=1}^{\infty}\left\{\frac{4\pi}{\omega}\cosh\frac{n\pi y'}{\omega} + \frac{4\alpha}{h_1\sqrt{1 - h_3^2}}\frac{\sinh\dfrac{n\pi y'}{\omega}}{n}\right\} \\ & \times \frac{q^n}{1 - q^{2n}}\sin\frac{n\pi}{2\omega}(\psi' + \beta_1),\end{aligned}$$

$$\frac{2}{\alpha h_3}(c_0 t + \beta_3') = \left\{-\frac{1}{h_3}\left[\zeta(z) - \frac{\eta z}{\omega}\right] + \frac{2h_3^2 - 3}{2(1 - h_3^2)^{3/2}}\left[\zeta(y') + \frac{\eta y'}{\omega}\right] + \frac{h_1}{2\alpha(h_3^2 - 1)}\left(\frac{\eta}{\omega} - \frac{1}{3}\right)\right\}(\psi' + \beta_1)$$

$$+ \sum_{n=1}^{\infty}\left\{\frac{4}{h_1}\frac{\sin\frac{n\pi z}{\omega}}{n} - \frac{2(h_3^2 - 3)}{(1 - h_3^2)^{3/2}}\frac{\sinh\frac{n\pi y'}{\omega}}{n} + \frac{2\pi h_1}{\alpha(h_3^2 - 1)\omega}\cosh\frac{n\pi y'}{\omega}\right\}$$

$$\times \frac{q^n}{1 - q^{2n}}\sin\frac{n\pi}{2\omega}(\psi' + \beta_1),$$

$$u - \frac{1}{3} = -\frac{\eta}{\omega} - 2\left(\frac{\pi}{\omega}\right)^2 \sum_{n=1}^{\infty}\frac{nq^n}{1 - q^{2n}}\cos\frac{n\pi}{2\omega}(\psi' + \beta_1).$$

There is no singularity in this motion. The motion has only points of libration, such as discussed in Section 2.4, and $e_3 \leqq x \leqq e_2$, or $\alpha/(e_2 + \frac{1}{3}) \leqq r \leqq \alpha/(e_3 + \frac{1}{3})$. The period of the motion with respect to $\psi$ is $4\omega$ but the period with respect to $t$ is

$$\frac{2\omega\alpha h_3}{c_0}\left\{-\frac{1}{h_3}\left[\zeta(z) - \frac{\eta z}{\omega}\right] + \frac{2h_3^2 - 3}{2(1 - h_3^2)^{3/2}}\left[\zeta(y') + \frac{y'\eta}{\omega}\right] + \frac{h_1}{2\alpha(h_3^2 - 1)}\left(\frac{\eta}{\omega} - \frac{1}{3}\right)\right\}.$$

The points of maximum and minimum $r$ rotate with a constant rate. This trajectory is called *quasi-elliptic.* When $4\omega < 2\pi$, the motion is the same in form as those of Einstein, Schwarzschild, and de Sitter. In general, the period for $\psi$ does not equal the period for $r$, the former being $4\omega$ and the latter being $2\pi$. When the ratio of these two periods is an irrational number, the motion is quasi-elliptic and the trajectory fills up everywhere densely the ring region bounded by the two circles, respectively, with the maximum and the minimum $r$ as the radii. It is what Poincaré called *quasi-ergodic.* Hence, it is *stable in Poisson's sense* (Poincaré, 1885, 1899; Hagihara, 1927), which will be discussed in a later chapter. The motion of the pericenter, the point of the minimum $r$, is approximately represented by Einstein's formula

$$\frac{24\pi^3 a^2}{c_0^2 T_0^2(1 - e^2)},$$

if $\pi/2\omega$ is approximately equal to 1. For a general value of $\pi/2\omega$ the period can be written

$$T_0 = \frac{2\sqrt{6}\pi\mathbf{a}}{c_0\left[1 - \left(\frac{\pi}{2\omega}\right)^2\right]^{1/2}}\left\{1 + \frac{\mathbf{e}^2}{1728}\,\frac{\Psi\left(\frac{2\omega}{\pi}\right)}{\left(\frac{\pi}{2\omega}\right)^2\left[2 + \left(\frac{\pi}{2\omega}\right)^2\right]\left[1 + 2\left(\frac{\pi}{2\omega}\right)^2\right]^2}\right\},$$

where

$$\Psi\left(\frac{\pi}{2\omega}\right) = 1685 + 6510\left(\frac{\pi}{2\omega}\right)^2 + 9369\left(\frac{\pi}{2\omega}\right)^2 + \cdots.$$

The third law of Kepler takes the form

$$T_0^2 = \frac{8\pi^2}{c_0^2}\frac{\mathbf{a}^3}{\alpha}\left\{1 + \frac{\alpha^2\mathbf{e}^2}{192\mathbf{a}^2}\,\frac{\left(\frac{\pi}{2\omega}\right)^2\Psi\left(\frac{\pi}{2\omega}\right)}{\left[1 - \left(\frac{\pi}{2\omega}\right)^2\right]\left[2 + \left(\frac{\pi}{2\omega}\right)^2\right]\left[1 + 2\left(\frac{\pi}{2\omega}\right)^2\right]^2}\right\}^2,$$

where **a** and **e** correspond, respectively, to the semimajor axis and the eccentricity of the orbit in the Keplerian motion and are similar to those in Gyldén's absolute orbit, which will be described later. Hagihara (1931) has extended the notion of Bessel functions to an infinite number of arguments and obtained the general solution of the relativistic trajectories in the form of infinite series. Similarly, for the Newtonian two-body problem, Hagihara obtained several theorems based on the expansion for a quasi-elliptic motion in the relativistic one-body problem (Section 5.13).

If two of the three roots $e_1$, $e_2$, $e_3$ coincide, then asymptotic motions appear, as in the limiting cases of libration discussed in Section 2.4, though more complicated as the motion is in two dimensions. It is known that elliptic functions degenerate to trigonometric or hyperbolic functions when two of $e_i$ coincide. If one root in the real domain coincides with another root, then the motion makes a spiral approach to the circle corresponding to these coincident roots. As the motion is asymptotic to the circle $r = \alpha$, the motion is called *pseudo-spiral*. If the two roots bounding the real domain coincide, then the motion is circular and is called *pseudo-circular*. If two of the $e_i$ coincide in Case II, then the circle with the coincident roots as the radius is an asymptotic orbit for $t \to -\infty$

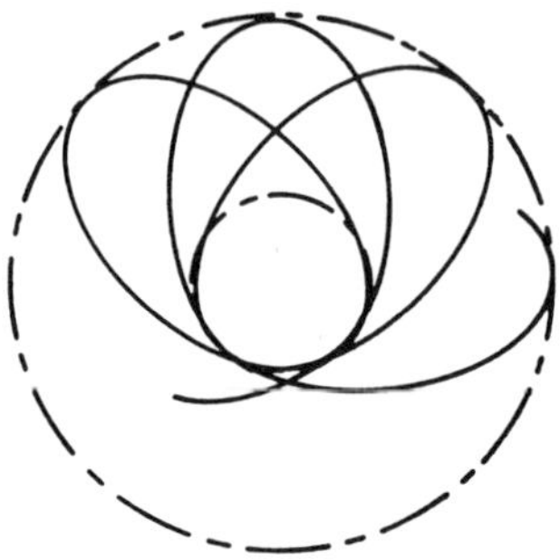

FIGURE 1

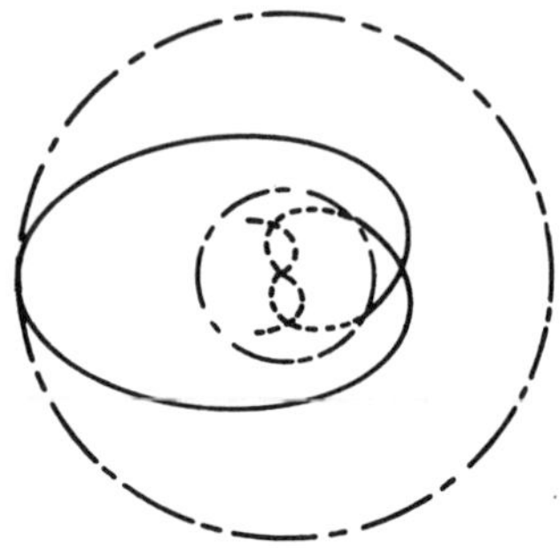

FIGURE 2

and for $t \to +\infty$. The circle itself being a pseudo-circle, the motion is what Poincaré (1899) called *doubly asymptotic* and the circle what he called the *limit cycle* (Poincaré, 1885); this will be described in a later chapter.

Summarizing:

If the three roots are all different, then the motions and their respective conditions are:

quasi-elliptic $\quad \dfrac{\alpha}{e_2 + \frac{1}{3}} \leqq r \leqq \dfrac{\alpha}{e_3 + \frac{1}{3}}$ (Figure 1)

quasi-parabolic $\quad \dfrac{\alpha}{e_2 + \frac{1}{3}} \leqq r \leqq \infty$

quasi-hyperbolic $\quad \dfrac{\alpha}{e_2 + \frac{1}{3}} \leqq r < \infty$

pseudo-elliptic $\quad \alpha < r \leqq \dfrac{\alpha}{e_1 + \frac{1}{3}}$ (Figure 2)

pseudo-parabolic $\quad \alpha < r < \infty$

pseudo-hyperbolic $\quad \alpha < r \leqq \infty.$

If two of the three roots coincide, then:

quasi-elliptic spiral $\quad \dfrac{\alpha}{e_A + \frac{1}{3}} < r < \dfrac{\alpha}{e_3 + \frac{1}{3}}, \qquad e_A = e_1 = e_2$

(doubly asymptotic motion, Figure 3)

quasi-parabolic spiral $\quad \dfrac{\alpha}{e_A + \frac{1}{3}} < r \leqq \infty$

quasi-hyperbolic spiral $\quad \dfrac{\alpha}{e_A + \frac{1}{3}} < r \leqq \infty$

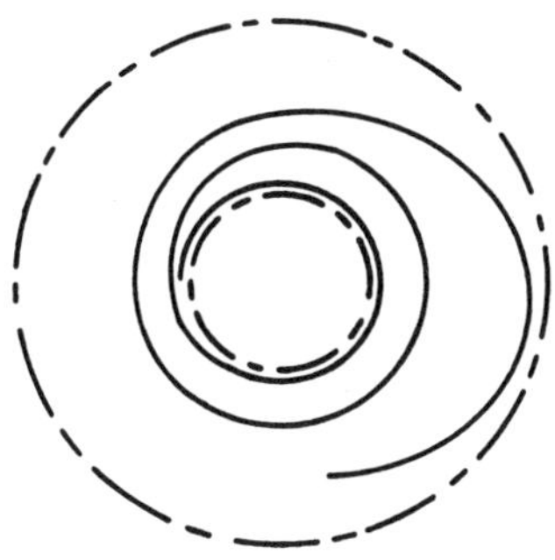

FIGURE 3

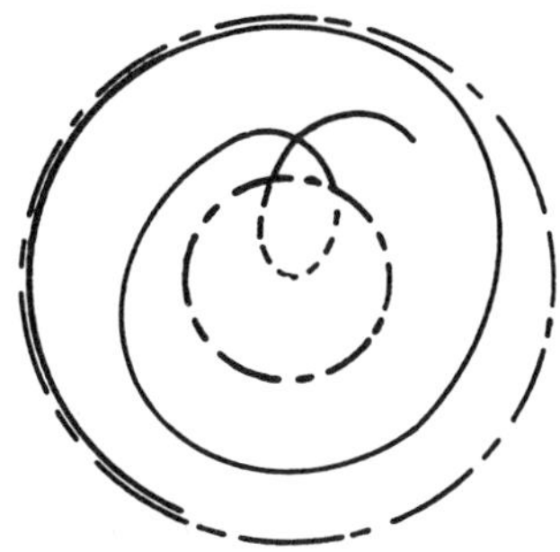

FIGURE 4

pseudo-spiral $\alpha < r < \dfrac{\alpha}{e_B + \frac{1}{3}}, \quad e_B = e_2 = e_3$ (Figure 4)

pseudo-circular $r = 3\alpha, \qquad e_1 = e_2 = e_3 = 0$

quasi-circular $r = \dfrac{\alpha}{e_B + \frac{1}{3}}.$

A more detailed discussion is found in Hagihara's paper (1931).

## 2.20 ALGEBRAIC INTEGRALS

The function $y$ defined by a rational polynomial

$$f(y, x) = A_0 y^n + A_1 y^{n-1} + \cdots + A_{n-1} y + A_n = 0 \tag{73}$$

with functions $A_0, A_1, \ldots, A_n$ of $x$ as coefficients is called an *algebraic function* of $x$. Consider an integral called an *Abelian integral*

$$\int_a^x R(y, x)\, dx \tag{74}$$

of a rational algebraic function $R(y, x)$ of $x$ and $y$ integrated along a path from a point $a$ to another point $x$ of a complex plane (Appell and Goursat, 1895; Forsyth, 1893; Stahl, 1896; Baker, 1897, 1907; Stahl and Noether, 1911; Picard, 1905, 1931; Osgood, 1932; Bliss, 1933; Conforto, 1956). As a simple example, consider the hyperelliptic case

$$y^2 = f(x), \qquad f(x) = \lambda_0 + \lambda_1 x + \cdots + \lambda_6 x^6. \tag{73a}$$

If $x$ is near $a$ and $a$ is not a root of (73a), then we can solve (73a) by substituting $x = a + t$, $y = P(t)$. If $P(t)$ is a power series arranged in positive powers of $t$, and $a$ is a branch point where $x$ and $y$ are both finite, then we solve (73a) by putting $x = a + t^2$, $y = P(t)$. We examine

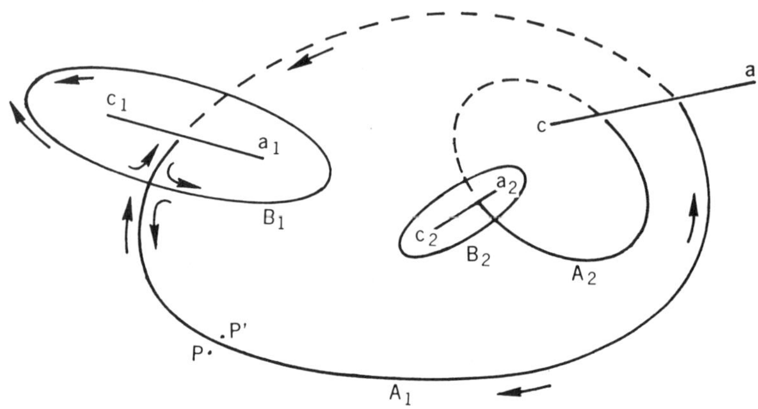


FIGURE 5

the behavior at the point at infinity by putting $x = t^{-1}$, $y = t^{-3}P(t)$. A rational function $R(y, x)$ of $x$ and $y$ can be written generally in the form $R(y, x) = t^{-m}Q(t)$ near a branch point, where $m$ is a positive or negative integer or zero, and $Q(t)$ does not vanish at $t = 0$. If $m < 0$, then $R(y, x)$ has a zero point of order $-m$ at $t = 0$. Then the Abelian integral (74) can be written $t^{-m}Q(t) + C \log t$ with a constant $C$. If $C$ is zero everywhere and $m < 0$, the integral is called an integral of the *first kind*. If $C = 0$ but $m > 0$ at any finite point, the integral possesses poles—that is, a finite number of algebraic infinity points—and is called an integral of the *second kind*. A function that possesses only poles is said to be *meromorphic*. It can be shown that there exist only a finite number of points at which $m > 0$ or $C \neq 0$, and that the sum of the values of $C$ for various points is zero. If $m \geqq 0$ and $C$ is nonzero at two or more points, then the integral is called an integral of the *third kind*.

Integrals of the first and the second kinds are not *uniform*, that is, they are not single-valued but multiform. To make the integral uniform (*uniformization*), we cut the Riemann surface along closed curves in the definite sense of description. Such a curve is called a *cross cut*. Figure 5 shows as an example the hyperelliptic case defined by (73a). Let the points $a, c; a_1, c_1; a_2, c_2$ be the branch points. Draw the curve segments connecting $a$ and $c$, $a_1$ and $c_1$, $a_2$ and $c_2$. These are called the *branch cuts* (*Verzweigungschnitt*). Two sheets of the Riemann surface intersect along these curve segments and the upper and the lower sheets interchange along these segments. In general, $m$ sheets branch out and interchange one after the other if $y^m = x$. Draw the closed curves $A_1$ and $A_2$, crossing the segments $ac$ and $a_1c_1$, and $ac$ and $a_2c_2$, respectively, and the closed curves $B_1$ and $B_2$ enclosing $a_1c_1$ and $a_2c_2$, respectively. After crossing the branch cut $ac$, $a_1c_1$, or $a_2c_2$, the curves $A_1$ and $A_2$ pass to the other sheet of the Riemann surface in order to complete the circuits. The cross cuts are $A_1$, $A_2$, $B_1$, $B_2$ in this case.

Let $A_1$ be the circuit round along the closed curve $A_1$ in the positive direction, and $A_1^{-1}$ be the circuit along $A_1$ in the negative direction. The path $A_1B_1A_1^{-1}B_1^{-1}$ forms the boundary of the Riemann surface cut along the cross cuts. Similarly, form the path $A_2B_2A_2^{-1}B_2^{-1}$. Along either of these two nonintersecting boundaries $A_1B_1A_1^{-1}B_1^{-1}$ and $A_2B_2A_2^{-1}B_2^{-1}$, the integrals (74) of the first and the second kinds are uniform. The value of the integral at a point $P$ just outside $A_1$ differs by $\Omega_1$ from its value at a point $P'$ opposite to $P$ and just inside $A_1$. This value $\Omega_1$ is the same all along the curve $A_1$. Similarly the value of integral (74) just outside $B_1$ differs by $\Omega_1'$ from its value at a point opposite and just inside $B_1$. Similar differences $\Omega_2$ and $\Omega_2'$ are obtained for $A_2$ and $B_2$. The four values $\Omega_1$, $\Omega_1'$, $\Omega_2$, $\Omega_2'$ are called the *periods* of the integral. Before the cut is made, the value of integral (74) is expressed by

$$\int_a^x R(y, x)\, dx + h_1\Omega_1 + h_2\Omega_2 + h_1'\Omega_1' + h_2'\Omega_2',$$

with integers $h_1$, $h_2$, $h_1'$, $h_2'$.

Since a term $C \log t$ appears in the expansion of an integral of the third kind, the value of the integral increases by $2\pi iC$ by going along a circuit round the point $t = 0$. If there are two logarithmic singularities with $C$'s of opposite sign, then the integral can be made uniform by drawing cross cuts connecting the two singular points without crossing any of the foregoing cross cuts. It can be seen that the sum of the values of $C$ is zero.

If a closed curve drawn on a surface can be contracted to a single point—without passing through any branch point—remaining always on the surface, then the surface is said to be *simply connected.* A Riemann surface can be made to consist of $n$ separate simply connected sheets by means of $q$ cross cuts. The difference $q - n$ is fixed for a given Riemann surface. The integer defined by $N = q - n + 2$ is called the *order of connectivity*; and the surface, *N-ply connected.* The surface of a sphere is simply connected, as $q = 0$, $n = 1$, $N = 1$. The surface of an open cylinder is made to be simply connected by cutting along a generator, so that $q = n = 1$, $N = 2$; thus the uncut surface is doubly connected. The surface of a torus is triply connected, as $q = 2$, $n = 1$, $N = 3$. A plane area with $m$ contours is $m$-ply connected. In general, a surface of the order of connectivity $N$ can be transformed into a simply connected

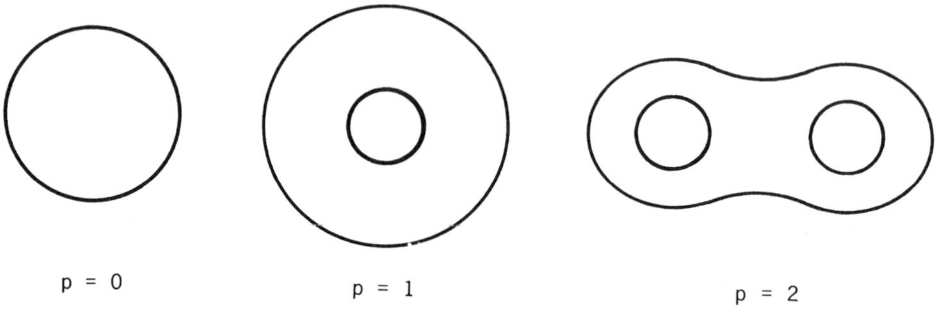


FIGURE 6

surface by means of $N - 1$ cuts. The number of the boundaries of a multiply connected surface increases or decreases by 1 by drawing a cross cut.

The boundary of a simply connected surface—or of a surface *made* simply connected by a suitable number of cross cuts—consists of one closed curve. The number of different boundaries of an $N$-ply connected surface is equal to $N$ less zero or an even integer. Hence the order of connectivity of a closed surface with only one boundary is equal to an odd integer. Write $N = 2p + 1$. The integer $p$ is called the *genus* (*Geschlecht*) of the surface. In the case of a spherical surface the genus $p = 0$. In the elliptic case $y^2 = 4x^3 - g_2x - g_3$, as a special case of (73), $p = 1$, and the surface is a torus (Figure 6). In the hyperelliptic case represented by (73a), $p = 2$.

If we take the upper and the lower limits of the integral (74) at two points on the opposite sides of a cross cut on a surface of genus $p$, then the value of the integral is constant along the cross cut until another cross cut is met; let this value be $\Omega$. The number of the different values of $\Omega$ equals the order of connectivity $N$. The value of the integral before the cross cuts are made can be written

$$\int_a^x R(y, x)\, dx + h_1\Omega_1 + h_2\Omega_2 + \cdots + h_N\Omega_N,$$

where $h_1, h_2, \ldots, h_N$ are positive or negative integers or zero.

A pair of corresponding values of $x$ and $y$ is called a *place*. Different values of $y$ may be related by (73) to the same value of $x$; they are located on different sheets of the Riemann surface. We call a *place* the corresponding points on different sheets having the same values of $x$ and $y$. There are $p$ linearly independent integrals for an integral of the first kind (which is finite everywhere). Let a value of the integral on the

Riemann surface be $W$; then the other values are expressed by $W + \sum_{r=1}^{2p} m_r\omega_r$ with integral values of $m_r$. There are $p + 1$ linearly independent integrals for an integral of the second kind (which has poles but no logarithmic singularity). All its values are expressed in the form $\Gamma + \sum_{r=1}^{2p} n_r\epsilon_r$ with integers $n_r$. Suppose that an integral of the third kind has $\mu$ logarithmic singularities in addition to poles, then the general value of the integral is represented by $\Pi + \sum_{r=1}^{2p} m_r\varpi_r + 2\pi i \sum_{q=1}^{\mu} m_q'C_q$, where $C_1 + C_2 + \cdots + C_\mu = 0$. The constants $\omega_r$, $\epsilon_r$, and $\varpi_r$ are called the *periods* with regard to the $r$th cross cut.

Since there are $p$ linearly independent functions for an integral of the first kind, any function of the first kind equals the sum of $\sum_{r=1}^{p} c_r w_r$ and a constant, where $c_1, c_2, \ldots, c_p$ are constants and $w_r$ are the $p$ linearly independent functions. Denote the periods along these $p$ cross cuts by $\omega_{r1}, \omega_{r2}, \ldots, \omega_{rp}$ $(r = 1, 2, \ldots, p)$, respectively. It can be shown that the determinant $\|\omega_{rs}\|$ is not zero. Hence, by suitably choosing $c_1, c_2, \ldots, c_p$, we can find linear combinations $W_1, W_2, \ldots, W_p$ of $\omega_1, \omega_2, \ldots, \omega_p$ such that the periods are transformed as shown in Table 1.

TABLE I

| | $\Omega_1$ | $\Omega_2$ | $\cdots$ | $\Omega_p$ | $\Omega_1'$ | $\Omega_2'$ | $\cdots$ | $\Omega_p'$ |
|---|---|---|---|---|---|---|---|---|
| $W_1$ | $\pi i$ | 0 | $\cdots$ | 0 | $a_{11}$ | $a_{12}$ | $\cdots$ | $a_{1p}$ |
| $W_2$ | 0 | $\pi i$ | $\cdots$ | 0 | $a_{21}$ | $a_{22}$ | $\cdots$ | $a_{2p}$ |
| $\vdots$ | | | | | | | | |
| $W_p$ | 0 | 0 | $\cdots$ | $\pi i$ | $a_{p1}$ | $a_{p2}$ | $\cdots$ | $a_{pp}$ |

This form of periods is called the *normal form* and the integrals $W_r$ the *normal integrals*. The constants $a_{rs}$ are not independent. First, $a_{rs} = a_{sr}$. Furthermore, if we put $a_{rs} = \rho_{rs} + i\sigma_{rs}$, then $\sigma_{rs} = \sigma_{sr}$, $\rho_{rs} = \rho_{sr}$ and, as can be proved, $\rho_{11}c_1^2 + 2\rho_{12}c_1c_2 + \rho_{22}c_{22}^2 + \cdots + \rho_{pp}c_p^2$ is negative unless all $c_i$ are zero at the same time.

Functions of the second kind are determined by the poles and the real parts of the periods. It can be shown that the periods can be so arranged that they are zero along the first $p$ cross cuts $A_1, A_2, \ldots, A_p$ but are algebraic functions at the poles along the remaining $p$ cross cuts $B_1, B_2, \ldots, B_p$; that is, the periods have the values $2dW_r/dx$ at simple poles, where $dW_r/dx$ are algebraic functions. Suppose that the function $\Gamma_a(x; c)$ has period zero along the cross cuts $A_1, A_2, \ldots, A_p$ of type $A$ on the Riemann surface cut by $2p$ cross cuts. Let the zero places of this function be $c_1, c_2, \ldots, c_p$, and the places of simple poles be $a$. These values are determined by the values of $x$ and $y$ satisfying (73). The places with these pairs of values $(x, y)$ are on each sheet of the Riemann surface. Let the periods with regard to the cross cuts $B_1, B_2, \ldots, B_p$ of type $B$ be $\Omega_1, \Omega_2, \ldots, \Omega_p$. Such a function $\Gamma_a(x; c)$ is a Riemann's normal integral of the second kind.

An integral of the third kind is uniform everywhere on the cut Riemann surface. It has periods along the cross cuts $B_1, B_2, \ldots, B_p$ of type $B$, but no period along the cross cuts $A_1, A_2, \ldots, A_p$ of type $A$. Denote by $\Pi_{x_1 x_2}(x; c)$ a function that is finite and continuous everywhere except for the logarithmic singularities at $(x_1, y_1)$ and $(x_2, y_2)$. It is proved that $\Pi_{x_1 x_2}(x; c) = \Pi_{xc}(x_1; x_2)$. This is a Riemann's normal integral of the third kind. A general integral of the third kind is expressed by

$$\Pi_{x_1 x_2}(x; c) + \sum_{i=1}^{p} \sum_{j=1}^{p} A_{ij} W_i(x; c) W_j(x_1; x_2), \qquad A_{ij} = A_{ji},$$

The periods are

$$0, 0, \ldots, 0, 2\pi i W_1(x_1; x_2), \ldots, 2\pi i W_p(x_1; x_2).$$

Consider two neighboring points $x_1$ and $x_1'$, and put $x_1' - x_1 = t_{x_1}^{m+1}$. Write the limit of

$$[\Pi_{x_1' x_2}(x; c) - \Pi_{x_1 x_2}(x; c)]/t_{x_1}$$

for $t_{x_1} \to 0$ as $D_{t_{x_1}} \Pi_{x_1 x_2}(x; c)$. Then we have

$$D_{t_{x_1}} \Pi_{x_1 x_2}(x; c) = \Gamma_{x_1}(x; c).$$

Thus an integral of the second kind is obtained from the derivative of an integral of the third kind.

The places on various branch sheets corresponding to a point $x = a$ are denoted by

$$x - a = t_l^{m_1 + 1}, \ldots, x - a = t_k^{m_k + 1}. \tag{73b}$$

By the *dimension* of a function $F$ is meant the order of infinity of $F$ at $x = a$; or, more precisely, the smallest positive integer $\sigma + 1$ for which $(x - a)^{\sigma + 1} F$ is finite at $x = a$ on all the sheets. Suppose that rational functions $h_1, h_2, \ldots, h_{n-1}$ become infinite only at $x = a$, and that $(x - a)^{\sigma_1 + 1} h_1, \ldots, (x - a)^{\sigma_{n-1} + 1} h_{n-1}$ are finite at $x = a$ and become infinite only at $x = \infty$. It can be shown that the dimension of $g_i \equiv (x - a)^{\sigma_i + 1} h_i$ is equal to the dimension of $h_i$. Let $(x, 1)_l$ be a polynomial of degree $l$ in $x$. An arbitrary *integral function*—that is, a function $F$ with no singularity other than that at $x = \infty$ — can be written

$$F = (1, x)_\mu + (1, x)_{\mu_1} g_1 + \cdots + (1, x)_{\mu_i - 1} g_{n-1}.$$

The sum of the dimensions of $g_1, g_2, \ldots, g_{n-1}$ is proved to be equal to $p + n - 1$. There exists an integral polynomial in $x, \beta_1, \ldots, \beta_{n-1}$ for which $\beta_i y^i$ are integral functions. By representing such a polynomial, in the same manner as $F$, as a function of linearly independent functions $g_1, g_2, \ldots, g_{n-1}$, we obtain in general

$$g_i(y, x) \equiv (\mu_{i,n-1}y^{n-1} + \cdots + \mu_{i,1}y + \mu_i)/D_i,$$

where $\mu_{i,n-1}, \ldots, \mu_{i,1}, \mu_i, D_i$ are integral polynomials in $x$.

Let the equation (73) defining the given Riemann surface be $f(y,x) = 0$. We write this in the form $\chi_n(y, x) = 0$, in which

$$\chi_i(y, x) \equiv A_0 y^i + A_1 y^{i-1} + \cdots + A_{i-1}y + A_i.$$

Furthermore, by equating various powers of $y$ on both sides, we define the functions $\phi_0, \phi_1, \ldots, \phi_{n-1}$, such that

$$\begin{aligned}\phi_0(y', x) &+ \phi_1'(y', x)g_1(y, x) + \cdots + \phi_{n-1}(y', x)g_{n-1}(y, x)\\ &= [f(y', x) - f(y, x)]/(y' - y)\\ &= y^{n-1}\chi_0 + y^{n-2}\chi_1(y', x) + \cdots + y\chi_{n-2}(y', x) + \chi_{n-1}(y', x).\end{aligned}$$

Then an integral of the third kind can be written

$$P_{x_1x_2}(x; c) = \int_c^x dx[(x, x_1) - (x, x_2)],$$

where

$$(x, \xi) = \frac{\phi_0(y, x) + \phi_1(y, x)g_1(y, \xi) + \cdots + \phi_{n-1}(y, x)g_{n-1}(y, \xi)}{(x - \xi)f'(y)}.$$

This expression for an integral of the third kind with logarithmic singularities at $(x_1, y_1)$ and $(x_2, y_2)$ is written in terms of the coefficients of (73).

Let $x = a$ be a zero point on the Riemann surface, and denote the linearly independent integrals of the first kind by $u_1(x; a), \ldots, u_p(x; a)$. Suppose that the derivatives $D_t u_1(x; a), \ldots, D_t u_p(x; a)$ are finite everywhere on the Riemann surface. Suppose, further, that we choose $p$ fixed places $c_1, c_2, \ldots, c_p$ on the Riemann surface such that the linear expression

$$\lambda_1 D_t u_1(x; a) + \cdots + \lambda_p D_t u_p(x; a)$$

with constant coefficients $\lambda_1, \lambda_2, \ldots, \lambda_p$ does not vanish at any place $c_1, c_2, \ldots, c_p$, but that one of such linear expressions does vanish for a pair of values of $\lambda$. Denote by $\omega_i(x)$ a linear expression that vanishes at all places $c_1, c_2, \ldots, c_p$ except a place $c_i$, where it equals 1. Write

$\mu_i(x)$ in place of $D_t u_i(x; a)$, and let $\nu_{ij}$ be the minor of the determinant formed of $\mu_j(c_i)$, then

$$\omega_i(x) = \nu_{i1}\mu_1(x) + \cdots + \nu_{ip}\mu_p(x) \qquad (i = 1, 2, \ldots, p).$$

The general integral of the first kind is written with this notation:

$$V_i(x; a) = \int_a^x \omega_i(t)\, dt_x,$$

where $t_x$ is defined as in (73b). Let

$$\Gamma_z(x; a) - [(z, x) - (z, a)]\frac{dz}{dt} \equiv H_z(x; a).$$

Then

$$H_z(x; a) = \sum_{i=1}^{p} \omega_i(z) H_{c_i}(x; a),$$

where $\Gamma_z(x; a)/(dz/dt)$ is a rational function of $z$ of the form

$$\Gamma_z(x; a) = [(z, x) - (z, a)]\frac{dz}{dt} + \sum_{i=1}^{p} \omega_i(z) H_{c_i}(x; a).$$

Integrate with respect to $z$; the function

$$\Pi_{z,c}(x; a) = P_{x,a}(z; c) + \sum_{i=1}^{p} V_i(z; c) H_{c_i}(x; a)$$

is an integral of the third kind, where $c$ is an arbitrary place. The general form of an integral of the second kind is

$$\Gamma_z(x; a) = D_z \Pi_{z,c}(x; a) = E_z(x; a) + \sum_{i=1}^{p} V_i(x; a) D_z H_{c_i}(z; c),$$

where $E_z(x; a)/(dz/dt)$ is a rational function of $z$. Denote by $\kappa$ an arbitrary symmetric matrix whose elements are $\kappa_{ij}$, and consider $p$ functions $L_i(x; a)$ defined by

$$L_i(x; a) = \sum_{j=1}^{p} \nu_{ji} H_{c_j}(x; a) - 2\sum_{j=1}^{p} \kappa_{ij} u_j(x; a).$$

This form is called the fundamental integral of the second kind associated with the integrals $u_1(x; a), \ldots, u_p(x; a)$. Multiply $L_i(x; a)$ by $\mu_i(z)$ and sum over $i$, then

$$\sum_{i=1}^{p} \mu_i(z) L_i(x; a) = H_z(x; a) - 2\sum_{i=1}^{p}\sum_{j=1}^{p} \kappa_{ij} u_i(x; a)\mu_j(z),$$

which is independent of the places $c_1, c_2, \ldots, c_p$. Substituting these $p$ independent places for $z$ successively, we obtain $p$ functions $L_i(x; a)$. In terms of these functions, the integrals of the second kind are represented by

$$\Gamma_z(x; a) = [(z, x) - (z, a)] \frac{dz}{dt} + \sum_{i=1}^{p} \mu_i(z) L_i(x; a)$$

$$+ 2 \sum_{i=1}^{p} \sum_{j=1}^{p} \kappa_{ij} u_i(x; a) u_j(z),$$

and the integrals of the third kind by

$$\Pi_{z,c}(x; a) = P_{z,c}(x; a) + \sum_{i=1}^{p} u_i(x; a) L_i(z; c)$$

$$+ 2 \sum \sum \kappa_{ij} u_i(x; a) u_j(z; c).$$

Let $w_1, w_2, \ldots, w_p$ be $p$ independent functions of the normal form of the first kind on the Riemann surface of genus $p$. Consider $q$ $(q \leqq p)$ independent variables $x_1, x_2, \ldots, x_q$ and construct the functions

$$v_r = w_r(x_1) + w_r(x_2) + \cdots + w_r(x_q) \qquad (r = 1, 2, \ldots, q).$$

The functions $v_r$ are multiform, that is, many-valued, with constant periods $\omega_{r1}, \omega_{r2}, \ldots, \omega_{r,2p}$ which are the same as those of $w_r$. Since $w_r(x_m)$ differs from $v_r$ by integral multiples of the periods, we can write $v_r$ in the general form

$$v_r + m_1 \omega_{r,1} + m_2 \omega_{r,2} + \cdots + m_{2p} \omega_{r,2p} \qquad (r = 1, 2, \ldots, p),$$

where the integers $m_s$ $(s = 1, 2, \ldots, 2p)$ are the same for the $q$ functions $v_r$. If $q = p$, then $r_1, r_2, \ldots, r_p$ reduce to uniform functions of the independent variables $x_1, x_2, \ldots, x_p$ when suitable integral multiples of the periods are added or subtracted. Conversely, $x_1, x_2, \ldots, x_p$, regarded as functions of $v_1, v_2, \ldots, v_p$, possess $2p$ pairs of periods $\omega_{1s}, \omega_{2s}, \ldots, \omega_{ps}$ $(s = 1, 2, \ldots, 2p)$. This procedure is called the *inversion* (*Umkehrung*) of Abelian integrals.

The fundamental theorem in the theory of algebraic functions (which Schwarz and Klein have proved in an elegant manner by referring to potential theory) states that there exists a function which is finite at all places on the Riemann surface and is uniform and continuous except along the cross cuts drawn to make the surface simply connected, and for which the real parts of the periods along the cross cuts are of the given constant values; and that there exists a function having the singularities of the given character at the given places on the Riemann surface and having the above-described properties at any other places. According to this fundamental existenc theorem, the functions with the properties in question exist; in particular, functions exist with the given poles, whose periods possess all zero real parts. We need the condition that the imaginary parts of the periods shall also be zero. Consider, for example, poles of the first order $c_1, c_2, \ldots, c_m$. Such functions are algebraic and finite, uniform and continuous, at all places except the poles. Let $Z_r$ be a function of the second kind with the place $c_r$ as the only pole. The condition for the vanishing of the periods along the cross cuts of types $A_1$ and $A_2$ is

$$y = \beta_1 Z_1 + \cdots + \beta_m Z_m + \beta_0.$$

The condition for the vanishing of the periods along the cross cuts of types $B_1$ and $B_2$ is

$$\beta_1 \phi_r(c_1) + \cdots + \beta_m \phi_r(c_m) = 0 \qquad (r = 1, 2, \ldots, p)$$

$$\phi_r(x) = \frac{dW_r}{dx}.$$

We must determine the $m + 1$ constants $\beta$ from these $m + 1$ equations. If $q$ functions $\phi_r$ in the last $p$ equations vanish at all places $c_1, c_2, \ldots, c_m$, then we are left with $p - q$ functions $\phi_r$. Since they contain $m$ constants $\beta$, we can express $p - q$ constants $\beta$ in terms of the remaining $m - p + q$ constants $\beta$. If the additive constant $\beta_0$ is taken into account, the function $y$ contains $m - p + q - 1$ arbitrary constants and is a linear combination of arbitrary multiples of $m - p + q$ functions, each with $p - q + 1$ poles of the first order, $p - q$ of which are common to all functions in the linear combination. This is the *Riemann-Roch theorem* and has important applications in algebraic geometry (Baker, 1897; Coble, 1929; van der Waerden, 1939; Hirzebruch, 1956).

Since $\phi_r(z) = dW_r/dx$ is the derivative of an integral of the first kind, it has neither any pole nor any logarithmic singularity except the branch points. It has a finite number of zero points and is finite and continuous at any other place. Hence it should be a polynomial in $x$. The $p - 1$ ratios of the coefficients $C$ in their linear combination

$$\phi(x) = C_1 \phi_1(x) + \cdots + C_p \phi_p(x)$$

can be determined so that $\phi(x)$ vanishes at the arbitrary $p - 1$ places; accordingly they can be determined by the $p - 1$ zero points except for a constant factor. We call this the $\phi$-polynomial.

Let $y$ be an algebraic function defined by the irreducible equation

$$f(y, x) = y^n + A_1 y^{n-1} + \cdots + A_{n-1} y + A_n = 0, \tag{75}$$

where $A_1, \ldots, A_n$ are polynomials in $x$. Let $R(y, x)$ be a rational function of $x$ and $y$, and the periods along the cross cuts $A_i$ and $B_i$ be respectively $\Omega_i$ and $\Omega_i'$ $(i = 1, 2, \ldots, p)$. Further denote by $\theta(y, x)$ a rational polynomial on the Riemann surface $f(y, x) = 0$ and let the zero points be the places $P_i(y_i, x_i)$ and the poles be the places $P_i'(y_i', x_i')$. Any one of these places does not coincide with a place of any kind of singularity of Abel's integral $\int R(y, x)\, dx$. Let $C$ be a simple closed curve on the Riemann surface which includes all places of singularities of this integral but contains neither the places of the zero points nor of the poles of $\theta(y, x)$. *Abel's theorem* states that

$$\sum_{i=1}^{q} \int_{(x_0,y_0)}^{(x_i,y_i)} R(y, x)\, dx - \sum_{i=1}^{q} \int_{(x_0,y_0)}^{(x_i',y_i')} R(y, x)\, dx$$

is equal to the sum of the residues of $R \log \theta$ inside $C$ added to

$$\sum_{k=1}^{p} (h_k \Omega_k + h_k' \Omega_k'),$$

where $h_k$ and $h_k'$ are integers.

In particular, if Abel's integral is respectively of the first, the second, and the third kinds, then this relation is written

$$\sum_{i=1}^{q} W(y_i, x_i) - \sum_{i=1}^{q} W(y_i', x_i') = \sum_{k=1}^{p} (h_k \Omega_k + h_k' \Omega_k'),$$

$$\sum_{i=1}^{q} E_\nu(y_i, x_i; P) - \sum_{i=1}^{q} E_\nu(y_i, x_i; P')$$

$$= -\left| \frac{d^{\nu+1} \log \theta}{dt^{\nu+1}} \right| + \sum_{k=1}^{p} (h_k \Omega_k + h_k' \Omega_k'),$$

and

$$\sum_{i=1}^{q} \Pi(y_i, x_i; P_1, P_2) - \sum_{i=1}^{q} \Pi(y_i', x_i'; P_1, P_2)$$

$$= \log \frac{\theta(P_1)}{\theta(P_2)} + \sum_{k=1}^{p} (h_k \Omega_k + h_k' \Omega_k').$$

Here it is supposed, as an example, that the integral of the second kind contains the poles at the place $P$ of the Riemann surface and can be expanded in the normal form $\nu!\, t^{-\nu-1} + c_0 + c_1 t + \cdots$ in the neighborhood of the poles. The integral of the third kind is supposed to have singularities at two places $P_1$ and $P_2$ with expansions of the form $\log t + c_0 + c_1 t + \cdots$ at $P_1$ and $-\log t + c_0' + c_2' t + \cdots$ at $P_2$.

## APPLICATION TO AN EARTH SATELLITE

### 2.21 EARTH POTENTIAL

The motion of an artificial satellite around the earth, whose dynamical model is taken as an oblate spheroid, can be discussed under suitable assumptions by Hamilton-Jacobi separation of variables.

The gravitational potential $V$ outside an axially symmetric oblate earth of mass $M$ (the symmetry axis being the polar axis) is expressed by

$$V(r, \theta) = -\frac{\mu}{r}\left[1 - \sum_{n=1}^{\infty} J_n\left(\frac{R}{r}\right)^n P_n(\sin\theta)\right], \tag{76}$$

where $R$ is the equatorial axis of the earth, $r$ is the radius vector, $\theta$ is the latitude of the satellite, $\mu = GM$ ($G$ is the gravitational constant), and $J_n$ is the coefficient of oblateness introduced by Merson and King-Hele (1958) (see Hagihara, 1962). If the earth is symmetrical with respect to the equator, then all odd harmonics vanish: $J_1 = J_3 = J_5 = \cdots = 0$. The coefficients $J_2$ and $J_4$ are given by

$$J_2 = \frac{2}{3}J = \frac{I_p - I_t}{MR^2}, \qquad J_4 = -\frac{8D}{35}, \tag{77}$$

where $J$ and $D$ were introduced by Jeffreys (1952) and $I_p$ and $I_t$ are the moments of inertia of the earth about its polar and transverse axes, respectively, through the earth center. Recently, asymmetry with respect to the equatorial plane has been observed. If the origin is taken at the earth center of mass on the polar axis, then $J_1$ vanishes; $J_3, J_5, \ldots$ do not vanish but are very small. On the other hand, if the origin is taken not coincident with the center of mass but with the displacement to the north of the center of mass from the origin, then $J_1 \neq 0$, and $J_3, J_5, \ldots$ still do not vanish.

Vinti (1959) derived a solution $V_L$ of Laplace's equation in oblate spheroidal coordinates leading to exact separability of the Hamilton-

Jacobi equation. It contains three adjustable constants $b_0$, $b_1$, and $c$, which may be so chosen that the expansion of $V_L$ is

$$V_L = -\frac{\mu}{r}\left[1 - J_2\left(\frac{R}{r}\right)^2 P_2(\sin\theta) + J_2^2\left(\frac{R}{r}\right)^4 P_4(\sin\theta) - J_2^3\left(\frac{R}{r}\right)^6 P_6(\sin\theta) + \cdots - J_1\left(\frac{R}{r}\right)P_1(\sin\theta) + J_1J_2\left(\frac{R}{r}\right)^3 P_3(\sin\theta) - J_1J_2\left(\frac{R}{r}\right)^5 P_5(\sin\theta) + \cdots\right], \tag{78}$$

where $J_2$ is given by (77), $J_1 = -\delta/R$ and $J_4 = -J_2^2$, $J_6 = J_2^3, \ldots,$ $J_{2n} = (-1)^{n+1}J_2^n$; $J_3 = -J_1J_2$, $J_5 = J_1J_2^2, \ldots, J_{2n+1} = (-1)^n J_1J_2^n$.

Kislik (1960, 1961) assumed a potential of the form

$$V_k = -\frac{\mu}{d}\frac{\sqrt{q_1}}{q_1 - q_2},$$

which is of Liouville's type of Section 1.15 with the curvilinear coordinates

$$q_1 = \frac{1}{2}\left[\left(\frac{r^2}{d^2} - 1\right) + \sqrt{\left(\frac{r^2}{d^2} - 1\right)^2 + 4\frac{r^2}{d^2}\sin^2\theta}\right],$$

$$q_2 = \frac{1}{2}\left[\left(\frac{r^2}{d^2} - 1\right) - \sqrt{\left(\frac{r^2}{d^2} - 1\right)^2 + 4\frac{r^2}{d^2}\sin^2\theta}\right]$$

$$(-1 \leqq q_2 \leqq 0 \leqq q_1 < +\infty),$$

with a disposable constant $d$. If we let $d = \sqrt{J_2}$, then this potential $V_k$ coincides with (76) up to the term $J_2$ and

$$V_k - V(r, \theta) = -\frac{\mu}{r^5}(J_2^2 - J_4)P_4(\sin\theta) + \cdots.$$

The corresponding Hamilton-Jacobi equation

$$\frac{2}{d^2}\left[\frac{q_1(1 + q_1)}{q_1 - q_2}\left(\frac{\partial S}{\partial q_1}\right)^2 + \frac{q_2(1 + q_2)}{q_2 - q_1}\left(\frac{\partial S}{\partial q_2}\right)^2 + \frac{1}{4(1 + q_1)(1 + q_2)}\left(\frac{\partial S}{\partial q_3}\right)^2\right] - \frac{\mu}{d}\frac{\sqrt{q_1}}{q_1 - q_2} = C$$

can be integrated by the separation of variables. Aksenov, Grebenikov, and Demin (1961, 1963) solved the problem by Zhongolovich's simplified potential (Zhongolovich, 1957):

$$V = -\frac{\mu}{r}\left[1 + \sum_{n=1}^{\infty}(-1)^n\left(\frac{R}{r}\right)^{2n} P_{2n}(\sin\theta)\right],$$

using the same method. By including the terms with $n = 1$ (see Gyldén's theory, Section 10.8), Aksenov, Grebenikov, and Demin (1963a) discussed the stability of particular solutions, such as a polar elliptic orbit, a circular equatorial orbit, and periplegmatic orbits with moving perigees on certain ellipsoids. By using the coordinates

$$r = R \cosh v \sin u, \qquad z = -R \sinh v \cos u,$$

$$d\tau = \frac{dt}{\sinh^2 v + \cos^2 u},$$

we may write the equations of motion in the form

$$\frac{d^2 v}{d\tau^2} = \frac{\mu}{R^3} \cosh v - h \sinh 2v - \frac{c_1^2 \sinh v}{\cosh^3 v},$$

$$\frac{d^2 u}{d\tau^2} = h \sin 2u - \frac{c_1^2 \cos u}{\sin^3 u},$$

where $h$ and $c_1$ are constants.

While Sterne (1957, 1958) referred to the separable disturbing potential

$$3k(\sin^2 \theta - \tfrac{1}{2} \sin^2 I)\left(\frac{1}{r^3} - \frac{1}{ar^2}\right)$$

with constants $k$ and $a$ and with the orbital inclination $I$, Garfinkel (1958, 1959) has based his intermediary orbit on the separable potential

$$V_0 = -\frac{1}{r} + 6k\left[\frac{c_1(\sin^2 \theta - c_2)}{2r^2} + \frac{c_3}{r} + \frac{c_4}{r^3}\right],$$

with

$$c_1 = \frac{1}{p} = \frac{1}{a(1 - e^2)}, \qquad c_2 = \cos^2 I,$$

$$c_3 = \frac{(1 - e^2)^{1/2}}{2p^2} P_2(\cos I), \quad c_4 = 0,$$

where $p$, $a$, and $e$ are the parameter, semimajor axis, and eccentricity, respectively, of the elliptic orbit. This potential gives the simplest possible orbit free from all secular parts by incorporating a major portion of the second spherical harmonics and leads to a closed solution with no secular variation of order $k$. Barrar (1961) simplified the solution of Vinti by reducing the four elliptic integrals of Vinti to two elliptic integrals as in Garfinkel's solution. Garfinkel (1964) further improved the constants

$c_1, c_2, c_3$ with the modification $c_1 + \delta_1, c_2 + \delta_2, c_3 + \delta_3$, where $\delta_1, \delta_2, \delta_3$ are determined so as to incorporate in the intermediary orbit the secular variations of the second order. In his later work, Garfinkel used instead of the Hamilton-Jacobi method the method of von Zeipel, which will be treated in Chapter 9.

On the other hand, Callender (1961) showed the motion of an artificial close satellite to be quasi-periodic by obtaining the bounds of the variation of the radial distance and the variation of the inclination.

Vinti's work (1959) is described in the following section.

## 2.22 OBLATE SPHEROIDAL COORDINATES

Take oblate spheroidal coordinates $\xi$, $\eta$, and $\phi$ such that

$$
\begin{aligned}
x &= c[(\xi^2 + 1)(1 - \eta^2)]^{1/2} \cos \phi \qquad (0 < \xi < \infty),\\
y &= c[(\xi^2 + 1)(1 - \eta^2)]^{1/2} \sin \phi \qquad (-1 \leqq \eta \leqq 1),\\
z &= c\xi\eta,
\end{aligned} \tag{79}
$$

where $c$ is a constant to be determined later. The surface $\xi$ = constant is an oblate spheroid. Its sections perpendicular to the $z$ axis are circles and its meridian sections are confocal ellipses of semiaxes $c(\xi^2 + 1)^{1/2}$ and $c\xi$, the focal separation being $2c$. The surface $\eta$ = constant is a hyperboloid of one sheet. The element of arc $ds^2$ is given by

$$
\begin{aligned}
ds^2 &= h_1^2\, d\xi^2 + h_2^2\, d\eta^2 + h_3^2\, d\phi^2,\\
h_1 &= c(\xi^2 + \eta^2)^{1/2}(\xi^2 + 1)^{-1/2},\\
h_2 &= c(\xi^2 + \eta^2)^{1/2}(1 - \eta^2)^{-1/2},\\
h_3 &= c(\xi^2 + 1)^{1/2}(1 - \eta^2)^{1/2}.
\end{aligned}
$$

Thus the kinetic energy is expressed by

$$
T = \frac{1}{2}\,(h_1^2\dot{\xi}^2 + h_2^2\dot{\eta}^2 + h_3^2\dot{\phi}^2) = \frac{1}{2}\left(\frac{p_\xi^2}{h_1^2} + \frac{p_\eta^2}{h_2^2} + \frac{p_\phi^2}{h_3^2}\right),
$$

where

$$
p_\xi = h_1^2\dot{\xi} = \frac{\partial S}{\partial \xi}, \qquad p_\eta = h_2^2\dot{\eta} = \frac{\partial S}{\partial \eta}, \qquad p_\phi = h_3^2\dot{\phi} = \frac{\partial S}{\partial \phi}.
$$

The Hamiltonian takes the form

$$
H = \frac{1}{2c^2}\left[\frac{\xi^2 + 1}{\xi^2 + \eta^2}\,p_\xi^2 + \frac{1 - \eta^2}{\xi^2 + \eta^2}\,p_\eta^2 + \frac{p_\phi^2}{(\xi^2 + 1)(1 - \eta^2)}\right] + V(\xi, \eta),
$$

and the Hamilton-Jacobi equation is

$$(\xi^2 + 1)\left(\frac{\partial S}{\partial \xi}\right)^2 + (1 - \eta^2)\left(\frac{\partial S}{\partial \eta}\right)^2 + \frac{\xi^2 + \eta^2}{(\xi^2 + 1)(1 - \eta^2)}\left(\frac{\partial S}{\partial \phi}\right)^2$$
$$+ 2c^2(\xi^2 + \eta^2)V(\xi, \eta) = 2c^2(\xi^2 + \eta^2)\alpha_1, \quad (80)$$

where $\alpha_1$ is the energy constant.

To separate the variables, write

$$S = S_1(\xi) + S_2(\eta) + S_3(\phi).$$

Then (80) takes the form

$$(\xi^2 + 1)\left(\frac{\partial S_1}{\partial \xi}\right)^2 + (1 - \eta^2)\left(\frac{\partial S_2}{\partial \eta}\right)^2 + \left(\frac{1}{1 - \eta^2} - \frac{1}{\xi^2 + 1}\right)\left(\frac{\partial S_3}{\partial \phi}\right)^2$$
$$+ 2c^2(\xi^2 + \eta^2)V(\xi, \eta) = 2c^2(\xi^2 + \eta^2)\alpha_1. \quad (81)$$

At first $\partial S_3/\partial \phi = \alpha_3 =$ constant. The elliptic coordinates $\xi$ and $\eta$ can be separated if and only if $V(\xi, \eta)$ is of the form

$$V(\xi, \eta) = \frac{1}{\xi^2 + \eta^2}[f(\xi) + g(\eta)]. \quad (82)$$

Inserting this in the Laplace equation

$$\frac{\partial}{\partial \xi}\left[(\xi^2 + 1)\frac{\partial V}{\partial \xi}\right] + \frac{\partial}{\partial \eta}\left[(1 - \eta^2)\frac{\partial V}{\partial \eta}\right] = 0 \quad (83)$$

shows that $f(\xi)$ and $g(\eta)$ must satisfy the equation

$$\xi^2(\xi^2 + 1)f''(\xi) - 2\xi(\xi^2 + 2)f'(\xi) + 2(\xi^2 + 2)f(\xi)$$
$$+ \eta^2(1 - \eta^2)g''(\eta) + 2\eta(\eta^2 - 2)g'(\eta)$$
$$- 2(\eta^2 - 2)g(\eta) + M(\xi, \eta) = 0, \quad (84)$$

where

$$\begin{aligned} M(\xi, \eta) &= \eta^2 F(\xi) - \xi^2 G(\eta), \\ F(\xi) &= (\xi^2 + 1)f''(\xi) + 2\xi f'(\xi) - 2f(\xi), \quad (85) \\ G(\eta) &= (\eta^2 - 1)g''(\eta) + 2\eta g'(\eta) - 2g(\eta). \end{aligned}$$

A solution of the Laplace equation (83) of form (82) can exist only if $M(\xi, \eta)$ is separable. The necessary and sufficient condition that $M(\xi, \eta)$ be separable is that

$$\begin{aligned} F(\xi) &= 4B + 4C\xi^2, \\ G(\eta) &= -4A + 4C\eta^2, \end{aligned} \quad (86)$$

where $A$, $B$, and $C$ are constants.

The condition is evidently sufficient. To show that the condition is also necessary when $M(\xi, \eta)$ is separable, note that separability means that

$$\eta^2 F(\xi) - \xi^2 G(\eta) = \psi(\eta) - \phi(\xi),$$

which, if $\eta$ or $\xi$ are equated to zero, gives either

$$\phi(\xi) = \psi(0) + G(0) \cdot \xi^2 \qquad (\eta = 0),$$

or

$$\psi(\eta) = \phi(0) + F(0) \cdot \eta^2 \qquad (\xi = 0).$$

Substituting the latter two equations gives

$$\eta^2[F(\xi) - F(0)] - \xi^2[G(\eta) - G(0)] = \phi(0) - \psi(0).$$

Let $\xi = 0$ or $\eta = 0$. Then, $\phi(0) = \psi(0)$. Hence,

$$\frac{1}{\xi^2}[F(\xi) - F(0)] = \frac{1}{\eta^2}[G(\eta) - G(0)] = 4C,$$

where $C$ is a constant. Thus we have

$$F(\xi) = F(0) + 4C\xi^2, \qquad G(\eta) = G(0) + 4C\eta^2.$$

The necessary condition is proved.

Comparing (85) with (86) we obtain

$$\begin{aligned} (\xi^2 + 1)f'' + 2\xi f' - 2f - 4C\xi^2 - 4B &= 0, \\ (\eta^2 - 1)g'' + 2\eta g' - 2g - 4C\eta^2 + 4A &= 0. \end{aligned} \tag{87}$$

Again, from (85) and (86),

$$M(\xi, \eta) = 4A\xi^2 + 4B\eta^2.$$

Substituting these in (84) gives the following equation, which $f$ and $g$ should satisfy with a constant $K$.

$$\begin{aligned} &\xi^2(\xi^2 + 1)f''(\xi) - 2\xi(\xi^2 + 2)f'(\xi) + 2(\xi^2 + 2)f(\xi) + 4A\xi^2 \\ &\quad = \eta^2(\eta^2 - 1)g''(\eta) - 2\eta(\eta^2 - 2)g'(\eta) + 2(\eta^2 - 2)g(\eta) - 4B\eta^2 \\ &\quad \equiv 4K. \end{aligned} \tag{88}$$

Multiply the first equation of (87) by $\xi^2$ and subtract it from the first equation of (88). Next multiply the second equation of (87) by $\eta^2$ and subtract it from the second equation of (88). Then

$$\begin{aligned} \xi f' - f &= (\xi^2 + 1)^{-1}[C\xi^4 + (A + B)\xi^2 - K], \\ \eta g' - g &= (\eta^2 - 1)^{-1}[C\eta^4 - (A + B)\eta^2 - K]. \end{aligned}$$

The most general solution of these equations is

$$f(\xi) = K + b_0\xi + C\xi^2 + (A + B - C + K)\xi \tan^{-1} \xi,$$

$$g(\eta) = -K - b_1\eta + C\eta^2 + \tfrac{1}{2}(A + B - C + K)\eta \log \frac{1+\eta}{1-\eta},$$

where $b_0$ and $b_1$ are the integration constants. Putting these expressions in (87) gives $B = A$. Hence,

$$f(\xi) = K + b_0\xi + C\xi^2 + (2A - C + K)\xi \tan^{-1} \xi,$$

$$g(\eta) = -K - b_1\eta + C\eta^2 + \tfrac{1}{2}(2A - C + K)\eta \log \frac{1+\eta}{1-\eta}.$$

Thus we have shown that all axially symmetric solutions of the form (82) of the Laplace equation in oblate spheroidal coordinates are expressible by

$$V(\xi, \eta) = C + \frac{1}{\xi^2 + \eta^2}\left[b_0\xi - b_1\eta + b_2\left(2\xi \tan^{-1} \xi + \eta \log \frac{1+\eta}{1-\eta}\right)\right],$$

where $\frac{1}{2}(2A - C + K) = b_2$. If $V$ is zero at infinity, then we must take $C = 0$ and we are left with three constants $b_0$, $b_1$, and $b_2$. Furthermore, since the polar axis $\eta = \pm 1$ cannot be a singularity, we must have $b_2 = 0$. Hence, a solution with two constants $b_0$ and $b_1$ of the Laplace equation that makes the Hamilton-Jacobi equation separable must be of the form

$$V = \frac{1}{\xi^2 + \eta^2}(b_0\xi - b_1\eta) = b_0 - \Re(\xi + i\eta)^{-1} + b_1\Im(\xi + i\eta)^{-1}. \quad (89)$$

Now, from transformation formulas (89),

$$(\xi + i\eta)^{-1} = cr^{-1}(1 - 2hs + h^2)^{-1/2}, \qquad s = \sin\theta, \qquad h = -\frac{ic}{r}.$$

If we choose $c \ll r$, then $|h| < 1$, and $(1 - 2hs + h^2)^{-1/2}$ is the generating function for the Legendre polynomials $P_n(s)$. Hence,

$$(\xi + i\eta)^{-1} = \frac{c}{r}\sum_{n=0}^{\infty} h^n P_n(s) = \frac{c}{r}\sum_{n=0}^{\infty} (-i)^n \frac{c^n}{r^n} P_n(s),$$

from which,

$$\frac{\xi}{\xi^2 + \eta^2} = \Re\left(\frac{1}{\xi + i\eta}\right) = \frac{c}{r}\left[1 - \frac{c^2}{r^2} P_2(\sin\theta) + \frac{c^4}{r^4} P_4(\sin\theta) - \cdots\right],$$

$$\frac{\eta}{\xi^2 + \eta^2} = -\Im\left(\frac{1}{\xi + i\eta}\right)$$

$$= \frac{c}{r}\left[\frac{c}{r} P_1(\sin\theta) - \frac{c^3}{r^3} P_3(\sin\theta) + \frac{c^5}{r^5} P_5(\sin\theta) - \cdots\right].$$

Combining this with (89) gives the potential in the form

$$V = b_0 \cdot \frac{c}{r}\left[1 - \frac{c^2}{r^2} P_2(\sin\theta) + \frac{c^4}{r^4} P_4(\sin\theta) - \cdots\right]$$
$$- b_1 \frac{c}{r}\left[\frac{c}{r} P_1(\sin\theta) - \frac{c^3}{r^3} P_3(\sin\theta) + \cdots\right]. \tag{90}$$

Comparing (90) with (76) shows that

$$b_0 c = -\mu, \qquad c^2 = R^2 J_2, \qquad b_1 c^2 = -\mu R J_1 = \mu\delta.$$

These are the flexible constants and the other $J_n$ can be expressed in terms of them. From (89),

$$V = -\frac{\mu}{c}\frac{1}{\xi^2 + \eta^2}\left(\xi + \eta\frac{\delta}{c}\right). \tag{91}$$

This Vinti's potential can be written in form (76) with the limitation that

$$J_{2m} = (-1)^{m+1} J_2^m, \quad J_{2m+1} = (-1)^m J_1 J_2^m \quad (m = 0, 1, 2, \ldots). \tag{92}$$

Substitute in (80) the expression (91) and the relation $\partial S_3/\partial\phi = \alpha_3$ obtained earlier; then

$$(\xi^2 + 1)\left(\frac{\partial S_1}{\partial\xi}\right)^2 - \frac{\alpha_3^2}{\xi^2 + 1} - 2\mu c\xi - 2\alpha_1 c^2 \xi^2$$
$$= -(1 - \eta^2)\left(\frac{\partial S_2}{\partial\eta}\right)^2 - \frac{\alpha_3^2}{1 - \eta^2} + 2\mu\delta\eta + 2\alpha_1 c^2 \eta$$
$$= -\alpha_2^2. \tag{93}$$

Here it is supposed that the plane $\eta = 0$ is intersected by the orbit and, hence, $-\alpha_2^2 \leqq \alpha_3^2$. Thus,

$$\frac{\partial S_1}{\partial\xi} = \pm\frac{1}{\xi^2 + 1}\left[\alpha_3^2 + (-\alpha_2^2 + 2\mu c\xi + 2\alpha_1 c^2\xi^2)(\xi^2 + 1)\right]^{1/2},$$
$$\frac{\partial S_2}{\partial\eta} = \pm\frac{1}{1 - \eta^2}\left[-\alpha_3^2 + (\alpha_2^2 + 2\mu\,\delta\eta + 2\alpha_1 c^2\eta^2)(1 - \eta^2)\right]^{1/2}, \tag{94}$$
$$\frac{\partial S_3}{\partial\phi} = \alpha_3.$$

If we let $\rho = c\xi$, $\rho_1 = c\xi_1$, and $\eta = \sigma$, the complete integral of the Hamilton-Jacobi equation becomes

$$S = \alpha_3\phi \pm \int_{\rho_1}^{\rho} \frac{1}{\rho^2 + c^2}[c^2\alpha_3^2 + (-\alpha_2^2 + 2\mu\rho + 2\alpha_1\rho^2)(\rho^2 + c^2)]^{1/2}\,d\rho$$
$$\pm \int_0^{\sigma} \frac{1}{1 - \sigma^2}[-\alpha_3^2 + (\alpha_2^2 + 2\mu\delta\sigma + 2\alpha_1 c^2\sigma^2)(1 - \sigma^2)]^{1/2}\,d\sigma. \tag{95}$$

The general solution of Hamilton's equations is thus

$$\frac{\partial S}{\partial \alpha_1} = t + \beta_1, \qquad \frac{\partial S}{\partial \alpha_2} = \beta_2, \qquad \frac{\partial S}{\partial \alpha_3} = \beta_3,$$

or

$$\pm\int_{\rho_1}^{\rho} \rho^2[c^2\alpha_3 + (-\alpha_2^2 + 2\mu\rho + 2\alpha_1\rho^2)(\rho^2 + c^2)]^{-1/2}\,d\rho$$
$$\pm c^2\int_0^{\sigma} \sigma^2[-\alpha_3^2 + (\alpha_2^2 + 2\mu\delta\sigma + 2\alpha_1c^2\sigma^2)(1 - \sigma^2)]^{-1/2}\,d\sigma = t + \beta_1,$$

$$\mp\alpha_2\int_{\rho_1}^{\rho} [c^2\alpha_3^2 + (-\alpha_2^2 + 2\mu\rho + 2\alpha_1\rho^2)(\rho^2 + c^2)]^{-1/2}\,d\rho$$
$$\pm \alpha_2\int_0^{\sigma} [-\alpha_3^2 + (\alpha_2^2 + 2\mu\,\delta\sigma + 2\alpha_1c^2\sigma^2)(1 - \sigma^2)]^{-1/2}\,d\sigma = \beta_2, \tag{96}$$

$$\phi \pm c^2\alpha_3\int_{\rho_1}^{\rho} \frac{1}{\rho^2 + c^2}[c^2\alpha_3^2 + (-\alpha_2^2 + 2\mu\rho + 2\alpha_1\rho^2)(\rho^2 + c^2)]^{-1/2}\,d\rho$$
$$\mp \alpha_3\int_0^{\sigma} \frac{1}{1 - \sigma^2}[-\alpha_3^2 + (\alpha_2^2 + 2\mu\,\delta\sigma + 2\alpha_1c^2\sigma^2)(1 - \sigma^2)]^{-1/2}\,d\sigma = \beta_3.$$

Vinti (1966a) considered further the third harmonic term. In the oblate spheroidal coordinates

$$x + \sqrt{-1}\,y = r\cos\theta\exp(\sqrt{-1}\,\phi) = [(\rho^2 + c^2)(1 - \eta^2)]^{1/2}\exp(\sqrt{-1}\,\phi),$$
$$z = r\sin\theta = -\delta + \rho\eta,$$

Vinti takes

$$c^2 = r_e^2J_2(1 - \tfrac{1}{4}J_3^2J_2^{-3}), \qquad \delta = -\tfrac{1}{2}r_eJ_2^{-1}J_3,$$

where $r_e$ is the equatorial radius. Then the gravitational potential

$$V = -\mu(\rho^2 + c^2\eta^2)^{-1}(\rho + \eta\delta)$$

helps to make the problem separable. According to Vinti, it gives about two-thirds of the fourth harmonic and negligible values, of order $J_2^3$ or smaller, for the higher harmonics. The inversion formulas for the elliptic integrals become very complicated, but Vinti succeeded in obtaining the solution. Vinti (1966b) studied the effect of a constant force on a Keplerian orbit — in an analogous manner to the theory of the Stark effect in atomic physics — using parabolic coordinates and applying them to the effect of solar radiation pressure on the motion of an artificial satellite.

## 2.23 INTEGRATION BY ELLIPTIC INTEGRALS

Izsak (1960) has worked out Vinti's idea in a complete manner and has taken the solution of Vinti's equations as the intermediary orbit of the motion, by including second-order terms. The solution has been obtained in a complete form, described in Section 2.7; the integrals have been transformed by a linear fractional transformation in the complex domain, referring to Weierstrass's theory of elliptic functions. The solution is expressed in Fourier series and the integration constants as functions of the Keplerian elements. Vinti (1961, 1961a, 1962, 1963) tried to simplify Izsak's procedure by expressing the solution in terms of the complete elliptic integrals, avoiding the complicated elliptic functions; that is, as the sum of integrals containing the quadrics as in (95), which are evaluated by convergent Fourier series after the independent variable is transformed into the eccentric anomaly for uniformization.

Following Izsak (1960), assume that $\delta = 0$ (the same as assuming that $J_1 = 0$) and let

$$\begin{aligned} P(\rho) &= 2\alpha_1\rho^4 + 2\mu\rho^3 - (\alpha_2^2 - 2\alpha_1c^2)\rho^2 + 2\mu c^2\rho - (\alpha_3^2 - \alpha_2^2), \\ Q(\sigma) &= -2\alpha_1c^2\sigma^4 + (2\alpha_1c^2 - \alpha_2^2)\sigma^2 + (\alpha_2^2 - \alpha_3^2). \end{aligned}$$

In equations (96), take the upper signs for $\pm$ or $\mp$, then

$$\begin{aligned} \frac{\partial S}{\partial \alpha_1} &= \int_{\rho_1}^{\rho} \frac{\rho^2\,d\rho}{\sqrt{P(\rho)}} + c^2\int_0^{\sigma} \frac{\sigma^2\,d\sigma}{\sqrt{Q(\sigma)}} = t + \beta_1, \\ \frac{\partial S}{\partial \alpha_2} &= -\alpha_2\int_{\rho_1}^{\rho} \frac{d\rho}{\sqrt{P(\rho)}} + \alpha_2\int_0^{\sigma} \frac{d\sigma}{\sqrt{Q(\sigma)}} = \beta_2, \\ \frac{\partial S}{\partial \alpha_3} &= \phi + c^2\alpha_3\int_{\rho_1}^{\rho} \frac{d\rho}{(\rho^2 + c^2)\sqrt{P(\rho)}} - \alpha_3\int_0^{\sigma} \frac{d\sigma}{(1 - \sigma^2)\sqrt{Q(\sigma)}} = \beta_3. \end{aligned} \tag{97}$$

Introduce $v$ with a constant $\Gamma$ (as yet indeterminate):

$$\Gamma\int_{\rho_1}^{\rho} \frac{d\rho}{\sqrt{P(\rho)}} = v; \tag{98}$$

then the second equation of (97) takes the form

$$\alpha_2 \int_0^\sigma \frac{d\sigma}{\sqrt{Q(\sigma)}} = \frac{\alpha_2}{\Gamma} v + \beta_2. \tag{99}$$

Suppose that $v = 0$ for $\rho = \rho_1$ and $v = v_0 = -\Gamma\beta_2/\alpha_2$ for $\sigma = 0$. The variables $\rho$ and $\sigma$ are obtained in terms of $v$ from these formulas (98) and (99). Then the first and third equations of (97) become

$$\phi - \beta_3 = \alpha_3 \int_{v_0}^{v} \frac{dv}{1 - \sigma^2} - \frac{c^2\alpha_3}{\Gamma} \int_0^v \frac{dv}{\rho^2 + c^2}, \tag{100}$$

$$t + \beta_1 = \frac{1}{\Gamma} \int_0^v \rho^2\, dv + \frac{c^2}{\Gamma} \int_{v_0}^{v} \sigma^2\, dv. \tag{101}$$

Izsak (1960) introduced a linear transformation with real coefficients of the complex $\rho$-plane (Section 2.4)

$$\rho = \frac{p + d_*\xi}{1 + e_*\xi}, \tag{102}$$

such that the roots $\rho_1, \rho_2, \rho_3, \rho_4$ of $P(\rho) = 0$ are mapped into the points $\xi_1 = 1$, $\xi_2 = -1$, $\xi_3 = +\sqrt{-1}/m$, $\xi_4 = -\sqrt{-1}/m$, respectively, of the $\xi$-plane. The constant $m$ is considered real or purely imaginary according as $\rho_3$ and $\rho_4$ are conjugate complex or real, and to be determined by

$$\frac{\rho_1 - \rho_3}{\rho_1 - \rho_4} : \frac{\rho_2 - \rho_3}{\rho_2 - \rho_4} = \frac{\xi_1 - \xi_3}{\xi_1 - \xi_4} : \frac{\xi_2 - \xi_3}{\xi_2 - \xi_4} = \left(\frac{1 + \sqrt{-1}\, m}{1 - \sqrt{-1}\, m}\right)^2.$$

In analogy to a Keplerian motion (see Sections 5.16–5.20), let

$$\begin{aligned} \rho_1 &= a(1 - e), & \rho_2 &= a(1 + e), \\ \rho_3 &= a(\kappa - \sqrt{-1}\,\lambda), & \rho_4 &= a(\kappa + \sqrt{-1}\,\lambda). \end{aligned} \tag{103}$$

The relations

$$\rho_j = \frac{p + d_*\xi_j}{1 + e_*\xi_j} \qquad (j = 1, 2, 3, 4),$$

with the notation (103) give

$$p = \frac{\rho_1 + \rho_2}{2} - \frac{\rho_2 - \rho_1}{2} e_* = a(1 - ee_*),$$

$$d_* = \frac{\rho_1 + \rho_2}{2} e_* - \frac{\rho_2 - \rho_1}{2} = a(e_* - e),$$

and

$$\frac{m}{1 - m^2} = \frac{e\lambda}{(1 - \kappa)^2 - (e^2 - \lambda^2)}, \tag{104}$$

$$\frac{e_*}{1 + e_*^2} = \frac{(1 - \kappa)e}{(1 - \kappa)^2 + (e^2 + \lambda^2)}, \tag{105}$$

where the values of $m$ and $e_*$ are taken so that they vanish for $e = 0$. By applying (103) to (98) we obtain, after some reduction,

$$-\frac{\Gamma}{A_*}\int_0^\xi \frac{d\xi}{(1-\xi^2)(1+m^2\xi^2)} = v,$$

$$A_* = \sqrt{\frac{-2\alpha_1\Delta^2}{(1-e_*^2)(e_*^2+m^2)}} \qquad \Delta = pe_* - d_* = ae(1-e_*^2).$$

Let

$$\xi = \sqrt{1-\eta^2}, \qquad A = \sqrt{\frac{-2\alpha_1\Delta^2(1+m^2)}{(1-e_*^2)(e_*^2+m^2)}};$$

then

$$\frac{\Gamma}{A}\int_0^\eta \frac{d\eta}{\sqrt{(1-\eta^2)(1-k^2\eta^2)}} = v,$$

and

$$\eta = \operatorname{sn}\left(\frac{Av}{\Gamma}\Big/k^2\right), \qquad \xi = \operatorname{cn}\left(\frac{Av}{\Gamma}\Big/k^2\right),$$

$$k = \frac{m}{\sqrt{1+m^2}}. \tag{106}$$

Try to determine $\Gamma$ so that the period of $\xi(v)$ is equal to $2\pi$. Since the real quarter-period of cn $(z/k^2)$ is known to be

$$K = \frac{\pi}{2}\left[1 + \left(\frac{1}{2}\right)^2 k^2 + \left(\frac{1\cdot 3}{2\cdot 4}\right)^2 k^4 + \cdots\right],$$

therefore

$$\xi = \operatorname{cn}\left[\left(\frac{2K}{\pi}v\right)\Big/k^2\right], \qquad \Gamma = \frac{\pi}{2k}A.$$

To transform the integral (99), write

$$Q(\sigma) = (\alpha_2^2 - \alpha_3^2)\left(1 - \frac{\sigma^2}{\sigma_1^2}\right)\left(1 - \frac{\sigma^2}{\sigma_2^2}\right),$$

$$\frac{\sigma}{\sigma_1} = \zeta, \qquad \frac{\sigma_1}{\sigma_2} = l;$$

then (99) becomes

$$\frac{\alpha_2\sigma_1}{\sqrt{\alpha_2^2-\alpha_3^2}}\int_0^\zeta \frac{d\zeta}{\sqrt{(1-\zeta^2)(1-l^2\zeta^2)}} = \frac{\alpha_2}{\Gamma}v + \beta_2,$$

or

$$\sigma = \sigma_1 \operatorname{sn}\left[\left(\frac{\sqrt{\alpha_2^2-\alpha_3^2}}{\Gamma\sigma_1}(v-v_0)\right)\Big/l^2\right], \qquad v_0 = -\frac{\Gamma\beta_2}{\alpha_2}.$$

Let

$$L = \frac{\pi}{2}\left[1 + \left(\frac{1}{2}\right)^2 l^2 + \left(\frac{1\cdot 3}{2\cdot 4}\right)^2 l^4 + \cdots\right],$$

$$\frac{2L}{\pi}(v + \varpi) = \frac{\sqrt{\alpha_2^2 - \alpha_3^2}}{\Gamma s}(v - v_0), \qquad \sigma_1 = s = \sin I,$$

$$\varpi = \epsilon v + \omega, \qquad \epsilon = \frac{\pi}{2L}\frac{\sqrt{\alpha_2^2 - \alpha_3^2}}{\Gamma s} - 1,$$

$$\omega = -\frac{\pi}{2L}\frac{\sqrt{\alpha_2^2 - \alpha_3^2}}{\Gamma s} v_0 = \frac{\pi}{2L}\frac{\sqrt{\alpha_2^2 - \alpha_3^2}}{\alpha_2 s}\beta_2,$$

where $I$ is the inclination of the orbital plane and $\epsilon$ is the motion of the perigee.

Thus we have obtained

$$\rho = \frac{p + d_* \operatorname{cn}\left[\left(\frac{2K}{\pi}v\right)\Big/k^2\right]}{1 + e_* \operatorname{cn}\left[\left(\frac{2K}{\pi}v\right)\Big/k^2\right]}, \tag{107}$$

$$\sigma = s \operatorname{sn}\left[\left(\frac{2L}{\pi}(v + \varpi)\right)\Big/l^2\right]. \tag{108}$$

These formulas correspond to those for a Keplerian motion, in which the radius vector $r$ and the declination $\delta$ are given by

$$r = \frac{p}{1 + e\cos v}, \qquad \sin\delta = s\sin(v + \omega).$$

The quantities $\kappa$ and $\lambda$ can be obtained from (104), (105), and (106) in terms of $a$, $e$, and $s$. The coefficients and the roots of $P(\rho) = 0$ are related to those of $Q(\sigma) = 0$ as follows:

$$\begin{aligned}
2a(1 + \kappa) &= -\frac{\mu}{\alpha_1},\\
a^2[(1 - e^2) + 4\kappa + (\kappa^2 + \lambda^2)] &= -\frac{\alpha_2^2 - 2c^2\alpha_1}{2\alpha_1} = c^2\left(s^2 + \frac{s^2}{e^2}\right),\\
2a^3[(1 - e^2)\kappa + (\kappa^2 + \lambda^2)] &= -\frac{c^2\mu}{\alpha_1},\\
\alpha^4(1 - e^2)(\kappa^2 + \lambda^2) &= -\frac{c^2(\alpha_2^2 - \alpha_3^2)}{2\alpha_1} = c^4 s^2\left(\frac{s^2}{l^2}\right).
\end{aligned} \tag{109}$$

Let $\nu = c/a$; then

$$\kappa = \frac{\nu^2(1 - s^2)(1 - e^2 - \nu^2 s^2)}{(1 - e^2 - \nu^2)(1 - e^2 - \nu^2 s^2) + 4\nu^2 s^2}, \tag{110}$$

$$\kappa^2 + \lambda^2 = \frac{\nu^2 s^2[(1 - e^2 - \nu^2)(1 - e^2 - \nu^2 s^2) + 4\nu^2]}{(1 - e^2 - \nu^2)(1 - e^2 - \nu^2 s^2) + 4\nu^2 s^2}, \tag{111}$$

or, from expansion in powers of $\nu^2$ (which is of the order $10^{-3}$ for artificial satellites with small eccentricities),

$$\kappa = \frac{\nu^2(1 - s^2)}{1 - e^2} + \frac{\nu^4(1 - s^2)}{(1 - e^2)^3}(1 - 4s^2 - e^2) + \cdots,$$

$$\lambda^2 = \nu^2 s^2 - \frac{\nu^4}{(1 - e^2)^2}(1 - s^2)(1 - 5s^2) + \cdots.$$

For a very low inclination, $\nu$ becomes imaginary so that $\rho_3$ and $\rho_4$ are real. The value of $I$ for $\lambda = 0$ is given approximately by $s^2 = \nu^2/(1 - e^2)^2$.

The last equation of (109) and equation (111) give

$$l^2 = \frac{\nu^2 s^2}{1 - e^2}\frac{(1 - e^2 - \nu^2)(1 - e^2 - \nu^2 s^2) + 4\nu^2 s^2}{(1 - e^2 - \nu^2)(1 - e^2 - \nu^2 s^2) + 4\nu^2}$$

$$= \frac{\nu^2 s^2}{1 - e^2}\frac{4\nu^4 s^2}{(1 - e^2)^3}(1 - s^2) + \cdots.$$

From (104), (105), (106), (110), and (111), Izsak obtained

$$e_* = e\left\{1 + \frac{\nu^2}{1 - e^2}(1 - 2s^2) + \frac{\nu^4}{(1 - e^2)^3}[(3 - 16s^2 + 14s^4) - 2(1 - s^2)^2 e^2] + \cdots\right\},$$

$$k^2 = \frac{\nu^2 e^2}{(1 - e^2)^2}s^2 - \frac{\nu^4 e^2}{(1 - e^2)^4}[(1 - 10s^2 + 11s^4) + s^4 e^2] + \cdots, \tag{112}$$

$$\epsilon = \frac{\nu^2}{4(1 - e^2)^2}(12 - 15s^2) + \frac{\nu^4}{64(1 - e^2)^4} \times [(288 - 1296s^2 + 1035s^4) - (144 + 288s^2 - 510s^4)e^2] + \cdots$$

Substituting (108) in (100) gives the first integral in the form

$$\frac{\alpha_3}{\Gamma(1 + \epsilon)}\int_0^w \frac{dw}{1 - s^2\,\mathrm{sn}^2\left[\left(\frac{2L}{\pi}w\right)\Big/l^2\right]}, \tag{113}$$

where

$$w = v + \varpi = (1 + \epsilon)v + \omega.$$

The integrand is doubly periodic with periods $2L$ and $2\sqrt{-1}\,L'$, and has two simple poles $z_1$ and $z_2$. For $l < s < 1$ these poles are

$$z_1 = L + \sqrt{-1}\,y,$$
$$z_2 = L + 2\sqrt{-1}\,L' - \sqrt{-1}\,y, \qquad (0 < y < L').$$

The residues at these poles computed by Izsak (1960) are respectively

$$\frac{-\sqrt{-1}/2}{\sqrt{1-s^2}\,\sqrt{1-l^2/s^2}}, \qquad \frac{\sqrt{-1}/2}{\sqrt{1-s^2}\,\sqrt{1-l^2/s^2}}.$$

The integrand of (113) is written

$$\frac{1}{1 - s^2\,\mathrm{sn}^2\left[\left(\frac{2L}{\pi}w\right)\Big/ l^2\right]} = a_0 + 2\sum_{n=1}^{\infty} a_{2n}\cos 2nw,$$

where

$$a_{2n} = \frac{1}{\pi}\int_0^{\pi} \frac{\exp\,(2\sqrt{-1}\,nw)}{1 - s^2\,\mathrm{sn}^2\left[\left(\frac{2L}{\pi}w\right)\Big/ l^2\right]}\,dw.$$

The integral over the entire period-parallelogram gives

$$\oint = \int_0^{\pi} + \int_{\pi}^{\pi+\pi\sqrt{-1}L'/L} + \int_{\pi+\pi\sqrt{-1}L'/L}^{\pi\sqrt{-1}L'/L} + \int_{\pi\sqrt{-1}L'/L}^{0} = (1-q^{2n})\int_0^{\pi}$$
$$= \pi\,(1-q^{2n})\,a_{2n},$$
$$q = \exp\left(-\frac{\pi L'}{L}\right).$$

Again, by computing the residues for the singularity of $\exp\,(2\sqrt{-1}\,nw)$ at infinity, Izsak obtained

$$a_{2n} = \frac{\pi}{2L}\,\frac{(-1)^n}{\sqrt{1-s^2}\,\sqrt{1-(l^2/s^2)}}\,\frac{\tau^{2n}-(q/\tau)^{2n}}{1-q^{2n}},$$

$$\tau = \exp\left(-\frac{\pi}{2L}y\right),$$

and

$$a_0 = 1 - 2\sum_{n=1}^{\infty} a_{2n}.$$

Let

$$\mathrm{sn}\left(\frac{2L}{\pi} w \Big/ l^2\right) = \sin \psi,$$

so that

$$\mathrm{cn}\left(\frac{2L}{\pi} w \Big/ l^2\right) = \cos \psi,$$

$$\mathrm{dn}\left(\frac{2L}{\pi} w \Big/ l^2\right) = \sqrt{1 - 2^2 \sin^2 \psi},$$

$$dw = \frac{\pi}{2L} \frac{d\psi}{\sqrt{1 - l^2 \sin^2 \psi}};$$

then (113) is transformed into

$$\frac{\alpha_3}{\Gamma(1 + \epsilon)} \frac{\pi}{2L} \int_0^{\psi} \frac{d\psi}{(1 - s^2 \sin^2 \psi)\sqrt{1 - l^2 \sin^2 \psi}}. \tag{114}$$

Inverting this integral gives

$$\psi = \mathrm{am}\left[\left(\frac{2L}{\pi} w\right) \Big/ l^2\right] = w + 2 \sum_{n=1}^{\infty} \frac{q^n \sin 2nw}{n(1 + q^{2n})},$$

$$q = \exp\left(-\frac{\pi L'}{L}\right) = \frac{l^2}{16}\left(1 + \frac{l^2}{2} + \frac{21 l^4}{64} + \cdots\right),$$

$$\psi = w + \frac{l^2}{8}\left(1 + \frac{l^2}{2}\right) \sin 2w + \frac{l^4}{256} \sin 4w + \cdots,$$

$$w = \psi - \frac{l^2}{8}\left(1 + \frac{l^2}{2}\right) \sin 2\psi + \frac{3 l^4}{256} \sin 4\psi + \cdots.$$

Thus (114) becomes

$$\int_0^{\psi} \frac{\sqrt{1 - s^2}\sqrt{1 - (l^2/s^2)}}{(1 - s^2 \sin^2 \psi)\sqrt{1 - l^2 \sin^2 \psi}} \, d\psi$$

$$= \int_0^{\psi} \frac{\sqrt{1 - s^2}\, d\psi}{1 - s^2 \sin^2 \psi} - \left(\frac{l^2}{2s^2} + \frac{l^4}{16 s^4}(2 + 3s^2)\right)\sqrt{1 - s^2}\, \psi$$

$$+ \frac{3 l^4}{32 s^2} \sqrt{1 - s^2} \sin 2\psi + \cdots.$$

The first integral on the right-hand side is familiar in Keplerian motion and equals $\tan^{-1}(\sqrt{1 - s^2} \tan \psi)$.

For the second integral of (100) write

$$\operatorname{sn}\left(\frac{2K}{\pi} v \Big/ k^2\right) = \sin \phi;$$

then

$$\int_0^v \frac{c^2\, dv}{\rho^2 + c^2} = \frac{\pi}{2K} \int_0^\phi \frac{c^2}{\rho^2 + c^2} \frac{d\phi}{\sqrt{1 - k^2 \sin^2 \phi}},$$

where

$$\rho = a \frac{(1 - ee_*) + (e_* - e) \cos \phi}{1 + e_* \cos \phi}. \tag{115}$$

Furthermore, letting

$$\nu_1 = \frac{\nu^2}{(1 - e^2)^2}, \qquad k_1 = \frac{k^2}{e^2} = \frac{\nu^2}{(1 - e^2)^2} s^2 + \cdots = \nu_1 s^2 + \cdots,$$

$$\delta_1 = \frac{e_* - e}{(1 - e^2)e} = \frac{\nu^2}{(1 - e^2)^3} (1 - 2s^2) + \cdots = \nu_1 (1 - 2s^2) + \cdots,$$

Izsak obtained

$$\int_0^\phi \frac{c^2\, d\phi}{(\rho^2 + c^2)\sqrt{1 - k^2 \sin^2 \phi}}$$
$$= \nu_1 \left[ (1 - \nu_1) + \left(\frac{1}{2} - 3\nu_1 + \delta_1 + \frac{k_1}{4}\right) e^2 - \left(\frac{3\nu_1}{8} - \frac{k_1}{16}\right) e^4 \right] \phi$$
$$+ \nu_1 \left[ (2 - 4\nu_1) - \left(3\nu_1 - \frac{\delta_1}{2} - \frac{k_1}{4}\right) e^2 \right] e \sin \phi + \cdots.$$

Likewise the integrals in (101) can be expanded:

$$\int_0^v \rho^2\, dw = \frac{\pi}{2K} a^2 \int_0^\phi \frac{(\rho/a)^2}{\sqrt{1 - k^2 \sin^2 \phi}}\, d\phi$$

with

$$\int_0^\phi \frac{(\rho/a)^2\, d\phi}{\sqrt{1 - k^2 \sin^2 \phi}} = \sqrt{1 - e_*^2} \left[ (1 - \delta_2^2) + \frac{k_2}{2} \left(1 - 4\delta_2 - \frac{9k_2}{4}\right) \right] E$$
$$- \sqrt{1 - e_*^2} \left[ (1 - \delta_2) - \frac{k_2}{2} \left(1 - \delta_2 - \frac{3k_2}{4}\right) \right] e \sin E$$
$$- \left[ \frac{k_2}{2} (1 - e_*^2) - \delta_2^2 - k_2 \left(\delta_2 + \frac{9k_2}{16}\right)(2 - e_*^2) \right] \phi$$
$$- k_2 \left(\delta_2 + \frac{3k_2}{4}\right) e \sin \phi + \cdots,$$

where

$$e = e_*(1 - \delta_2), \qquad \cos E = \frac{e_* + \cos\phi}{1 + e_* \cos\phi}, \qquad \sin E = \frac{\sqrt{1 - e_*^2}\,\sin\phi}{1 + e_*\cos\phi},$$

$$\begin{aligned}\delta_2 &= \frac{e_* - e}{e_*} \\ &= \frac{\nu^2}{1 - s^2}(1 - 2s^2) + \frac{\nu^4}{(1 - e^2)^3}[(2 - 12s^2 + 10s^4) - (1 - 2s^4)e^2] \\ &\quad + \cdots, \\ k_2 &= \frac{k^2(1 - e_*^2)}{e_*^2} \\ &= \frac{\nu^2}{1 - e^2}s^2 - \frac{\nu^4}{(1 - e^2)^3}[(1 - 8s^2 + 7s^4) + s^4e^2] + \cdots.\end{aligned}$$

Now (115) can be written in the form

$$\rho = a(1 - e\cos E).$$

This is familiar in a Keplerian motion where $\rho$ and $E$ denote the radius vector and the eccentric anomaly, respectively.

Thus we have obtained

$$\begin{aligned}\phi - \beta_3 &= \int_0^\psi \frac{\sqrt{1 - s^2}\,\sqrt{1 - (l/s)^2}\,d\psi}{(1 - s^2\sin^2\psi)\sqrt{1 - l^2\sin^2\psi}} \\ &\quad - \frac{\alpha_3}{A}\int_0^\phi \frac{c^2\,d\phi}{(\rho^2 + c^2)\sqrt{1 - k^2\sin^2\phi}}, \\ t + \beta_1 &= \frac{a^2}{A}\left[\int_0^\phi \frac{(\rho/a)^2\,d\phi}{\sqrt{1 - k^2\sin^2\phi}} + \frac{K/L}{1 + \epsilon}\nu^2 s^2 \int_0^\psi \frac{\sin^2\psi\,d\psi}{\sqrt{1 - l^2\sin^2\psi}}\right],\end{aligned}$$

where the expansions are substituted for the integrals. The constants are given by

$$\begin{aligned}\hat{n}\frac{a^2}{A} &= \frac{1}{\sqrt{1 - e_*^2}}\frac{\sqrt{1 + k_2}}{(1 + \kappa)(1 - \delta_2)}, \\ \hat{n} &= \sqrt{\frac{\mu}{a^3}}\left\{1 - \frac{3\nu^2(1 - s^2)}{2(1 - e^2)} + \frac{3\nu^4(1 - s^2)}{8(1 - e^2)^3}\right. \\ &\qquad \left.\times [(1 + 11s^2) - (1 - 5s^2)e^2] - \cdots\right\},\end{aligned}$$

$$\frac{\sqrt{1+k_2}}{(1+\kappa)(1-\delta_2)} = 1 - \frac{\nu^2 s^2}{2(1-e^2)} + \frac{\nu^4}{8(1-e^2)^3}$$

$$\times [(12 - 48s^2 + 39s^4) - (8 - 16s^2 + 7s^4)e^2] + \cdots,$$

$$\frac{\alpha_3}{A} = \sqrt{1-s^2}\left\{1 + \frac{\nu^2}{2(1-e^2)^2}\right.$$

$$\left.\times [(5 - 7s^2) + (1 - s^2)e^4 + \cdots]\right\},$$

$$\frac{K/L}{1+\epsilon} = 1 - \frac{\nu^2}{2(1-e^2)^2}[(6 - 7s^2) - s^2e^2] + \cdots.$$

The final expressions for the radius vector and the declination are

$$r = \sqrt{(1 - e\cos E)^2 + \nu^2(1 - s^2\sin^2\psi)},$$

$$\sin\delta = \frac{s\sin\psi}{\sqrt{1 + \nu^2\dfrac{1 - s^2\sin^2\psi}{(1 - e\cos E)^2}}}.$$

The orbital elements are $a$, $e$, $s = \sin I$, $\beta_1$, $\beta_3$, and $\omega$. Other important constants are $e_*$, $k^2$, $\epsilon$, given by (112). The mean motion in the anomaly is $(1+\epsilon)$ times the mean motion in space. The motion of the node is represented by

$$-\frac{3\nu^2\sqrt{1-s^2}}{2(1-e^2)^2} + \frac{3\nu^4\sqrt{1-s^2}}{16(1-e^2)^4}[(18 - 13s^2) + 24s^2e^2] + \cdots.$$

Allen and Knolle (1964, 1965) derived differential corrections for the Izsak solution, and showed that the Izsak intermediary orbit specified by only six parameters was accurate enough, as compared with the rotating-precessing ellipse which requires eight parameters. Borchers (1963), by extending the result of Izsak on the second-order solution of Vinti's problem, has presented a procedure for computing the coordinates and velocity of a satellite from the initial values of the position coordinates and velocity components. Borchers also derived the exact expression for the velocity in Vinti's intermediary orbit, using Izsak's orbital elements, for comparison with other methods of perturbation by means of the so-called Newton-Raphson iteration method (Scarborough, 1958). One drawback of the ordinary perturbation methods is that they necessitate frequent multiplications of Fourier series, inevitably accompanied by the question of convergence or divergence of the expansions.

### 2.24 REDUCTION TO THE TWO-FIXED-CENTER PROBLEM

Move the coordinate origin (0, 0, 0) to (0, 0, $\sqrt{2k}$) and write the new rectangular coordinates by $(x, y, z')$, where $z' = z + \sqrt{2k}$,

$$z' = r_2 \sin \theta', \qquad r_2^2 = x^2 + y^2 + z'^2,$$

$$z = r \sin \theta, \qquad r^2 = x^2 + y^2 + z^2.$$

Then it can be shown (Barrar, 1961) that

$$-\frac{1}{r_2} + \frac{\sqrt{2k}\, P_1(\sin \theta')}{r_2^2} = -\frac{1}{r} + \frac{2kP_2(\sin \theta)}{r^3} + \sum_{n=3}^{\infty} (n - 1)(2k)^{n/2} \frac{P_n(\sin \theta)}{r^{n+1}}.$$

Since

$$\frac{1}{r_2} = \sum_{n=0}^{\infty} c^n \frac{P_n(\sin \theta)}{r^{n+1}}, \qquad \frac{P_n(\sin \theta)}{r^{n+1}} = \frac{(-1)^n}{n!} \frac{\partial^n}{\partial z^n} \left(\frac{1}{r}\right),$$

therefore

$$\frac{1}{r_2} = \sum_{n=0}^{\infty} c^n \frac{(-1)^n}{n!} \frac{\partial^n}{\partial z^n} \left(\frac{1}{r}\right).$$

Partial differentiation of both sides with respect to $z$ and $z'$, respectively, gives

$$\frac{P_1(\sin \theta')}{r_2^2} = -\frac{\partial}{\partial z'} \left(\frac{1}{r_2}\right) = \sum_{n=0}^{\infty} c^n \frac{(-1)^{n+1}}{n!} \frac{\partial^{n+1}}{\partial z^{n+1}} \left(\frac{1}{r}\right)$$

$$= \sum_{n=0}^{\infty} c^n (n + 1) \frac{P_{n+1}(\sin \theta)}{r^{n+2}}.$$

This potential is written in the form

$$-\frac{\mu}{r} \left[1 - \frac{2kP_2(\sin \theta)}{r^2} - k^{3/2} O\left(\frac{1}{r^3}\right)\right],$$

of which the first two terms coincide with those of Vinti's potential

$$-\frac{\mu}{r} \sum_{n=0}^{\infty} (-1)^n (2k)^n \frac{P_{2n}(\sin \theta)}{r^{2n}},$$

but differ from those of the exact potential by $k^{1/2}/r$ times the second term. The equations of motion can be integrated with this potential by the separation of variables in the Hamilton-Jacobi partial differential equation (Barrar, 1961).

Consider two fixed centers with mass $M_1$ and $M_2$ at respective distances $a_1 = aM_2/M$ and $a_2 = -aM_1/M$ $(M = M_1 + M_2)$ from the center of mass of these two centers (Aksenov, Grebenikov, Demin, 1963b; Marchal, 1964, 1966). The problem of two fixed centers can be solved by the separation of variables (Charlier, 1906). Let

$$V = -f\left(\frac{M_1}{r_1} + \frac{M_2}{r_2}\right), \qquad r^2 = x^2 + y^2 + z^2,$$

$$r_1^2 = x^2 + y^2 + (z - a_1)^2, \qquad r_2^2 = x^2 + y^2 + (z - a_2)^2.$$

Then

$$\frac{1}{r_1} = \frac{1}{r}\sum_{n=0}^{\infty}\left(\frac{a_1}{r}\right)^n P_n\left(\frac{z}{r}\right), \qquad \frac{1}{r_2} = \frac{1}{r}\sum_{n=0}^{\infty}\left(\frac{a_2}{r}\right)^n P_n\left(\frac{z}{r}\right).$$

Hence

$$V = -\frac{fM}{r}\left[1 + \sum_{n=0}^{\infty}\frac{\gamma_n}{r^n}P_n\left(\frac{z}{r}\right)\right],$$
$$\gamma_1 = 0, \qquad \gamma_n = \frac{M_1a_1^n + M_2a_2^n}{M} \qquad (n > 1). \tag{116}$$

In order that $V$ be real it is necessary that $M$ and $\gamma_n$, $n = 2, 3, \ldots$, be real; besides, $\gamma_1$ must be zero, since the center of mass is taken as origin; that is,

$$\gamma_1 = 0, \qquad \Im(\gamma_n) = \Im(M) = 0, \qquad n = 2, 3, \ldots. \tag{117}$$

Following Yarov-Yarovoy (1962), let

$$M_1 = \rho_1(\cos\psi_1 + \sqrt{-1}\sin\psi_1), \qquad M_2 = \rho_2(\cos\psi_2 + \sqrt{-1}\sin\psi_2),$$

$$a_1 = R_1(\cos\phi_1 + \sqrt{-1}\sin\phi_1), \qquad a_2 = R_2(\cos\phi_2 + \sqrt{-1}\sin\phi_2).$$

Then the first condition of (117) gives

$$\rho_1R_1\cos(\psi_1 + \phi_1) + \rho_2R_2\cos(\psi_2 + \phi_2) = 0,$$

$$\rho_1R_1\sin(\psi_1 + \phi_1) + \rho_2R_2\sin(\psi_2 + \phi_2) = 0,$$

that is,

$$\tan(\psi_1 + \phi_1) = \tan(\psi_2 + \phi_2),$$

or

$$\psi_1 + \phi_1 = \psi_2 + \phi_2 + s\pi,$$

where $s$ is an integer. If $\cos(\psi_1 + \phi_1) \neq 0$, then we should have

$$\frac{\rho_2}{\rho_1} = \frac{R_1}{R_2}. \tag{118}$$

The second condition of (117) gives

$$\sin(\psi_1 + n\phi_1) + \frac{\rho_2}{\rho_1}\left(\frac{R_2}{R_1}\right)^n \sin(\psi_2 + n\phi_2) = 0,$$

or, by (118),

$$\sin(\psi_1 + n\phi_1) + \left(\frac{R_2}{R_1}\right)^{n-1} \sin(\psi_2 + n\phi_2) = 0. \tag{119}$$

Suppose that $\phi_1 \neq 0, \pi$, $\psi_1 \neq 0, \pi$, $\phi_2 \neq 0, \pi$, $\psi_2 \neq 0, \pi$; then the only solution of (119) that holds for all values of $n$ is $R_1 = R_2$. Then (119) becomes

$$\tan(\psi_1 + \phi_1) = \cot\left[\frac{(n-1)}{2}(\psi_1 + \phi_1)\right].$$

The only solution that holds for all values of $n$ is $\phi_1 + \psi_1 = 0,\ 2\pi$. Hence

$$\psi_1 + \phi_1 = \pi/2 + s_1\pi,$$

$$\psi_2 + \phi_2 = \pi/2 + s_2\pi \qquad (s_1, s_2 = 0, 1, 2, \ldots).$$

From these,

$$\psi_1 + \psi_2 = s_3\pi \qquad (s_3 = 0, 1, 2, \ldots).$$

These relations show that $M_1$ and $M_2$ on one hand, and $a_1$ and $a_2$ on the other, are pairs of conjugate complex quantities.

Next suppose that $\phi_1 = 0, \pi$, $\psi_1 = 0, \pi$, $\phi_2 = 0, \pi$, $\psi_2 = 0, \pi$. Then $M_1$, $M_2$ and $a_1$, $a_2$ should be all real.

Thus there are only two cases in which $U$ is real. In one case, $M_1$ and $M_2$, $a_1$ and $a_2$ are conjugate complex; in the other case, $M_1$, $M_2$, $a_1$, $a_2$ are all real. The third condition of (117) is naturally satisfied.

In the first case,

$$M_1 = \frac{M}{2}(1 + \sqrt{-1}\,\sigma), \qquad M_2 = \frac{M}{2}(1 - \sqrt{-1}\,\sigma),$$

$$a_1 = c(\sigma + \sqrt{-1}), \qquad a_2 = c(\sigma - \sqrt{-1}),$$

and the potential is represented by

$$V = -\frac{\mu}{2}\left(\frac{1 + \sqrt{-1}\,\sigma}{r_1} + \frac{1 - \sqrt{-1}\,\sigma}{r_2}\right), \qquad \mu = fM,$$

$$r_1^2 = x^2 + y^2 + [z - c(\sigma + \sqrt{-1})]^2,$$

$$r_2^2 = x^2 + y^2 + [z - c(\sigma - \sqrt{-1})]^2,$$

or

$$V = -\frac{\mu}{r}\left[1 + \sum_{n=2}^{\infty} \frac{\gamma_n}{r^n} P_n\left(\frac{z}{r}\right)\right],$$

$$\gamma_n = \frac{c^n}{2}[(1 + \sqrt{-1}\,\sigma)(\sigma + \sqrt{-1})^n + (1 - \sqrt{-1}\,\sigma)(\sigma - \sqrt{-1})^n]. \tag{120}$$

This potential coincides as far as the third term with the exact potential (76), which (if we put $z/r$ for $\sin\theta$) is

$$V = -\frac{\mu}{r}\left[1 - \sum_{n=2}^{\infty} J_n\left(\frac{R}{r}\right)^n P_n\left(\frac{z}{r}\right)\right],$$

if $M$ is the earth's mass and $\sigma$ and $c$ are so chosen that the second and third terms of these two expansions should be equal, that is, if

$$c^2(1 + \sigma^2) = J_2, \qquad 2\sigma c^3(1 + \sigma^2) = J_3,$$

or if

$$c = R\sqrt{J_2 - (J_3/2J_2)^2}, \qquad \sigma = -\frac{J_3}{2J_2}\frac{1}{\sqrt{J_2 - (J_3/2J_2)^2}},$$

$$\gamma_4 = c^4(1 + \sigma^2)(1 - 3\sigma^2).$$

In the second case,

$$M_1 = M(1 - \gamma), \qquad M_2 = M\gamma,$$

$$a_1 = a\gamma, \qquad a_2 = -a(1 - \gamma),$$

and the potential is

$$V = -fM\left\{\frac{1 - \gamma}{\sqrt{x^2 + y^2 + (z - a\gamma)^2}} + \frac{\gamma}{\sqrt{x^2 + y^2 + [z - a(1 - \gamma)]^2}}\right\},$$

or

$$V = -\frac{\mu}{r}\left[1 + \sum_{n=2}^{\infty} \frac{\bar{\gamma}_n}{r^n} P_n\left(\frac{z}{r}\right)\right], \tag{121}$$

$$\bar{\gamma}_n = a^n\gamma(1 - \gamma)[\gamma^{n-1} + (-1)^n(1 - \gamma)^{n-1}].$$

The second term coincides with the second term of the exact potential if

$$\bar{\gamma}_2 = -J_2R^2.$$

It is not possible to make the third terms coincide.

Barrar's potential is obtained by expanding

$$V = -\mu\left[\frac{1-\gamma}{\sqrt{x^2+y^2+(z-\delta+\delta-a_1)^2}} + \frac{\gamma}{\sqrt{x^2+y^2+(z-\delta+\delta-a_2)^2}}\right]$$

in the form

$$V = -\frac{\mu}{\rho}\left[1-\frac{\delta(z-\delta)}{\rho^2}\right],$$

$$\rho = \sqrt{x^2+y^2+(z-\delta)^2}$$

with

$$\delta = a\sqrt{\gamma(1-\gamma)} = R\sqrt{J_2}.$$

Vinti's potential is obtained if the coordinate origin is displaced by $c\sigma$ along the rotation axis of the oblate spheroid. Let

$$r_1 = \sqrt{\rho^2 - 2(\sqrt{-1}\,c)\zeta + (\sqrt{-1}\,c)^2},$$

$$r_2 = \sqrt{\rho^2 - 2(-\sqrt{-1}\,c)\zeta + (\sqrt{-1}\,\zeta)^2}$$

$$\rho = \sqrt{x^2+y^2+(z-c\sigma)^2}, \qquad \zeta = z - c\sigma,$$

and expand $1/r_1$ and $1/r_2$ in powers of $\sqrt{-1}\,c/\rho$; then,

$$V = -\frac{\mu}{\rho}\left[1+\sum_{n=1}^{\infty}(-1)^n c^{2n}\left(\frac{1}{\rho}\right)^{2n}P_{2n}\left(\frac{\zeta}{\rho}\right) - c\sigma\sum_{n=0}^{\infty}(-1)^n c^{2n}\left(\frac{1}{\rho}\right)^{2n+1}P_{2n+1}\left(\frac{\zeta}{\rho}\right)\right],$$

which is Vinti's potential. Thus the physical significance of Vinti's potential is the translation of the origin of the spheroidal coordinates by a distance $\sigma = -J_3/2J_2$ times the equatorial radius. This distance, amounting to 7 km in the case of the earth, is larger than any change produced by $J_3$. According to Vinti (1966), it is related to the long-period terms of perturbation theory and turns out to be equal to the displacement by $J_3$ of the plane of symmetry of those exactly elliptical polar orbits that are possible solutions with the spheroidal model.

The integration can be carried out by the transformation

$$x = c\cosh v \sin u \cos w, \qquad y = c \cosh v \sin u \sin w,$$

$$z = c\sigma + c \sinh v \cos u, \qquad dt = I\,d\tau = (\sinh^2 v + \cos^2 u)\,d\tau.$$

The kinetic energy $T$ and the potential $V$ take the form

$$T = \frac{c^2}{2}\left\{\left[\left(\frac{du}{dt}\right)^2 + \left(\frac{dv}{dt}\right)^2\right]I + \left(\frac{dw}{dt}\right)^2 \cosh^2 v \sin^2 u\right\},$$

or

$$T = \frac{1}{2c^2}\left[\frac{1}{I}(p_u^2 + p_v^2) + \frac{p_w^2}{\cosh^2 v \sin^2 u}\right],$$

and

$$V = -\frac{\mu}{c}\frac{\sinh v - \sigma \cos u}{\sinh^2 v + \cos^2 u}.$$

The corresponding Hamilton-Jacobi equation is

$$\left(\frac{\partial S}{\partial u}\right)^2 + \left(\frac{\partial S}{\partial v}\right)^2 + \left(\frac{1}{\sin^2 u} - \frac{1}{\cosh^2 v}\right)\left(\frac{\partial S}{\partial w}\right)^2$$
$$- 2\mu c(\sinh v - \sigma \cos u) = 2c^2 h(\sinh^2 v + \cos^2 u).$$

Separating the variables gives

$$\left(\frac{\partial S}{\partial u}\right)^2 = c_2 - \frac{c_1^2}{\sin^2 u} - 2\mu c\sigma \cos u + 2c^2 h \cos^2 u,$$

$$\left(\frac{\partial S}{\partial v}\right)^2 = -c_2 + \frac{c_1^2}{\cosh^2 v} + 2\mu c \sinh v + 2c^2 h \sinh^2 v,$$

$$\left(\frac{\partial S}{\partial w}\right)^2 = c_1^2,$$

where $h$, $c_1$, $c_2$ are integration constants. The integration can be carried out by using elliptic integrals.

Timoshkova (1968) wrote

$$U = \frac{k^2 M}{r}\left[1 + \sum_{j=2}^{\infty} \frac{J_j}{r^j} P_j\left(\frac{z}{r}\right)\right],$$

$$J_j = \frac{c^j}{2}[(1 + \sqrt{-1}\,\sigma)(\sigma + \sqrt{-1})^j + (1 - \sqrt{-1}\,\sigma)(\sigma - \sqrt{-1})^j],$$

$$c = R\left[-J_2 - \left(\frac{J_3}{2J_2}\right)^2\right]^{1/2}, \qquad \sigma = \frac{J_3}{2J_2}\left[-J_2 - \left(\frac{J_3}{2J_2}\right)^2\right]^{-1/2},$$

and transformed the coordinates

$$x = \sqrt{(\xi^2 + c^2)(1 - \eta^2)} \cdot \cos w, \qquad y = \sqrt{(\xi^2 + c^2)(1 - \eta^2)} \cdot \sin w,$$

$$z = c\sigma + \xi\eta \qquad (-1 \leqq \eta \leqq 1, \xi > 0, \xi^2 + c^2 > R_{\text{pole}}^2),$$

and the time

$$dt = (\xi^2 + \eta^2)\, d\tau.$$

Andrade (1890) solved the two-fixed-center problem, that is, the motion of a particle attracted by two fixed centers, in complete detail by referring to elliptic functions. Recently the discussion has been revived by Sémirot (1949), Deprit (1962), Alekseev (1965), Degtjarow and Ewdokimowa (1966), and Marchal (1966); some have classified various kinds of possible real motions in three dimensions.

## REFERENCES


Ahrens, W.

1895 *Ueber eine Gattung n-fach periodischer Functionen von n reelen Veränderlichen,* Dissertation, Rostock.

Aksenov, E. P., Grebenikov, E. A., and Demin, V. G.

1961 *Iskusstvenye Sputniki Zemli,* **8**, 64.

1963 *Ibid.,* **16**, 173.

1963a *Ibid.,* **16**, 163.

1963b *Astron. Zh.,* **40**, 363; trans. *Sov. Astron.–A.J.,* **7**, 276, 1963.

Alexeyev, V. M.

1965 *Byull. Inst. Teor. Astron.* (Akad. Nauk SSSR), **10**, no. 4, 241.

Allen, W. A., and Knolle, W. E.

1964 *Astron. J.,* **69**, 393.

1965 *Ibid.,* **70**, 152.

Andersen, W.

1929 *Z. Physik,* **56**, 854.

1930 *Ibid.,* **66**, 280.

Andrade, J.

1890 *J. Ecole Polytech. Paris, cahier 60,* 1.

Appell, P., and Goursat, H.

1895 *Théorie des Fonctions Algébriques et de leurs Intégrales,* I, Gauthier-Villars, Paris.

Appell, P., and Lacour, E.

1922 *Principes de la Théorie des Fonctions Elliptiques et Applications,* Gauthier-Villars, Paris.

Bacon, H. M.

1934 *Annals of Math.,* **35**, 776.

Baire, R.

1925 *Leçons sur les Fonctions Discontinues,* Gauthier-Villars, Paris.

Baker, H. F.

1897 *Abel's Theorem and the Allied Theory including the Theta Functions,* Cambridge University Press, London.

1907 *Multiply-Periodic Functions,* Cambridge University Press, London.

Barrar, R. B.

1961 *Astron. J.,* **66**, 11.

Bernstein, S. N.
1926 *Leçons sur les Propriétés Extrémales et les Meilleurs Approximations d'une Variable Réelle*, Gauthier-Villars, Paris.

Besicovitch, A. S.
1932 *Almost-Periodic Functions*, Cambridge University Press, London.

Bliss, G. A.
1933 *Algebraic Functions*, Am. Math. Soc., New York.

Bochner, S.
1925 *Compt. Rend.*, **180**, 1156.
1927 *Math. Annalen*, **96**, 119, 338.
1933 *Acta Math.*, **61**, 149.

Bohl, P.
1893 Dissertation, Tartu. Published as *Über die Darstellung von Functionen einer Variabeln durch Trigonometrische Reihen mit mehreren einer Variabeln proportionalen Argumenten*, C. Mattiesen, Tartu, 1893.
1906 *J. reine angew. Math.*, **131**, 268.

Bohr, H.
1923 *Proc. London Math. Soc.*, ser. 2, **21**, 315.
1923a *Compt. Rend.*, **177**, 737, 1090.
1925 *Acta Math.*, **45**, 29; **46**, 101.
1925a *Math. Z.*, **28**, 38.
1925b *Compt. Rend.*, **181**, 766.
1926 *Acta Math.*, **47**, 237.
1932 *Fast-periodische Funktion*, Springer, Berlin.

Bolza, O.
1909 *Vorlesungen über Variationsrechnung*, Teubner, Leipzig.

Borchers, R. V.
1963 *NASA Technical Note D-1539*, National Aeronautics and Space Administration, Washington, D.C.

Born, M.
1925 *Vorlesungen über Atommechanik*, Springer, Berlin.

Bourbaki, N.
1963 *Intégration*, Hermann, Paris.

Brown, E. W.
1932 *Elements of the Theory of Resonance*, Cambridge University Press, London.

Brunt, D.
1917 *Combination of Observations*, Cambridge University Press, London.

Callender, E. D.
1961 *Astron. J.*, **66**, 134.

Charlier, C. V. L.
1906 *Mechanik des Himmels*, I, Gruyter, Leipzig.

Coble, A. B.
1929 *Algebraic Geometry and Theta Functions*, Am. Math. Soc., New York.

Conforto, F.
1956 *Algebraische Funktionen und algebraische Geometrie*, Springer, Berlin.

de Donder, Th.
1930 *Applications de la Gravifique Einsteinienne*, Gauthier-Villars, Paris.

Degtjarow, W. G., and Ewdokimowa, L. S.
1966 *Byull. Inst. Teor. Astron.* (Akad. Nauk SSSR), **10**, 510.
Denjoy, A.
1941 *Leçons sur le Calcul des Coefficients d'une Série Trigonométrique*, I–III, Gauthier-Villars, Paris.
1949 *Ibid.*, IV.
Deprit, A.
1962 *Bull. Soc. Math. Belg.*, **14**, 12.
De Sitter, W.
1916 *Monthly Notices Roy. Astron. Soc.*, **76**, 699.
1917 *Proc. Acad. Sci. Amsterdam*, **19**, 367.
Eddington, A. S.
1923 *The Mathematical Theory of Relativity*, Cambridge University Press, London.
Eddington, A. S., and Clark, G. L.
1938 *Proc. Roy. Soc. London*, ser. A, **166**, 465.
Einstein, A.
1914 *Sitzber. preuss. Akad. Wiss. Berlin*, **1914**, 1030.
1915 *Ibid.*, **1915**, 315, 831.
1916 *Ibid.*, **1916**, 688.
1916a *Ann. Physik*, **49**, 769.
Einstein, A., and Infeld, L.
1940 *Annals of Math.*, **41**, 455.
Einstein, A., Infeld, L., and Hoffmann, B.
1938 *Annals of Math.*, **39**, 65.
Esclangon, E.
1902 *Compt. Rend.*, **135**, 891.
1903 *Ibid.*, **137**, 30.
1904 Thesis, Paris. Published as *Les Fonctions Quasi-Périodiques*, Gauthier-Villars, Paris, 1904.
1916 *Ann. Obs. Bordeaux*, **16**, 53.
Favard, J.
1928 *Acta Math.*, **51**, 31.
1933 *Leçons sur les Fonctions Presque-Périodiques*, Gauthier-Villars, Paris.
Forsyth, A. R.
1893 *Theory of Functions of a Complex Variable*, Cambridge University Press, London.
Fues, E.
1925 *Z. Physik*, **34**, 788.
Garfinkel, B.
1958 *Astron. J.*, **63**, 88.
1959 *Ibid.*, **64**, 353.
1964 *Ibid.*, **69**, 223.
Goursat, H.
1918 *Cours d'Analyse Mathématique*, II, 123.
Haag, J.
1931 *Le Probléme de Schwarzschild*, Gauthier-Villars, Paris.

Hagihara, Y.
1927 *Japan. J. Astron. Geophys.*, **5**, 1.
1931 *Ibid.*, **8**, 67.
1944 *Ibid.*, **21**, 29.
1962 *Astron. J.*, **67**, 108.
Hahn, H.
1921 *Theorie des reellen Funktionen*, Springer, Berlin.
Halmos, P. R.
1950 *Measure Theory*, Van Nostrand, New York.
Halphen, G. H.
1886 *Traité des Fonctions Elliptiques*, I, 408, 426, 428, Gauthier-Villars, Paris.
Hardy, G. H., and Littlewood, J. E.
1914 *Acta Math.*, **37**, 156.
Hardy, G. H., and Rogosinski, W. W.
1950 *Fourier Series*, Cambridge University Press, London.
Hilb, E. and Riesz, F.
1924 *Neuere Untersuchungen über trigonometrischen Reihen*, Teubner, Leipzig.
Hirzebruch, F.
1956 *Neue Topologische Methoden in der algebraischen Geometrie*, Springer, Berlin.
Hobson, E. W.
1926 *Theory of Functions of a Real Variable*, II, Cambridge University Press, London.
Infeld, L.
1931 *Phys. Z.*, **32**, 110.
Izsak, I. G.
1960 *Smithsonian Astrophys. Obs. Spec. Rep. No. 52.*
Jeffreys, H.
1952 *The Earth*, 3rd ed., Cambridge University Press, London.
Kaczmarz, S., and Steinhaus, H.
1935 *Theorie der Orthogonalreihen*, Monografje Matematycze, Warsaw.
Kestelman, H.
1937 *Modern Theories of Integration*, Clarendon, Oxford.
Kislik, M. D.
1960 *Iskusstvenye Sputniki Zemli*, **4**, 3; translated in *Artificial Earth Satellite*, L. V. Kurnosova, ed., Plenum Press, New York, 1963.
1961 *Iskusstvenye Sputniki Zemli*, **13**, 23.
Knopp, K.
1931 *Theorie und Anwendung der unendlichen Reihen*, Springer, Berlin.
Koksma, J. F.
1936 *Diophantische Approximationen*, Springer, Berlin.
Kolmogorov, A. N., and Fomin, S. V.
1961 *Measure, Lebesgue Integrals and Hilbert Space*, Academic Press, New York.
König, D., and Szücs, A.
1913 *Rendiconti Cir. Mat. Palermo*, **36**, 79.
Kronecker, L.
1899 *Leopold Kronecker's Werke*, III, Part 1, 49, Teubner, Leipzig.

Landau, E.
1927 *Vorlesungen über Zahlentheorie*, I–III, Hirzel, Leipzig.
1929 *Math. Annalen*, **102**, 177.
1935 *Math. Z.*, **39**, 745.
Lebesgue, H.
1906 *Leçons sur les Séries Trigonométriques*, Gauthier-Villars, Paris.
1930 *Leçons sur l'Intégration et la Recherches des Fonctions Primitives*, Gauthier-Villars, Paris.
Lettenmeyer, F.
1923 *Proc. London Math. Soc.*, ser. 2, **21**, 306.
Levi-Civita, T.
1933 *Z. angew. Math. Mech.*, **13**, 112.
Maak, W.
1950 *Fastperiodische Funktionen*, Springer, Berlin.
Marchal, C.
1964 *Actes XV$^e$ Congrès Intern. d'Astronautique* (Warsaw), I, 159.
1966 *Bull. Astron.*, ser. 3, **1**, fasc. 3.
McShane, E. J.
1947 *Integration*, Princeton University Press, Princeton, N.J.
Merson, R. H., and King-Hele, D. G.
1958 *Nature*, **182**, 640.
Mineur, H.
1936 *J. Math. pures appl.*, ser. 9, **15**, 385.
Minkowski, H.
1907 *Diophantische Approximation*, Teubner, Leipzig.
Montel, P.
1927 *Leçons sur les Familles Normales des Fonctions Analytiques et leurs Applications*, Gauthier-Villars, Paris.
Natanson, I. P.
1955 *Theory of Functions of a Real Variable*, I, Frederick Ungar, New York.
1960 *Ibid.*, II.
Oppenheimer, J. R., and Volkoff, G. H.
1939 *Phys. Rev.*, **55**, 374.
Osgood, W. F.
1924 *Lehrbuch der Funktionentheorie*, II, Part 1, Teubner, Leipzig. Reprinted by Chelsea Publishing, Bronx, N.Y., 1965.
1932 *Ibid.*, II, Part 2.
Picard, É.
1905 *Traité d'Analyse*, II, Chap. 14, Gauthier-Villars, Paris.
1931 *Quelques Applications Analytiques de la théorie des Courbes et des Surfaces Algébriques*, Gauthier-Villars, Paris.
Pilowski, G. I.
1928 *Z. Physik*, **49**, 588.
Poincaré, H.
1885 *J. Math. pures appl.*, ser. 4, **1**, 167. Reprinted in *Oeuvres de Henri Poincaré*, I, 90, Gauthier-Villars, Paris, 1928.
1899 *Méthodes Nouvelles de la Mécanique Céleste*, III, Gauthier-Villars, Paris.

Roberts, W. R. W., and Hartford, R. R.
1938 *Elliptic and Hyperelliptic Integrals and Allied Theory*, Cambridge University Press, London.
Robertson, H. P.
1938 *Annals of Math.*, **39**, 101.
Rogosinski, W. W.
1930 *Fouriersche Reihen*, Vieweg, Brunswick.
Saks, S.
1937 *Theory of Integrals*, Lwów, Warsaw.
Scarborough, J. B.
1958 *Numerical Mathematical Analysis*, Johns Hopkins Press, Baltimore, Md.
Schlesinger, L., and Plessner, A.
1926 *Lebesguesche Integrale und Fouriersche Reihe*, Gruyter, Berlin.
Schönberg, I.
1928 *Math. Z.*, **28**, 171.
Schwarzschild, K.
1916 *Sitzber. preuss. Akad. Wiss. Berlin*, **1916**, 189, 424.
Sémirot, P.
1949 *Bull. Astron.*, ser. 2, **14**, 37.
Sierpinski, W.
1928 *Leçons sur les Nombres Transfinis*, Lwów, Warsaw.
1934 *L'Hypothèse du Continu*, Lwów, Warsaw.
Stäckel, P.
1891 *J. reine angew. Math.*, **107**, 319.
1893 *Math. Annalen*, **42**, 537, 563.
1912 *J. reine angew. Math.*, **128**, 222.
Stahl, H.
1896 *Theorie der Abel'schen Funktionen*, Teubner, Leipzig.
Stahl, H., and Noether, M.
1911 *Abriss einer Theorie der algebraischen Funktionen einer Veränderlichen in neuer Fassung*, Teubner, Leipzig.
Staude, O.
1887 *Math. Annalen*, **29**, 468.
1889 *J. reine angew. Math.*, **105**, 298.
Sterne, T. E.
1957 *Astron. J.*, **62**, 96.
1958 *Ibid.*, **63**, 28.
Stoner, E. C.
1929 *Phil. Mag.*, **7**, 63.
1930 *Ibid.*, **9**, 951.
Stumpff, K.
1937 *Grundlagen und Methoden der Periodenforschung*, Springer, Berlin.
Synge, J. L.
1928 *Math. Annalen*, **99**, 738.
Thielman, H. P.
1953 *Theory of Functions of Real Variables*, Butterworth, London.
Thomas, L. H.
1924 *Proc. Cambridge Phil. Soc.*, **22**, 886.

Timoshkova, E. I.
1968 *Byull. Inst. Teor. Astron.* (Akad. Nauk SSSR), **11**, 465.
Tonelli, L.
1928 *Serie Trigonometriche*, Zanichelli, Bologna.
Vallée Poussin, C. de la
1916 *Intégrales de Lebesgue, Fonctions d'Ensemble, Classe de Baire*, Gauthier-Villars, Paris.
1919 *Leçons sur l'Approximation des Fonctions d'une Variable Réelle*, Gauthier-Villars, Paris.
van der Corput, J. G.
1921 *Math. Annalen*, **84**, 53.
1931 *Acta Math.*, **56**, 373.
1932 *Ibid.*, **59**, 209.
van der Waerden, B. L.
1937 *Moderne Algebra*, I, Springer, Berlin.
1939 *Einführung in die algebraische Geometrie*, Springer, Berlin.
Vinti, J. P.
1959 *J. Res. Natl. Bur. Standards*, **63B**, 105.
1961 *Astron. J.*, **66**, 514.
1961a *J. Res. Natl. Bur. Standards*, **65B**, 169.
1962 *Ibid.*, **66B**, 5.
1963 *Ibid.*, **67B**, 191.
1966 *Ibid.*, **70B**, 1.
1966a *Ibid.*, **70B**, 17.
1966b *Proc. IAU Symp. No. 25, Salonika*, G. Contopoulos, ed., Academic Press, New York.
Weierstrass, K.
1866 *Monatshefte Berliner Akad. Wiss.*, **1866**, 97.
1895 *Mathematische Werke*, II, 125, Mayer u. Müller, Berlin.
Weyl, H.
1916 *Math. Annalen*, **77**, 313.
1919 *Phys. Z.*, **20**, 31.
1919a *Ann. Physik*, **59**, 185.
Whitney, H.
1957 *Geometric Integration Theory*, Princeton mathematical series, no. 21, Princeton University Press, Princeton, N.J.
Whittaker, E. T., and Watson, G. N.
1924 *Modern Analysis*, Cambridge University Press, London.
Yarov-Yaravoy, M. S.
1962 *Byull. Inst. Teor. Astron.* (Akad. Nauk SSSR), **8**, 647.
Zhongolovich, I. D.
1957 *Byull. Inst. Teor. Astron.* (Akad. Nauk SSSR), **6**, 507.
Zygmund, A.
1935 *Trigonometrical Series*, Monografje Matematycze, Warsaw.

# Particular Solutions of the Many-Body Problem

## HOMOGRAPHIC MOTION

### 3.1 FUNDAMENTAL EQUATIONS

Before we attempt a general solution to the problem of many bodies we discuss its particular solutions, that is, the rigorous solutions of the problem when the initial motion of the bodies obeys certain conditions. In the case of three bodies Euler discovered a particular solution in which the three bodies always lie on a straight line, their positions being determined by a quintic equation with functions of the three masses as coefficients. This straight line passes through the center of mass of the three bodies, and the bodies individually perform Keplerian motions with the center of mass as one of the foci. Lagrange has found a particular solution in which each of the three bodies lies at a vertex of an equilateral triangle, irrespective of the masses. Each radius vector from the center of mass of the system to each of the masses moves in the same way as the radius vector of a Keplerian motion around the focus. At each instant the three bodies form an equilateral triangle. Lagrange reduced the equations of motion of the three bodies to a system of two differential equations of order 2 and one differential equation of order 3, that is, to a system of differential equations of order 7, and obtained from the system the particular solutions of Euler and of Lagrange. In a special case when each of the bodies performs a circular motion, the particular solution of Euler is called the *straight-line equilibrium-point solution* and the particular solution of Lagrange the *equilateral-triangular equilibrium-point solution*. These definitions refer to points in equilibrium, that is, points which are at rest with respect to rotating axes. These solutions are the only particular solutions (apart from the trivial solution in which the three bodies move along a single straight line fixed in space).

When the resultant force acting on each particle in the many-body problem is directed to the center of mass of the system and is of magnitude proportional to the distance of the particle from the center of mass, and when the initial velocity of each particle is in the direction making a fixed angle (the same for all the particles) with the radius vector to the particle drawn from the center of mass and is of magnitude proportional to the magnitude of the radius vector, then, according to Laplace, there exist possible motions in which the radius vectors of the particles at any given instant are mutually in a constant ratio independent of time, and the angular velocities of the particles at any given instant are equal. Such motions are called *homographic* (Lehmann-Filhés, 1891; Dziobek, 1900; Seydler, 1886; Moulton, 1900, 1904; Pizzetti, 1904; Banachiewicz, 1906; Andoyer, 1906; Longley, 1907; Woronetz, 1907; Brehm, 1908; Zapp, 1908; Bilimovitch, 1911; Chazy, 1918, 1927; Müntz, 1922; Lindow, 1922, 1923, 1924, 1926, 1928; Schaub, 1929; MacMillan and Bartkey, 1932; Meyer, 1933; Williams, 1938; Meffroy, 1962, 1967). Dziobek proved in the case of $n$ particles mutually attracted according to Newtonian law the possibility of their homographic motion. Such motions are a generalization of the particular solutions of Euler and Lagrange for the three-body problem.

In the generalized particular solution of Lagrange, the configuration of $n$ particles all in one plane in their homographic motion is called a *central figure*. The generalized collinear particular solution of Euler, in which all $n$ particles lie on the same straight line at each instant, is also possible. When $n \geqq 4$, there may occur homographic motions in three dimensions. The motion of each particle is along the corresponding straight line fixed in space, all lines pass through the center of mass; the configuration of the $n$ bodies at each instant remains similar, and the corresponding sides remain parallel and proportional in length. Such a configuration is called *homothetic*. A homographic motion for which the total angular momentum of the particles is zero forms a homothetic configuration, while a homographic motion for which the total angular momentum is not zero occurs either on a straight line or on a plane and in no other way. As will be shown later, if all particles in the problems of three bodies and of many bodies collide at the same point at the same instant, the angular momentum of the whole system is necessarily zero and the particles tend to one of such homographic configurations at the collisions. We propose to discuss such particular solutions of the $n$-body problem in the most general manner (Meyer, 1933).

Suppose that the center of mass in the $n$-body problem is fixed in space, and take it as the coordinate origin. (Because, if there is no external force acting on the particle system, the center of mass is at rest.) In order to discuss such homographic motions, take coordinate axes that rotate around the origin relative to the axes fixed in space, and that at the same time expand or contract. Denote the coordinates of the $n$ particles at the initial epoch by $x_i, y_i, z_i$ $(i = i, 2, \ldots, n)$, and at any instant $t$ by

$$\xi_i = \lambda x_i, \qquad \eta_i = \lambda y_i, \qquad \zeta_i = \lambda z_i \qquad (i = 1, 2, \ldots, n), \tag{1}$$

with respect to the rotating and expanding or contracting axes. For a homographic motion, the coefficient $\lambda$ is a function of $t$ only. The configuration is supposed to have expanded in the ratio $\lambda$ since the initial epoch; in other words, every length is supposed to have expanded in all directions in the ratio $\lambda$. Further, denote by $\omega_1, \omega_2, \omega_3$ the angular velocities of the rotating axes around the instantaneous coordinate axes. Call these rotating and expanding axes the $(\xi, \eta, \zeta)$-axes.

The rate of variation with time of a vector $\mathfrak{V}(u, v, w)$, referred to the $(\xi, \eta, \zeta)$-axes, is

$$\frac{du}{dt} - v\omega_3 + w\omega_2, \qquad \frac{dv}{dt} - w\omega_1 + u\omega_3, \qquad \frac{dw}{dt} - u\omega_2 + v\omega_1,$$

because the increments for each component of $\mathfrak{V}$ in $dt$ are

$$\begin{array}{llll} \dfrac{du}{dt}\,dt, & \omega_3 u\,dt, & -\omega_2 u\,dt, & \text{for } u, \\[2ex] -\omega_3 v\,dt, & \dfrac{dv}{dt}\,dt, & \omega_1 v\,dt, & \text{for } v, \\[2ex] \omega_2 w\,dt, & -\omega_1 w\,dt, & \dfrac{dw}{dt}\,dt, & \text{for } w. \end{array}$$

Substitute for $\mathfrak{V}$ the coordinates of the $i$th particles $(\xi_i, \eta_i, \zeta_i)$ referred to the $(\xi, \eta, \zeta)$-axes. This gives the velocity of the $i$th particle relative to the $(\xi, \eta, \zeta)$-axes in the form

$$V_{\xi_i} = \frac{d\xi_i}{dt} - \eta_i\omega_3 + \zeta_i\omega_2 = -\lambda(\omega_3 y_i - \omega_2 z_i) + \frac{d\lambda}{dt} x_i,$$

$$V_{\eta_i} = \frac{d\eta_i}{dt} - \zeta_i\omega_1 + \xi_i\omega_3 = -\lambda(\omega_1 z_i - \omega_3 x_i) + \frac{d\lambda}{dt} y_i,$$

$$V_{\zeta_i} = \frac{d\zeta_i}{dt} - \xi_i\omega_2 + \eta_i\omega_1 = -\lambda(\omega_2 x_i - \omega_1 y_i) + \frac{d\lambda}{dt} z_i,$$

and the kinetic energy of the whole system as

$$T = \frac{1}{2}\sum_{i=1}^{n} [V_{\xi_i}^2 + V_{\eta_i}^2 + V_{\zeta}^2].$$

Denote the mutual distance of the $k$th and the $i$th particles at the initial epoch by $r_{ki}$ and at an arbitrary epoch $t$ by $\rho_{ki}$; then

$$\rho_{ki}^2 = (\xi_k - \xi_i)^2 + (\eta_k - \eta_i)^2 + (\zeta_k - \zeta_i)^2,$$
$$r_{ki}^2 = (x_k - x_i)^2 + (y_k - y_i)^2 + (z_k - z_i)^2,$$
$$\rho_{ki} = \lambda r_{ki}.$$

Let the mass of the $i$th particle be $m_i$, and choose the units so that the constant of gravitation is equal to 1; then the force function is expressed by

$$U = \sum_i \sum_k{}' \frac{m_k}{\rho_{ki}} = \frac{1}{\lambda}\sum_i \sum_k{}' \frac{m_k}{r_{ki}},$$

where $\sum'$ represents the summation over $k = 1, 2, \ldots, n$, excluding $i = k$.

Substitute these expressions for $T$ and $U$ in the Lagrangian equations

$$\frac{d}{dt}\left(\frac{\partial T}{\partial \dot{\xi}_i}\right) - \frac{\partial T}{\partial \xi_i} = \frac{\partial U}{\partial \xi_i},$$

$$\frac{d}{dt}\left(\frac{\partial T}{\partial \dot{\eta}_i}\right) - \frac{\partial T}{\partial \eta_i} = \frac{\partial U}{\partial \eta_i},$$

$$\frac{d}{dt}\left(\frac{\partial T}{\partial \dot{\zeta}_i}\right) - \frac{\partial T}{\partial \zeta_i} = \frac{\partial U}{\partial \zeta_i} \qquad (i = 1, 2, \ldots, n).$$

This gives the equations of motion of the $i$th particle:

$$x_i\left[\lambda\frac{d^2\lambda}{dt^2} - \lambda^2(\omega_2^2 + \omega_3^2)\right] + y_i\left[\lambda^2\omega_1\omega_2 - \frac{d}{dt}(\lambda^2\omega_3)\right]$$
$$+ z_i\left[\lambda^2\omega_1\omega_3 + \frac{d}{dt}(\lambda^2\omega_2)\right] = \frac{1}{\lambda}\sum_k{}' m_k\frac{x_k - x_i}{r_{ik}^3},$$

$$x_i\left[\lambda^2\omega_1\omega_2 + \frac{d}{dt}(\lambda^2\omega_3)\right] + y_i\left[\lambda\frac{d^2\lambda}{dt^2} - \lambda^2(\omega_2^3 + \omega_1^2)\right]$$
$$+ z_i\left[\lambda^2\omega_2\omega_3 - \frac{d}{dt}(\lambda^2\omega_1)\right] = \frac{1}{\lambda}\sum_k{}' m_k\frac{y_k - y_i}{r_{ik}^3}, \qquad (2)$$

$$x_i\left[\lambda^2\omega_1\omega_3 - \frac{d}{dt}(\lambda^2\omega_2)\right] + y_i\left[\lambda^2\omega_2\omega_3 + \frac{d}{dt}(\lambda^2\omega_1)\right]$$
$$+ z_i\left[\lambda\frac{d^2\lambda}{dt^2} - \lambda^2(\omega_1^2 + \omega_2^2)\right] = \frac{1}{\lambda}\sum_k{}' m_k\frac{z_k - z_i}{r_{ki}^3}$$
$$(i = 1, 2, \ldots, n).$$

These equations of motion reduce to those of Andoyer (1906) for $\lambda = 1$. Brehm (1908) has deduced the same equations by a different method. From these equations it looks as though Laplace's assumption that the net force on a particle is directed to the center of mass in a homographic motion, were not necessary. In the first equation of (2), for instance, it looks as though the components of force were not proportional to $x_i$, since $\omega_1$, $\omega_2$, $\omega_3$ are not all zero. But it can be shown from the properties of homographic motions, as will be seen immediately, that the velocity of each particle should be proportional to its distance from the center of masses both at the initial epoch and at any time following, and that the angle between the radius vector and the velocity should be the same for all particles at any given instant.

We will discuss the solutions of (2), distinguishing the cases of straight-line configurations, plane configurations, and space configurations. Homographic motions forming these configurations are called the straight-line, plane, and space homographic motions, respectively. The converse is also true. That is, take as the $\xi$-axis in the relation (1) the straight line forming the line configuration at a given instant; then $\eta_i = \zeta_i = 0$ $(i = 1, 2, \ldots, n)$, and $y_i = z_i = 0$ $(i = 1, 2, \ldots, n)$, since $\lambda \neq 0$; and hence $\eta_i = \lambda y_i = 0$, $\zeta_i = \lambda z_i = 0$. In the plane configuration, take the plane $\zeta = 0$; then $\zeta_i = 0$ $(i = 1, 2, \ldots, n)$ and $z_i = 0$ $(i = 1, 2, \ldots, n)$; hence, $\zeta_i = \lambda z_i = 0$ $(i = 1, 2, \ldots, n)$. If we exclude the case of simultaneous collisions of all $n$ bodies, then all the $x_i, y_i, z_i$ are not simultaneously zero, if all of the $\xi_i, \eta_i, \zeta_i$ are not simultaneously zero at some instant. Hence, $\xi_i, \eta_i, \zeta_i$ are never simultaneously zero.

## STRAIGHT-LINE HOMOGRAPHIC MOTIONS

### 3.2 STRAIGHT-LINE CONFIGURATION

When all $n$ particles lie on a straight line, we take the straight line as our $\xi$-axis and $y_i = z_i = 0$ $(i = 1, 2, \ldots, n)$. We can choose the $\eta$-axis so that $\omega_2 = 0$. Then equations (2) take the form

$$\begin{aligned} \frac{d^2\lambda}{dt^2} - \lambda\omega_3^2 &= \frac{1}{\lambda^2}\frac{1}{x_i}\sum_k{}' m_k \frac{x_k - x_i}{r_{ik}^3}, \\ \frac{d}{dt}(\lambda^2\omega_3) = 0, &\qquad \lambda^2\omega_1\omega_3 = 0. \end{aligned} \tag{3}$$

The last two equations give either $\omega_3 = 0$, $\omega_1 \neq 0$, or $\omega_1 = 0$, $\omega_3 \neq 0$. In the first case, the $\xi$-axis is fixed and the motion is on the straight line fixed in space, the angular momentum of the system being zero. In the second case, $\lambda^2\omega_3$ = constant = $h$ and the $\xi$-axis rotates around the $\eta$-axis. This is a plane motion, and the angular momentum is not zero. We suppose that $h = 0$ in the first case and $h \neq 0$ in the second case; then

$$\frac{d^2\lambda}{dt^2} - \frac{h^2}{\lambda^3} = \frac{1}{\lambda^2}\frac{1}{x_i}\sum_k{}' m_k \frac{x_k - x_i}{r_{ik}^3}. \tag{4}$$

The right-hand side of (4) depends on the $i$th particle, while the left-hand side is independent of $i$. Hence, both sides should equal $-(k/\lambda)^2$, where $k$ is a constant. (The sign is minus because the force is one of attraction.) Thus,

$$\frac{d^2\lambda}{dt^2} = -\frac{k^2}{\lambda^2} + \frac{h^2}{\lambda^3}, \qquad \lambda^2\omega_3 = h. \tag{5}$$

Multiplying both sides by $d\lambda/dt$ and integrating, we obtain

$$\frac{1}{2}\left(\frac{d\lambda}{dt}\right)^2 = \frac{k^2}{\lambda} - \frac{h^2}{2\lambda^2} + \gamma$$

$$= \frac{1}{2\lambda^2}(2\gamma\lambda^2 + 2k^2\lambda - h^2) \qquad (\gamma = \text{constant}).$$

We see that this equation is identical with the energy integral in the Keplerian motion

$$\frac{1}{2}\left(\frac{d\lambda}{dt}\right)^2 + \frac{1}{2}\lambda^2\omega_3^2 - \frac{k^2}{\lambda} = \text{constant},$$

if we notice that the second equation of (5) represents the constancy of the areal velocity. Thus (5) represents the equations for the Keplerian motion, as will be seen in Sections 5.9–5.15. If $\lambda$ = constant, then $\omega_3$ = constant and the $n$ particles are at rest with respect to the rotating axes. This is called the *straight-line equilibrium point.* In particular, if $h = 0$, then the motion is along a fixed straight line and ends at a simultaneous collision of all the $n$ particles at the center of mass of the system.

The condition for the possibility of the motion is that the $n$ equations

$$X_i \equiv \sum_k{}' m_k \frac{x_k - x_i}{r_{ki}^3} + k^2 x_i = 0 \qquad (i = 1, 2, \ldots, n), \tag{6}$$

obtained by making the right-hand member of (4) equal to $-(k/\lambda)^2$, should hold for all values of $i$. We have the relation

$$\sum_{i=1}^{n} m_i X_i = 0, \tag{7}$$

among the $n$ equations (6); hence, there are only $n - 1$ independent equations among them.

Write $\sum''$ for the summation over $k = 1, 2, \ldots, n$ by excluding $k = i$ and $k = j$. Then, from (6),

$$X_i - X_j \equiv (x_i - x_j)\left(k^2 - \frac{m_i + m_j}{r_{ij}^3}\right) + \sum_k'' m_k \frac{x_k - x_i}{r_{ki}^3} - \sum_k'' m_k \frac{x_k - x_j}{r_{kj}^3} = 0. \tag{8}$$

Equations similar to (7) and (8) were obtained by Meffroy (1962).

First suppose that $n = 2$. Then (6) becomes

$$m_2 \frac{x_2 - x_1}{r_{21}^3} + k^2 x_1 = 0, \qquad m_1 \frac{x_1 - x_2}{r_{12}^3} + k^2 x_2 = 0.$$

By solving these equations we obtain

$$x_1 = -\frac{m_2 r_{12}}{m_1 + m_2}, \qquad x_2 = \frac{m_1 r_{12}}{m_1 + m_2}, \qquad k^2 = \frac{m_1 + m_2}{r_{12}^3};$$

that is, the motion is Keplerian around the center of masses, and $\omega_3$ is its angular velocity; furthermore, the areal velocity is constant, as the second equation of (5) shows.

Next, suppose that $n = 3$, and $x_1 < x_2 < x_3$. Since $r_{12} = x_2 - x_1$, $r_{13} = x_3 - x_1$, $r_{23} = x_3 - x_2$; then we may define the ratios $r_{12}/r_{23} = \alpha$, $r_{31}/r_{21} = (1 + \alpha)/\alpha$, $r_{31}/r_{32} = 1 + \alpha$, where $\alpha$ is a constant. Hence, we may write (8) in the form

$$k^2 = \frac{m_1 + m_2}{r_{12}^3} + \frac{m_3}{r_{12}^3}\left(\frac{\alpha}{1 + \alpha}\right)^2 - \frac{m_3}{r_{12}^3}\alpha^2,$$

$$k^2 = \frac{m_1 + m_3}{r_{12}^3}\left(\frac{\alpha}{1 + \alpha}\right)^3 + \frac{m_2}{r_{12}^3}\left(\frac{\alpha}{1 + \alpha}\right) + \frac{m_2}{r_{12}^3}\left(\frac{\alpha^3}{1 + \alpha}\right),$$

$$k^2 = \frac{m_2 + m_3}{r_{12}^3}\alpha^3 - \frac{m_1}{r_{12}^3}\alpha + \frac{m_1}{r_{12}^3}\left(\frac{\alpha^3}{(1 + \alpha)^2}\right).$$

We add these three equations after multiplying them by $\alpha$, $-(1 + \alpha)$, 1, respectively; this gives an identity $0 \equiv 0$. Hence, the three equations are not independent. From the first and the third we derive a quintic equation

$$(m_2 + m_3)\alpha^5 + (3m_3 + 2m_2)\alpha^4 + (3m_3 + m_2)\alpha^3 - (3m_1 + m_2)\alpha^2 - (3m_1 + 2m_2)\alpha - (m_1 + m_2) = 0.$$

By Descartes' rule in the theory of algebraic equations, this must have at least one positive root. Hence, there are homographic motions of the type that we consider. This algebraic equation is the quintic equation of Lagrange in the three-body problem.

In the case of the sun, the earth, and the moon, the ratio $\alpha$ is about 1/100. According to whether the moon is situated between the other two or the earth is between the other two, there is generally either a new or a full moon. Actually, if the motion were strictly homographic, the three bodies would be on a straight line, and there would always be either a solar or a lunar eclipse; that is, the three bodies would always be in syzygy.

Andoyer (1906) discussed the case $n = 4$. Lehmann-Filhés (1891) and Moulton (1900, 1904) discussed a more general case. Recently Brumberg (1958) has obtained the equilibrium configurations in the way described above and has shown the instability of the motion. The motion of each particle is Keplerian around the center of mass. All particles are on a straight line at any instant and the relative positions of the particles on the straight line are given by (6) and are independent of time; hence the configuration is similar at all times.

The relative positions of the $n$ bodies in such cases are the same for $h = 0$ and $h \neq 0$. In particular, if $h = 0$, that is, if the total angular momentum of the system is zero, then each of the $n$ bodies moves on the corresponding straight line fixed in space through the center of mass, and the motion is homothetic. If $h \neq 0$, then homothetic motions are not possible. The motion in this case is a plane motion, and the straight line on which the $n$ bodies are located rotates in that plane. If $\lambda =$ constant, then the positions of the $n$ bodies are the straight-line equilibrium points. As there are $n!/2$ ways in which the $n$ bodies interchange their relative positions, there are $(n!/2)$-line homographic motions.

Suppose that we know the motion of $n$ bodies with finite masses. What is the motion, under the attraction of these $n$ bodies, of the $(n + 1)$th body with an infinitesimal mass? By infinitesimal we mean so vanishingly small as not to affect the motion of the first $n$ bodies. This restricted problem of $(n + 1)$ bodies is called the "$(n + 1)$-body problem."

The first $n$ bodies can lie on a straight line in $n!/2$ different ways. Suppose that we have proved the existence of homographic motions for the $n$ bodies and consider the motion of the $(n + 1)$th body with $m_{n+1} = 0$. The equations (6) are divided into two sets. The first $n$ equations are the set that determine the motion of the first $n$ bodies. Such homographic motions are $n!/2$ in number, by assumption, and are supposed to be known. The second set of equations is

$$X_{n+1} = m_1 \frac{x_1 - x_{n+1}}{r_{1,n+1}^3} + m_2 \frac{x_2 - x_{n+1}}{r_{2,n+1}^3} + \cdots + m_n \frac{x_n - x_{n+1}}{r_{n,n+1}^3} + k^2 x_{n+1} = 0.$$

The unknown quantity $x_{n+1}$ is determined when the values of all the other $x$ are known. Let a solution of the first set of equations be

$$x_1^0, x_2^0, \ldots, x_n^0,$$

such that $x_1^0 < x_2^0 < \cdots < x_n^0$, and suppose that $\partial X_{n+1}/\partial x_{n+1} > 0$. Then there is one and only one real root $x_{n+1}$ in each of the $n + 1$ regions bounded by $-\infty, x_1^0, x_2^0, \ldots, x_n^0, +\infty$. Thus there exist $(n + 1)!/2$ solutions in all.

Take one such solution, for example, $x_1^0 < x_2^0 < \cdots < x_n^0 < x_{n+1}^0$. As one of the masses $m_k$ starts with the value zero and increases continuously through positive values, the quantities $x_1, \ldots, x_{n+1}$ vary from their initial values in a continuous manner. They cease to be continuous either when one of the masses $m_k$ becomes infinite or when two of the masses coincide. On the other hand, a root of an algebraic equation with real coefficients changes from a real number to a complex number either when it becomes infinite or when it is one of two roots which coincide to become a double root.

Suppose at first that $x_i$ and $x_j$ $(i < j)$ are finite, and $r_{ij} \to 0$. As the term of (6) that contains $r_{ij}$ becomes infinite for $r_{ij} \to 0$, the values of other $r_{ig}$ $(g < i)$ should become zero and the corresponding terms should be $-\infty$, in order that (6) shall remain valid. Hence, we should have $r_{ig} \to 0$. However, the summation in (6) for $i = 1$ consists of the sum of positive terms. This would mean that infinity equals zero — hence, the two finite roots $x_i$ and $x_j$ cannot coincide.

If $x_i \to -\infty$, then we have $x_1 \to -\infty$; then $x_{n+1} \to +\infty$ by (7). According to (6), we also have $x_2 \to -\infty$, because $X_1 = 0$. Hence, $x_2 - x_1 \to 0$ and $k_2 x_2(x_2 - x_1)/r_{12}^3 \to -\infty$. Then, $x_3 \to -\infty$ from $X_2 = 0$ in (6). If we repeat the argument, we obtain $x_{n+1} \to -\infty$. This, again, is contradictory. Hence, no root $x_i$ can become infinite.

The equations (6) admit double roots only when

$$\Delta = \left\| \frac{\partial(X_1, X_2, \ldots, X_{n+1})}{\partial(x_1, x_2, \ldots, x_{n+1})} \right\| = 0.$$

If none of the $m_i$ are zero, then, multiplying each row of this determinant by $\sqrt{m_1}, \sqrt{m_2}, \ldots$, respectively, and dividing each column by the same quantities leads to

$$\Delta = \left\| \begin{matrix} k^2 + \sum' \dfrac{2m_j}{r_{1j}^3} & -\dfrac{2\sqrt{m_1 m_2}}{r_{12}^3} & \cdots \\ -\dfrac{2\sqrt{m_1 m_2}}{r_{12}^3} & k^2 + \sum_j' \dfrac{2m_j}{r_{2j}^3} & \cdots \\ \cdots & \cdots & \cdots \end{matrix} \right\|.$$

This is equivalent to the determinant equation for the principal axes transformation

$$\left\| \frac{\partial^2 Q}{\partial \alpha_i \, \partial \alpha_j} - \delta_{ij} S \right\| = 0,$$

of the quadratic form

$$Q = \frac{2(\sqrt{m_2}\,\alpha_1 - \sqrt{m_1}\,\alpha_2)^2}{r_{12}^3} + \frac{2(\sqrt{m_3}\,\alpha_1 - \sqrt{m_1}\,\alpha_3)^2}{r_{13}^3} + \cdots = 0$$

for an unknown quantity $S = -k^2$. But this quadratic form consists of only positive terms. The roots $S$ are all real and cannot be negative. Hence, $\Delta > 0$. But, if $m_k = 0$, then the determinant equals the product of a determinant of similar form and a positive factor. Hence $\Delta \neq 0$, and (6) cannot have any double root.

Thus, when $m_k$, starting with the value zero, varies continuously through positive values, any root of (6) is a continuous function of $m_k$ and cannot change from a real number to a complex number. Hence the $(n + 1)!/2$ solutions obtained for $m_k = 0$ exist for all positive values of $m_k$. Thus the existence of homographic motions for any $n$ has been proved.

## PLANE HOMOGRAPHIC MOTIONS

### 3·3 CONDITION FOR PLANE CONFIGURATIONS

When the $n$ bodies form a plane configuration, we take that plane as our coordinate plane $\zeta = 0$. Equations (2) of Section 3.1 become

$$x_i\left[\lambda \frac{d^2\lambda}{dt^2} - \lambda^2(\omega_2^2 + \omega_3^2)\right] + y_i\left[\lambda^2\omega_1\omega_2 - \frac{d}{dt}(\lambda^2\omega_3)\right] = \frac{1}{\lambda}\sum_k{}' m_k \frac{x_k - x_i}{r_{ki}^3},$$

$$x_i\left[\lambda^2\omega_1\omega_2 + \frac{d}{dt}(\lambda^2\omega_3)\right] + y_i\left[\lambda \frac{d^2\lambda}{dt^2} - \lambda^2(\omega_1^2 + \omega_3^2)\right] = \frac{1}{\lambda}\sum_k{}' m_k \frac{y_k - y_i}{r_{ki}^3}, \quad (9)$$

$$x_i\left[\lambda^2\omega_1\omega_3 - \frac{d}{dt}(\lambda^2\omega_2)\right] + y_i\left[\lambda^2\omega_2\omega_3 + \frac{d}{dt}(\lambda^2\omega_1)\right] = 0,$$

$$(i = 1, 2, \ldots, n).$$

The third equation gives

$$-\frac{\lambda^2\omega_1\omega_3 - d(\lambda^2\omega_2)/dt}{\lambda^2\omega_2\omega_3 + d(\lambda^2\omega_1)/dt} = \frac{x_1}{y_1} = \frac{x_2}{y_2} = \cdots = \frac{x_n}{y_n}.$$

This states that the $n$ bodies are on a straight line and the motion is straight-line homographic, as discussed earlier. We exclude this case. Then there remains

$$\lambda^2\omega_1\omega_3 - \frac{d}{dt}(\lambda^2\omega_2) = \lambda^2\omega_2\omega_3 + \frac{d}{dt}(\lambda^2\omega_1) = 0,$$

or

$$\lambda^2\omega_2\frac{d}{dt}(\lambda^2\omega_2) + \lambda^2\omega_1\frac{d}{dt}(\lambda^2\omega_1) = 0.$$

Thus,

$$\lambda^4(\omega_1^2 + \omega_2^2) = \text{constant}.$$

From the first two equations of (9), we see that

$$\lambda\frac{d^2\lambda}{dt^2} - \lambda^2(\omega_2^2 + \omega_3^2), \qquad \lambda\frac{d^2\lambda}{dt^2} - \lambda^2(\omega_1^2 + \omega_3^2),$$

$$\lambda^2\omega_1\omega_2, \qquad \frac{d}{dt}(\lambda^2\omega_3)$$

are all of the form (constant/$\lambda$). Subtracting these four equations shows that $\lambda^2(\omega_2^2 - \omega_1^2)$ and $\lambda^2(\omega_1^2 + \omega_2^2)$ are also of the same form. Thus, $\lambda^3(\omega_1^2 + \omega_2^2) = \text{constant}$. But we know that $\lambda^4(\omega_1^2 + \omega_2^2) = \text{constant}$. Hence, either $\omega_1 = \omega_2 = 0$, or $\lambda = \text{constant}$. In the latter case,

$$\omega_1^2 + \omega_2^2, \qquad \omega_2^2 + \omega_3^2, \qquad \omega_1^2 + \omega_3^2$$

are constant; thus, $\omega_1\omega_3 = \omega_2\omega_3 = 0$, from the third equation of (9). Hence, either $\omega_1 = \omega_2 = 0$, or $\omega_3 = 0$. The case $\omega_1 = \omega_2 = 0$ reduces to the former case.

Suppose that $\omega_3 = 0$. Then the motion is a rotation around a straight line on the plane $\zeta = 0$. Take the straight line as the $\xi$-axis, for which $\omega_2 = 0$. Then the condition for a homographic motion is

$$0 = \sum_k{}' m_k\frac{x_k - x_i}{r_{ki}^3}, \qquad -\omega_1^2 y_i = \frac{1}{\lambda^3}\cdot\sum_k{}' m_k\frac{y_k - y_i}{r_{ki}^3}.$$

From the first equation we form $\sum_i m_i x_i \cdot \sum_k' m_k(x_k - x_i)/r_{ki}^3$; this equals $\frac{1}{2}\sum_{i\neq j}' m_i m_j(x_i - x_j)^2/r_{ij}^3$. The required condition is that this expression should be zero but this is a contradiction because it states that the sum of positive quantities should be zero. Hence, $\omega_3 = 0$ cannot occur. Consequently, it is necessary to have a rotation round the $\zeta$-axis in this case; that is, the angular momentum of the system cannot be zero; thus the motion cannot be homothetic. That is, in a plane homographic motion the angular momentum cannot be zero. Furthermore, since $\omega_1 = \omega_2 = 0$, and $\omega_3 \neq 0$, the motion is a plane motion that takes place always in one plane.

We return to the former case, $\omega_1 = \omega_2 = 0$. Here we should have $\omega_3 \neq 0$, and a rotation should take place around the $\zeta$-axis. Then

$$\begin{aligned} x_i\left(\lambda \frac{d^2\lambda}{dt^2} - \lambda^2\omega_3^2\right) - y_i \frac{d}{dt}(\lambda^2\omega_3) &= \frac{1}{\lambda}\sum_k{}' m_k \frac{x_k - x_i}{r_{ki}^3}, \\ x_i \frac{d}{dt}(\lambda^2\omega_3) + y_i\left(\lambda \frac{d^2\lambda}{dt^2} - \lambda^2\omega_3^2\right) &= \frac{1}{\lambda}\sum_k{}' m_k \frac{y_k - y_i}{r_{ki}^3}. \end{aligned} \tag{10}$$

We multiply these equations by $-y_i$ and $x_i$, respectively, and take the sum over $i$; then

$$(x_i^2 + y_i^2)\frac{d}{dt}(\lambda^2\omega_3) = \frac{1}{\lambda}\sum_k{}' m_k \frac{x_i y_k - x_k y_i}{r_{ik}^3}.$$

The right-hand member becomes zero when multiplied by $m_i$ and summed over $i$, that is,

$$\sum_i m_i(x_i^2 + y_i^2)\frac{d}{dt}(\lambda^2\omega_3) = 0.$$

Hence,

$$\frac{d}{dt}(\lambda^2\omega_3) = 0,$$

that is,

$$\lambda^2\omega_3 = h = \text{constant}.$$

From (10) we obtain

$$\lambda^2\frac{d^2\lambda}{dt^2} - \frac{h^2}{\lambda} = \frac{1}{x_i}\sum_k{}' m_k \frac{x_k - x_i}{r_{ik}^3} = \frac{1}{y_i}\sum_k{}' m_k \frac{y_k - y_i}{r_{ik}^3}.$$

This holds for all values of $i$; so the quantity on the left is equal to a constant $-k^2$, independent of $i$. (The sign is negative because the force is one of attraction.) Thus,

$$\frac{d^2\lambda}{dt^2} = -\frac{k^2}{\lambda^2} + \frac{h^2}{\lambda^3}, \qquad \lambda^2\omega_3 = h. \tag{11}$$

This shows that the particle is performing a Keplerian elliptic motion in the $(\xi, \eta)$-plane. The center of mass of the system is the focus. The configuration of the $n$ bodies at any instant is similar to that at any other instant. This circumstance prevails whether $\lambda$ is a constant or is variable, and the same equation applies in both cases. In particular, if $\lambda$ is constant then so is $\omega_3$, and the motion is Keplerian and circular and the moving points are at rest relative to the rotating coordinate axes. These points that are fixed with respect to the rotating frame are called the *plane equilibrium points*. For $n = 3$ they are the *Langrangian equilateral-triangular equilibrium points*.

An equilibrium configuration of the $n$ bodies in a plane is called a *central figure*. For such a motion to be possible, the initial condition

$$\begin{aligned} X_i &\equiv \sum_j{}' m_j \frac{x_j - x_i}{r_{ji}^3} + k^2 x_i = 0 \\ Y_i &\equiv \sum_j{}' m_j \frac{y_j - y_i}{r_{ji}^3} + k^2 y_i = 0 \\ &\qquad (i = 1, 2, \ldots, n) \end{aligned} \tag{12}$$

should be satisfied, whether $\lambda$ is variable or constant. But the origin is the center of mass:

$$\sum_i m_i x_i = \sum_i m_i y_i = 0.$$

Furthermore, since

$$\sum_i m_i \sum_j{}' m_j \frac{x_j - x_i}{r_{ji}^3} = \sum_i m_i \sum_j{}' m_j \frac{y_j - y_i}{r_{ji}^3} = 0,$$

we obtain the conditions

$$\sum_i m_i X_i = \sum_i m_i Y_i = 0. \tag{13}$$

Again, since

$$\begin{aligned} \sum_i m_i y_i \sum_j{}' m_j \frac{x_j - x_i}{r_{ji}^3} &= \sum_i \sum_j{}' \frac{m_i m_j (x_i - x_j)(y_i - y_j)}{2 r_{ji}^3} \\ &= \sum_i m_i x_i \sum_j{}' m_j \frac{y_j - y_i}{r_{ji}^3}, \end{aligned}$$

therefore

$$\sum_i m_i (x_i Y_i - y_i X_i) = 0, \tag{14}$$

even though (12) is not satisfied. Thus there are $2n$ equations (12). But, since there are three relations existing between (12), given by (13) and (14), the equations (12) give only $2n - 3$ independent conditions in all.

Rewrite equations (12). Let

$$(i,j)_x \quad \text{represent} \quad x_j - x_i,$$

$$(i,j)_y \quad \text{represent} \quad y_j - y_i,$$

$$(i,j,k) \quad \text{represent} \quad \begin{vmatrix} x_i & y_i & 1 \\ x_j & y_j & 1 \\ x_k & y_k & 1 \end{vmatrix}.$$

Then,

$$(j,k) = (i,k) - (i,j),$$

$$(i,j,k) = (i,j)_x(i,k)_y - (i,j)_y(i,k)_x. \tag{15}$$

The symbol $(i,j,k)$ indicates twice the area of the triangle formed by the $i$th, $j$th, and $k$th bodies, in which the sign of $(i,j,k)$ is taken into account. Then from (12) we obtain

$$X_i - X_j = (i,j)_x\left[k^2 + \frac{m_i + m_j}{r_{ij}^3} + \sum_k{}'' \frac{m_k}{r_{jk}^3}\right]$$

$$+ \sum_k{}'' m_k(i,k)_x\left(\frac{1}{r_{ik}^3} - \frac{1}{r_{jk}^3}\right) = 0,$$

$$Y_i - Y_j = (i,j)_y\left[k^2 + \frac{m_i + m_j}{r_{ij}^3} + \sum_k{}'' \frac{m_k}{r_{jk}^3}\right]$$

$$+ \sum_k{}'' m_k(i,k)_y\left(\frac{1}{r_{ik}^3} - \frac{1}{r_{jk}^3}\right) = 0,$$

where $\sum''$ denotes the sum over $k$ obtained by excluding $k = i$ and $k = j$. Multiply these equations respectively by $(i,j)_y$ and $(i,j)_x$ and subtract. Then we have, taking (15) into consideration,

$$\sum_k{}'' m_k(i,j,k)\left(\frac{1}{r_{ik}^3} - \frac{1}{r_{jk}^3}\right) = 0. \tag{16}$$

As many plane homographic motions are possible as the number of distinct solutions of this equation.

### 3.4 CENTRAL FIGURES

Denote the total mass of the $n$ bodies by $M$; then

$$\sum_j{}' m_j(x_i - x_j) = \sum_j m_j(x_i - x_j) = \sum_j m_j x_i = M x_i,$$

and

$$\frac{\partial}{\partial x_i}\sum_i\sum_j{}' \frac{m_i m_j}{r_{ij}} = m_i \sum_j{}' m_j \frac{x_j - x_i}{r_{ji}^3},$$

$$\frac{\partial}{\partial x_i}\left(\sum_i\sum_j{}' m_i m_j r_{ij}^2\right) = 2m_i \sum_j{}' m_j(x_i - x_j).$$

Hence, (12) can be written

$$\frac{\partial U}{\partial x_i} = 0, \qquad \frac{\partial U}{\partial y_i} = 0 \qquad (i = 1, 2, \ldots, n),$$

where

$$U = \sum_i\sum_j{}' \frac{m_i m_j}{r_{ij}} + \frac{k^2}{2M}\sum_i\sum_j{}' m_i m_j r_{ij}^2,$$

Let

$$U_1 = \sum_i\sum_j{}' \frac{m_i m_j}{r_{ij}}, \qquad U_2 = -\frac{k^2}{2M}\sum_i\sum_j{}' m_i m_j r_{ij}^2. \tag{17}$$

Then (12) takes the form

$$\frac{\partial U_1}{\partial x_i} = \frac{\partial U_2}{\partial x_i}, \qquad \frac{\partial U_1}{\partial y_i} = \frac{\partial U_2}{\partial y_i}, \qquad (i = 1, 2, \ldots, n). \tag{18}$$

The motion under the force function $U_2$ is determined by

$$m_i \frac{d^2 x_i}{dt^2} = \frac{\partial U_2}{\partial x_i} = -\sum_j{}' \frac{k^2}{2M} m_i m_j (x_i - x_j) = -\frac{k^2}{2} m_i x_i,$$

and is a simple harmonic motion around the system center of mass with period $2\sqrt{2}\,\pi/k$. Each particle describes its own distinct orbit, which is an ellipse around the center of mass as focus. Hence, if in the motion of $n$ particles all on one plane the composite force acting on each particle is directed to the center of mass and its magnitude is proportional to its distance from the center of mass and, furthermore, if the initial velocity of each particle is proportional to its distance from the center of mass and is in a direction that makes with its radius vector the same angle for all $n$ particles, then the system of $n$ particles performs a homographic motion.

This theorem is due to Laplace. We have obtained (18) by seeking the condition for homographic motions. Thus, this initial condition is necessary and sufficient for a homographic motion, that is, for the motion to be represented in the form (1) of Section 3.1. In the case of $n$ particles attracting each other according to Newtonian law, Dziobek (1900) proved the existence of homographic motions under the action of the same force exerted on each particle as in Laplace's case. Equations (18) show that this is the case.

If there exists a $U_2$ in (18) such that

$$U_2 = -C = \text{constant}, \tag{19}$$

or if there exist configurations of $n$ particles such that $U_2 = -C$, then there is at least one configuration for which $U_1$ is a relative minimum. This is one of the central figures. This theorem is due to Dziobek (1900).

Because of (19), $m_i m_j r_{ij}^2 < C$; therefore

$$U_1 = \sum_i \sum_j{}' \frac{m_i m_j}{r_{ij}} > \frac{1}{\sqrt{C}} \sum_i \sum_j{}' m_j^{3/2} m_i^{3/2}.$$

If $r_{ij} = 0$, then this equation is satisfied for $U_1 = \infty$. Otherwise $U_1$ is finite and continuous, along with its first-order derivatives. Hence, there must be at least one configuration with a minimum $U_1$ among the family of configurations $U_2 = -C$. The configuration with a minimum $U_1$ satisfies

$$\frac{\partial U_1}{\partial x_i} = \frac{\partial U_1}{\partial y_i} = 0,$$

and

$$\frac{\partial U_1}{\partial x_i} : \frac{\partial U_2}{\partial x_i} = \frac{\partial U_1}{\partial y_i} : \frac{\partial U_2}{\partial y_i}.$$

Thus, the existence of homographic motions in Newtonian mechanics has been proved. Note that, although the theorem has only been proved for $U_1 =$ minimum, it is also true for $U_1 =$ maximum or minimax.

In the foregoing proof we have assumed that the masses are all finite. If one of the masses is infinitesimal, we must proceed in a different manner. Consider the $(n + 1)$-body problem in which $m_{n+1} = 0$. Let

$$U_1' = \sum_{i=1}^{n} \frac{m_i}{r_{i,n+1}}, \qquad U_2' = \sum_{i=1}^{n} m_i r_{i,n+1}^2.$$

Then,

$$\frac{\partial U_1'}{\partial x_{n+1}} : \frac{\partial U_2'}{\partial x_{n+1}} = \frac{\partial U_1'}{\partial y_{n+1}} : \frac{\partial U_2'}{\partial y_{n+1}},$$

for the mass $m_{n+1}$. Because of (17), the coefficient $k^2/2M$ in $U_2'$ contains only $m_1, m_2, \ldots, m_n$ but not $m_{n+1}$. Hence, from (18),

$$\frac{\partial U'}{\partial x_{n+1}} = \frac{\partial U'}{\partial y_{n+1}} = 0,$$

where

$$U_1' + \frac{k^2}{2M} U_2' = U'.$$

$U_1'$ becomes infinite when $m_{n+1}$ approaches any one of the $n$ masses. Otherwise, $U'$ is finite and continuous, as are its first-order derivatives. Moreover, $U_2'$ becomes infinite as $x_{n+1}$, $y_{n+1}$ are infinite. Hence, $U'$ should have at least one minimum in between. As $C'$, by beginning with the value $+\infty$, approaches this minimum, the family of curves $U' = C'$ for a very large value of $C'$ consists of $n$ small circles around each of the $n$ masses $m_1, m_2, \ldots, m_n$ and a very large circle enclosing all the $n$ masses. As $C'$ decreases from this value, the curve $U' = C'$ should lie in the $(n + 1)$-ply connected region lying between this large circle and these $n$ small circles. Such a value of $C'$ must be smaller than its value in the outside region of the large circle around infinity and also smaller than its value in the inside region of each of the $n$ small circles. As $C'$ decreases further, two or more of these circles coalesce. There should be at least $n$ points at which the large circle coalesces with each of the $n$ small circles in turn. Such points of coalescence correspond to the extrema in question.

When we increase $m_{n+1}$, starting with the value $m_{n+1} = 0$, circumstances should vary continuously. The existence of extrema in this case can be proved by the principle of analytical continuation in the theory of analytic functions. But this analytical continuation becomes impossible, as we have shown in the case of straight-line equilibrium configurations, either when $m_{n+1} \to \infty$, or when two of the points of coalescence coincide and then become complex. Since equations (12) are of the same form as (6) of Section 3.2, it can be proved in a similar manner that such a coincidence can never occur.

Take an example. Put $\lambda = 1$; then $\omega_3 =$ constant and

$$U_2' = \sum_{i=1}^{n} m_i r_{i,n+1}^2 = \sum_{i=1}^{n} m_i r_i^2 + M r_{n+1}^2 = Mg^2 + M r_{n+1}^2,$$

$$r_j^2 = x_j^2 + y_j^2 \qquad (j = 1, 2, \ldots, n + 1),$$

where $g$ is a constant representing the radius of gyration of the whole system of the $n$ particles. Hence,

$$U' = \sum_{i=1}^{n} \frac{m_i}{r_{i,n+1}} + \frac{k^2}{2M} M(x_{n+1}^2 + y_{n+1}^2) + \text{constant}$$

$$= \sum_{i=1}^{n} \frac{m_i}{r_{i,n+1}} + \frac{\omega_3}{2} (x_{n+1}^2 + y_{n+1}^2) + \text{constant.}$$

This is called *Jacobi's integral* and will be discussed later. It is an integral for the motion of the $(n + 1)$th particle of vanishingly small mass when the $n$ particles perform a uniform circular motion. The family of curves $U' = C'$ in the case is shown in Figure 1 for $n = 2$, $m_1 \neq m_2$, $m_3 = 0$. The five points in the figure indicated by the letter $L$ represent the extrema. They are sometimes called the *libration points*, but it is not necessary to have librations around these points. $L_4$ and $L_5$ correspond to minima. Each of the other three points corresponds to a minimax, that is, to a minimum for any displacement in the direction connecting the two finite masses but to a maximum for any displacement perpendicular to it.

The case $n = 3$, $m_1 = m_2 = m_3 = 1$, $m_4 = 0$, is illustrated in Figures 2 and 3. Figure 2 shows the three bodies on a straight line; Figure 3 shows them at each vertex of an equilateral triangle. Both are the central figures for $n = 3$ (Lindow, 1922, 1923, 1924, 1926, 1928; Schaub, 1929; Hütterhain, 1933, 1935; see also Section 14.24).

The case $n = 4$, $m_1 = m_2 = m_3 = m_4 = 1$, $m_5 = 0$, is shown likewise in Figures 4 and 5. It is the $(4 + 1)$-body problem.

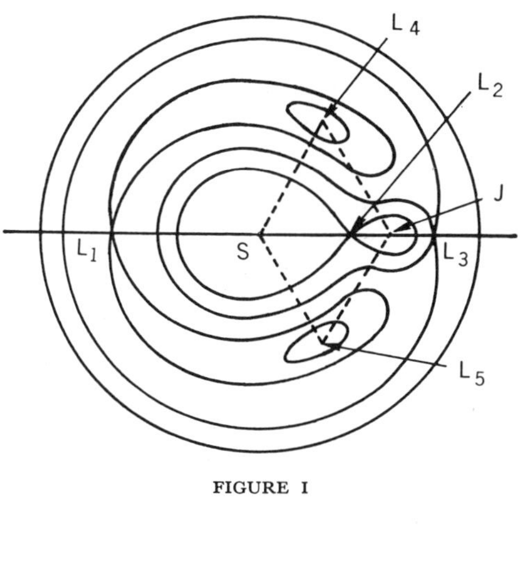


FIGURE I

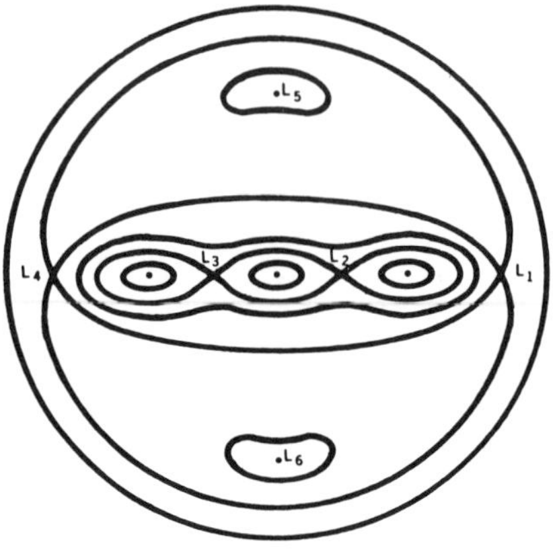


FIGURE 2

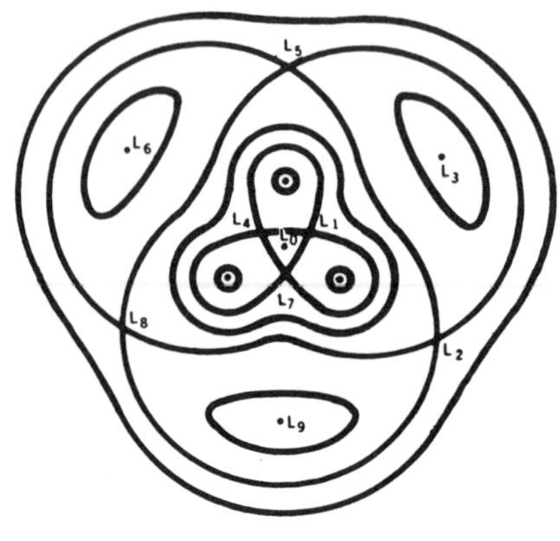


FIGURE 3

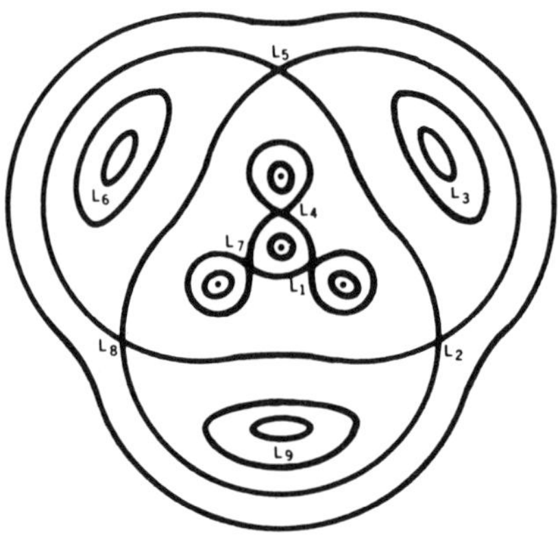


FIGURE 4

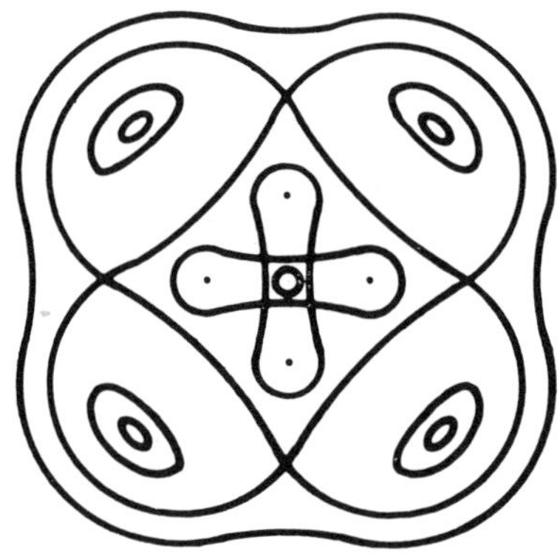

FIGURE 5

### 3.5 PROPERTIES OF CENTRAL FIGURES

There are $n(n-1)/2$ relations of the form (16) in the case of $n$ finite masses. These relations determine $n(n-1)/2$ mutual distances $r_{ij}$. But there are $(n-2)(n-3)/2$ relations expressing the condition that these $n(n-1)/2$ mutual distances should be all in one plane.

In the case $n = 4$, there is only one condition specifying that the tetrahedron formed by those four points be of volume zero. Suppose now that the conditions for $n$ points to be in one and the same plane are $(n-2)(n-3)/2$ in number. Then we proceed to prove by mathematical induction that the conditions for $n + 1$ points to be in one and the same plane are $(n-1)(n-2)/2$ in number.

If we connect each of the first $n$ points with the new $(n + 1)$th point then we obtain $n$ new distances $r_{ij}$. By equating to zero the volume of the tetrahedron formed by two points $A$ and $B$ among the first $n$ points, the $(n + 1)$th point $P$, and one of the remaining $n - 2$ points among the first $n$ points, we obtain $n - 2$ conditions. These are new conditions, and there is no other condition when we increase the number of points from $n$ to $n + 1$. Hence there are

$$[(n - 2)(n - 3)/2] + (n - 2) = (n - 1)(n - 2)/2$$

conditions.

Thus there should be $(n - 2)(n - 3)/2$ relations among the $n(n - 1)/2$ mutual distances $r_{ij}$ in order that the points shall be in one and the same plane.

Now we express the condition for four points $(x_i, y_i, z_i)$ $(i = 1, 2, 3, 4)$ to be in one and the same plane, in terms of the mutual distances $r_{ij}$. Take a new set of four points $(x'_i, y'_i, z'_i)$ $(i = 1, 2, 3, 4)$, and denote the volumes of the two tetrahedrons by $V$ and $V'$, respectively.

$$6V = \begin{vmatrix} x_1 & x_2 & x_3 & x_4 & 0 \\ y_1 & y_2 & y_3 & y_4 & 0 \\ z_1 & z_2 & z_3 & z_4 & 0 \\ 1 & 1 & 1 & 1 & 0 \\ 0 & 0 & 0 & 0 & 1 \end{vmatrix}, \qquad 6V' = -\begin{vmatrix} x'_1 & y'_1 & z'_1 & 0 & 1 \\ x'_2 & y'_2 & z'_2 & 0 & 1 \\ x'_3 & y'_3 & z'_3 & 0 & 1 \\ x'_4 & y'_4 & z'_4 & 0 & 1 \\ 0 & 0 & 0 & 1 & 0 \end{vmatrix}.$$

Let $(i, j) = (x_i - x'_j)^2 + (y_i - y'_j)^2 + (z_i - z'_j)^2$. Then,

$$-\tfrac{1}{2}(i, j) = (x_i x'_j + y_i y'_j + z_i z'_j) - \tfrac{1}{2}(x_i^2 + y_i^2 + z_i^2) - \tfrac{1}{2}(x_j'^2 + y_j'^2 + z_j'^2).$$

Form the product of the two determinants $36VV'$. Subtract the last row multiplied by $\frac{1}{2}(x_1^2 + y_1^2 + z_1^2)$ from each row, and subtract the last column multiplied by $\frac{1}{2}(x_j'^2 + y_j'^2 + z_j'^2)$ from each column; then

$$288VV' = \begin{vmatrix} (11) & (12) & (13) & (14) & 1 \\ (21) & (22) & (23) & (24) & 1 \\ (31) & (32) & (33) & (34) & 1 \\ (41) & (42) & (43) & (44) & 1 \\ 1 & 1 & 1 & 1 & 0 \end{vmatrix}.$$

If these two tetrahedrons coincide, then $(i, j) = 0$ and

$$288V^2 = \begin{vmatrix} 0 & r_{12}^2 & r_{13}^2 & r_{14}^2 & 1 \\ r_{21}^2 & 0 & r_{23}^2 & r_{24}^2 & 1 \\ r_{31}^2 & r_{32}^2 & 0 & r_{34}^2 & 1 \\ r_{41}^2 & r_{42}^2 & r_{43}^2 & 0 & 1 \\ 1 & 1 & 1 & 1 & 0 \end{vmatrix} \equiv F. \tag{20}$$

Hence, $F = 0$ expresses the condition for four points to be in one and the same plane, in terms of the mutual distances.

Take two triangles $(x_i, y_i)$ and $(x_i', y_i')$ $(i = 1, 2, 3)$, in one plane and denote the areas by $\Delta$ and $\Delta'$, respectively. Then,

$$2\Delta = \begin{vmatrix} x_1 & x_2 & x_3 & 0 \\ y_1 & y_2 & y_3 & 0 \\ 1 & 1 & 1 & 0 \\ 0 & 0 & 0 & 1 \end{vmatrix}, \qquad 2\Delta' = -\begin{vmatrix} x_1' & x_1' & 0 & 1 \\ x_2' & x_2' & 0 & 1 \\ x_3' & y_3' & 0 & 1 \\ 0 & 0 & 1 & 0 \end{vmatrix}.$$

Let $d_{ij}^2 = (x_i^2 + y_i^2) + (x_j'^2 + y_j'^2) - 2(x_i x_j' + y_i y_j')$; then,

$$32\Delta\Delta' = \begin{vmatrix} d_{11}^2 & d_{12}^2 & d_{13}^2 & 1 \\ d_{21}^2 & d_{22}^2 & d_{23}^2 & 1 \\ d_{31}^2 & d_{32}^2 & d_{33}^2 & 1 \\ 1 & 1 & 1 & 0 \end{vmatrix}.$$

Suppose now that $\Delta$ is the area of the triangle formed by three points (1, 2, 3), and that $\Delta'$ is the area of the triangle formed by (2, 3, 4), and denote them by $\Delta_1$ and $\Delta_2$ respectively. Then,

$$32\Delta_1\Delta_2 = \begin{vmatrix} r_{12}^2 & r_{13}^2 & r_{14}^2 & 1 \\ r_{32}^2 & 0 & r_{34}^2 & 1 \\ r_{42}^2 & r_{43}^2 & 0 & 1 \\ 1 & 1 & 1 & 0 \end{vmatrix} = \frac{\partial F}{\partial (r_{21}^2)}.$$

Hence,

$$\frac{\partial F}{\partial r_{21}^2} = 32\Delta_1\Delta_2.$$

or, more generally,

$$\frac{\partial F}{\partial r_{ij}^2} = 32\Delta_i\Delta_j. \tag{21}$$

Our problem is to find the extremum of $U$ under the $(n-2)(n-3)/2$ conditions $F = 0$. It is sufficient to find the extremum of

$$U + \rho_1 F_1 + \rho_2 F_2 + \cdots + \rho_l F_l, \qquad l = (n-2)(n-3)/2, \quad (22)$$

with Lagrange's indeterminate multipliers $\rho_i$ by the calculus of variations. We obtain the solution by supposing that $n(n-1)/2$ of the mutual distances $r_{ij}$ are independent increments and by eliminating the $(n-2)(n-3)/2$ indeterminate multipliers $\rho_i$ from the $(n-2)(n-3)/2$ conditions $F = 0$ (Carathéodory, 1933; Pylarinos, 1937, 1941).

There is no such condition for $n = 3$. Hence (16) can be written

$$m_3(123)(r_{13}^{-3} - r_{23}^{-3}) = 0,$$
$$m_1(231)(r_{21}^{-3} - r_{31}^{-3}) = 0,$$
$$m_2(312)(r_{32}^{-3} - r_{12}^{-3}) = 0.$$

The solution is either $(123) = 0$ or $r_{12} = r_{23} = r_{31}$, where there is no restriction on $m_1$, $m_2$, $m_3$. In the first case the area of the triangle is zero and the three points are on one straight line, as already discussed in Section 3.2. The case is trivial when the three points coincide in one point. In the second case the three points form an equilateral triangle, that is, the equilateral-triangular equilibrium points (Bilimovitch and Petronievics, 1936).

For $n = 4$ we have six equations

$$\begin{aligned}
m_3(123)(r_{13}^{-3} - r_{23}^{-3}) + m_4(124)(r_{14}^{-3} - r_{24}^{-3}) &= 0,\\
m_2(132)(r_{12}^{-3} - r_{23}^{-3}) + m_4(134)(r_{14}^{-3} - r_{34}^{-3}) &= 0,\\
m_2(142)(r_{12}^{-3} - r_{24}^{-3}) + m_3(143)(r_{13}^{-3} - r_{34}^{-3}) &= 0, \qquad (23)\\
m_1(231)(r_{12}^{-3} - r_{13}^{-3}) + m_4(234)(r_{24}^{-3} - r_{34}^{-3}) &= 0,\\
m_1(241)(r_{12}^{-3} - r_{14}^{-3}) + m_3(243)(r_{23}^{-3} - r_{34}^{-3}) &= 0,\\
m_1(341)(r_{13}^{-3} - r_{14}^{-3}) + m_2(342)(r_{23}^{-3} - r_{24}^{-3}) &= 0.
\end{aligned}$$

Suppose at first that $m_1 = m_2 = m_3 = m_4$, then the conditions are satisfied when the four points are at each vertex of a regular quadrangle (Figure 6). Next we transform the first three equations of (23) and suppose that neither of the masses is zero and neither of the triangles is of area zero; this gives the condition

$$\begin{aligned}
&(r_{13}^{-3} - r_{23}^{-3})(r_{12}^{-3} - r_{24}^{-3})(r_{14}^{-3} - r_{34}^{-3})\\
&\qquad = (r_{14}^{-3} - r_{24}^{-3})(r_{13}^{-3} - r_{34}^{-3})(r_{12}^{-3} - r_{23}^{-3}). \qquad (24)
\end{aligned}$$

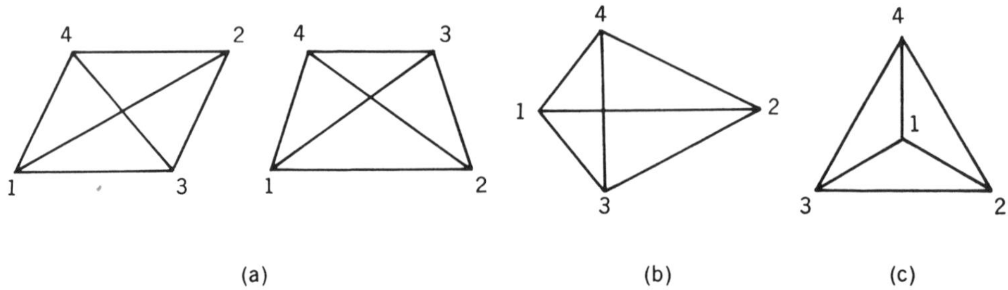


FIGURE 6

This relation was deduced by Dziobek (1900). Let us consider a few examples satisfying this relation.

The first example is a parallelogram or a trapezoid with $r_{13} = r_{24}$, $r_{14} = r_{23}$. For a parallelogram, $-(234) = (134) = (124) = -(123)$. For a trapezoid, $(234) = (134)$, $(124) = (123)$. From (23),

$$m_3(123) - m_4(124) = 0, \qquad m_1(134) - m_2(234) = 0.$$

Hence, there is no equilibrium configuration with $m_1, m_2, m_3, m_4 > 0$ for a parallelogram. On the other hand, the relations $m_1 = m_2 > 0$, $m_3 = m_4 > 0$ hold for a trapezoid. In particular, for a square:

$$r_{12} = r_{23} = r_{14} = r_{34}, \qquad r_{13} = \sqrt{2}\, r_{12}, \qquad m_1 = m_2 = m_3 = m_4$$

and

$$r_{12}^3 \omega_3^2 = m_1(2 + 1/\sqrt{2}).$$

The next example is a rhombus with $r_{13} = r_{14}$, $r_{23} = r_{24}$. We have $(124) + (123) = 0$, and should have $m_3 = m_4$. In particular, when there is a mass $m_1$ at the center of an equilateral triangle, $m_2 = m_3 = m_4$, $r_{12} = r_{13} = r_{14}$, $r_{34} = r_{42} = r_{23}$, and $\omega_3^2 = m_1/r_{12}^3 + m_2/r_{23}^3$.

This has been similarly discussed by Brumberg (1958). He shows that each of these configurations is unstable.

In order to obtain a general conclusion for $n = 4$, Dziobek (1900) transformed (22) by writing $4\rho$ for $\rho_1$ in the form

$$m_i m_j \left( -\frac{1}{r_{ij}^2} + \frac{k^2}{2M} r_{ij} \right) + \rho r_{ij}\, \Delta_i \Delta_j = 0,$$

or

$$m_i m_j \left( \frac{1}{r_{ij}^3} - \delta \right) + \epsilon \Delta_i \Delta_j = 0 \qquad (i, j = 1, 2, 3, 4), \tag{25}$$

where $\Delta_1 = (234)$, $\Delta_2 = (431)$, $\Delta_3 = (241)$, $\Delta_4 = (321)$. From (25),

$$\begin{aligned}(r_{12}^{-3} - \delta)(r_{34}^{-3} - \delta) &= (r_{13}^{-3} - \delta)(r_{24}^{-3} - \delta) \\ &= (r_{14}^{-3} - \delta)(r_{23}^{-3} - \delta),\end{aligned} \tag{26}$$

and the condition for a central figure takes the form

$$\begin{vmatrix} 1 & r_{12}^{-3} + r_{34}^{-3} & r_{12}^{-3}r_{34}^{-3} \\ 1 & r_{13}^{-3} + r_{24}^{-3} & r_{13}^{-3}r_{24}^{-3} \\ 1 & r_{14}^{-3} + r_{23}^{-3} & r_{14}^{-3}r_{23}^{-3} \end{vmatrix} = 0. \tag{27}$$

Combining any two of (26) gives

$$\delta = \frac{r_{12}^{-3}r_{34}^{-3} - r_{13}^{-3}r_{42}^{-3}}{r_{12}^{-3} + r_{34}^{-3} - r_{31}^{-3} - r_{42}^{-3}} = \frac{r_{13}^{-3}r_{42}^{-3} - r_{14}^{-3}r_{23}^{-3}}{r_{13}^{-3} + r_{42}^{-3} - r_{14}^{-3} - r_{23}^{-3}}$$

$$= \frac{r_{14}^{-3}r_{23}^{-3} - r_{13}^{-3}r_{42}^{-3}}{r_{14}^{-3} + r_{23}^{-3} - r_{31}^{-3} - r_{42}^{-3}}.$$

Let

$$N = r_{12}^{-3} + r_{34}^{-3} - r_{13}^{-3} - r_{42}^{-3},$$

$$Z_1 = (r_{12}^{-3} - r_{13}^{-3})(r_{12}^{-3} - r_{42}^{-3}),$$

$$Z_2 = (r_{12}^{-3} - r_{13}^{-3})(r_{13}^{-3} - r_{34}^{-3});$$

then, from (25),

$$m_1m_2Z_1 = -\epsilon\Delta_1\Delta_2N, \qquad m_1m_3Z_2 = -\epsilon\Delta_1\Delta_3N.$$

Hence

$$\frac{m_1}{m_3} = \frac{\Delta_2}{\Delta_3}\frac{Z_2}{Z_1} = \frac{\Delta_2}{\Delta_3}\frac{r_{13}^{-3} - r_{34}^{-3}}{r_{12}^{-3} - r_{42}^{-3}}.$$

We draw the following conclusion:

Suppose that the four points are each at a vertex of a convex quadrangle and in the order $m_1, m_2, m_3, m_4$; then $\Delta_2/\Delta_3 < 0$. Hence, in order that $m_2/m_3 > 0$, both $(r_{21}^{-3} - r_{24}^{-3})$ and $(r_{34}^{-9} - r_{31}^{-3})$ must have the same sign; that is, the four points $P_1, P_2, P_3, P_4$ should be such that

$$P_2P_4 - P_2P_1 \quad \text{and} \quad P_3P_1 - P_3P_4$$

have the same sign. But, as $P_2P_4$ and $P_3P_1$ are the diagonals, we should have $P_2P_4 > P_2P_1$, $P_3P_1 > P_3P_4$. We obtain similar conditions for $m_3/m_4$, $m_4/m_1$, $m_1/m_2$. When the four points form a convex quadrangle, there exist central figures with positive masses if both diagonals are larger than any of the four sides.

Next suppose that one of the four points, $P_4$ for example, lies inside the triangle $P_1P_2P_3$. A similar discussion shows that, in order to have central figures with positive masses, the sides $P_2P_3$, $P_3P_1$, $P_1P_2$ in each of the triangles $\Delta P_2P_3P_4$, $\Delta P_3P_1P_4$, $\Delta P_1P_2P_4$, respectively, are the largest, and that $P_4P_1 > P_4P_2 > P_4P_3$ if $P_2P_3 > P_3P_1 > P_1P_2$.

MacMillan and Bartkey (1932) and Brumberg (1958) considered also concave quadrangles. (For convex quadrangles, $r_{12}, r_{23}, r_{34}, r_{41} \leqq r_{13}, r_{24}$, where $r_{13}$, $r_{24}$ are the diagonals, and the ratio $r_{13}{:}r_{24}$ is between $1/\sqrt{3}$ and $\sqrt{3}$. For concave quadrangles, $r_{12}, r_{23}, r_{13} \leqq r_{34}, r_{41}, r_{24}$, for example.) They showed the existence of at least one permissible quadrangle in each case.

Williams (1938) studied the five-body problem in which one of the five bodies is at the center of a square or a rhombus. If $m_5$ is at the center of a square, then, he states

$$0.439 < \frac{m_3}{m_4} = \frac{m_5}{m_4} = 1.414.$$

If $m_4$ is at the center of a rhombus formed of $m_1$, $m_2$, $m_3$, $m_5$, then

$$\frac{r_{23}}{r_{12}} = \frac{r_{35}}{r_{12}} = \frac{r_{15}}{r_{12}} = 1, \qquad \frac{r_{14}}{r_{12}} = \frac{r_{34}}{r_{12}} = 0.781,$$

$$\frac{r_{24}}{r_{12}} = \frac{r_{45}}{r_{12}} = 0.625, \qquad \frac{r_{25}}{r_{15}} = 1.249 \qquad \frac{r_{13}}{r_{12}} = 1.562,$$

and

$$\frac{m_1}{m_3} = 1, \qquad \frac{m_2}{m_3} = \frac{m_5}{m_3} = \frac{m_5}{m_1} = 2.617,$$

$$\frac{m_4}{m_3} = 0.536, \qquad \frac{m_5}{m_4} = 4.878, \qquad \frac{m_2}{m_5} = 1.$$

The above discussion on the existence of central figures is not based on algebraic proof for any arbitrary masses. However, their existence has been proved by our analytical method; indeed, they exist whether the parameter $\lambda$ of Section 3.1 is constant or variable.

Brumberg (1958) discussed the stability of circular solutions in the four-body problem. Let

$$u_i = x_i + \sqrt{-1}\, y_i, \qquad v_i = x_i - \sqrt{-1}\, y_i, \qquad i = 1, 2, 3, 4;$$

let $u_i = a_i$, $v_i = b_i$, $z_i = 0$, $i = 1, 2, 3, 4$, be the relative equilibrium positions. The equations for variations for $\xi_i$, $\eta_i$, $\zeta_i$, such that $u_i = a_i + \xi_i$, $v_i = b_i + \eta_i$, $z_i = \zeta_i$, are

$$\ddot{\xi}_i + 2\omega_3\sqrt{-1}\,\dot{\xi}_i = \omega_3^2\xi_i + \frac{1}{2}\sum_j \xi_j P_{ij} + \frac{3}{2}\sum_j \eta_j P'_{ij},$$

$$\ddot{\eta}_i - 2\omega_3\sqrt{-1}\,\dot{\eta}_i = \omega_3^2\eta_i + \frac{3}{2}\sum_j \xi_j P''_{ij} + \frac{1}{2}\sum_j \eta_j P_{ij},$$

$$\ddot{\zeta}_i + \sum_j \zeta_j P_{ij} = 0,$$

where

$$a_{ij} = [(a_i - a_j)(b_i - b_j)]^{-3/2},$$

$$P_{ii} = \sum_j{}' m_j a_{ij}, \qquad P_{ij} = -m_j a_{ij}, \qquad \sum_j P_{ij} = 0,$$

$$P'_{ii} = \sum_j{}' m_j a_{ij} \frac{a_i - a_j}{b_i - b_j}, \qquad P'_{ij} = -m_j a_{ij} \frac{a_i - a_j}{b_i - b_j}, \qquad \sum_j P'_{ij} = 0,$$

$$P''_{ii} = \sum_j{}' m_j a_{ij} \frac{b_i - b_j}{a_i - a_j}, \qquad P''_{ij} = -m_j a_{ij} \frac{b_i - b_j}{a_i - a_j}, \qquad \sum_j P''_{ij} = 0.$$

The condition for relative equilibrium gives

$$\omega_3^2 a_i = \sum_j a_j P_{ij} = \sum_j b_j P'_{ij}, \qquad \omega_3^2 b_i = \sum_j b_j P_{ij} = \sum_j a_j P''_{ij}.$$

The $\zeta$-equation can be solved by substituting $\zeta_i = C_i e^{ht}$, where $h$ is determined by

$$\Delta(h^2) = \begin{vmatrix} P_{11} + h^2 & P_{12} & P_{13} & P_{14} \\ P_{21} & P_{22} + h^2 & P_{23} & P_{24} \\ P_{31} & P_{32} & P_{33} + h^2 & P_{34} \\ P_{41} & P_{42} & P_{43} & P_{44} + h^2 \end{vmatrix} = 0.$$

The equation $\Delta(h^2) = 0$ admits one zero root and three negative roots. The zero root corresponds to the solution of Lagrange. The root $h^2 = -\omega_3^2$ corresponds to the straight-line solution of Euler. Thus, for any mass and any initial perturbations perpendicular to the plane of the unperturbed motion, the circular solutions are stable to the first approximation and the perturbed motion along the $z$-axis is periodic.

Put $\xi_i = A_i e^{\sqrt{-1}\,\lambda t}$, $\eta_i = B_i e^{\sqrt{-1}\,\lambda t}$ in the equations for $\xi_i$, $\eta_i$; then,

$$A_i(\lambda + \omega_3)^2 + \frac{1}{2}\sum_j A_j P_{ij} + \frac{3}{2}\sum_j B_j P'_{ij} = 0,$$

$$B_i(\lambda - \omega_3)^2 + \frac{3}{2}\sum_j A_j P''_{ij} + \frac{1}{2}\sum_j B_j P_{ij} = 0.$$

The determinant equation $\Delta(\lambda^2) = 0$ is of the eighth degree in $\lambda^2$ with real coefficients. The root $\lambda^2 = \omega_3^2$ is triple, and the root $\lambda^2 = 0$ is simple. Put $x_i = x_i^0 + \phi_i$, $y_i = y_i^0 + \psi_i$, $z_i = z_i^0 + \zeta_i$, where $x_i^0 = \text{const.}$, $y_i^0 = \text{const.}$, $z_i^0 = 0$ are the equilibrium positions of Lagrange. From the equations for variations with respect to $\phi_i$, $\psi_i$, $\zeta_i$ we obtain the determinant equation

$$\Delta(g^2) = g^2(g^2 + \omega_3^2)^3(g^8 + ag^6 + bg^4 + cg^2 + d) = 0,$$

$$a = 5\omega_3^2 - (P_{11} + P_{22} + P_{33} + P_{44}).$$

The roots of

$$a_0x^4 + 4a_1x^3 + 6a_2x^2 + 4a_3x + a_4 = 0$$

are all real, negative, and different, when $a_0 > 0$, $a_1 > 0$, $a_2 > 0$, $a_3 > 0$, $a_4 > 0$, $a_1^2 - a_0a_2 > 0$, $12(a_1^2 - a_0a_2)^2 - a_0^2 f_1 > 0$, and

$$D = f_1^3 - 27f_2^2 > 0,$$

where

$$f_1 = a_0a_4 - 4a_1a_3 + 3a_2^2,$$

$$f_2 = a_0a_2a_4 + 2a_1a_2a_3 - a_0a_3^2 - a_1^2a_4 - a_2^3.$$

For $D = 0$ this algebraic quartic equation admits multiple roots (Serret, 1885; Halphen, 1891).

By applying this principle, Brumberg (1958) has shown that the square configuration, the rhombus, and the equilateral triangle with one of the masses at the center are all unstable. On the other hand, Hütterhain (1933, 1935) proved by the same method that the motion of the infinitesimal mass in the (3 + 1)-body problem with finite masses $m_1$, $m_2$, $m_3 = m_1$ in the straight-line case is stable and periodic if $m_2/m_1 > 11.72138$ around the equilibrium points $L_5$ and $L_6$ (Figure 2). Brumberg identified the configuration of bright stars in the Trapezium of Orion with a tetrahedron.

Longley (1907) proved the existence of more complicated configurations as shown in Figure 7. The masses on each circle are equal and the

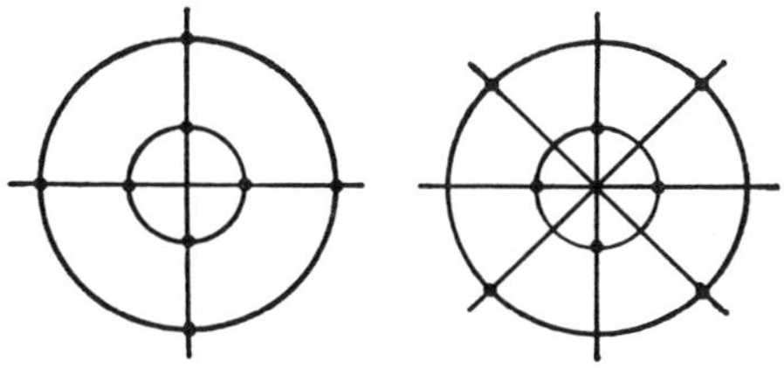

FIGURE 7

ratio of the masses on these two concentric circles is determined by the radii of the circles. Bilimovitch (1927, 1929) discussed more complicated configurations.

Recently Klemperer (1962) has considered, by extending the works of Dziobek and Bilimovitch, equilibrium configurations for limited ranges of rhombic configurations and of arrays in the shape of hexagonal and octagonal regular rosettes. Such arrangements comprise alternating heavier bodies, all alike, and an equal number of lighter bodies, also all alike, in regularly alternating fashion. See Figure 8.

When three finite masses form the equilibrium figures of Lagrange or Euler, the position of the fourth body for equilibrium is discussed as the (3 + 1)-body problem. We have already treated it analytically; it will be discussed algebraically in a later chapter with the use of elliptic functions.

When all $n$ masses are equal, that is, $m_1 = m_2 = \cdots = m_n = 1/n$, there can occur a homographic motion of each mass at each vertex of a regular $n$-sided polygon. If the radius of the circumscribed circle be taken as the unit of measurement, then (12) is satisfied by

$$x_i = \cos \frac{2i\pi}{n}, \qquad y_i = \sin \frac{2i\pi}{n} \qquad (i = 0, 1, \ldots, n - 1),$$

$$r_{ij} = 2 \sin (i - j) \frac{\pi}{n} \qquad (i > j),$$

$$k^2 = \frac{1}{4n} \sum_{i=1}^{n-1} \frac{1}{\sin (i\pi/n)}.$$

By supposing that $m_{n+1} = 0$, we can discuss the $(n + 1)$-body problem in this case.

Summarizing, we see that a homographic motion of all particles in one and the same plane is at any instant a plane motion in that same plane and can occur only when the total angular momentum is not zero. If each of the masses describes a circular motion around the center of

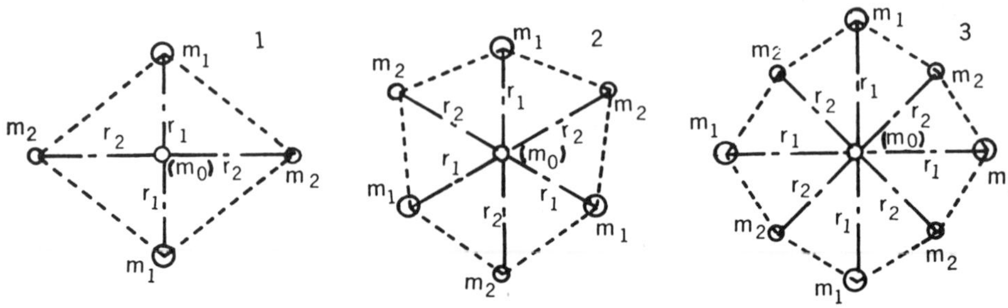


FIGURE 8

mass, the particles form an equilibrium configuration of a central figure referred to the rotating coordinate axes. If each mass describes an elliptic motion, the particles form similar configurations at each instant referred to the rotating coordinate axes.

## SPACE HOMOGRAPHIC MOTION

### 3.6 CONDITION FOR SPACE CONFIGURATIONS

Suppose that $n$ bodies are not in one plane at a certain instant. Let the instantaneous rotation axis be the $\zeta$-axis; then $\omega_1 = \omega_2 = 0$, $\omega_3 \neq 0$ and equations (2) of Section 3.1 are now

$$x_i\left(\lambda \frac{d^2\lambda}{dt^2} - \lambda^2\omega_3^2\right) - y_i \frac{d}{dt}(\lambda^2\omega_3) = \frac{1}{\lambda}\sum_j{}' m_j \frac{x_j - x_i}{r_{ij}^3},$$

$$x_i \frac{d}{dt}(\lambda^2\omega_3) + y_i\left(\lambda \frac{d^2\lambda}{dt^2} - \lambda^2\omega_3^2\right) = \frac{1}{\lambda}\sum_j{}' m_j \frac{y_j - y_i}{r_{ij}^3}, \tag{28}$$

$$z_i\lambda \frac{d^2\lambda}{dt^2} = \frac{1}{\lambda}\sum_j{}' m_j \frac{z_j - z_i}{r_{ij}^3}.$$

Multiply the first equation by $-m_iy_i$, and the second equation by $m_ix_i$, and add; then the sum of the right-hand members is zero, and

$$\sum_i m_i(x_i^2 + y_i^2) \frac{d}{dt}(\lambda^2\omega_3) = 0.$$

Hence,

$$\lambda^2\omega_3 = \text{constant}. \tag{29}$$

Similarly to the plane homographic motion, there is a relation

$$\lambda \frac{d^2\lambda}{dt^2} - \lambda^2\omega_3^2 = \frac{1}{\lambda} \times \text{constant};$$

then the third equation of (28) gives

$$\lambda^3\omega_3^2 = \text{constant}. \tag{30}$$

From (29) and (30), either $\lambda = \text{constant}$, or $\omega_3 = 0$. In the first case $\omega_3$ is also constant; the third equation of (28) gives

$$\sum_j{}' m_j \frac{z_j - z_i}{r_{ij}^3} = 0 \qquad (i = 1, 2, \ldots, n).$$

Multiply each of these equations by $m_i z_i$ and sum over $i$; then

$$\sum_i \sum_j{}' \frac{m_i m_j}{r_{ij}} = 0,$$

which is impossible. Hence $\omega_3 = 0$. In the second case, in addition,

$$\lambda^2 \frac{d^2\lambda}{dt^2} = -K^2 = \text{constant}.$$

Each particle performs a Keplerian motion on a straight line through the center of mass and fixed in space, that is, the motion is homothetic with $n \geqq 4$. The total angular momentum is zero. The configuration of the masses at each instant remains always similar and similarly situated with regard to the center of mass.

To summarize the line and the plane homographic motions: the homographic motion with zero total angular momentum is a homothetic motion on fixed straight lines through the center of mass, each mass at a vertex of a similar and similarly situated polyhedron at each instant, and the homographic motion with nonvanishing total angular momentum is either the plane or the line homographic motion, each mass lying on a straight line or forming a central figure. Homographic motions in which all the masses are on one and the same plane at any instant are plane homographic motions. Homographic motions in which not all of the masses are on one and the same plane at an instant are homothetic motions on fixed straight lines through the center of mass; the masses form a similar and similarly situated polyhedron at each instant, and the total angular momentum is zero. Such a motion is a special case of the line homographic motion in which the total angular momentum is zero. Central figures can occur only when the total angular momentum is not zero.

The condition for space homographic motion is

$$\begin{aligned} &\sum_j{}' m_j \frac{x_j - x_i}{r_{ij}^3} + K^2 x_i = 0 \\ &\sum_j{}' m_j \frac{y_j - y_i}{r_{ij}^3} + K^2 y_i = 0 \qquad (i = 1, 2, \ldots, n). \\ &\sum_j{}' m_j \frac{z_j - z_i}{r_{ij}^3} + K^2 z_i = 0 \end{aligned} \tag{31}$$

With the same definition of $X_i$, $Y_i$, $Z_i$ as in Section 3.2

$$\sum_i m_i x_i = \sum_i m_i y_i = \sum_i m_i z_i = 0,$$

$$\sum_i m_i(y_i X_i - x_i Y_i) = \sum_i m_i(z_i Y_i - y_i Z_i) = \sum_i m_i(x_i Z_i - z_i X_i) = 0.$$

Hence, (31) gives only $(3n - 6)$ independent equations. From these,

$$\sum_k{}''' m_k(i, j, h, k)(r_{ik}^{-3} - r_{jk}^{-3}) = 0 \qquad (i, j, h, k = 1, 2, \ldots, n), \tag{32}$$

where $\sum'''$ denotes the sum over $k$ equal to neither $i$, nor $j$, nor $h$, and $(i, j, h, k)$ denotes six times the volume of the tetrahedron formed by $m_i$, $m_j$, $m_h$, $m_k$. From (31) we see that

$$U = \sum_i \sum_j{}' \frac{m_i m_j}{r_{ij}} + \frac{K^2}{2M} \sum_i \sum_j{}' m_i m_j r_{ij}^2 \tag{33}$$

should be an extremum. Hence we can prove Laplace's theorem of Section 3.4 for such space homographic motions. The theorem is also true for line homographic motions.

Lehmann-Filhés (1891) and Bilimovitch (1911) proved that the condition is satisfied for $n = 4$ when all mutual distances $r_{ij}$ are equal, that is, when all masses form a regular tetrahedron. Recently Brumberg (1958) has again shown the existence of such a configuration.

For $n = 5$, form (32) and put $r_{13} = r_{14} = a$,

$$m_1(2341)(r_{12}^{-3} - r_{13}^{-3}) + m_5(2345)(r_{25}^{-3} - r_{35}^{-3}) = 0,$$

$$m_1(2351)(r_{12}^{-3} - r_{13}^{-3}) + m_4(2354)(r_{24}^{-3} - r_{34}^{-3}) = 0.$$

From these, either $(2345) = 0$ or $r_{24} = r_{34} = b$, $r_{25} = r_{35} = c$. In the first case, the four points 2, 3, 4, 5, are all in one and the same plane and $r_{12} = r_{13} = r_{14} = r_{15} = a$. In the second case, either $m_2 = m_3$ or $a = b = c$; in the latter event, either $(1245) = (1345) = 0$ or

$$r_{14} = r_{15} = r_{45}.$$

This last condition is satisfied either for $m_2 = m_3$ or for $m_1 = m_4 = m_5$. Such a configuration has been obtained by Brehm (1908).

If the four masses form a regular tetrahedron at one instant and approach to or recede from the center of masses with velocities proportional to their respective distances from the center of masses, then the condition (33) is satisfied, as in the case of plane homographic motion. This theorem was proved by Lehmann-Filhés (1891) and Dziobek (1900).

Bilimovitch (1927, 1929) proved the existence of homographic motions for these two cases: (i) when two sets of equal masses are situated at the vertices of the corresponding two similar and similarly situated regular polyhedrons, and (ii) when one set of equal masses are at the vertices of a regular polyhedron and the other set of equal masses are at the vertices of its conjugate regular polyhedron, that is, a polyhedron obtained by joining the middle points of the surfaces of the original polyhedron (for example, a regular tetrahedron and a regular octahedron, a regular hexahedron and a regular octahedron). Furthermore, the existence of homographic motions was proved for two regular polyhedrons, the second being obtained by joining the middle points of the edges of the original polyhedron (for example, a regular tetrahedron and a regular octahedron). These are not equilibrium configurations.

There seem to be no existing celestial bodies with such space homographic motions. But it has been proved that, in the immediate neighborhood of simultaneous collisions of all $n$ bodies at a point, or at infinity when all $n$ bodies recede from the finite region, the $n$ bodies tend indefinitely to one of such homographic configurations. In such cases the total angular momentum is zero, as will be proved later. Sundman proved that in two-body collisions the motion can be analytically continued after the collisions, but in collisions of three or more bodies the motion may become complex if the masses and the initial conditions do not satisfy certain restrictions, and the motion cannot be analytically continued after the collisions. Such studies have been carried out by Sundman, Block, Freundlich, Chazy, Merman, and others, as will be discussed in a later chapter.

## ISOSCELES TRIANGULAR SOLUTIONS

### 3·7 REGULAR AND IRREGULAR SINGULARITIES

Suppose that the coefficients $p_0(x), p_1(x), \ldots, p_n(x)$ of the differential equation

$$p_0(x) \frac{d_n y}{dx^n} + p_1(x) \frac{d^{n-1} y}{dx^{n-1}} + \cdots + p_n(x) y = 0 \tag{34}$$

are uniform analytic functions with only isolated singularities in the complex plane. The equation is said to be analytic if the coefficients are analytic. A point $x = \xi$ where all coefficients are holomorphic, that is, single-valued and continuous, is called an *ordinary point*; all other points are called *singular*. At an ordinary point, the coefficients are expanded in the form

$$p_i(x) = \sum_{k=0}^{\infty} p_i^{(k)}(x - \xi)^k \qquad (i = 0, 1, 2, \ldots, n),$$

convergent for $|x - \xi| < d(\xi)$, where $d(\xi)$ is the shortest distance between the point $x = \xi$ and the nearest singular point. At an ordinary point $x = \xi$, the analytic linear differential equation (34) is satisfied by the power series

$$y(x) = \phi(x) = \sum_{k=0}^{\infty} c^{(k)}(x - \xi)^k, \tag{35}$$

which is uniquely determined by the initial condition. This series is convergent for $|x - \xi| < d(\xi)$. Every solution $y = \phi(x)$ is analytic. Its only singularities in the finite part of the complex plane are due to the singularities of the coefficients.

The necessary and sufficient condition that an isolated singularity of equation (34) of canonical form can be regular is that it should be at most a pole of order $r$ of the coefficient $p_r(x)$ of $d^{n-r}y/dx^{n-r}$ in the canonical form in which the highest order coefficient is $p_0(x) \equiv 1$. This is called Fuchs's condition (Heffter, 1894; Schlesinger, 1908, 1922; Ince, 1927; Poole, 1936). A differential equation satisfying Fuchs's condition is said to be of *Fuchsian type*.

At an ordinary point $x = \xi$ we can construct $n$ independent solutions, with arbitrary initial values that are not linearly connected, such that any other solution can be written

$$y = \sum_{i=1}^{n} c_i \phi_i(x),$$

with constants $c_i$. Any such system of $n$ solutions is called a *fundamental system* of solutions. The solutions $\phi_1(x), \ldots, \phi_n(x)$ are said to be *linearly connected* if there exist constants $c_1, \ldots, c_n$, not all zero, such that

$$\sum_{i=1}^{n} c_i \phi_i(x) \equiv 0.$$

If there are $n$ functions that are differentiable $n - 1$ times, then,

$$\sum_{i=1}^{n} c_i \frac{d^{j-1}\phi_i(x)}{dx^{j-1}} = 0 \qquad (j = 1, 2, \ldots, n).$$

Eliminating $c_1, \ldots, c_n$, we obtain

$$W(\phi_1, \ldots, \phi_n) \equiv \begin{vmatrix} \phi_1 & \phi_2 & \cdots & \phi_n \\ \dfrac{d\phi_1}{dx} & \dfrac{d\phi_2}{dx} & \cdots & \dfrac{d\phi_n}{dx} \\ \vdots & & & \\ \dfrac{d^{n-1}\phi_1}{dx^{n-1}} & \dfrac{d^{n-1}\phi_2}{dx^{n-1}} & \cdots & \dfrac{d^{n-1}\phi_n}{dx^{n-1}} \end{vmatrix} = 0.$$

This is a Wronskian and is a necessary condition for the functions $\phi_1, \ldots, \phi_n$ to be linearly connected. If $n$ functions $\phi_1(x), \ldots, \phi_n(x)$ are differentiable $n - 1$ times and if $W(\phi_1, \ldots, \phi_n) \equiv 0$ but

$$W(\phi_1, \ldots, \phi_{n-1}) \neq 0,$$

then the functions are connected by a linear relation with constant coefficients. Let $p_0(x) \equiv 1$; then

$$\begin{aligned} \frac{dW}{dx} &= \sum_{j=1}^{n} \left[\sum_{i=1}^{n} \frac{\partial W}{\partial \phi_i^{(j-1)}} \phi_i^{(j)}(x)\right] = \sum_{i=1}^{n} \frac{\partial W}{\partial \phi_i^{(n-1)}} \phi_i^{(n)}(x) \\ &= -p_1(x)\left[\sum_{i=1}^{n} \frac{\partial W}{\partial \phi_i^{(n-1)}} \phi_i^{(n-1)}(x)\right] = -p_1(x) \cdot W, \end{aligned}$$

where $\phi_i^{(j)}$ denotes the $j$th derivative of $\phi_i$. Hence,

$$W(x) = W(\xi) \exp\left[-\int_\xi^x p_1(t)\, dt\right]. \qquad (35a)$$

This relation is due to Abel and Liouville. If $p_1(x)$ is continuous, then $W(\phi_1, \ldots, \phi_n)$ vanishes either identically or not at all. The necessary and sufficient condition for $n$ solutions $\phi_1(x), \ldots, \phi_n(x)$ to form a fundamental system of solutions of (34) with $p_0(x) \equiv 1$ is that their Wronskian should not vanish at an arbitrary chosen point $x = \xi$.

Consider a linear transformation

$$\Phi_i(x) = \sum_{j=1}^{n} a_{ij}\phi_j(x) \qquad (i = 1, 2, \ldots, n)$$

corresponding to any selected circuit $\Gamma$ on the complex plane. Consider also

$$y = c_1\phi_1(x) + \cdots + c_n\phi_n(x) \qquad (36)$$

such that

$$\left[\sum_{i=1}^{n} c_i\Phi_i(x)\right] = \lambda\left[\sum_{i=1}^{n} c_i\phi_i(x)\right].$$

Since $\phi_1(x), \ldots, \phi_n(x)$ are linearly independent,

$$\sum_{i=1}^{n} c_i a_{ij} = \lambda c_j \qquad (j = 1, 2, \ldots, n).$$

If this system of equations for $c_1, \ldots, c_n$ is compatible, then $\lambda$ must satisfy

$$\Delta_n(\lambda) = \begin{vmatrix} a_{11} - \lambda & a_{21} & \cdots & a_{n1} \\ a_{12} & a_{22} - \lambda & \cdots & a_{n2} \\ \vdots & & & \\ a_{1n} & a_{2n} & \cdots & a_{nn} - \lambda \end{vmatrix} = 0, \tag{37}$$

which is called the characteristic equation for the matrix $(a_{ij})$. No root can be zero, since $\|a_{ij}\| \neq 0$. Each root has at least one linear form (36). When the characteristic equation admits $n$ distinct roots, then there are $n$ linearly independent combinations of $\phi_1, \ldots, \phi_n$, which form a new fundamental system obeying the canonical transformation $Y_i = \lambda_i y_i$ $(i = 1, 2, \ldots, n)$. The root $\lambda$ should be independent of the choice of the original fundamental system $\phi_1(x), \ldots, \phi_n(x)$. For a canonical solution, $y_i$ is provided with the same multiplier $\lambda_j$, whether it is expressed in terms of $\phi_1(x), \ldots, \phi_n(x)$ or of some other fundamental system

$$\psi_1(x), \ldots, \psi_n(x)$$

admitting the transformation with constant coefficients $b_{ij}$:

$$\Psi_i(x) = \sum_{j=1}^{n} b_{ij}\psi_j(x) \qquad (i = 1, 2, \ldots, n).$$

Suppose that $x = 0$ is a singularity where Fuchs's condition is satisfied. Consider the normal form

$$x^n \frac{d^n y}{dx^n} + P_1(x)x^{n-1}\frac{d^{n-1}y}{dx^{n-1}} + \cdots + P_n(x)y = 0, \tag{38}$$

where $P_1(x), \ldots, P_n(x)$ are holomorphic at $x = 0$. Let $\delta \equiv x(d/dx)$; then

$$x^m \frac{d^m}{dx^m} = \delta(\delta - 1)\cdots(\delta - m + 1),$$

so that (38) becomes

$$F(x, \delta)y \equiv \delta^n y + Q_1(x)\,\delta^{n-1}y + \cdots + Q_n(x)y = 0. \tag{38a}$$

The coefficients $Q_1, \ldots, Q_n$ are linear functions of $P_1, \ldots, P_n$. Let

$$Q_i(x) = \sum_{j=0}^{\infty} Q_{ij}x^j \qquad (i = 1, 2, \ldots, n),$$

$$F_0(\delta) = \delta^n + Q_{10}\delta^{n-1} + Q_{20}\delta^{n-2} + \cdots + Q_{n0},$$

$$F_j(\delta) = Q_{1j}\delta^{n-1} + Q_{2j}\delta^{n-2} + \cdots + Q_{nj};$$

then (38a) becomes

$$F(x, \delta)y = \sum_{j=0}^{\infty} x^j F_j(\delta)y = 0.$$

Substituting a solution

$$y = x^\rho(c_0 + c_1x + c_2x^2 + \cdots) \tag{39}$$

gives

$$F(x, \delta)y = \sum_{i=0}^{\infty}\sum_{j=0}^{\infty} x^j F_j(\delta)c_i x^{\rho+i} = \sum_{i=0}^{\infty}\sum_{j=0}^{\infty} x^{\rho+i+j}F_j(\rho + i)c_i$$

$$= \sum_{k=0}^{\infty} x^{\rho+k}\left[\sum_{i=0}^{k} F_{k-i}(\rho + i)c_i\right] = 0.$$

This vanishes identically if each coefficient is zero, or if $c_0, c_1, \ldots$ satisfy

$$\sum_{i=0}^{k} F_{k-i}(\rho + i)c_i = 0 \qquad (k = 0, 1, 2, \ldots). \tag{40}$$

This is called the *indicial equation* or the fundamental characteristic equation, and the roots $\rho$ are the *exponents* of the given singularity. If the solution in the form (39) contains only a finite number or none of negative powers of $x$ at a singular point ($x = 0$ in the example), the singularity is called *regular*. Otherwise it is *irregular*.

The number of linearly independent regular solutions (39) cannot exceed the number of distinct roots of the indicial equation. One distinct power series (39) at most is associated with each distinct exponent.

If $F_0(\rho) = 0$, $F_0(\rho + k) \neq 0$, $(k = 1, 2, \ldots)$, then we may choose $c_0 \neq 0$; all subsequent coefficients $c_1, c_2, \ldots$ are uniquely determined. If several exponents differ by integers, then (40) may or may not be compatible and the number of distinct series (39) may be smaller than the number of distinct roots. If the indicial equation has multiple roots, then the deficiency of regular solutions of the type (39) is supplied by solutions involving logarithms.

Let

$$\lambda = \exp(2\pi i\rho)$$

and

$$y = x^{\rho}(u_0 + u_1x + u_2x^2 + \cdots),$$

where $u_0, u_1, u_2, \ldots$ are polynomials in $\log x$ of degree lower than $n$. Then

$$F(x, \delta)y = \sum_{i=0}^{\infty}\sum_{j=0}^{\infty} x^j F_j(\delta)x^{\rho+i}u_i = \sum_{i=0}^{\infty}\sum_{j=0}^{\infty} x^{\rho+i+j}F_j(\delta + \rho + i)u_i$$

$$= \sum_{k=0}^{\infty} x^{\rho+k}\left[\sum_{i=0}^{k} F_{k-i}(\delta + \rho + i)u_i\right],$$

which vanishes identically if $u_0, u_1, u_2, \ldots$ satisfy

$$\sum_{i=0}^{k} F_{k-i}(\delta + \rho + i)u_i = 0 \qquad (k = 0, 1, 2, \ldots).$$

This is a system of linear differential equations with constant coefficients in the independent variable $t = \log x_1$.

The equation for $k = 0$ is written

$$F_0(\delta + \rho)u_0 \equiv F_0(\rho)u_0 + \frac{1}{1!}F_0'(\rho)\,\delta u_0 + \frac{1}{2!}F_0''(\rho)\,\delta^2 u_0 + \cdots = 0.$$

If $u_0$ is a polynomial in $t$ not identically zero, then this expression is a polynomial of the same degree unless $F_0(\rho) = 0$. For a solution, $\rho$ must satisfy the indicial equation. If we identify $\rho$ with the lowest root of the set $\rho = \rho_1$, then the series belonging to the higher roots appears. Since $F_0(\delta + \rho_1)u_0 = G_1(\delta)\,\delta^{\kappa_1}u_0$, where the polynomial $G_1(0) \neq 0$, the polynomial $u_0$ must satisfy $\delta^{\kappa_1}u_0 = 0$. The structure of the solution is determined by the first $N + 1$ relations, where $N$ is the difference between the highest and the lowest exponents of the set.

For the point at infinity, put $x = 1/z$, $\delta' = z(d/dz) = -x(d/dx) - \delta$ in (38a); then

$$\delta'^n y - Q_1(1/z)\delta'^{n-1}y + \cdots + (-1)^n Q_n(1/z)y = 0.$$

The necessary and sufficient condition in order that $z = 0$, i.e. $x = \infty$, shall be regular, is that $Q_1(\infty), Q_2(\infty), \ldots$ be finite. In the canonical form (34) with $p_0(x) = 1$, this means that $xp_1(x), x^2p_2(x), \ldots, x^np_n(x)$ should remain finite for $x \to \infty$.

Let

$$\psi(x) \equiv (x - a_1)(x - a_2)\cdots(x - a_m),$$

where $a_1, a_2, \ldots, a_m$ are the singularities at finite distances from the origin. The necessary and sufficient condition for these to be regular is that $\psi^r p_r(x)$ $(r = 1, 2, \ldots, n)$ should be holomorphic for all finite values of $x$. These expressions are then integral functions of $x$. But, since $x^r p_r(x)$ $(r = 1, 2, \ldots, n)$ are finite at infinity,

$$[\psi(x)]^r p_r(x) = \Pi_{(m-1)r}(x) \qquad (r = 1, 2, \ldots, n),$$

where $\Pi_\alpha(x)$ means a polynomial of degree $\alpha$.

The equation can then be written

$$\psi^n \frac{d^n y}{dx^n} + \Pi_{m-1}\psi^{n-1}\frac{d^{n-1}y}{dx^{n-1}} + \cdots + \Pi_{mn-n}y = 0$$

with polynomial coefficients. If $n = 1$, that is, if the equation is linear, then

$$p_1(x) \equiv \frac{\Pi_{m-1}(x)}{\psi(x)} = \sum_{i=1}^{n} \frac{A_i}{(x - a_i)},$$

and the indicial equation at $x = a_i$ is

$$\rho(\rho - 1)\cdots(\rho - n + 1) + A_i\rho(\rho - 1)\cdots(\rho - n + 2) + \cdots = 0,$$

so that the sum of the exponents at $x = a_i$ is $[n(n - 1)/2 - A_i]$. The indicial equation at infinity is

$$\rho(\rho + 1)\cdots(\rho + n - 1) - \Big(\sum_i A_i\Big)\rho(\rho + 1)\cdots(\rho + n - 2) + \cdots = 0.$$

Hence the sum of the exponents at infinity is $\sum_i A_i - n(n - 1)/2$. Thus the sum of all $(m + 1)n$ exponents is

$$\sum_{i=1}^{m+1}\sum_{j=1}^{n} \rho_{ij} = (m - 1)n(n - 1)/2,$$

where $\rho_{ij}$ denotes the $j$th root at the singularity $x = a_i$.

The most general second-order equation of the Fuchsian type is

$$\frac{d^2y}{dx^2} + \frac{\Pi_{m-1}}{\psi}\frac{dy}{dx} + \frac{\Pi_{2m-2}}{\psi^2}y = 0,$$

or

$$\frac{d^2y}{dx^2} + \left[\sum_{i=1}^{m} \frac{A_i}{(x - a_i)}\right]\frac{dy}{dx} + \left[\sum_{i=1}^{m} \frac{B_i}{(x - a_i)^2} + \sum_{i=1}^{m} \frac{C_i}{(x - a_i)}\right]y = 0,$$

where

$$\sum_i C_i = 0.$$

At $x = a_i$ the indicial equation is

$$\rho(\rho - 1) + A_i\rho + B_i = 0.$$

Let the roots be $\rho = \rho_i$ and $\rho = \rho_i'$; then the equation is of the canonical form

$$\frac{d^2y}{dx^2} + \left[\sum_{i=1}^{m} \left(\frac{1 - \rho_i - \rho_i'}{x - a_i}\right)\right]\frac{dy}{dx} + \left[\sum_{i=1}^{m} \frac{\rho_i\rho_i'}{(x - a_i)^2} + \frac{\Pi_{2m-2}}{\psi}\right]y = 0.$$

The indicial equation at infinity is

$$\rho(\rho + 1) - \left(\sum_{i=1}^{m} A_i\right)\rho + \sum_{i=1}^{m} (B_i + C_ia_i) = 0.$$

Fuchs's condition is

$$\sum_{i=1}^{m} (1 - \rho_i - \rho_i') + (1 - \rho_\infty - \rho_\infty') = 2.$$

The equations of Gauss, Lamé, Legendre, and Bessel, that appear in mathematical physics, are special cases of such Fuchsian-type equations. (Klein, 1894, 1933; Hagihara, 1970).

In order to solve the nonhomogeneous equation

$$\frac{d^ny}{dx^n} + p_1(x)\frac{d^{n-1}y}{dx^{n-1}} + \cdots + p_n(x)y = f(x), \tag{34a}$$

let

$$\frac{d^{j-1}y}{dx^{j-1}} = \sum_{i=1}^{n} u_i \frac{d^{j-1}\phi_i(x)}{dx^{j-1}} \qquad (j = 1, 2, \ldots, n).$$

This equation determines uniquely the functions $u_1, u_2, \ldots, u_n$, since $W(\phi_1, \ldots, \phi_n) \neq 0$. We also have

$$0 = \sum_{i=1}^{n} \frac{du_i}{dx} \frac{d^{j-1}\phi_i(x)}{dx^{j-1}} \qquad (j = 1, 2, \ldots, n-1),$$

$$\frac{d^n y}{dx^n} = \sum_{i=1}^{n} \left[ u_i \frac{d^n \phi_i(x)}{dx^n} + \frac{du_i}{dx} \frac{d^{n-1}\phi_i(x)}{dx^{n-1}} \right].$$

Substituting these equations in (34) with $p_0(x) = 1$, we obtain

$$\sum_{i=1}^{n} \frac{du_i}{dx} \frac{d^{n-1}\phi_i(x)}{dx^{n-1}} = f(x).$$

Hence

$$\frac{du_i}{dx} = f(x) \frac{1}{W(x)} \frac{\partial W(x)}{\partial \phi_i^{(n-1)}} \qquad (i = 1, 2, \ldots, n),$$

where $W(x)$ denotes the Wronskian in (35a). If the principal solution at $x = \xi$, say $y = \Phi(x; \xi)$, is determined by the condition

$$\frac{d^{j-1}y}{dx^{j-1}} = 0 \qquad (j = 1, 2, \ldots, n)$$

for $x = \xi$, then $u_1 = u_2 = \cdots = u_n = 0$ for $x = \xi$. Hence

$$u_i = \int_\xi^x \frac{f(t)}{W(t)} \frac{\partial W(t)}{\partial \phi_i^{(n-1)}} \, dt,$$

$$\Phi(x; \xi) = \sum_{i=1}^{n} \phi_i(x) \int_\xi^x \frac{f(t)}{W(t)} \frac{\partial W(t)}{\partial \phi_i^{(n-1)}} \, dt.$$

If $t$ is a constant, then

$$\phi(x; t) \equiv \sum_{i=1}^{n} \frac{\phi_i(x)}{W(t)} \frac{\partial W(t)}{\partial \phi_i^{(n-1)}}$$

gives the solution of the homogeneous equation determined by the initial condition

$$\frac{d^{j-1}y}{dx^{j-1}} = 0 \qquad (j = 1, 2, \ldots, n-1), \qquad \frac{d^{n-1}y}{dx^{n-1}} = 1$$

for $x = t$. With this definition of $\phi(x; t)$, the formula for the principal solution is

$$\Phi(x; \xi) = \int_\xi^x \phi(x; t) f(t) \, dt.$$

Let

$$Fy \equiv p_0 \frac{d^n y}{dx^n} + p_1(x) \frac{d^{n-1} y}{dx^{n-1}} + \cdots + p_n(x) y,$$

$$F^* z \equiv (-1)^n \frac{d^n(p_0 z)}{dx^n} + (-1)^{n-1} \frac{d^{n-1}(p_1 z)}{dx^{n-1}} + \cdots + p_n z.$$

The operators $F$ and $F^*$ are said to be *adjoint* to each other. We have an identity

$$zFy - yF^*z \equiv D\Psi(y, z),$$

where

$$\Psi(y, z) \equiv p_0 z \frac{d^{n-1} y}{dx^{n-1}} + \left[p_1 z - \frac{d}{dx}(p_0 z)\right] \frac{d^{n-2} y}{dx^{n-2}} + \cdots$$

$$+ \left[p_{n-1} z - \frac{d}{dx}(p_{n-2} z) + \cdots + (-1)^{n-1} \frac{d^{n-1}(p_0 z)}{dx^{n-1}}\right] y$$

is called the *bilinear concomitant.*

### 3.8 EXISTENCE OF THE SOLUTIONS

So far we have discussed the particular solution of the three-body problem in which the masses are all quite arbitrary. Now we consider a particular solution obtained by restricting the masses, and examine whether there is a solution where the masses form an isosceles triangle. Fransèn (1895) discussed three kinds of *isosceles-triangular solutions:* one with an axis of symmetry, one with a plane of symmetry, and one with both an axis of symmetry and a plane of symmetry — the axis of symmetry being on the plane of symmetry. These are particular cases for equal masses in Euler's solution and in Lagrange's equilateral-triangular solution. Pavanini (1907) and MacMillan (1913) calculated the solution for the case with an axis of symmetry, that is, when two of the bodies move in circular orbits on a plane perpendicular to the axis of symmetry and the third body (with infinitesimal mass) moves along the axis of symmetry. Buchanan (1920) obtained periodic solutions and asymptotic solutions in their neighborhood when the two bodies with equal masses move in circular orbits and in elliptic orbits, according to Poincaré's theory of periodic solutions. Wilczynski (1913) applied projective differential geometry (Wilczynski, 1906; Fubini and Čech,

1926) to a generalization of Euler's and Lagrange's solutions. Supposing the center of mass of the three bodies to be at rest, he considered the properties of the trajectories of each of the three bodies. Two of these trajectories lie upon the ruled surface described by the straight line joining two of the three bodies, and all three trajectories lie upon the conical surface enveloping the instantaneous orbital planes of the three bodies (Salmon, 1912). If the velocities of the three bodies are in the plane of the three bodies at each instant, then the orbits are all plane curves and the surface in question coincides with the common plane of the three orbital planes. Wilczynski proved that, except when at least one of the three bodies has a plane curve as its orbit, the case where one of the ruled surfaces is a developable surface or a conical surface occurs only when two sides of the triangle formed by the three bodies are equal. Furthermore, the case where the curve described by one of the three bodies is asymptotic to one of the ruled surfaces occurs only for an isosceles triangle. Seeking thc condition under which the ruled surface describes a conical surface, he showed that the vertex of the conical surface recedes to infinity and the surface becomes a cylinder. Then two sides of the triangle become equal and the base of the triangle describes a cylindrical surface which is the ruled surface in question. The ruled surface may sometimes degenerate into a plane or a straight line. From this it is evident that, if two sides of the triangle are equal, then the mass at the vertex of the triangle describes a plane curve passing through the center of mass. This is the case with a plane of symmetry. In other cases where one of the masses describes a curve asymptotic to the ruled surface, the two sides of the triangle meeting at the center of mass should be equal. This is the case with an axis of symmetry. Thus Wilczynski proved the existence of particular solutions for an isosceles triangle. Wilczynski (1913) and MacMillan (1913), by studying the zero points and the singular points of the functions appearing in the discussion on the basis of function theory, proved that there is no solution of the three-body problem other than Lagrange's equilateral-triangular solution if two sides of the triangle formed by the three bodies are equal and if at the same time the masses situated at the two ends of the base of the triangle are not equal. We base our discussion of this property on Fuchs's theory of linear differential equations according to Chazy (1921). This particular solution is important in discussing the collisions of two bodies in the three-body problem, that is, in discussing under what conditions the third mutual distance may become infinitely small without being equal to zero if the other two mutual distances of the three bodies are always larger than a finite length.

Let the two masses on the base of the triangle be $m_1$ and $m_2$ and the third mass be $m_3$. Let the rectangular coordinates of $m_2$ relative to $m_1$ be $x$, $y$, $z$ and the coordinates of $m_3$ relative to the center of masses of $m_1$ and $m_2$ be $x_1$, $y_1$, $z_1$. Define $r$ and $r_1$ such that $r^2 = x^2 + y^2 + z^2$, $r_1^2 = x_1^2 + y_1^2 + z_1^2$. Let $R$ be the equal distances of $m_1$ and $m_2$ from $m_3$. The equations of motion in the Jacobian canonical coordinate system are

$$\frac{d^2x}{dt^2} = -Ax, \qquad \frac{d^2y}{dt^2} = -Ay, \qquad \frac{d^2z}{dt^2} = -Az,$$
$$\frac{d^2x_1}{dt^2} = -A_1x_1, \qquad \frac{d^2y_1}{dt^2} = -A_1y_1, \qquad \frac{d^2z_1}{dt^2} = -A_1z_1, \tag{41}$$

where

$$A = \frac{m_1 + m_2}{r^3} + \frac{m_3}{R^3}, \qquad A_1 = \frac{m_1 + m_2 + m_3}{R^3}.$$

The integrals of angular momentum are

$$y\frac{dz}{dt} - z\frac{dy}{dt} = a, \qquad z\frac{dx}{dt} - x\frac{dz}{dt} = b, \qquad x\frac{dy}{dt} - y\frac{dx}{dt} = c,$$
$$y_1\frac{dz_1}{dt} - z_1\frac{dy_1}{dt} = a_1, \qquad z_1\frac{dx_1}{dt} - x_1\frac{dz_1}{dt} = b_1, \qquad x_1\frac{dy_1}{dt} - y_1\frac{dx_1}{dt} = c_1; \tag{42}$$

hence, each mass moves in its own plane. By our assumption, the distances of $m_3$ from $m_1$ and $m_2$ are equal.

$$\left(x_1 - \frac{m_1}{m_1 + m_2}x\right)^2 + \left(y_1 - \frac{m_1}{m_1 + m_2}y\right)^2 + \left(z_1 - \frac{m_1}{m_1 + m_2}z\right)^2$$
$$= \left(x_1 + \frac{m_2}{m_1 + m_2}x\right)^2 + \left(y_1 + \frac{m_2}{m_1 + m_2}y\right)^2 + \left(z_1 + \frac{m_2}{m_1 + m_2}z\right)^2, \tag{43}$$

or

$$xx_1 + yy_1 + zz_1 = \frac{1}{2}\frac{m_1 - m_2}{m_1 + m_2}r^2. \tag{43a}$$

If $m_1 = m_2$, then the right-hand member is zero. If

$$A - A_1 = (m_1 + m_2)\left(\frac{1}{r^3} - \frac{1}{R^3}\right),$$

then the three masses form an equilateral triangle of Lagrange.

From (42),

$$(xx_1 + yy_1 + zz_1)\left(\frac{dx}{dt}\frac{dx_1}{dt} + \frac{dy}{dt}\frac{dy_1}{dt} + \frac{dz}{dt}\frac{dz_1}{dt}\right)$$

$$- \left(\frac{dx}{dt}x_1 + \frac{dy}{dt}y_1 + \frac{dz}{dt}z_1\right)\left(x\frac{dx_1}{dt} + y\frac{dy_1}{dt} + z\frac{dz_1}{dt}\right)$$

$$= aa_1 + bb_1 + cc_1. \quad (44)$$

Furthermore, from (41) and (43),

$$p + q = \frac{m_1 - m_2}{m_1 + m_2} r \frac{dr}{dt}, \qquad \frac{dp}{dt} - \frac{dq}{dt} = \frac{1}{2}\frac{m_1 - m_2}{m_1 + m_2} r^2, \quad (45)$$

where

$$p = \frac{dx}{dt}x_1 + \frac{dy}{dt}y_1 + \frac{dz}{dt}z_1, \qquad q = x\frac{dx_1}{dt} + y\frac{dy_1}{dt} + z\frac{dz_1}{dt}.$$

Expand

$$\frac{dp}{dt} + \frac{dq}{dt}$$

and substitute (41) and (44); then

$$\frac{1}{4}\left(\frac{m_1 - m_2}{m_1 + m_2}\right)^2\left[r\frac{d^2r}{dt^2} + \left(\frac{dr}{dt}\right)^2 + \frac{1}{2}(A + A_1)r^2\right] - pq$$

$$= aa_1 + bb_1 + cc_1. \quad (46)$$

Differentiate $(dr/dt)^2$ and combine the result with (42) and the identity

$$(x^2 + y^2 + z^2)\left[\left(\frac{dx}{dt}\right)^2 + \left(\frac{dy}{dt}\right)^2 + \left(\frac{dz}{dt}\right)^2\right]$$

$$\equiv \left(x\frac{dx}{dt} + y\frac{dy}{dt} + z\frac{dz}{dt}\right)^2 + \left(y\frac{dz}{dt} - z\frac{dy}{dt}\right)^2$$

$$+ \left(z\frac{dx}{dt} - x\frac{dz}{dt}\right)^2 + \left(x\frac{dy}{dt} - y\frac{dx}{dt}\right)^2;$$

then,

$$r\frac{d^2r}{dt^2} = -Ar^2 + (a^2 + b^2 + c^2)\left(\frac{dr}{dt}\right)^{-2}. \quad (47)$$

Substituting these in (46) gives

$$\frac{1}{8}\left(\frac{m_1 - m_2}{m_1 + m_2}\right)^2 (A_1 - A)r^4 + \frac{1}{4}(p - q)^2 + \frac{1}{4}\left(\frac{m_1 - m_2}{m_1 + m_2}\right)^2 (a^2 + b^2 + c^2) = aa_1 + bb_1 + cc_1. \quad (48)$$

Put

$$R^2 = r_1^2 + \frac{m_1 m_2}{(m_1 + m_2)^2} r^2; \quad (49)$$

then (43) and (43a) give the equation corresponding to (47):

$$r_1 \frac{d^2 r_1}{dt^2} = -A_1 r_1^2 + \frac{1}{r_1^2}(a_1^2 + b_1^2 + c_1^2). \quad (50)$$

When we substitute $A$ and $A_1$ from (41), we obtain the relations (45), (47), (48), (49), (50) among the unknown quantities $p - q$, $r$, $R$, $r_1$.

First suppose that $m_1 \neq m_2$. There are now five relations among four quantities. We propose that these relations can only be satisfied by Lagrange's solution. For proof, apply Fuchs's theory on the singularities of a linear differential equation. First, take $u = -\int r^{-2}\,dt$ as the independent variable instead of $t$. This is a regularizing transformation, to be discussed in a later chapter; we only note here that $r = 0$ if $x = y = z = 0$, that is, if the two $m_1$ and $m_2$ bodies collide. This case is trivial; it reduces to the two-body problem. Put

$$p - q = \frac{m_1 - m_2}{m_1 + m_2} v. \quad (50a)$$

From (45) and (48),

$$\frac{dv}{du} + v^2 + (a^2 + b^2 + c^2) - 4\left(\frac{m_1 + m_2}{m_1 - m_2}\right)^2 (aa_1 + bb_1 + cc_1) = 0. \quad (51)$$

Then (45) and (47) are transformed to

$$\frac{1}{R^3} = \frac{1}{r^3} + \frac{2}{m_1 + m_2}\frac{1}{r^4}\frac{dv}{du}, \quad (52)$$

$$\frac{d^2}{du^2}\frac{1}{r} = Ar^3 - \frac{1}{r}(a^2 + b^2 + c^2). \quad (53)$$

Referring to (52) gives the following linear differential equation for $1/r$:

$$\frac{d^2}{du^2}\frac{1}{r} + \left(-\frac{2m_3}{m_1 + m_2}\frac{dv}{du} + a^2 + b^2 + c^2\right)\frac{1}{r} = m_1 + m_2 + m_3. \quad (54)$$

Equation (50) takes the form

$$\frac{d}{du}\left(\frac{1}{r^2}\frac{dr_1}{du}\right) + A_1 r^2 r_1 - (a_1^2 + b_1^2 + c_1^2)\frac{r^2}{r_1^3} = 0, \tag{55}$$

or

$$\frac{d}{du}\left(\frac{1}{r^2}\frac{dr_1}{du}\right) + (m_1 + m_2 + m_3)\frac{r^2 r_1}{R^3} - (a_1^2 + b_1^2 + c_1^2)\frac{r^2}{r_1^3} = 0. \tag{56}$$

Thus it is sufficient to prove that the five relations (49), (51), (52), (54), and (56) among four functions $v$, $r$, $R$, and $r_1$, all considered as functions of $u$, cannot be satisfied simultaneously except in the Lagrange case.

Suppose that $u$ is a complex variable. If these five equations admit a common solution as $u$ moves on the real axis of the complex plane of $u$, then the same five equations are satisfied in the whole complex plane of $u$ by the principle of analytical continuation. Hence, we consider the common solution in the whole plane of $u$. At first we see that (51) contains only $v$. If the solution be a constant or zero, then $1/r^3 - 1/R^3$ is also zero, by (52), and the case reduces to Lagrange's. If $v(u)$ is neither a constant nor zero, then by (51) we can continue analytically to the whole plane of $u$, and $v(u)$ has either a simple pole with residue 1 or an infinite number of such simple poles. Referring to (54) shows that $1/r$ can have no singularity other than the singularities of $v$ and can be analytically continued to the whole plane of $u$. From (52) and (49), we see that $R$ and $r_1$ can also be analytically continued to the whole plane of $u$.

For the purpose of studying the behavior of $v$, $r$, $R$, $r_1$ in the neighborhood of the poles of $v(u)$, take a pole of $v$ as the origin $u = 0$. By (51), the expansion of $u$ in the neighborhood of $u = 0$ is of the form

$$v(u) = \frac{1}{u} + \alpha_1 u + \alpha_3 u^3 + \cdots.$$

Hence, substitution in (54) gives

$$\frac{d^2}{du^2}\frac{1}{r} + \left(\frac{2m_3}{m_1 + m_2}\frac{1}{u^2} + \beta_0 + \beta_2 u^2 + \cdots\right)\frac{1}{r} = m_1 + m_2 + m_3, \tag{57}$$

where $\beta_0, \beta_2, \ldots$ are constants. This is a linear differential equation of Fuchsian type. The fundamental characteristic equation, or indicial equation, in the Fuchsian theory in the neighborhood of a singularity is obtained by replacing $1/r$ by $u^\lambda(1 + r_1 u + \cdots)$ with constants $r_1, \ldots,$

and comparing the coefficients of $u^{\lambda-2}$ by the methods of Section 3.7, that is,

$$\lambda(\lambda - 1) + \frac{2m_3}{m_1 + m_2} = 0.$$

Denote the two roots of this equation by $\lambda_1$ and $\lambda_2$. But (57) has a particular solution $1/r = f(u)$, which is regular even at $u = 0$ and can be expanded in an ascending power series of $u$ beginning with the first term $(m_1 + m_2)u^2/2$.

First suppose that $2m_3/(m_1 + m_2) < 1/4$; then $\lambda_1$ and $\lambda_2$ are both real and distinct, and $0 < \lambda_1 < \frac{1}{2}$, $\frac{1}{2} < \lambda_2 < 1$. Hence, the general solution of (57) in the neighborhood of $u = 0$ is of the form

$$\frac{1}{r} = C_1 u^{\lambda_1}\phi_1(u) + C_2 u^{\lambda_2}\phi_2(u) + f(u), \tag{58}$$

where $C_1$ and $C_2$ are the integration constants and $\phi_1$ and $\phi_2$ are integral power series each with the constant term equal to 1.

Next suppose that $2m_3/(m_1 + m_2) > \frac{1}{4}$; then $\lambda_1$ and $\lambda_2$ are both conjugate complex with real parts $\frac{1}{2}$. Hence, the general solution of (57) is of the form

$$\frac{1}{r} = u^{1/2}\{C_1 u^{i\mu}\phi_1(u) + C_2 u^{-i\mu}\phi_2(u)\} + f(u), \tag{59}$$

where $\mu$ is a positive constant depending only on the masses.

If $2m_3/(m_1 + m_2) = \frac{1}{4}$, then $\lambda_1 = \lambda_2 = \frac{1}{2}$ and the general solution of (57) is of the form

$$\frac{1}{r} = u^{1/2}(C_1 + C_2 \log u)\phi_1(u) + C_2 u^{1/2}\phi_2(u) + f(u), \tag{60}$$

where the constant term of $\phi_2$ is zero.

Assume now that $1/r$ is expressed by a regular integral $f(u)$. Corresponding to this solution, the functions $r$, $1/R^2$, $R$, and $r_1$, being analytically continued to the same contour of integration, are then expanded in ascending power series of $u$ in the neighborhood of $u = 0$. The first terms of these expansions, that is, the principal values, will be, according to (49) and (52), $u^{6+n}$ for $1/R^3$, $u^{-2-n/3}$ for $R$, $u^{-2}$ for $r$, and $u^{-2-n/3}$ for $r_1$, where $n$ is an even positive integer. If we substitute these expansions for $r$, $R$, and $r_1$ in (56), it becomes an identity. But the orders of the pole at $u = 0$ in these three expansions are $-n/3$, $2n/3$, and $2 + n$, respectively. Hence, none of them can be an identity for any choice of the even positive integer $n$. Hence, $1/r$ cannot be expressed by $f(u)$, contrary to our assumption.

Next assume that $1/r$ as a function of $u$ is expanded in the form (58), with particular values of $C_1$ and $C_2$, of which at least one is different from zero, or in the form (60) with $C_2 = 0$, $C_1 \neq 0$. As before, $1/r$, $r$, $1/R^3$, $R$, $r_1$ can be expanded in ascending power series of $u$ in the neighborhood of $u = 0$, where the exponents may be irrational. The first terms of these expansions for $1/r$ and $r$ are $Cu^\lambda$ and $C^{-1}u^{-\lambda}$, respectively. Here it is supposed that $C$ is a constant different from zero. $\lambda$ is equal to $\lambda_1$ or $\lambda_2$, or equal to $\frac{1}{2}$. According to (52) the principal value of $1/R^3$ is

$$-\frac{2C^4}{m_1 + m_2}\, u^{4\lambda - 2},$$

and that of $R$ is

$$\left(-\frac{m_1 + m_2}{2C^4}\, u^{2 - 4\lambda}\right)^{1/3}.$$

Thus in the neighborhood of $u = 0$ the function $r$ is more important than $R$. By (49), the principal value of $r_1$ is equal to that of $kr$, where

$$k = \frac{\sqrt{-m_1 m_2}}{m_1 + m_2}.$$

By continuing such arguments we see that (55) cannot be an identity unless $C = 0$. Thus we are led again to a contradiction.

Finally, assume that the expansion of $1/r$ in the neighborhood of $u = 0$ is of the form (59), where at least one of $C_1$ and $C_2$ differs from zero, or of the form (60) with $C_2 \neq 0$. Then $1/r$, $r$, $1/R^3$, $R$, and $r_1$ are expanded in ascending fractional and integral powers of $u$; the coefficients (except the coefficient of $\log u$ if it is contained) are bounded as $u$ approaches $u = 0$ along a suitable path of integration. Hence, it is sufficient that $C_1 u^{i\mu} + C_2 u^{-i\mu}$ in (59), or $C_1 + C_2 \log u$ in (60), is larger than a positive number in absolute value. The condition is satisfied for $C_1 + C_2 \log u$ along any path of integration passing through $u = 0$, because the expression tends to infinity. On the other hand, the expression $C_1 u^{i\mu} + C_2 u^{-i\mu}$, if none of $C_1$ and $C_2$ is zero, has an infinite number of zero points on a semi-infinite line starting from $u = 0$, with their distances from $u = 0$ in geometrical progression. The required condition is satisfied if $u$ does not enter the inside of the angle formed by two semi-infinite lines starting from $u = 0$. If either $C_1$ or $C_2$ is zero, then it is sufficient that the path of integration does not rotate an infinite number of times around $u = 0$. In any case, the principal value of the expansion of $r_1$ is equal to that of $kr$; by substituting on the left-hand side of (55), we see that it cannot be an identity unless both $C_1$ and $C_2$ are zero. Hence, this last assumption, too, should be rejected.

Thus we have proved the theorem of Wilczynski and MacMillan that, if $m_1 \neq m_2$, then the isosceles triangle in question is the equilateral triangle of Lagrange.

### 3.9 NATURE OF THE ISOSCELES-TRIANGULAR SOLUTIONS

Suppose that $m_1 = m_2$. Equations (45) and (48) are now

$$p + q = 0 \quad \text{and} \quad \tfrac{1}{4}(p - q)^2 = aa_1 + bb_1 + cc_1, \tag{61}$$

where $p$ and $q$ are now constants. Putting $m_1 = m_2$ in (42) or (43) and in the definition of $p$ gives

$$ax + by + cz = 0, \qquad xx_1 + yy_1 + zz_1 = 0,$$

$$\frac{dx_1}{dt}x + \frac{dy_1}{dt}y + \frac{dz_1}{dt}z = 0. \tag{62}$$

Solving these three equations, we obtain

$$\begin{aligned}(aa_1 + bb_1 + cc_1)x &= p(bz_1 - cy_1),\\ (aa_1 + bb_1 + cc_1)y &= p(cx_1 - az_1),\\ (aa_1 + bb_1 + cc_1)z &= p(ay_1 - bx_1).\end{aligned} \tag{63}$$

Equations (63) give a unique solution if $aa_1 + bb_1 + cc_1 \neq 0$. If we differentiate each of (63) twice with respect to $t$, then the left-hand side is multiplied by $-A$, and the right-hand side by $-A_1$ and $p$ is a constant by (61). Hence both sides of (63) should be identically zero if $A \neq 0$. (We have excluded the trivial case in which all of $x$, $y$, and $z$ are zero.) Since at least one of $x, y$, or $z$ is not zero, $aa_1 + bb_1 + cc_1 = 0$. Therefore by the second equation of (61), $p = q = 0$. Thus (62) reduces to

$$xx_1 + yy_1 + zz_1 = 0, \qquad x\frac{dx_1}{dt} + y\frac{dy_1}{dt} + z\frac{dz_1}{dt} = 0. \tag{64}$$

For our discussion we classify four cases:

First suppose that (64) is satisfied for $x_1 = y_1 = z_1 = 0$ by being reduced to an identity for $x$, $y$, $z$. In this case the three masses are on a straight line and $m_3$ is situated at the middle point of $m_1$ and $m_2$. This is a special case of Euler's straight-line equilibrium solution.

Next suppose that the coefficients of $x$, $y$, $z$ are in proportion. Then $a_1 = b_1 = c_1 = 0$, by (42). The position vector $(x_1, y_1, z_1)$ has a fixed direction and position vector $(x, y, z)$ is perpendicular to that direction by the first equation of (64). The motion has an axis of symmetry; $m_3$ moves on the axis of symmetry, and $m_1$ and $m_2$ move symmetrically with respect to this axis.

Next suppose that $a_1$, $b_1$, $c_1$ are not zero. Then $x:a_1 = y:b_1 = z:c_1$, from (64); position vector $(x, y, z)$ has a fixed direction and the vector $(x_1, y_1, z_1)$ is perpendicular to it. The motion has a plane of symmetry; $m_3$ moves on the plane of symmetry, and $m_1$ and $m_2$ move symmetrically with respect to this plane.

Last, suppose that the two preceding cases coincide, that is, position vectors $(x, y, z)$ and $(x_1, y_1, z_1)$ have their own fixed directions and are mutually perpendicular. The motion is a plane motion and has an axis of symmetry lying on that plane; $m_3$ moves on the axis of symmetry.

Let us consider the last three of these four cases in greater detail: I, with an axis of symmetry; II, with a plane of symmetry; and III, with an axis of symmetry together with a plane of symmetry. (These three cases correspond respectively to Fransèn's three kinds of solution — see beginning of Section 3.8.

I. Consider an isosceles triangle with an axis of symmetry. Take the origin $O$ at the center of masses $m_1 = m_2 = m$, and the $Oz$-axis along the axis of symmetry. Then $z = 0$, $x_1 = y_1 = 0$ in (41) and the coordinates of $m_1$, $m_2$, and $m_3$ are, respectively, $(-x, -y, 0)$, $(x, y, 0)$, and $(0, 0, z_1)$; and $r = \sqrt{x^2 + y^2}$. The equations of motion are

$$\frac{d^2x}{dt^2} = -\left[\frac{m}{4r^3} + \frac{m_3}{(r^2 + z_1^2)^{3/2}}\right]x,$$

$$\frac{d^2y}{dt^2} = -\left[\frac{m}{4r^3} + \frac{m_3}{(r^2 + z_1^2)^{3/2}}\right]y, \tag{65}$$

$$\frac{d^2z_1}{dt^2} = -\frac{(2m + m_3)z_1}{(r^2 + z_1^2)^{3/2}}.$$

Let $x = r\cos\theta$, $y = r\sin\theta$; then the integral of (65), with the constants of integration $C$ and $h$, is given by

$$\left(\frac{dr}{dt}\right)^2 + \frac{m_3}{2m + m_3}\left(\frac{dz_1}{dt}\right)^2 = -\frac{C^2}{r^2} + \frac{m}{2r} + \frac{2m_3}{\sqrt{r^2 + z_1^2}} + h,$$

$$r^2\frac{d\theta}{dt} = C. \tag{66}$$

If $C = 0$, then the motion reduces to case III. Suppose that $C \neq 0$. Since the right-hand side of (63) is positive or zero, then

$$-\frac{C^2}{r^2} + \frac{m + 4m_3}{2r} + h \geqq 0. \tag{67}$$

Hence, if $h \geqq 0$, $r$ cannot be smaller than a positive root of this equation for $r$. If $h < 0$, $r$ is in the interval between the two roots. Let the positive root or the smaller of the two roots be $l$; then the three mutual distances are less than or equal to $l$. There is no collision. Hence, the motion can be traced from $t = -\infty$ to $t = +\infty$. For $h < 0$ the mass $m_3$ oscillates on the $Oz$-axis, its motion being symmetric about the origin $O$. The areal velocity for the motion of either $m_1$ or $m_2$ is invariant, but the motion is not Keplerian, because the second term of the right-hand member in the first or second equation of (65) is a perturbation.

If the motion of $m_1$ and $m_2$ is a revolution in the plane $(x, y)$ around $O$ with the same period as the motion of $m_3$, then the motion is confined to periodic orbits. For $h \geqq 0$, the mass $m_3$ recedes to infinity, and consequently the motion of $m_1$ and $m_2$ tends asymptotically to a Keplerian motion in the $(x, y)$-plane around $O$. For $m_3 \to 0$ the motion of $m_1$ and $m_2$ also tends to the Keplerian. If we look at these motions from the center of mass of the three bodies, then the origin $O$ together with the plane $Oxy$ moves in the direction of $Oz$ with respect to the center of mass. The straight line joining $m_1$ and $m_2$ describes a ruled surface in space. Its asymptote is the axis of symmetry $Oz$ on which $m_3$ moves. This is the case of Wilczynski's second category.

II. Consider the case with a plane of symmetry. Take the plane of symmetry as the $Oxy$ plane; then $x = y = 0$, $z_1 = 0$. The coordinates of $m_1$, $m_2$, and $m_3$ are respectively, $(0, 0, -z)$, $(0, 0, z)$, and $(x_1, y_1, 0)$. Suppose that $z \geqq 0$. The equations of motion, if we write $r_1^2 = x_1^2 + y_1^2$, are

$$\frac{d^2 z}{dt^2} = -\left[\frac{m}{4z^3} + \frac{m_3}{(z^2 + r_1^2)^{3/2}}\right] z,$$

$$\frac{d^2 x_1}{dt^2} = -\frac{(2m + m_3)x_1}{(z^2 + r_1^2)^{3/2}}, \tag{68}$$

$$\frac{d^2 y_1}{dt^2} = -\frac{(2m + m_3)y_1}{(z^2 + r_1^2)^{3/2}}.$$

Put $x_1 = r_1 \cos \theta_1$, $y_1 = r_1 \sin \theta_1$; then

$$\left(\frac{dz}{dt}\right)^2 + \frac{m_3}{2m + m_3}\left[\left(\frac{dr_1}{dt}\right)^2 + \frac{C^2}{r_1^2}\right] = \frac{m}{2z} + \frac{2m_3}{\sqrt{z^2 + r_1^2}} + h,$$

$$r_1^2 \frac{d\theta_1}{dt} = C. \tag{69}$$

For $C = 0$ the case reduces to case III (Wilczynski's third category). Suppose that $C \neq 0$. As we have taken $z \geqq 0$, we have $d^2z/dt^2 < 0$ and $d^2z/dt^2 = -\infty$ at $z = 0$. Hence, $z$ cannot take any minimum value other than 0. The masses $m_1$ and $m_2$ collide at the origin $O$. But, as $C \neq 0$, the mass $m_3$ revolves on the plane of symmetry $xy$ around $O$, and the areal velocity is invariant. The motion tends asymptotically to a Keplerian motion for $z \to \infty$.

The center of mass $G$ of the three bodies lies always in the $xy$-plane. Seen from $G$, the straight line joining $m_1$ and $m_2$ describes a cylindrical surface with the straight line passing through $G$ and parallel to the $Oz$-axis as the axis of the cylinder. This is the ruled surface of Wilczynski, and the motion is of his second category. The trajectories of $m_1$ and $m_2$ lie on the $z$-axis and are mirror images of one another; $m_1$ and $m_2$ collide at $O$ in a finite time $t_1$ or $-t_1$. But, as will be shown in a later chapter, the motion, according to Sundman, can be analytically continued after any two-body collision. In this sense the motion of $m_1$ and $m_2$ can be traced from $t = -\infty$ to $t = +\infty$. For $m_3 \to 0$ the motion tends to a straight-line Keplerian motion.

III. Consider the case of plane motion with an axis of symmetry and a plane of symmetry; the axis lies on the plane. We take the axis of symmetry as the $x$-axis and the plane of symmetry as the plane $z = 0$. The coordinates of $m_1$, $m_2$, and $m_3$ are, respectively, $(0, -y, 0)$, $(0, y, 0)$, $(x_1, 0, 0)$, with $y \geqq 0$. The equations of motion are

$$\frac{d^2y}{dt^2} = -\left[\frac{m}{4y^3} + \frac{m_3}{(x_1^2 + y^2)^{3/2}}\right]y,$$
$$\frac{d^2x_1}{dt^2} = -\frac{(2m + m_3)x_1}{(x_1^2 + y^2)^{3/2}}. \tag{70}$$

As $y$ has no minimum other than 0, the masses $m_1$ and $m_2$ collide at $O$ in a finite time $t_1$ or $-t_1$. If owing to the initial conditions $x_1$ and $y$ do not become zero simultaneously, then only $m_1$ and $m_2$ collide and the motion can be analytically continued beyond the two-body collision, as in case II. The angular momentum is zero and the energy integral is

$$\left(\frac{dy}{dt}\right)^2 + \frac{m_3}{2m + m_3}\left(\frac{dx_1}{dt}\right)^2 = \frac{m}{2y} + \frac{2m_3}{\sqrt{y^2 + x_1^2}} + h. \tag{71}$$

For $h > 0$ either all of the three masses $m_1$, $m_2$, $m_3$ or two masses $m_1$ and $m_2$, or only the single mass $m_3$ can recede to infinity. If we continue analytically after the two-body collision of $m_1$ and $m_2$, then there may be periodic solutions as in case II. There may occur a three-body collision at $O$. We do not touch on the question as to whether or not repeated two-body collisions can cause a three-body collision. After a three-body collision, the analytical continuation is impossible except for some special values of the masses and the initial conditions, as will be shown in a later chapter. The possibility of collisions for the periodic orbits in cases I and II will be discussed in a later chapter.

We now discuss the motion of an infinitesimal mass $m_3$ in case I — the case with the axis of symmetry. We suppose that $m_1$ and $m_2$ move in a common circular orbit and that $m_3$ moves on the straight line passing normally through the center of this circle. This is a special case of the restricted three-body problem. Denote the radius of the circle by $a$, and the integration constant by $\gamma$; then, from (65) and (66),

$$\frac{d^2 z_1}{dt^2} = -\frac{2mz_1}{(a^2 + z_1^2)^{3/2}}$$

or

$$\left(\frac{dz_1}{dt}\right)^2 = \frac{4m}{\sqrt{a^2 + z_1^2}} - 4\gamma.$$

For $\gamma < 0$, the variable $z_1$ reaches $\infty$ and $dz_1/dt = 2\sqrt{-\gamma}$ at $z_1 = \infty$. For $\gamma > 0$, the variable $z_1$ is at a standstill, $dz_1/dt = 0$ at

$$z_1^2 = \zeta^2 = \frac{m^2}{\gamma^2} - a^2,$$

and $z_1$ oscillates between $-\zeta \leqq z_1 \leqq \zeta$ through $z_1 = 0$. This is a periodic solution. But, unless $m/\gamma \geqq a$, the velocity $dz_1/dt$ becomes imaginary and the motion is imaginary. In the case $m/\gamma = a$, the moving point is always at rest at the origin; this is the case of Euler.

Let $1/w = \sqrt{a^2 + z_1^2}$; then $w$ vanishes only when $z_1$ becomes infinite. The variable $w$ does not vanish for $\gamma > 0$, and it never becomes infinite. Then the energy integral is

$$\left(\frac{dw}{dt}\right)^2 = 4w^4(1 - a^2w^2)(mw - \gamma).$$

We take $u$ (defined by $w^2\,dt = du$) as the new independent variable instead of $t$. Then

$$\left(\frac{dw}{du}\right)^2 = 4(1 - a^2w^2)(mw - \gamma).$$

This is transformed to

$$\left(\frac{dv}{du}\right)^2 = 4v^3 - g_2 v - g_3,$$

where

$$w = -\frac{v}{(a^2m)^{1/3}} + \frac{\gamma}{3m}.$$

Denote the three roots of this equation by $e_1 > e_2 > e_3$. The solution can be written by Weierstrass's elliptic function $v = \wp(u)$ and

$$dt = [-(a^2m)^{-1/3}\wp(u) + \gamma m^{-1}]^{-2}\, du.$$

If $\gamma > 0$, $\gamma/m < 1/a$; then

$$(a^2m)^{-1/3}e_1 = \frac{\gamma}{3m} + \frac{1}{a} < 0, \qquad (a^2m)^{-1/3}e_2 = -\frac{2\gamma}{3m} < 0,$$

$$(a^2m)^{-1/3}e_3 = \frac{\gamma}{3m} - \frac{1}{a} < 0.$$

The corresponding values of $w$ are $w_1 = -1/a$, $w_2 = \gamma/m$, $w_3 = 1/a$; $w_1 < w_2 < w_3$. Let the periods of the elliptic functions be $2\omega$ and $2\omega'$; then as $u$ varies from $\omega + \omega'$ to $\omega'$, the function $\wp(u)$ decreases from $e_2$ to $e_3$ and $\wp'(u)$ is real and positive, $w$ increases from $\gamma/m$ to $1/a$, and $\sqrt{a^2 + z_1^2}$ decreases from $m/\gamma$ to $a$. Since there is no zero point of $w$, that is, no point at which $z_1 = \infty$, the relation between $dt$ and $du$ is regular. The variable $u$ is also regular for finite values of $t$. Hence $w$, and accordingly $z_1$, is regular. The quantity $\sqrt{a^2 + z_1^2}$ oscillates between $m/\gamma$ and $a$; it corresponds to a periodic solution. The coordinate $z_1$ oscillates between $\pm\sqrt{(m/\gamma)^2 - a^2}$, situated symmetrically about $O$.

If $\gamma > 0$, $\gamma/m > 1/a$; then $e_2$ and $e_3$ interchange. But the motion is imaginary in this case.

If $\gamma < 0$, then the values of $w$ corresponding to $e_1 > e_2 > e_3$ are either $w_1 = -1/a$, $w_2 = \gamma/m$, $w_3 = 1/a$, or $w_1 = \gamma/m$, $w_2 = -1/a$, $w_3 = 1/a$; $w_1 < w_2 < w_3$. As $1/a$ is the only positive root among the three, there is a zero point for $w$ between $w_2$ and $w_3$. $t$ cannot be related to $u$ regularly at $w = 0$ in the transformation $w^2\, dt = du$. In fact, $z_1 = \infty$ at $w = 0$. Hence, the real motion occurs only between $w = 0$ and $w = w_3$, that is, between $v = (\gamma/3m)(a^2m)^{1/3}$ and $e_3$. At $w = 0$, that is, at $z_1 = \infty$, the motion recedes to infinity.

The relation between $t$ and $u$ is obtained from the formulas for elliptic functions

$$\frac{d}{dx}\zeta(x) = -\wp(x),$$

$$\frac{d}{dx}\log\sigma(x) = \zeta(x),$$

$$\zeta(x+y) - \zeta(x-y) - 2\zeta(y) = -\frac{\wp'(y)}{\wp(x) - \wp(y)},$$

$$\wp(x+y) + \wp(x-y) - 2\wp(y) = \frac{[\wp'(y)]^2}{[\wp(x) - \wp(y)]^2} + \frac{\wp''(y)}{\wp(x) - \wp(y)};$$

that is,

$$[\wp'(y)]^2 \int \frac{dx}{[\wp(x) - \wp(y)]^2} = -\zeta(x+y) - \zeta(x-y) - 2x\wp(y)$$

$$+ \frac{\wp''(y)}{\wp'(y)}\left[\log\frac{\sigma(x+y)}{\sigma(x-y)} - 2x\zeta(y)\right].$$

Let $\wp(\alpha) = \gamma a^{2/3}m^{-2/3}/3$; then,

$$dt = a^{4/3}m^{2/3}[\wp(u) - \wp(\alpha)]^{-2}\,du;$$

or, if we integrate with the integration constant $t_0$,

$$\left[\frac{\wp'(\alpha)}{a^{2/3}m^{1/3}}\right]^2(t - t_0) = -\zeta(u+\alpha) - \zeta(u-\alpha) - 2u\wp(\alpha)$$

$$+ \frac{\wp'(\alpha)}{\wp''(\alpha)}\left[\log\frac{\sigma(u+\alpha)}{\sigma(u-\alpha)} - 2u\zeta(\alpha)\right].$$

Write $q = e^{-\pi\omega'/i\omega}$; then, using the formulas of elliptic functions,

$$\wp(w) = -\frac{\eta}{\omega} + \left(\frac{\pi}{2\omega}\right)^2 \frac{1}{\sin^2\frac{\pi w}{2\omega}} - 2\left(\frac{\pi}{\omega}\right)^2 \sum_{n=1}^{\infty} \frac{nq^{2n}}{1 - q^{2n}}\cos\frac{n\pi w}{\omega},$$

$$\zeta(w) = \frac{\eta w}{\omega} + \frac{\pi}{2\omega}\cot\frac{\pi w}{2\omega} + \frac{2\pi}{\omega}\sum_{n=1}^{\infty}\frac{q^{2n}}{1 - q^{2n}}\sin\frac{n\pi w}{\omega},$$

$$\log\sigma(w) = \frac{\eta w^2}{2\omega} + \log\left(\frac{2\omega}{\pi}\sin\frac{\pi w}{2\omega}\right) + 2\sum_{n=1}^{\infty}\frac{q^{2n}}{1 - q^{2n}}\,\frac{1 - \cos\frac{n\pi w}{\omega}}{n},$$

we obtain

$$w = \frac{\gamma}{3m} - a^{-2/3}m^{-1/3}\left[-\frac{\eta}{\omega} + \left(\frac{\pi}{2\omega}\right)^2\left(\sin^2\frac{\pi u}{2\omega}\right)^{-2} - 2\left(\frac{\pi}{\omega}\right)^2\sum_{n=1}^{\infty}\frac{nq^{2n}}{1-q^{2n}}\cos\frac{n\pi u}{\omega}\right],$$

$$\left[\frac{\wp'(\alpha)}{a^{2/3}m^{1/3}}\right]^2\cdot(t-t_0)$$

$$= -\frac{2qu}{\omega} - 4u\wp'(\alpha) - 4u\wp(\alpha)\cos\frac{\pi u}{\omega}\operatorname{cosec}\frac{\pi(u+\alpha)}{2\omega}\operatorname{cosec}\frac{\pi(u-\alpha)}{2\omega}$$

$$+ \frac{\wp''(\alpha)}{\wp'(\alpha)}\left\{\frac{2\alpha u}{\omega} - 4u\zeta(\alpha) + \log\left[\frac{2\omega}{\pi}\sin\frac{\pi(u+\alpha)}{2\omega}\right] - \log\left[\frac{2\omega}{\pi}\sin\frac{\pi(u-\alpha)}{2\omega}\right] + 4\sum_{n=1}^{\infty}\frac{q^{2n}}{1-q^{2n}}\frac{1}{n}\sin\frac{n\pi u}{\omega}\sin\frac{n\pi\alpha}{\omega}\right\}.$$

As we see from these expansions, the variable $w$ has no singularity, as $u$ varies from $\omega + \omega'$ to $\omega'$. But $t - t_0$ has a singularity, indeed a logarithmic singularity, at $u = \alpha$; this corresponds to $w = 0$, that is, to $z_1 = \infty$. For $\gamma > 0$ this singularity does not lie in the interval between $v = e_2$ and $v = e_3$. Hence, the growth of $t - t_0$ consists of a uniform component, superposed on periodic oscillations in $u$ of period $4\omega$. The variable $w$, and accordingly $z_1$, oscillates with period $4\omega$. Thus, if $\gamma > 0$, $\gamma/m < 1/a$, then there exist period solutions of period $4\omega$. The period can be computed by

$$\sqrt{\frac{2\omega}{\pi}} = \frac{2}{(e_1-e_3)^{1/4} + (e_1-e_2)^{1/4}}(1 + 2h^4 + 2h^{16} + \cdots),$$

where

$$h = \tfrac{1}{2}l + 2(\tfrac{1}{2}l)^5 + 15(\tfrac{1}{2}l)^9 + \cdots, \qquad l = \frac{(e_1-e_3)^{1/4} - (e_1-e_2)^{1/4}}{(e_1-e_3)^{1/4} + (e_1-e_2)^{1/4}}.$$

The period with regard to $t$ is

$$4\omega\left[\frac{\wp'(\alpha)}{a^{2/3}m^{1/3}}\right]^2\left\{-\frac{2\eta}{\omega} + \frac{\wp''(\alpha)}{\wp'(\alpha)}\left[\frac{2\eta\alpha}{\omega} - 4\zeta(\alpha)\right]\right\}.$$

Compare this with the two-body problem; we see that the case $\gamma > 0$ corresponds to elliptic motions and the case $\gamma < 0$ to hyperbolic motions.

When $m_1$ and $m_2$ move in elliptic orbits, we may be able to make a similar transformation but still cannot carry the computation through. Buchanan (1920) computed the periodic solutions in the sense of Poincaré, but we postpone consideration of this to a later chapter, along with the other classes of isosceles-triangular solutions, and the discussion of their collisions.

Various attempts have been made to attack the three-body problem by introducing new ideas. Sokolov (1945) introduced a new independent variable in the plane problem and derived the known configurations of Lagrange and Euler (Sections 3.2, 3.7). Pylarinos (1941) introduced a rigid triangle associated with the triangle formed by the three bodies. Egervary, according to Meffroy, used the analogy of the motion of a top. Recently Meffroy (1962, 1968) studied the circular cone with one of the three bodies as vertex and a fixed direction in space as axis. Let the vertex angle of the cone be fixed at $\pi - 2\theta$. The plane $P$ containing the three masses rolls around this cone, the polar coordinates of the plane being $\pi/2 - \theta$ and $\phi$. The angle $\theta$ is fixed because of the areal velocity integrals and the energy integral. If the plane $P$ rotates around a fixed direction on the plane $P$ which passes through one of the three masses, then either the motion is Lagrange's or in the limit, two of the three bodies collide. In a special case $\theta = \pi/2$, we determine the relative motion from a differential equation of the third order,

$$\lambda = \left\{2\frac{d\phi}{dt}\frac{d^3\phi}{dt^3} - 3\left(\frac{d\phi}{dt}\right)^2 + 4\left(\frac{d\phi}{dt}\right)^4\right\}\Big/\left\{4k^2\left(\frac{d\phi}{dt}\right)^{7/2}\right\},$$

by putting $\lambda = 0$ or $\lambda = \infty$ as is appropriate. Otherwise the plane $P$ cannot rotate around any direction fixed on $P$ that passes through one of the bodies. If one of the bodies is of mass zero, then the differential equations are reduced to a complicated one of second order.

## REFERENCES

Andoyer, H.
1906 *Bull. Astron.*, **25**, 50.

Banachiewicz, T.
1906 *Compt. Rend.*, **142**, 510.

Bilimovitch, A.
1911 *Astron. Nachr.*, **189**, 181.
1927 *Berichte serbische Akad. Wiss. Belgrad*, **59**, 17.
1929 *Ibid.*, **63**, 1.

Bilimovitch, A., and Petronievics, B.
1936 *Astron. Nachr.*, **258**, 209.
Brehm, E.
1908 *Partikuläre Integrale des Problems der drei Körper*, Dissertation, Berlin.
Brumberg, V. A.
1958 *Astron. Zh.*, **34**, 55.
Buchanan, D.
1920 *Periodic Orbits*, F. R. Moulton, ed., Carnegie Institution, Washington.
Carathéodory, C.
1933 *Sitzber. bayer. Akad. Wiss.*, **1933**, 257.
Chazy, J.
1918 *Bull. Astron.*, **35**, 321.
1921 *Ibid.*, ser. 2, **1**, 171.
1927 *Bull. Soc. Math. France*, **55**, 222.
Dziobek, O.
1900 *Astron. Nachr.*, **152**, 33.
Fransèn, A. E.
1895 *Öfversigt af. kongl. Vetenskaps Akad. Handl.*, **52**, 788.
Fubini, G. and Čech, E.
1926 *Geometria Proiettiva Differenziale*, I, II, Zanichelli, Bologna.
Hagihara, Y.
1970 *Theories of Equilibrium Figures of a Rotating Liquid Mass*, National Aeronautics and Space Administration, Washington, D.C.
Halphen, G. H.
1891 *Traité des Fonctions Elliptiques et de leurs Applications*, III, Gauthier-Villars, Paris.
Heffter, L.
1894 *Einführung in die Theorie der linearen Differentialgleichungen*, Teubner, Leipzig.
Hütterhain, E.
1933 *Astron. Nachr.*, **250**, 297.
1935 *Astron. Nachr.*, **254**, 281.
Ince, E. L.
1927 *Ordinary Differential Equations*, Longmans, London.
Klein, F.
1894 *Über lineare Differentialgleichungen zweiter Ordnung, Vorlesung gehalten in Sommersemester 1894, ausgearbeitet von R. Ritter*, Göttingen (mimeograph).
1933 *Vorlesungen über hypergeometrische Funktionen*, Springer, Berlin.
Klemperer, W. B.
1962 *Astron. J.*, **67**, 162.
Lehmann-Filhés, R.
1891 *Astron. Nachr.*, **127**, 137.
Lindow, M.
1922 *Astron. Nachr.*, **216**, 389.
1923 *Ibid.*, **219**, 141.
1924 *Ibid.*, **220**, 369.
1926 *Ibid.*, **228**, 233.

1928 *Ibid.*, **233**, 177.
Longley, W. R.
1907 *Bull. Am. Math. Soc.*, **13**, 324.
MacMillan, W. D.
1913 *Astron. J.*, **27**, 11.
MacMillan, W. D., and Bartkey, W.
1932 *Trans. Am. Math. Soc.*, **34**, 838.
Meffroy, J.
1962 *Compt. Rend.*, **255**, 477.
1967 *Icarus*, **7**, 351.
1968 *Icarus*, **8**, 196.
Meyer, O.
1933 *Ann. Obs. Bordeaux*, **17**, 75.
Moulton, F. R.
1900 *Trans. Am. Math. Soc.*, **1**, 510.
1904 *Quart. J. Math.*, **35**, 116.
Müntz, Ch. H.
1922 *Math. Z.*, **15**, 169.
Pavanini, G.
1907 *Annali Mat. pura appl.*, ser. 3, **13**, 179.
Pizzetti, P.
1904 *Rendiconti Accad. Lincei*, ser. 5, **13**, fasc. 1, 17.
Poole, E. G. C.
1936 *Introduction to the Theory of Differential Equations*, Clarendon Press, Oxford.
Pylarinos, O.
1937 *Math. Annalen*, **114**, 150.
1941 *Math. Z.*, **47**, 357.
Salmon, G.
1912 *A Treatise on the Analytical Geometry of Three Dimensions*, I, 5th ed., Longmans, London.
Schaub, W.
1929 *Astron. Nachr.*, **236**, 33.
Schlesinger, L.
1908 *Einführung in die Theorie der gewöhnlichen Differentialgleichungen auf funktionentheoretischen Grundlagen*, Gruyter, Leipzig.
1922 *Vorlesungen über linearen Differentialgleichungen*, Teubner, Leipzig.
Serret, J. A.
1885 *Cours d'Algèbre Supérieure*, **I**, Gauthier-Villars, Paris.
Seydler, A.
1886 *Abhandl. Math.-Naturwiss. Kl. Gesellschaft Wien*, ser. 7, **1**, 1.
Sokolev, B.
1945 *Dokl. Akad. Nauk SSSR*, **46**, 95, 98.
Wilczynski, E. J.
1906 *Projective Differential Geometry of Curves and Ruled Surfaces*, Teubner, Leipzig.
1913 *Annali Mat. pura appl.*, ser. 3, **21**, 1.
Williams, W. L.
1938 *Trans. Am. Math. Soc.*, **44**, 563.

Woronetz, P.
1907 *Math. Annalen*, **63**, 387.
Zapp, E.
1908 *Untersuchung eines speciellen Falles des Drei- und Vierkörperproblems*, Dissertation, Munich.

# Part II
# Transformation Theory

# Continuous Groups of Transformations

## INFINITESIMAL TRANSFORMATIONS

### 4.1 THE FIRST INTEGRALS

Consider a system of differential equations

$$\frac{dx_1}{X_1} = \frac{dx_2}{X_2} = \cdots = \frac{dx_n}{X_n} = dt. \tag{1}$$

Under certain conditions which must be satisfied by the functions $X_i$, this system admits the *first integrals*, which are regular in $t, x_1, x_2, \ldots, x_n$,

$$f_i(t, x_1, x_2, \ldots, x_n) = c_i \qquad (i = 1, 2, \ldots, n), \tag{2}$$

with integration constants $c_i$. These integrals are independent if the Jacobian $\partial(f_1, f_2, \ldots, f_n)/\partial(x_1, x_2, \ldots, x_n)$ does not vanish. Solving (2) in that case gives *general integrals*

$$x_i = \phi_i(t, c_1, c_2, \ldots, c_n) \qquad (i = 1, 2, \ldots, n). \tag{3}$$

If $f$ is an integral of (1), then $f$ satisfies a partial differential equation of the first order. We consider $x_1, x_2, \ldots, x_n$ as the coordinates for an integral curve in an $n$-dimensional space, which are represented as functions of a parameter $t$. Since the function $f$ becomes a constant when the variables $x_i$ in $f$ are expressed as functions of $t$ we have $df = 0$. In other words, the function $f$ is invariant under a transformation

$$t' = t + dt, \qquad x_i' = x_i + X_i\, dt \qquad (i = 1, 2, \ldots, n) \tag{4}$$

as the transformed point moves along the integral curve. This transformation (4) is called an *infinitesimal transformation*, which was studied by

Lie (Lie, 1888, 1890, 1891, 1893, 1893a, 1896; Kowalewski, 1931; Eisenhardt, 1933; Campbell, 1903; Engel and Faber, 1932; Hoheisel, 1928; Cartan, 1933, 1937; Vivanti, 1918; Bianchi, 1918; Potron, 1936; Cerf, 1929; Buhl, 1928; Carathéodory, 1933; Pontrjagin, 1939; Chevalley 1946, 1951). We say then that the system of differential equations (1) *admits* an infinitesimal transformation (4), or, replacing $dx_i$ in $df$ by $X_i\,dt$ according to (1), we have

$$X(f) = \frac{\partial f}{\partial t} + X_1 \frac{\partial f}{\partial x_1} + X_2 \frac{\partial f}{\partial x_2} + \cdots + X_n \frac{\partial f}{\partial x_n} = 0. \tag{5}$$

$X(f)$ is the symbol for an infinitesimal transformation; an infinitesimal transformation (5) is said to *generate* (2). Any first integral (2) satisfies (5). Conversely, any function $f$ satisfying (5) is a first integral of (1), because substituting (1) for $X_i$ of (5) gives $df = 0$. Any function $F$ of $f_1, f_2, \ldots, f_n$ satisfies (5). If we eliminate the coefficients $1, X_1, \ldots, X_n$ from these equations, we obtain the Jacobian

$$\partial(F, f_1, \ldots, f_n)/\partial(t, x_1, \ldots, x_n) = 0,$$

which shows that $F$ is a function of $f_1, \ldots, f_n$.

Thus an integral of (5) is an integral of (1), and vice versa.

Next, consider the problem in the reverse order. Does any solution of a partial differential equation of the first order become an integral of an ordinary differential equation? Consider a partial differential equation

$$F(x_1, \ldots, x_n, z, p_1, \ldots, p_n) = 0, \tag{6}$$

where $p_i = \partial z/\partial x_i$. Suppose that we know an arbitrary solution of this equation $z = \Phi(x_1, x_2, \ldots, x_n)$. Letting $X_i = \partial F/\partial x_i$, $P_k = \partial F/\partial p_k$, $Z = \partial F/\partial z$, we construct

$$\frac{dx_1}{P_1} = \frac{dx_2}{P_2} = \cdots = \frac{dx_n}{P_n}.$$

The $P_i$ are known functions of $x_1, x_2, \ldots, x_n$. As these equations constitute $n - 1$ relations among $n$ variables $x_1, x_2, \ldots, x_n$, they represent a curve on an $n$-dimensional hypersurface $z = \Phi$ embedded in an $(n + 1)$-dimensional space. The given differential equations represent relations among infinitesimals. By giving arbitrary values to the integration constants we obtain a family of curves on that hypersurface. Through any point of an integral hypersurface $z = \Phi(x_1, x_2, \ldots, x_n)$ there passes one and only one curve belonging to this family. Let a set $(p_1, p_2, \ldots, p_n)$ correspond to each point $z = \Phi(x_1, x_2, \ldots, x_n)$; that is, if we restrict the direction to be $p_1, p_2, \ldots, p_n$ from that point, we obtain $\infty^1$ elements.

Any set of $2n + 1$ values for $z$, $x_i$, $p_i$ is called an *element*. The curve obtained in this manner is called the *characteristic* curve of Cauchy.

If we do not know the function $\Phi$, we proceed as follows. Suppose that $z$, $x_i$, $p_i$ satisfies $F = 0$ along the characteristic curve starting at the elements $(z^0, x_i^0, p_i^0)$, that is, if we set $p_{ik} = \partial^2 z/\partial x_i\, \partial x_k$, then

$$dz = p_1\, dx_1 + \cdots + p_n\, dx_n, \qquad dp_i = p_{i1}\, dx_1 + \cdots + p_{in}\, dx_n.$$

Partial differentiation of $F = 0$ with respect to $x_i$ gives

$$X_i + p_i Z + P_1 p_{i1} + \cdots + P_n p_{in} = 0.$$

Or, if this equation is written in the form

$$\frac{dx_1}{P_1} = \frac{dx_2}{P_2} = \cdots = \frac{dx_n}{P_n} = du,$$

we obtain

$$(X_i + p_i Z)\, du + p_{i1}\, dx_1 + \cdots + p_{in}\, dx_n = 0,$$

that is,

$$(X_i + p_i Z)\, du + dp_i = 0.$$

Hence

$$\frac{dx_i}{P_i} = \frac{dz}{P_1 p_1 + \cdots + P_n p_n} = \frac{-dp_k}{X_k + p_k Z} = du \qquad (i, k = 1, 2, \ldots, n). \tag{7}$$

This equation does not depend on $\Phi$. We can solve these ordinary differential equations for $z$, $x_i$, $p_i$ if we know $z^0$, $x_i^0$, $p_i^0$.

When the elements $z$, $x_i$, $p_i$ are connected by relations of the form $dz = \sum_{i=1}^{n} p_i\, dx_i$, they are called, following Lie, a *family of elements* (*Element-verein*). This is an $n$-dimensional continuum. According to Lie, the method of Lagrange and Charpit for solving a partial differential equation is equivalent to that of finding the family of elements of the highest order. Lie called it a *family of integrals*. Let a family of *complete integrals*, that is, an integral with $n$ arbitrary constants $\alpha_i$ of a partial differential equation (6), be

$$V(z, x_i, \alpha_i) = 0. \tag{8}$$

Since $z$ should satisfy the partial differential equation with any set of constants $\alpha_i$, therefore

$$\frac{\partial V}{\partial x_i} + p_i \frac{\partial V}{\partial z} = 0 \qquad (i = 1, 2, \ldots, n). \tag{9}$$

If we eliminate $\alpha_i$ from the $(n + 1)$ equations (8) and (9), we obtain the given partial differential equation. Hence, our problem is to find a function $z$ of $x_i$ and $\alpha_i$ that satisfies these equations, that is,

$$z = f_0(x_i), \qquad \alpha_i = f_i(x_j).$$

Differentiate the complete integral with regard to $x_i$ and refer to (9); then

$$\sum_{j=1}^{n} \frac{\partial V}{\partial \alpha_j} \frac{\partial \alpha_j}{\partial x_i} = 0 \qquad (i = 1, 2, \ldots, n). \tag{10}$$

Solving (8) and (9) is equivalent to solving (8) and (10). In order to satisfy (8) and (9), put

$$V = 0, \qquad \frac{\partial V}{\partial \alpha_i} = 0 \qquad (i = 1, 2, \ldots, n).$$

In this case $z = f_0(x_i)$ is called a *singular integral*. If the $\partial V/\partial \alpha_i$ are not all zero, then (10) gives

$$\frac{\partial(\alpha_1, \ldots, \alpha_n)}{\partial(x_1, \ldots, x_n)} = 0.$$

This shows that there is a relation among $\alpha_1, \alpha_2, \ldots, \alpha_n$. Let the relation be $\phi = 0$. The differentials $d\alpha_1, d\alpha_2, \ldots, d\alpha_n$ satisfy

$$d\phi = 0. \tag{11}$$

Then

$$\frac{\partial V/\partial \alpha_1}{\partial \phi/\partial \alpha_1} = \frac{\partial V/\partial \alpha_2}{\partial \phi/\partial \alpha_2} = \cdots = \frac{\partial V/\partial \alpha_n}{\partial \phi/\partial \alpha_n} = \lambda. \tag{12}$$

Equations (8), (11), and (12) are $n + 2$ equations with $z, \alpha_1, \ldots, \alpha_n, \lambda$ as the unknown quantities. Eliminate $\lambda, \alpha_1, \ldots, \alpha_n$ from these equations; then we obtain a relation that gives an integral $z$ containing an arbitrary function $\phi$. This is called the *general integral.*

The equations (1) are of the first order with regard to each of the $n$ unknown functions. They are said to be of *order n*, as they are equivalent to a differential equation of order $n$ for a single unknown function.

In fact, first suppose that $n = 2$. Combining

$$\frac{dx_1}{dt} = X_1(x_1, x_2) \qquad \text{and} \qquad \frac{dx_2}{dt} = X_2(x_1, x_2),$$

we obtain

$$\frac{dx_2}{dt} = f\left(x_1, \frac{dx_1}{dt}\right).$$

If we put $x_2 = dx_1/dt$, we obtain a differential equation of the second order with respect to $x_1$.

Next suppose that this theorem is true when the number of equations is equal to $n - 1$; we shall prove by mathematical induction that it is true when the number of equations is $n$. From the first $n - 1$ equations, we obtain

$$\frac{dx_i}{dt} = x_{i+1} \qquad (i = 1, 2, \ldots, n - 2),$$

where $x_n$ is taken as a parameter. Solving these $n - 2$ equations gives $x_n$ as a function of $x_1$ and its $n - 1$ derivatives up to the order $n - 1$. Substitute these in the $n$th equation of (1); this gives $dx_n/dt$ in terms of $x_1$ and its $n - 1$ derivatives. Put $x_n = dx_{n-1}/dt$; this gives a differential equation of the $n$th order for $x_1$. Conversely, a differential equation of the $n$th order

$$\frac{d^n x}{dt^n} = F\left(\frac{d^{n-1}x}{dt^{n-1}}, \ldots, \frac{dx}{dt}, x, t\right)$$

can be transformed into

$$\frac{dx_1}{x_2} = \frac{dx_2}{x_3} = \cdots = \frac{dx_{n-1}}{x_n} = \frac{dx_n}{F(x_n, x_{n-1}, \ldots, x, t)} = dt$$

if we let $x = x_1$, $dx_i/dt = x_{i+1}$ $(i = 1, 2, \ldots, n - 1)$. Hence, (1) is equivalent to a differential equation of the $n$th order with respect to a single variable $x_1$; it is of order $n$.

If (1) has $p$ $(p \leqq n)$ independent integrals, then the order of the differential equation is reduced to $n - p$. Suppose the $p$ integrals to be

$$f_1 = C_1, \qquad f_2 = C_2, \ldots, \qquad f_p = C_p.$$

We express $x_1, x_2, \ldots, x_p$ in terms of $t$ and $x_{p+1}, \ldots, x_n$. Denote the result of substituting these values of $x_1, x_2, \ldots, x_p$ in $X_{p+1}, \ldots, X_n$ by $[X_{p+1}], \ldots, [X_n]$. Then $[X_{p+1}], \ldots, [X_n]$ are functions of

$$t, x_{p+1}, \ldots, x_n$$

only. From this arises a system of differential equations of order $n - p$,

$$\frac{dx_{p+1}}{[X_{p+1}]} = \cdots = \frac{dx_n}{[X_n]} = dt.$$

Thus, if we can find $n$ independent integrals, we can reduce the order of the differential equation to zero, and the equations will no longer contain any derivative; that is, we can solve the equations without integration. If we are able to do this, we say that the equations have been solved by *quadrature*. To find the $n$ independent integrals, we use Mayer's important method of solving a differential equation of order $n$. The differential equations for the motion of masses in the general three-body problem are of order 18. The equations would be solved if we could find 18 independent integrals. As will be shown later, only 12 independent integrals are known at present. A certain type of integral has even been proved nonexistent. Solving the general three-body problem evidently calls for some transcendental process.

## 4.2 INFINITESIMAL TRANSFORMATIONS

If there is a relation giving the third element from two other elements in a set $G$ of objects, such as $AB = C$, if the relation obeys an associative law $(AB)C = A(BC)$, and if $G$ contains a unit element $I$ independent of any element of $G$ such that $IA = AI = A$ and also an inverse element such as $A^{-1}A = I$, then $G$ is called a *group*. If $AB = BA$, then it is a *commutative group* or an *Abelian group*. For an Abelian group we sometimes use $A + B$ instead of $AB$ and call it an *additive group*, or a *modulus* (Van der Waerden, 1955, 1959).

Let $\epsilon$ be an infinitesimal quantity independent of $x_1, x_2, \ldots, x_n$. A transformation of the form

$$x_i' = x_i + \alpha_i(x_1, x_2, \ldots, x_n)\epsilon \qquad (i = 1, 2, \ldots, n)$$

is called an *infinitesimal transformation* and is written $\Delta x_i = \alpha_i\epsilon$. If $f$ is a function of $x_i$, then,

$$f(x_i') = f(x_i + \alpha_i\epsilon) = f(x_i) + \sum_{i=1}^{n} \frac{\partial f}{\partial x_i} \Delta x_i,$$

or

$$\Delta f = \sum_{i=1}^{n} \alpha_i \frac{\partial f}{\partial x_i} \epsilon.$$

Such a transformation forms a *group*. Two infinitesimal transformations represented by $\epsilon_1$ and $\epsilon_2$ operated successively are equivalent to a single transformation belonging to the group; $\epsilon = 0$ is an identity transformation; this group also contains an inverse transformation corresponding to $-\epsilon$. This is a one-parameter group. We denote an infinitesimal transformation by an operator $A$:

$$A = \sum_{i=1}^{n} \alpha_i \frac{\partial}{\partial x_i}.$$

This means

$$A(f) = \sum_{i=1}^{n} \alpha_i \frac{\partial f}{\partial x_i}.$$

Let $c$ be a constant and $F$ be a function of $f_1, f_2, \ldots, f_m$; then

$$A(f_1 + f_2) = A(f_1) + A(f_2), \qquad A(cf) = cA(f),$$

$$A(f_1 f_2) = f_1 A(f_2) + f_2 A(f_1), \qquad A(F(f_1, \ldots, f_m)) = \sum_{j=1}^{m} \frac{\partial F}{\partial f_j} A(f_j).$$

Thus $A$ is a *linear operator*. Furthermore, let

$$B = \sum_{i=1}^{n} \beta_i \frac{\partial}{\partial x_i};$$

then the compound operator

$$(A, B) \equiv AB - BA = \sum_{i=1}^{n} (A\beta_i - B\alpha_i) \frac{\partial}{\partial x_i} \tag{13}$$

is called a *commutator* (*Klammerausdruck*). Since it can be shown that

$$(A, B)f = A(Bf) - B(Af),$$

this gives

$$(A, B) = -(B, A), \qquad (A, A) = 0.$$

Hence, a commutator is an infinitesimal transformation. Since it contains the inverse transformation that can be obtained by putting $-\epsilon$ in place of $\epsilon$, such infinitesimal transformations form a group. Lie called this group of infinitesimal transformations a *continuous transformation group*, as $\epsilon$ can vary continuously.

If there is no independent relation of the form

$$\sum_{r=1}^{m} \lambda_r A_r f \equiv 0$$

among the $m$ infinitesimal transformations

$$A_r = \sum_{i=1}^{n} \alpha_i^{(r)} \frac{\partial}{\partial x_i} \qquad (r = 1, 2, \ldots, m)$$

with constants $\lambda_i$, then the $A_r f$ are said to be *linearly independent* infinitesimal transformations, $m$ in number. A solution common to all partial differential equations

$$A_r f = 0 \qquad (r = 1, 2, \ldots, m)$$

satisfies

$$(A_r, A_s) f = 0 \qquad (r, s = 1, 2, \ldots, m).$$

Hence, $(A_r, A_s) f = 0$ is a consequence of

$$A_1 f = 0, A_2 f = 0, \ldots, A_m f = 0.$$

In other words,

$$(A_r, A_s) f = \sum_{t=1}^{m} c_{rst}(x_1, x_2, \ldots, x_n) A_t f.$$

In this case, Clebsch called the system $A_1 f = 0, A_2 f = 0, \ldots, A_m f = 0$ an *m-parameter complete system.* Lie showed that $c_{rst}$ is a constant; it is called the *structure constant.* This is the second fundamental theorem on the Lie groups (Section 4.26). Lie (1888, 1890, 1893, 1893a, 1896) has made a detailed investigation of the continuous transformation groups, and his work has been applied to the theory of differential equations and to geometry.

In order to solve a partial differential equation of the first order

$$\frac{\partial f}{\partial t} + X_1 \frac{\partial f}{\partial x_1} + \cdots + X_n \frac{\partial f}{\partial x_n} = 0, \qquad X_i = X_i(t, x_1, \ldots, x_n), \quad (14)$$

for $f(t, x_1, \ldots, x_n)$, it is necessary to find $n$ independent functions $f_1, f_2, \ldots, f_n$, that fulfill the identities

$$\frac{\partial f_i}{\partial t} + X_1 \frac{\partial f_i}{\partial x_1} + \cdots + X_n \frac{\partial f_i}{\partial x_n} \equiv 0 \qquad (i = 1, 2, \ldots, n).$$

The most general solution is $F(f_1, f_2, \ldots, f_n)$ with an arbitrary function $F$.

If we let each of these $n$ functions $f_i$ equal a constant, then each of these equations is an integral of the system of ordinary differential equations

$$\frac{dx_i}{dt} = X_i \qquad (i = 1, 2, \ldots, n). \quad (15)$$

By an infinitesimal transformation

$$t' = t + \Delta t, \qquad x_i' = x_i + X_i \Delta t \qquad (i = 1, 2, \ldots, n), \quad (16)$$

such a function $f_i$ is unaltered, that is, because of (14),

$$f_i(t', x_1', \ldots, x_n') = f_i(t, x_1, \ldots, x_n) + \left(\frac{\partial f_i}{\partial t} + \sum_{j=1}^{n} X_j \frac{\partial f_i}{\partial x_j}\right) \Delta t$$

$$= f_i(t, x_1, \ldots, x_n).$$

Hence, an integral of the system of ordinary differential equations is *invariant* with respect to an infinitesimal transformation admitted by the system. $\Delta t$ is what we have denoted by $\epsilon$.

If a system of ordinary differential equations (15) admits an infinitesimal transformation $Af$ and if it has an integral $y$ = constant, then $Ay = 0$. Actually, $y$ has the same value at each point of the integral curve. At the point transformed from a point on this curve by this infinitesimal transformation, the value of $y$ increases by $A(y)\,\Delta t$. Along the integral curve the quantity $y$ must have the same value whatever $\Delta t$ is. Hence, $Ay = 0$.

Conversely, if the results of operating with $A$ on the $n$ independent integrals $y_i$ are all zero, then the system of differential equations admits an infinitesimal transformation $Af$. At the point transformed from a point on the integral curve, the independent integrals $y_1, y_2, \ldots, y_n$ take the values $y_1 + A(y_1)\,\Delta t, y_2 + A(y_2)\,\Delta t, \ldots, y_n + A(y_n)\,\Delta t$, respectively. Since they are also the integrals, they are equal to constants and do not depend on the values initially taken; in other words, the points obtained by the infinitesimal transformation form an integral curve. Hence, the differential equations admit an infinitesimal transformation $Af$. Thus, if there exists a function $f$ for which $Af$ becomes zero, then $f$ is an integral. This is the property that we apply to obtain the integrals of the $n$-body problem in the following section.

When an infinitesimal transformation is a contact transformation, it is an infinitesimal contact transformation and naturally forms a group. Let

$$\Delta q_i = \phi_i(q, p)\,\Delta t, \qquad \Delta p_i = \psi_i(q, p)\,\Delta t,$$

$$q_i' = q_i + \Delta q_i, \qquad p_i' = p_i + \Delta p_i.$$

Then

$$\sum_{i=1}^{n} (p_i'\,dq_i' - p_i\,dq_i) = dW(q, p)$$

is a perfect differential, since it is a contact transformation. Thus

$$\Delta t \sum_{i=1}^{n} (\psi_i \, dq_i + p_i \, d\phi_i) = dW,$$

and

$$\sum_{i=1}^{n} (\psi_i \, dq_i + p_i \, d\phi_i) = dU,$$

where

$$dW = U(q, p) \, \Delta t.$$

Hence

$$\sum_{i=1}^{n} (\psi_i \, dq_i - \phi_i \, dp_i) = d\Big(U - \sum_{i=1}^{n} p_i \phi_i\Big) = -dH(q, p).$$

Thus the most general infinitesimal contact transformation is expressed by

$$q_i' = q_i + \frac{\partial H}{\partial p_i} \Delta t, \qquad p_i' = p_i - \frac{\partial H}{\partial q_i} \Delta t, \qquad (i = 1, 2, \ldots, n),$$

where $H$ is an arbitrary function of $q_i$ and $p_i$, and $\Delta t$ is an increment independent of $q_i$, $p_i$. This infinitesimal transformation is admitted by a system of differential equations

$$\frac{dq_i}{dt} = \frac{\partial H}{\partial p_i}, \qquad \frac{dp_i}{dt} = -\frac{\partial H}{\partial q_i} \qquad (i = 1, 2, \ldots, n). \tag{17}$$

Hence, the study of a Hamiltonian system of canonical differential equations is reduced to the study of this infinitesimal contact transformation. By repeating such infinitesimal transformations we can finally obtain the whole trajectory, if we start at the initial values of $q_i$, $p_i$ of the dynamical system.

Denote two independent infinitesimal increments of an infinitesimal contact transformation by $\delta$ and $\Delta$. We have proved in Section 1.8 that a bilinear covariant is invariant under a contact transformation, that is,

$$\sum_{i=1}^{n} (\Delta q_i \, \delta p_i - \delta q_i \, \Delta p_i) = \text{invariant},$$

or

$$\frac{d}{dt} \sum_{i=1}^{n} (\Delta q_i \, \delta p_i - \delta q_i \, \Delta p_i) = 0.$$

Let $u_1 = a_1, u_2 = a_2, \ldots, u_{2n} = a_{2n}$ be the integrals in question and $u_r$

be a function of $q_i$, $p_i$, $t$. Let $\Delta$ be an increment due to the variation of $a_r$, and $\delta$ be an increment due to the variation of $a_s$. Then

$$\frac{d}{dt}\sum_{i=1}^{n}\left(\frac{\partial q_i}{\partial a_r}\frac{\partial p_i}{\partial a_s} - \frac{\partial q_i}{\partial a_s}\frac{\partial p_i}{\partial a_r}\right) = 0,$$

or, by definition of the Lagrangian brackets,

$$\frac{d}{dt}[a_r, a_s] = 0 \qquad (r, s = 1, 2, \ldots, n).$$

Hence, the Lagrangian brackets are invariant during the motion. This is called Lagrange's theorem. We have proved in Section 1.9 that the Lagrangian and Poisson brackets are reciprocal. Hence, the Lagrangian brackets and Poisson brackets formed of $2n$ independent constants of the integrals are both invariant by any infinitesimal contact transformation, that is, have the same values throughout the whole domain of the motion expressed by a Hamiltonian system of canonical equations.

## 4.3 POISSON'S THEOREM

We have seen that the function $f$, for which the ordinary differential equation (15) has an integral $f(x_i, t)$ = constant, satisfies the first-order partial differential equation (14). Let $F(x_i, t)$ = constant be an integral of a Hamiltonian system of differential equations (17); then (14) can be written in the form

$$\frac{\partial F}{\partial t} + (H, F) = 0, \tag{18}$$

where

$$(H, F) = \sum_{i=1}^{n}\left(\frac{\partial H}{\partial p_i}\frac{\partial F}{\partial q_i} - \frac{\partial H}{\partial q_i}\frac{\partial F}{\partial p_i}\right)$$

is a Poisson bracket. In other words, an integral $F$ = constant of a system of canonical equations should satisfy equation (18). Conversely, as we have proved, a solution, $F$ = constant, of a partial differential equation (18) is an integral of the Hamiltonian system of canonical equations (17). We call this *Poisson's first theorem.*

From this theorem with a Poisson operator

$$\sum_{i=1}^{n}\left(\frac{\partial H}{\partial p_i}\frac{\partial}{\partial q_i} - \frac{\partial H}{\partial q_i}\frac{\partial}{\partial p_i}\right)$$

we desire

$$(H, f_1 + f_2) = (H, f_1) + (H, f_2), \qquad (H, cf) = c(H, f),$$
$$(H, f_1 f_2) = f_1(H, f_2) + f_2(H, f_1).$$

Generally,

$$(H, F) = \sum_{j=1}^{n} (H, f_j) \frac{\partial F}{\partial f_j}.$$

Hence, a Poisson operator is a linear operator and a special type of commutator.

Here we have an important identity

$$(u, (v, w)) + (v, (w, u)) + (w, (u, v)) \equiv 0. \tag{19}$$

This is called *Poisson's identity.*

By our construction, the left-hand member of (19) is the sum of products of three derivatives of $u$, $v$, $w$. Two factors come from the first-order derivatives of two of $u$, $v$, $w$; the third factor is a second-order derivative of the remaining third variable. If we can show that the sum does not contain any second-order derivative, then the theorem will be proved. Consider two Poisson operators $Af = (v, f)$ and $Bf = (w, f)$ operating on $f$. Then the left-hand side of (19) becomes

$$(u, (v, w)) + ABu - BAu = (u, (v, w)) + (A, B)u,$$

where $(A, B)$ is a commutator operator. If the sum contains a second-order derivative of $u$, then such a derivative should be contained in $(A, B)u$, because $(u, (v, w))$ contains only first-order derivatives of $u$. But by (13), $(A, B)$ contains only first-order derivatives.

Next consider *Poisson's second theorem.* If $f_1$ and $f_2$ are two independent solutions of canonical equations (17), then a Poisson bracket $(f_1, f_2)$ is also an independent solution of (17). In other words, if

$$\frac{\partial f_1}{\partial t} + (H, f_1) = 0, \qquad \frac{\partial f_2}{\partial t} + (H, f_2) = 0,$$

then

$$\frac{\partial (f_1, f_2)}{\partial t} + (H, (f_1, f_2)) = 0.$$

By Poisson's identity (19),

$$(f_1, (f_2, H)) + (f_2, (H, f_1)) + (H, (f_1, f_2)) \equiv 0.$$

From our assumption, we obtain

$$\left(f_1, \frac{\partial f_2}{\partial t}\right) - \left(f_2, \frac{\partial f_1}{\partial t}\right) + (H, (f_1, f_2)) \equiv 0.$$

But we know that

$$\frac{\partial}{\partial t}(f_1, f_2) = \left(f_1, \frac{\partial f_2}{\partial t}\right) - \left(f_2, \frac{\partial f_1}{\partial t}\right).$$

This proves the theorem.

## LIE'S THEOREM

### 4.4 ELIMINATION OF TIME

Suppose that in the canonical equations

$$\frac{\partial q_i}{dt} = \frac{\partial H}{\partial p_i}, \qquad \frac{\partial p_i}{dt} = -\frac{\partial H}{\partial q_i} \qquad (i = 1, 2, \ldots, n), \tag{20}$$

$H$ does not contain $m$ coordinates $q_i$. Then these equations have the integrals

$$p_r = \text{constant} = p_r^{(0)} \qquad (r = 1, 2, \ldots, m),$$

and the equations split into two sets of equations: the first set is

$$\frac{dq_{m+j}}{dt} = \frac{\partial H}{\partial p_{m+j}}, \qquad \frac{dp_{m+j}}{dt} = -\frac{\partial H}{\partial q_{m+j}} \qquad (j = 1, 2, \ldots, n - m),$$

and the second set is

$$\frac{dq_r}{dt} = \frac{\partial H}{\partial p_r} \qquad (r = 1, 2, \ldots, m).$$

In the first set of equations, $H$ does not contain $q_1, q_2, \ldots, q_m$. This set can be solved as $2(n - m)$ equations by putting $p_r = p_r^{(0)}$, that is, a system of equations of order $2(n - m)$. The general integral contains $2(n - m)$ constants. Substitute this solution in the second set of equations and solve for $q_1, q_2, \ldots, q_m$. There are $m$ integration constants $q_1^{(0)}, q_2^{(0)}, \ldots, q_m^{(0)}$, making $2n$ integration constants altogether. Hence, if $H$ does not contain the $q_i$, $m$ in number, the order of differential equations (20) is reduced by $2m$.

Next suppose that $H$ does not contain $t$ explicitly, then the integral

$$H = h = \text{constant},$$

is obtained, which is the energy integral. In this case we can reduce the order of equations (20) by two. In fact, if we put aside a solution such as that in which $q_i$ and $p_i$ are all constant, then each pair $q_i, p_i$ depends on $t$. Let one of the latter pairs be $q_n, p_n$. Since $dq_n/dt = 0$, therefore $\partial H/\partial p_n \neq 0$. Now we can solve $H = h$ for $p_n$,

$$p_n = K(q_1, \ldots, q_n, p_1, \ldots, p_{n-1}, h).$$

Divide each of the first $2(n-1)$ equations of (20) side by side by $dq_n/dt = \partial H/\partial p_n$. Let $u$ be any one of $q_1, \ldots, q_{n-1}, p_1, \ldots, p_{n-1}$. Since

$$\frac{\partial K}{\partial u} = -\frac{\partial H/\partial u}{\partial H/\partial p_n}$$

we obtain

$$\frac{dq_i}{dq_n} = -\frac{\partial K}{\partial p_i}, \qquad \frac{dp_i}{dq_n} = \frac{\partial K}{\partial q_i} \qquad (i = 1, 2, \ldots, n-1). \tag{21}$$

The function $K$ does not contain $p_n$. Hence the order of differential equations is reduced to $2(n-1)$. After solving these equations we obtain $t$ from

$$dt = \frac{dq_n}{\partial H/\partial p_n} \tag{22}$$

without solving any differential equation. We call this procedure the *elimination of time.*

## 4.5 INVOLUTION

If among the conjugate variables $q_i$, $p_i$ $(i = 1, 2, \ldots, n)$, of (20), there exist $m$ $(m < n)$, relations

$$f_r(q, p) = 0 \qquad (r = 1, 2, \ldots, m), \tag{23}$$

from which we can derive Poisson's equations

$$(f_r, f_s) = 0 \qquad (r, s = 1, 2, \ldots, m), \tag{24}$$

and if (23) can be solved for $m$ variables $p_i$, for example, for $p_1, p_2, \ldots, p_m$, that is, if $\partial(f_1, f_2, \ldots, f_m)/\partial(p_1, p_2, \ldots, p_m) \neq 0$, then (23) is said to be in *involution.*

If (23) is solved for $p_1, p_2, \ldots, p_m$ and is reduced to the form

$$p_i = \phi_i(p_{m+1}, \ldots, p_n, q_1, q_2, \ldots, q_n) \qquad (i = 1, 2, \ldots, m), \tag{25}$$

then from (24) we can derive

$$(p_i - \phi_i, p_i - \phi_j) = 0 \qquad (i, j = 1, 2, \ldots, m). \tag{26}$$

In other words, (25) is also in involution. This theorem is called *Jacobi's lemma.*

Put $u$ for any one of the variables $q_i$, $p_i$ that is not $p_1, p_2, \ldots, p_m$. From (23),

$$\frac{\partial f_r}{\partial u} + \sum_{i=1}^{m} \frac{\partial f_r}{\partial p_i}\frac{\partial \phi_i}{\partial u} = 0 \qquad (r = 1, 2, \ldots, m).$$

Hence,

$$\frac{\partial f_r}{\partial u} = \sum_{i=1}^{m} \frac{\partial f_r}{\partial p_i} \frac{\partial(p_i - \phi_i)}{\partial u} \qquad (r = 1, 2, \ldots, m).$$

If $u$ is any one of $p_1, p_2, \ldots, p_m$, then it is an identity. Thus,

$$\frac{\partial f_r}{\partial q_k} \frac{\partial f_s}{\partial p_k} = \sum_{i,j}^{1,\ldots,m} \frac{\partial f_r}{\partial p_i} \frac{\partial f_s}{\partial p_j} \frac{\partial(p_i - \phi_i)}{\partial q_k} \frac{\partial(p_j - \phi_j)}{\partial p_k}.$$

From this,

$$(f_r, f_s) = \sum_{i,j}^{1,\ldots,m} \frac{\partial f_r}{\partial p_i} \frac{\partial f_s}{\partial q_j} (p_i - \phi_i, p_j - \phi_j).$$

From our assumption (24), we obtain

$$\sum_{i=1}^{m} \frac{\partial f_r}{\partial p_j} \left[ \sum_{j=1}^{m} \frac{\partial f_s}{\partial p_i} (p_i - \phi_i, p_j - \phi_j) \right] = 0 \qquad (r, s = 1, 2, \ldots, m).$$

Consider these equations as a system of $m$ simultaneous linear equations for the $m$ unknown functions in the parentheses. Then

$$\sum_{j=1}^{m} \frac{\partial f_s}{\partial p_j} (p_i - \phi_i, p_j - \phi_j) = 0 \qquad (i, s = 1, 2, \ldots, m),$$

because $\partial(f_1, f_2, \ldots, f_m)/\partial(p_1, p_2, \ldots, p_m) \neq 0$. By a similar argument,

$$(p_i - \phi_i, p_j - \phi_j) = 0 \qquad (i, j = 1, 2, \ldots, m),$$

and the lemma is proved.

Einstein in a 1929 paper applied the idea of involution to the unified field theory of general relativity (Cartan, 1931).

## 4.6 INVARIANTIVE RELATIONS

We say that

$$f(t, x_1, x_2, \ldots, x_n) = 0 \tag{27}$$

is an *invariantive relation* with regard to

$$\frac{dx_i}{dt} = X_i(t, x_1, x_2, \ldots, x_n) \qquad (i = 1, 2, \ldots, n), \tag{28}$$

when the relation (27) satisfying the differential equations (28) for a particular value of $t$ causes it to satisfy them for all values of $t$. An invariantive relation does not contain any integration constants (unlike integrals of the equations).

If $f = 0$ is an invariantive relation with regard to (28), then

$$\frac{df}{dt} = \frac{\partial f}{\partial t} + \sum_{j=1}^{n} X_j \frac{\partial f}{\partial x_j} = 0, \tag{29}$$

or, if $\lambda$ denotes a function of $t, x_1, x_2, \ldots, x_n$,

$$\frac{df}{dt} = \lambda f. \tag{30}$$

Equation (30) is the necessary and sufficient condition for (27) to be an invariantive relation with regard to (28). Indeed, $df/dt$ is a function of $t, x_i$. If we substitute for $x_n$ its value obtained from $f = 0$, then $df/dt$ becomes a function $G(x_1, x_2, \ldots, x_{n-1}, f, t)$ of $x_1, x_2, \ldots, x_{n-1}, f, t$. Expanding this function with respect to $f$ gives

$$\frac{df}{dt} = G(x_1, \ldots, x_{n-1}, 0, t) + \left(\frac{\partial G}{\partial f}\right)_{f=0} f + \cdots.$$

If $f = 0$, then $df/dt = 0$. Hence, $G(x_1, \ldots, x_{n-1}, 0, t) = 0$. Denote the remaining terms by $\lambda$; this gives (30). Conversely, if $f = 0$ in (30), then $df/dt = 0$.

In general, if there are $m + 1$ independent relations

$$f_r(t, x_1, \ldots, x_n) = 0 \qquad (r = 0, 1, \ldots, m), \tag{31}$$

the necessary and sufficient condition for (31) to be invariantive with regard to (28) is that

$$\frac{df_r}{dt} = \sum_{s=0}^{m} \lambda_{rs} f_s \qquad (r = 0, 1, \ldots, m), \tag{32}$$

where $\lambda_{rs}$ are functions of $t, x_1, \ldots, x_n$ and $d/dt$ has the meaning given in (29).

Now suppose that we know the solution $x_i(t)$ of (28) and that $x_i(t) + \delta x_i$ also satisfies (28). $\delta x_r$ is a function of $t$ and $\delta$ obeys (27). We consider $\delta$ to be simultaneous displacements, that is, $\delta t = 0$. Then

$$\frac{d\delta x_i}{dt} = \delta X_i = \sum_{j=1}^{n} \frac{\partial X_i}{\partial x_j} \delta x_j \tag{33}$$

represents the difference between the varied motion with a modified initial condition and the original motion as a function of $t$. Suppose that the variables $x_i$ in $\partial X_i/\partial x_j$ are replaced by the known functions $x_i(t)$ of $t$.

Poincaré called (33) the *variational equation* (*équation aux variations*) for (28). From (33),

$$\frac{d}{dt}\,\delta x_i = \delta\,\frac{dx_i}{dt}.$$

This expresses the commutativity of $\delta$ and $d/dt$. More generally, the relation

$$\frac{d}{dt}\,\delta f = \delta\,\frac{df}{dt}$$

holds true for any function $f$ of $t, x_1, x_2, \ldots, x_n$. $\delta$ represents the variation obeying (28) and $d/dt$ has the meaning given in (29).

Consider the variation

$$\delta f = 0 \tag{34}$$

together with the invariantive relation (27) with regard to the canonical equations (20). Since the initial displacement $\delta x_i$ of (33) is arbitrary, (34) is reduced to

$$\frac{\partial f}{\partial x_i} = 0 \qquad (i = 1, 2, \ldots, n). \tag{35}$$

If the combination of (35) and $f = 0$ gives compatible relations among the $n$ variables $x_1, x_2, \ldots, x_n$, then (35) is an invariantive relation. Combining (30) with

$$\delta f = \sum_{j=1}^{n} \frac{\partial f}{\partial x_j}\,\delta x_j \qquad \text{and} \qquad \frac{d}{dt}\,\delta f = \delta\,\frac{df}{dt},$$

we obtain

$$\sum_{j=1}^{n}\left(\frac{d}{dt}\frac{\partial f}{\partial x_j}\right)\delta x_j = \sum_{i=1}^{n}\frac{\partial f}{\partial x_i}\left(\lambda\delta x_i - \frac{d\delta x_i}{dt}\right) + f\,\delta\lambda.$$

Referring to (27) and (35) and noting that $\delta x_j$ is arbitrary, we see that

$$\frac{d}{dt}\frac{\partial f}{\partial x_j} = 0 \qquad (j = 1, 2, \ldots, m).$$

Hence, $\partial f/\partial x_j = 0$ is an invariantive relation. The relations (35) obtained from (34) are called the *stationarity condition.*

More generally, $m + 1$ independent invariantive relations of form (31) with regard to differential equations (28) should satisfy the relations of form (32). Assume that the system of differential equations representing the stationarity condition that one of the invariantive relations

$$\delta f_0 = 0 \tag{36}$$

should be stationary for all possible virtual displacements satisfies not only $f = 0$ but also

$$f_1 = 0, f_2 = 0, \ldots, f_m = 0. \tag{37}$$

Then we can express (37) in terms of $x_{m+1}, \ldots, x_n$ by solving them with respect to $x_1, x_2, \ldots, x_m$. Use $f_1, \ldots, f_m$ as new variables in place of $x_1, \ldots, x_m$. Then (28) can be written

$$\begin{aligned} \frac{df_r}{dt} &= \Xi_r \qquad (r = 1, 2, \ldots, m), \\ \frac{dx_i}{dt} &= \Xi_i \qquad (i = m + 1, \ldots, n). \end{aligned} \tag{38}$$

The function $\Xi$ is the result of substituting $f_1, \ldots, f_m$ for $x_1, \ldots, x_m$, respectively, in $X$, and is a function of $f_1, \ldots, f_m, x_{m+1}, \ldots, x_n, t$.

From the first equation of (32) and the last $n - m$ equations of (38) we obtain

$$\frac{\partial f_0}{\partial t} + \sum_{r=1}^{m} \frac{\partial f_0}{\partial f_r}\frac{df_r}{dt} + \sum_{i=m+1}^{n} \frac{\partial f_0}{\partial x_i}\Xi_i = \sum_{s=0}^{m} \lambda_{0s} f_s.$$

Putting $f_1 = \cdots = f_m = 0$ in (32) gives

$$\overline{\left(\frac{df_r}{dt}\right)} = \bar{\lambda}_{r0}\bar{f}_0 \qquad (r = 1, 2, \ldots, m),$$

where, in general, $\bar{F}$ is the result of substituting $f_1, \ldots, f_m$ for $x_1, \ldots, x_m$, respectively in $F(x_1, \ldots, x_n, t)$ according to (37). Again, since $t, x_{m+1}, \ldots, x_n$ do not depend on $f_1, \ldots, f_m$, therefore

$$\overline{\left(\frac{\partial f_0}{\partial t}\right)} = \frac{\partial \bar{f}_0}{\partial t}, \qquad \overline{\left(\frac{\partial f_0}{\partial x_r}\right)} = \frac{\partial \bar{f}_0}{\partial x_r} \qquad (r = m + 1, \ldots, n),$$

and

$$\frac{\partial \bar{f}_0}{\partial t} + \sum_{i=m+1}^{n} \frac{\partial \bar{f}_0}{\partial x_i}\Xi_i = \left[\bar{\lambda}_{00} - \sum_{r=1}^{m} \overline{\left(\frac{\partial f_0}{\partial f_r}\right)}\bar{\lambda}_{r0}\right]\bar{f}_0.$$

The left-hand member of this equation is a perfect differential of $\bar{f}_0$ and the right-hand side is of the form $\lambda\bar{f}_0$. Since the equation is of the form $d\bar{f}_0/dt = \lambda\bar{f}_0$, the equation $\bar{f}_0 = 0$ is an invariantive relation with regard to

$$\frac{dx_i}{dt} = \Xi_i \qquad (i = m + 1, \ldots, n).$$

Thus, if there exists an invariantive relation (37), then there exists also an invariantive relation $f = 0$, in the form of the variational equation (36), with regard to equations (38) which were obtained by eliminating some of the variables.

We now apply this theorem to the canonical equations (20). Suppose that $H$ does not contain $t$, and that there exist $m$ invariantive relations

$$f_r(q, p) = 0 \qquad (r = 1, 2, \ldots, m), \tag{39}$$

which do not contain $t$, are independent of $H =$ constant, and are mutually in involution. According to Jacobi's lemma the equations

$$p_i - \phi_i(p_{m+1}, \ldots, p_n, q_1, \ldots, q_n) = 0 \qquad (i = 1, 2, \ldots, m), \tag{40}$$

obtained by solving (39) are also in involution. Write

$$\{u, v\} = \sum_{j=1}^{n-m} \left(\frac{\partial u}{\partial p_{m+j}} \frac{\partial v}{\partial q_{m+j}} - \frac{\partial u}{\partial q_{m+j}} \frac{\partial v}{\partial p_{m+j}}\right).$$

Then the condition for involution $(p_i - \phi_i, p_j - \phi_j) = 0$ is written

$$\frac{\partial \phi_i}{\partial q_j} - \frac{\partial \phi_j}{\partial q_i} + \{\phi_i, \phi_j\} = 0 \qquad (i, j = 1, 2, \ldots, m), \tag{41}$$

because the right-hand side of (41) does not depend on $p_j$ $(j = 1, 2, \ldots, m)$, according to (40).

We now express the fact that (40) are invariantive relations with regard to (20): differentiate (40) with respect to $t$ and refer to (20); this gives

$$\frac{\partial H}{\partial q_i} + \{H, \phi_i\} + \sum_{j=1}^{m} \frac{\partial H}{\partial p_j} \frac{\partial \phi_j}{\partial q_i} = 0 \qquad (i = 1, 2, \ldots, m). \tag{42}$$

Let $\bar{H}$ be the result of substituting (40) in $H$; then the differential coefficients of $\bar{H}(p_{m+1}, \ldots, p_n, q_1, \ldots, q_n)$ are

$$\frac{\partial \bar{H}}{\partial p_{m+j}} = \frac{\partial H}{\partial p_{m+j}} + \sum_{i=1}^{m} \frac{\partial H}{\partial p_i} \frac{\partial \phi_i}{\partial p_{m+j}},$$

$$\frac{\partial \bar{H}}{\partial q_{m+j}} = \frac{\partial H}{\partial q_{m+j}} + \sum_{i=1}^{m} \frac{\partial H}{\partial p_i} \frac{\partial \phi_i}{\partial q_{m+j}} \qquad (j = 1, 2, \ldots, n - m), \tag{43}$$

$$\frac{\partial \bar{H}}{\partial q_r} = \frac{\partial H}{\partial q_r} + \sum_{i=1}^{m} \frac{\partial H}{\partial p_i} \frac{\partial \phi_i}{\partial q_r} \qquad (r = 1, 2, \ldots, m). \tag{44}$$

From (43),

$$\{\bar{H}, \phi_r\} = \{H, \phi_r\} - \sum_{i=1}^{m} \frac{\partial H}{\partial p_i} \{\phi_r, \phi_i\} \qquad (r = 1, 2, \ldots, m).$$

Adding (44) and this equation side by side gives

$$\frac{\partial \bar{H}}{\partial q_r} + \{\bar{H}, \phi_r\} = \frac{\partial H}{\partial q_r} + \{H, \phi_r\} + \sum_{i=1}^{m} \frac{\partial H}{\partial p_i} \left( \frac{\partial \phi_i}{\partial q_r} - \{\phi_r, \phi_i\} \right)$$

$$(r = 1, 2, \ldots, m).$$

Finally, referring to (41) and (42), we obtain

$$\frac{\partial \bar{H}}{\partial q_r} + \{\bar{H}, \phi_r\} = 0 \qquad (r = 1, 2, \ldots, m). \tag{45}$$

Now, as $H$ is an integral of (20), the stationarity condition for $H$, that is, $\delta H = 0$, is satisfied for all possible displacements fulfilling the conditions (45). As we have proved, the condition is equivalent to $\delta \bar{H} = 0$, or to

$$\frac{\partial \bar{H}}{\partial p_{m+j}} = 0, \qquad \frac{\partial \bar{H}}{\partial q_{m+j}} = 0 \qquad (j = 1, 2, \ldots, n - m),$$

$$\frac{\partial \bar{H}}{\partial q_r} = 0 \qquad (r = 1, 2, \ldots, m).$$

But the last $m$ equations can be derived from the first $n - m$ equations by means of (40). Hence, even if there exist $m$ relations of the form (29) or (40), the stationarity condition $\delta H = 0$ gives only $2(n - m)$ relations. This is called *Levi-Civita's theorem.* This discussion is based on those particular solutions of differential equations that satisfy the stationarity condition. We shall now prove that even in the general solution the order of canonical differential equations is reduced by $2m$ when $m$ integrals are mutually in involution.

## 4.7 LIOUVILLE'S THEOREM

If a canonical system of $2n$th-order equations admits $n$ integrals that are independent, in involution, and solvable for $p$, then the differential equations can be solved by quadrature alone. This is called *Liouville's theorem.*

Suppose that we know $n$ mutually independent integrals

$$f_r(q, p, t) = c_r = \text{constant} \qquad (r = 1, 2, \ldots, n),$$

in involution and solvable for $p$. It is sufficient to prove by the Hamilton-Jacobi theorem that we can obtain a complete integral $W(q, c, t)$ of

$$\frac{\partial W}{\partial t} + H = 0$$

by mere quadrature. Since $f_r$ are integrals of the canonical equations, then

$$\frac{\partial f_r}{\partial t} + (H, f_r) = 0 \qquad (r = 1, 2, \ldots, n) \tag{46}$$

by Poisson's first theorem and, moreover,

$$(f_r, f_s) = 0 \qquad (r, s = 1, 2, \ldots, n). \tag{47}$$

Take a new pair of variables $q_0 = t$ and $p_0$ besides $q_i, p_i$ $(i = 1, 2, \ldots, n)$, and suppose that $p_0$ is an auxiliary parameter contained neither in $H$ nor in $f$. Denote these $2(n + 1)$ canonical variables by $u$, $v$ and the Poisson brackets for these variables by $((u, v))$. Then,

$$((u, v)) = \frac{\partial u}{\partial p_0}\frac{\partial v}{\partial t} - \frac{\partial u}{\partial t}\frac{\partial v}{\partial p_0} + (u, v).$$

When neither $u$ nor $v$ contains $p_0$, then $((u, v)) = (u, v)$ and

$$((p_0 + H, f_r)) = \frac{\partial f_r}{\partial t} + (H, f_r) \qquad (r = 1, 2, \ldots, n).$$

Hence, conditions (46), (47) become

$$((p_0 + H, f_r)) = 0, \quad ((f_r, f_s)) = 0 \qquad (r, s = 1, 2, \ldots, n).$$

These equations show that $p_0 + H$ and $f_r$ are in involution for the $n + 1$ pairs of $q$ and $p$. Since we can solve equations

$$f_r = c_r, \quad p_0 + H = 0 \qquad (r = 1, 2, \ldots, n)$$

for $p_0, p_1, \ldots, p_n$, we can express $p_0$ as a function of $q_i$, $c_i$, $t$ by the second equation. Let them be expressed by

$$p_r - \phi_r(q, c, t) = 0 \qquad (r = 1, 2, \ldots, n),$$

$$p_0 - \phi_0(q, c, t) = 0.$$

Hence, $\phi_0$ is the result of substituting $\phi_r$ for $p_r$ in $H$. From Jacobi's lemma, we obtain

$$((p_r - \phi_r, p_s - \phi_s)) = 0 \qquad (r, s = 0, 1, 2, \ldots, n).$$

The functions $\phi$ do not contain $p$ in these equations, therefore

$$\frac{\partial \phi_r}{\partial q_s} = \frac{\partial \phi_s}{\partial q_r} \qquad (r, s = 0, 1, 2, \ldots, n).$$

This shows that there exists a function $W$ such that

$$\frac{\partial W}{\partial t} = \phi_0, \qquad \frac{\partial W}{\partial q_r} = \phi_r \qquad (r = 1, 2, \ldots, n).$$

Hence we can obtain $W$ from these equations by mere quadrature and Liouville's theorem is proved.

## 4.8 LIE'S THEOREM

If in a canonical system of $2n$th-order differential equations we know $m$ integrals that are independent, in involution, and solvable for $p$, then we can reduce the order of the equations to $2(n - m)$. This is called *Lie's theorem.*

Define $q_0 = t$ and $p_0$ as in the proof of Liouville's theorem; this gives integrals

$$p_0 + H = 0,$$

$$f_r(q, p, t) - c_r = \text{constant} \qquad (r = 1, 2, \ldots, m), \tag{48}$$

that are independent, in involution, and solvable for $p$. Hence they can be written

$$p_r - \phi_r(q, c, t) = 0 \qquad (r = 1, 2, \ldots, m), \tag{49}$$

$$p_0 - \phi_0(q, c, t) = 0,$$

and

$$((p_r - \phi_r, p_s - \phi_s)) = 0 \qquad (r, s = 0, 1, \ldots, m). \tag{50}$$

Also, the $n + 1$ integrals (49) are mutually in involution. As in Liouville's theorem, the canonical equations (20) can be written in one of the forms

$$\frac{\partial W}{\partial t} + H = 0, \qquad p_0 + H = 0,$$

$$\frac{\partial W}{\partial q_0} + H\left(q_0, q_1, \ldots, q_n, \frac{\partial W}{\partial q_1}, \ldots, \frac{\partial W}{\partial q_n}\right) = 0.$$

The $m$ known integrals are written in the form

$$\frac{\partial W}{\partial q_r} - \phi_r(q_0, q_1, \ldots, q_n, c_1, \ldots, c_m) = 0 \qquad (r = 1, 2, \ldots, m).$$

Thus the canonical equations (20) can be replaced by this system of $m + 1$ partial differential equations of $n + 1$ variables. Hence, to have integrals that are independent, in involution, and solvable for $p$ in a partial differential equation of $n + 1$ variables, is equivalent to having a system of $m + 1$ partial differential equations for $n + 1$ variables that are in involution. It is sufficient to prove that this system of $m + 1$ partial differential equations can be reduced to a partial differential equation of $n + 1 - m$ variables.

Let the system of $m + 1$ partial differential equations be

$$\frac{\partial W}{\partial q_i} = \phi_i\left(q_0, q_1, \ldots, q_n, \frac{\partial W}{\partial q_{m+1}}, \ldots, \frac{\partial W}{\partial q_n}\right) \qquad (i = 0, 1, \ldots, m). \tag{51}$$

This is obtained by substituting (49) in

$$p_r = \frac{\partial W}{\partial q_r} \qquad (r = 0, 1, 2, \ldots, m),$$

and eliminating $p_0, p_1, \ldots, p_m$. It also satisfies (50). To integrate (51): according to Jacobi's lemma, find integrals (not containing $p_1, p_2, \ldots, p_m$) of the equations

$$(p_i - \phi_i, f) = 0 \qquad (i = 0, 1, \ldots, m), \tag{52}$$

or of

$$\frac{\partial f}{\partial q_i} + \sum_{j=m+1}^{n} \left(\frac{\partial \phi_i}{\partial q_j}\frac{\partial f}{\partial p_j} - \frac{\partial \phi_i}{\partial p_j}\frac{\partial f}{\partial q_j}\right) = 0 \qquad (i = 0, 1, \ldots, m).$$

For that purpose, take $x_0, x_1, \ldots, x_m$ as new variables after carrying out Mayer's transformation

$$q_0 = q_0^{(0)} + x_0, \qquad q_1 = q_1^{(0)} + x_0 x_1, \qquad \ldots, \qquad q_m = q_m^{(0)} + x_0 x_m.$$

Then (52) can be written

$$\frac{\partial f}{\partial x_i} + \sum_{j=m+1}^{n} \left(\frac{\partial \Phi_i}{\partial q_j}\frac{\partial f}{\partial p_j} - \frac{\partial \Phi_i}{\partial p_j}\frac{\partial f}{\partial q_j}\right) = 0 \qquad (i = 0, 1, \ldots, m). \tag{53}$$

Here $\Phi_i$ $(i = 1, 2, \ldots, m)$ and $\Phi_0$ are the results of performing Mayer's transformation on $\phi_i$ $(i = 1, \ldots, m)$ and on

$$\phi_0 + x_1\phi_1 + \cdots + x_m\phi_m,$$

respectively. We have

$$\frac{\partial W}{\partial q_0} = \Phi_0, \qquad \frac{\partial W}{\partial q_r} = x_0\Phi_r \qquad (r = 1, 2, \ldots, m) \tag{54}$$

after Mayer's transformation on (51). Thus the relation between (51) and (52) corresponds to the relation between (53) and (54). Hence, integrating either both or the first equation (54) is equivalent to integrating a linear differential equation

$$(p_0 - \Phi_0, f) = 0,$$

or integrating the first equation of (53).

Our next step is to prove that, if we can integrate an equation of $n + 1 - m$ variables

$$\frac{\partial W}{\partial q_0} - \Phi_0 = 0,$$

then we can obtain a complete integral of (54). This equation can be written

$$\frac{\partial W}{\partial x_0} = \Phi_0\left(x_0, x_1, \ldots, x_m, q_{m+1}, \ldots, q_n, \frac{\partial W}{\partial q_{m+1}}, \ldots, \frac{\partial W}{\partial q_n}\right).$$

By the Hamilton-Jacobi theorem, this is equivalent to integrating

$$\frac{dq_i}{dx_0} = -\frac{\partial \Phi_0}{\partial p_i}, \qquad \frac{dp_i}{dx_0} = \frac{\partial \Phi_0}{\partial q_i} \qquad (i = m + 1, \ldots, n). \tag{55}$$

Then integrals are obtained in the form

$$\begin{aligned} q_i &= q_i(x_0, x_1, \ldots, x_m, a_{m+1}, \ldots, a_n, b_{m+1}, \ldots, b_n), \\ p_i &= p_i(x_0, x_1, \ldots, x_m, a_{m+1}, \ldots, a_n, b_{m+1}, \ldots, b_n), \end{aligned} \tag{56}$$

$$(i = m + 1, \ldots, n).$$

Here the values of $p_{m+1}, \ldots, p_n, q_{m+1}, \ldots, q_n$ for $x_0 = 0$ are denoted by $a_{m+1}, \ldots, a_n, b_{m+1}, \ldots, b_n$, respectively. Then

$$W = \sum_{i=m+1}^{n} a_i b_i + \int_0^{x_0} \left(\Phi_0 - \sum_{i=m+1}^{n} p_i \frac{\partial \Phi_0}{\partial p_i}\right) dx_0 \tag{57}$$

is an integral of $\partial W/\partial x_0 = \Phi_0$ that becomes

$$a_{m+1}q_{m+1} + \cdots + a_n q_n \qquad \text{for} \quad x_0 = 0.$$

Since the solution is

$$p_i = \frac{\partial W}{\partial q_i}, \qquad b_i = \frac{\partial W}{\partial a_i}, \qquad (i = m + 1, \ldots, n),$$

we have

$$\sum_{i=m+1}^{n} \frac{\partial W}{\partial q_i} dq_i = \sum_{i=m+1}^{n} p_i \, dq_i = \sum_{i=m+1}^{n} \left(-p_i \frac{\partial \Phi_0}{\partial p_i}\right) dx_0.$$

Furthermore, $W$ contains $x_0$, and $\partial W/\partial x_0 = \Phi_0$. Hence,

$$W = \int_0^{x_0} \Phi_0 \, dx_0.$$

Adding these two equations gives (57).

Substitute in $W$ the values of $b_{m+1}, \ldots, b_n$ obtained from (56), giving $\overline{W}$. We now prove that $\overline{W}$ satisfies (51), if put in the place of $W$.

Substitute the values from (56) for $q_{m+1}, \ldots, q_n$ in $W$. This gives $\overline{W}$, according to our definition. Consequently,

$$\frac{\partial W}{\partial x_i} = \frac{\partial \overline{W}}{\partial x_i} + \sum_{j=m+1}^{n} \frac{\partial \overline{W}}{\partial q_j} \frac{\partial q_j}{\partial x_i}.$$

On the other hand, $W$ is a function of

$$x_0, x_1, \ldots, x_m, a_{m+1}, \ldots, a_n, b_{m+1}, \ldots, b_n.$$

Hence,

$$\begin{aligned} \frac{\partial W}{\partial x_i} &= \int_0^{x_0} \left\{ \frac{\partial \Phi_0}{\partial x_i} + \sum_{j=m+1}^{n} \left[ \frac{\partial \Phi_0}{\partial q_j} \frac{\partial q_j}{\partial x_i} - p_j \frac{\partial}{\partial x_i} \left( \frac{\partial \Phi_0}{\partial p_j} \right) \right] \right\} dx_0 \\ &= \int_0^{x_0} \frac{\partial \Phi_0}{\partial x_i} \, dx_0 + \int_0^{x_0} \sum_{j=m+1}^{n} \left( \frac{dp_j}{dx_0} \frac{\partial q_j}{\partial x_i} + p_j \frac{\partial^2 q_j}{\partial x_0 \, \partial x_i} \right) dx_0 \\ &= \int_0^{x_0} \frac{\partial \Phi_0}{\partial x_i} \, dx_0 + \sum_{j=m+1}^{n} \left[ p_j \frac{\partial q_j}{\partial x_i} \right]_0^{x_0}. \end{aligned}$$

Since $q_j = b_j$ for $x_0 = 0$, therefore

$$\frac{\partial \overline{W}}{\partial x_i} = \int_0^{x_0} \frac{\partial \Phi_0}{\partial x_i} \, dx_0 + \sum_{j=m+1}^{n} \left( p_j - \frac{\partial \overline{W}}{\partial q_j} \right) \frac{\partial q_j}{\partial x_i}.$$

But, as $\overline{W}$ is an integral of the first equation of (54), we have $\partial \overline{W}/\partial q_j = p_j$. Thus

$$\frac{\partial \overline{W}}{\partial x_i} = \int_0^{x_0} \frac{\partial \Phi_0}{\partial x_i} \, dx_0.$$

Now, since (54) is in involution,

$$((p_0 - \Phi_0, p_i - x_0 \Phi_i)) = 0, \qquad x_0 \Phi_i = \phi_i,$$

that is,

$$\frac{\partial \Phi_0}{\partial x_i} - \frac{\partial \phi_i}{\partial x_0} + \sum_{j=m+1}^{n} \left( \frac{\partial \phi_i}{\partial q_j} \frac{\partial \Phi_0}{\partial p_j} - \frac{\partial \Phi_0}{\partial q_j} \frac{\partial \phi_i}{\partial p_j} \right) = 0.$$

Substituting $q_j$, $p_j$ from (56), which are themselves integrals of (55), gives

$$\frac{\partial \Phi_0}{\partial x_i} = \frac{\partial \phi_i}{\partial x_0} + \sum_{j=m+1}^{n} \left( \frac{\partial \phi_i}{\partial q_j} \frac{dq_j}{dx_0} + \frac{\partial \phi_i}{\partial p_j} \frac{dp_j}{dx_0} \right) = \frac{d\phi_i}{dx_0}$$

and

$$\frac{\partial \overline{W}}{\partial x_i} = \int_0^{x_0} \frac{d\phi_i}{dx_0}\, dx_0 = [\phi_i]_0^{x_0} = x_0 \Phi_i.$$

Hence, $\overline{W}$ satisfies (54).

Thus we have proved that, if we can integrate the single first equation of (54)—which is nothing but a partial differential equation of $n + 1 - m$ variables—then we can integrate the entire system of partial differential equations (54). Equations (54) are equivalent to the system of partial differential equations (51). Hence, if we integrate a single partial differential equation of $n + 1 - m$ variables, or (going back to our original variables before we have adopted $q_0$, $p_0$) if we integrate a Hamilton-Jacobi partial differential equation of $n - m$ variables, then our problem is completely solved. This is the case of canonical equations of $2(n - m)$ canonically conjugate variables. Thus, the order of the given differential equations is reduced to $2(n - m)$.

Van Kampen and Wintner (1938) proved by their method of parametrization that the order of the differential equations can be reduced even if there exist $m$ invariantive relations instead of $m$ integrals (Section 4.14).

## FUNCTION GROUPS

### 4.9 COMPLETE SYSTEMS

Suppose that $a_i^{(j)}$ in the linear differential operators

$$A_j = \sum_{i=1}^{n} a_i^{(j)} \frac{\partial}{\partial x_i}$$

are functions of $x_1, x_2, \ldots, x_n$ and that there is no linear homogeneous relation of the form

$$\lambda_1 A_1 f + \lambda_2 A_2 f + \cdots + \lambda_m A_m f = 0$$

among the operators

$$A_1 f, A_2 f, \ldots, A_m f$$

with nonvanishing functions $\lambda$ of $x_1, x_2, \ldots, x_n$. Then $A_1f, A_2f, \ldots, A_mf$ are said to be *independent*. Let $X_i$ be $m$ independent linear operators and $f$ be an integral of a system of partial differential equations of $n$ variables $x_1, x_2, \ldots, x_n$:

$$X_if = 0 \qquad (i = 1, 2, \ldots, m). \tag{58}$$

Then, since both $X_\alpha f = 0$ and $X_\beta f = 0$ $(\alpha, \beta = 1, 2, \ldots, m)$, we have $X_\alpha X_\beta f = 0$, $X_\beta X_\alpha f = 0$; hence

$$(X_\alpha, X_\beta)f = (X_\alpha X_\beta - X_\beta X_\alpha)f = 0.$$

In general, a commutator $(X_\alpha, X_\beta)$ can be expressed linearly in terms of $X_1, X_2, \ldots, X_m$; that is, it is a linear combination of $X_1, X_2, \ldots, X_m$. Suppose that we acquire $s$ new independent operators by the operations $(X_\alpha, X_\beta)$, making altogether $m + s$ operators $X_i$. If $m + s = n$, then there is no integral other than $f =$ constant, because this is a system of $n$ simultaneous linear homogeneous partial differential equations of $n$ independent variables. If $m + s < n$, we carry out a similar operation again. Thus, either we obtain $n$ independent operators $X_i$ and come to the above case, or we obtain $r < n$ independent operators. In the latter case we repeat the operations. Finally, we come to the stage where all combinations $(X_i, X_j)f$ are linear combinations of $X_1f, X_2f, \ldots, X_rf$. Clebsch calls such a system of partial differential equations a *complete system*. The case for $r = n$ is special. Equations (51) form a complete system. Thus the integration of a system of linear simultaneous partial differential equations $X_1f = 0, X_2f = 0, \ldots, X_mf = 0$ is reduced to the integration of a complete system.

We now announce four propositions.

I. Any transformation of the independent variables in a complete system leads to a complete system.

When we make a transformation $x_i = \phi_i(y_1, y_2, \ldots, y_n)$ to (58), then

$$Xf = Yf = \sum_{i=1}^{n} b_i \frac{\partial f}{\partial y_i},$$

where $b_i$ are functions of $y_1, y_2, \ldots, y_n$. From $X_if = 0$ we obtain $Y_jf = 0$. Since $X_if = Y_if$, and

$$(X_i, X_j)f = \lambda_1X_1f + \cdots + \lambda_rX_rf,$$

therefore,

$$(Y_i, Y_j)f = \lambda_1'Y_1f + \cdots + \lambda_r'Y_rf.$$

The coefficients $\lambda_j'$ are the result of substituting $y$ for $x$ in $\lambda_j$. Hence, the operators $Y_j f$ also form a complete system.

II. If $Z_1, Z_2, \ldots, Z_r$ are linear combinations of $X_1, X_2, \ldots, X_r$, and, conversely, if $X_1, X_2, \ldots, X_r$ are linear combinations of $Z_1, Z_2, \ldots, Z_r$, then the system

$$Z_1 f = 0, Z_2 f = 0, \ldots, Z_r f = 0$$

is said to be *equivalent to* the system

$$X_1 f = 0, X_2 f = 0, \ldots, X_r f = 0.$$

A system of partial differential equations that is equivalent to a complete system is itself a complete system. A complete system can be replaced by an infinite number of complete systems that are equivalent to the first complete system. A complete system among these, which is such that all combinations $(X_i, X_j) = 0$ $(i, j = 1, 2, \ldots, r)$, is called a *Jacobian system.* An infinitesimal transformation group in such a case is *commutative,* that is, an Abelian group, and forms a subgroup of the original infinitesimal transformation group.

III. Any complete system is equivalent to a Jacobian system. We solve a complete system $X_1 f = 0, X_2 f = 0, \ldots, X_r f = 0$ with respect to $\partial f / \partial x_i$ and denote the result by

$$Z_i f = \frac{\partial f}{\partial x_i} + \sum_{j=1}^{n-r} b_{ij} \frac{\partial f}{\partial x_{r+j}} = 0 \qquad (i = 1, 2, \ldots, r). \tag{59}$$

This is a complete system equivalent to the original system. As $(Z_i, Z_j) f$ contains only

$$\frac{\partial f}{\partial x_{r+1}}, \ldots, \frac{\partial f}{\partial x_n},$$

then $(Z_i, Z_j) f = 0$ if $(Z_i, Z_j) f$ is to be a linear combination of $Z_1 f, Z_2 f, \ldots, Z_r f$. Thus we obtain a Jacobian system.

IV. A complete system of $r$ partial differential equations of $n$ independent variables is reduced, if we know an integral of the system, to a complete system of $r - 1$ partial differential equations of $n - 1$ independent variables.

Suppose that we can integrate $X_1 f = 0$ and choose $y_1, y_2, \ldots, y_n$ so that its $n - 1$ integrals are $y_2, \ldots, y_n$. By this transformation the first equation $X_1 f = 0$ of the system $X_1 f = 0, \ldots, X_r f = 0$ becomes

$\partial f/\partial y_1 = 0$. Solve the remaining $r - 1$ equations with respect to $\partial f/\partial y_2, \ldots, \partial f/\partial y_r$ and let the solution be

$$Y_1 f = \frac{\partial f}{\partial y_1} = 0,$$

$$Y_i f = \frac{\partial f}{\partial y_i} + \sum_{j=1}^{n-r} c_{ij} \frac{\partial f}{\partial y_{r+j}} = 0 \qquad (i = 2, 3, \ldots, r). \tag{60}$$

This is a Jacobian system, as it is of the form just given. Because of the relation

$$(Y_1, Y_i)f = \sum_{j=1}^{n-r} \frac{\partial c_{ij}}{\partial y_1} \frac{\partial f}{\partial y_{r+j}} \equiv 0,$$

the $c_{ij}$ do not depend on $y_1$. Furthermore, $(Y_i, Y_j)f = 0$ for $i > 1, j > 1$. Hence, $r - 1$ equations

$$Y_2 f = 0, \ldots, Y_r f = 0$$

form a system of $r - 1$ partial differential equations of $n - 1$ independent variables.

By repeating this process we can reduce the given complete system to a partial differential equation of $n - r + 1$ variables. This gives us the following theorem: a complete system of $r$ partial differential equations of $n$ independent variables admits $n - r$ independent integrals; the general integral is an arbitrary function of these integrals.

If we call the integration of a system consisting of $n$ ordinary differential equations an *operation*, then we need $n - r, n - r - 1, \ldots, 2, 1$ operations in the above case. The last operation is a mere quadrature. When we know $p$ independent integrals, we need $n - p - r, n - p - r - 1, \ldots$ operations. In particular, if the given equations are canonical with $n$ pairs of conjugate variables $q_i, p_i$, and if we know $r$ independent integrals that are in involution, then it is sufficient to carry out only $2(n - r + 1)$ operations, according to Lie's theorem. A method of solving a Hamilton-Jacobi equation

$$H_0(q_1, \ldots, q_n, p_1, \ldots, p_n) = a_0$$

is: first find a solution $H_1$ that is independent of $H_0$ and satisfies $(H_0, H_1) = 0$. We need $2n - 1$ operations for this purpose, as we already know an integral $H_0 = a_0$. Next, find a solution $H$ of a complete system $(H_0, H) = 0$, $(H_1, H_0) = 0$, $(H_1, H) = 0$. We need

$$2(n - 1) - 1 = 2n - 3$$

operations for this purpose. Thus the number of operations is reduced by two.

## 4.10 FUNCTION GROUPS

Let $u_1, u_2, \ldots, u_r$ be $r$ independent functions of $2n$ variables

$$q_1, q_2, \ldots, q_n, p_1, p_2, \ldots, p_n.$$

If all the Poisson brackets $(u_i, u_j)$ are expressed in terms of $u_1, u_2, \ldots, u_r$ and nothing else, that is, if

$$(u_i, u_j) = f_{ij}(u_1, \ldots, u_r) \qquad (i, j = 1, 2, \ldots, r), \tag{61}$$

then $u_1, u_2, \ldots, u_r$ are said to form a *function group*. The functions $u_1, u_2, \ldots, u_r$ are called its *base*, and $r$ the *order* of the function group. It is an $r$-parameter group. In particular, if $f_{ij} = 0$, then this function group is in involution, that is, it is an Abelian group (Lie, 1929, 1888, 1890, 1893, 1893a, 1896; Goursat, 1921).

When $u_1, u_2, \ldots, u_r$ form a function group, any function

$$F(u_1, u_2, \ldots, u_r)$$

belongs to that group. When $F(u_1, u_2, \ldots, u_r)$ and $\Phi(u_1, u_2, \ldots, u_r)$ belong to the same group, then $(F, \Phi)$ also belongs to the same group. This is because

$$\begin{aligned}(F, \Phi) &= \sum_{i,j} \frac{\partial F}{\partial u_i} \frac{\partial \Phi}{\partial u_j} (u_i, u_j) \\ &= \sum_{i,j} \frac{\partial F}{\partial u_i} \frac{\partial \Phi}{\partial u_j} f_{ij}(u_1, \ldots, u_r).\end{aligned}$$

When we take $r$ independent variables $v_1, v_2, \ldots, v_r$ of $u_1, u_2, \ldots, u_r$, then $(v_i, v_j)$ can be expressed in terms of $v_1, v_2, \ldots, v_r$. These two function groups $(u_1, u_2, \ldots, u_r)$ and $(v_1, v_2, \ldots, v_r)$ are considered two different forms of the same function group. Each of them is a base of that group.

If $r$ independent functions $u_1, u_2, \ldots, u_r$ of $2n$ variables

$$q_1, q_2, \ldots, q_n, p_1, p_2, \ldots, p_n$$

form a function group of order $r$, then $r$ linear expressions

$$(u_1, f) = 0, (u_2, f) = 0, \ldots, (u_r, f) = 0 \tag{61a}$$

form a complete system.

In proving this theorem, it is seen at once that $u_1, u_2, \ldots, u_r$ are independent if the $r$ equations (61a) are independent—that is, if not all of the minor determinants of order $r$ formed from the determinant

$\|\partial u_i/\partial q_j\|$ are zero, that is, if the rank of the determinant is $r$. Let $(u_i, f) = A_i f$. From Jacobi's identity,

$$((u_i, u_j), f) + ((u_j, f), u_i) + ((f, u_i), u_j) \equiv 0,$$

we obtain

$$(A_i, A_j)f \equiv A_i(A_j f) - A_j(A_i f) = ((u_i, u_j), f).$$

Writing $(u_i, u_j) = f_{ij}(u_1, u_2, \ldots, u_r)$, we obtain

$$A_j(A_i f) - A_i(A_j f) = \frac{\partial f_{ij}}{\partial u_1} A_1 f + \cdots + \frac{\partial f_{ij}}{\partial u_r} A_r f.$$

Hence, the operators $A_r f$ form a complete system.

Conversely, if the functions $\phi_1, \phi_2, \ldots, \phi_r$ of $q_i, p_i$ are independent, it is necessary and sufficient that $\phi_i$ should form a function group of order $r$ in order that

$$(\phi_1, f) = 0, \ldots, (\phi_r, f) = 0$$

form a complete system.

From the above identity, $((\phi_i, \phi_j), f) = 0$. If the system is complete, then this formula should be derivable from the original system of equations. This happens only when the Poisson brackets $(\phi_i, \phi_j)$ are functions of $\phi_1, \phi_2, \ldots, \phi_r$. This proves the converse theorem.

Now the system of equations

$$(u_1, f) = 0, (u_2, f) = 0, \ldots, (u_r, f) = 0 \tag{62}$$

admits $2n - r$ independent integrals $v_1, v_2, \ldots, v_{2n-r}$ according to the theorem stated at the end of Section 4.9, and other integrals can be expressed in terms of these $2n - r$ integrals. Poisson's theorem states that $(v_i, v_j)$ are also integrals of this system and accordingly are expressed in terms of $v_1, v_2, \ldots, v_{2n-r}$. Hence, $v_1, v_2, \ldots, v_{2n-r}$ form a function group of order $2n - r$. Each function of this group is in involution with any function of the original function group $u_1, u_2, \ldots, u_r$. The new group is called the *polar group* of the original function group.

The system

$$(v_1, f) = 0, (v_2, f) = 0, \ldots, (v_{2n-r}, f) = 0 \tag{63}$$

is a complete system with $u_1, u_2, \ldots, u_r$ as the $r$ independent integrals; all other integrals are expressed in terms of $u_1, u_2, \ldots, u_r$. Thus,

$$(u_1, u_2, \ldots, u_r) \qquad \text{and} \qquad (v_1, v_2, \ldots, v_{2n-r})$$

are in reciprocal relationship. Hence, a polar group is also a *reciprocal group*.

If any function $U$ belonging to a function group $u_1, u_2, \ldots, u_r$ is in involution with any other function in the same function group, then $U$ is called a *distinguished function* (*ausgezeichnete Funktion*). A distinguished function is a function common to a function group and its polar group. Denote the polar group of $(u_1, u_2, \ldots, u_r)$ by $(v_1, v_2, \ldots, v_{2n-r})$ and the distinguished function by $F(q_1, q_2, \ldots, q_n, p_1, p_2, \ldots, p_n)$; then

$$F(q_1, q_2, \ldots, q_n, p_1, p_2, \ldots, p_n) = U(u_1, u_2, \ldots, u_r)$$
$$= V(v_1, v_2, \ldots, v_{2n-r}). \qquad (64)$$

If there are $m$ distinguished functions, then

$$U_i(u_1, u_2, \ldots, u_r) = V_i(v_1, v_2, \ldots, v_{2n-r}) \qquad (i = 1, 2, \ldots, m).$$

Let $u_1, u_2, \ldots, u_r, v_1, v_2, \ldots, v_{2n-r}$ be a group $G$. The functions $u_1, u_2, \ldots, u_r$ and $v_1, v_2, \ldots, v_{2n-r}$ form respectively the subgroups $U$ and $V$ of $G$. A distinguished function belongs to a subgroup $A$ of each of $U$ and $V$. The Poisson brackets $(u_i, u_j)$ form a coset (Nebengruppe). As $vU$, that is, $(u_i, v_j)$, and $Uv$, that is, $(v_i, u_j)$, are both zero, the right coset coincides with the left coset and we have $UA = AU$; in other words this is an Abelian group. As $U^{-1}AU = A$, the group is also self-conjugate. Moreover, $V^{-1}AV = A$. We have $G^{-1}AG = A$; that is, $A$ is invariant. Hence, $A$ is an invariant subgroup of $G$.

We can determine the distinguished functions belonging to a function group of order $r$ in the following manner. By definition, the number of distinguished functions belonging to the given function group equals the number of such independent integrals of the complete system (62) that are functions of $u_1, u_2, \ldots, u_r$ only.

Write

$$f = U(u_1, u_2, \ldots, u_r) \qquad (u_i, f) = A_i f.$$

Then, (62) yields $r$ equations for $u_1, u_2, \ldots, u_r$:

$$A_i U = \sum_{j=1}^{r} \frac{\partial U}{\partial u_j}(u_i, u_j) = 0 \qquad (i = 1, 2, \ldots, r), \qquad (65)$$

because $(u_i, u_j)$ can be expressed in terms of $u_1, u_2, \ldots, u_r$, according to our assumption. Suppose that these $r$ equations are reduced to $r - m$ independent equations

$$A_1 U = 0, A_2 U = 0, \ldots, A_{r-m} U = 0. \qquad (66)$$

This is a complete system, because, by putting $(u_i, u_j) = f_{ij}(u_1, u_2, \ldots, u_r)$, we obtain

$$A_iA_jU - A_jA_iU = \frac{\partial f_{ij}}{\partial u_1} A_1U + \cdots + \frac{\partial f_{ij}}{\partial u_r} A_rU$$

from Jacobi's identity. Since this is a complete system of $r - m$ equations for $r$ variables, it should have $m$ integrals. Hence the number of the distinguished functions is equal to the number of linearly independent integrals of (65). Let the order of the first nonzero minor of the determinant $\Delta = \|(u_i, u_j)\|$ $(i, j = 1, 2, \ldots, r)$ be $r - m$. Then $m$ is the number of distinguished functions. Take $r - m$ linearly independent equations (66) out of the system (65), and look for $m$ integrals of (66). As this determinant is skew-symmetric, it is zero if its order is odd. Hence there must be at least one distinguished function in this case.

## 4.11 CANONICAL FORM OF A FUNCTION GROUP

If a function group is in involution, then it is in involution in whatever form it is written. A group in involution is the simplest form of a function group and can be considered to be of a canonical form.

Consider a function group that is not in involution. A function group of order $r$ that is not in involution cannot contain more than $r - 2$ independent distinguished functions. Suppose that there were $r - 1$ independent distinguished functions. Let them be denoted by $U_1, U_2, \ldots, U_{r-1}$; and the remaining functions belonging to the group be denoted by $V$. Then, we should have

$$(U_1, V) = 0, (U_2, V) = 0, \ldots, (U_{r-1}, V) = 0,$$
$$(U_i, U_j) = 0 \qquad (i, j = 1, 2, \ldots, r - 1),$$

that is, this function group would be in involution, contrary to our assumption. Hence, there cannot be more than $r - 2$ independent distinguished functions.

Consider a function group $(u_1, u_2, \ldots, u_r)$ that is not in involution, and suppose that $u_1$ is not a distinguished function of this group. Then at least one of $(u_1, u_2), \ldots, (u_1, u_r)$ is not zero. We can find such a function $F(u_1, u_2, \ldots, u_r)$, belonging to this function group, that satisfies $(u_1, F) = 1$. Let such a function $F$ be $u_2$, that is, a function $F$ that satisfies the partial differential equations

$$(u_1, u_2) \frac{\partial F}{\partial u_2} + \cdots + (u_1, u_r) \frac{\partial F}{\partial u_r} = 1, \tag{67}$$

and is a function of $u_1, u_2, \ldots, u_r$. Then $(u_1, u_2) = 1$.

Next, the two equations $(u_1, F) = 0$, $(u_2, F) = 0$, or

$$\begin{aligned} \frac{\partial F}{\partial u_2} + (u_1, u_3)\frac{\partial F}{\partial u_3} + \cdots + (u_1, u_r)\frac{\partial F}{\partial u_r} &= 0, \\ -\frac{\partial F}{\partial u_1} + (u_2, u_3)\frac{\partial F}{\partial u_3} + \cdots + (u_2, u_r)\frac{\partial F}{\partial u_r} &= 0, \end{aligned} \tag{67a}$$

are independent and form a complete system. By Jacobi's identity, we may write

$$(u_1, (u_2, F)) - (u_2, (u_1, F)) = ((u_1, u_2), F) = (1, F) \equiv 0.$$

The coefficients $(u_i, u_j)$ of this complete system are functions of $u_1, u_2, \ldots, u_r$. Let the $r - 2$ independent integrals of these partial differential equations be $u'_1, u'_2, \ldots, u'_{r-2}$. These functions form a function group of order $r - 2$. The functions $u_1, u_2, u'_1, \ldots, u'_{r-2}$ are independent. If we assume that there exists a relation among them, then it should contain either $u_1$ or $u_2$, because otherwise it would be contrary to our hypothesis that $u'_1, u'_2, \ldots, u'_{r-2}$ are independent integrals of (67a). Now suppose that the relation contains $u_1$, and let it be

$$\psi(u_1, u_2, u'_1, \ldots, u'_{r-2}) = 0.$$

Then

$$(u_2, \psi) = \frac{\partial\psi}{\partial u_1}(u_2, u_1) + \frac{\partial\psi}{\partial u'_1}(u_2, u'_1) + \cdots + \frac{\partial\psi}{\partial u'_{r-2}}(u_2, u'_{r-2}) = 0.$$

As $u'_1, u'_2, \ldots, u'_{r-2}$ are solutions of (67), therefore

$$(u_2, u'_1) = \cdots = (u_2, u'_{r-2}) = 0.$$

We have taken $(u_1, u_2) = 1$. Hence $(u_2, \psi) = -\partial\psi/\partial u_1 = 0$. This contradicts our hypothesis. Therefore $u_1, u_2, u'_1, \ldots, u'_{r-2}$ are independent.

Thus, if $(u_1, u_2, \ldots, u_r)$ are not in involution, first take $(u_1, u_2) = 1$; another set $u'_1, u'_2, \ldots, u'_{r-2}$ is now in involution with $u_1$ and $u_2$. If these $u'_1, u'_2, \ldots, u'_{r-2}$ are in involution, write $u_2 = Q_1$, $u_1 = P_1$; the new set $Q_1, P_1, u'_1, \ldots, u'_{r-2}$ satisfies

$$(P_1, Q_1) = 0, (P_1, u'_i) = 0, (Q_1, u'_i) = 0, (u'_i, u'_j) = 0 \qquad (i, j = 1, 2, \ldots, r - 2). \tag{68}$$

If $u'_1, u'_2, \ldots, u'_{r-2}$ are not in involution, repeat this operation; this gives the set $P_{,2}\, Q_2, u''_1, \ldots, u''_{r-4}$, which has the same property. Repeating this process leads to the following

*Theorem:* Any function group can be put in the form

$$Q_1, P_1, \ldots, Q_q, P_q, Q_{q+1}, \ldots, Q_{q+m},$$

such that

$$(P_i, Q_i) = 1 \qquad (i = 1, 2, \ldots, q),$$

and the other Poisson brackets all vanish. In such a case the distinguished functions are functions of $Q_{q+1}, \ldots, Q_{q+m}$ only.

The second part of this theorem is proved as follows: Let a distinguished function be $\Pi(Q_1, \ldots, Q_{q+m}, P_1, \ldots, P_q)$. Then the equations for determining the distinguished functions are

$$(u_i, \Pi) = \frac{\partial \Pi}{\partial u_1}(u_i, u_1) + \frac{\partial \Pi}{\partial u_2}(u_i, u_2) + \cdots + \frac{\partial \Pi}{\partial u_r}(u_i, u_r) = 0$$

$$(i = 1, 2, \ldots, r),$$

and they can be written

$$(Q_i, \Pi) = \frac{\partial \Pi}{\partial P_i} = 0, \qquad (P_i, \Pi) = \frac{\partial \Pi}{\partial Q_i} = 0 \qquad (i = 1, 2, \ldots, q).$$

Thus the distinguished functions are functions of $Q_{q+1}, \ldots, Q_{q+m}$ only. As they are in involution with all $P_i$, $Q_i$ $(i = 1, 2, \ldots, q)$, the functions $Q_{q+1}, \ldots, Q_{q+m}$ are themselves the distinguished functions, $m$ in number. Hence, the difference between the number $2q + m$ of the functions belonging to a function group of order $2q + m$ and the number $m$ of the distinguished functions is $2q$ and therefore always even.

Suppose that the functions in the system

$$f_1 = c_1, f_2 = c_2, \ldots, f_q = c_q \tag{69}$$

of $n$ variables are mutually in involution, and that we know $r - q$ independent integrals $f_{q+1}, \ldots, f_r$ of a complete system

$$(f_1, f) = 0, (f_2, f) = 0, \ldots, (f_q, f) = 0.$$

Suppose that we cannot obtain any new integral from the Poisson brackets formed of $f_{q+1}, \ldots, f_r$. These functions $f_1, \ldots, f_r$ form a function group. If $r = 2n - q$, then we can integrate the system by quadrature. Suppose that $r < 2n - q$. Suppose that there are $m$ distinguished functions that are independent of the $q$ distinguished functions $f_1, f_2, \ldots, f_q$ among $f_1, f_2, \ldots, f_r$. Then, $r = q + m + 2\nu$. As this function group contains a function group of order $q + m + \nu$ which is in involution, we must integrate a system of $q + m + \nu$ equations in involution to determine these functions. This is done in

$2n - 2q - 2m - 2\nu$, $2n - 2q - 2m - 2\nu - 2, \ldots, 2$ operations. This method of integration is due to Jacobi-Mayer.

Without leading to the integration of a system of differential equations in involution we may consider $X_j f = 0$ $(j = 1, 2, \ldots, q)$, such that

$$(X_i X_j - X_j X_i) f = \sum_{k=1}^{j} \lambda_{ij}^{k} X_k f \qquad (i = 1, 2, \ldots, j - 1; j = 1, 2, \ldots, q).$$

A collection of such $X_j f = 0$ $(j = 1, 2, \ldots, q)$, is called a *successively complete* system and is used in Pfeiffer's (1933) method of integration. A generalization of Jacobi's method has been treated by de Donder (1927) and Saltykov (1934, 1935).

Now we prove the following important

*Theorem:* The number of independent functions in involution in a given function group is $(r + m)/2$, where $r$ is the order of the function group and $m$ is the number of the distinguished functions.

When reduced to a canonical form according to the process just described, any function group of order $r$ can be written in the form

$$P_1, \ldots, P_q, \quad Q_1, \ldots, Q_q, \quad Q_{q+1}, \ldots, Q_{q+m} \qquad (2q + m = r), \qquad (70)$$

such that $(P_i, Q_i) = 1$ $(i = 1, 2, \ldots, q)$; the other Poisson brackets all vanish. Hence, $q + m = (r + m)/2$ functions

$$Q_1, \ldots, Q_q; \quad Q_{q+1}, \ldots, Q_{q+m}$$

are in involution. We proceed to prove that there cannot be more than $q + m$ functions in involution.

Assume tentatively that there are $\nu > q$ functions. Let them be $Q_1, \ldots, Q_\nu$. Then we should be able to determine independent functions $P_{q+1}, \ldots, P_{q+m}, Q_{q+m+1}, \ldots, Q_n$ other than those in (70), so that $P_1, \ldots, P_n, Q_1, \ldots, Q_n$ would form a canonical function group of order $2n$. If we eliminate one of the distinguished functions $Q_{q+1}, \ldots, Q_{q+m}$ (for example, $Q_{q+1}$), the polar group of a function group of order $2q + m - 1$

$$P_1, \ldots, P_q, \quad Q_1, \ldots, Q_q, \ *, Q_{q+2}, \ldots, Q_{q+m} \qquad (71)$$

in which the asterisk denotes the absence of $Q_{q+1}$, will have the form

$$Q_{q+1}, U_1, U_2, \ldots, U_{2n-2q-m},$$

because $Q_{q+1}$ is in involution with the $2q + m - 1$ functions of (71). Hence, in the same manner in which we constructed a canonical function group, we can determine $P_{q+1}$ so that it satisfies

$$(P_{q+1}, Q_{q+1}) = 1.$$

The functions $P_1, \ldots, P_q, P_{q+1}, Q_1, \ldots, Q_{q+m}$ form a canonical function group and contain the functions (70). The group's distinguished functions are $Q_{q+2}, \ldots, Q_{q+m}$; they are mutually independent. By repeating similar processes we obtain

$$P_1, \ldots, P_{q+m}, \qquad Q_1, \ldots, Q_{q+m}, \tag{72}$$

but these do not contain distinguished functions. We add to these functions the polar group of $2(n - q - m)$ functions belonging to (72). As the function group (72) does not contain distinguished functions, its polar group does not contain any distinguished function. Hence, by treating this polar function group similarly, we obtain both a canonical function group of order $2n$, and, from (72), the polar group belonging to it. The canonical function group that we obtain is

$$P_1, \ldots, P_q, \ P_{q+1}, \ldots, P_n, \ \ Q_1, \ldots, Q_q, \ \ Q_{q+1}, \ldots, Q_{q+m}, \ldots, Q_n. \tag{73}$$

Thus, $n - q - m$ functions $Q_{q+m+1}, \ldots, Q_n$ are mutually in involution and also in involution with the functions $P_1, \ldots, P_q, Q_1, \ldots, Q_{q+m}$ of (70). Hence, they are in involution with the functions $\phi_1, \phi_2, \ldots, \phi_q$ contained in (70). By hypothesis, $\phi_1, \phi_2, \ldots, \phi_\nu$ are mutually in involution. Hence, we have altogether a function group consisting of $n - q - m + \nu$ functions:

$$Q_{q+m+1}, \ldots, Q_n, \qquad \phi_1, \phi_2, \ldots, \phi_\nu, \tag{74}$$

which are mutually in involution. But the order of the polar group of a given function group of order $r$ is $2n - r$. If $r$ functions of a function group of order $r$ are all mutually in involution, then a complete system

$$(u_1, f) = 0, \ldots, (u_r, f) = 0$$

admits $r$ independent integrals $u_1, \ldots, u_r$. As these functions of the polar group should satisfy these $r$ equations, the polar group must contain these $r$ functions. Hence, $r \leqq 2n - r$, that is, $r \leqq n$.

Going back to our function group (74), we have

$$n - q - m + \nu \leqq n.$$

Hence

$$\nu \leqq q + m = \tfrac{1}{2}(r + m),$$

since $n - q - m + \nu$ functions of $u$ are mutually in involution. Thus, we arrive at the conclusion that the number of functions mutually in involution among functions (70) is $(r + m)/2$.

In the course of our proof it is seen that a general function group $Q_1, \ldots, Q_n, P_1, \ldots, P_n$ of order $2n$, that are not in involution and, hence, have no distinguished function, are functions of our original independent variables $q_1, \ldots, q_n, p_1, \ldots, p_n$ and are such that

$$(P_i, Q_j) = \delta_{ij},$$
$$(P_i, P_j) = (Q_i, Q_j) = 0 \qquad (i, j = 1, 2, \ldots, n).$$

Hence, by the theorem in Section 1.9, the transformation from $(q_i, p_i)$ to $(Q_i, P_i)$ is a contact transformation.

## INTEGRALS OF THE $n$-BODY PROBLEM

### 4.12 EULERIAN INTEGRALS

Consider $n$ particles moving in an Euclidean three-dimensional space according to Newton's gravitational law. Following Poincaré (1897, 1905) we denote the coordinates of the first particle by $x_1$, $x_2$, $x_3$ and its mass by $m_1 = m_2 = m_3$, the second particle by $x_4$, $x_5$, $x_6$ and its mass by $m_4 = m_5 = m_6$, and so on, finally the coordinates of the $j$th particle by $x_{3j-2}, x_{3j-1}, x_{3j}$ and its mass by $m_{3j-2} = m_{3j-1} = m_{3j}$ $(j = 1, 2, \ldots, n)$. The corresponding momenta are

$$y_i = m_i \frac{dx_i}{dt} \qquad (i = 1, 2, \ldots, 3n).$$

The total kinetic energy of the system is

$$T = \frac{1}{2} \sum_{i=1}^{3n} m_i \left(\frac{dx_i}{dt}\right)^2 = \frac{1}{2} \sum_{i=1}^{3n} \frac{y_i^2}{m_i}, \tag{75}$$

and the force function is

$$U = \sum_{i,j} \frac{m_{3i} m_{3j}}{\Delta_{3i,3j}} \qquad (i, j = 1, 2, \ldots, n), \tag{76}$$

where $\Delta_{3i,3j}$ is the mutual distance between the $i$th and $j$th particles. Here the units are so chosen that the constant of universal gravitation equals 1.

Let

$$H = T - U, \tag{77}$$

and the equations of motion of the dynamical system be

$$\frac{dx_i}{dt} = \frac{\partial H}{\partial y_i}, \qquad \frac{dy_i}{dt} = -\frac{\partial H}{\partial x_i} \qquad (i = 1, 2, \ldots, 3n). \tag{78}$$

The order of this system of equations is $6n$.

Now, since the function $H$ of (77) does not contain $t$ explicitly, the differential equations remain unaltered when we make the transformation $t' = t + \epsilon$. Hence, (78) admits the infinitesimal transformation

$$A_0 f = \frac{\partial f}{\partial t}.$$

Corresponding to this transformation, we have $dH/dt = \partial H/\partial t = 0$, which gives the integral $H =$ constant, the energy integral.

Differential equations (78) remain unaltered when we translate the coordinate origin $x'_{3j-2} = x_{3j-2} + \epsilon$ $(j = 1, 2, \ldots, n)$ in the $x_1$ direction. The force function $U$ depends only on the distances between the particles, and the kinetic energy $T$ depends only on the momenta. Hence, the system (78) admits the infinitesimal transformation

$$A_1 f = \sum_{j=1}^{n} \frac{\partial f}{\partial x_{3j-2}}.$$

Likewise, (78) admits

$$A_2 f = \sum_{j=1}^{n} \frac{\partial f}{\partial x_{3j-1}}, \qquad A_3 f = \sum_{j=1}^{n} \frac{\partial f}{\partial x_{3j}}.$$

Now the result of applying to the integral $H =$ constant the infinitesimal transformations admitted by the system (78) is also an integral. Thus

$$A_1 H = \sum_{j=1}^{n} \frac{\partial H}{\partial x_{3j-2}} = 0.$$

This last transformation has been set equal to zero because $T$ does not contain the coordinates and $U$ depends only on the mutual distances.

$$\sum_{j=1}^{n} \frac{\partial H}{\partial x_{3j-2}} = 0.$$

Hence, we have from (78) the equation

$$\sum_{j=1}^{n} \frac{dy_{3j-2}}{dt} = 0,$$

which gives the integral

$$H_1 = -\sum_{j=1}^{n} y_{3j-2} = \text{constant}.$$

Similarly, we have the integrals

$$H_2 = -\sum_{j=1}^{n} y_{3j-1} = \text{constant}, \qquad H_3 = -\sum_{j=1}^{n} y_{3j} = \text{constant},$$

or

$$\sum_{j=1}^{n} m_{3j-2} \frac{dx_{3j-2}}{dt} = \text{constant}, \qquad \sum_{j=1}^{n} m_{3j-1} \frac{dx_{3j-1}}{dt} = \text{constant},$$

$$\sum_{j=1}^{n} m_{3j} \frac{dx_{3j}}{dt} = \text{constant}.$$

These three integrals are what we call the *first* integrals of the center of mass.

Since $U$ depends only on the differences of coordinates and $T$ on the momenta, (78) remains unaltered when we rotate the coordinate axes. Let the rotation be

$$x'_{3j-1} = x_{3j-1} \cos \epsilon + x_{3j} \sin \epsilon \approx x_{3j-1} + x_{3j}\epsilon,$$

$$x'_{3j} = -x_{3j-1} \sin \epsilon + x_{3j} \cos \epsilon \approx -x_{3j-1}\epsilon + x_{3j},$$

Then

$$f(x'_{3j-1}, x'_{3j}) - f(x_{3j-1}, x_{3j}) = \epsilon x_{3j} \frac{\partial f}{\partial x_{3j-1}} - \epsilon x_{3j-1} \frac{\partial f}{\partial x_{3j}}.$$

Equations (78) admit the infinitesimal transformation

$$A_4 f = \sum_{j=1}^{n} \left( -x_{3j} \frac{\partial f}{\partial x_{3j-1}} + x_{3j-1} \frac{\partial f}{\partial x_{3j}} \right).$$

Similarly, they admit

$$A_5 f = \sum_{j=1}^{n} \left( -x_{3j-2} \frac{\partial f}{\partial x_{3j}} + x_{3j} \frac{\partial f}{\partial x_{3j-2}} \right),$$

$$A_6 f = \sum_{j=1}^{n} \left( -x_{3j-1} \frac{\partial f}{\partial x_{3j-2}} + x_{3j-2} \frac{\partial f}{\partial x_{3j-1}} \right).$$

When we apply these transformations to the integral $H =$ constant, then, from the expression $\partial H/\partial x_i$ in (78), we obtain

$$\sum_{j=1}^{n}\left(x_{3j}\frac{dy_{3j-1}}{dt} - x_{3j-1}\frac{dy_{3j}}{dt}\right) = 0,$$

that is,

$$H_4 = \sum_{j=1}^{n}(x_{3j}y_{3j-1} - x_{3j-1}y_{3j}) = \text{constant}.$$

Similarly,

$$H_5 = \sum_{j=1}^{n}(x_{3j-2}y_{3j} - x_{3j}y_{3j-2}) = \text{constant},$$

and

$$H_6 = \sum_{j=1}^{n}(x_{3j-1}y_{3j-2} - x_{3j-2}y_{3j-1}) = \text{constant}.$$

These are the integrals of the momenta of momentum, or the integrals of angular momentum, or the integrals of areal velocity.

Finally, we apply the transformation

$$x'_{3j-2} = x_{3j-2} + t\epsilon \qquad (j = 1, 2, \ldots, n).$$

Then $U$ remains unaltered, as it depends only on the coordinate differences. Differentiating this and multiplying the result by $m_{3j-2}$ gives

$$y'_{3j-2} = y_{3j-2} + m_{3j-2}\epsilon \qquad (j = 1, 2, \ldots, n).$$

The kinetic energy $T$ increases by the amount

$$\sum_{j=1}^{n}\frac{1}{2m_{3j-2}}2m_{3j-2}\epsilon y_{3j-2},$$

if we neglect quantities of order $\epsilon^2$. The derivative $dx_{3j-2}/dt$ increases by $\epsilon$. Hence, the equation

$$\frac{dx_{3j-2}}{dt} = \frac{\partial H}{\partial y_{3j-2}}$$

remains unaltered by this transformation. Similarly, $\partial H/\partial x_{3j-2}$ remains unaltered, $dy_{3j-2}/dt$ remains unaltered, and

$$\frac{dy_{3j-2}}{dt} = -\frac{\partial H}{\partial x_{3j-2}}$$

remains unaltered. Thus equations (78) admit the infinitesimal transformation

$$A_7 f = \sum_{j=1}^{n} \left( m_{3j-2} \frac{\partial f}{\partial y_{3j-2}} + t \frac{\partial f}{\partial x_{3j-2}} \right).$$

Similarly, they admit

$$A_8 f = \sum_{j=1}^{n} \left( m_{3j-1} \frac{\partial f}{\partial y_{3j-1}} + t \frac{\partial f}{\partial x_{3j-1}} \right),$$

$$A_9 f = \sum_{j=1}^{n} \left( m_{3j} \frac{\partial f}{\partial y_{3j}} + t \frac{\partial f}{\partial x_{3j}} \right).$$

Apply the transformation $A_7 f$ to $H =$ constant. By (75), we have

$$\sum_{j=1}^{n} y_{3j-2} = \text{constant} = \alpha_1,$$

as $H$ depends only on the coordinate differences and $y$ is contained only in $T$. Further integration gives

$$\sum_{j=1}^{n} m_{3j-2} x_{3j-2} = \alpha_1 t + \beta_1, \qquad \beta_1 = \text{constant},$$

that is,

$$H_7 = \sum_{j=1}^{n} (m_{3j-2} x_{3j-2} - y_{3j-2} t) = \text{constant}.$$

Similarly,

$$H_8 = \sum_{j=1}^{n} (m_{3j-1} x_{3j-1} - y_{3j-1} t) = \text{constant},$$

and

$$H_9 = \sum_{j=1}^{n} (m_{3j} x_{3j} - y_{3j} t) = \text{constant}.$$

These are the *second* integrals of the center of mass.

Thus, we have obtained ten integrals, that is, one integral of the energy, $H$, six integrals of the center of mass, $H_1$, $H_2$, $H_3$, $H_7$, $H_8$, $H_9$, and three integrals of angular momentum, $H_4$, $H_5$, $H_6$. The corresponding infinitesimal transformations are $A_0 f$, $A_1 f$, $A_2 f$, $A_3 f$, $A_7 f$, $A_8 f$, $A_9 f$, $A_4 f$, $A_5 f$, $A_6 f$; they constitute a function group of ten members, as will be shown immediately. These integrals are sometimes called the *classical Eulerian integrals*, or — from a transformation-group viewpoint — a *function group of Galilei* of order ten. It is to be noted that these ten integrals are all algebraic with respect to the coordinates and the momenta (Saltykov, 1950).

## 4.13 DISTINGUISHED FUNCTIONS

TABLE 1

| | $H$ | $H_1$ | $H_2$ | $H_3$ | $H_4$ | $H_5$ | $H_6$ | $H_7'$ | $H_8'$ | $H_9'$ |
|---|---|---|---|---|---|---|---|---|---|---|
| $H$ | 0 | 0 | 0 | 0 | 0 | 0 | 0 | 0 | 0 | 0 |
| $H_1$ | 0 | 0 | 0 | 0 | 0 | $H_3$ | $-H_2$ | 0 | $MH_3$ | $-MH_2$ |
| $H_2$ | 0 | 0 | 0 | 0 | $-H_3$ | 0 | $H_1$ | $-MH_3$ | 0 | $MH_1$ |
| $H_3$ | 0 | 0 | 0 | 0 | $H_2$ | $-H_1$ | 0 | $MH_2$ | $-MH_1$ | 0 |
| $H_4$ | 0 | 0 | $H_3$ | $-H_2$ | 0 | $H_6$ | $-H_5$ | 0 | $H_9'$ | $-H_8'$ |
| $H_5$ | 0 | $-H_3$ | 0 | $H_1$ | $-H_6$ | 0 | $H_4$ | $-H_9'$ | 0 | $H_7'$ |
| $H_6$ | 0 | $H_2$ | $-H_1$ | 0 | $H_5$ | $-H_4$ | 0 | $H_8'$ | $-H_7'$ | 0 |
| $H_7'$ | 0 | 0 | $MH_3$ | $-MH_2$ | 0 | $H_9'$ | $-H_8'$ | 0 | $MH_9'$ | $-MH_8'$ |
| $H_8'$ | 0 | $-MH_3$ | 0 | $MH_1$ | $-H_9'$ | 0 | $H_7'$ | $-MH_9'$ | 0 | $MH_7'$ |
| $H_9'$ | 0 | $MH_2$ | $-MH_1$ | 0 | $H_8'$ | $-H_7'$ | 0 | $MH_8'$ | $-MH_7'$ | 0 |

Table 1 shows the Poisson brackets

$$(f, g) = \sum_{i=1}^{3n} \left( \frac{\partial f}{\partial x_i} \frac{\partial g}{\partial y_i} - \frac{\partial f}{\partial y_i} \frac{\partial g}{\partial x_i} \right)$$

formed from the ten classical integrals of Euler, where

$$M = \sum_{j=1}^{n} m_{3j-2} = \sum_{j=1}^{n} m_{3j-1} = \sum_{j=1}^{n} m_{3j}$$

is the total mass of the system. As it is inconvenient to have $t$ explicitly in the integrals, we use

$$H_7' = H_3H_8 - H_2H_9, \qquad H_8' = H_1H_9 - H_3H_7, \qquad H_9' = H_2H_7 - H_1H_8$$

instead of $H_7$, $H_8$, $H_9$. This leads to the identity

$$H_1H_7' + H_2H_8' + H_3H_9' \equiv 0. \tag{79}$$

Hence, nine of the ten integrals $H$, $H_1$, $H_2$, $H_3$, $H_4$, $H_5$, $H_6$, $H_7'$, $H_8'$, $H_9'$ are independent and form a ninth-order function group. Table 1 lists the structure constants for the Poisson operators; in other words, it is a *representation* of a function group, that is, it is equivalent to a representation of this group by the matrix of a linear homogeneous transformation.

Now we form the ninth-order determinant. Any combination with $H$ in the Poisson brackets is zero, as $H$ is in involution with any of the integrals. The first nonzero minor of this determinant is of order six; that is, the order of this determinant is six. Put $r = 9$, $m = 3$ in the theorem of Section 4.11. Because of identity (79), the number of distinguished functions is three and the number of functions mutually in involution is $(r + m)/2 = 6$. $H$ is included among these functions. For the moment, we put aside any $H$ which is in involution with any other function. As the Poisson brackets among the three functions $H_1$, $H_2$, $H_3$ are all zero, these three functions constitute a function group of order three that are mutually in involution. Denote a function in involution with $H_1, H_2, H_3$ and $H_4, H_5, H_6$ by $\Phi(H_1, H_2, H_3, H_4, H_5, H_6)$ and look for such a function that satisfies the requirement

$$(H_1, \Phi) = 0, \qquad (H_2, \Phi) = 0, \qquad (H_3, \Phi) = 0.$$

Take $F(f_1, f_2, \ldots, f_m)$ and put it in the definition of a Poisson bracket; then,

$$(K, F) = \sum_{j=1}^{m} (K, f_j) \frac{\partial F}{\partial f_j}.$$

We substitute each $H$ in succession for one of $K$ and $f_j$ and, bearing in mind the identity

$$H_1(H_1, \Phi) + H_2(H_2, \Phi) + H_3(H_3, \Phi) \equiv 0$$

among these equations, we see that only two of the Poisson brackets are independent. We thus have two independent equations

$$H_3 \frac{\partial \Phi}{\partial H_5} - H_2 \frac{\partial \Phi}{\partial H_6} = 0, \qquad -H_3 \frac{\partial \Phi}{\partial H_4} + H_1 \frac{\partial \Phi}{\partial H_6} = 0.$$

Integrating these equations, we obtain

$$\Phi = H_1 H_4 + H_2 H_5 + H_3 H_6.$$

Thus we obtain five functions $H, H_1, H_2, H_3, \Phi$ that are mutually in involution (Lie, 1874, 1929; Engel, 1916, 1917; Engel and Faber, 1932; Cartan, 1922; Goursat, 1921, 1923).

Next we seek a function $\Pi$ of ten integrals $H$ that is in involution with $H, H_1, H_2, H_3, \Phi$. The required equations are

$$(H, \Pi) = (H_1, \Pi) = (H_2, \Pi) = (H_3, \Pi) = (\Phi, \Pi) = 0.$$

Such a function is found to be

$$\Pi = (MH_4 - H_7')^2 + (MH_5 - H_8')^2 + (MH_6 - H_9')^2.$$

Thus we have six functions

$$H, H_1, H_2, H_3, \Phi, \Pi$$

that are mutually in involution.

A distinguished function should be in involution with any function belonging to the same group. Let $f$ be such a function. It should satisfy

$$(H_4,f) = (H_5,f) = (H_6,f) = (H_7',f) = (H_8',f) = (H_9',f) = 0.$$

We have two functions

$$f = \Psi = \tfrac{1}{2}(H_1^2 + H_2^2 + H_3^2) \qquad \text{and}$$

$$f = \Pi = (MH_4 - H_7')^2 + (MH_5 - H_8')^2 + (MH_6 - H_9')^2$$

satisfying these equations. The two functions $\Psi$ and $\Pi$ together with $H$ are the required three distinguished functions. Hence, the six functions mutually in involution are $H, H_1, H_2, H_3, \Psi, \Pi$ (Hagihara, 1938).

On the other hand, according to Lie's theorem, the order of a system of differential equations of order $2m$ is reduced (when there are $p$ integrals mutually in involution) to order $2(m - p)$. Hence, the order of the differential equations of the $n$-body problem is reduced to $6n - 12$, because there exist six integrals mutually in involution. Note that the number of the known Eulerian integrals is ten. This difference between twelve and ten is explained by the existence of two other integrals: the integral for the elimination of time (Section 4.4), and the integral for the elimination of nodes discovered by Jacobi (1842, 1844; see also Section 5.4).

In the plane $n$-body problem, the order of the differential equations is $4n$. There are six corresponding classical integrals:

$$H = T - U, \qquad H_1 = -\sum_{j=1}^{n} y_{2j-1}, \qquad H_2 = -\sum_{j=1}^{n} y_{2j},$$

$$H_6 = \sum_{j=1}^{n} (x_{2j}y_{2j-1} - x_{2j-1}y_{2j}),$$

$$H_7 = \sum_{j=1}^{n} (m_{2j-1}x_{2j-1} - y_{2j-1}t),$$

$$H_8 = \sum_{j=1}^{n} (m_{2j}x_{2j} - y_{2j}t).$$

The function group is of order $r = 6$. Table 2 gives the matrix representation of this function group.

TABLE 2

| | $H$ | $H_1$ | $H_2$ | $H_6$ | $H_7$ | $H_8$ |
|---|---|---|---|---|---|---|
| $H$ | 0 | 0 | 0 | 0 | 0 | 0 |
| $H_1$ | 0 | 0 | 0 | $-H_2$ | $-M$ | 0 |
| $H_2$ | 0 | 0 | 0 | $H_1$ | 0 | $-M$ |
| $H_6$ | 0 | $H_2$ | $-H_1$ | 0 | $H_8$ | $-H_7$ |
| $H_7$ | 0 | $M$ | 0 | $-H_8$ | 0 | 0 |
| $H_8$ | 0 | 0 | $M$ | $H_7$ | 0 | 0 |

The order of the determinant formed of this matrix is four; therefore $m = 2$. That is, there are four independent functions mutually in involution and two distinguished functions. The functions mutually in involution are $H$, $H_1$, $H_2$, $X$; and the distinguished functions are $H$ and $X$, where

$$X = MH_6 - H_2H_7 + H_1H_8.$$

Thus, the order of the differential equations is reduced to $4n - 8$. The order is reduced by eight, but there are only six known classical integrals; this difference is due to two integrals: the integral for the elimination of time, and the integral expressing the fact that the longitude of perihelion appears only in the difference of longitudes (Hagihara, 1944).

It is known that with cyclic coordinates, a Hamiltonian system possesses an integral common to the whole class of potential functions. Schmeidler (1963) obtained a general criterion for the existence of such an integral, on the basis of Lie's idea of transformation groups and function groups. He applied this criterion to problems of stellar dynamics and showed that the integral could not exist (see Chapter 14).

## 4.14 PARAMETRIZATION

The classical reduction of the differential equations for a dynamical system described in the foregoing sections can be carried out owing to the existence of a function group of the first integrals (Section 4.10). By using the Hamiltonian matrices (Sections 1.18–1.22) van Kampen and Wintner (1938) treated a more general problem — in which known functions or hypersurfaces in the phase space cannot be represented by first integrals — by substituting invariantive relations for the first integrals (Section 4.6). In particular, the known hypersurfaces need not constitute a continuous family but may be isolated. The restricted three-body problem can be reduced by substituting for these first integrals certain combinations of complicated invariantive-relations systems. This substitution is based on a distinct idea of the *parametrization* of an invariantive-relations system. The parametrization is symmetric with respect to the coordinates and momenta. Thus the reduction has a formal advantage when an invariantive relation is a first integral; it can be carried out by canonical transformations, as has been shown in Sections 1.18–1.22.

A matrix with $k$ rows and $l$ columns is called a $(k, l)$-matrix. The product $CD$ of a $(k, l)$-matrix $C$ and an $(l, m)$-matrix $D$ is a $(k, m)$-matrix. The transposed matrix of a $(k, l)$-matrix is an $(l, k)$-matrix. A $(k, 1)$-matrix is a $k$-vector. A $(k, l)$-matrix with all its $kl$ elements zero is denoted by $0_l^k$. A zero $k$-vector $0_1^k$ is written as $0^k$. Total differentiation with respect to $t$ is denoted by a dot; partial differentiation with respect to $t$ is denoted by a subscript $t$. If a scalar $c = c(Z, t)$ is a function of $t$ and a $k$-vector $Z = Z(t)$, then

$$\dot{c} = c_t + \dot{Z}' \operatorname{grad}_Z c,$$

where the prime denotes the transposed matrix.

Consider

$$G = \begin{pmatrix} 0_n^n & -E^n \\ E^n & 0_n^n \end{pmatrix}, \qquad G = -G' = -G^{-1},$$

where $E^n$ is a unit matrix with $n$ rows and columns, and a canonical dynamical system with $n$ degrees of freedom

$$G\dot{X} = \operatorname{grad}_X h, \qquad h = h(q_1, \ldots, q_n, p_1, \ldots, p_n, t), \tag{80}$$

where

$$X = (x_1, \ldots, x_n, x_{n+1}, \ldots, x_{2n}) = \begin{pmatrix} Q \\ P \end{pmatrix}, \tag{81}$$

$$Q = \begin{pmatrix} q_1 \\ \vdots \\ q_n \end{pmatrix} = \begin{pmatrix} x_1 \\ \vdots \\ x_n \end{pmatrix}, \qquad P = \begin{pmatrix} p_1 \\ \vdots \\ p_n \end{pmatrix} = \begin{pmatrix} x_{n+1} \\ \vdots \\ x_{2n} \end{pmatrix}.$$

Suppose that $K(X, t)$ is a $(2n - m)$-vector, where $m \leqq 2n$, in the $(2n + 1)$-dimensional $(X, t)$-domain; then $2n - m$ is called the *rank* of $\operatorname{grad}_X K$. Then

$$K = K(X, t) = 0^{2n-m} \tag{82}$$

determines a nonempty set $\Lambda(t)$ in the $2n$-dimensional $X$-space for a fixed $t$. Thus $\Lambda(t)$ is a locally $m$-dimensional manifold for a fixed $t$.

We choose a $2n$-vector $F = F(Y, t)$ dependent on $t$, and an $m$-vector $Y = (y_1, \ldots, y_m)$, such that

$$X = F(Y, t) \tag{83}$$

is a local parametrization of the manifold $\Lambda(t)$ and $m$ is the rank of a $(2n, m)$-matrix

$$J = J(Y, t) = \operatorname{grad}_Y F.$$

A curve

$$X(t) = F(Y(t), t) \tag{83a}$$

in the $2n$-dimensional phase space obeys the relation

$$\dot{x} = F_t + J\dot{Y}$$

(where $J = J(Y, t) = \operatorname{grad}_Y F$ is the Jacobian), as an identity in $t$. Thus, if every solution $X(t)$ of (80) that lies on $\Lambda(t)$ for a *single* $t$ lies on $\Lambda(t)$ for *every* $t$, then the $2n - m$ relation (82) is called an invariantive system of the differential equations (80). The parametrization (83) of (80) is called an invariantive system of (80). An invariantive system does not determine $F = F(Y, t)$ uniquely, since $m =$ rank of the Jacobian $J$ is the only restriction imposed on the choice of the parametrization (83).

If a given solution $X(t)$ satisfies (82), then there exists just one path $Y = Y(t)$, such that (83a) is satisfied in the space of the parameter vector $Y$. This unique $Y(t)$ satisfies

$$GJ\dot{Y} = -GF_t + \operatorname{grad}_X h \tag{84}$$

identically when (83) is substituted. Hence a unique vector $L = L(Y, t)$ in the $(m + 1)$-dimensional $(Y, t)$-space—defined so that

$$GJL = -GF_t + \operatorname{grad}_X h$$

is satisfied identically by (83) — belongs to this parametrization. Thus $Y(t)$ and $L(Y, t)$ are uniquely determined, and

$$\dot{Y} = L(Y, t). \tag{85}$$

If this system of equations has a solution $Y(t)$ such as (83a), then and only then is $X(t)$ the solution of the given system of equations satisfying the invariantive system (82).

### 4.15 SEPARATING PARAMETRIZATION

Define an $m$-vector $R$ and a scalar $r$ by

$$R = R(Y, t) = \tfrac{1}{2}J'GF, \qquad r = r(Y, t) = \tfrac{1}{2}F_t'GF; \tag{86}$$

then it can be proved (van Kampen and Wintner, 1938) that a matrix function $J'GJ$ in $(Y, t)$ is represented in the $(Y, t)$-domain by an alternating derivative, that is, the curl

$$J'GJ = (\text{grad}_Y R) - (\text{grad}_Y R)',$$

and a vector function $JGF_t$ in $(Y, t)$ is represented by

$$J'GF_t = R_t - \text{grad}_Y r.$$

Since the Jacobian $J$ is a $(2n, m)$-matrix, $J'GJ$ is a skew-symmetric $(m, m)$-matrix, and $2l$ is the rank of $J'GJ$ for an integer $l$ such that $2l \leqq m$. This integer $l$ is determined uniquely by the invariantive system (82), because the matrix $N = N(X, t) = \text{grad}_X K$ is a $(2n - m, 2n)$-matrix, and $NGN'$ is a skew-symmetric $(2n - m, 2n - m)$-matrix.

The $m$-vector $Y = (y_1, \ldots, y_m)$ can be split into a $2l$-vector $Y^0$ and an $(m - 2l)$-vector $Y^*$ such that

$$(y_1, \ldots, y_m) = Y = \begin{pmatrix} Y^0 \\ Y^* \end{pmatrix},$$

$$Y^0 = (y_1, \ldots, y_{2l}), \qquad Y^* = (y_{2l+1}, \ldots, y_m). \tag{87}$$

If there exists a $(2l, 2l)$-matrix $S^0 = S^0(Y, t)$ such that the $(m, m)$-matrix $J'GJ$ is represented in the form

$$J'GJ = \begin{pmatrix} S^0 & 0^{2l}_{m-2l} \\ 0^{m-2l}_{2l} & 0^{m-2l}_{2l} \end{pmatrix}, \tag{88}$$

then the parametrization (83) is called a *separating parametrization.*

If the parametrization (83) of the invariantive system (82) of the differential equation (80) is a separating parametrization, then the first $2l$ of the $m$ differential equations (85) are of the form

$$\dot{Y}^0 = L^0(Y^0, t); \tag{89}$$

that is, these $2l$ differential equations are separated from the remaining $m - 2l$ differential equations defining the components $Y^*$ of the $m$-vector $Y = (y_1, \ldots, y_{2l}, y_{2l+1}, \ldots, y_m)$. The differential equations are of the form

$$\dot{Y}^* = L^*(Y^*, Y^0, t).$$

When the solution $Y^0(t)$ of the equations (89) is known, the corresponding function $Y^*(t)$ is the solution of

$$\dot{Y}^* = L^*(Y^*, t), \tag{90}$$

where

$$L^*(Y^*, t) = L^*(Y^*, Y^0(t), t)$$

is a known function of $t$ and the $(m - 2l)$-vector $Y^*$. Thus the equation (85) is split into two separate systems (89) and (90). The $m$-vector $Y(t)$ determined by $Y^0(t)$ and $Y^*(t)$ gives the solution (83a) of equation (80). In the case of separating parametrization, the $2l$-vector $Y^*$ is determined by the system of $2l$ differential equations forming a non-degenerate Pfaffian dynamical system (Section 1.22) of $l$ degrees of freedom in the sense used by Birkhoff (1927).

If $S^0$ exists in (88), then $\det S^0 \neq 0$ and $S^0 = -(S^0)'$, since $J'GJ$ is skew-symmetric and $2l$ is its rank. In such a case $S^0$ can be represented by the alternating derivative

$$S^0(Y^0, t) \equiv S^0 = (\text{grad}_{Y^0} R^0) - (\text{grad}_{Y^0} R^0)',$$

and

$$J'GF_t = \text{grad}_Y h + \begin{pmatrix} R_t^0 & -\text{grad}_{Y^0} h^0 \\ 0^{m-2l} & 0^{m-2l} \end{pmatrix},$$

where

$$R^0 = R^0(Y^0, t), \qquad h^0 = h^0(Y^0, t)$$

are both $2l$-vectors. In the particular case of separating parametrization, the functions $J'GJ$ and $J'GF_t - \text{grad}_Y h$ (which are functions of $t$ and the $m$-vector $Y$) are independent of the $(m - 2l)$-vector $Y^*$ of (87).

It can be proved (van Kampen and Wintner, 1938) that there always exists a separating parametrization of an invariantive system. Since we know

$$J'GJ\dot{Y} + J'GF_t - \text{grad}_Y h = 0^m,$$

we obtain

$$[(\text{grad}_Y R) - (\text{grad}_Y R)']\dot{Y} + R_t - \text{grad}_Y (r + h) = 0^m, \qquad (91)$$

where $R$ and $r$ are given by (86) and

$$h = h(Y, t) = h(F(Y, t), t).$$

Let $\omega$ be an $(m + 1)$-ary Pfaffian

$$\omega = R'dY + (r + h)\, dt;$$

then the foregoing equation (91) is equivalent to the Euler-Lagrange equations belonging to the variational problem with fixed end points

$$\delta \int \omega \equiv \delta \int \{[R(Y, t)]'\, dY + [r(Y, t) + h(Y, t)]\, dt\} = 0.$$

This can be expressed by

$$\delta \int [-\tfrac{1}{2}Y^{0\prime}G^0\, dY^0 + h^0(Y^0, Y^*, t)\, dt] = 0,$$

where the Pfaffian $\omega$ is represented in the $(2l + 1)$-ary normal form

$$-\tfrac{1}{2}Y^{0\prime}G^0\, dY^0 + h^0\, dt, \tag{92}$$

and

$$G^0 = \begin{pmatrix} 0_l^l & -E_l \\ E_l & 0_l^l \end{pmatrix} = -(G^0) = -(G^0)^{-1}.$$

The Euler-Lagrange equation belonging to the new variational problem is

$$\begin{pmatrix} G^0 & 0_{m-2l}^{2l} \\ 0_{2l}^{m-2l} & 0_{m-2l}^{m-2l} \end{pmatrix}\begin{pmatrix} \dot{Y}^0 \\ \dot{Y}^* \end{pmatrix} - \begin{pmatrix} \operatorname{grad}_{Y^0} h^0 \\ \operatorname{grad}_{Y^*} h^0 \end{pmatrix} = 0^m.$$

Hence

$$J'GJ = \begin{pmatrix} G^0 & 0_{m-2l}^{2l} \\ 0_{2l}^{m-2l} & 0_{m-2l}^{m-2l} \end{pmatrix}.$$

Thus, the condition (88) characterizing the separating parametrization is satisfied.

A canonical parametrization is a separating parametrization for which the corresponding separated system of $2l$ differential equations — that is, the system (89) — is a canonical dynamical system of $l$ degrees of freedom; accordingly the equation

$$[(\operatorname{grad}_{Y^0} R^0) - (\operatorname{grad}_{Y^0} R^0)']\dot{Y}^0 = -R_t^0 + \operatorname{grad}_{Y^0} h^0$$

is of the particular form

$$G^0\dot{Y}^0 = \operatorname{grad}_{Y^0} h^0, \tag{93}$$

where

$$\operatorname{grad}_{Y^*} h^0 = 0^{m-2l}.$$

Thus, the invariantive system (82) of equations (80) always admits a canonical parametrization. The construction of such canonical parametrization of a given invariantive system depends on the point transformation that transforms the Pfaffian $\omega$ into its normal form (92). This point transformation is determined by the solution of a complete system of partial differential equations (see Section 4.9). In general, the solution can be obtained not by mere quadrature but by successive approximations.

## 4.16 FIRST INTEGRALS

A scalar function $g = g(X, t)$ is a first integral of the system (80), if and only if $\dot{g} = 0$ is satisfied identically by this system of differential equations. If the equation

$$K(X, t) = C \tag{94}$$

with a constant $(2n - m)$-vector $C$ is an invariantive system for any $C$ in the $(2n - m)$-dimensional domain, then $\dot{K} = 0^{2n-m}$ is an identity because of the system of differential equations. When the invariantive system consists of only first integrals, if the $2n - m$ first integrals represented by (94), which are mutually independent, form a function group (see Section 4.10) in the sense of Lie, then this function group contains $n - l$ independent first integrals in involution and nothing more (see Section 4.5; also see van Kampen and Wintner, 1938; Williamson, 1940). Denote by $k$ the rank of $NGN'$, where $N = \text{grad}_X K$. It can be proved that, if (94) consists of $2n - m$ first integrals of the canonical system (80) of $n$ degrees of freedom, and if these first integrals are independent and form a function group, then the maximum number $j$ of the first integrals in involution that belong to the function group is equal to $2n - m - k$ (see Sections 4.10, 4.11).

Consider the three-body problem in which $\xi_i$, $\eta_i$, $\zeta_i$ $(i = 1, 2, 3)$ are the rectangular coordinates of the mass $\mu_i$ referred to the inertial system, and $\Xi_i$, $\text{H}_i$, $\text{Z}_i$ are the conjugate momenta $\mu_i\dot{\xi}_i$, $\mu_i\dot{\eta}_i$, $\mu_i\dot{\zeta}_i$. The system of differential equations of motion admits, besides the energy integral, nine classical integrals

$$f_1 = \sum_{i=1}^{3} (\text{H}_i\zeta_i - \text{Z}_i\eta_i), \qquad f_2 = \sum_{i=1}^{3} (\text{Z}_i\xi_i - \Xi_i\zeta_i),$$

$$f_3 = \sum_{i=1}^{3} (\Xi_i\eta_i - \text{H}_i\xi_i), \qquad f_4 = \sum_{i=1}^{3} \Xi_i,$$

$$f_5 = \sum_{i=1}^{3} \text{H}_i, \qquad f_6 = \sum_{i=1}^{3} \text{Z}_i,$$

$$f_7 = \sum_{i=1}^{3} (\mu_i\xi_i - t\Xi_i), \qquad f_8 = \sum_{i=1}^{3} (\mu_i\eta_i - t\text{H}_i),$$

$$f_9 = \sum_{i=1}^{3} (\mu_i\zeta_i - t\text{Z}_i).$$

The $(2n - m)$-vector $K$ defines a 9-vector with components $f_1, \ldots, f_9$. Thus $n = 9$, $m = 9$; hence $l = k$, $j = 9 - k$, where $k = 4$. Hence $l = 4$ and $j = 5$; these are respectively the degree of freedom and the maximum number of the first integrals in involution contained in the function group generated by $f_1, \ldots, f_9$ of the separated system (93). The Poisson brackets (see Sections 4.12, 4.13) are represented by

$$(f_1, f_2) = f_3, \qquad (f_2, f_4) = -f_6, \qquad (f_3, f_4) = f_5,$$

$$(f_2, f_7) = -f_9, \qquad (f_3, f_7) = f_8, \qquad (f_4, f_7) = \sum_{i=1}^{3} \mu_i.$$

Thus,

$$\|(f_\rho, f_\sigma)\| = \begin{pmatrix} F_1 & F_2 & F_3 \\ F_2 & 0 & \sum_{i=1}^{3} \mu_i E^3 \\ F_3 & -\sum_{i=1}^{3} \mu_i E^3 & O \end{pmatrix},$$

where

$$F_i = \begin{pmatrix} O & f_{3i} & -f_{3i-1} \\ -f_{3i} & O & f_{3i-2} \\ f_{3i-1} & -f_{3i-2} & O \end{pmatrix} \qquad (i = 1, 2, 3).$$

In the plane problem, $\zeta_i = Z_i = 0$ $(i = 1, 2, 3)$. Hence we have five integrals $f_3, f_4, f_5, f_7, f_8$. Thus, $n = 6$, $2n - m = 5$, and $l = k + 1$, $j = 5 - k$.

The 5-rowed skew-symmetric matrix is of rank 4; and $k = 2$, $l = 3$, $j = 3$.

## INTEGRAL INVARIANTS

### 4.17 INTEGRALS AND INTEGRAL INVARIANTS

Suppose that, in the system of differential equations

$$\frac{dx_1}{X_1} = \frac{dx_2}{X_2} = \cdots = \frac{dx_n}{X_n} = dt, \tag{95}$$

the functions $X_i$, together with their derivatives, are single-valued and continuous, and do not contain $t$ explicitly. The trajectory $x_i = f_i(t)$ $(i = 1, 2, \ldots, n)$ in the $n$-dimensional space $(x_1, x_2, \ldots, x_n)$ represents the solution of (95) that passes through $x_i = x_i^0$ $(i = 1, 2, \ldots, n)$ at $t = t_0$. Take a subspace $E_p^{(0)}$ of $p$ dimensions in this space and suppose that the trajectory that was in this subspace $E_p^{(0)}$ at $t = t_0$ is transferred to a $p$-dimensional subspace $E_p$ at time $t$. A homogeneous expression of degree $p$ with regard to the differentials

$$\omega = \sum_{\alpha_1, \alpha_2, \cdots, \alpha_p} A_{\alpha_1 \alpha_2 \cdots \alpha_p} \, dx_{\alpha_1} \, dx_{\alpha_2} \cdots dx_{\alpha_p} \tag{96}$$

is called a *Pfaffian* of degree $p$ (Section 1.7). If the integral $I_p = \int \omega$ taken over this Pfaffian has the same value for any $E_p$ and $E_p^{(0)}$, then $I_p$ is called an *integral invariant* of degree $p$ (Section 1.2). Here an integral invariant is meant in the sense used by Poincaré. The index $p$ can take any value between 1 and $n$ (Poincaré, 1889, 1899; Weber, 1900; Goursat, 1908, 1912, 1915, 1921; de Donder, 1901, 1902, 1927; Cartan, 1922; Schouten and van der Kulk, 1949).

It is evident from this definition that an integral invariant remains an integral invariant after any transformation of the variables. Let $f_1, f_2, \ldots, f_{n-1}$ be $n - 1$ independent integrals of

$$X(f) = X_1 \frac{\partial f}{\partial x_1} + X_2 \frac{\partial f}{\partial x_2} + \cdots + X_n \frac{\partial f}{\partial x_n} = 0.$$

Then we have $n - 1$ integrals of (95): $f_i = c_i$ $(i = 1, 2, \ldots, n - 1)$ and $t = f_n + c_n$, where the $c_i$ are constants and $f_i$ depend only on $x_1, x_2, \ldots, x_n$. Take new variables $y_1 = f_1, y_2 = f_2, \ldots, y_{n-1} = f_{n-1}$, $z = f_n$; then (95) is transformed into the system

$$\frac{dy_1}{0} = \frac{dy_2}{0} = \cdots = \frac{dy_{n-1}}{0} = \frac{dz}{1} = dt, \tag{97}$$

whose general integrals are

$$y_1 = c_1, \qquad y_2 = c_2, \ldots, \qquad y_{n-1} = c_{n-1}, \qquad z = t + c_n.$$

The original Pfaffian $\omega$ is changed to a Pfaffian $\omega$ of $dy_1, dy_2, \ldots, dy_{n-1}$, $dz$; the coefficients $A$ of the differentials in $\omega$ are now functions of $y_1, y_2, \ldots, y_{n-1}, z$. Hence, an integral invariant $I_p = \int \omega$ is transformed to an integral invariant $\int \varpi$ with regard to (97). If we increase $z$ by $t - t_0$ by keeping $y_1, y_2, \ldots, y_{n-1}$ unaltered in the subspace $E_p^{(0)}$, then the points in $E_p^{(0)}$ are displaced to the points in $E_p$. In order that $\int \varpi$ may not depend on $t$—that is, in order that $\int \varpi$ may be an integral invariant with regard to (97)—it is necessary and sufficient that $\varpi$ should be invariant under the change of variable from $z$ to $z + t$ in $\varpi$, in other words, that the coefficients of $\varpi$ should not depend on $z$. But the system (97) does not alter its form even if we take $n - 1$ arbitrary independent functions of $y_1, y_2, \ldots, y_{n-1}$ as our new variables. Hence, there must be an infinite number of such integral invariants of degree $p$.

As a corollary, consider $f_1, f_2, \ldots, f_r$ $(r < n)$, which are integrals of (95) independent of $t$, and a Pfaffian $\omega$ of degree $p$, in which the coefficients of $df_1, df_2, \ldots, df_r$ are functions of $f_1, f_2, \ldots, f_r$ only. Then $\int \omega$ is an integral invariant of degree $r$. If, for example, $f$ is an integral of (97), then $\int df$ is an integral invariant. Further, suppose that

$$\int \Big(\sum_{i,j} B_{ij}\, dx_i\, dx_j\Big)^{1/2}$$

becomes

$$\int \Big(\sum_{i,j} B'_{ij}\, dy_i\, dy_j\Big)^{1/2}$$

by this transformation and that $B'_{ij}$ do not depend on $z$. Then this is an integral invariant. Let $\int \omega_p$ and $\int \omega_q$ be two integral invariants with regard to (95). In the transformation from (95) to (97), the coefficients in these integral invariants do not contain $z$. Hence, the situation is the same for the product, and $\int \omega_p \omega_q$ is also an integral invariant.

Denote the differential of (96) formally by

$$\omega' = \sum dA_{\alpha_1\alpha_2\cdots\alpha_p}\, dx_{\alpha_1}\, dx_{\alpha_2} \ldots dx_{\alpha_p}$$

and call it a differential form. Put it in another form

$$\omega' = \sum B_{\alpha_1\alpha_2\cdots\alpha_p\alpha_{p+1}}\, dx_{\alpha_1}\, dx_{\alpha_2} \cdots dx_{\alpha_p}\, dx_{\alpha_{p+1}}.$$

If $p$ is even, then

$$B_{\alpha_1\alpha_2\cdots\alpha_p} = \frac{\partial A_{\alpha_1\alpha_2\cdots\alpha_p}}{\partial x_{\alpha_{p+1}}} + \frac{\partial A_{\alpha_2\cdots\alpha_{p+1}}}{\partial x_{\alpha_1}} + \cdots + \frac{\partial A_{\alpha_{p+1}\alpha_1\cdots\alpha_{p-1}}}{\partial x_{\alpha_p}}, \tag{98}$$

and if $p$ is odd, then

$$B_{\alpha_1\cdots\alpha_{p+1}} = \frac{\partial A_{\alpha_1\alpha_2\cdots\alpha_p}}{\partial x_{\alpha_{p+1}}} - \frac{\partial A_{\alpha_2\cdots\alpha_{p+1}}}{\partial x_{\alpha_1}} + \cdots - \frac{\partial A_{\alpha_{p+1}\alpha_1\cdots\alpha_{p-1}}}{\partial x_{\alpha_p}}. \tag{99}$$

In particular, if $\omega$ is a Pfaffian of degree one,

$$\omega = A_1\, dx_1 + A_2\, dx_2 + \cdots + A_n\, dx_n,$$

then

$$\omega' = \sum_{i,j} \left(\frac{\partial A_i}{\partial x_j} - \frac{\partial A_j}{\partial x_i}\right) dx_i\, dx_j.$$

Exchanging $dx_i\,dx_j$ with $dx_j\,\delta x_i - \delta x_j\,dx_i$ gives the bilinear covariant of Section 1.4,

$$\omega' = \sum_{i,j} \left(\frac{\partial A_i}{\partial x_j} - \frac{\partial A_j}{\partial x_i}\right)(dx_j\,\delta x_i - \delta x_j\,dx_i).$$

By the above theorem, it is necessary and sufficient that $\partial A_i/\partial x_j - \partial A_j/\partial x_i$ should be independent of $z$ for this expression $\omega'$ to be an integral invariant.

Let the points on a trajectory, which starts at time $t_0$ from the subspace $E_{p+1}^{(0)}$, embedded in an $n$-dimensional space, occupy the subspace $E_{p+1}$ at time $t$. The subspaces $E_{p+1}^{(0)}$ and $E_{p+1}$ are bounded by the subspaces $E_p^{(0)}$ and $E_p$, respectively. By a generalization of Stokes's theorem, the integrals $\int \omega'$ extended over $E_{p+1}^{(0)}$ and $E_{p+1}$ are equal to $\int \omega$ extended over $E_p^{(0)}$ and $E_p$, respectively. If $\int \omega$ is an integral invariant, then the integrals $\int \omega$ extended over $E_p^{(0)}$ and $E_p$ have the same value. Hence, the integrals $\int \omega'$ extended over $E_{p+1}^{(0)}$ and $E_{p+1}$ have the same value; that is, they are integral invariants. Thus, a new integral invariant $I' = \int \omega'$ is derived from an integral invariant with regard to (95).

The integral $\int \omega$ extended over $E_p$ is called an *absolute integral invariant.* When and only when $E_p$ is a closed space, an integral invariant over it is said to be a *relative integral invariant* and written as $J$. The above integrals $\int \omega'$, extended over the spaces $E_{p+1}^{(0)}$ and $E_{p+1}$, respectively, are equal to $\int \omega$ extended over the closed subspaces $E_p^{(0)}$ and $E_p$ bounding $E_{p+1}^{(0)}$ and $E_{p+1}$, respectively. Since $\int \omega$ is a relative integral invariant by definition, $\int \omega'$ is an absolute integral invariant. The necessary and sufficient condition that $\int \omega$ be a relative but not an absolute integral invariant is that $\int \omega'$ should be an absolute integral invariant. Conversely, if $\int \omega'$ is an absolute integral invariant, then $\int \omega$ is a relative or an absolute integral invariant. A perfect differential is zero when integrated over a closed space; therefore, an integral of the sum of the integrand $\omega$ of a relative integral invariant and any perfect differential of the same degree extended over a closed space is also a relative integral invariant.

### 4.18 CONDITION FOR AN INTEGRAL INVARIANT

Let us find the condition for (96) to be an absolute integral invariant with regard to (95). Denote the integral $\int \omega$ extended over $E_p$ by $I_p(t)$. It is required that $dI_p/dt = 0$ in order that $\int \omega$ shall be an integral invariant. As $t$ increases by $\delta t$, the variable $x_i$ increases to

$$x_i' = x_i + X_i(x_1, \ldots, x_n)\,\delta t \qquad (i = 1, 2, \ldots, n).$$

By taking $p$ independent variables $u_1, u_2, \ldots, u_p$ we can write

$$I_p(t) = \int \sum A_{\alpha_1 \cdots \alpha_p} \frac{\partial x_{\alpha_1}}{\partial u_1} \cdots \frac{\partial x_{\alpha_p}}{\partial u_p} du_1 \cdots du_p,$$

$$I_p(t + \delta t) = \int \sum A'_{\alpha_1 \cdots \alpha_p} \frac{\partial x'_{\alpha_1}}{\partial u_1} \cdots \frac{\partial x'_{\alpha_p}}{\partial u_p} du_1 \cdots du_p.$$

$A'$ is the result of substituting $x'_i$ for $x_i$ in $A_i$. We also have

$$A'_{\beta_1 \cdots \beta_p} = A_{\beta_1 \cdots \beta_p} + X(A_{\beta_1 \cdots \beta_p})\, \delta t,$$

$$\frac{\partial x'_\alpha}{\partial u_j} = \frac{\partial x_\alpha}{\partial u_j} + \delta t \sum_i \frac{\partial X_\alpha}{\partial x_i} \frac{\partial x_i}{\partial u_j} \qquad (\alpha, j = 1, \ldots, p),$$

where

$$X(f) = X_1 \frac{\partial f}{\partial x_1} + X_2 \frac{\partial f}{\partial x_2} + \cdots + X_n \frac{\partial f}{\partial x_n}.$$

Hence, the coefficient of

$$\frac{\partial x_{\alpha_1}}{\partial u_1} \cdots \frac{\partial x_{\alpha_p}}{\partial u_p} \delta t$$

in $I_p(t + \delta t)$ is

$$X(A_{\alpha_1 \cdots \alpha_p}) + \sum_i \left( A_{i\alpha_2 \cdots \alpha_p} \frac{\partial X_i}{\partial x_{\alpha_1}} + A_{\alpha_1 i \cdots \alpha_p} \frac{\partial X_i}{\partial x_{\alpha_2}} + \cdots \right.$$
$$\left. + A_{\alpha_1 \cdots \alpha_{p-1} i} \frac{\partial X_i}{\partial x_{\alpha_p}} \right). \qquad (100)$$

In order that $I_p(t + \delta t) - I_p(t)$ shall be zero, it is necessary and sufficient that the quantity (100) be zero for any combination of the indices $\alpha_1, \ldots, \alpha_p$. In order that $\int \omega$ shall be a relative integral invariant with regard to (95), it is necessary that the expression (100) integrated over a closed space of $p$ dimensions be zero.

From the integral invariant of order $n$,

$$I_n = \int M(x_1, x_2, \ldots, x_n)\, dx_1\, dx_2 \cdots dx_n, \qquad (101)$$

we form the expression

$$X(M) + \sum_{i=1}^{n} M \frac{\partial X_i}{\partial x_i} = 0,$$

or

$$\frac{\partial}{\partial x_1}(X_1 M) + \frac{\partial}{\partial x_2}(X_2 M) + \cdots + \frac{\partial}{\partial x_n}(X_n M) = 0. \qquad (102)$$

Hence, for (101) to be an integral invariant with regard to (95) it is necessary and sufficient for $M$ to be the *last multiplier* of the differential equations (95). That is, (95) becomes a perfect differential when multiplied by $M$ on the left-hand side (Poincaré, 1899).

If

$$\int M\, dx_1\, dx_2 \cdots dx_n, \qquad \int M'\, dx_1\, dx_2 \cdots dx_n,$$

are two integral invariants, then $M/M'$ is an integral of (95). For, we obtain

$$\frac{\partial}{\partial x_1}\left(X_1 \frac{M}{M'}\right) + \frac{\partial}{\partial x_2}\left(X_2 \frac{M}{M'}\right) + \cdots + \frac{\partial}{\partial x_n}\left(X_n \frac{M}{M'}\right) = 0$$

from

$$\frac{\partial}{\partial x_1}(X_1 M) + \cdots + \frac{\partial}{\partial x_n}(X_n M) = 0,$$

$$\frac{\partial}{\partial x_1}(X_1 M') + \cdots + \frac{\partial}{\partial x_n}(X_n M') = 0.$$

In particular, if $\sum_{i=1}^{n} \partial X_i/\partial x_i = 0$, then letting $M = 1$ gives the $n$-dimensional volume $\int dx_1\, dx_2 \cdots dx_n$ as an integral invariant. A dynamical system with such a relation is called a *Liouville dynamical system.* Since the volume is invariant, the dynamical system is represented by an incompressible fluid motion

$$\sum_{i=1}^{n} \frac{\partial X_i}{\partial x_i} = \operatorname{div} \mathbf{X} = 0,$$

where $\mathbf{X}$ is a vector with components $X_1, X_2, \ldots, X_n$.

When $X_i$ does not contain $t$ in the Pfaffian of order $n - 1$, we have

$$\Omega_{n-1} = X_1\, dx_2\, dx_3 \cdots dx_n \pm X_2\, dx_3 \cdots dx_n\, dx_1 + \cdots$$
$$\pm X_n\, dx_1\, dx_2 \cdots dx_{n-1},$$

where the plus sign is taken for $n$ odd and the minus sign for $n$ even. An integral of $X(f) = 0$ is also an integral of $\Omega_{n-1}\, df = 0$, since

$$\Omega_{n-1}\, df = \pm X(f)\, dx_1\, dx_2 \cdots dx_n.$$

Conversely, an integral of $\Omega_{n-1}\, df = 0$ is also an integral of $X(f) = 0$. Suppose that transforming from (95) to (97) gives

$$\Omega_{n-1} = K\, dy_1\, dy_2 \cdots dy_{n-1}.$$

This occurs when and only when $MK$ does not depend on $z$, or when $\int M\Omega_{n-1}$ is an integral invariant if $M\Omega_{n-1}$ is a differential; that is, $\int M\Omega_{n-1}$ is an integral invariant if and only if

$$\frac{\partial(MX_1)}{\partial x_1} + \frac{\partial(MX_2)}{\partial x_2} + \cdots + \frac{\partial(MX_n)}{\partial x_n} = 0.$$

Hence, we can obtain an integral invariant $\int M\Omega_{n-1}$ from the last multiplier of (95). Here it is assumed that $M$ and $\Omega_{n-1}$ contain neither $z$ nor $dz$ after the transformation.

Conversely, by constructing an expression analogous to (100) from

$$\omega_{n-1} = A_1\,dx_2\,dx_3\cdots dx_n \pm A_2\,dx_3\cdots dx_n\,dx_1 + \cdots \pm A_n\,dx_1\,dx_2\cdots dx_{n-1},$$

we obtain

$$\begin{aligned}
&X(A_1)\,dx_2\cdots dx_n \pm \cdots \pm X(A_n)\,dx_1\cdots dx_{n-1}\\
&+ A_1[dX_2\,dx_3\cdots dx_n + dx_2\,dX_3\cdots dx_n + \cdots]\\
&\pm A_2[dX_3\cdots dx_n\,dx_1 + dx_3\,dX_4\cdots dx_1 + \cdots]\\
&\vdots\\
&\pm A_n[dX_1\,dx_2\cdots dx_{n-1} + dx_1\,dX_2\cdots dx_{n-1} + \cdots],
\end{aligned}$$

where the coefficient of $dx_2\,dx_3\cdots dx_n$, for example, is

$$X(A_1) - A(X_1) + A_1H,$$

with

$$A = \sum_{i=1}^{n} A_i \frac{\partial}{\partial x_i}, \qquad H = \sum_{i=1}^{n} \frac{\partial X_i}{\partial x_i}.$$

Letting all coefficients equal zero, we see that the necessary and sufficient condition for $\int \omega_{n-1}$ to be an integral invariant is that

$$X(A(f)) - A(X(f)) + HA(f) = 0 \tag{103}$$

for any function $f$.

Furthermore,

$$X(f) = 0, \qquad A(f) + Mf = 0 \qquad \text{with} \qquad M = \sum_{i=1}^{n} \frac{\partial A_i}{\partial x_i}$$

form a system in involution, since the necessary and sufficient condition is that

$$X(A(f) + Mf) - A(X(f)) - MX(f) = 0$$

should be a consequence of $X(f) = 0$, $A(f) + Mf = 0$. From (103), this becomes

$$X(M)f - HA(f) = [X(M) + HM]f = 0.$$

The relation $X(M) + HM = 0$ expresses that $M$ is a last multiplier, since the derived form of $\omega_{n-1}$ is $M\,dx_1\,dx_2 \cdots dx_n$.

The system $X(f) = 0$, $A(f) + Mf = 0$ is obtained by finding those integrals of $X(f) = 0$ that are also the multipliers for $\omega_{n-1}$. We have

$$(\omega_{n-1}f)' = \omega'_{n-1}f + \omega_{n-1}\,df = [A(f) + Mf]\,dx_1 \cdots dx_n.$$

We obtain the equations $X(f) = 0$, $A(f) = 0$ at the same time by finding those integrals of $X(f) = 0$ that satisfy also the relation $\omega_{n-1}\,df = 0$. Hence, if the integral $\int \omega_{n-1}$ is an integral invariant, the linear equations $(\omega_{n-1}f)' = 0$, $X(f) = 0$ form a system in involution, and consequently the equations $\omega_{n-1}\,df = 0$, $X(f) = 0$ form a complete system of linear homogeneous equations.

## 4.19 INTEGRAL INVARIANTS ALONG TRAJECTORIES

Write equations (95) in the form $dx_i = X_i\,dt$ $(i = 1, 2, \ldots, n)$, where $X_i$ contains $x_1, x_2, \ldots, x_n$ but not $t$. Consider the second system

$$dx_i = \lambda X_i\,dt \qquad (i = 1, 2, \ldots, n),$$

where $\lambda$ is a function of $x_1, x_2, \ldots, x_n$. The integrals $f_1, f_2, \ldots, f_{n-1}$ of the first system are also the integrals of the second system. Only the $n$th integral $f_n$ contains $t$. Hence, the $n$th integral of the second system contains $\lambda$. The trajectory has the same form in both systems, but there is a difference in the way it is described in terms of $t$. If the coefficients in a Pfaffian $\omega$ of order $p$ with respect to $df_1, df_2, \ldots, df_{n-1}$ are functions of $f_1, f_2, \ldots, f_{n-1}$ only, then $\int \omega$ is an integral invariant whatever $\lambda$ may be. Such an integral invariant is called an *integral invariant along the trajectory* and denoted by $I_p^l$. We have seen that the necessary and sufficient condition for $\int \Omega$ to be an integral invariant in the transformation from (95) to (97) is that the coefficients of $\Omega$ do not depend on $z$. Also, whatever the factor $\lambda$ is, $\Omega$ must be independent of $z$ and $dz$ for $\int \Omega$ to be an integral invariant. That is, $\Omega$ must contain neither $z$ nor $dz$, for $\int \Omega$ to be an integral invariant along the trajectory.

In order that an integral invariant $\int \Omega$ with regard to (95) shall be also an integral invariant along the trajectory, it is necessary and sufficient that $\int \Omega$ should still be an integral invariant even if we write $\lambda X_i$ for $X_i$, whatever may be the value of $\lambda$. Write $\lambda X_i$ for $X_i$ in the condition (100).

In order that this rewritten condition hold for any function $\lambda$, it is necessary and sufficient to have

$$\sum_{i=1}^{n} A_{\alpha_1 \cdots \alpha_{p-1} i} X_i = 0 \tag{104}$$

for all combinations of $p - 1$ indices taken arbitrarily out of the $n$ indices. When we take $dx_{\alpha_1} \cdots dx_{\alpha_{p-1}}$ out of each term of

$$A_{\alpha_1 \cdots \alpha_{p-1} i} \, dx_{\alpha_1} \cdots dx_{\alpha_{p-1}} \, dx_i,$$

we obtain a Pfaffian of order one

$$\omega_{\alpha_1 \cdots \alpha_{p-1}} = \sum_{i=1}^{n} A_{\alpha_1 \cdots \alpha_{p-1} i} \, dx_i.$$

By taking all combinations of $p - 1$ out of the $n$ indices we obtain the Pfaffian of order one associated with

$$\Omega = \sum A_{\alpha_1 \cdots \alpha_p} \, dx_{\alpha_1} \cdots dx_{\alpha_p}.$$

When we substitute (95) in these last two expressions, they are reduced to zero by reason of (104); that is, such equations are satisfied by a function of one variable satisfying $n - 1$ integrals $f_i = c_i$ $(i = 1, 2, \ldots, n - 1)$, of the system (95). Hence, $\omega_{\alpha_1 \cdots \alpha_{p-1}} = 0$ is a result of

$$df_1 = 0, \ldots, df_{n-1} = 0,$$

and the Pfaffian of order one associated with $\Omega$ is a linear homogeneous expression with respect to $df_1, \ldots, df_{n-1}$. Also, (104) is the condition for an integral invariant along the trajectory, and (100) gives the condition for an integral invariant in general. It can be proved, as in the general case, that if $\int \Omega$ is an integral invariant along the trajectory then so is $\int \Omega'$.

By applying Poincaré's method we can obtain an integral invariant of order $p - 1$ along the trajectory from a general integral invariant of order $p$ not necessarily along the trajectory. Let an integral invariant in general, not along the trajectory, be

$$I_p = \int \omega_p = \int \sum A_{\alpha_1 \cdots \alpha_p} \, dx_{\alpha_1} \cdots dx_{\alpha_p},$$

and put

$$C_{\alpha_1 \cdots \alpha_{p-1}} = \sum_{i=1}^{n} A_{\alpha_1 \cdots \alpha_{p-1} i} X_i$$

then it does not necessarily follow that all $C_{\alpha_1 \cdots \alpha_{p-1}}$ are zero. If we write

$$\omega_{p-1} = \sum C_{\alpha_1 \cdots \alpha_{p-1}}\, dx_{\alpha_1} \cdots dx_{\alpha_{p-1}},$$

then $\int \omega_{p-1}$ is an integral invariant along the trajectory. That this is an integral invariant follows from the fact that the expression (100) formed for $\omega_{p-1}$,

$$X(C_{\alpha_1 \cdots \alpha_{p-1}}) + \sum_{i=1}^{n} \left( C_{i\alpha_2 \cdots \alpha_{p-1}} \frac{\partial X_i}{\partial x_{\alpha_1}} + \cdots + C_{\alpha_1 \cdots \alpha_{p-2} i} \frac{\partial X_i}{\partial x_{\alpha_{p-1}}} \right) = 0$$

is the result of equating the expression (100) formed for $\omega_p$ to zero. It is

$$X(C_{\alpha_1 \cdots \alpha_{p-1}}) = \sum_{i} X_i X(A_{\alpha_1 \cdots \alpha_{p-1} i}) + \sum_{i} \sum_{j} A_{\alpha_1 \cdots \alpha_{p-1} i} X_j \frac{\partial X_i}{\partial x_j}.$$

This can be proved if we substitute in $X(A_{\alpha_1 \cdots \alpha_{p-1} i})$ the expression obtained by equating the expression (100) formed for $\omega_p$ to zero, and if at the same time we substitute in $C_{\alpha_1 \cdots \alpha_{p-1}}$ the expression in terms of such $A_{\alpha_1 \cdots \alpha_{p-1} i}$. Hence, $\int \omega_{p-1}$ is an integral invariant. That it is also an integral invariant along the trajectory can be proved by the fact that the sum in

$$\sum_{i=1}^{n} C_{\alpha_1 \cdots \alpha_{p-2} i} X_i = \sum_{i} \sum_{j} A_{\alpha_1 \cdots \alpha_{p-2} ji} X_i X_j$$

becomes zero by exchanging $i$ and $j$ on the right-hand side, and equation (104) is then satisfied.

De Donder (1901) applied the theory of integral invariants to the theory of turbulence.

## 4.20 VARIATIONAL EQUATIONS

Let $x_i = x_i^{(0)}(t)$ $(i = 1, 2, \ldots, n)$ be a particular solution of the system of differential equations

$$\frac{dx_i}{dt} = X_i \qquad (i = 1, 2, \ldots, n). \tag{105}$$

Put $x_i = x_i^{(0)} + \xi_i$ $(i = 1, 2, \ldots, n)$, and substitute in (105); then we have the differential equations for the variation $\xi_i$. The system of differential equations obtained by neglecting higher-order terms with regard to $\xi_i$,

$$\frac{d\xi_i}{dt} = \sum_{j=1}^{n} \left.\frac{\partial X_i}{\partial x_j}\right|_{x=x^{(0)}} \xi_j \qquad (i = 1, 2, \ldots, n), \tag{106}$$

are called the *variational equations*. Let a solution of (105) be

$$x_i = \phi_i(t, c_1, c_2, \ldots, c_p),$$

and call a solution corresponding to $x_i = x_i^{(0)}$ with

$$c_1 = c_2 = \cdots = c_p = 0$$

the *generating solution*; then (106) has particular solutions

$$\xi_i = \frac{\partial \phi_i}{\partial c_j} \qquad (i = 1, 2, \ldots, n; j = 1, 2, \ldots, p).$$

Thus, when we know solutions of (105) containing integration constants, we can obtain a particular solution of (106). If we know an integral of (105), $F(x_1, x_2, \ldots, x_n) =$ constant, then we have the integrals

$$F[\phi_1(t), \phi_2(t), \ldots, \phi_n(t)] = \text{constant}$$

for (105), and

$$F[\phi_1(t) + \xi_1, \phi_2(t) + \xi_2, \ldots, \phi_n(t) + \xi_n] = \text{constant}$$

for (106). From these two integrals we obtain an integral of (106)

$$\left.\frac{\partial F}{\partial x_1}\right|_{x=x^{(0)}} \xi_1 + \cdots + \left.\frac{\partial F}{\partial x_n}\right|_{x=x^{(0)}} \xi_n = \text{constant},$$

where $\partial F/\partial x_i|_{x=x^{(0)}}$ is the result of substituting $x_i^{(0)} = \phi_i(t)$ after the differentiation.

Poincaré (1899) has found a close connection between an integral of (106) and an integral invariant with regard to (105) in general. Let $F(\xi_1, \xi_2, \ldots, \xi_n) =$ constant be an integral of (106). It is a homogeneous function with respect to $\xi_i$ and depends on $x_i$. $F$ can be considered linear homogeneous with respect to $\xi_i$; if not, we can still construct a linear homogeneous expression by forming a power of expression $F$ with a suitable index. Thus,

$$\int F(dx_1, dx_2, \ldots, dx_n) \tag{107}$$

is an integral invariant with regard to (105). In fact, the integrand is supposed to be infinitely small and linear homogeneous with respect to

$dx_i$. Let $F_0$ be the infinitely short arc of a curve with the values of $(x_1, x_2, \ldots, x_n)$ and $(x_1 + \xi_1, x_2 + \xi_2, \ldots, x_n + \xi_n)$ at $t = 0$ as the coordinates of the two ends, then the integral (107) extended over this short arc is nothing but $F(\xi_1, \xi_2, \ldots, \xi_n)$. Since this expression is an integral, its values on $F_0$ and on a curve $F$ consisting of the points at time $t$ should be equal. If $F_0$ and $F$ are both finite, we divide these curves into infinitely short arcs and extend (107) over the corresponding arcs on each of the curves. Then the integral (107) extended over an arc of $F$ equals the integral (107) extended over the corresponding arc of $F_0$. Hence, (107) is an integral invariant.

Conversely, if (107) is an integral invariant of order one, then $F(\xi_1, \xi_2, \ldots, \xi_n)$ is an integral of (106). Indeed, the integral (107), being of equal values for $F_0$ and for $F$, has the same value $F(\xi_1, \xi_2, \ldots, \xi_n)$ when the ends of $F_0$ are $x_i$ and $x_i + \xi_i$. But, as this is an integral invariant, it is a constant. Hence, it is an integral of (106).

Now we extend this notion to the integral invariants of order two. Take two particular solutions $\xi_i$ and $\xi_i'$ $(i = 1, 2, \ldots, n)$ of the variational equations (106). There exists a function $F(x_i, \xi_i, \xi_i')$ that is independent of $t$ whatever the two particular solutions are; in other words, $F$ is an integral of the systems of differential equations

$$\frac{d\xi_j}{dt} = \sum_{i=1}^{n} \left.\frac{\partial X_j}{\partial x_i}\right|_{x=x^{(0)}} \xi_i, \qquad \frac{d\xi_j'}{dt} = \sum_{i=1}^{n} \left.\frac{\partial X_j}{\partial x_i}\right|_{x=x^{(0)}} \xi_i', \tag{108}$$

for $\xi_i$ and $\xi_i'$, respectively. In particular, if we take $F$ to be of the form

$$\sum_{i,j} A_{ij}(\xi_i\xi_j' - \xi_j\xi_i'),$$

where $A_{ij}$ are functions of $x$ only, then the double integral

$$I = \iint \sum A_{ij}\, dx_i\, dx_j$$

is an integral invariant with regard to (105).

To prove this theorem, construct for $F_0$ an infinitely small parallelogram with its vertices at the points $x_i$, $x_i + \xi_i$, $x_i + \xi_i'$, $x_i + \xi_i + \xi_i'$ at time $t = 0$. For $F$ we construct another infinitely small parallelogram with its vertices at the same coordinates at time $t$. Then $I$ is equal to

$$\sum_{i,j} A_{ij}(\xi_i\xi_j' - \xi_j\xi_i')$$

and, being an integral of (108) by our assumption, should have equal values for $F$ and $F_0$. If $F$ and $F_0$ have finite areas, we divide them into infinitely small parts and proceed in the same way as in the above case of integral invariants of order one. The value of $I$ extended over any one of the parts of $F_0$ equals the value of $I$ extended over the corresponding part of $F$. Thus, its values for $F_0$ and $F$ are equal. Hence, $I$ is an integral invariant. The converse of this theorem can be proved in the same way as in the foregoing case.

Integral invariants of order three can be obtained from three particular solutions $\xi_i$, $\xi_i'$, $\xi_i''$, respectively, of the three systems of variational equations

$$\frac{d\xi_j}{dt} = \sum_i \frac{\partial X_j}{\partial x_i}\bigg|_{x=x^{(0)}} \xi_i, \qquad \frac{d\xi_j'}{dt} = \sum_i \frac{\partial X_j}{\partial x_i}\bigg|_{x=x^{(0)}} \xi_i',$$

$$\frac{d\xi_j''}{dt} = \sum_i \frac{\partial X_j}{\partial x_i}\bigg|_{x=x^{(0)}} \xi_i''.$$

Since these systems admit the integral

$$\sum_{i,j,k} A_{ijk} \begin{vmatrix} \xi_i & \xi_i' & \xi_i'' \\ \xi_j & \xi_j' & \xi_j'' \\ \xi_k & \xi_k' & \xi_k'' \end{vmatrix},$$

it follows that the integral

$$\iiint \sum_{i,j,k} A_{ijk}\, dx_i\, dx_j\, dx_k,$$

where $A_{ijk}$ are functions of $x$ only, is an integral invariant of (105).

In general, let there be $p$ ($p < n$) integrals of (106): $\Phi_1, \Phi_2, \ldots, \Phi_p$; suppose that they are functions of $x$ only. Denote $q$ first-order integral invariants with regard to (105) by $\int F_1(dx_i), \ldots, \int F_q(dx_i)$. Then $F_1(\xi_i), \ldots, F_q(\xi_i)$ are integrals of (106), linear homogeneous with respect to $\xi_i$. Take a function

$$\Theta(\Phi_1, \ldots, \Phi_p, F_1, \ldots, F_q) \equiv \Theta[\Phi_k, F_l],$$

that is linear homogeneous with respect to $F$; then $\Theta[\Phi_k, F_l(\xi_i)]$ is a new integral of (106), linear homogeneous with respect to $\xi_i$. Hence, $\int \Theta[\Phi_k, F_l(dx_i)]$ is a first-order integral invariant with regard to (105).

Similarly, let

$$\iint F_1(dx_i\, dx_j), \ldots, \iint F_q(dx_i\, dx_j)$$

be $q$ second-order integral invariants and $\Theta[\Phi_k, F_l]$ be a function of $\Phi$ and $F$, and linear homogeneous with respect to $F$. Then

$$\Theta[\Phi_k, F_l(\xi_i\xi_j' - \xi_j\xi_i')]$$

is an integral of (106) and

$$\iint \Theta[\Phi_k, F_l(dx_i\, dx_j)]$$

is a second-order integral invariant with regard to (105).

More generally, let $\int F_1$ and $\int F_2$ be integral invariants of order $p$ and $q$, respectively, and let $\xi_i^{(1)}, \ldots, \xi_i^{(p+q)}$ be $p + q$ integrals of (106). Denote by $F_1'$ the result of substituting for the determinant formed of $p$ differentials in $F_1$ the determinant formed of $p$ solutions $\xi_i^{(1)}, \ldots, \xi_i^{(p)}$. Similarly, denote by $F_2'$ the result of substituting, for $q$ differentials in $F_2$, $q$ solutions $\xi_i^{(p+1)}, \ldots, \xi_i^{(p+q)}$. Then $F_1' \cdot F_2'$ is an integral of (106). We make arbitrary permutations of the $p + q$ solutions $\xi_i^{(j)}$. If we assign a plus or minus sign according to the odd or even number of the permutations and add the result, then the sum is also an integral of (106). There are $(p + q)!/(p!\, q!)$ such independent permutations. This is a determinant of order $p + q$ formed of $p + q$ solutions $\xi_i^{(1)}, \ldots, \xi_i^{(p+q)}$, and its integral is linear homogeneous. The result is an integral invariant of order $p + q$, with regard to (105). This is how Poincaré constructed integral invariants of various orders.

Consider, instead of (105), the system of canonical equations

$$\frac{dx_i}{dt} = \frac{\partial F}{\partial y_i}, \qquad \frac{dy_i}{dt} = -\frac{\partial F}{\partial x_i} \qquad (i = 1, 2, \ldots, n). \tag{109}$$

The variational equations are now

$$\begin{aligned} \frac{d\xi_i}{dt} &= \sum_j \frac{\partial^2 F}{\partial y_i\, \partial x_j}\, \xi_j + \sum_j \frac{\partial^2 F}{\partial y_i\, \partial y_j}\, \eta_j, \\ \frac{d\eta_i}{dt} &= -\sum_j \frac{\partial^2 F}{\partial x_i\, \partial x_j}\, \xi_j - \sum_j \frac{\partial^2 F}{\partial x_i\, \partial y_j}\, \eta_j \qquad (i = 1, 2, \ldots, n). \end{aligned} \tag{110}$$

Denote another solution of (110) by $\xi_i', \eta_i'$; then,

$$\sum_{i=0}^{n} \frac{d}{dt}(\eta_i'\xi_i - \xi_i'\eta_i) = \sum_{i=1}^{n} \left(\eta_i' \frac{d\xi_i}{dt} - \xi_i' \frac{d\eta_i}{dt} - \eta_i \frac{d\xi_i'}{dt} + \xi_i \frac{d\eta_i'}{dt}\right) = 0.$$

Hence, we have an integral of (110) in the form

$$\sum_{i=1}^{n} (\xi_i\eta_i' - \eta_i\xi_i') = \text{constant}. \tag{111}$$

If $\xi_i$ and $\eta_i$ are integrals of (110), then so are $d\xi_i/dt$ and $d\eta_i/dt$. The integral invariant obtained in this way,

$$I_1 = \int (dx_1\,dy_1 + \cdots + dx_n\,dy_n), \tag{112}$$

is of order two and has a very important application.

The integral formed of four sets of particular solutions

$$\sum_{i,j} \begin{vmatrix} \xi_i & \xi_i' & \xi_i'' & \xi_i''' \\ \eta_i & \eta_i' & \eta_i'' & \eta_i''' \\ \xi_j & \xi_j' & \xi_j'' & \xi_j''' \\ \eta_j & \eta_j' & \eta_j'' & \eta_j''' \end{vmatrix} = \text{constant}$$

is also an integral of (110); the corresponding integral invariant with regard to (109),

$$I_2 = \int \sum_{i,j} dx_i\,dy_i\,dx_j\,dy_j,$$

is of order four, where the sum is extended over the $n(n-1)/2$ combinations of $i$ and $j$. Similarly,

$$I_3 = \int \sum_{i,j,k} dx_i\,dy_i\,dx_j\,dy_j\,dx_k\,dy_k$$

extended over the $n(n-1)(n-2)/6$ combinations of $i$, $j$, $k$ is a sixth-order integral invariant. In addition, the equations (106) admit the integral expressed by the sum of determinants

$$\sum_{\alpha_1,\cdots,\alpha_p} \|\xi_{\alpha_1}\eta_{\alpha_1}\xi_{\alpha_2}\eta_{\alpha_2}\cdots\xi_{\alpha_p}\eta_{\alpha_p}\| = \text{constant}$$

$$(\alpha_1, \alpha_2, \ldots, \alpha_p = 1, 2, \ldots, n).$$

Hence, we can construct $n$ similar integral invariants $I_1, I_2, \ldots, I_n$ of order twice the index; they are not necessarily independent. This is how Poincaré constructed integral invariants of higher orders.

Let an integral of (105) be $\Phi(x_i, y_i, t) = \text{constant}$. Then the equation

$$\sum_i \frac{\partial\Phi}{\partial x_i}\xi_i + \sum_i \frac{\partial\Phi}{\partial y_i}\eta_i = \text{constant}$$

is an integral of (105). Since, identically,

$$\sum_i \frac{d}{dt}\left(\frac{\partial \Phi}{\partial x_i}\right)\xi_i + \sum_i \frac{d}{dt}\left(\frac{\partial \Phi}{\partial y_i}\right)\eta_i + \sum_i \frac{\partial \Phi}{\partial x_i}\left(\sum_j \frac{\partial^2 F}{\partial y_i\, \partial x_j}\,\xi_j + \sum_j \frac{\partial^2 F}{\partial y_i\, \partial y_j}\,\eta_j\right)$$

$$- \sum_i \frac{\partial \Phi}{\partial y_i}\left(\sum_j \frac{\partial^2 F}{\partial x_i\, \partial x_j}\,\xi_j + \sum_j \frac{\partial^2 F}{\partial x_i\, \partial y_j}\,\eta_j\right) = 0,$$

therefore,

$$\frac{d}{dt}\left(\frac{\partial \Phi}{\partial x_i}\right) = -\sum_j \frac{\partial^2 F}{\partial y_i\, \partial x_j}\frac{\partial \Phi}{\partial x_j} + \sum_j \frac{\partial^2 F}{\partial x_i\, \partial x_j}\frac{\partial \Phi}{\partial y_j},$$

$$\frac{d}{dt}\left(\frac{\partial \Phi}{\partial y_i}\right) = -\sum_j \frac{\partial^2 F}{\partial y_i\, \partial y_j}\frac{\partial \Phi}{\partial x_j} + \sum_j \frac{\partial^2 F}{\partial x_i\, \partial y_j}\frac{\partial \Phi}{\partial y_j}.$$

Hence, $\xi_i = \partial\Phi/\partial y_i$, $\eta_i = -\partial\Phi/\partial x_i$ is a solution of (106). Furthermore, if $\Phi =$ constant, $\Phi_1 =$ constant are both solutions of (105), then, by substituting the expressions for $\xi_i$, $\eta_i$ just obtained in

$$\sum_i \frac{\partial \Phi}{\partial x_i}\,\xi_i + \sum_i \frac{\partial \Phi}{\partial y_i}\,\eta_i = \text{constant},$$

we obtain

$$(\Phi, \Phi_1) \equiv \sum_i \left(\frac{\partial \Phi}{\partial x_i}\frac{\partial \Phi_1}{\partial y_i} - \frac{\partial \Phi}{\partial y_i}\frac{\partial \Phi_1}{\partial x_i}\right) = \text{constant}.$$

This is Poisson's theorem. Since $\xi_i' = \partial\Phi_1/\partial y_i$, $\eta_i' = -\partial\Phi_1/\partial x_i$ constitute a particular solution of (106), there follows the integral

$$\sum_i (\xi_i\eta_i' - \xi_i'\eta_i) = \text{constant}.$$

This is the formula (111) which we have just obtained. The integral invariant $I_1$ of (112) is merely another form of Poisson's theorem.

Let $\Phi$, $\Phi_1$, $\Phi_2$, $\Phi_3$ be four integrals of (105). Let their Jacobian with respect to $x_i$, $x_j$, $y_i$, $y_j$ be $\Delta_{ij}$. The sum $\sum_{i,j}\Delta_{ij}$ of $\Delta_{ij}$ with all possible combinations of $i$, $j$ is also an integral of (106), that is, the integral invariant $I_2$. By a similar correspondence we can generalize Poisson's theorem to $I_2, I_3, \ldots, I_n$.

### 4.21 INTEGRAL INVARIANTS OF THE $n$-BODY PROBLEM

The energy integral of the differential equations for the $n$-body problem

$$\frac{dx_i}{dt} = \frac{\partial H}{\partial y_i}, \qquad \frac{dy_i}{dt} = -\frac{\partial H}{\partial x_i} \qquad (i = 1, 2, \ldots, 3n) \tag{113}$$

is

$$H = \sum_{i=1}^{3n} \frac{1}{2m_i} y_i^2 - U = \text{constant}. \tag{114}$$

Let a solution of (113) be

$$x_i = \phi_i(t), \qquad y_i = m_i\phi_i'(t). \tag{115}$$

Then another solution is written with an arbitrary constant $c$ in the form

$$x_i = \phi_i(t + c), \qquad y_i = m_i\phi_i'(t + c).$$

A particular solution of the variational equations is

$$\xi_i = c\phi_i'(t) = c\frac{y_i}{m_i}, \qquad \eta_i = cm_i\phi_i''(t) = c\frac{\partial U}{\partial x_i}.$$

Hence

$$\xi_i = \frac{y_i}{m_i}, \qquad \eta_i = \frac{\partial U}{\partial x_i} \qquad (i = 1, 2, \ldots, 3n),$$

constitute a particular solution of the variational equations, and the variational equations admit an integral

$$\sum_i \frac{y_i\eta_i}{m_i} - \sum_i \frac{\partial U}{\partial x_i}\xi_i = \text{constant},$$

corresponding to (114). The corresponding integral invariant is

$$\int \omega_0 = \int \sum_i \left(\frac{1}{m_i} y_i\, dy_i - \frac{\partial U}{\partial x_i} dx_i\right).$$

Similarly, the particular solutions of the variational equations corresponding to the first integrals of the center of mass, i.e.,

$$H_1 = -\sum_{j=1}^{n} y_{3j-2}, \qquad H_2 = -\sum_{j=1}^{n} y_{3j-1}, \qquad H_3 = -\sum_{j=1}^{n} y_{3j},$$

are

$$-\sum_{j=1}^{n} \eta_{3j-2}, \qquad -\sum_{j=1}^{n} \eta_{3j-1}, \qquad -\sum_{j=1}^{n} \eta_{3j}.$$

The corresponding integral invariants are

$$\int \omega_1 = -\int \sum_{j=1}^{n} dy_{3j-2}, \qquad \int \omega_2 = -\int \sum_{j=1}^{n} dy_{3j-1},$$

$$\int \omega_3 = -\int \sum_{j=1}^{n} dy_{3j}.$$

Corresponding to the integral of angular momentum

$$H_4 = \sum_{j=1}^{n} (x_{3j}y_{3j-1} - x_{3j-1}y_{3j}),$$

the particular solution of the variational equations is

$$\sum_{j=1}^{n} (x_{3j}\eta_{3j-1} - y_{3j}\xi_{3j-1} - x_{3j-1}\eta_{3j} + y_{3j-1}\xi_{3j}),$$

and the corresponding integral invariant is

$$\int \omega_4 = \int \sum_{j=1}^{n} (x_{3j}\, dy_{3j-1} - y_{3j}\, dx_{3j-1} - x_{3j-1}\, dy_{3j} + y_{3j-1}\, dx_{3j}).$$

Similarly

$$\int \omega_5 = \int \sum_{j=1}^{n} (x_{3j-2}\, dy_{3j} - y_{3j-2}\, dx_{3j} - x_{3j}\, dy_{3j-2} + y_{3j}\, dx_{3j-2}),$$

$$\int \omega_6 = \int \sum_{j=1}^{n} (x_{3j-1}\, dy_{3j-2} - y_{3j-1}\, dx_{3j-2} - x_{3j-2}\, dy_{3j-1} + y_{3j-2}\, dx_{3j-1}).$$

We also obtain from (115), putting $i \equiv \alpha \pmod 3$, another particular solution

$$x_i = \phi_i(t) + c_\alpha + k_\alpha t, \qquad y_i = m_i\phi_i'(t) + m_i k_\alpha \qquad (i = 1, 2, \ldots, 3n),$$

$$c_0^2 + c_1^2 + c_2^2 = c^2, \qquad k_0^2 + k_1^2 + k_2^2 = k^2.$$

Supposing the constants $c$ and $k$ to be infinitely small, we obtain the solutions of the variational equations

$$\xi_\alpha = 1, \qquad \eta_\alpha = 0,$$

and

$$\xi_\alpha = t, \qquad \eta_\alpha = m_\alpha \qquad (\alpha = 3j - 2, 3j - 1, 3j).$$

Corresponding to the integrals

$$\sum_{j=1}^{n} \eta_{3j-2} = \text{constant}, \qquad \sum_{j=1}^{n} \eta_{3j-1} = \text{constant}, \qquad \sum_{j=1}^{n} \eta_{3j} = \text{constant},$$

$$\sum_{j=1}^{n} \eta_{3j-2} t - \sum_{j=1}^{n} m_{3j-2}\xi_{3j-2} = \text{constant},$$

$$\sum_{j=1}^{n} \eta_{3j-1} t - \sum_{j=1}^{n} m_{3j-1}\xi_{3j-1} = \text{constant},$$

$$\sum_{j=1}^{n} \eta_{3j} t - \sum_{j=1}^{n} m_{3j}\xi_{3j} = \text{constant},$$

of the variational equations, we obtain the respective integral invariants

$$\int \omega_7 = \int \sum_{j=1}^{n} m_{3j-2}(\delta x_{3j-2} - t\delta y_{3j-2} - y_{3j-2}\,\delta t),$$

$$\int \omega_8 = \int \sum_{j=1}^{n} m_{3j-1}(\delta x_{3j-1} - t\,\delta y_{3j-1} - y_{3j-1}\,\delta t),$$

$$\int \omega_9 = \int \sum_{j=1}^{n} m_{3j}(\delta x_{3j} - t\,\delta y_{3j} - y_{3j}\,\delta t).$$

A very interesting integral invariant is obtained when $U$ is homogeneous, of degree $p$: the Newtonian law of gravitation corresponds to $p = -1$. Equations (113) remain unaltered when we multiply $t$ by $\lambda^{2-p}$, $x_i$ by $\lambda^2$, $y_i$ by $\lambda^p$, where $\lambda$ is a constant. If we take $\lambda$ to be equal to $1 + \epsilon$ (very close to 1), then we have a particular solution of the variational equations

$$\xi_i = 2\phi_i - (2 - p)t\phi_i', \qquad \eta_i = pm_i\phi_i' - (2 - p)m_i t\phi_i',$$

or

$$\xi_i = 2x_i - (2 - p)t\frac{y_i}{m_i}, \qquad \eta_i = py_i - (2 - p)t\frac{\partial U}{\partial x_i}.$$

From these we derive the particular solution

$$\sum_{i=1}^{3n} (2x_i\eta_i - py_i\xi_i) = (2 - p)t\sum_{i=1}^{3n}\left(\frac{y_i\eta_i}{m_i} - \frac{\partial U}{\partial x_i}\xi_i\right)$$

$$= \text{constant}$$

of the variational equations. This solution cannot be obtained by differentiating any known integral of (113). The corresponding integral invariant is

$$\int \omega_{10} = \int \sum_{i=1}^{3n} (2x_i\, dy_i - py_i\, dx_i) + (p-2)t \int \sum_{i=1}^{3n} \left(\frac{y_i\, dy_i}{m_i} - \frac{\partial U}{\partial x_i}\, dx_i\right).$$

As this expression contains $t$ explicitly, this integral invariant is different in character from those already found. The second integral is written

$$\int d\left[\sum_{i=1}^{3n} \left(\frac{y_i^2}{2m_i} - U\right)\right];$$

there is a perfect differential under the integral sign. The sum is equal to the energy constant, by (114). $\int \omega_{10}$ is linear and is an integral extended over an arbitrary arc of a curve. Let the values of the energy integral at the two ends of the arc be $C_0$ and $C_1$, respectively. This arc is what we have called $F_0$ in the foregoing discussion; since it is deformed continuously to $F$, the values of $C_0$ and $C_1$ do not alter. Hence

$$\omega_{10} = \int \sum_{i=1}^{3n} (2x_i\, dy_i - py_i\, dx_i) - (p-2)t(C_1 - C_0).$$

The first term on the right-hand side varies as $F$ is deformed, but the variation is proportional to $t$. If $C_0 = C_1$ at both ends of this curve, then $\int \omega_{10}$ is a constant. If this curve is closed, then the same situation occurs. Hence, this is a relative integral invariant. But in the case of a closed curve the integral remains unaltered when we put a perfect differential under the integral sign. Adding $\sum_{i=1}^{3n} (x_i\, dy_i + y_i\, dx_i)$ after multiplying by a constant factor, we see that $\int \sum_{i=1}^{3n} y_i\, dx_i$ and $\int \sum_{i=1}^{3n} x_i\, dy_i$ are both relative integral invariants. We have shown how we can get an absolute integral invariant of order two from a relative integral invariant of order one. The former invariant in the present case is

$$I_1 = \iint \sum_{i=1}^{3n} dx_i\, dy_i,$$

which we have already derived. The differential $\sum_{i=1}^{3n} (2x_i\, dy_i - py_i\, dx_i)$ is a perfect differential when $p = -2$, that is, when the attractive force is inversely proportional to the cube of the mutual distance (instead of the square as in Newtonian law), and the invariant takes the form

$$\int \sum_{i=1}^{3n} (2x_i\, dy_i - py_i\, dx_i) = \sum_{i=1}^{3n} 2x_i y_i.$$

Hence, $\sum_{i=1}^{3n} 2x_i y_i$ is linear with respect to $t$. But this is equal to

$$\frac{d}{dt}\sum_{i=1}^{3n} m_i x_i^2,$$

and

$$\sum_{i=1}^{3n} m_i x_i^2$$

is quadratic with respect to $t$, as was found by Jacobi. In general,

$$\sum_{i=1}^{3n} (2x_i\, dy_i - py_i\, dx_i)$$

is not a perfect differential; it equals

$$\int \sum_{i=1}^{3n} (2x_i\, dy_i + y_i\, dx_i) - 3t(C_1 - C_0)$$

in the Newtonian case, $p = -1$. Even if we form Poisson's brackets using this new integral and the classical ten integrals, there occurs no additional new integral, as can be seen from a simple computation. Indeed, this integral is not in involution with any known integral except the energy integral $H$. Hence, the order of the differential equations cannot be reduced by such an integral.

## 4.22 CARTAN'S EXTERIOR MULTIPLICATION

Cartan (1922, 1945) used the notation

$$\begin{vmatrix} u_i & u_j \\ v_i & v_j \end{vmatrix} = [u_i \quad u_j], \qquad \begin{vmatrix} u_1 & u_2 & u_3 \\ v_1 & v_2 & v_3 \\ w_1 & w_2 & w_3 \end{vmatrix} = [u_1 \quad u_2 \quad u_3], \ldots,$$

and called these (following Grassmann and Plücker) the *exterior multiplication forms*. With the linear forms

$$f(u) = \sum_{i=1}^{n} a_i u_i, \qquad f'(u) = \sum_{i=1}^{n} a_i' u_i,$$

we obtain

$$[f(u), f'(u)] = \begin{vmatrix} f(u) & f'(u) \\ f(v) & f'(v) \end{vmatrix} = \sum_{i,j} a_i a_j' \begin{vmatrix} u_i & u_j \\ v_i & v_j \end{vmatrix} = \sum_{i,j} a_i a_j' [u_i u_j],$$

if we regard this quantity as a product of two linear forms where the order of the factors in the partial products has not been altered, and such that the product of two identical variables is zero and the product of two different variables changes sign when the order of the factors is reversed. Thus, to any alternating bilinear form corresponds a form of the second degree with a single series of variables but with exterior multiplication; conversely, to any quadratic form with exterior multiplication there corresponds an alternating bilinear form.

For the case of three variables,

$$[u_1u_2u_3] = -[u_2u_1u_3] = -[u_1u_3u_2] = -[u_3u_2u_1]$$
$$= [u_2u_3u_1] = [u_3u_1u_2].$$

In general, an exterior multiplication form of order $n$ is written

$$A[u_1u_2\cdots u_n].$$

An exterior multiplication form $[F\Phi\Psi]$ can be defined from exterior multiplication forms $F$, $\Phi$, $\Psi$ of arbitrary orders, the degree of the product being the sum of the degrees of the factors. The product is zero if the sum of the degrees exceeds $n$. For example, from

$$F = [u_1u_2] + [u_3u_4] + \cdots + [u_{2s-1}u_{2s}]$$

we obtain

$$\frac{1}{2!}[F^2] = [u_1u_2u_3u_4] + [u_1u_2u_5u_6] + \cdots + [u_{2s-3}u_{2s-2}u_{2s-1}u_{2s}],$$

$$\frac{1}{3!}[F^3] = [u_1u_2u_3u_4u_5u_6] + \cdots,$$

$$\vdots$$

$$\frac{1}{s!}[F^s] = [u_1u_2u_3\cdots u_{2s-1}u_{2s}],$$

$$\frac{1}{(s+1)!}[F^{s+1}] = 0.$$

From $\omega_\delta = \sum_i a_i\,\delta x_i$ we derive

$$\omega'_\delta = \sum_i [\delta a_i\,\delta x_i] = \sum_{i,j}\left(\frac{\partial a_j}{\partial x_i} - \frac{\partial a_i}{\partial x_j}\right)[\delta x_i\,\delta x_j].$$

This is called the *exterior derivative* of the Pfaffian $\omega_\delta$. From Stokes's formula in the present notation,

$$\int_C \omega_\delta = \iint_S \omega'_\delta,$$

the necessary and sufficient condition for $\omega' = 0$ is that $\omega$ should be a perfect differential.

To a bilinear form

$$\Omega = \sum_{i,j} a_{ij}[\delta x_i \; \delta x_j]$$

corresponds a bilinear alternating form

$$\sum_{i,j} a_{ij}(\delta x_i \; \delta x'_j - \delta x_j \; \delta x'_i).$$

Corresponding to a trilinear form

$$\Omega'(\delta \; \delta' \; \delta'') = \sum_{i,j} [\delta a_{ij}(\delta' x_i \; \delta'' x_j - \delta' x_j \; \delta'' x_i) - \delta' a_{ij}(\delta x_i \; \delta'' x_j - \delta x_j \; \delta'' x_i) + \delta'' a_{ij}(\delta x_i \; \delta' x_j - \delta x_j \; \delta' x_i)],$$

there is the trilinear alternating exterior derived form:

$$\Omega'_\delta = \sum_{i,j} [\delta a_{ij} \; \delta x_i \; \delta x_j] = \sum_{i,j,k} \left(\frac{\partial a_{ij}}{\partial x_k} + \frac{\partial a_{jk}}{\partial x_i} + \frac{\partial a_{ki}}{\partial x_j}\right)[\delta x_i \; \delta x_j \; \delta x_k].$$

This is the exterior differential of $\Omega$. If an exterior multiplication form of any degree is exteriorly derived, then the degree increases by one. A term $A[\delta x_i \; \delta x_j \cdots \delta x_l]$ generates a term $[\delta A \; \delta x_i \; \delta x_j \cdots \delta x_l]$ by exterior derivation. From a constant $m$ and an exterior multiplication form $\Omega$, we obtain $(m\Omega)' = [dm \cdot \Omega] + m\Omega'$. From two exterior multiplication forms $\Omega$ and $\Pi$ we obtain

$$[\Omega\Pi]' = [\Omega'\Pi] + [\Omega\Pi']$$

if the degree of $\Omega$ is even, and

$$[\Omega\Pi]' = [\Omega'\Pi] - [\Omega\Pi']$$

if the degree is odd. In particular,

$$[\Omega^p]' = p[\Omega^{p-1}\Omega']$$

if the degree of $\Omega$ is even.

The derivative of the exterior derivative $\Omega'$ of an exterior multiplication form is identically zero. Cartan (1945) called this Poincaré's theorem. The term of $\Omega'$ corresponding to an arbitrary term $a[\delta x_1 \cdots \delta x_p]$ of $\Omega$ is $[\delta a \; \delta x_1 \cdots \delta x_p]$. If $a$ depends only on $x_1, \ldots, x_p$, then the last term and its derivatives are zero. If $a$ does not depend on $x_1, \ldots, x_p$, then — by a change of variables so that $a$ becomes equal to $x_{p+1}$ — the exterior derivative of a term $[\delta x_{p+1}\, \delta x_1 \cdots \delta x_p]$ is zero, because, the coefficient of this term being unity, the derivative contributes nothing toward the formation of the exterior derivative.

Conversely, if the derivative of a differential form $\Omega$ is zero, then the form $\Omega$ can be regarded as the derivative of a form $\Pi$ whose degree is one less than the degree of $\Omega$. To prove this theorem we refer to the lemma: if the derivative of a form $\Omega$ is zero and this form does not contain the differential $\delta x_n$, then the coefficients are all independent of $x_n$. In fact a term such as $A[\delta x_1\, \delta x_2 \cdots \delta x_p]$ generates in the derivation a term $[\delta A\, \delta x_1\, \delta x_2 \cdots \delta x_p]$, or, by expansion, several terms one of which is $(\partial A/\partial x_n)[\delta x_n\, \delta x_1 \cdots \delta x_p]$. This last term cannot cancel by combining with another term, since no term of $\Omega$ contains $\delta x_n$. Hence $\partial A/\partial x_n = 0$, because $\Omega' = 0$.

To prove the theorem, let $\Omega_0$ be the result of substituting $x_n = x_n^0$, $\delta x_n = 0$ in $\Omega$. Evidently $\Omega_0' = 0$ if $\Omega' = 0$. Suppose that the theorem is true for $n - 1$ variables and that it is possible to find a form $\Pi_0$ constructed with the variables $x_1, \ldots, x_{n-1}$ such that $\Pi_0' = \Omega_0$. Divide $\Omega$ and $\Pi$ into the parts containing $\delta x_n$ and the parts independent of $\delta x_n$, so that

$$\Omega = \Omega_1 + [\delta x_n \Omega_2], \qquad \Pi = \Pi_1 + [\delta x_n \Pi_2].$$

The term containing $\delta x_n$ in $\Pi'$ is

$$\Pi' = \left[\delta x_n \frac{\partial \Pi_1}{\partial x_n}\right] - [\delta x_n \Pi_2'] + \cdots.$$

Choose $\Pi_2$ arbitrarily and determine $\Pi_1$ by the condition that (*i*) $\Pi_1$ reduces to $\Pi_0$ for $x_n = x_n^0$, and (*ii*) $\partial \Pi_1/\partial x_n = \Pi_2' + \Omega_2$, so that $\Pi_1$ can be evaluated by quadrature. The form $\Pi' - \Omega$ does not contain $\delta x_n$ after similar terms are cancelled and reduces to zero if we put $x_n = x_n^0$ in the coefficients. Because of the lemma, all coefficients are independent of $x_n$. Hence the difference $\Pi' - \Omega$ should be identically zero.

This theorem shows that if the exterior derivative of a linear form $\Omega$ is zero then $\Omega$ is an exact differential. If $\Omega$ is a quadratic form with three variables

$$\Omega = P[\delta y\, \delta z] + Q[\delta z\, \delta x] + R[\delta x\, \delta y],$$

then the condition

$$\frac{\partial P}{\partial x} + \frac{\partial Q}{\partial y} + \frac{\partial R}{\partial z} = 0$$

is necessary and sufficient in order that $\Omega$ shall be regarded as the derivative of a linear form, that is, in order that we can find three functions $A$, $B$, $C$ satisfying

$$\frac{\partial C}{\partial y} - \frac{\partial B}{\partial z} = P, \qquad \frac{\partial A}{\partial z} - \frac{\partial C}{\partial x} = Q, \qquad \frac{\partial B}{\partial x} - \frac{\partial A}{\partial y} = R.$$

By the *rank* of a quadratic form we mean the number of quadratic terms necessary to express the quadratic form as their sum. To find the rank of a general quadratic form we need only see how many independent equations the actual equations would reduce to, if we equated all the partial derivatives of the quadratic form to zero. Similarly the rank is defined for an exterior multiplication form $F = \sum_{i,j} a_{ij}[u_i u_j]$. The equations obtained by equating all its partial derivatives to zero are called the *associated system*. Note that

$$\frac{\partial [u_1 u_2]}{\partial u_1} = u_2, \qquad \frac{\partial [u_1 u_2]}{\partial u_2} = -u_1, \qquad \frac{\partial [u_1 u_2]}{\partial u_i} = 0$$

$$(i = 3, 4, \ldots, n).$$

If $a_{12} \neq 0$ in the coefficients of $F(u)$, then

$$\frac{1}{a_{12}}\left[\frac{\partial F}{\partial u_1}\frac{\partial F}{\partial u_2}\right] = \frac{1}{a_{12}}\,[(a_{12}u_2 + a_{13}u_3 + \cdots + a_{1n}u_n)$$

$$\times\ (a_{21}u_1 + a_{23}u_3 + \cdots + a_{2n}u_n)]$$

has the same coefficients as $F$ in the terms combining $u_1$ or $u_2$, so that

$$F(u) - \frac{1}{a_{12}}\left[\frac{\partial F}{\partial u_1}\frac{\partial F}{\partial u_2}\right] \equiv F'(u)$$

contains only $u_3, \ldots, u_n$. If in addition $a'_{34} \neq 0$, then

$$F'(u) - \frac{1}{a'_{34}}\left[\frac{\partial F'}{\partial u_3}\frac{\partial F'}{\partial u_4}\right] = F''(u)$$

contains only $u_5, \ldots, u_n$. Continuing this process, we obtain

$$F(u) = \frac{1}{a_{12}}\left[\frac{\partial F}{\partial u_1}\frac{\partial F}{\partial u_2}\right] + \frac{1}{a'_{34}}\left[\frac{\partial F'}{\partial u_3}\frac{\partial F'}{\partial u_4}\right] + \frac{1}{a''_{56}}\left[\frac{\partial F''}{\partial u_5}\frac{\partial F''}{\partial u_6}\right] + \cdots,$$

or

$$F(U) = [U_1U_2] + [U_3U_4] + \cdots + [U_{2s-1}U_{2s}] \qquad (2s \leqq n),$$

where

$$\frac{\partial F}{\partial u_1} = U_1, \qquad \frac{\partial F}{\partial u_2} = a_{12}U_2, \qquad \frac{\partial F'}{\partial u_3} = U_3, \qquad \frac{\partial F'}{\partial u_4} = a'_{34}U_4,$$

$$\frac{\partial F''}{\partial u_5} = U_5, \qquad \frac{\partial F''}{\partial u_6} = a''_{56}U_6, \ldots.$$

This is called the *canonical form* of $F$. We see that the rank of $F$ should always be even. The associated system is $U_1 = U_2 = \cdots = U_{2s} = 0$.

If a given exterior multiplication form $\Omega$ is invariant with regard to the system of differential equations

$$\frac{dx_1}{X_1} = \frac{dx_2}{X_2} = \cdots = \frac{dx_n}{X_n},$$

then $\Omega$ can be expressed by $n - 1$ independent integrals $y_1, y_2, \ldots, y_{n-1}$ and their differentials, by the theorem of Section 4.17. The situation is the same for the exterior derivative $\Omega'$. The system of linear differential equations associated with two exterior multiplication forms $\Omega$ and $\Omega'$ is a consequence of the equations

$$dy_1 = 0, dy_2 = 0, \ldots, dy_{n-1} = 0, \tag{117}$$

and hence a consequence of (116). In other words, in order that (116) shall admit $\Omega(\delta)$ as an invariant form, it is necessary that the system associated with $\Omega$ and $\Omega'$ be satisfied when the variables $\delta x_1, \ldots, \delta x_n$ are replaced by $X_1, \ldots, X_n$, respectively. The system of equations expressed by perfect differentials in the form (117) is called the *characteristic* system with respect to (116).

Conversely, suppose that the above condition for perfect differentials is satisfied. Since the system associated with $\Omega(d)$ is satisfied when (116) is taken into account, it is satisfied when the equivalent system (117) is taken into account. Hence $\Omega(d)$, considered as an exterior multiplication form of $dx_i$, can be expressed uniquely by $dy_1, \ldots, dy_{n-1}$, the coefficients being functions of $x$ that can be expressed in terms of $y_1, \ldots, y_{n-1}$ and $x_n$, if $X_n \neq 0$. Thus

$$\Omega = \sum A_{i_1 \cdots i_p}[dy_{i_1} \cdots dy_{i_p}].$$

Only the term in $[dx_n\, dy_{i_1} \cdots dy_{i_p}]$ has the coefficient $\partial A_{i_1 \cdots i_p}/\partial x_n$. By hypothesis $\Omega'$ should be expressed by means of the variables $dy_i$ only. Hence

$$\frac{\partial A_{i_1 \cdots i_p}}{\partial x_n} = 0,$$

that is, $\Omega$ is expressed by the first integrals of the given system and their differentials. This is an invariant form. Thus the equations of the characteristic system of $\Omega$ reduce to the equations of the associated system of $\Omega$ joined with the equations of the associated system of $\Omega'$.

In particular, if $\Omega$ is a perfect differential, that is, if $\Omega' = 0$, then the characteristic equation of a differential form $\Omega$ coincides with the associated system of form $\Omega$.

The characteristic equation of a relative integral invariant $\int \Omega$ coincides with the associated system of the derived form $\Omega'$, because a relative invariant leads to an absolute invariant $\int \Omega'$, where $\Omega'$ is a perfect differential. From the linear integral invariant in dynamics

$$\int \omega_\delta = \int \sum_i p_i \, \delta q_i - H \, \delta t,$$

we obtain

$$\omega'_\delta = \sum_i [\delta p_i \, \delta q_i] - [\delta H \, \delta t].$$

The associated system of form $\omega'$ is

$$\frac{\partial \omega'}{\partial(\delta q_i)} = 0, \qquad \frac{\partial \omega'}{\partial(\delta p_i)} = 0, \qquad \frac{\partial \omega'}{\partial(\delta t)} = 0;$$

that is, replacing $\delta$ by $d$,

$$-\,dp_i - \frac{\partial H}{\partial q_i}\, dt = 0, \qquad dq_i - \frac{\partial H}{\partial p_i}\, dt = 0, \qquad dH - \frac{\partial H}{\partial t}\, dt = 0.$$

This is the result already obtained in Chapter 1.

An electromagnetic field defined by the exterior form

$$\begin{aligned}\Omega = {} & H_x[\delta y \, \delta z] + H_y[\delta z \, \delta x] + H_z[\delta x \, \delta y] + E_x[\delta x \, \delta t] \\ & + E_y[\delta y \, \delta t] + E_z[\delta z \, \delta t]\end{aligned}$$

satisfies the condition that $\Omega'$ is zero. Thus,

$$\begin{aligned}\Omega' = {} & \left(\frac{\partial H_x}{\partial x} + \frac{\partial H_y}{\partial y} + \frac{\partial H_z}{\partial z}\right)[\delta x \, \delta y \, \delta z] + \left(\frac{\partial H_x}{\partial t} + \frac{\partial E_z}{\partial y} - \frac{\partial E_y}{\partial z}\right)[\delta y \, \delta z \, \delta t] \\ & + \left(\frac{\partial H_y}{\partial t} + \frac{\partial E_x}{\partial z} - \frac{\partial E_z}{\partial x}\right)[\delta z \, \delta x \, \delta t] + \left(\frac{\partial H_z}{\partial t} + \frac{\partial E_y}{\partial x} - \frac{\partial E_x}{\partial y}\right)[\delta x \, \delta y \, \delta t].\end{aligned}$$

Equating to zero the four coefficients of $\Omega'$, we obtain the classical four equations of Maxwell

$$\operatorname{div} H = 0, \qquad \frac{\partial H}{\partial t} + \operatorname{curl} E = 0,$$

If the exterior differential of a form $\omega$ of degree $p$ is zero, then the integral $\int \omega$ extended over the boundary of an arbitrary domain of $(p + 1)$ dimensions is zero. In this case the form is said to be *closed*. Let $\theta_\alpha \equiv A_{\alpha_i}\, dx_i = 0$ $(\alpha = 1, 2, \ldots, r)$ be a differential system of linear equations where $\theta_\alpha$ are supposed to be independent. The set of exterior forms

$$\Phi = [F_1\phi_1] + [F_2\phi_2] + \cdots + [F_h\phi_h]$$

— where $\phi_i$ are homogeneous exterior forms under the condition that the sum of the degrees of two factors in any term $[F_i\phi_i]$ is the same for all terms on the right-hand side — is called the *ring* (Van der Waerden, 1955) determined by the $h$ homogeneous exterior forms $F_1, F_2, \ldots, F_h$ of $n$ variables. Any form $\Phi$ of the ring determined by $h$ given forms

$$F_1, F_2, \ldots, F_h$$

vanishes for any solution of the system $F_1 = 0, F_2 = 0, \ldots, F_h = 0$. But in order that an exterior form $\Phi$ may vanish for all solutions of the system of linear independent equations $F_1 = 0, F_2 = 0, \ldots, F_h = 0$, it is necessary and sufficient that $\Phi$ belongs to the ring determined by the forms $F_i$. The necessary and sufficient condition for $\Phi$ to belong to the ring is that the form $[F_1F_2\cdots F_h\Phi]$ be zero.

In order that a system of $r$ linear differential equations shall be completely integrable, it is necessary and sufficient that in the neighborhood of any point in space the exact differential $d\theta_\alpha$ belong to the ring determined by the forms $\theta_\alpha$. A differential system is said to be *closed* if the exterior differentials $d\theta_\alpha$ belong to the ring determined by $\theta_\alpha$. The necessary and sufficient condition for the complete integrability of the system

$$\theta_\alpha \equiv A_{\alpha_i}\, dx_i = 0 \qquad (\alpha = 1, 2, \ldots, r)$$

is that

$$[\theta_1\theta_2\cdots\theta_r\, d\theta_\alpha] = 0 \qquad (\alpha = 1, 2, \ldots, r).$$

Attempts to solve differential systems lead to the study of the solutions of closed differential systems.

The exterior product of two exterior invariant forms is evidently an invariant form. Hence, if an invariant exterior form $\Omega$ is known, we can derive a series of other invariant forms, that is, $\Omega'$ and all forms deducible from $\Omega$ and $\Omega'$, by exterior multiplication.

If $\Omega$ is an absolute invariant form of an even degree, then two series of absolute invariant forms

$$[\Omega^p], \qquad [\Omega^{p-1}\Omega'] \qquad (p = 1, 2, \ldots),$$

can be deduced. The derivative of a form of the first series is equal to a form of the second series. The derivative of a form of the second series is zero. If $\Omega$ is an absolute invariant of an odd degree, then we obtain two series

$$[\Omega'^p], \qquad [\Omega\Omega'^{p-1}] \qquad (p = 1, 2, \ldots).$$

The derivative of a form of the first series is zero, and the derivative of a form of the second series is equal to a form of the first series.

Consider a relative integral invariant $\int \Omega$ and suppose that $\Omega$ is of an even degree. Then only one new invariant, that is, the absolute invariant $\int \Omega'$, can be derived. If $\Omega$ is of an odd degree, then we obtain a series of relative integral invariants $\int \Omega\Omega'^{p-1}$ and a series of absolute integral invariants $\int \Omega'^p$. The relative invariant $\int \Omega\Omega'^{p-1}$ leads to the absolute invariant $\int \Omega'^p$ by derivation.

For example, take the relative invariant of dynamics

$$\int \omega = \int \sum_{i=1}^{n} p_i \, \delta q_i - H \, \delta t.$$

The relative integral invariants derivable from this invariant are $\int \omega\omega'^{p-1}$ $(p = 1, 2, \ldots, n)$, and the absolute integral invariants are $\int \omega'^p$ $(p = 1, 2, \ldots, n)$. Hence there exists an invariant, relative or absolute, of any given degree, lower than or equal to $2n$.

Consider a form $F$ of degree $p$. Denote by $\xi, \xi', \ldots, \xi^{(n-p+1)}$ the linear forms with indeterminate coefficients

$$\xi = \sum_{i=1}^{n} \xi_i u_i, \quad \xi' = \sum_{i=1}^{n} \xi'_i u_i, \quad \ldots.$$

The exterior product $[F\xi\xi' \cdots \xi^{(n-p+1)}]$ is of degree $n$ and hence of the form $\Phi[u_1 u_2 \cdots u_n]$. The coefficient $\Phi$ is linear with respect to each series of coefficients $\xi$ and is alternating. It corresponds to an exterior form of degree $n - p$ with respect to the variables $\xi_1, \xi_2, \ldots, \xi_n$. This is defined as the *adjoint* form of $F$.

If we apply a linear substitution to the variables $u$ and a linear substitution to the variables $\xi$ so that the expression

$$\xi_1 u_1 + \xi_2 u_2 + \cdots + \xi_n u_n$$

is conserved, then the expression $\Phi[u_1u_2\cdots u_n]$ is conserved. In other words, the adjoint form is multiplied by the determinant of the linear substitution applied to the variables $u$.

An exterior form of degree $p$ is called a *monomial* if it can be expressed as an exterior product of $p$ linear forms. Consider the system of linear equations obtained by equating to zero all partial derivatives of order $p - 1$ of that exterior form, that is, the associated system. If a linear substitution with nonvanishing determinant is applied to the form $F(u_1, u_2, \ldots, u_p)$ to transform it into the form $\Phi(u_1, u_2, \ldots, u_p)$, then the same process transforms the associated system of $F$ into the associated system of $\Phi$. If an exterior monomial form is *linearly* transformed so that it contains only a single term $a[u_1u_2\cdots u_p]$, then the associated system consists of $p$ independent equations $u_1 = 0, u_2 = 0, \ldots, u_p = 0$. Conversely, if the associated system of a form $F$ of degree $p$ is reduced to $p$ independent equations, then the associated system can take the form $u_1 = 0, u_2 = 0, \ldots, u_p = 0$ by a linear substitution, but none of the nonvanishing coefficients of the form can contain indices other than $1, 2, \ldots, p$. Hence the form should be $a[u_1u_2\cdots u_p]$, a monomial. The necessary and sufficient condition for an exterior form of degree $p$ to be a monomial is that the associated system be of rank $p$.

The adjoint form of $F = F_1 + F_2$ is the sum of the adjoint forms $\Phi_1$ and $\Phi_2$. The adjoint form of $aF$ is $a\Phi$, where $a$ is a numerical constant. To calculate the adjoint form it is sufficient to calculate the adjoint form of a monomial such as $F = [u_{\alpha_1}u_{\alpha_2}\cdots u_{\alpha_p}]$. Then $\Phi = [\xi_{\alpha_{p+1}}\cdots\xi_{\alpha_n}]$, where the indices $\alpha_{p+1}, \ldots, \alpha_n$ are those indices $1, 2, \ldots, n$ that do not appear in the sequence $\alpha_1, \alpha_2, \ldots, \alpha_p$, and these indices are arranged in an order such that the total sequence $\alpha_1, \ldots, \alpha_p, \alpha_{p+1}, \ldots, \alpha_n$ is even.

If $F$ is of degree $n - 1$, then the adjoint form is of the first degree. We can suppose that the adjoint form is $\xi_n$ so that $F$ can be put in the form $F = [u_1u_2\cdots u_{n-1}]$. If $F$ is of degree $n - 2$, then $\Phi$ is of the second degree, and is given by $\Phi = [\xi_1\xi_2] + \cdots + [\xi_{2s-1}\xi_{2s}]$.

Thus,

$$F = [u_3u_4u_5\cdots u_n] + [u_1u_2u_5u_6\cdots u_n] + \cdots + [u_1u_2\cdots u_{2s-2}u_{2s+1}\cdots u_n].$$

Suppose that $f_1, f_2, \ldots, f_h$ are $h$ independent linear forms. Let $F$ be an arbitrary exterior form. In order that $F$ may vanish when there are relations $f_1 = 0, f_2 = 0, \ldots, f_h = 0$ among the variables, it is necessary and sufficient that $[Ff_1f_2\cdots f_h] = 0$. For, if any term of $F$ contains at least one of the variables $u_1, u_2, \ldots, u_h$ as a factor, then the product $[Fu_1\cdots u_h]$ is zero. Conversely, if this product is zero, then an arbitrary term of $F$ must contain at least one of the variables $u_1, u_2, \ldots, u_h$, otherwise the multiplication of the term by $[u_1u_2\cdots u_h]$ will give a nonzero product that we cannot cancel by combining it with any other product.

Consider an exterior quadratic form $F$ in $n$ variables $u_1, \ldots, u_n$. Suppose that the variables are connected by a linear relation

$$f \equiv a_1u_1 + a_2u_2 + \cdots + a_nu_n = 0.$$

The form $F$, where $u_n$ is expressed in terms of $u_1, \ldots, u_{n-1}$ by means of this linear relation, is of a certain *rank* corresponding to the number of linearly independent equations of the associated system, namely

$$\frac{\partial F}{\partial u_1} - \frac{a_1}{a_n}\frac{\partial F}{\partial u_n} = 0, \ldots, \qquad \frac{\partial F}{\partial u_{n-1}} - \frac{a_{n-1}}{a_n}\frac{\partial F}{\partial u_n} = 0, \qquad f = 0.$$

If there are $p$ linearly independent relations

$$f_1 = \sum_{i=1}^{n} a_iu_i = 0, \qquad f_2 = \sum_{i=1}^{n} a_i'u_i = 0, \ldots, \qquad f_p = \sum_{i=1}^{n} a_i^{(p-1)}u_i = 0,$$

then the associated system of $F$ is

$$\begin{Vmatrix} \dfrac{\partial F}{\partial u_1} & \dfrac{\partial F}{\partial u_2} & \cdots & \dfrac{\partial F}{\partial u_n} \\ a_1 & a_2 & \cdots & a_n \\ a_1' & a_2' & \cdots & a_n' \\ \vdots & & & \\ a_1^{(p-1)} & a_2^{(p-1)} & \cdots & a_n^{(p-1)} \end{Vmatrix} = 0, \quad f_1 = 0, \quad f_2 = 0, \ldots, \quad f_p = 0.$$

The rank $2s'$ of the form $F$, when the variables are connected by the $p$ linearly independent relations, is twice the largest exponent for which the form $[f_1f_2\cdots f_pF^{s'}]$ is not zero.

If we define the bilinear alternating form

$$\Phi(\xi, \xi') = \sum_{i,j} a_{ij}\xi_i\xi_j'$$

by the equation

$$s[F^{s-1}(\xi_1u_1 + \cdots + \xi_nu_n)(\xi_1'u_1 + \cdots + \xi_n'u_n)] = \Phi(\xi, \xi')[F^s],$$

then the exterior quadratic form $\Phi(\xi) = \sum_{i,j} a_{ij}[\xi_i\xi_j]$ is equal to the adjoint form of $F^{s-1}/(s-1)!$, multiplied by a constant factor. It is an absolute covariant of $F$ in the sense that if we apply a linear substitution to the variables $u_1, \ldots, u_n$ and a linear substitution to the variables $\xi_1, \ldots, \xi_n$ such that $\xi_1u_1 + \cdots + \xi_nu_n$ is conserved, and if $F(u)$

and $\Phi(\xi)$ become respectively equal to $\bar{F}(\bar{u})$ and $\bar{\Phi}(\bar{\xi})$ through these two substitutions, then

$$s[\bar{F}^{s-1}(\bar{\xi}_1\bar{u}_1 + \cdots + \bar{\xi}_n\bar{u}_n)(\bar{\xi}'_1\bar{u}_1 + \cdots + \bar{\xi}'_n\bar{u}_n)] = \bar{\Phi}(\bar{\xi}\bar{\xi}')[\bar{F}^s].$$

In particular, if $F$ is of the canonical form

$$F = [u_1u_2] + \cdots + [u_{2s-1}u_{2s}],$$

then

$$\Phi = [\xi_1\xi_2] + \cdots + [\xi_{2s-1}\xi_{2s}].$$

This leads to the general identity

$$\left[\frac{F^{s-p}}{(s-p)!}(\xi_1u_1 + \cdots + \xi_nu_n)(\xi'_1u_1 + \cdots + \xi'_nu_n)\cdots\right.$$

$$\left.(\xi_1^{(2p-1)}u_1 + \cdots + \xi_n^{(2p-1)}u_n)\right] = \frac{\Phi^{(p)}(\xi, \xi', \ldots, \xi^{(2p-1)})}{p!}\left[\frac{F^s}{s!}\right], \quad (118)$$

where the exterior form of degree $p$ corresponding to the multilinear alternating form $\Phi^{(p)}$ is equal to $[\Phi^{(p)}(\xi)]$. If $p = 2$, then, by taking the terms in $[\xi_i\xi'_j\xi''_k\xi'''_l]$, we obtain

$$\left[\frac{F^{s-2}}{(s-2)!}u_iu_ju_ku_l\right] = (a_{ij}a_{kl} + a_{ik}a_{lj} + a_{il}a_{jk})\left[\frac{F^s}{s!}\right], \quad (119)$$

where $a_{ij}$ are defined by

$$\left[\frac{F^{s-1}}{(s-1)!}u_iu_j\right] = a_{ij}\left[\frac{F^s}{s!}\right].$$

Consider a form of degree $2s - 1$:

$$\left[\frac{F^{s-2}}{(s-2)!}u_iu_ju_k\right] - \left[\frac{F^{s-1}}{(s-1)!}(a_{ij}u_k + a_{jk}u_i + a_{ki}u_j)\right].$$

Multiply exteriorly by any one of the variables $u_1, \ldots, u_{2s}$, for example by $u_l$; then, by (119), the product is zero. Hence the form itself is identically zero. Since $u_i$, $u_j$, $u_k$ can be replaced by any three linear forms of the variables, we obtain the following

*Theorem:* Consider linear forms $f_1, \ldots, f_p$ and put

$$\left[\frac{F^{s-1}}{(s-1)!}f_if_j\right] = a_{ij}\left[\frac{F^s}{s!}\right] \qquad (i, j = 1, 2, \ldots, p),$$

then,

$$\left[\frac{F^{s-2}}{(s-2)!}f_if_jf_k\right] = \left[\frac{F^{s-1}}{(s-1)!}(a_{ij}f_k + a_{jk}f_i + a_{ki}f_j)\right]. \quad (120)$$

Consider $p$ linearly independent relations $f_1 = 0, f_2 = 0, \ldots, f_p = 0$. We can suppose that they are reduced to the forms

$$u_1 = 0, u_2 = 0, \ldots, u_p = 0,$$

on the single condition that the first $p$ variables $u_1, \ldots, u_p$ can be exchanged among themselves. Thus an arbitrary linear substitution can be applied to the variables $\xi$ on the single condition that the last $2s - p$ variables $\xi_{p+1}, \ldots, \xi_{2s}$ can be exchanged among themselves. Put

$$\left[\frac{F^{s-1}}{(s-1)!} u_i u_j\right] = a_{ij}\left[\frac{F^s}{s!}\right] \qquad (i, j = 1, 2, \ldots, p) \tag{121}$$

and drop the terms in $\xi_{p+1}, \ldots, \xi_{2s}$ in $\Phi$; then,

$$\overline{\Phi} = \sum_{i,j=1}^{p} a_{ij}[\xi_i \xi_j].$$

Let $2q$ be the rank of the form $\overline{\Phi}$. By suitable linear substitution on $\xi_1, \ldots, \xi_p$, we can reduce $\overline{\Phi}$ to

$$\Phi = [\xi_1 \xi_2] + \cdots + [\xi_{2q-1} \xi_{2q}].$$

Replacing $\xi_1, \ldots, \xi_p$ by linear combinations of $\xi_{p+1}, \ldots, \xi_{2s}$, we obtain $\Phi$ in the form

$$\Phi = [\xi_1 \xi_2] + \cdots + [\xi_{2q-1} \xi_{2q}] + [\xi_{2q+1} \xi_{p+1}] + \cdots + [\xi_p \xi_{2p-2q}] \\ + [\xi_{2p-2q+1} \xi_{2p-2q+2}] + \cdots + [\xi_{2s-1} \xi_{2s}].$$

Then $F$ takes the form

$$F = [u_1 u_2] + \cdots + [u_{2q-1} u_{2q}] + [u_{2q+1} u_{p+1}] + \cdots + [u_p u_{2p-2q}] \\ + \cdots + [u_{2s-1} u_{2s}].$$

Hence, if there are $p$ relations $u_1 = 0, u_2 = 0, \ldots, u_p = 0$, then the rank of $F$ is reduced by $2p - 2q$. From this we obtain the following

*Theorem:* Consider linearly independent forms $f_1, f_2, \ldots, f_p$, and $\frac{1}{2}p(p-1)$ quantities $a_{ij}$ defined by

$$\left[\frac{F^{s-1}}{(s-1)!} f_i f_j\right] = a_{ij}\left[\frac{F^s}{s!}\right], \tag{121a}$$

and the exterior quadratic form of $p$ variables $\xi_1, \ldots, \xi_p$,

$$\Phi(\xi) = \sum_{i,j=1}^{p} a_{ij}[\xi_i \xi_j]. \tag{122}$$

If the rank of this form is $2q$, then the rank of the form $F$ is reduced by $2p - 2q$ when the variables are connected by $p$ relations

$$f_1 = 0, f_2 = 0, \ldots, f_p = 0.$$

If, in addition, we apply to the $p$ given linear forms a linear substitution such that $\Phi$ reduces to the canonical form

$$\Phi = [\xi_1\xi_2] + \cdots + [\xi_{2q-1}\xi_{2q}],$$

then the form $F$ reduces to the canonical form

$$F = [f_1f_2] + \cdots + [f_{2q-1}f_{2q}] + [f_{2q+1}f_{p+1}] + \cdots + [f_pf_{2p-2q}] + \cdots + [f_{2s-1}f_{2s}],$$

where $f_{p+1}, \ldots, f_{2s}$ denote the new linear forms conveniently chosen so that they are independent of one another and of the given forms.

Suppose that a system of differential equations

$$\frac{dx_1}{dt} = X_1, \ldots, \frac{dx_n}{dt} = X_n \tag{123}$$

admits an integral invariant of the maximum degree $n$

$$\Omega = M[(\delta x_1 - X_1\,\delta t)(\delta x_2 - X_2\,\delta t)\cdots(\delta x_n - X_n\,\delta t)].$$

The required condition is that $\Omega' = 0$, that is,

$$\frac{\partial M}{\partial t} + \frac{\partial(MX_1)}{\partial x_1} + \cdots + \frac{\partial(MX_n)}{\partial x_n} = 0.$$

The coefficient $M$ is called the *Jacobi last multiplier* (Section 4.18). This condition indicates that the form $\Omega$ can be expressed by means of $n$ independent first integrals $y_1, y_2, \ldots, y_n$ of (123) and their differentials; that is,

$$M[(\delta x_1 - X_1\,\delta t)\cdots(\delta x_n - X_n\,\delta t)] = H(y_1, \ldots, y_n)[\delta y_1\cdots\delta y_n].$$

Let $\omega_1 = 0, \ldots, \omega_r = 0$ be a system of Pfaffians completely integrable, of which we know an invariant form of the maximum degree $r$, i.e., $\Omega = M[\omega_1\omega_2\cdots\omega_r]$. If we know $r - 1$ first integrals $y_1, \ldots, y_{r-1}$ of the system, then, by equating $y_1, \ldots, y_{r-1}$ each to an arbitrary constant, we can replace the given system by a single system $\omega_r = 0$, and obtain

$$\Omega = N[\delta y_1\cdots\delta y_{r-1}\omega_r].$$

The coefficient $N$ can be calculated from $M$. Thus $N\omega_r$ is an invariant form for a single equation $\omega_r = 0$; that is, $N\omega_r$ is a perfect differential, and the integration can be carried out by quadrature.

If this Pfaffian system admits an infinitesimal transformation

$$Af = \xi_1 \frac{\partial f}{\partial x_1} + \cdots + \xi_n \frac{\partial f}{\partial x_n},$$

then we can deduce from an invariant differential form $\Omega$ another invariant form $A(\Omega)$. Let $\Omega = [\omega_1 \cdots \omega_r]$. We construct

$$\omega_1(A), \omega_2(A), \ldots, \omega_r(A)$$

and suppose that $\omega_1(A) = 1, \omega_2(A) = \cdots = \omega_r(A) = 0$, are the results of a suitable transformation. Suppose that $\Omega$ is of the third degree, and consider the corresponding trilinear differential form $\Omega(\delta, \delta', \delta'')$. Replace $\delta$ by the symbol $A$ for the infinitesimal transformation; then, we obtain $\Omega(A, \delta', \delta'')$ — with two series of differentials $\delta'$ and $\delta''$ — to which corresponds an exterior quadratic form $\Omega(A, \delta)$. If $\Omega$ is expressed in terms of the first integrals $y_i$ and their differentials, then $\Omega(A, \delta)$ is expressed in terms of $y_i$ and $\delta y_i$. Thus, from any invariant form we can derive another invariant form whose degree is diminished by one:

$$\Omega(A, \delta) = \xi_1 \frac{\partial \Omega}{\partial(\delta x_1)} + \cdots + \xi_n \frac{\partial \Omega}{\partial(\delta x_n)}.$$

Hence, if we know an infinitesimal transformation, we can deduce from the invariant form $\Omega(\delta)$ a new invariant form $\Omega(A, \delta)$ by means of this formula.

Let $2n$ be the rank of the exterior derivative $\omega'$, and $f$ and $g$ be two first integrals of its characteristic system. The two differential forms $[\omega'^{n-1}\, df\, dg]$ and $[\omega'^n]$ are invariants of the maximum degree $2n$. Hence these two forms differ only by a factor that is a first integral. Put

$$\frac{1}{(n-1)!}[\omega'^{n-1}\, df\, dg] = \frac{1}{n!}(f, g)[\omega'^n], \tag{124}$$

where $(f, g)$ is a Poisson bracket and is an alternating bilinear form with respect to the partial derivatives of $f$ and $g$.

If the condition $(f, g) = 0$ is satisfied, then the rank of $\omega'$ is $2n - 4$. In this case we say that $f$ and $g$ are in *involution*. If this condition is not satisfied, then, by the definition of $(f, g)$, the form $\omega' - [df\, dg]/(f, g)$ is of rank $2n - 2$, and its $n$th power is

$$[\omega'^n] - \frac{n}{(f, g)}[\omega'^{n-1}\, df\, dg] = 0.$$

If (124) is applied to $\omega' = \sum_i [\delta p_i\, \delta q_i] - [\delta H\, \delta t]$ in dynamics, and if the terms $[\delta p_1\, \delta q_1\, \delta p_2 \cdots \delta q_n]$ are equated on both sides, then we obtain

$$(f, g) = \sum_i \left(\frac{\partial f}{\partial p_j}\frac{\partial g}{\partial q_i} - \frac{\partial f}{\partial q_j}\frac{\partial g}{\partial p_i}\right),$$

as in Chapter 1.

From (120) we obtain the identity

$$\frac{1}{(n-1)!}[\omega'^{n-1}\{(f,g)\,dh + (g,h)\,df + (h,f)\,dg\}]$$

$$= \frac{1}{(n-2)!}[\omega'^{n-2}\,df\,dg\,dh].$$

By exterior derivation, and noting that the exterior derivative of the right-hand member is zero, we see that

$$[\omega'^{n-1}\,d(f,g)\,dh] + [\omega'^{n-1}\,d(g,h)\,df] + [\omega'^{n-1}\,d(h,f)\,dg] = 0.$$

This is Jacobi's identity (Section 4.3).

If there are two linear invariant forms $\varpi_1$ and $\varpi_2$, then the quantity $\alpha$ defined by

$$n[\omega'^{n-1}\varpi_1\varpi_2] = \alpha[\omega'^n] \tag{125}$$

is a first integral. It reduces to $(y_1, y_2)$ if $\varpi_1$ and $\varpi_2$ are differentials of two first integrals $y_1$ and $y_2$.

Suppose that the characteristic equations for $\omega'$ admit two infinitesimal transformations $A_1f$ and $A_2f$. Then

$$\varpi_1 = \omega'(A_1, \delta), \qquad \varpi_2 = \omega'(A_2, \delta).$$

Apply to both sides of (125) the operation that carries an invariant form $\Omega(\delta)$ to the form $\Omega(A_1, \delta)$. Then,

$$n(n-1)[\omega'^{n-2}\varpi_1\varpi_2] - n[\omega'^{n-1}\varpi_1]\varpi_2(A_1) = n\alpha[\omega'^{n-1}\varpi_1].$$

Since $[\omega'^{n-1}\varpi_1]$ is not zero,

$$\alpha = -\varpi_2(A_1) = \omega'(A_1, A_2) = \varpi_1(A_2). \tag{126}$$

Thus, if we apply the generalized theorem of Poisson-Jacobi to the two invariant forms $\omega'(A_1, \delta)$ and $\omega'(A_2, \delta)$, we can derive the first integral $\omega'(A_1, A_2)$ by twice performing on $\omega'$ the operation corresponding to the infinitesimal transformations $A_1f$ and $A_2f$.

After these general considerations we now apply the theory to the reduction of the differential equations for the $n$-body problem.

Suppose that we reduce the equations to the order $6n - 6$ by using the six integrals for the center of mass. The equations then admit four infinitesimal transformations $A_if$ $(i = 0, 4, 5, 6)$. Corresponding to these transformations are four integrals $H_i$ $(i = 0, 4, 5, 6)$. The infinitesimal transformation corresponding to the integral invariant $\omega_{10}$ (Section 4.21) is

$$A_{10}f = \sum_{i=1}^{3n} \left[x_i \frac{\partial f}{\partial x_i} - \frac{p}{2} y_i \frac{\partial f}{\partial y_i}\right] + \left[1 + \frac{p}{2}\right] t \frac{\partial f}{\partial t}.$$

The corresponding linear invariants $\omega_i = \omega'(A_i, \delta)$ are

$$\omega_0 = \delta H_0, \qquad \omega_4 = \delta H_4, \qquad \omega_5 = \delta H_5, \qquad \omega_6 = \delta H_6,$$

$$\omega_{10} = -\sum_{i=1}^{3n} m_i\left(x_i\, \delta y_i + \frac{p}{2} y_i\, \delta x_i\right) + \left(1 + \frac{p}{2}\right) t\, \delta H_0 + pH_0\, \delta t.$$

We have

$$\omega_{10}' = A_{10}(\omega') = \left(1 - \frac{p}{2}\right)\omega'.$$

If we compute $a_{ij} = \omega'(A_i, A_j)$ by (125) and (126), then we obtain

$$\Phi(\xi) = pH_0[\xi_{10}\xi_0] + \frac{p-2}{2}[\xi_{10}(H_4\xi_4 + H_5\xi_5 + H_6\xi_6)]$$
$$+ H_4[\xi_5\xi_6] + H_5[\xi_6\xi_4] + H_6[\xi_4\xi_5].$$

The rank is four. Let

$$\xi_0' = pH_0\xi_0 + \frac{p-2}{2}(H_4\xi_4 + H_5\xi_5 + H_6\xi_6).$$

$$\xi_4' = \alpha_1\xi_4 + \alpha_2\xi_5 + \alpha_3\xi_6, \qquad \xi_5' = \beta_1\xi_4 + \beta_2\xi_5 + \beta_3\xi_6,$$

$$\xi_6' = \xi_0, \qquad \xi_{10}' = \xi_{10}.$$

Then we obtain the canonical form

$$\Phi = [\xi_{10}'\xi_0'] + [\xi_4'\xi_5'],$$

where $\alpha$ and $\beta$ are chosen so that

$$\alpha_2\beta_3 - \alpha_3\beta_2 = H_4, \qquad \alpha_3\beta_1 - \alpha_1\beta_3 = H_5, \qquad \alpha_1\beta_2 - \alpha_2\beta_1 = H_6,$$

and with further conditions

$$\alpha_1\beta_1 + \alpha_2\beta_2 + \alpha_3\beta_3 = 0,$$

$$\alpha_1^2 + \alpha_2^2 + \alpha_3^2 = \beta_1^2 + \beta_2^2 + \beta_3^2 = \sqrt{H_4^2 + H_5^2 + H_6^2}.$$

If we define five linear forms $\varpi_0, \varpi_4, \varpi_5, \varpi_6, \varpi_{10}$, by

$$\xi_0\omega_0 + \xi_4\omega_4 + \xi_5\omega_5 + \xi_6\omega_6 + \xi_{10}\omega_{10}$$
$$= \xi_0'\varpi_0 + \xi_4'\varpi_4 + \xi_5'\varpi_5 + \xi_6'\varpi_6 + \xi_{10}'\varpi_{10},$$

then we obtain

$$\varpi_0 = \frac{2}{p-2}\frac{H_4\,dH_4 + H_5\,dH_5 + H_6\,dH_6}{H_4^2 + H_5^2 + H_6^2},$$

$$\varpi_4 = \frac{\alpha_1\,dH_4 + \alpha_2\,dH_5 + \alpha_3\,dH_6}{\sqrt{H_4^2 + H_5^2 + H_6^2}},$$

$$\varpi_5 = \frac{\beta_1\,dH_4 + \beta_2\,dH_5 + \beta_3\,dH_6}{\sqrt{H_4^2 + H_5^2 + H_6^2}},$$

$$\varpi_6 = dH_0 - \frac{2pH_0}{p-2}\frac{H_4\,dH_4 + H_5\,dH_5 + H_6\,dH_6}{H_4^2 + H_5^2 + H_6^2},$$

$$\varpi_{10} = \omega_{10},$$

and

$$\omega' = [\varpi_{10}\varpi_0] + [\varpi_4\varpi_5] + [\omega_1\varpi_6] + [\omega_2\omega_3] + \cdots.$$

Hence

$$\omega' = \frac{2}{p-2}\left[\omega_{10}\frac{H_4\,dH_4 + H_5\,dH_5 + H_6\,dH_6}{H_4^2 + H_5^2 + H_6^2}\right]$$

$$+ \frac{H_4[dH_5\,dH_6] + H_5[dH_6\,dH_4] + H_6[dH_4\,dH_5]}{H_4^2 + H_5^2 + H_6^2} + \Omega, \quad (127)$$

where

$$\Omega = \left[\omega_1\left(dH_0 - \frac{2pH_0}{p-2}\frac{H_4\,dH_4 + H_5\,dH_5 + H_6\,dH_6}{H_4^2 + H_5^2 + H_6^2}\right)\right]$$

$$+ (\omega_2\omega_3) + \cdots.$$

By the theorem relative to (121), the rank of $\Phi$ is four. Since there are five integral invariants, the rank of $\omega'$ is reduced by $2 \times 5 - 4 = 6$, and the order of the differential equations is reduced from $6n - 6$ to $6n - 12$; that is, we obtain the integrals by putting each of $H_0$, $H_4$, $H_5$, $H_6$ equal to arbitrary constants. However, the corresponding characteristic system contains arbitrary parameters.

There is a process for reducing the number of the degree of freedom by avoiding the introduction of arbitrary parameters. Equate the exterior derivative of the right-hand member of (127) to zero; then we obtain

$$\Omega' = \left[\frac{H_4\,dH_4 + H_5\,dH_5 + H_6\,dH_6}{H_4^2 + H_5^2 + H_6^2}\,\Omega\right],$$

by referring to $\omega'_{10} = \frac{1}{2}(2 - p)\omega'$. This shows that the exterior derivative of a quadratic form

$$\frac{1}{\sqrt{H_4^2 + H_5^2 + H_6^2}}\,\Omega = \frac{2}{2-p}\left(\frac{1}{\sqrt{H_4^2 + H_5^2 + H_6^2}}\,\omega_{10}\right)' - \frac{H_4[dH_5\,dH_6] + H_5[dH_6\,dH_4] + H_6[dH_4\,dH_5]}{(H_4^2 + H_5^2 + H_6^2)^{3/2}} \tag{127a}$$

is zero. The first term on the right-hand side is zero because it is an exact exterior derivative. For the second term, consider a vector $OS$ of length $\gamma = (H_4^2 + H_5^2 + H_6^2)^{1/2}$. This is the angular momentum of the whole system with respect to the center of mass. Let $d\sigma$ be the surface element described by the point $S$, and $\alpha_1$, $\alpha_2$, $\alpha_3$ be the direction cosines of the normal to this surface element. Then

$$[dH_5\,dH_6] = \alpha_1\,d\sigma, \qquad [dH_6\,dH_4] = \alpha_2\,d\sigma, \qquad [dH_4\,dH_5] = \alpha_3\,d\sigma.$$

The second term on the right-hand side of (127a) becomes

$$-\frac{\alpha_1 H_4 + \alpha_2 H_5 + \alpha_3 H_6}{\gamma^3}\,d\sigma = -\frac{\cos\psi\,d\sigma}{\gamma}, \tag{127b}$$

where $\psi$ is the angle between the normal and the vector $OS$. The right-hand member of (127b) equals the solid angle subtended by $d\sigma$ at the center of mass. Denote by $\theta$ and $\phi$ the colatitude and the longitude of $S$; then this right-hand member becomes $-\sin\theta[d\theta\,d\phi]$, that is, the exterior derivative of $\cos\theta\,d\phi$. Let

$$\varpi = \frac{2}{2-p}\frac{\omega_{10}}{\gamma} + \cos\theta\,d\phi;$$

then the characteristic equation of the relative integral invariant $\int \varpi$ is

$$dH_0 - \frac{2p}{p-2}H_0\frac{d\gamma}{\gamma} = 0, \qquad \omega_4 = 0, \qquad \omega_5 = 0, \qquad \omega_6 = 0,$$
$$\omega_1 = 0, \ldots, \qquad \omega_7 = 0, \ldots, \qquad \omega_{10} = 0.$$

## ADIABATIC INVARIANTS

### 4.23 ADIABATIC INVARIANTS

Ehrenfest (1917) and Burgers (1917) showed that the mean value of the variation of the phase integral (Section 2.7)

$$J = \oint p\,dq$$

becomes zero in multiply-periodic motions of atomic systems in classical quantum theory, when the system is submitted to slowly varying perturbations. This invariance of the phase integral for infinitely slow variation has been called the *adiabatic invariance* by these two workers, and also by Born (1926). Sommerfeld (1922) assumed that the time variation of the mechanical system is nonsystematic, rather than assuming zero time-average. But a slow variation does not necessarily satisfy this assumption. Kneser (1924) supposed that the mechanical system contains a slowly varying parameter $a$, such that $a = f(t)$ for $0 < t < T$, and, by extending the interval to $0 < t < lT$, proved that the total variation of the phase integral $J(t = lT) - J(t = 0)$ tends to zero as $l \to \infty$, for the case of one degree of freedom. The proof of Fermi and Persico (1926) is not satisfactory. Born and Fock (1928), using Schrödinger's equation in quantum mechanics as a basis, estimated the order for the deviation from invariance which tends to zero asymptotically for $T \to \infty$.

Let the eigenfunctions of the operator $H(s)$ be $\phi_1(q, s), \phi_2(q, s), \ldots$, which satisfy

$$H(s)\phi_n(q, s) = W_n(s)\phi_n(q, s) \qquad (n = 1, 2, \ldots).$$

The Schrödinger equation

$$H\psi_n + \frac{h}{2\pi i}\frac{\partial \psi_n}{\partial t} = 0 \qquad (n = 1, 2, \ldots)$$

is satisfied by the complete, normalized orthogonal system of functions $\psi_1(q, t), \psi_2(q, t), \ldots$, with the initial condition

$$\psi_n(q, 0) = \phi_n(q, 0) \qquad (n = 1, 2, \ldots).$$

We have

$$\dot{q} = \frac{2\pi i}{h}(Hq - qH), \qquad \dot{p} = \frac{2\pi i}{h}(Hp - pH).$$

These equations are satisfied by

$$q = U^\dagger q^0 U, \qquad p = U^\dagger p^0 U, \qquad H = U^\dagger H^0 U,$$

with unitary matrix $U$, where $U^\dagger$ denotes the transposed conjugate matrix of $U$, and $H^0$, $q^0$, $p^0$ are the initial values of $H$, $q$, $p$, respectively. $U$ should satisfy

$$\frac{h}{2\pi i}\dot{U} + H^0 U = 0.$$

The solution is of the form $U = VY$ with unitary matrices $V$ and $Y$. Let $s = t/T$; then,

$$\frac{dY}{ds} = -\frac{2\pi i T}{h} WY + iQY,$$

$$Q = -i\frac{dV^\dagger}{ds} V = iV^\dagger \frac{dV}{ds}.$$

Referring to the matrix elements

$$U_{mn} = \int \phi_m^*(q, 0)\psi_n(q, t)\rho \, dq, \qquad V_{mn} = \int \phi_m^*(q, 0)\phi_n(q, s)\rho \, dq,$$

$$Y_{mn} = \int \phi_m^*(q, s)\psi_n(q, t)\rho \, dq,$$

where a superscript * means the conjugate complex and $\rho$ denotes the density function in $q$-configuration space, we obtain

$$\frac{dY_{mn}}{ds} = -\frac{2\pi i T}{h} W_m Y_{mn} + i \sum_k{}' Q_{mk} Y_{kn}.$$

Let

$$\omega_k(s) = \frac{2\pi}{h}\int_0^s W_k(s)\, ds,$$

$$c_{mn}(s) = Y_{mn} e^{iT\omega_m}, \qquad c_{mn}(0) = \delta_{mn}.$$

Assume that $d(\omega_m - \omega_n)/ds = 2\pi\nu_{mn}(s)$ admits only a finite number of zero points of the $r$th order such that

$$\frac{1}{|2\pi\nu_{mn}(s)|} < \frac{A}{|s - s_0|^r}$$

with a finite positive constant $A$ in the neighborhood of a zero point $s_0$, that $|Q_{mn}(s)| \leqq M_{mn}$ with a finite positive constant matrix $M_{mn}$, and that the real and the imaginary parts of $Q_{mn}(s)/\nu_{mn}(s)$ are piecewise monotonic with a finite number of discontinuity points. Then Born and Fock proved that $c_{mn} - \delta_{mn}$ is of order $T^{-1/(r+1)}$ for a finite $s$ and an infinite $T$ such that

$$c_{mn} = \delta_{mn} + O(T^{-1/(r+1)}).$$

Suppose, with Levi-Civita (1928), that the system of differential equations

$$\frac{dx_i}{dt} = X_i(x_1, x_2, \ldots, x_n, a_1, a_2, \ldots, a_r) \qquad (i = 1, 2, \ldots, n), \quad (128)$$

containing the parameters $a_1, a_2, \ldots, a_r$, but not $t$ explicitly, admits the integrals

$$
\begin{aligned}
&f_j(x_1, x_2, \ldots, x_n, a_1, a_2, \ldots, a_r) = c_j \qquad (j = 1, 2, \ldots, n-1),\\
&f_n(x_1, x_2, \ldots, x_n, a_1, a_2, \ldots, a_r) = t - t_0,
\end{aligned}
\tag{129}
$$

where $a_\rho$ are all constant. An integral curve of (128) is the trajectory of a point moving in the $n$-dimensional space $R_n$, and the trajectory is represented by the intersection of the $n - 1$ hypersurfaces, that is, the $(n-1)$-dimensional continuum represented by the first set of $n-1$ equations of (129) embedded in $R_n$.

When the parameter $a_\rho$ varies as a function of $t$, the variables

$$t_0, c_1, \ldots, c_{n-1}$$

of (129) are functions of $t$. By the principle of the variation of constants we must integrate

$$
\begin{aligned}
\frac{dc_j}{dt} &= \sum_{\rho=1}^{r} \frac{\partial f_j}{\partial a_\rho} \frac{da_\rho}{dt}\\
\frac{dt_0}{dt} &= -\sum_{\rho=1}^{r} \frac{\partial f_n}{\partial a_\rho} \frac{da_\rho}{dt} \qquad (j = 1, 2, \ldots, n-1).
\end{aligned}
\tag{130}
$$

For the coefficients of the time derivatives of $a_\rho$ we substitute the solution of (128), and the result is of the form

$$\frac{\partial f_i}{\partial a_\rho} = \phi_{i\rho}(t - t_0, c_1, \ldots, c_{n-1}, a_1, \ldots, a_r)$$

$$(i = 1, 2, \ldots, n;\ \rho = 1, 2, \ldots, r);$$

also, (130) is transformed to

$$
\begin{aligned}
dc_j &= \sum_{\rho=1}^{r} \phi_{i\rho}\, da_\rho,\\
dt_0 &= -\sum_{\rho=1}^{r} \phi_{n\rho}\, da_\rho \qquad (j = 1, 2, \ldots, n-1).
\end{aligned}
\tag{131}
$$

If we know $a_\rho(t)$, then, when (130) is solved, $c_j(t)$ should be a *functional* of $a_\rho(t)$. Let the domain of $t$ be $T$, and the two ends be $t_0 + T_0$ and $t_0 + T_1$. Take a distribution function

$$F(t - t_0, c_1, \ldots, c_{n-1}, a_1, \ldots, a_r)$$

and consider the mean value of an arbitrary function

$$J(t - t_0, c_1, \ldots, c_{n-1}, a_1, \ldots, a_r):$$

$$\bar{J} = \int_{t_0+T_0}^{t_0+T_1} JF\,dt \Big/ \int_{t_0+T_0}^{t_0+T_1} F\,dt, \tag{132}$$

where the integration is to be carried out with regard to $t$ appearing explicitly in the integrands. By applying this operation to (131) we obtain

$$dc_j = \sum_{\rho=1}^{r} \bar{\phi}_{j\rho}\,da_\rho, \qquad dt_0 = -\sum_{\rho=1}^{r} \bar{\phi}_{n\rho}\,da_\rho; \tag{133}$$

the formulas still contain $t_0, c_1, \ldots, c_{n-1}, a_1, \ldots, a_r$, but the parameters $a_1, \ldots, a_r$ are now considered independent variables. This happens when the parameters $a_1, \ldots, a_r$ vary slowly. An invariant expression of the system of differential equations of the form (133) which were obtained by taking the mean values is called an *adiabatic invariant* (Levi-Civita, 1928, 1928a; Kneser, 1924; Graffi, 1936; Mineur, 1937; Mattioli, 1932; Geppert, 1928, 1929; Lenard, 1959). In order that

$$J(t_0, c_1, \ldots, c_{n-1}, a_1, \ldots, a_r)$$

shall be an adiabatic invariant, it must satisfy the condition

$$\sum_{j=1}^{n-1} \frac{\partial J}{\partial c_j}\bar{\phi}_{j\rho} - \frac{\partial J}{\partial t_0}\bar{\phi}_{n\rho} + \frac{\partial J}{\partial a_\rho} = 0 \qquad (\rho = 1, 2, \ldots, r). \tag{134}$$

If — referring to (134) — we can show that the differential coefficients of all orders with respect to $t$ of the mean values defined by (132) are zero, then the function is said to be *invariant in the mean* with regard to (128), where the word "in the mean" is taken in the sense of Section 2.2. According to the definition commonly accepted by physicists an adiabatic invariant is a function of the integration constants and the parameters of (128) that is invariant in the mean for all values of

$$da_\rho/dt \qquad (\rho = 1, 2, \ldots, r).$$

From such a definition, we obtain,

$$\frac{dJ}{dt} = \sum_{\rho=1}^{r} \left( \sum_{i=1}^{n-1} \frac{\partial J}{\partial c_i}\phi_{i\rho} - \frac{\partial J}{\partial t_0}\phi_{n\rho} + \frac{\partial J}{\partial a_\rho} \right) \frac{da_\rho}{dt}.$$

As $J$ does not depend on $t$ explicitly, this definition requires the validity of (134). This notion was conceived by Ehrenfest (1917) and developed by Burgers (1917). It deals with invariance in the mean rather than ordinary invariance. An external electromagnetic field, for example, in quantum mechanics varies very slowly and slightly compared to variations within the atom itself, so we can quantize with respect to action variables which are considered to be adiabatically invariant.

The mean value defined by (132) depends on $t_0$. Now suppose that it does not. Then the domain $T$, also, does not depend on $t_0$. We consider with Levi-Civita (1928) the mean value taken over the infinite space

$$[J] = \lim_{T \to \infty} \left\{ \int_{t_0}^{t_0+T} JF\, dt \Big/ \int_{t_0}^{t_0+T} F\, dt \right\}. \tag{135}$$

The definition of adiabatic invariants is narrower, and the adiabatic invariant $J$ must identically satisfy the condition

$$\sum_{i=1}^{n-1} \frac{\partial J}{\partial c_i} [\phi_{i\rho}] + \frac{\partial J}{\partial a_\rho} = 0 \qquad (\rho = 1, 2, \ldots, r). \tag{136}$$

As any function of one or more than one adiabatic invariants is also an adiabatic invariant, our problem is to find independent particular solutions of the equations (136). The number of such solutions is at most $n - 1$ and they express the relation between the parameters and

$$c_1, \ldots, c_{n-1}.$$

In particular, an integral $f_i(x_1, \ldots, x_n) = c_i$ independent of the parameters is an adiabatic invariant.

The first $n - 1$ integrals (129) determine the trajectory $S$ in the $n$-dimensional space. Suppose that $f_i$ is single-valued or at least a finite-number-valued function along $S_i$, and suppose in addition that, if the domain $T$ of $t$ over which to take the mean extends to infinity, $f_i$ has the property of *ergodicity* — because we have no information as to its behavior in the infinite domain. That is, *almost all* trajectories defined by (128) fill up *densely* a certain continuum $\Phi$. The words "almost all" are used (in the sense of point-set theory) to mean that we exclude such values of the integration constants as the Lebesgue measure zero (Sections 2.9, 2.19). As we shall see later, Poincaré and Carathéodory proved that such a situation occurs when the trajectories are in a domain of $R_n$ with finite measure, and the last multiplier $M$ for (128) is either positive or zero (at most) at points of measure zero in that domain. This is called the *quasi-ergodic theorem* (see Dragoni, 1934, 1934a).

If we make this additional assumption, then the trajectory fills up densely the $(n - m)$-dimensional continuum $\Phi$ if $m$ of the $n$ equations (129) are single-valued (or, at most, finite-number-valued) along the whole trajectory and the remaining $n - m$ equations are infinite-number-valued. For convenience we assume that $\Phi$ is a closed continuum. The number $n$ is called the *primitivity order*. When $m = 0$, the space $\Phi$ is called *primitive*. By this assumption of quasi-ergodicity in topology, we can replace the operations extending over $t$ of (135) by the operations with respect to the space $\Phi$.

Consider the case $m = 1$. There is only one integral,

$$f(x_1, \ldots, x_n, a_1, \ldots, a_r) = c,$$

and it determines an $(n - 1)$-dimensional hypersurface $\Phi$. Assume that $F = $ constant. Suppose that an element $d\Phi_0$ of $\Phi$ at $t = t_0$ is displaced to $d\Phi_1$ at $t = t_1$, and define a single-valued density function $\kappa$ such that $\kappa_0 d\Phi_0 = \kappa_1 d\Phi_1$. The quantity $\kappa_0$ is the value of $\kappa$ where $t = t_0$, and $\kappa_1$ is the value where $t = t_1$. There is one and only one function $\kappa$ with this property. For, if there were another single-valued function $\lambda$ with this property, then $\kappa d\Phi$ and $\lambda d\Phi$ would be invariant because of (128). On the assumption that $f$ is a finite-number-valued function, the ratio $\kappa : \lambda$, being an integral of (128), should be a function of $f$, that is, should be equal to a constant on the hypersurface $\Phi$. Hence, there is only one such $\kappa$.

Now consider besides $f = c$ another hypersurface $f = c + dc$ and call it $\Phi'$. Let $P'$ be the point of intersection of the normal to $\Phi$ at a point $P$ of $\Phi$ with the hypersurface $\Phi'$. The direction cosines of the normal are

$$\frac{1}{\Gamma_{1n}} \frac{\partial f}{\partial x_i} \qquad (i = 1, 2, \ldots, n),$$

$$\Gamma_{1n}^2 = \sum_{i=1}^{n} \left(\frac{\partial f}{\partial x_i}\right)^2.$$

If we denote a line element of the normal by $d'n$, then $\Gamma_{1n}\, d'n = dc$, and the components of $d'n$ are

$$d'x_i = \frac{1}{\Gamma_{1n}^2} \frac{\partial f}{\partial x_i}\, dc \qquad (i = 1, 2, \ldots, n).$$

Let the volume element situated between $\Phi$ and $\Phi'$ be $d\tau = d\Phi\, d'n$. As the trajectories are on this hypersurface, by (128), then

$$d\tau = d\Phi_1\, d'n_1.$$

Hence,

$$\kappa\, d\Phi\, dc = \kappa \Gamma_{1n}\, d\tau = \kappa_1\, d\Phi_1\, dc = [\kappa \Gamma_{1n}]_1\, d\tau_1,$$

that is, $M\,d\tau = M_1\,d\tau_1$, where $M = \kappa\Gamma_{1n}$. In other words, $\int M\,d\tau$ is an integral invariant with regard to (128). Hence, $M$ is the last multiplier by the theorem in Section 4.18 due to Poincaré and must satisfy

$$\sum_{i=1}^{n} \frac{\partial M}{\partial x_i} X_i + M \sum_{i=1}^{n} \frac{\partial X_i}{\partial x_i} = 0. \tag{137}$$

Hence, if $M$ is the last multiplier of (128), then $\kappa = M/\Gamma_{1n}$.

Assume here that the solution of (137) is finite-number-valued and does not become zero on $\Phi$ except at a finite number of points and that its sign is positive; then, instead of (135), we have

$$[J] = \int_\Phi J \frac{M}{\Gamma_{1n}}\,d\Phi \Big/ \int_\Phi \frac{M}{\Gamma_{1n}}\,d\Phi.$$

If $F \not\equiv 1$, then

$$[J] = \int_\Phi JF \frac{M}{\Gamma_{1n}}\,d\Phi \Big/ \int_\Phi F \frac{M}{\Gamma_{1n}}\,d\Phi.$$

Hence, the adiabatic invariant (136) can be written

$$\frac{\partial J}{\partial c} \int_\Phi \frac{\partial f}{\partial a_\rho} F \frac{M}{\Gamma_{1n}}\,d\Phi + \frac{\partial J}{\partial a_\rho} \int_\Phi F \frac{M}{\Gamma_{1n}}\,d\Phi = 0 \tag{138}$$

$$(\rho = 1, 2, \ldots, r).$$

Geppert (1928, 1929), by studying the condition for integrability of this equation, derived the necessary and sufficient condition for the existence of adiabatic invariants in a dynamical system of primitivity order $m = 1$ to be

$$\left[\frac{\partial f}{\partial a_\rho} \frac{\partial \log MF}{\partial a_\lambda}\right] - \left[\frac{\partial f}{\partial a_\rho}\right]\left[\frac{\partial \log MF}{\partial a_\lambda}\right]$$

$$= \left[\frac{\partial f}{\partial a_\lambda} \frac{\partial \log MF}{\partial a_\rho}\right] - \left[\frac{\partial f}{\partial a_\lambda}\right]\left[\frac{\partial \log MF}{\partial a_\rho}\right]$$

$$(\lambda, \rho = 1, 2, \ldots, r). \tag{139}$$

In particular, if $\sum_{i=1}^{n} \partial X_i/\partial x_i = 1$, the dynamical system is called *Liouville's*; the $n$-dimensional volume $V = \int dx_1, \ldots, dx_n$ is now an integral invariant of the system of differential equations. It occurs in the motion of an incompressible fluid. In this case $M =$ constant and $F = 1$. Such a dynamical system has been studied by Gibbs and Hertz in analytical dynamics and statistical mechanics. We can prove that an adiabatic invariant exists by taking the mean value. In general, when $MF$ does not depend on the parameters $a_1, \ldots, a_r$ we call it the *general Liouville case*; there then exists an adiabatic invariant. Hence, there exists an integral invariant

$$V = \int_{\Phi} MF\, dx_1 \cdots dx_n \tag{140}$$

extended over the inside of the hypersurface $\Phi$. In the general Liouville case, because of (140), equations (138) can be written in the form

$$\frac{\partial J}{\partial c}\frac{\partial V}{\partial a_\rho} - \frac{\partial J}{\partial a_\rho}\frac{\partial V}{\partial c} = 0 \qquad (\rho = 1, 2, \ldots, r).$$

Thus

$$J = V = \int MF\, dx_1 \cdots dx_n$$

is an adiabatic invariant in the general Liouville case.

For a general value of $m$, if there exists an independent integral

$$f_i(x_1, \ldots, x_n; a_1, \ldots, a_r) = c_i \qquad (i = 1, 2, \ldots, m),$$

then, by a similar argument, the density function $\kappa$ is $M/\Gamma_{mn}$, where $\Gamma_{mn}^2$ is the square of the matrix $(\partial(f_1, \ldots, f_m)/\partial(x_1, \ldots, x_m))$; and the differential equations for the adiabatic invariants corresponding to (138) are written

$$\sum_{i=1}^{m} \frac{\partial J}{\partial c_i} \int_{\Phi} \frac{\partial f_i}{\partial a_\rho} F \frac{M}{\Gamma_{mn}}\, d\Phi + \frac{\partial J}{\partial a_\rho} \int_{\Phi} \frac{M}{\Gamma_{mn}}\, d\Phi = 0$$

$$(\rho = 1, 2, \ldots, r). \tag{141}$$

Now consider the Hamiltonian canonical equations

$$\frac{dq_i}{dt} = \frac{\partial H}{\partial p_i}, \qquad \frac{dp_i}{dt} = -\frac{\partial H}{\partial q_i} \qquad (i = 1, 2, \ldots, n), \tag{142}$$

where $H$ does not contain $t$ explicitly. There exists the integral $H = h$. Assume that there exists no integral other than $H = h$ and that the hypersurface $H = h$ is closed and that the dynamical system is quasi-ergodic (as is evident from topological transformations of the motion of molecules in a cube — see Section 2.9). Then the integral $\int FM\, dx_1 \cdots dx_n$ extended over the volume in the phase space $\Phi_{2n}$ bounded by $H = h$ is an adiabatic invariant; in particular, if $F =$ constant in a Liouville dynamical system, then the volume is an integral invariant. If, in addition, (142) has $m$ integrals $f_i = c_i$ mutually in involution, then

$$(f_\alpha, f_\beta) = 0 \qquad (\alpha, \beta = 1, 2, \ldots, m),$$

and

$$(f_\alpha, H) = 0 \qquad (\alpha = 1, 2, \ldots, m).$$

Let the $m$ solutions of $f_i = c_i$ with respect to the $p_\alpha$ be

$$p_\alpha = p_\alpha(p_{m+1}, \ldots, p_n, q_1, \ldots, q_n, c_1, \ldots, c_m) \qquad (\alpha = 1, 2, \ldots, m)$$

and the result of the substitution of these equations in $H$ be $\mathfrak{H}$. By Lie's theorem (Section 4.8), the canonical equations (142) are reduced to the differential equations of order $2(n - m)$:

$$\frac{dq_i}{dt} = \frac{\partial \mathfrak{H}}{\partial p_i}, \qquad \frac{dp_i}{dt} = -\frac{\partial \mathfrak{H}}{\partial q_i} \qquad (i = m + 1, \ldots, n),$$

and $q_\alpha$ are obtained from

$$\frac{dq_\alpha}{dt} = \frac{\partial \mathfrak{H}}{\partial p_\alpha} \qquad (\alpha = 1, 2, \ldots, m).$$

We call such a $2(n - m)$-dimensional space the reduced phase space. If the reduced phase space is closed and the dynamical system is Liouville's, then the volume $W(h, c_1, \ldots, c_m)$ of this phase space is an integral invariant. If $H$ and $f_\alpha$ both contain the parameters $a_1, \ldots, a_r$, and hence, $\mathfrak{H}$ contains $2m + r$ parameters $q_1, \ldots, q_m; c_1, \ldots, c_m; a_1, \ldots, a_r$, all of which vary slowly, and if the trajectory is quasi-ergodic in the $2(n - m)$-dimensional closed space, then the volume $W(h, c_1, \ldots, c_r)$ of the reduced phase space is an adiabatic invariant. Since $H = h$, we rewrite it $f_0 = c_0$; then

$$(f_\alpha, f_\beta) = 0 \qquad (\alpha, \beta = 0, 1, 2, \ldots, m).$$

Let the result of substitution of these $m$ integrals for $p_1, \ldots, p_m$ in the $m + 1$ functions $f_\alpha$ be $F_\alpha$; then,

$$\frac{dq_i}{dt} = \frac{\partial F_\alpha}{\partial p_i}, \qquad \frac{dp_i}{dt} = -\frac{\partial F_\alpha}{\partial q_i} \qquad (i = m + 1, \ldots, n; \alpha = 0, 1, \ldots, m).$$

All these equations give the same trajectory. There exist adiabatic invariants $W_1, \ldots, W_m$, which are the respective $2(n - m)$-dimensional volumes corresponding to $F_1, \ldots, F_m$ taken for $F_0$ or $\mathfrak{H}$ in the above discussion. Hence, such a dynamical system given by canonical equations of primitivity order $m + 1$ has $m + 1$ adiabatic invariants, if the $m$ independent integrals are mutually in involution.

A special case $n = m + 1$ contains the case in which the solution can be obtained by separation of the variables according to Stäckel's theorem. Suppose that the functions $\phi_{\alpha\beta}$ (which appeared in our proof of Stäckel's theorem in Section 1.16) are functions of $q_\beta$ only. Then we have $n$ integrals

$$F_\alpha = \sum_{\beta=0}^{n-1} \phi^{\alpha\beta}(\tfrac{1}{2}p_\beta^2 - U_\beta) \qquad (\alpha = 0, 1, \ldots, n-1),$$

where $U_\beta$ is a function of a single variable $q_\beta$ only, and this function $F_\alpha$ is a function of a single variable $q_\alpha$ only. Hence the $F_\alpha$ are mutually in involution. The reduced phase space is now a plane, and there exist $n$ adiabatic invariants

$$W_i = \int p_i \, dq_i \qquad (i = 0, 1, \ldots, n-1).$$

This proves that the action in quantum mechanics is an adiabatic invariant (Section 2.7). The proof does not necessitate our assuming that the periods of the component oscillations are mutually in irrational ratios as in the case studied by Burgers (1917). Thus we have proved the fundamental theorem of quantum mechanics. We have assumed quasi-ergodicity in our proof; this assumed property is valid in any Liouville dynamical system, as will be shown in a later chapter on the fundamentals of the statistical mechanics for Boltzmann's theory. We have proved that the coordinates and the momenta in Stäckel's case are expressed by quasi-periodic functions (Section 2.10). Hence, the motion in this case is quasi-periodic.

Lenard (1959) has shown that a positive constant $M > 0$ can be found such that, for a change $\Delta Q$ of an adiabatic invariant $Q$, $|\Delta Q| < \epsilon^n M$ for all small values of $\epsilon$; he states that the adiabatic invariant is of the $n$th order. There are many such invariants that are known to be constant to a low order but are actually so to all orders. Kulsrud (1957) proved this proposition for the harmonic oscillator and Lenard (1959) for non-linear oscillators.

The theory of adiabatic invariants has been applied to the fall of meteoric swarms or cosmic dust, and to the motion of celestial bodies of varying mass as affected by the ejection or accretion of their masses (Levi-Civita, 1934).

## 4.24 ADIABATIC APPROXIMATION

A function $J(p, q, \lambda)$ was called by Arnold (1963) an adiabatic invariant of a dynamical system with the Hamiltonian

$$H(p, q, \lambda) \qquad (p = p_1, p_2, \ldots, p_n;\ q = q_1, q_2, \ldots, q_n),$$

on the following condition: for any $\kappa > 0$ there exists a $\mu_0 > 0$, such that, if $0 < \mu < \mu_0$, then for any $t$ in the interval $0 < t < 1/\mu$ there exists the inequality $|J(t) - J(0)| < \kappa$.

The linear oscillatory system occupies an exceptional position: the frequency of its oscillation does not depend on the amplitude, and there are well-defined resonant frequencies. In nonlinear systems the frequency changes with amplitude, and oscillations do not grow without limit as in the linear case; that is, the condition of resonance in the nonlinear case changes as the amplitude changes.

Consider a nonlinear oscillatory system with one degree of freedom with the analytic Hamiltonian function $H(p, q, \lambda)$ depending on slow time $\lambda = \mu t$ periodically, that is, $H(p, q, \lambda + 2\pi) = H(p, q, \lambda)$. In the zero-th approximation, $\lambda =$ constant and the system is autonomous, i.e., independent of an explicit appearance of $t$. Also, the system is integrable. Periodic motions of the system can be described by the action $I$ and angle $w$. We have

$$p = \frac{\partial S}{\partial q}, \qquad w = \frac{\partial S}{\partial I},$$

with the generating function

$$S(I, q) = \int_{H=h}^{q} p\, dq,$$

where $H_0(I) = h$ is the inverse function of

$$I(h) = \frac{1}{2\pi} \int_{H=h} p\, dq,$$

and $p$ is the value derived from $H(p, q) = h$. The variation of $I$ and $w$ with time is given by

$$\dot{w} = \omega(I) = \frac{\partial H_0}{\partial I}, \qquad I = \text{constant}.$$

This approximation is valid during the time $t \sim 1$.

The transformation $p, q \rightarrow I, w$ being canonical, the change of $I$ and $w$ is determined by the canonical equations with the Hamiltonian

$$H(I, w, \lambda) = H_0(I) + \mu H_1(I, w, \lambda), \tag{143}$$

where $H_1 = \partial S/\partial \lambda$ is a single-valued function of period $2\pi$ with respect to $w$ and $\lambda$. In the space of $p, q, \lambda$ the equation $I =$ constant determines a two-dimensional torus and the angular variables $w$ and $\lambda$ are the latitude and longitude on this torus (Figure 1). In the conservative approximation

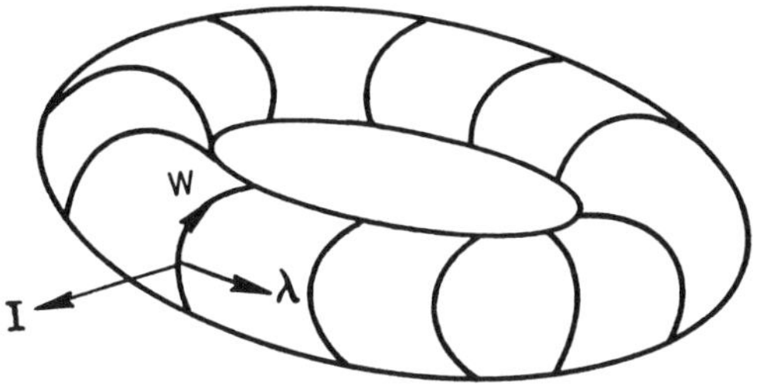


FIGURE I

$\mu = 0$, every point of the torus moves along the meridian $\lambda$ = constant with angular velocity $\omega(I, \lambda)$ depending on the longitude. For $\mu \neq 0$, a slow motion $\lambda = \mu$ is added across the meridian; the motion now has two frequencies. But in the approximation the phase point remains on the invariant torus $I$ = constant. This approximation is called adiabatic. The true motion is near the adiabatic approximation during $t \sim 1/\mu$.

In the linear system, $\omega(I)$ = constant, $H_0(I) = I\omega$. In the nonlinear system,

$$\bar{\omega}(I) = \frac{1}{2\pi}\oint \omega(I, \lambda)\, d\lambda = \frac{d\bar{H}_0}{dI}, \qquad \bar{H}_0(I) = \frac{1}{2\pi}\oint H_0(I, \lambda)\, d\lambda.$$

On the assumption that

$$\frac{d^2\bar{H}_0}{dI^2} = \frac{d\bar{\omega}}{dI} \neq 0 \tag{144}$$

we can prove the *perpetual* adiabatic invariance of $I$ (Arnold, 1963).

In the case of the perturbed Hamiltonian (143) we obtain many invariant tori, real but not approximate, near the tori $I$ = constant, for small $\mu$. They are two-dimensional and divide the three-dimensional phase space $p$, $q$, $\lambda$ into thin layers. It can be proved that, if the initial point $(p_0, q_0, \lambda)$ lies between two such tori $T_1$ and $T_2$, then the entire trajectory $p(t)$, $q(t)$, $\lambda(t)$ lies wholly between $T_1$ and $T_2$; more precisely, we can find a constant $\mu_0 > 0$ for any $\kappa > 0$ such that, if $|\mu| < \mu_0$, then any point $(p_0, q_0, \lambda_0)$ lies between two invariant tori $T_1$ and $T_2$, and

$$|I(p_1, q_1, \lambda_1) - I(p_2, q_2, \lambda_2)| < \kappa$$

for $p_1$, $q_1$, $\lambda_1$ belonging to $T_1$ and $p_2$, $q_2$, $\lambda_2$ belonging to $T_2$.

Let us introduce a new time, $T = w$. Then the integral curves of the Hamiltonian system in space $I$, $w$, $\lambda$ are invariantly connected with the differential form

$$I\, dw - H(I, w, \mu t)\, dt = -\frac{1}{\mu}(H\, d\lambda - \mu I\, dw).$$

Consider $H$, $\lambda$, $w$ instead of $I$, $w$, $t$ as the independent variables. Solving (143) with respect to $I$, we obtain

$$I(H, \lambda, w) = I_0(H, \lambda) + \mu I_1(H, \lambda, w) + \cdots.$$

Put

$$p' = H, \qquad q' = \lambda, \qquad T = w, \qquad K = \mu I.$$

Then,

$$\begin{aligned} K(p', q', T) &= \mu I_0(p', q') + \mu^2 I_1(p', q', T) + \cdots, \\ H\, d\lambda - \mu I\, dw &= p'\, dq' - K(p', q', T)\, dT, \end{aligned} \tag{145}$$

so that the systems with the Hamiltonians (143) and (145) are equivalent. The variables $P$ and $Q$ are introduced by the canonical transformation

$$p' = \frac{\partial S}{\partial q'}, \qquad Q = \frac{\partial S}{\partial P},$$

with the generating function

$$S(P, q') = \int^{q'} H_0(I, \lambda)\, d\lambda.$$

The Hamiltonian is transformed to

$$k(P, Q, T) = \mu k_0(P) + \mu^2 k_1(P, Q, T) + \cdots, \tag{146}$$

where $I = k_0(P)$ is a function inverse to $H_0(I)$.

Hence, if the frequency $\omega(I, \lambda) = \partial H_0/\partial I$ does not vanish, there exist analytic functions $P$, $Q$, $T$ of the variables $I$, $w$, $\lambda$ independent of $\mu$, such that (i) $P$, $Q - \lambda$, $T - w$ are periodic with period $2\pi$ with respect to $w$ and $\lambda$; (ii) the canonical equations with the Hamiltonian (143) are equivalent to the equations

$$\frac{dP}{dT} = -\frac{\partial k}{\partial Q}, \qquad \frac{dQ}{dT} = \frac{\partial k}{\partial P},$$

with the Hamiltonian (146) which is of period $2\pi$ with respect to both $Q$ and $T$ and analytic in the complex neighborhood of the toroidal layer $|P - I_0| \leqq r$; and (iii) the function $k_0(P)$ of (146) is inverse to $\overline{H}_0(I)$, so that $\overline{H}_0(k_0(P)) = P$.

From this theorem we can show the perpetual adiabatic invariance of action.

*Theorem:* Let an oscillatory system be described by the analytic Hamiltonian (143), in terms of action-angle variables. Suppose that

$\partial H_0/\partial I \neq 0$ and $\partial^2 H_0/\partial I^2 \neq 0$ in the region $|I - I_0| \leqq r$ everywhere. If for any $\kappa > 0$ we can find a constant $\mu_0 > 0$ such that $|\mu| < \mu_0$ and $|I(0) - I_0| \leqq r - \kappa$, then $|I(t) - I(0)| < \kappa$ will hold for

$$-\infty < t < +\infty.$$

In fact, the system with the Hamiltonian (146) can be reduced to a set of canonical equations with two degrees of freedom

$$\frac{dP}{d\tau} = -\frac{\partial k}{\partial Q}, \qquad \frac{dR}{d\tau} = -\frac{\partial k}{\partial T}, \qquad \frac{dQ}{d\tau} = \frac{\partial k}{\partial P}, \qquad \frac{dT}{d\tau} = 1,$$

with the Hamiltonian

$$R + k(P, Q, T) = R + \mu k_0(P) + \mu^2 k_1(P, Q, T) \cdots.$$

The Hamiltonian is now of the degenerate form

$$H(p, q) = H_0(p_0) + \mu H_1(p, q) + \cdots, \tag{147}$$

where $p = (p_0, p_1)$ with a vector $p_0$ of $n_0$ dimensions (corresponding to the fast-varying coordinates $q_0$) and a vector $p_1$ of $n_1$ dimensions, $n = n_0 + n_1$ (corresponding to the slow-varying coordinates $q_1$). The solution with this Hamiltonian (147) in the zero-th approximation is

$$\dot{p}_0 = 0, \qquad \dot{q}_0 = \omega_0(p_0) = \frac{\partial H_0}{\partial p_0},$$

$$\dot{p}_1 = 0, \qquad \dot{q}_1 = \mu \xi_1(p) = \mu \frac{\partial \bar{H}_1}{\partial p_1}, \tag{148}$$

where $\omega_1 = \mu \xi_1$ for the slow frequencies.

Assume that the secular part of the perturbation averaged over the fast motion in the coordinates $q_0$,

$$\bar{H}_1 = \frac{1}{(2\pi)^{n_0}} \oint H_1 \, dq_0,$$

is independent of the phase $q_1$ of the slow motion $\bar{H}_1 = \bar{H}_1(p_1)$, and that the determinants of the frequencies with respect to momenta $p_0$ and $p_1$ do not become identically zero; that is,

$$\left\| \frac{\partial \omega_0}{\partial p_0} \right\| = \left\| \frac{\partial^2 H_0}{\partial p_0^2} \right\| \not\equiv 0, \qquad \left\| \frac{\partial \xi_1}{\partial p_1} \right\| = \left\| \frac{\partial^2 \bar{H}_1}{\partial p_1^2} \right\| \not\equiv 0.$$

Arnold (1963) has proved that for sufficiently small $|\mu|$ for the majority of initial conditions — that is, all except for sets of Lebesgue measure zero — the motion determined by the canonical equations with the

Hamiltonian (147) is conditionally periodic (Section 2.7) and, for $-\infty < t < +\infty$, is near the conditionally periodic motion (148). This is, according to Arnold, because almost every vector

$$\omega = (\omega_1, \omega_2, \ldots, \omega_n)$$

satisfies the inequalities

$$|(\omega, k)| \geqq K|k|^{-\nu} \qquad (|k| = |k_1| + \cdots + |k_n|; \nu = n + 1)$$

for all integers $k \neq 0$ for some $K(\omega) > 0$, as will be discussed in a later chapter.

Thus we come to the following conclusion: If the function (146) is analytic for $|\mu| < \bar{\mu}$ in the region

$$|\Im Q, T| \leqq \rho, \qquad |P - P_0| \leqq r$$

and in this region

$$|k_0| \leqq M, \qquad |k_1 + (\mu)\cdots| \leqq M, \qquad \left|\frac{d^2k_0}{dP_0^2}\right| \geqq \theta > 0,$$

then for any $\kappa > 0$ we can find a constant $\mu_0(\kappa, \bar{\mu}, r, \rho, M, \theta) > 0$ with the property that if $|\mu| < \mu_0$, then the real toroidal layer $F$, where $|P - P_0| \leqq r$, is filled with invariant tori with accuracy up to the remainder of measure smaller than $\kappa m(F)$, and each of these invariant tori is located at a distance less than $\kappa$ from some torus $P =$ constant. Hence the theorem on the perpetual adiabatic invariance of action is proved.

We confine ourselves for the moment to a conservative dynamical system with two degrees of freedom $X$, $Y$. Assume that the change of the coordinate $X$ only slightly affects the state of the system; then the system can be regarded as having approximately one degree of freedom but depending on the slowly varying parameter $\mu X$. The action variable $I_Y$ takes the part of an adiabatic invariant.

*Lemma:* Let $X$, $Y$, $P_X$, $P_Y$ be canonically conjugate variables, and consider the Hamiltonian $H(\mu X, Y, P_X, P_Y)$. If $\mu$ is small, then the state of the system changes little during the variation of $X$ by a first-order quantity. If we fix $X$ and $P_X$, then the Hamiltonian is of the form $H_Y(Y, P_Y)$, and the action variable is $I_Y(Y, P_Y)$. The value of $I_Y$ depends also on $\mu X$ and $P_X$. For the Hamiltonian $H(\mu X, Y; P_X, P_Y)$, the action $I_Y$ is an adiabatic invariant in the sense that the change of

$$I_Y[\mu X(t), Y(t), P_X(t), P_Y(t)]$$

during the time $t \sim 1/\mu$ is small, and of order $\mu$. This Hamiltonian can be written $H = H_X(\mu X, P_X, I_Y)$. For an approximation the system is described by the Hamiltonian $H_X(\mu X, P_X)$. If it is periodic with one degree of freedom, then we introduce the action variable $I_X \sim 1/\mu$. In the adiabatic approximation, the motion consists of fast oscillations of $Y$ with frequence $\omega_Y$ and slow oscillations of $X$ with frequency $\omega_X$ of order $\mu$, in which $I_Y$ and $\mu I_X$ are conserved.

Suppose that the Hamiltonian is reduced to the form (145). That $I_Y$ and $\mu I_X$ are perpetual adiabatic invariants is easily concluded from the existence of two-dimensional invariant tori, since these tori divide the three-dimensional energy level $H = h$.

Let the Hamiltonian function $H(x, Y, P_X, P_Y)$, where $x = \mu X$, be analytic and, for a fixed $x$, let the momenta $P_X$ determine the oscillatory system with the variables $I_Y(x, P_X, h)$ and $w_Y(x, Y, P_X, P_Y)$. Then there exists an analytic substitution, changing $x$, $Y$, $P_X$, $P_Y$ to

$$x', w', P', I'$$

such that: (i) the functions of $x$, $Y$, $P_X$, $P_Y$ admit period $2\pi$ with regard to $w$, and for $\mu \to 0$ the variables $x'$, $w'$, $P'$, $I'$ are transformed to $x$, $w_Y$, $P_X$, $I_Y$; (ii) along the integral curves of the canonical equations

$$\frac{dP_X}{dt} = -\frac{\partial H}{\partial X}, \qquad \frac{dP_Y}{dt} = -\frac{\partial H}{\partial Y}, \qquad \frac{dX}{dt} = \frac{\partial H}{\partial P_X}, \qquad \frac{dY}{dt} = \frac{\partial H}{\partial P_Y},$$

the canonical equations

$$\frac{dP'}{dw'} = -\frac{\partial K}{\partial x'}, \qquad \frac{dx'}{dw'} = \frac{\partial K}{\partial P'} \tag{149}$$

are satisfied with the Hamiltonian $K(P', x', w', h)$ depending on the parameter $h$; and (iii) the function $K$ is of the form $K = -\mu I'$, where

$$I' = I_0(P', x', h) + \mu I_1(P', x', w', h) + \cdots \tag{150}$$

is analytic with period $2\pi$ with respect to $w'$ and

$$I_0(P', x', h) = I_Y(x', P', h).$$

From $H(x, Y, P_X, P_Y) = h$, we have

$$P_Y = P_Y(x, P_X, Y, h).$$

Assume that $2\pi I_Y(x, P_X, h) = \oint P_Y \, dX$; then this relation determines $h(x, P_X, I_Y)$. The generating function $P'X + S(x, P', Y, I')$, where

$$S = \int^Y P_Y[x, P', Y, h(x, P', I')] \, dY,$$

gives the canonical transformation $X, Y, P_X, P_Y \to X', w', P', I'$, according to the formulas

$$X' = X + \frac{\partial S}{\partial P'}, \qquad w' = \frac{\partial S}{\partial I'}, \qquad P_X = P' + \mu \frac{\partial S}{\partial x}, \qquad P_Y = \frac{\partial S}{\partial Y}.$$

Assume that $x' = \mu X'$. Since

$$H\{x, Y, P', P_Y[x, P', Y, h(x, P', I')]\} = h(x, P', I'),$$

then

$$H\{x, Y, P_X, P_Y[x, P', Y, h(x, P', I')]\} = h(x, P', I') + \mu \frac{\partial S}{\partial x} \frac{\partial H}{\partial P_X} + \cdots.$$

Substituting the new variables for $X$ and $Y$, we obtain

$$H(x, Y, P_X, P_Y) = h(x', P', I') + \mu \frac{\partial S}{\partial x} \frac{\partial H}{\partial P_X} - \mu \frac{\partial S}{\partial P'} \frac{\partial H}{\partial x} + \cdots,$$

or

$$\begin{aligned} H(x, Y, P_X, P_Y) &= H'(x', P', I', w') \\ &= h_0(x', P', I') + \mu h_1(x', P', I', w') + \cdots, \end{aligned} \tag{151}$$

where $\mu h_1 + \cdots$ is analytic, with period $2\pi$ with respect to $w'$.

Taking $P', X', -H, t, w'$, instead of $P', X', I', w', t$ as the new independent variables, and taking $h$ as a parameter and $w'$ as time, gives the new Hamiltonian as $I'(X', P', w', h)$, where $I'$ is determined by

$$H'(\mu X', P', I', w') = h. \tag{152}$$

The coordinate $t$ is cyclical. By dropping $-H$ and $t$ we obtain a nonautonomous system with one degree of freedom. Multiply $X'$ and $-I'$ by the constant $\mu$, giving $\mu X' = x'$ and $-\mu I' = K$. The derivatives of $x'$ and $P'$ with respect to $w'$ are determined by the canonical equations (149) with the Hamiltonian $K$.

These forms of $H$ and $H'$ show that $I'$ is of the form (150); thus the lemma is proved.

According to (150), the Hamiltonian $K$ is of the form (145). By our assumption, the inequalities (144) are satisfied. Consequently the argument in the proof of perpetual adiabatic invariance can be applied. Hence the existence of the invariant tori and the perpetual adiabatic invariance of $I_Y$ and $\mu I_X$ are proved, on the assumption that the mean value $\overline{(\omega_Y/\omega_X)}$ of the ratio $\omega_Y/\omega_X$ depends on $I_Y$ for a fixed total energy $h$.

Thus the method of adiabatic approximation is justified in this case of two degrees of freedom.

Arnold applied the idea to the problem of magnetic traps, that is, the motion of a charged particle in a magnetic field. Assume that the instantaneous radius of the helix along which the particle moves is small compared with the distance at which the field changes by a noticeable amount. Arnold has shown that in axially-symmetric magnetic traps the adiabatic invariant $W_\perp/B$ is conserved perpetually, where $B$ is the magnetic field intensity and $W_\perp$ is the kinetic energy of the motion transverse to the field.

Suppose that the magnetic field $B$ is derived from the vector potential $A$, which has only one component $A_\phi = A(r, z)$ in polar coordinates. Then the intensity is

$$B_r = -\frac{\partial A}{\partial z}, \qquad B_z = \frac{1}{r}\frac{\partial(rA)}{\partial r},$$

and the lines of force are given by $rA$ = constant and $\phi$ = constant. The Lagrangian function for unit charge and unit mass is

$$L = \tfrac{1}{2}(\dot{r}^2 + \dot{z}^2 + r^2\dot{\phi}^2) + rA\dot{\phi},$$

and the Hamiltonian is

$$H = \frac{1}{2}\left[p_r^2 + p_z^2 + \frac{(p_\phi - rA)^2}{r^2}\right].$$

Since $\phi$ is cyclical, the value of $p_\phi = M$ is constant. There remains to be studied the plane motion in the field with potential

$$U(r, z) = \frac{(M - rA)^2}{2r^2}.$$

This potential represents a potential ditch with zero bottom along the line of force $rA = M$. In the neighborhood of this line we have

$$U(r, z) = \tfrac{1}{2}B^2y^2 + \cdots,$$

where $y$ is the distance to a line of force.

We introduce the coordinates $x$ and $y$ by

$$dr^2 + dz^2 = [1 + yk(x)]^2\, dx^2 + dy^2,$$

where $k(x)$ is the curvature of the line of force at the point $(x, 0)$. The new Hamiltonian is

$$H = \frac{1}{2}\left\{\frac{p_x^2}{[1 + yk(x)]^2} + p_y^2\right\} + U,$$

$$U(x, y) = \tfrac{1}{2}B(x)y^2 + \cdots.$$

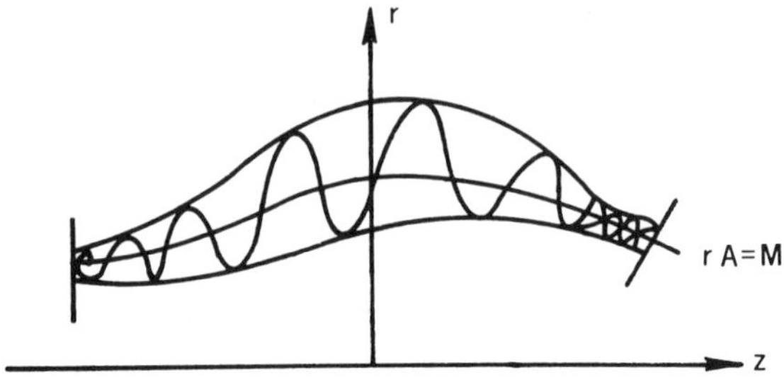


FIGURE 2

We are interested in the case in which the radius of the Larmor spiral described by the charge around the lines of force is very small compared with the characteristic dimension of the field; that is, we consider the values of $M$ and the total energy $h$ such that $U \leqq h$ determines a strip of width $\sim\mu$ around the line of force $rA = M$ (Figure 2). Introduce new variables $X$, $Y$, $P_X$, $P_Y$ by $x = \mu X$, $y = \mu Y$, $p_x = \mu P_X$, $p_y = \mu P_Y$; then the Hamiltonian becomes

$$H' = \frac{1}{2}\left\{\frac{P_X^2}{[1 + \mu Yk(\mu X)]^2} + P_Y^2\right\} + U',$$

$$U'(\mu X, Y) = \tfrac{1}{2}B^2(\mu X)Y^2 + \cdots.$$

or

$$H' = \frac{P_X^2 + P_Y^2 + B^2(x^2)Y^2}{2} + \mu H_1(x, Y, P_X, P_Y) + \cdots$$

of the form similar to the foregoing.

We select the initial values of $y$, $\dot{x}$, $\dot{y}$ in such a manner that $H \sim \mu^2$, $\mu \ll 1$, and $y, \dot{x}, \dot{y} \sim \mu$. We fix $Y$, $\dot{X}$, $\dot{Y}$, choose $y = \mu Y$, $\dot{x} = \mu\dot{X}$, $\dot{y} = \mu\dot{Y}$, and put $\mu$ equal to zero. If the magnetic field is analytic and $B(x) > 0$, $B(x) \to \infty$ for $|x| \to \infty$, then for any $\kappa > 0$ we have $\mu_0 > 0$ such that, if $|\mu| < \mu_0$, then $|I_Y(t) - I_Y(0)| < \kappa$ for all $-\infty < t < +\infty$, where $I_Y = (P_Y^2 + B^2Y^2)/(2B)$. Hence, a particle for which $I_Y \neq 0$ is locked in a bounded region if the field $B$ grows indefinitely as $x \to \infty$. This is the phenomenon of the trap with magnetic mirrors. We see that when $Y = 0$, then $\dot{\phi} = 0$ and $\dot{y}^2/2 = W_\perp$, so that $\mu^2 I_Y = W_\perp/B$. Hence, in an axially-symmetric magnetic trap, the magnetic moment $W_\perp/B$ is a perpetual adiabatic invariant.

This picture of the magnetic trap applies to the motion of charged particles around the geomagnetic lines of force in the Van Allen belt, or that of charged particles along the magnetic lines of force in the solar chromosphere between pairs of sunspots.

The generalization of the theorem to multiparametric problems can be made on the assumption that

$$|(k, \xi)| > K|k|^{-\nu}, \qquad |k| > 0, \qquad \nu = n + 1, \qquad K > 0.$$

The extension to the case of many degrees of freedom has yet to be accomplished.

## 4.25 EXISTENCE OF ADIABATIC INVARIANTS

Kasuga (1961) has given a precise formulation of the adiabatic theorem in the general case and proved it by the operator method in the classical mechanics of von Neumann (1932) and Koopman (1931).

According to Kasuga a quantity $I(p, q, s)$, which is an integral of

$$\frac{dp}{dt} = -\frac{\partial H}{\partial q}, \qquad \frac{dq}{dt} = \frac{\partial H}{\partial p}, \qquad H = H(p, q, s), \tag{153}$$

where $p = (p_1, p_2, \ldots, p_n)$, $q = (q_1, q_2, \ldots, q_n)$, for every fixed $s$, $(0 \leqq s \leqq 1)$, is called an adiabatic invariant of (153) if $I(p, q, t/\lambda)$ is conserved, asymptotically for $\lambda \rightarrow +\infty$, along almost all trajectories of the system

$$\frac{dp}{dt} = -\frac{\partial H}{\partial q}, \qquad \frac{dq}{dt} = \frac{\partial H}{\partial p}, \qquad H = H(p, q, t/\lambda); \tag{154}$$

that is (if we let $t' = t/\lambda$), of the system

$$\frac{dp}{dt'} = -\lambda\frac{\partial H}{\partial q}, \qquad \frac{dq}{dt'} = \lambda\frac{\partial H}{\partial p}, \qquad H = H(p, q, t'), \tag{154a}$$

in the whole interval of time $0 \leqq t \leqq \lambda$, that is, $0 \leqq t' \leqq 1$. The fundamental case of the adiabatic theorem in celestial mechanics is the case in which (153) admits no single-valued Lebesgue measurable integral other than the energy integral for almost all $s$, $0 \leqq s \leqq 1$. In this case the phase volume

$$\mathfrak{J}(p, q, s) = \iint dp\, dq,$$

where the integral extended over the domain $I_{E,s}$ in the phase space enclosed by the energy surface $S_{E,s} = \{(p, q) \mid E(p, q, s) = E\}$ of (153) passing through $(p, q)$, is the only adiabatic invariant.

Kasuga first determined the region $D$ in a $(2n + 1)$-dimensional space $R^{2n+1}$ such that the adiabatic theorem can be stated, and studied the properties of the phase volume function $\mathfrak{J}(p, q, s)$ in $D$.

Let $B$ be a subset of a Euclidian space or a subset of a Hilbert space (von Neumann 1929; Stone, 1932; Hille, 1948); then the closure of $B$

in the space is denoted by $\bar{B}$, and the set of all inner points of $B$ by $B^0$. A subset of $B$ is called *relatively open in B* if it can be considered as an intersection of an open set in the Euclidian space with $B$. We denote by $C^m(B)$ the set of all $m$-times continuously differentiable functions on $B$ — noting that $\bar{B}^0 \supset B$ — and by $C_0^m(B)$ the set of all $m$-times continuously differentiable functions on $B$ vanishing outside compact sets contained in $B$. By $K$ we denote the point set

$$\{(p, q, s) \mid -\infty < p_i < +\infty, -\infty < q_i < +\infty, i = 1, 2, \ldots, n, \\ a \leqq s \leqq b\}$$

in $R^{2n+1}$ with two fixed numbers $a$ and $b$, $a < b$.

Assume that the point set in $R^{2n+1}$, where the real-valued function $H(p, q, s)$ is defined, contains a subset $G$ of $K$ relatively open in $K$, satisfying

$$\sum_{i=1}^{n} \left[\left(\frac{\partial H}{\partial p_i}\right)^2 + \left(\frac{\partial H}{\partial q_i}\right)^2\right] \neq 0$$

on $G$ and the following condition: For each $s$, $a \leqq s \leqq b$, there is an open interval

$$\Gamma_s = \{E \mid E_s < E < E_s'\}, \qquad -\infty \leqq E_s < E_s' = +\infty,$$

on the $E$-line such that the set

$$S_{E,s} = \{(p, q) \mid H(p, q, s) = E, (p, q) \in G_s\}$$

in the open set $G_s = \{(p, q) \mid (p, q, s) \in G\}$ is nonempty and is a closed $(2n - 1)$-dimensional manifold of $R^{2n}$ enclosing a domain $I_{E,s}$ in $R^{2n}$ for each $E \in \Gamma_s$ and $G_s = \mathbf{U}_{E \in \Gamma_s}$ (the neighborhood of $S_{E,s}$ for which $E \in \Gamma_s$).

Assume further that there are two numbers $J_1^*$ and $J_2^*$ independent of $s$ such that $J_s < J_1^* < J_2^* < J_s'$ for all $s$ in the interval $a \leqq s \leqq b$.

Put

$$S(J, s) = \{(p, q) \mid (p, q) \in G_s, \bar{\mathfrak{J}}(p, q, s) = J\}$$

and

$$I(s) = I(J_1^*, J_2^*, s) = \{(p, q) \mid (p, q) \in G_s, J_1^* < \bar{\mathfrak{J}}(p, q, s) < J_2^*\}.$$

It can be seen that

$$\mathfrak{J}(E, s) = \int_{I_{E,s}} dp\, dq \in C'(\Lambda), \qquad \bar{\mathfrak{J}}(p, q, s) = \mathfrak{J}(H(p, q, s), s),$$

and $\partial\mathfrak{J}/\partial E > 0$ for $(E, s) \in \Lambda$, and that there is a unique measure $m_{J,s}$ defined for all Borel subsets (Section 2.1) of each $S(J, s)$, such that

$$\frac{d}{dJ}\int_{I(J_1^*,J,s)} f(p, q)\, dp\, dq = \int_{S(J,s)} f(p, q)\, dm_{J,s},$$

for every function $f(p, q) \in C_0^0[I(s)]$, where $m_{J,s}$ denotes the Lebesgue measure of $S(J, s)$. For any integrable function $f(p, q)$ on $I(s)$, the function $f(p, q)$ is integrable on the measure space $(S(J, s), m_{J,s})$ for almost all $J$ $(J_1^* < J < J_2^*)$, and

$$\int_{S(J,s)} f(p, q)\, dm_{J,s}$$

is integrable with respect to $J$ in the interval $J_1^* < J < J_2^*$, and

$$\int_{I(s)} f(p, q)\, dp\, dq = \int_{J_1^*}^{J_2^*} \left(\int_{S(J,s)} f(p, q)\, dm_{J,s}\right) dJ,$$

$$\int_{S(J,s)} dm_{J,s} = 1.$$

Kasuga then proved the theorem that

$$\int_{S(J,s)} \frac{\partial\bar{\mathfrak{J}}}{\partial s}(p, q, s)\, dm_{J,s} = 0$$

for $J_1^* < J < J_2^*$ and $a \leqq s \leqq b$.

He then studied the flows defined by (153) and the infinitesimal generator $iA$ of the unitary group $\{U_t \mid -\infty < t < +\infty\}$ associated with one of the flows, and proved the adiabatic theorem for a multidimensional and multiparametric case.

Let $(X, m)$ be a measure space where $m$ is a finite, separable, and complete measure (Section 2.1) defined on a Borel field in $X$. A one-parameter group $\{\mathfrak{T}_t \mid -\infty < t < +\infty\}$ of a one-to-one mapping $\mathfrak{T}_t$ of $X$ onto $X$ is called a *flow* on $(X, m)$. A measurable function $f(P)$ on $(X, m)$ is called an *invariant* function of a flow $\{\mathfrak{T}_t\}$ on $(X, m)$ if

$$f(\mathfrak{T}_t(P)) = f(P)$$

almost everywhere on $(X, m)$ for every fixed $t$. It is called a *strictly invariant* function of a flow $\{\mathfrak{T}_t\}$ on $(X, m)$ if it is defined everywhere on $X$ and if $f(\mathfrak{T}_t(P)) = f(P)$ for all $(P, t)$ such that $P \in X$, $-\infty < t < +\infty$. A measure-preserving and measurable flow $\{\mathfrak{T}_t\}$ on $(X, m)$ is ergodic in the sense used by von Neumann (1932) if and only if all its invariant functions are equivalent, that is, coincident almost everywhere, to constants on $(X, m)$. If a flow $\{\mathfrak{T}_t\}$ on $(X, m)$ is measure-preserving and measurable, then we can associate with it a one-parameter group $\{\mathfrak{U}_t \mid -\infty < t < +\infty\}$ of unitary transformations (Section 1.18) $\mathfrak{U}_t$ on $L^2(X, m)$ by

$$\mathfrak{U}_t f(P) = f(\mathfrak{T}_t(P)), \qquad f \in L^2(X, m), \qquad P \in X;$$

and $\mathfrak{U}_t$ is a continuous function of $t$ in the strong topology of $\mathfrak{U}_t$. (See a later chapter on the ergodic theorem.)

If $X$ is a Lebesgue measurable subset of a Euclidian space $R^r$, and $m$ is the Lebesgue measure of $R^r$ defined for all Lebesgue measurable subsets of $X$, a flow on $(X, m)$ is called a flow on $X$.

Consider the Hamiltonian system with a parameter $s$. The canonical equations are

$$\frac{dp}{dt} = -\frac{\partial H}{\partial q}, \qquad \frac{dq}{dt} = \frac{\partial H}{\partial p}, \qquad H = H(p, q, s).$$

By assumption, the solution of this system,

$$p = p(t, p^0, q^0, s), \qquad q = q(t, p^0, q^0, s),$$

in the open set $I(s)$ for a fixed $s$ $(a \leqq s \leqq b)$ with the initial conditions

$$(p, q) = (p^0, q^0), \qquad (p^0, q^0) \in I(s), \qquad \text{at } t = 0$$

can be uniquely prolonged for the whole interval $-\infty < t < +\infty$ and

$$(p(t, p^0, q^0, s), q(t, p^0, q^0, s)) \in S(J, s) \qquad \text{for } -\infty < t < +\infty$$

if $(p^0, q^0) \in S(J, s)$.

Denote by $T_t^{(s)}$ the one-to-one mapping of $I(s)$ onto $I(s)$:

$$(p^0, q^0) \to (p(t, p^0, q^0, s), q(t, p^0, q^0, s));$$

then $\{T_t^{(s)} \mid -\infty < t < +\infty\}$ constitutes a flow $F_s$ on $I(s)$ for $a \leqq s \leqq b$. The flow $F_s$ on $I(s)$ is measure-preserving and measurable. Since $T_t^{(s)}$ transforms $S(J, s)$ onto $S(J, s)$, the mapping $T_t^{(s)}$ induces a one-to-one mapping $T_t^{(J,s)}$ of $S(J, s)$ onto $S(J, s)$, such that

$$\{T_t^{(J,s)} \mid -\infty < t < +\infty\}$$

constitutes a flow $F_{J,s}$, which is measure-preserving and measurable, on the measure space $(S(J, s), m_{J,s})$ for $J_1^* < J < J_2^*$, $a \leqq s \leqq b$. Kasuga proved the equivalence for any fixed $s$ of two conditions: (i) that every invariant function $f(p, q)$ of the flow on $I(s)$ is equivalent on $I(s)$ to a function of the form $\phi(H(p, q, s))$, where $\phi(E)$ is a measurable function of $E$ for the interval $\mathfrak{E}(J_1^*, s) < E < \mathfrak{E}(J_2^*, s)$, and (ii) that the flow $F_{J,s}$ on $(S(J, s), m_{J,s})$ is ergodic for almost all $J$ in the interval

$$J_1^* < J < J_2^*.$$

Suppose that either of these two conditions is satisfied for almost all $s$ in $a \leqq s \leqq b$. If an invariant function $f(p, q, s)$ of the flow $F$ on $D$ belongs to $L^2(D)$, then

$$\int_D \overline{f(p, q, s)}\, \frac{\partial \mathfrak{J}}{\partial s}\, (p, q, s)\, dp\, dq\, ds = 0.$$

Now consider the one-parameter group $\{U_t\}$ of unitary transformations on $L^2(D)$ associated with the flow $F$ on $D$. Define an operator $Af$ by

$$\left\| \frac{U_t f - f}{it} - Af \right\|_D \to 0 \qquad (t \to 0)$$

for all $f \in L^2(D)$ for which such an $Af$ exists. Then $A$ is a self-adjoint operator in the sense used by von Neumann on $L^2(D)$, and $U_t = e^{iAt}$ in the sense of the operator calculus (see a later chapter on the ergodic theorem; Stone, 1932). Denote the domain of $A$ by $\mathfrak{D}(A)$. If $f \in C_0^1(D^0)$, we define $f = 0$ on $D - D^0$. If $f \in C_0^1(D^0)$, then $f \in \mathfrak{D}(A)$ and

$$Af = -i(f, H) = i(H, f),$$

where $(H, f)$ is the Poisson bracket.

Denote by $N$ the set of all invariant functions of the flow $F$ on $D$ belonging to $L^2(D)$. The set $N$ coincides with the set

$$\{f \,|\, f \in L^2(D), f = U_t f, -\infty < t < +\infty\} = \{f \,|\, f \in \mathfrak{D}(A), Af = 0\}.$$

Hence $N$ is a closed linear subspace of $L^2(D)$. Denote the orthogonal complement of $N$ in $L^2(D)$ by $N^\perp$. Then it can be shown that $A[C_0^1(D^0)]$, the image of $C_0^1(D^0)$ by $A$, is contained in $N^\perp$ and is dense in $N^\perp$. For any $\epsilon > 0$, there is a function $f_0(p, q, s) \in C_0^1(D^0)$ such that

$$\left\| (H, f_0) - \frac{\partial \mathfrak{J}}{\partial s} \right\|_D < \epsilon.$$

Let $a'$ and $b'$ be two numbers such that $a \leqq a' < b' \leqq b$ and

$$(b' - a') < (J_2^* - J_1^*)/(2M),$$

and put

$$J_2 = J_2^* - M(b' - a'), \qquad J_1 = J_1^* + M(b' - a').$$

Then the solution of (154a) passing through $(p^0, q^0, a')$, where

$$(p^0, q^0) \in I(J_1, J_2, a')$$

can be prolonged in $D$ to the interval $a' \leqq t \leqq b'$ for every $\lambda > 0$.

When $a \leqq a' < b' \leqq b$, $(p^0, q^0) \in I(a')$, and $\lambda > 0$, we define

$$\Delta(a', b', p', q', \lambda) = \max_{a' \leqq t \leqq b'} |\mathfrak{J}\{\tilde{p}(t, p^0, q^0, a', \lambda), \tilde{q}(t, p^0, q^0, a', \lambda), t\} - \mathfrak{J}(p^0, q^0, a')|,$$

if the solution $p = \tilde{p}(t, p^0, q^0, a', \lambda)$, $q = \tilde{q}(t, p^0, q^0, a', \lambda)$ of (154a) can be continued to $t = b'$ in $D$; we put

$$\Delta(a', b', p^0, q^0, \lambda) = +\infty$$

if this solution cannot be continued to $t = b'$ in $D$. When

$$a \leqq a' < b' \leqq b, \qquad \lambda > 0, \delta > 0,$$

we denote the subset of $I(a')$,

$$\{(p^0, q^0) \mid (p^0, q^0) \in I(a'), \Delta(a', b', p^0, q^0, \lambda) < \delta\},$$

by $L(a', b', \lambda, \delta)$. The subset $L(a', b', \lambda, \delta)$ is an open set in $R^{2n}$.

Denote the $m$-dimensional Lebesgue measure of a measurable set in $R^m$ by $\mu_m[\ ]$. Then, for any fixed $\delta(>0)$, we have

$$\mu_{2n}[I(J_1, J_2, a') - L(a', b', \lambda, \delta)] \to 0 \qquad (\lambda \to +\infty).$$

The adiabatic theorem of Kasuga states that

$$\mu_{2n}[I(J_1, J_2, a) - L(a, b, \lambda, \delta)] \to 0 \qquad (\lambda \to +\infty),$$

for any fixed $J_1, J_2, \delta$ such that $J_1^* < J_1 < J_2 < J_2^*$ and $\delta > 0$.

## LIE GROUPS

### 4.26 LIE'S FUNDAMENTAL THEOREMS

Consider a set of independent functions

$$f_i(x_1, x_2, \ldots, x_n; a_1, a_2, \ldots, a_r) \qquad (i = 1, 2, \ldots, n),$$

of $n$ independent variables $x_1, x_2, \ldots, x_n$ containing $r$ parameters $a_1, a_2, \ldots, a_r$, continuous of class $C^2$ in $x$ and in $a$. By assumption, the Jacobian of $f_i$ with respect to $x$ is not zero. If there is no relation of the form

$$\sum_{\sigma=1}^{r} \phi_\sigma(a_1, a_2, \ldots, a_r) \frac{\partial f_i}{\partial a_\sigma} = 0 \qquad (i = 1, 2, \ldots, n)$$

with nonzero functions $\phi_\sigma(a_1, a_2, \ldots, a_r)$, then and only then are $a_1, a_2, \ldots, a_r$ said to be the *essential parameters*. Any transformation by an $r$-parameter group

$$x'_i = f_i(x_1, x_2, \ldots, x_n; a_1, a_2, \ldots, a_r) \qquad (i = 1, 2, \ldots, n) \quad (155)$$

transforms a point $P(x_1, \ldots, x_n)$ of an $n$-dimensional manifold to a point $P'(x'_1, \ldots, x'_n)$; that is, $T_a x = x'$, $T_a^{-1}x' = x$. The composite transformation $T_b T_a = T_c$ is determined uniquely; thus,

$$T_c(T_b T_a) = (T_c T_b) T_a.$$

A system with these properties is called a *semi-group*. If it contains an identity such as $T_a I = I T_a = T_a$ and an inverse element such as $T_a^{-1} T_a = T_a T_a^{-1} = I$, then it is called a *group*. From (155) we obtain

$$x_j = f'_j(x'_i, \ldots, x'_n, a_1, \ldots, a_r) \qquad (j = 1, 2, \ldots, n).$$

The transformations (155) define a *continuous transformation group* $T$ (Lie, 1888, 1890, 1891, 1893, 1893a, 1896).

From (155),

$$\delta x'_i = \sum_\rho f_{i\rho}(x, a)\, \delta a_\rho \qquad (i = 1, 2, \ldots, n),$$

or

$$\delta x'_i = \sum_\rho f_{i\rho}\{x', a\}\, \delta a_\rho \qquad (i = 1, 2, \ldots, n). \quad (156)$$

This is an infinitesimal transformation $T_a^{-1} T_{a+\delta a}$. According to Lie it is written

$$\sum_\rho \left( \sum_i f_{i\rho}\{x', a\} \frac{\partial f'}{\partial x'_i} \right) \frac{\delta a_\rho}{\delta t}. \quad (156a)$$

We write the relation $T_b T_a = T_c$ in the form

$$c_\rho = \psi_\rho(a, b) \qquad (\rho = 1, 2, \ldots, r). \quad (157)$$

Since $T_b T_{a+\delta a} = T_{c+\delta c}$, we have

$$T_c^{-1} T_{c+\delta c} = T_a^{-1} T_b^{-1} T_b T_{a+\delta a} = T_a^{-1} T_{a+\delta a}.$$

Thus, $T_a^{-1} T_{a+\delta a}$ is invariant for the transformation (157). This leads to the relation

$$\sum_\rho \left( \sum_i f_{i\rho}\{x', a\} \frac{\partial f'}{\partial x'_i} \right) \frac{\delta a_\rho}{\delta t} = \sum_\sigma \left( \sum_i f_{i\sigma}\{x', c\} \frac{\partial f'}{\partial x'_i} \right) \frac{\delta c_\sigma}{\delta t}. \quad (156b)$$

From (157) we have

$$\delta c_\sigma = \sum_\rho \phi_{\sigma\rho}(b, a)\, \delta a_\rho = \sum_\rho \phi_{\sigma\rho}\{a, c\}\, \delta a_\rho.$$

Substituting this expression for $\delta c_\sigma$ in (156b) makes the infinitesimal transformation (156a) become

$$\sum_\rho \left(\sum_i f_{i\rho}\{x', a\} \frac{\partial f'}{\partial x'_i}\right) \frac{\delta a_\rho}{\delta t} = \sum_\sigma \left[\left(\sum_i f_{i\sigma}\{x', c\} \frac{\partial f'}{\partial x'_i}\right) \cdot \sum_\rho \phi_{\sigma\rho}\{a, c\} \frac{\delta a_\rho}{\delta t}\right]. \tag{156c}$$

We fix $c$. Then $b$ is given by (157), and the $f_{i\sigma}\{x', c\}$ become functions of $x'$ only. We can put

$$f_{i\sigma}\{x', c\} = \xi_{\sigma i}(x'), \qquad \phi_{\sigma\rho}\{a, c\} = \psi_{\sigma\rho}(a),$$

$$\sum_i \xi_{\sigma i}(x') \frac{\partial f'}{\partial x'_i} = X'_\sigma f'. \tag{158}$$

This gives an important relation:

$$\sum_\rho \left(\sum_i f_{i\rho}\{x', a\} \frac{\partial f'}{\partial x'_i}\right) \frac{\delta a_\rho}{\delta t} = \sum_\sigma \left(\sum_\rho \psi_{\sigma\rho}(a) \frac{\delta a_\rho}{\delta t}\right) X'_\sigma f'. \tag{158a}$$

These infinitesimal transformations $X_1 f, \ldots, X_r f$ constitute a ring with the operations of addition and multiplication, which is called an *infinitesimal ring.*

We have $(x') = (x)S_t$, $(x) = (x')S_t^{-1}$; accordingly

$$(x' + \delta x') = (x')S_t^{-1} S_{t+\delta t}.$$

Hence, the infinitesimal transformation $T_a^{-1} T_{a+\delta a}$ can be formed from the basic transformations $X_1 f, \ldots, X_r f$, linearly with the coefficients

$$\sum_\tau \psi_{1\tau}(a)\, \delta a_\tau, \ldots, \sum_\tau \psi_{r\tau}(a)\, \delta a_\tau.$$

Write (157) in the form $a'_\rho = \psi_\rho(a, b)$ and let $T_b T_a = T_{a'}$. If $T_c T_{a'} = T_{a''}$, then $T_c T_b T_a = T_{a''}$. But since $T_c T_b = T_d$, therefore, $T_d T_a = T_{a''}$. This is called a left-hand group. It is the second parameter group of Lie. Write (157) in the form $a'_\rho = \psi_\rho(b, a)$ and let $T_a T_b = T_{a'}$. If $T_{a'} T_c = T_{a''}$ then $T_a T_b T_c = T_{a''}$. Since $T_b T_c = T_d$, therefore $T_a T_d = T_{a''}$. This is Lie's first parameter group; it is a right-hand group. The parameter $b$ is essential in these groups. Cartan (1933, 1937) introduced the notion of *equipollence* by means of a geometrical illustration. Changing $T_a^{-1}T_{a+\delta a}$ to $T_{a'}^{-1}T_{a'+\delta a'}$ by means of (158), we obtain

$$\sum_\rho \Big(\sum_\tau \psi_{\rho\tau}(a)\ \delta a_\tau\Big) X_\rho f = \sum_\rho \Big(\sum_\tau \psi_{\rho\tau}(a')\ \delta a'_\tau\Big) X_\rho f$$

for the second parameter group. Since the basic infinitesimal transformations are linearly independent, therefore

$$\sum_\tau \psi_{\rho\tau}(a)\ \delta a_\tau = \sum_\tau \psi_{\rho\tau}(a')\ \delta a'_\tau \qquad (\rho = 1, 2, \ldots, r).$$

That is, the $r$ Pfaffians $\sum_\tau \psi_{\rho\tau}(a)\ \delta a_\tau$ are invariant for the second parameter group.

Construct an $r$-dimensional parameter space with coordinates $a_1, a_2, \ldots, a_r$. Put $\sum_\tau \psi_{\rho'\tau}(a)\ \delta a_\tau = \epsilon_{\rho\rho'}\ \delta t$, where $\epsilon_{\rho\rho} = 1$, $\epsilon_{\rho\rho'} = 0$ $(\rho \neq \rho')$. Such a vector field is invariant for the second parameter group; it is the field of the infinitesimal increment defined by the infinitesimal transformations

$$A_\rho f = \sum_\tau \alpha_{\rho\tau}(a) \frac{\partial f}{\partial a_\tau} \qquad (\rho = 1, 2, \ldots, r), \tag{159}$$

with the determinant $\|\psi_{\sigma\tau}\| \neq 0$, where $\alpha_{\rho\tau}(a)$ is the minor of this determinant. Since any infinitesimal transformation $Af$ is formed linearly with constant coefficients from $A_1 f, \ldots, A_r f$, we obtain

$$(A_\sigma, A_\rho) f = \sum_\tau c_{\sigma\rho\tau} A_\tau f \qquad (\sigma, \rho = 1, 2, \ldots, r) \tag{160}$$

with the commutator $(A_\sigma, A_\rho)$ (Section 4.2).

If the second parameter group $T_c T_{a'} = T_{a''}$ is applied to the first parameter group $T_a T_b = T_{a'}$, then $T_c T_a T_b = T_{a''}$. Let $T_c T_a = T_{a_1}$, $T_{a_1} T_b = T_{a_2}$; then $T_c T_a T_b = T_{a_2}$. Hence, if transformations from the first and the second parameter groups are denoted by $A_b$ and $B_c$, respectively, then $A_b B_c = B_c A_b$. Thus, $B_c = A_b^{-1} B_c A_b$, $A_b = B_c^{-1} A_b B_c$; that is, any transformation of one of the parameter groups is invariant for transformations of the other parameter group. Hence, $A_b^{-1} A_{b+\delta b}$ is invariant for $B_c$. We see that the infinitesimal transformations of the first parameter group are formed by linear combinations with constant coefficients from $A_1 f, \ldots, A_r f$. Let an infinitesimal transformation of the second parameter group be $Bf$, and let $Bf$ operate on $A_\rho f$; then we obtain $A_\rho f + (A_\rho, B)\,\delta t$. Since this is invariant for the second parameter group, therefore $(A_\rho, B_\sigma) = 0$ $(\rho, \sigma = 1, 2, \ldots, r)$.

Let $A_\rho f$ be a basic infinitesimal transformation of the first parameter group, and let $\delta_\rho$ be the displacement due to $A_\rho f$; then, by (159) and (160),

$$(A_\sigma, A_\tau) = \sum_\rho \left(\frac{\delta_\sigma \alpha_{\tau\rho}}{\delta t} - \frac{\delta_\tau \alpha_{\sigma\rho}}{\delta t}\right) \frac{\delta f}{\delta \alpha_\rho} = \sum_\rho \sum_\kappa c_{\sigma\tau\kappa} \alpha_{\kappa\rho} \frac{\partial f}{\partial \alpha_\rho}.$$

Multiply by $\psi_{\lambda\rho}$ and sum over $\rho$; then, remembering that the matrix $\alpha$ is the reciprocal of the matrix $\psi$, we obtain

$$\sum_\rho \left(\psi_{\lambda\rho} \frac{\delta_\sigma \alpha_{\tau\rho}}{\delta t} - \psi_{\lambda\rho} \frac{\delta_\tau \alpha_{\sigma\rho}}{\delta t}\right) = c_{\sigma\tau\lambda}.$$

With $\sum_\rho \psi_{\lambda\rho}\alpha_{\tau\rho} = \epsilon_{\lambda\tau}$, $\sum_\rho \psi_{\lambda\rho}\alpha_{\sigma\rho} = \epsilon_{\lambda\sigma}$, we have

$$\sum_\rho \left(\alpha_{\sigma\rho} \frac{\delta_\tau \psi_{\lambda\rho}}{\delta t} - \alpha_{\tau\rho} \frac{\delta_\sigma \psi_{\lambda\rho}}{\delta t}\right) = \sum_\rho \sum_\kappa \left(\alpha_{\sigma\rho}\alpha_{\tau\kappa} \frac{\partial \psi_{\lambda\rho}}{\partial a_\kappa} - \alpha_{\tau\rho}\alpha_{\sigma\kappa} \frac{\partial \psi_{\lambda\rho}}{\partial a_\kappa}\right)$$

$$= c_{\sigma\tau\lambda} = \sum_\rho \sum_\kappa \alpha_{\sigma\rho}\alpha_{\tau\kappa} \left(\frac{\partial \psi_{\lambda\rho}}{\partial a_\kappa} - \frac{\partial \psi_{\lambda\kappa}}{\partial a_\rho}\right).$$

Hence

$$\frac{\partial \psi_{\lambda\rho}}{\partial a_\sigma} - \frac{\partial \psi_{\lambda\sigma}}{\partial a_\rho} = \sum_\kappa \sum_\tau c_{\kappa\tau\lambda} \psi_{\kappa\rho} \psi_{\tau\sigma}. \tag{161}$$

This is called *Maurer's relation.*

If the group contains at least a transformation (155) that maps a point $(x_1, \ldots, x_n)$ of the $n$-dimensional manifold to any other point $(x_1', \ldots, x_n')$ of that manifold, then the group is said to be *transitive.* When there are exactly $n$ independent equations among

$$X_1 f = 0, \ldots, X_r f = 0,$$

then the transformation of the point $(x_1, \ldots, x_n)$ is transitive. Otherwise it is *intransitive.* If $n = r$, then the system of $x$ can be transformed to the system of $a$ under certain conditions. Such a group is said to be *simply* transitive.

By combining $P_\rho = \sum_\tau \psi_{\rho\tau}(a)\,\delta a_\tau$ $(\rho = 1, 2, \ldots, r)$ and the basic infinitesimal transformations

$$X_\rho' f' = \sum_i \xi_{\rho i}(x') \frac{\partial f'}{\partial x_i'},$$

we have seen that the transformation $T_a^{-1} T_{a+\delta a}$ satisfies

$$\delta x_i'' = \sum_\rho P_\rho \xi_{\rho i}(x') \qquad (i = 1, 2, \ldots, n).$$

If $x'$ is regarded as a function of $a$, then it should satisfy

$$dx'_i - \sum_{\rho,\sigma} \psi_{\sigma\rho}(a)\xi_{\sigma i}(x')da_\rho = 0 \qquad (i = 1, 2, \ldots, n), \tag{162}$$

or

$$\frac{\partial x'_i}{\partial a_\rho} = \sum_\sigma \psi_{\sigma\rho}(a)\xi_{\sigma i}(x') \qquad (i = 1, 2, \ldots, n;\ \rho = 1, 2, \ldots, r). \tag{163}$$

Hence, a transformation of an $r$-parameter group should satisfy this system (163) of partial differential equations. This is the *first fundamental theorem of Lie.*

Substituting (162) in the relation

$$dF = \sum_i \frac{\partial F}{\partial x'_i} dx'_i + \sum_\rho \frac{\partial F}{\partial a_\rho} da_\rho$$

we obtain

$$\frac{\partial F}{\partial a_\rho} + \sum_{i,\sigma} \psi_{\sigma\rho}(a)\xi_{\sigma i}(x') \frac{\partial F}{\partial x'_i} = 0 \qquad (\rho = 1, 2, \ldots, r), \tag{164}$$

or, with the definition of the matrix $\alpha$,

$$\sum_i \xi_{\rho i}(x') \frac{\partial F}{\partial x'_i} + \sum_\sigma \alpha_{\rho\sigma}(a) \frac{\partial F}{\partial a_\sigma} = 0 \qquad (\rho = 1, 2, \ldots, r). \tag{165}$$

Since $x'_i = f_i(x, a)$ satisfies (162) with the integration constants $a$, the relation $x_i = f'_i(x', a)$ satisfies (165). As these $n$ functions $f'_i$ are independent with regard to $x'$, the equations (165) admit $n$ independent integrals $f'_i$. There are $r$ independent equations for $n + r$ variables in the form (165). Hence, (165) is a complete system and can be integrated completely.

Conversely, it can be shown that if the transformation (155) containing the identical elements $a_1^0, \ldots, a_r^0$ of the parameters satisfies the Pfaffian system (162) which is completely integrable with respect to $x'$ and $a$, such that $\|\psi_{\sigma\rho}(a^0)\| \neq 0$, then the transformation gives a transformation group.

According to (162) the transformation $T_a^{-1}T_{a+\delta a}$ coincides with the infinitesimal transformation $\sum_\sigma P_\sigma(a)X'_\sigma$. Let $(a_1, \ldots, a_r)$ represent a point $A$ and $(a_1^0, \ldots, a_r^0)$ a point $A^0$ of the parameter space. Divide the arc $AA^0$ joining $A^0$ to $A$ into segments by the points $A^1, A^2, \ldots, A^{n-1}$ on the arc. Then $T_{A^0}^{-1}T_A$ is divided into factors

$$T_{A^0}^{-1}T_{A^1},\ T_{A^1}^{-1}T_{A^2},\ \ldots,\ T_{A^{n-1}}^{-1}T_{A^n}.$$

Each of these factors can become an infinitesimal transformation, if we increase the number of these segments indefinitely. If the arc is written

$$P_\sigma(a) = c_\sigma\, dt \qquad (\sigma = 1, 2, \ldots, r), \tag{166}$$

then $T_{A^0}^{-1} T_A$ is said to be *generated* (*erzeugt*) by the infinitesimal transformation $\sum_\sigma c_\sigma X_\sigma$. Referring to the definition of $P_\rho$, we obtain, from (166),

$$\frac{da_\rho}{dt} = \sum_\sigma c_\sigma \alpha_{\sigma\rho}(a) \qquad (\rho = 1, 2, \ldots, r). \tag{167}$$

The integral curve of these equations is a trajectory that admits the infinitesimal transformation

$$\sum_\rho \sum_\sigma c_\sigma a_{\sigma\rho}(a) \frac{\partial f}{\partial a_\rho} = \sum_\sigma c_\sigma A_\sigma f,$$

and passes through $A^0$. Thus $A^0$ is transformed to $A$ by the transformation $\sum_\sigma c_\sigma A_\sigma$ operating in the parameter space. At the same time the infinitesimal transformation $\sum_\sigma c_\sigma X_\sigma$ should operate in the $x$-manifold, where

$$X_\sigma f = \sum_i \xi_{\sigma i}(x) \frac{\partial f}{\partial x_i}, \qquad \delta x_i = \delta t \cdot \sum_\rho c_\rho \xi_{\rho i}(x) \qquad (i = 1, 2, \ldots, n).$$

Formula (163) shows that

$$\sum_\rho \psi_{i\rho}(a)\, \delta a_\rho = c_i\, \delta t,$$

which is derived from (155).

Suppose that the points $A^0, B^0, C^0, \ldots$ are transformed respectively to $A, B, C, \ldots$ by $\sum_\sigma c_\sigma A_\sigma$ for the same variation of $t$. Then the transformations

$$T_{A^0}^{-1} T_A, \qquad T_{B^0}^{-1} T_B. \qquad T_{C^0}^{-1} T_C, \ldots \tag{168}$$

all coincide and generate the infinitesimal transformation $\sum_\sigma c_\sigma X_\sigma$. Denote by $A^0, B^0, C^0, \ldots$ the successive images of the point $(a^0)$ corresponding to the identical element, and by $A, B, C, \ldots$ the corresponding images of the point $(a)$. Then $T_{a^0}^{-1} T_a$ or $T_a$ should be among the transformations (168). Since $T_{A^0}^{-1} T_A = T_a$, therefore $T_A = T_{A^0} T_a$. Thus, from two transformations $T_{A^0}$ and $T_a$ belonging to (155) we obtain the transformation $T_A$ belonging to (155). The point $A^0$ is an arbitrary point near the point $(a^0)$, and $(a)$ is obtained from $(a^0)$ by the transformation

$\sum_\sigma c_\sigma A_\sigma$. Hence, by choosing $c_\sigma$ suitably, we can reach any point $(a)$ in the neighborhood of $(a^0)$. For, integrating (167) from the point $(a^0)$ gives

$$a_\rho = a_p^0 + \sum_\sigma tc_\sigma \alpha_{\sigma\rho}(a^0) + \cdots \qquad (\rho = 1, 2, \ldots, r), \qquad (169)$$

Thus $(a_\rho)$ is a function of $\lambda_\sigma = tc_\sigma$. The Jacobian of $a_\rho$ with respect to $\lambda_\sigma$ at $\lambda_\sigma = 0$ is $\|\alpha_{\sigma\rho}(a^0)\|$ and does not vanish. By solving these series we obtain

$$tc_\sigma = \sum_\rho \psi_{\sigma\rho}(a^0)(a_\rho - a_\rho^0) + \cdots \qquad (\sigma = 1, 2, \ldots, r).$$

Hence $T_a$ is generated by the infinitesimal transformation $\sum_\sigma c_\sigma X_\sigma$ if $(a)$ is in the neighborhood of $(a^0)$.

Since by definition $\delta f' = X'f'\delta t$, $X'f' = \sum_i \xi_{\sigma i}(x')\, \partial f'/\partial x_i'$, therefore $df'/dt = X'f'$, $d^2f'/dt^2 = X'X'f' = X'^2f', \ldots$. Denote the values of $f', df'/dt, d^2f'/dt^2, \ldots$ at $t = 0$ by $f, Xf, X^2f, \ldots$, respectively; then

$$f' = f + \frac{t}{1!} Xf + \frac{t^2}{2!} X^2f + \cdots.$$

If the infinitesimal transformations $X_1f, \ldots, X_rf$ belonging to the foregoing group satisfy the conditions of Lie's theorem, then any transformation belonging to the group in the neighborhood of the identical element is generated by

$$Xf = e_1X_1f + \cdots + e_rX_rf.$$

Let $t = 1$; then such a transformation is written

$$f(x') = f(x) + \frac{1}{1!} \sum_\rho e_\rho X_\rho f + \frac{1}{2!} \sum_{\rho,\sigma} e_\rho e_\sigma X_\rho X_\sigma f + \cdots. \qquad (170)$$

The $r$ constants $e_1, \ldots, e_r$ are called the *canonical parameters*. If $f(x')$ satisfies

$$\frac{df(x')}{dt} = \sum_\rho \lambda_\rho X_\rho' f(x'),$$

without $\delta t$ being put equal to zero, then the infinitesimal transformation $(\lambda_1 X_1 f + \cdots + \lambda_r X_r f)\delta t$ generates a finite transformation

$$f(x') = f(x) + \frac{1}{1!} t \sum_\rho \lambda_\rho X_\rho f + \frac{1}{2!} t^2 \sum_{\rho,\sigma} \lambda_\rho \lambda_\sigma X_\rho X_\sigma f + \cdots. \qquad (171)$$

Here $X_1f, \ldots, X_rf$ are called the *generators* of the infinitesimal transformations.

The transformation (155) is represented in general by

$$X_\rho f = \sum_{i=1}^{n} \xi_{\rho i}(x) \frac{\partial f}{\partial x_i} \qquad (\rho = 1, 2, \ldots, r). \tag{172}$$

Integrating the differential equations

$$\frac{dx_1'}{\sum_{\rho=1}^{r} \lambda_\rho \xi_{\rho 1}(x')} = \cdots = \frac{dx_n'}{\sum_{\rho=1}^{r} \lambda_\rho \xi_{\rho n}(x')} = dt \tag{173}$$

from $x_i' = x_i$ for $t = 0$, we obtain

$$x_i' = \phi_i(x_1, \ldots, x_n, \lambda_1 t, \ldots, \lambda_r t) \qquad (i = 1, 2, \ldots, n).$$

This is in the required canonical form, and is written

$$x_i' = \phi_i(x_1, \ldots, x_n, u_1, \ldots, u_r) \qquad (i = 1, 2, \ldots, r),$$

where $\lambda_\rho t = u_\rho$.

The equations (165) can be put in the form $X_\sigma' F + A_\sigma F = 0$, where

$$\sum_\rho \alpha_{\sigma\rho}(a) \frac{\partial F}{\partial x_\rho} = A_\sigma F,$$

and are satisfied by the transformation (155). $A_\sigma F$ is the first parameter group. Since this group is a complete system (Section 4.9), the commutator $(X_\sigma' F + A_\sigma F, X_\tau' F + A_\tau F)$ is a linear combination of

$$X_\rho' F + A_\rho F \qquad (\rho = 1, 2, \ldots, r),$$

that is,

$$(X_\sigma' F + A_\sigma F, X_\tau' F + A_\tau F) = \sum_\rho \phi_{\sigma\tau\rho}(X_\rho' F + A_\rho F).$$

Since

$$(X_\sigma' F + A_\sigma F, X_\tau' F + A_\tau F) = (X_\sigma' F, X_\tau' F) + (A_\sigma F, A_\tau F),$$

the commutator is divided into

$$(X_\sigma', X_\tau') = \sum_\rho \phi_{\sigma\tau\rho} X_\rho' F, \qquad (A_\sigma, A_\tau) = \sum_\rho \phi_{\sigma\tau\rho} A_\rho F$$
$$(\sigma, \tau = 1, 2, \ldots, r).$$

Since $\|\alpha_{\sigma\rho}(a)\| \neq 0$, the second relation determines $\phi_{\sigma\tau\rho}$ uniquely as functions of $a$. Differentiating the first relation with respect to $a_\lambda$, we obtain

$$\sum_\rho \frac{\partial \phi_{\sigma\tau\rho}}{\partial a_\lambda} X_\rho' F = 0.$$

Since the $X'F$ are linearly independent, we should have

$$\frac{\partial \phi_{\sigma\tau\rho}}{\partial a_\lambda} = 0, \quad \text{that is,} \quad \phi_{\sigma\tau\rho} = \text{constant}.$$

Thus

$$(X_\sigma, X_\tau)f = \sum_\rho c_{\sigma\tau\rho} X_\rho f \qquad (\sigma, \tau = 1, 2, \ldots, r). \tag{174}$$

This is the *second fundamental theorem of Lie*; $c_{\sigma\tau\rho}$ are the *structure constants.*

Conversely, if $r$ independent infinitesimal transformations

$$X_1 f, \ldots, X_r f$$

satisfy (174), then these transformations belong to the $r$-parameter group, and any transformation of this group in the neighborhood of the identical element is generated by $\sum_\rho c_\rho X_\rho f$.

There are relations among the $r^3$ structure constants. Since

$$(X_\sigma, X_\tau) + (X_\tau, X_\sigma) = 0,$$

we have

$$\sum_\rho (c_{\sigma\tau\rho} + c_{\tau\sigma\rho}) X_\rho f = 0.$$

Since $X_\rho f$ are linearly independent, we also have

$$c_{\sigma\tau\rho} + c_{\tau\sigma\rho} = 0. \tag{175}$$

Substituting $X_\sigma f$, $X_\tau f$, $X_\rho f$ in Poisson's identity (Section 4.3), we obtain

$$((X_\sigma, X_\tau), X_\rho) + ((X_\tau, X_\rho), X_\sigma) + ((X_\rho, X_\sigma), X_\tau) \equiv 0,$$

or

$$\sum_\lambda \sum_\kappa [(c_{\sigma\tau\lambda} c_{\lambda\rho\kappa} + c_{\tau\rho\lambda} c_{\lambda\sigma\kappa} + c_{\rho\sigma\lambda} c_{\lambda\tau\kappa})] \equiv 0 \tag{176}$$

$$(\rho, \sigma, \tau = 1, 2, \ldots, r).$$

Equations (175) represent a plane manifold in the $r^3$-dimensional space, while the equations (176) represent a quadratic manifold. Each point on the intersection $\mathfrak{M}$ of these two manifolds corresponds to the structure constant of the group. Substituting $\bar{X}_1 f, \ldots, \bar{X}_r f$ for

$$X_1 f, \ldots, X_r f$$

by means of

$$\bar{X}_\rho f = a_{\rho 1} X_1 f + \cdots + a_{\rho r} X_r f \qquad (\rho = 1, 2, \ldots, r), \tag{177}$$

we obtain

$$X_\sigma f = \bar{a}_{1\sigma} \bar{X}_1 f + \cdots + \bar{a}_{r\sigma} \bar{X}_r f \qquad (\sigma = 1, 2, \ldots, r),$$

since the $X_\rho f$ are mutually independent. From the relations

$$\begin{aligned}(\bar{X}_\tau, \bar{X}_\kappa)f &= \sum_{\lambda,\mu,\nu} a_{\tau\lambda}a_{\kappa\mu}c_{\lambda\mu\nu}\bar{X}_\nu f \\ &= \sum_{\lambda,\mu,\nu,\rho} a_{\tau\lambda}a_{\kappa\mu}\bar{a}_{\rho\nu}c_{\lambda\mu\nu}\bar{X}_\rho f = \sum_\rho \bar{c}_{\tau\kappa\rho}\bar{X}_\rho f,\end{aligned}$$

we obtain

$$\bar{c}_{\tau\kappa\rho} = \sum_{\lambda,\mu,\nu} a_{\tau\lambda}a_{\kappa\mu}\bar{a}_{\rho\nu}c_{\lambda\mu\nu} \qquad (\rho, \tau, \kappa = 1, 2, \ldots, r). \tag{178}$$

Thus to each linear transformation (177) there corresponds a linear transformation of $c_{\tau\kappa\rho}$ of (178). Conversely, by solving (178) we obtain

$$c_{\tau\kappa\rho} = \sum_{\lambda,\mu,\nu} \bar{a}_{\lambda\tau}a_{\mu\kappa}\bar{a}_{\nu\rho}c_{\lambda\mu\nu} \qquad (\rho, \tau, \kappa = 1, 2, \ldots, r). \tag{179}$$

Since (177) constitute a group, the transformations (178) also constitute a group. This group of $c_{\tau\kappa\rho}$ is said to be *induced* by the group $X_\sigma f$. The constants $\bar{c}_{\tau\kappa\rho}$ and $c_{\lambda\mu\nu}$ related through (178) are not essentially different. A point of $\mathfrak{M}$ is simply transformed to another point of $\mathfrak{M}$. Two groups with the same structure constants are said to be *equivalent*.

For proving the converse of this theorem — that is, that there exists an $r$-parameter group (174) for the $r^3$ constants satisfying (175) and (176) — it is sufficient to show the existence of $\psi_{\sigma\rho}(a)$, such that $\|\psi_{\sigma\rho}(a)\| \neq 0$ satisfies Maurer's relations (161)

$$\frac{\partial\psi_{\rho\sigma}}{\partial a_\tau} - \frac{\partial\psi_{\rho\tau}}{\partial a_\sigma} = \sum_{\lambda,\kappa} c_{\lambda\kappa\rho}\psi_{\lambda\sigma}\psi_{\kappa\tau} \qquad (\rho, \sigma, \tau = 1, 2, \ldots, r).$$

This relation is merely the integrability condition for the Pfaffian (162), which can be written, similarly to (158a),

$$df(x') = \sum_\rho P_\rho X'_\rho f(x'), \qquad P_\rho = \sum_\sigma \psi_{\rho\sigma}(a)\, da_\sigma. \tag{180}$$

Let $a_1, \ldots, a_r$ be the canonical parameters. The finite transformations of the corresponding group are obtained by the infinitesimal transformation $(\sum_i a_i X_i f)\delta t$ from $t = 0$ to $t = 1$. If we continue it from $t = 1$ to $t = 1 + \delta t$, then we obtain the finite transformations with the canonical parameters $a_1(1 + \delta t), \ldots, a_r(1 + \delta t)$. From (180) we obtain

$$\delta f(x') = \sum_{\rho,\sigma} \psi_{\rho\sigma}(a)a_\sigma X'_\rho f(x')\delta t.$$

On the other hand,

$$\delta f(x') = \sum_{\rho} a_\rho X'_\rho f(x')\,\delta t.$$

Hence

$$\sum_{\sigma} \psi_{\rho\sigma}(a) a_\sigma = a_\rho \qquad (\rho = 1, 2, \ldots, r). \tag{181}$$

Multiply Maurer's relation by $a_\tau$ and sum over $\tau$; then, referring to (181), we obtain

$$\sum_{\tau} \frac{\partial \psi_{\rho\sigma}}{\partial a_\tau} a_\tau - \sum_{\tau} \frac{\partial \psi_{\rho\tau}}{\partial a_\sigma} a_\tau = \sum_{\lambda,\kappa} c_{\lambda\kappa\rho} \psi_{\lambda\sigma} a_\kappa. \tag{182}$$

Write

$$Uf = a_1 \frac{\partial f}{\partial a_1} + \cdots + a_r \frac{\partial f}{\partial a_r};$$

then, from (181),

$$U\psi_{\rho\sigma} = \epsilon_{\rho\sigma} - \psi_{\rho\sigma} + \sum_{\lambda,\kappa} c_{\lambda\kappa\rho} \psi_{\lambda\sigma} a_\kappa \qquad (\rho, \sigma = 1, 2, \ldots, r).$$

Put $a_\rho = \lambda_\rho t$; then (182) becomes

$$\frac{\partial (t\psi_{\rho\sigma})}{\partial t} = \epsilon_{\rho\sigma} + \sum_{\lambda,\kappa} c_{\lambda\kappa\rho} (t\psi_{\lambda\sigma}) \lambda_\kappa,$$

since

$$\frac{\partial \psi_{\rho\sigma}}{\partial t} = \sum_{\tau} \frac{\partial \psi_{\rho\sigma}}{\partial a_\tau} \lambda_\tau = t^{-1} U\psi_{\rho\sigma}.$$

Denote by $L$ the matrix $(l_{\rho\lambda})$, such that $\sum_\kappa c_{\lambda\kappa\rho}\lambda_\kappa = l_{\rho\lambda}$, and by $\Psi = t\psi$ the matrix $(\Psi_{\rho\sigma}) = t(\psi_{\rho\sigma})$; then the $r^3$ equations become

$$\frac{\partial \Psi}{\partial t} = L\Psi + I, \tag{183}$$

where $\Psi$ reduces to a null matrix for $t = 0$. If $\lambda_1, \ldots, \lambda_r$ are given, then this matrix equation determines $\Psi$ with the initial condition that $\Psi = 0$ for $t = 0$ in the form

$$\Psi = tI + \frac{1}{2!} t^2 L + \frac{1}{3!} t^3 L^2 + \cdots.$$

Hence,

$$\psi = I + \frac{t}{2!} L + \frac{t^2}{3!} L^2 + \cdots.$$

If we write $\sum_\kappa c_{\sigma\kappa\rho} a_\kappa = A_{\rho\sigma}$ and denote the matrix of $A_{\rho\sigma} = tl_{\rho\sigma}$ by $A$, then the $\psi_{\rho\sigma}$ are given as functions of $a$ by means of

$$\psi = I + \frac{1}{1!} A + \frac{1}{2!} A^2 + \cdots, \qquad A^m = t^m L^m. \tag{184}$$

It can be shown that this matrix $\psi$ satisfies Maurer's relation (161).

Denote the matrices of $\psi_{\rho\sigma}$ and $\alpha_{\rho\sigma}$ by $\psi$ and $\alpha$, respectively. By definition, $\psi_\alpha = I$. Then (183) is written in the form

$$\psi = \frac{e^{tL} - 1}{tL}$$

and the transposed reciprocal in the form

$$\alpha' = \frac{tL}{e^{tL} - 1}$$

in the neighborhood of $t = 0$. With Bernoulli's numbers $B_1, B_2, \ldots$, it is known that

$$\frac{x}{e^x - 1} = 1 - \frac{x}{2} + \frac{1}{2!} B_1 x^2 - \frac{1}{4!} B_2 x^4 + \cdots.$$

Thus

$$\alpha' = I - \frac{1}{2} A + \frac{1}{2!} B_1 A^2 - \frac{1}{4!} B_2 A^4 + \cdots.$$

Since the elements of the transposed matrix are $A'_{\rho\sigma} = -\sum_\tau a_\tau c_{\tau\rho\sigma}$, we have

$$\alpha = I - \frac{1}{2} A' + \frac{1}{2!} B_1 A'^2 - \frac{1}{4!} B_2 A'^4 + \cdots, \tag{185}$$

or, by writing the matrix of $c_{\tau\rho\sigma}$ as $C_\tau$,

$$\alpha = I + \frac{1}{2} \sum_\tau a_\tau C_\tau + \frac{1}{2!} B_1 \Big(\sum_\tau a_\tau C_\tau\Big)^2 - \frac{1}{4!} B_2 \Big(\sum_\tau a_\tau C_\tau\Big)^4 + \cdots,$$

where

$$A' = -\sum_\tau a_\tau C_\tau, \qquad C_\tau = (c_{\tau\rho\sigma}),$$

with fixed $\tau$. Thus, the infinitesimal transformations $A_1 f, \ldots, A_r f$ are determined from the matrix $\alpha$.

This proof is due to Schur (1891). This is the foundation of Baker's method of solving a system of linear differential equations with variable coefficients (Baker, 1902, 1904; Hagihara, 1940), which will be applied in Chapter 12 (Volume 3) to several problems in celestial mechanics. It is to be noticed that the series (184) is convergent if $A$, and accordingly $t$, is sufficiently small. Since the elements $\psi_{\rho\sigma}$ are convergent series arranged in powers of $a_1, \ldots, a_r$, so are the elements $\alpha_{\rho\sigma}$ of the reciprocal matrix.

Lie's series are applied by Gröbner (1960) to the numerical solution of differential equations, and by Knapp (1964) to the computation of perturbation (Section 10.44).

## 4.27 ADJOINT GROUPS

A finite continuous group containing at least one transformation that transforms every point of a region in a space $R_n(x_1, \ldots, x_n)$ to another point of the space is said to be *transitive*. If for a given set of values of $x_1, \ldots, x_n, x'_1, \ldots, x'_n$ there is at least one set of values of $a_1, \ldots, a_r$ satisfying (155), then the transformation group (155) is transitive. Otherwise it is *intransitive*. Hence, a transitive group should contain at least $n$ essential parameters. For $n = r$ there exists in general a transformation that transforms one arbitrary point to another arbitrary point. Thus, there is no transformation that makes a general point invariant, except the identity transformation. Such a group is called *simply* transitive. (For the same group definitions in another context, see discussion preceding the first fundamental theorem of Lie, Section 4.26.) Since the finite group of $r$ independent infinitesimal transformations

$$X_\rho f = \sum_{i=1}^{n} \xi_{\rho i}(x) \frac{\partial f}{\partial x_i} \qquad (\rho = 1, 2, \ldots, r)$$

can be written

$$x'_i = x_i + \sum_{\rho=1}^{r} e_\rho \xi_{\rho i}(x) + \cdots,$$

it is sufficient for this group to be transitive that none of the $n$th order determinants $\|\xi_{\rho i}\|$ be zero. Denote the rank of the matrix $(\xi_{\rho i})$ by $m < n$; then $m$ transformations out of $X_\rho f$ are independent and the remaining transformations are linear combinations of these. These $m$ transformations form a complete system (Section 4.9). Hence the system admits $n - m$ independent solutions $\Omega_1(x), \ldots, \Omega_{n-m}(x)$. Thus there are $n - m$ relations between $x$ and $x'$; that is,

$$\Omega_s(x') = \Omega_s(x) \qquad (s = 1, 2, \ldots, n - m).$$

Hence, in order for the group to be transitive, it is necessary and sufficient that $m$ of $X_1 f = 0, \ldots, X_r f = 0$ be independent. If the group is transitive it contains no invariant. If it is intransitive, then the solutions common to the $r$ transformations $X_\rho f = 0$ are the only invariants, and

$$\Omega_1(x) = C_1, \ldots, \Omega_{n-m}(x) = C_{n-m} \tag{186}$$

divide the $n$-dimensional space into $m$-dimensional regions $M_m$ that are invariant for $X_1 f = 0, \ldots, X_r f = 0$. This invariant manifold $M_m$ can be determined by eliminating the given transformation group from the finite equations (155).

Since a point of the space $R_n$ can belong to one and only one of the $\infty^{n-m}$ manifolds (186), we say that any manifold *decomposes* (*zerlegen*) the space $R_n$; this decomposition is invariant for any transformation of $X_1 f = 0, \ldots, X_r f = 0$. Even a transitive group can decompose $R_n$ into $\infty^{n-m}$ manifolds of the form (186). Such manifolds are invariant for the transformation group as a whole, but each individual manifold may be interchanged by a single transformation of the group. If a group decomposes $R_n$ invariantly into $\infty^{n-m}$ $m$-dimensional manifolds $M_m$ in this manner, then the group is called *imprimitive*. Otherwise it is *primitive*. An intransitive group is imprimitive. A primitive group is transitive.

The condition for imprimitivity is that the complete system of $m$ independent equations $Y_1 f = 0, \ldots, Y_m f = 0$ admits all transformations of the group. This system of equations defines an $m$-dimensional manifold $M_m$. If the system admits all transformations of the group, then $M_m$ is invariant for those transformations. The integer $m$ is called the *order* of imprimitivity (Section 4.23). The necessary and sufficient condition for the order of imprimitivity to be $m$ is that the complete system of equations $X_1 f = 0, \ldots, X_m f = 0$ admits $n - m$ independent solutions.

Consider in $m$ independent infinitesimal transformations

$$X_\rho f = \sum_{i=1}^{n} \xi_{\rho i} \frac{\partial f}{\partial x_i} \qquad (\rho = 1, 2, \ldots, m),$$

a set of $\infty^{n-m}$ infinitesimal transformations represented by

$$e_1 X_1 f + \cdots + e_m X_m f$$

with $m$ arbitrary constants $e_1, \ldots, e_m$. The transformed set can be written

$$\sum_{\rho=1}^{m} e_\rho \sum_{i=1}^{n} \xi_{\rho i}(x) \frac{\partial f}{\partial x_i} = \sum_{\rho=1}^{m} e'_\rho \sum_{i=1}^{n} \xi_{\rho i}(x') \frac{\partial f}{\partial x'_i}.$$

Hence, if

$$\sum_{\rho=1}^{m} e_\rho X_\rho f = \sum_{\sigma=1}^{m} e'_\sigma X'_\sigma f, \tag{187}$$

then the set of transformations $\sum_{\rho=1}^{m} e_\rho X_\rho f$ is invariant for the transformation of $x$ to $x'$; that is, $ST_eS^{-1} = T_{e'}$, or $ST_e = T_{e'}S$.

Let $Yf$ be a one-parameter group for the transformation of the parameter $t$ from $x$ to $x'_i = f_i(x_1, \ldots, x_n, t)$. Then $x'_i = x_i + tY + \cdots$, and $X'_\rho f = X_\rho f + t(Y, X_\rho)f + \cdots$. In order that this transformation may belong to the set of those transformations, we should have

$$(Y, X_\rho)f = \sum_{j=1}^{m} g_{\rho j} X_j f \qquad (\rho = 1, 2, \ldots, m).$$

Hence

$$\frac{\partial(X'_\rho f)}{\partial t} = \sum_{j=1}^{m} g_{\rho j} X'_j f \qquad (\rho = 1, 2, \ldots, m).$$

Since

$$\begin{aligned}\frac{\partial}{\partial t}\sum_\rho e'_\rho X'_\rho f &= \sum_\rho \frac{de'_\rho}{dt} X'_\rho f + \sum_\rho e'_\rho \sum_j g_{\rho j} X'_j f \\ &= \sum_\rho \left\{\frac{de'_\rho}{dt} + \sum_j g_{j\rho} e'_j\right\} X'_\rho f,\end{aligned}$$

therefore

$$\frac{de'_\rho}{dt} + \sum_j g_{j\rho} e'_j = 0. \tag{188}$$

The values of $e_\rho$ are determined by integrating (188) from $e'_\rho = e_\rho$ at $t = 0$.

If $X_1 f, \ldots, X_r f$ satisfies (188), that is, if $\sum_\rho e_\rho X_\rho f$ constitutes an $r$-parameter group and all transformations of this set of transformation groups make the one-parameter group $Yf$ invariant, then

$$(Y, X_\rho)f = \sum_{\sigma=1}^{r} g_{\rho\sigma} X_\sigma f \qquad (\rho = 1, 2, \ldots, r). \tag{189}$$

The coefficients $g_{\rho\sigma}$ are given by (188) in terms of $e_\rho$. Then $r + 1$ infinitesimal transformations $X_1 f, \ldots, X_r f$, $Yf$ constitute an $(r + 1)$-parameter group, and the $r$ transformations $X_1 f, \ldots, X_r f$ constitute its $r$-parameter subgroup. Let $x'_i = \psi_i(x_1, \ldots, x_n; a_1, \ldots, a_{r+1})$ be the finite transformations of the $(r + 1)$-parameter group; then the set of transformations $\sum_\rho e_\rho X_\rho f$ is invariant for the change of $x$ to $x'$.

A similar situation exists for infinitesimal transformations $Y_1f, \ldots, Y_mf$ instead of a single $Yf$. If $X_1f, \ldots, X_rf$ are invariant for all transformations of an $(r + m)$-parameter group, then they constitute an *invariant subgroup* (Lie, 1888). Then (189) holds for all such $Y_if$. Hence, if $X_1f, \ldots, X_rf$ are contained in $X_1f, \ldots, X_rf, Y_1f, \ldots, Y_mf$, and all commutators $(Y_i, Y_k)f$ are linear combinations of $X_1f, \ldots, X_rf$ with constant coefficients, then $X_1f, \ldots, X_rf$ constitute an invariant subgroup. In particular, if the transformation group $(X_\sigma, X_\rho)$ is an $r$-parameter group, then the group is the same as the group $X_1f, \ldots, X_rf$. If it contains less than $r$ parameters, then the group is an invariant subgroup of $X_1f, \ldots, X_rf$. It can be shown that the group $X_1f, \ldots, X_rf$ constitutes an invariant subgroup if mutually independent infinitesimal transformations $e_1X_1f + \cdots + e_rX_rf$, also independent of all commutators $(X_\rho, X_\sigma)$, are added to $(X_\rho, X_\sigma)$. Actually, $(X_\sigma, X_\rho)f$ contains at most $r$-parameter groups, as it belongs to $X_1f, \ldots, X_rf$. Since

$$((X_\sigma, X_\rho), (X_\lambda, X_\mu)) = \sum_{\tau,\kappa} c_{\sigma\rho\tau}c_{\lambda\mu\kappa}(X_\tau, X_\kappa)$$

from (182), the transformations $(X_\sigma, X_\rho)$ constitute a group. Since

$$(X_\tau, (X_\rho, X_\sigma)) = \sum_{\lambda} c_{\rho\sigma\lambda}(X_\tau, X_\lambda),$$

these transformations are invariant for $X_1f, \ldots, X_rf$. This theorem will be applied in Section 4.28 for simple groups.

If infinitesimal transformations $Y_1f, \ldots, Y_pf$ and $Z_1f, \ldots, Z_qf$ of two invariant subgroups of the same group $G_r$ have no infinitesimal transformation in common, then the $(Y_i, Z_j)$ should all vanish identically. If they have any common transformation, then the common infinitesimal transformation generates invariant subgroups of $G_r$. Actually, the $(Y_i, Z_j)$ are linear combinations of the infinitesimal transformations of the invariant subgroup with constant coefficients. If they have no transformation in common, then these combinations should be zero. If the $(Y_i, Z_j)$ are all zero, then the $Y_i$ and $Z_j$ are said to be *permutable* (*vertauschbar*). This theorem will be applied also in Section 4.28 for semisimple groups.

Let the independent transformations $Y_1f, \ldots, Y_rf$ belonging to the group $X_1f, \ldots, X_rf$ be such that

$$(Y_\sigma, Y_\rho)f = \sum_{\tau} c'_{\sigma\rho\tau}Y_\tau f, \qquad (X_\sigma, X_\rho)f = \sum_{\tau} c_{\sigma\rho\tau}X_\tau f,$$

$$Y_\sigma f = \sum_{\rho} h_{\sigma\rho}X_\rho f.$$

Then,

$$(Y_\sigma, Y_\rho)f = \sum_{\lambda,\mu,\tau} h_{\sigma\lambda}h_{\rho\mu}c_{\lambda\mu\tau}X_\tau f = \sum_{\lambda,\tau} h_{\lambda\tau}c'_{\sigma\rho\lambda}X_\tau f.$$

Hence

$$c'_{\sigma\rho\tau} = \frac{1}{D}\sum_{\lambda}\left\{\frac{\partial D}{\partial h_{\tau\lambda}}\sum_{\mu,\nu} h_{\sigma\mu}h_{\rho\nu}c_{\mu\nu\lambda}\right\}, \qquad D = \|h_{\rho\sigma}\|.$$

If arbitrary $Z_1f, \ldots, Z_rf$ are such that $(Z_\sigma, Z_\rho)f = \sum_\tau c'_{\sigma\rho\tau}Z_\tau f$ with the constants $c'_{\sigma\rho\tau}$ satisfying these relations, then $Z_\rho f$ and $X_\sigma f$ are said to be of *equal structure* (*gleichzusammengesetzt*). If $Y_1f, \ldots, Y_{r-q}f$ can be chosen so that, for the infinitesimal transformations

$$Z_\rho f = \sum_{\sigma=1}^{r-q} h_{\rho\sigma}Y_\sigma f \qquad (\rho = 1, 2, \ldots, r)$$

of the $(r-q)$-parameter group, none of the minor determinants of the $(r-q)$th order of $\|h_{\rho\sigma}\|$ can be made zero, and, at the same time, chosen so that $(Z_\sigma, Z_\rho) = \sum_{\tau=1}^{r} c_{\sigma\rho\tau}Z_\tau f$ are satisfied identically, then the group $X_1f, \ldots, X_rf$ is said to be *isomorphic* for $q = 0$, and *homomorphic* for $q > 0$, to $Y_1f, \ldots, Y_{r-q}f$. Two isomorphic or homomorphic groups are of equal structure, that is, have the same structure constants.

If $\sum_\rho c_\rho X_\rho f$ formed of the infinitesimal transformations $X_\rho f$ of an $r$-parameter group $x'_i = f_i(x_1, \ldots, x_n; a_1, \ldots, a_r)$ can be written in the form (187) for a transformation of $x$ to $x'$ as new variables, then

$$e'_\tau = \sum_\sigma \rho_{\tau\sigma}(a_1, \ldots, a_r)e_\sigma \qquad (\tau = 1, 2, \ldots, r), \tag{190}$$

where $e'$ is independent of $x$. Substituting $x''_i = f_i(x', b)$ gives

$$\sum_\rho e'_\rho X'_\rho f = \sum_\sigma e''_\rho X''_\rho f, \qquad e''_\sigma = \sum_\tau \rho_{\sigma\tau}(b_1, \ldots, b_r)e'_\tau.$$

Since $x'_i = f_i(x, a)$ form a group, we can write

$$x''_i = f_i(x, c), \qquad c_\rho = \phi_\rho(a_1, \ldots, a_r, b_1, \ldots, b_r).$$

Hence,

$$\sum_\rho e_\rho X_\rho f = \sum_\sigma e''_\sigma X''_\sigma f,$$

and

$$e''_\sigma = \sum_\tau \rho_{\sigma\tau}(c_1, \ldots, c_r)e_\tau = \sum_\tau \rho_{\sigma\tau}(\phi_1(a, b), \ldots, \phi_r(a, b))e_\tau.$$

Thus $e'_\sigma = \sum_\tau \rho_{\sigma\tau}(a)e_\tau$ form a group, and

$$e''_\sigma = \sum_{\tau,\kappa} \rho_{\sigma\tau}(b_1, \ldots, b_r)\rho_{\tau\kappa}(a_1, \ldots, a_r)e_\kappa,$$

$$\rho_{\sigma\tau}(\phi_1(a, b), \ldots, \phi_r(a, b)) \equiv \sum_\kappa \rho_{\kappa\tau}(a_1, \ldots, a_r)\rho_{\sigma\kappa}(b_1, \ldots, b_r).$$

Hence, the group (190) belongs to a group $x'_i = f_i(x, a)$ or to a group of infinitesimal transformations $X_1 f, \ldots, X_r f$. This last group is called the *adjoint group*.

In other words, let $ST_eS^{-1} = T_{e'}$ be the first transformation, and $S_1T_{e'}S_1^{-1} = T_{e''}$ be the second. Then

$$T_{e''} = S_1ST_eS^{-1}S_1^{-1} = S_2T_eS_2^{-1},$$

with $S_2 = S_1S$. If $ST_eS^{-1} = T_{e'}$, $ST_{e'}S^{-1} = T_{e''}$; then

$$T_{e''}T_{e'} = ST_{e'}S^{-1}ST_eS^{-1} = ST_{e'}T_eS^{-1};$$

that is, $T_e$ is in *automorphism*. Thus $T_e$ constitutes a group, and every transformation of the adjoint group is homomorphic to the given group. Hence, if a canonical parameter $e_\rho$ admits the transformations of an adjoint group, then the structure constants are invariant. Consequently, if the infinitesimal transformations of an adjoint group are written in the form

$$E_\mu f = \sum_{\lambda,\kappa} c_{\lambda\mu\kappa}e_\lambda \frac{\partial f}{\partial e_\kappa} \qquad (\mu = 1, 2, \ldots, r), \tag{191}$$

then $\lambda_1E_1f + \cdots + \lambda_rE_rf$ is a general infinitesimal transformation of the adjoint group and satisfies the relation

$$(E_\mu, E_\nu)f = \sum_\rho c_{\mu\nu\rho}E_\rho f \qquad (\mu, \nu = 1, 2, \ldots, r). \tag{192}$$

If $E_1f, \ldots, E_rf$ are not independent, but $\sum_\rho g_\rho E_\rho f \equiv 0$, then, from

$$\sum_\rho g_\rho \sum_{\sigma,\tau} c_{\sigma\rho\tau}e_\sigma \frac{\partial f}{\partial e_\tau} = 0,$$

we obtain

$$\sum_\rho g_\rho c_{\sigma\rho\tau} = 0 \qquad (\sigma, \tau = 1, 2, \ldots, r).$$

Hence

$$\left(X_\sigma f, \sum_\rho g_\rho X_\rho f\right) = \sum_\tau \left(\sum_\rho g_\rho c_{\sigma\rho\tau}\right)X_\tau f \equiv 0.$$

Thus, $\sum_\rho g_\rho X_\rho f$ is permutable with all infinitesimal transformations $X_\rho f$. Conversely, if $X_1 f, \ldots, X_r f$ contain all infinitesimal transformations that are permutable with all $X_\sigma f$ in $\sum_\rho g_\rho X_\rho f$, then $\sum_\rho g_\rho E_\rho f \equiv 0$. Such a $\sum_\rho g_\rho X_\rho f$ is called the *distinguished* (*ausgezeichnete*) infinitesimal transformation of the group. A distinguished infinitesimal transformation is invariant for the transformation $x_i' = f_i(x, a)$; that is,

$$\sum_\rho g_\rho X_\rho f = \sum_\rho g_\rho' X_\rho' f.$$

In the corresponding adjoint group, $\sum_\rho g_\rho E_\rho f = 0$. Hence, if there are just $m$ such distinguished transformations among $X_\rho f$ of the transformation group $G_r$, then $r - m$ among $E_\sigma f$ are not independent. The adjoint group has $r - m$ essential parameters. Thus, since $\sum_\rho g_\rho X_\rho f$ is invariant for the group $G_r$, this group $G_r$ is imprimitive. Conversely, if $G_r$ is primitive in $x_1, \ldots, x_n$, then $G_r$ does not contain any distinguished infinitesimal transformation. The adjoint group of an $r$-parameter primitive group contains $r$ essential parameters. Conversely, $G_r$ has a primitive group, that is, an adjoint group.

The transformation (155) with $r$ essential parameters satisfies

$$\frac{\partial x_i'}{\partial a_\rho} = \sum_\sigma \psi_{\sigma\rho}(a)\xi_{\sigma i}(x'). \tag{193}$$

Since $\|\psi_{\sigma\rho}\| \neq 0$, these equations can be solved in the form

$$\xi_{\sigma i}(x') = \sum_\rho \alpha_{\sigma\rho}(a)\frac{\partial x_i'}{\partial a_\rho}, \tag{194}$$

and the transformation (155) is generated by $r$ independent infinitesimal transformations

$$X_\rho f = \sum_{i=1}^n \xi_{\rho i}(x_1, \ldots, x_n)\frac{\partial f}{\partial x_i} \qquad (\rho = 1, 2, \ldots, r).$$

The $r$ infinitesimal transformations of the corresponding parameter group

$$A_\rho f = \sum_\sigma \alpha_{\rho\sigma}(a)\frac{\partial f}{\partial a_\sigma}$$

are independent. It has been seen that their structure constants are the same as those of $X_\rho f$; that is, if

$$(X_\rho, X_\sigma)f = \sum_\tau c_{\rho\sigma\tau} X_\tau f,$$

then

$$(A_\rho, A_\sigma)f = \sum_\tau c_{\rho\sigma\tau} A_\tau f.$$

We have seen also that the finite transformation

$$x'_i = F_i(x_1, \ldots, x_n, \lambda_1 t, \ldots, \lambda_r t) \tag{195}$$

with the canonical form

$$x'_i = x_i + \sum_{\sigma} e_\sigma \xi_{\sigma i}(x) + \cdots$$

of the transformation (155) is obtained by integrating

$$\frac{dx'_1}{\sum_\rho \lambda_\rho \xi'_{\rho 1}} = \cdots = \frac{dx'_n}{\sum_\rho \lambda_\rho \xi'_{\rho n}} = dt,$$

with the initial condition $x'_1 = x_1, \ldots, x'_n = x_n$ for $t = 0$. Multiply (194) by $\lambda_\sigma$ and sum over $\sigma$, then from these differential equations we obtain

$$\frac{dx'_i}{dt} = \sum_{\rho,\sigma} \lambda_\rho \alpha_{\rho\sigma}(a) \frac{\partial x'_i}{\partial a_\sigma}.$$

Hence, regarded as functions of $t$, the parameters $a_1, \ldots, a_r$ must satisfy

$$\frac{da_\rho}{dt} = \sum_{\sigma} \lambda_\sigma \alpha_{\sigma\rho}(a). \tag{196}$$

The initial condition is that $a$ should be $a^0$, obtained by putting $x = x'$ for $t = 0$ in the equations (155). Integrating (196) with this initial condition, we obtain

$$a_\rho = \Phi_\rho(\lambda_1 t, \ldots, \lambda_r t). \tag{197}$$

Putting these in (155), we obtain the required canonical form (195). Write this as

$$a_\rho = \Phi_\rho(u_1, \ldots, u_r), \qquad u = \lambda t.$$

Then the canonical transformation is of the form

$$x'_i = f_i(x_1, \ldots, x_n; \Phi_1(u), \ldots, \Phi_r(u)).$$

In other words, the system of equations in $a_1, \ldots, a_r, t$ admits the infinitesimal transformation

$$Af = \lambda_1 A_1 f + \cdots + \lambda_r A_r f + \frac{\partial f}{\partial t},$$

and can be integrated with respect to $a$ with the initial condition that $a = a^0$ for $t = 0$, where it is assumed that $\|\psi_{\rho\sigma}\| \neq 0$ for $a = a^0$.

The computation is simplified if we use the adjoint group (Lie, 1893, 1893a). Substituting $x_i' = f_i(x)$ in

$$\sum_\rho e_\rho \sum_i \xi_{\rho i}(x) \frac{\partial f}{\partial x_i} = \sum_\rho e_\rho X_\rho f,$$

we obtain

$$\sum_\rho e_\rho' X_\rho' f = \sum_\sigma e_\sigma X_\sigma f.$$

Accordingly

$$e_\sigma' = \sum_\tau \rho_{\sigma\tau}(a) e_\tau \tag{198}$$

defines the adjoint group. If the integral (197) of the system (196) is known, then the canonical form of the adjoint group is obtained by substituting (197) in (198). The result is of the form

$$e_\sigma' = \sum_\tau \rho_{\sigma\tau}(\Phi_1(\lambda t), \ldots, \Phi_r(\lambda t)) e_\tau. \tag{198a}$$

If the integral (197) is not known, then we determine the canonical form by solving

$$\frac{de_1'}{\sum_\rho \lambda_\rho e_{\rho 1}'} = \cdots = \frac{de_r'}{\sum_\rho \lambda_\rho e_{\rho r}'} = dt, \tag{199}$$

with the initial condition that $e' = e$ for $t = 0$, because the adjoint group is generated by

$$E_\rho f = \sum_{\sigma,\tau} c_{\sigma\rho\tau} e_\sigma \frac{\partial f}{\partial e_\tau}, \qquad e_{\rho\sigma}' = \sum_\tau c_{\tau\rho\sigma} e_\tau'$$

and satisfies

$$(E_\sigma, E_\rho) f = \sum_\tau c_{\sigma\rho\tau} E_\tau f.$$

Equations (199) can be solved by mere algebraic operations, since they have constant coefficients. Thus the canonical form of the adjoint group is determined in the form

$$e_\mu' = \sum_\nu \sigma_{\mu\nu}(\lambda_1 t, \ldots, \lambda_r t) e_\nu,$$

where $\sigma_{\mu\nu}$ contain $\lambda t$ only in the combination $\lambda t = u$. This will be identical with (198) if we substitute (197); that is,

$$\rho_{\mu\nu}(a_1, \ldots, a_r) = \sigma_{\mu\nu}(\lambda_1 t, \ldots, \lambda_r t)$$

is an identity. This is the integral of (196). It satisfies (197) identically for $a = a^0$, $t = 0$. It is when the $r$-parameter group does not contain the distinguished infinitesimal transformation that the equations (199) can be solved algebraically. If the group contains $l$ independent distinguished transformations, that is, if $X_1f, \ldots, X_rf$ contain $l$ transformations $Zf = \sum_i \xi_i(x)\, \partial f/\partial x_i$ that are permutable with each of $X_1f, \ldots, X_rf$, then just $l$ independent integrations should be carried out besides algebraic operations (Lie, 1927). The adjoint group constitutes, then, an $(r - l)$-parameter group and obeys the relation (195). The structure constants $c_{\rho\sigma\tau}$ should satisfy Lie's third fundamental theorem. Conversely, if $r^3$ structure constants $c_{\rho\sigma\tau}$ are given, then $X_1f, \ldots, X_rf$ need not be considered, since the structure constants give the $r$-parameter group.

Baker's matrix method of integrating a system of linear differential equations is founded on this use of the adjoint group (Baker, 1902, 1904; Hagihara, 1940).

## 4.28 SIMPLE AND SEMISIMPLE GROUPS

A group that does not contain any invariant subgroup is called *simple*. Otherwise it is *nonsimple* (Lie, 1893, 1927; Cartan, 1930, 1933; Chevalley, 1951; see also Weyl, 1923, 1923a, 1924, 1931, 1939; van der Waerden, 1955, 1959; Speiser, 1923; Pontrjagin, 1939; Gröbner, 1960). A simple group contains only one homomorphic subgroup, namely, the identity transformation.

Consider a nonsimple group. We choose the infinitesimal transformations of the group $G_r$ so that $X_{m+1}f, \ldots, X_rf$ generate an $(r - m)$-parameter group. Then the structure of $G_r$ is such that

$$(X_\mu, X_\nu)f = \sum_{\sigma=1}^{m} c_{\mu\nu\sigma} X_\sigma f + \sum_{\tau=1}^{r-m} c_{\mu,\nu,m+\tau} X_{m+\tau} f \qquad (\mu, \nu = 1, 2, \ldots, m),$$

$$(X_\mu, X_{m+k})f = \sum_{\tau=1}^{r-m} c_{\mu,m+k,m+\tau} X_{m+\tau} f,$$

$$(X_{m+k}, X_{m+j})f = \sum_{\tau=1}^{r-m} c_{m+k,m+j,m+\tau} X_{m+\tau} f \qquad (k, j = 1, 2, \ldots, r - m).$$

Let the $m$-parameter group homomorphic to $G_r$ determined by the invariant subgroup be

$$(\bar{X}_\mu, \bar{X}_\nu)f = \sum_{\sigma=1}^{m} c_{\mu\nu\sigma} \bar{X}_\sigma f \qquad (\mu, \nu = 1, 2, \ldots, m),$$

and suppose that $\bar{X}_{q+1}f, \ldots, \bar{X}_m f$ $(0 < q < m)$ generate an invariant subgroup composed of other transformations besides the identity transformation. Then $\bar{X}_{q+1}f, \ldots, \bar{X}_m f, X_{m+1}f, \ldots, X_r f$ generate an $(r - q)$-parameter invariant subgroup $G_{r-q}$ containing the $(r - m)$ parameter invariant subgroup $G_{r-m}$. Hence, in order that the $m$-parameter group $G_m$, homomorphic to the given $r$-parameter group $G_r$ $(r > m)$, shall be simple, it is necessary and sufficient that it be possible to obtain the structure of $G_r$ so that the invariant subgroup of the highest order $r - m$ corresponds to the identity transformation.

We can derive from the structure of the $r$-parameter group the structure of a sequence of simple groups by choosing the subgroups $\bar{G}_1, \bar{G}_2, \ldots$ of $G_r$ so that $\bar{G}_1$ is invariant in $G_r$ but is not contained in any larger invariant subgroup of $\bar{G}_k$, and so that $\bar{G}_{k+1}$ is invariant in $\bar{G}_k$ but is not contained in any larger invariant subgroup of $\bar{G}_k$. Then $\bar{G}_1$ gives the structure of the simple group that is isomorphic to $G_r$, and $\bar{G}_2$ gives the structure of the simple group that is isomorphic to $\bar{G}_1$, and so on. The numbers of the essential parameters of such simple groups in this sequence are $r - r_1, r_1 - r_2, r_2 - r_3, \ldots$, respectively. These are called the *indices* of the simple groups; since each index is smaller than the preceding one, the process will end at the stage where $\bar{G}_m$, for example, consists of only the identity transformation and $\bar{G}_{m-1}$ is simple. Thus we get $m$ such subgroups and the number of the essential parameters is $r$.

This sequence of $G_r, \bar{G}_1, \bar{G}_2, \ldots, \bar{G}_m$ of the group $G_r$ is called the *composition series* (*Kompositionsreihe*). The sequence of invariant subgroups only is called the *normal series* (*Normalreihe*). If no other subgroup can be inserted between any two of the subgroups in a sequence, that is, if no *refinement* (*Verfeinerung*) can be made, then the sequence is a composition series. Two normal series are isomorphic and can be refined. This is *Schreier's theorem* (Schreier, 1926, 1927). Since $\bar{G}_1$ is not necessarily the only subgroup in $G_r$ with the same property, there may be more than one composition series — not necessarily of the same property.

Let the differences between the essential parameters of $G_r$ and those of the largest subgroup containing $\bar{G}_1$ (but contained in $G_r$) be $l_1$. In general, let the differences between the essential parameters of $\bar{G}_{k-1}$ and those of the largest subgroup containing $\bar{G}_k$ but contained in $\bar{G}_{k-1}$ be $l_k$. The numbers $l_1, l_2, \ldots$ are the same (if we disregard the order) for any composition series $G_r, \bar{G}_1, \bar{G}_2, \ldots$. This is *Lie-Jordan's theorem.*

Let $G_m$ be the invariant subgroup of $G_r$, $m < r$, and its basic infinitesimal transformations be $X_1 f, \ldots, X_m f = 0$. In general,

$$c_{\tau\lambda\kappa} = 0 \qquad (\tau = 1, 2, \ldots, m;\ \lambda = 1, 2, \ldots, r;\ \kappa = m + 1, \ldots, r).$$

Substituting in (176), we obtain

$$c_{\rho\sigma\lambda}c_{\lambda\mu\tau} + c_{\sigma\mu\lambda}c_{\lambda\rho\tau} + c_{\mu\rho\lambda}c_{\lambda\sigma\tau} = 0$$

$$(\rho, \sigma, \tau, \lambda, \mu = m + 1, \ldots, r)$$

Hence, there exists the group $H_{r-m}$ of these structure constants, according to Lie's third fundamental theorem. Since these groups are all isomorphic, each of them is called the *factor group* of $G_r$ determined by the invariant subgroup $G_m$, and is written $F_{r-m} = G_r/G_m$. The subgroup $G_m$ is called the *normal divisor* (*Normalteiler*). Let its basic infinitesimal transformations be $Y_{m+1}f, \ldots, Y_rf$, and add $Y_1f = 0, \ldots, Y_mf = 0$ to them; then the groups $G_r$ and $F_{r-m}$ are homomorphic and $G_m$ corresponds to the identity transformation in $F_{r-m}$. Conversely, if $F_{r-m}$ is homomorphic to $G_r$, then $F_{r-m}$ is the factor group of $G_r$ determined by the invariant subgroup $G_r$ that corresponds to the identity transformation in $F_{r-m}$. If $F_s$ is the largest invariant subgroup of $F_{r-m}$, then and only then is the corresponding invariant subgroup $G_{s+m}$ of $G_r$ the largest. If $F_{r-m}$ is simple, then and only then is $G_m$ the largest invariant subgroup. Hence, the necessary and sufficient condition for the factor group $G_r/G_m$ to be simple is that $G_m$ should be the largest invariant subgroup of $G_r$. If $F_{r-m}$ is not simple, then it contains the composition series

$$F_{r-m}, F_{r-m_1}, \ldots, F_{r-m_q};$$

the corresponding subgroups $G_{r-m_1+m}, \ldots, G_{r-m_1+m_q}, G_m, 1$ constitute the composition series. Hence the invariant subgroup of $G_r$ is one of the composition series of $G_r$.

Suppose that $G_r$ contains the composition series $G_r, \bar{G}_1, \bar{G}_2, \ldots, \bar{G}_\rho, 1$, and that the order of $\bar{G}_\alpha$ is $r_\alpha$. As has just been seen, $F_\alpha = \bar{G}_\alpha/\bar{G}_{\alpha+1}$ is simple. Thus there are simple factor groups $F_{r-r_1}, F_{r_1-r_2}, \ldots, F_{r_\rho}$ that are, respectively, of order $r - r_1, r_1 - r_2, \ldots, r_{\rho-1} - r_\rho, r_\rho$. Let the second composition series of $G_r$ be $G_r, \bar{G}'_1, \bar{G}'_2, \ldots, \bar{G}'_\rho, 1$; then the numbers of elements are equal for these two composition series, according to Lie-Jordan's theorem. Hence, if $H_m$ is the common subgroup of $\bar{G}_1$ and $\bar{G}'_1$, it must be the largest invariant subgroup of the two groups. If the orders of the subgroups in the second composition series are $r'_1, \ldots, r'_\rho$, then $r = r_1 + r'_1 - m$. Suppose that the basic infinitesimal transformations of $G_r$ are so chosen that the basic infinitesimal transformations of $H_m$ are $X_1f, \ldots, X_mf$. Then $X_1f, \ldots, X_mf, X_{m+1}f, \ldots, X_{r_1}f$ are the base of $\bar{G}_1$; and $X_1f, \ldots, X_mf, X_{r_1+1}f, \ldots, X_rf$ are the base of $\bar{G}'_1$. The infinitesimal transformations

$$X_1f, \ldots, X_mf, X_{m+1}f, \ldots, X_{r_1}f, X_{r_1+1}f, \ldots, X_rf$$

are the base of $G_r$. However, the three indices of the structure constants of the factor group $G_r/\bar{G}_1$ are $m + 1, \ldots, r$, as has just been seen, and those of the factor group $\bar{G}_1/H_m$ are $m + 1, \ldots, r_1$. The factor groups $\bar{G}'_1/H_m$ and $G_r/\bar{G}'_1$ must have the same structure constants as the former ones. Thus, if $\bar{G}_1$ and $\bar{G}'_1$ are the largest invariant subgroups of $G_r$ with the common invariant subgroup $H_m$, then the factor groups $G_r/G_1$ and $\bar{G}'_1/H_m$ are isomorphic, and hence $G_r/\bar{G}'_1$ and $\bar{G}_1/H_m$ are also isomorphic.

It can be shown that, if $G_r$ has two composition series, then the simple factor groups of these two composition series are isomorphic if we disregard the order. Indeed, let $H$ be the largest invariant subgroup of $G'_r$, $\bar{G}_1$, $\bar{G}'_1$, and let the order be $h$. If $H$, $H_1$, $H_2, \ldots$ are a composition series of subgroups of $H$, then $G_r$, $\bar{G}_1$, $H$, $H_1, \ldots$ and $G_r$, $\bar{G}'_1$, $H$, $H_1, \ldots$ are two composition series of subgroups of $G_r$. These two composition series have the same structure constants, as just stated. Next,

$$G_r, \bar{G}_1, \bar{G}_2, \ldots \qquad \text{and} \qquad G_r, \bar{G}_1, H, H_1, \ldots$$

give the same series of simple groups. Similarly, $G_r$, $\bar{G}'_1$, $\bar{G}'_2, \ldots$ and $G_r$, $\bar{G}'_1$, $H$, $H_1, \ldots$ give the same series of simple groups. Hence, the simple factor groups of the two composition series in question are isomorphic as a whole. This is *Hölder-Engel's theorem.*

The subgroups of an Abelian group are normal divisors. If the structure is written by means of addition, then the group and the subgroups are called *modulus,* and the factor group the *remainder class modulus* (*Restklassenmodul*).

If the group $G_r$ consisting of $X_1f, \ldots, X_rf$ is an $r$-parameter group, then $(X_\rho, X_\sigma)f$ is the invariant subgroup of $G_r$, according to Section 4.27. Hence, in the adjoint group $E_r$ consisting of $E_1f, \ldots, E_rf$, the commutator $(E_\rho, E_\sigma)f$ generates the invariant subgroup of the adjoint group. This invariant subgroup is called the *derived group* of $G_r$. A derived group of $G_r$ includes also a derived group of itself. Such a derived group is called the second derived group. In general, the $m$th derived group can be defined in a similar manner. All derived groups are invariant for $G_r$. If the first derived group of $G_r$ coincides with itself, then all derived groups coincide. In such a case they are said to be *perfect* in the sense of point-set theory (Section 2.1). The order of the $m$th derived group cannot be higher than $m - 1$. Hence, the order of the $q$th derived group may coincide with that of the $(q + 1)$th. Then the $q$th derived group may or may not consist of only the identity transformation. A group $G_r$ whose derived group consists only of the identity transformation is called *integrable* (*auflösbar*). A group that has no such derived group is called *non-integrable* (*unauflösbar*). Every subgroup of an integrable group is integrable.

An $(r - m)$-parameter group homomorphic to an $r$-parameter integrable group is itself integrable. Hence, the adjoint group of an integrable group is integrable.

The difference between the order of $G_r$ and the order of its derived group is equal to the number of independent invariants of the adjoint group of $G_r$. For, let $e_1$ be an invariant of the adjoint group; then

$$E_\mu e_1 = \sum_\sigma e_\sigma c_{\sigma\mu 1} = 0.$$

Conversely, if the derived group is determined by $X_{m+1}f, \ldots, X_r f$, then $c_{\sigma\mu\tau} = 0$ $(\sigma, \mu = 1, 2, \ldots, r; \tau = 1, 2, \ldots, m)$, where $e_1, e_2, \ldots, e_m$ are the invariants of the adjoint group. An integrable group $G_r$ of order $r$ contains an invariant subgroup $G_{r-1}$ of order $r - 1$; $G_{r-1}$ contains an invariant subgroup $G_{r-2}$ of order $r - 2$; in general, an invariant subgroup of order $r - r_1$ contains an invariant subgroup of order $r - r_1 - 1$. Thus the infinitesimal transformations of $G_r$ can be written

$$(X_\sigma, X_{\sigma+\rho})f = \sum_{\tau=1}^{\sigma} c_{\sigma,\sigma+\rho,\tau} X_\tau f$$

$$(\sigma = 1, 2, \ldots, r;\ \rho = 1, 2, \ldots, r - \sigma). \quad (200)$$

We propose to determine all second-order groups that contain a transformation $\sum_\tau e_\tau X_\tau f$ of $G_r$. We search for $\lambda$ such that, with two constants $\rho$ and $\omega$, we may form

$$\Big(\sum_\tau e_\tau X_\tau, \sum_\sigma \lambda_\sigma X_\sigma\Big)f = \rho \sum_\tau e_\tau X_\tau f + \omega \sum_\sigma \lambda_\sigma X_\sigma f.$$

Let $X_1 f$ be the required group, then, for $\lambda_1 = 0$,

$$\Big(X_1, \sum_{\sigma=2}^{r} \lambda_\sigma X_\sigma\Big)f = \sum_{\sigma,\tau} \lambda_\sigma c_{1\sigma\tau} X_\tau f = \rho X_1 f + \omega \sum_{\tau=2}^{r} \lambda_\sigma X_\sigma f. \quad (201)$$

Hence, $\lambda$ can be obtained from

$$\sum_{\sigma=2}^{r} c_{1\sigma\tau}\lambda_\sigma = \omega\lambda_\tau,$$

with the roots $\omega$ of $\|c_{1\sigma\tau} - \omega\| = 0$. Then $\rho$ is given by

$$\rho = \sum_{\sigma=2}^{r} \lambda_\sigma c_{1\sigma 1}.$$

If $\rho = 0$ in particular, that is, if we search for $\sum_\sigma \lambda_\sigma X_\sigma$ invariant for $\sum_\tau e_\tau X_\tau$, then $\omega$ is determined by the *characteristic equation* or *Killing's equation*

$$\Delta(\omega) = \left\|\sum_{\tau=1}^{r} e_\tau c_{\tau\lambda\kappa} - \omega\right\| = 0.$$

Write this equation in the form

$$(-1)^r\Delta(\omega) = \omega^r - \psi_1(e)\omega^{r-1} + \psi_2(e)\omega^{r-2} + \cdots + (-1)^{r-1}\psi_{r-1}(e)\omega.$$

Then $\psi_i(e)$ is a homogeneous polynomial of degree $i$ in $e_1, \ldots, e_r$, such that

$$\psi_1(e) = \sum_{\sigma,\tau} e_\sigma c_{\sigma\tau\tau},$$

$$[\psi_1(e)]^2 - 2\psi_2(e) = \sum_{\sigma,\tau,\lambda,\mu} e_\sigma e_\tau c_{\sigma\lambda\mu} c_{\tau\mu\lambda},$$

$$\cdots\cdot$$

If we transform by $X'_\rho f = \sum_\sigma h_{\rho\sigma} X_\sigma f$, such that $e_\sigma = \sum_\rho h_{\rho\sigma} e'_\rho$, $(X'_\rho, X'_\sigma)f = \sum_\tau c'_{\rho\sigma\tau} X'_\tau f$ then we obtain

$$\sum_\lambda h_{\lambda\sigma} c'_{\rho\tau\lambda} = \sum_{\mu,\nu} h_{\rho\mu} h_{\tau\nu} c_{\mu\nu\sigma} \qquad (\sigma, \rho, \tau = 1, 2, \ldots, r).$$

If the new characteristic equation is $\Delta'(\omega)$, then

$$\Delta(\omega) = \Delta'(\omega),$$

and

$$\psi'_i(e_1, \ldots, e_r) = \psi_i\left(\sum_\rho h_{\rho 1} e'_1, \ldots, \sum_\rho h_{\rho r} e'_r\right)$$

It can be shown by the third fundamental theorem of Lie that $\psi_1(e), \ldots, \psi_{r-1}(e)$ are invariant for the transformations of the adjoint group.

Killing called the number of independent functions among

$$\psi_1(e), \ldots, \psi_{r-1}(e)$$

the *rank* of $G_r$. The function $\psi_i(e)$ is the sum of the principal minor determinants of order $r - i$ of the characteristic equation for $\omega = 0$, that is, of $\Delta(0)$. Hence the rank is either equal to or lower than the order of the principal minor of $\Delta(0)$. It equals at most the number of roots that are identically zero of the characteristic equation. The nonzero roots of the characteristic equation, as functions of $e_1, \ldots, e_r$, are the invariants of the adjoint group; and the number of the nonzero roots that are functionally independent is the rank of the group, because the number of independent functions $\psi_i$ equals the number of independent nonzero roots.

The root of $\psi_1(e)$ of the characteristic equation is always zero. Since the Poisson bracket of (201) is of the form $(S, T) = ST - TS$ and the traces of $ST$ and $TS$ are equal, the trace of $ST - TS$ is zero. A *trace* (*spur*) is the sum of the elements in the principal diagonal of a matrix. Corresponding to a root $\omega^{(\alpha)}$ of $\omega$ there is always a root $-\omega^{(\alpha)}$. For the zero roots of $\omega$ the group $X_1 f, \ldots, X_r f$ is Abelian.

The characteristic equation of an integrable group derived from (200) is written

$$\Delta(\omega) = -\omega \prod_{\sigma=1}^{r-1} \left( \sum_{\tau=1}^{r} e_\tau c_{\tau\sigma\sigma} - \omega \right),$$

and is decomposed into linear factors in $\omega, e_1, \ldots, e_r$. Since $\sum_\sigma e_\sigma C_{\sigma ii}$ $(i = 1, 2, \ldots, r)$ are the invariants of the adjoint group, they are equal to zero for a derived group. Hence the rank of the derived group of an integrable group is zero. If the rank of a derived group is zero, then and only then is the original group integrable.

Cartan (1933) proved that the necessary and sufficient condition for $G_r$ to be integrable is that the transformations of the derived group make identically zero the coefficient $\psi_2(e)$ of the characteristic equation. The group is integrable if

$$\sum_\tau c_{\rho\sigma\tau} \frac{\partial \psi_2}{\partial e_\tau} \equiv 0 \qquad (\rho, \sigma = 1, 2, \ldots, r).$$

In other words, the necessary and sufficient condition for integrability is derived by substituting this relationship in the equation for $\psi_2$, that is,

$$\sum_{\lambda,\mu,\nu} c_{\rho\lambda\mu} c_{\sigma\mu\nu} c_{\tau\nu\lambda} = \sum_{\lambda,\mu,\nu} c_{\rho\mu\lambda} c_{\sigma\nu\mu} c_{\tau\lambda\nu} \qquad (\rho, \sigma, \tau = 1, 2, \ldots, r).$$

A group that contains no invariant subgroup other than the identical element is called *semisimple*. Since a simple group contains no invariant subgroup, it is also semisimple.

In order that a group be semisimple, it is necessary and sufficient that $\partial\psi_2/\partial e_\tau = 0$ shall always be associated with $e_1 = \cdots = e_r = 0$. Hence, $\psi_2(e)$ should be written as the sum of $r$ squares. For a group to be semisimple, it is therefore necessary and sufficient that the discriminant of the quadratic term $\psi_2(e)$ shall not be zero; in other words, the rank of the determinant formed by the coefficients of $e_1, \ldots, e_r$ of the quadratic form must be $r$.

Suppose that a semisimple group $G_r$ contains an invariant subgroup $g_m$. Since $\psi_2$ is a function of $e_1, \ldots, e_m$ only, the equations

$$\frac{\partial\psi_2}{\partial e_1} = \cdots = \frac{\partial\psi_2}{\partial e_m} = 0$$

determine another invariant subgroup $g'$ or $G_r$. The transformations common to $g_m$ and $g'$ determine the integrable invariant subgroup of $G_r$. Hence $e_1 = \cdots = e_r = 0$ should be derived from

$$\frac{\partial\psi_2}{\partial e_1} = \cdots = \frac{\partial\psi_2}{\partial e_m} = e_{m+1} = \cdots = e_r = 0.$$

Thus $\partial\psi_2/\partial e_1, \ldots, \partial\psi_2/\partial e_m$ are independent of $e_1, \ldots, e_m$. The subgroup $g'$ contains $r - m$ infinitesimal transformations $X'_{m+1}f, \ldots, X'_rf$. Hence it is written in the form

$$\psi_2 = \phi(e'_1, \ldots, e'_m) + \chi(e'_{m+1}, \ldots, e'_r),$$

where $\phi$ and $\chi$ are both quadratic forms with nonzero discriminants. Thus the two invariant subgroups $g_m$ and $g'$ are semisimple, and the order of $g'$ is $r - m$.

Let $(e_1, \ldots, e_r)$ be a vector defined in a vector space, and consider the matrix of the coefficients of the quadratic form

$$\psi_2(e) = -\tfrac{1}{2} \sum_{\sigma,\tau,\lambda,\mu} e_\sigma e_\tau c_{\sigma\lambda\mu} c_{\tau\mu\lambda} = \sum_{\sigma,\tau} g_{\sigma\tau} e_\sigma e_\tau,$$

$$g_{\sigma\tau} = -\tfrac{1}{2} \sum_{\lambda,\mu} c_{\sigma\lambda\mu} c_{\tau\mu\lambda},$$

as the matrix of the linear transformation

$$e'_\rho = \sum_\sigma g_{\rho\sigma} e_\sigma$$

from a vector $(e_1, \ldots, e_r)$ to a vector $(e'_1, \ldots, e'_r)$. The vector $(e_1, \ldots, e_r)$ corresponding to a root of the characteristic equation defines *the root space* (*Wurzelraum*) $E_r$. A group is integrable if and only if the quadratic form $\psi_2(e)$ is zero for each vector $(e_1, \ldots, e_r)$ defining the derived group. If an arbitrary vector of a group becomes zero for a root of the characteristic equation, then the group is integrable. If two vectors $(e_1, \ldots, e_r)$ and $(\epsilon_1, \ldots, \epsilon_r)$ are such that $\sum_{\sigma,\tau} g_{\sigma\tau} e_\sigma \epsilon_\tau = 0$ for each pair of $\sigma$ and $\tau$, then they are said to be *orthogonal*. For a simple group, any vector of the root space $E_0$ corresponding to a zero root of the characteristic equation is orthogonal to any vector of the other root space $E_\alpha$. In fact,

$$\sum_{\sigma,\tau} g_{\sigma\tau} e_\sigma^{(\alpha)} \epsilon_\tau = 0$$

for an $e^{(\alpha)}$ corresponding to the root space $E_\alpha$. Denote the vector in $E_0$ by $e_1^{(0)}, \ldots, e_r^{(0)}$; then $\sum_\sigma e_\sigma^{(0)} X_\sigma f$ constitute a subgroup of $G_r$. Denote it by $U$. It can be shown that such a subgroup $U$ is Abelian for a semi-simple group (Cartan, 1914, 1914a, 1926, 1927, 1927a, 1927b, 1929, 1930, 1933, 1936, 1938; Weyl, 1924, 1925, 1926, 1926a, 1931, 1939; Chevalley, 1946, 1951).

### 4.29 REPRESENTATION OF A LIE GROUP

A finite and continuous abstract group is called a *Lie group* if in a sufficiently small neighborhood of the unit element we can find the coordinates or the parameters $a_1, \ldots, a_r$ such that—with the given $A$ with parameters $a_i$ as coordinates, and $B$ with parameters $b_i$ as coordinates—the product $C = AB$ with parameters $c_1, \ldots, c_r$ is of class $C^2$ and is represented by $c_i = \phi_i(a, b)$. The operations on the parameter group are generated by the linearly independent infinitesimal transformations $X_1 f, \ldots, X_r f$. They are such that $(X_i, X_j) = \sum_s c_{ijs} X_s$, and $c_{ijs}$ satisfy the requirements of the third fundamental theorem of Lie. In the place of $X$ we take the notation $T$ and regard it as a vector in an $r$-dimensional space. An arbitrary vector of this space is written $\bar{A} = \sum_\mu \alpha^\mu T_\mu$, where the coefficients $\alpha^\mu$ are the components of vector $\bar{A}$. From $\bar{A}$ we can construct a *ring*, which by definition does not obey the permutation law. In contrast to a group, we call it the *infinitesimal ring* of the group. A ring is an aggregate of elements obeying two operations—addition and multiplication—where the commutative law is not obeyed for multiplication. The distributive law is obeyed by both a ring and a group, that is, if $\bar{A} = \sum_\mu \alpha^\mu T_\mu$, $\bar{B} = \sum_\mu \beta^\mu T_\mu$, then $(\bar{A}\bar{B}) = \bar{A} \times \bar{B}$ and

$$A \times B = \sum_{\kappa,\lambda} \alpha^\kappa \beta^\lambda (T_\kappa, T_\lambda) = \sum_{\kappa,\lambda,\mu} \alpha^\kappa \beta^\lambda c_{\kappa\lambda\mu} T_\mu,$$

$$A + B = \sum_\mu \alpha^\mu T_\mu + \sum_\mu \beta^\mu T_\mu = \sum_\mu (\alpha^\mu + \beta^\mu) T_\mu = \bar{B} + \bar{A},$$

and

$$(\bar{A} + \bar{B}) \times \bar{C} = \bar{A} \times \bar{C} + \bar{B} \times \bar{C},$$
$$\bar{A} \times (\bar{B} + \bar{C}) = \bar{A} \times \bar{B} + \bar{A} \times \bar{C},$$
$$(k\bar{A}) \times \bar{B} = k(\bar{A} \times \bar{B}) = \bar{A} \times (k\bar{B})$$

with a scalar $k$. Such operations are called *linear* (Van der Waerden, 1955, 1959; Weyl, 1926, 1931, 1939). We have $\bar{A} \times \bar{B} = -\bar{B} \times \bar{A}$ instead of the commutative law, and

$$(\bar{A} \times \bar{B}) \times \bar{C} + (\bar{B} \times \bar{C}) \times \bar{A} + (\bar{C} \times \bar{A}) \times \bar{B} \equiv 0$$

(let us call this the *combination law*) instead of the associative law. This is determined by the third fundamental theorem of Lie. If these rules are satisfied, then Lie's $r$-parameter group, that is, an $r$-parameter *group germ*, belongs to it. By Schreier's (1926, 1927) definition: if the (new) combination law in the neighborhood of the unit element $I$ of a local continuous group can be applied only for the elements sufficiently near the center $I$, then this part of space is called the *group germ* (*Gruppenkeim*). This combination law is the same as the law for the continuous groups. The algebra in which this combination law obeys Lie's fundamental theorems, that is, the Lie algebra—an algebra in which we restrict ourselves only in such infinitesimal elements of a group—is called the *kernel* (*Kern*). Such an algebra determines the group germ on the kernel of such infinitesimal elements.

With a group or a ring, or an algebra $\mathfrak{A}$ in general, it is possible to associate a linear transformation $(a)$: $x \to x' = ax$ that operates on the varying element $x$ of the algebra $\mathfrak{A}$ (Schur, 1905, 1906, 1924; Frobenius and Schur, 1906; Brauer and Schur, 1930; von Neumann, 1929; Van der Waerden, 1933, 1935; Chevalley, 1951). Then an algebra $\mathfrak{A}$ of order $n$ can be regarded either as an aggregate of element $a$ corresponding to the transformation $(a)$, or as an $n$-dimensional vector space $\Omega$ in which this transformation operates. This homomorphic correspondence $(\mathfrak{A}): a \to (a)$ defines the *representation* (*Darstellung*); that is, to any element $s$ of a group there corresponds its image $U(s)$ in the $n$-dimensional vector space $\mathfrak{R} = \mathfrak{R}_n$, such that $U(s)U(t) = U(st)$. In this case, $s$ is said to *induce* (*induzieren*) the *mapping image* or simply *the image* $U(s)$ in the representation space $\mathfrak{R}$. If it is isomorphic instead of homomorphic, then it is the *true representation* (Murnaghan, 1938). Introducing the coordinates in $\mathfrak{R}$, we represent the mapping $U(s)$ by a matrix with a nonzero determinant consisting of $n$ rows and $n$ columns. If $A$ is a transformation of this coordinate to another, then the image of the mapping $U(s)$ is represented by the matrix $AU(s)A^{-1}$.

Let the correspondence $s \to U(s)$ be the representation $\mathfrak{H}$; then the correspondence $s \to AU(s)A^{-1}$ is also the representation $\mathfrak{H}$. These two representations are *equivalent*, only the coordinate system being differently chosen. The trace, that is, the sum of the elements on the principal diagonal of the matrix $A$, is not affected by the choice of such coordinates.

In the representation $\mathfrak{H}$ of a group $G$, the trace $\chi(s)$ of the image $U(s)$ corresponding to the element $s$ of the group depends on $s$. This is called the *character* of the representation (Littlewood, 1940). Equivalent representations have the same character. Since

$$U(tst^{-1}) = U(t)U(s)U(t^{-1}) = U(t)U(s)U^{-1}(t),$$

the images $U(s)$ and $U(tst^{-1})$ differ by position but have the same trace, i.e., $\chi(tst^{-1}) = \chi(s)$.

A representation, if transitive, is called *irreducible*; otherwise, *reducible*. A reducible representation may sometimes be regarded as the sum of irreducible representations. The matrix of an irreducible representation does not possess any invariant submatrix.

Let $U_\mu$ correspond uniquely to Lie's infinitesimal transformation $T_\mu$, and $\sum_\mu \alpha^\mu U_\mu$ to $\sum_\mu \alpha^\mu T_\mu$. Put $\sum_\mu \alpha^\mu U_\mu = A$, $\sum_\mu \beta^\mu U_\mu = B$; then

$$(AB) = \sum_{\kappa,\lambda,\mu} \alpha^\kappa \beta^\lambda c_{\kappa\lambda\mu} U_\mu, \qquad \sum_\mu \delta^\mu U_\mu = A \times B,$$

$$\delta^\mu = \sum_{\kappa,\lambda} \alpha^\kappa \beta^\lambda c_{\kappa\lambda\mu}.$$

Thus, to $\sum_\mu \delta^\mu U_\mu$ corresponds the parameter group $\sum_\mu \delta^\mu T_\mu$. The coefficients $\delta^\mu$ are the same for both. To

$$\bar{A} = \sum_\mu \alpha^\mu T_\mu, \qquad \bar{B} = \sum_\mu \beta^\mu T_\mu$$

corresponds the transformation $(\bar{A}\bar{B}) \to (AB)$. Thus to each vector $\bar{A}$ of an infinitesimal ring corresponds uniquely the infinitesimal transformation $A$, such that $(\bar{A}\bar{B})$ corresponds to $(AB)$, and $\sum_\mu \delta^\mu U_\mu$ to $\sum_\mu \delta^\mu T_\mu$. If the representation of an infinitesimal ring by means of linear infinitesimal transformations

$$A^i = \sum_j a^i_j x^j, \qquad B^i = \sum_j b^i_j x^j,$$

is such that the matrix $A$ corresponds to $\bar{A}$ and the operator

$$(AB) = AB - BA$$

corresponds to $\bar{A} \times \bar{B}$, then the representation of Lie groups by means of linear infinitesimal transformations becomes a problem in algebra.

An adjoint group can be the representation of the original group by means of $C_\kappa = (c^\mu_{\kappa\lambda})$, where the structure constants $c_{\kappa\lambda\mu}$ of the previous sections are now denoted by $c^{\mu}_{\kappa\lambda}$. In order to see this possibility, let $\bar{B} = \sum_{\mu} \beta^{\mu} T_{\mu}$ be an element of the infinitesimal ring. Then

$$C_{\kappa}\bar{B} = \sum_{\kappa,\mu} c^{\mu}_{\kappa\lambda}\beta^{\lambda} T_{\mu} = T_{\kappa} \times \bar{B};$$

and this means that the linear transformation $C_{\kappa}$ generates $T_{\kappa}$. Similarly, for an arbitrary infinitesimal transformation $A_a = \sum_{\kappa} \alpha^{\kappa} C_{\kappa}$ of an adjoint group,

$$A_a\bar{B} = \sum_{\kappa} \alpha^{\kappa} C_{\kappa}\bar{B} = \sum_{\kappa} \alpha^{\kappa}(T_{\kappa} \times \bar{B}) = \bar{A} \times \bar{B}.$$

Thus a linear transformation $A_a$ means multiplication by $\bar{A}$. It is known from the correspondence $\bar{A} \times \bar{B} \rightarrow A_aB_a - B_aA_a$ that $A_a$ of the adjoint group gives the representation of $\bar{A}$. By applying this to an arbitrary vector $\bar{C}$ we obtain

$$\begin{aligned}(\bar{A} \times \bar{B}) \times \bar{C} &= (A_aB_a - B_aA_a)\bar{C} = A_a(B_a\bar{C}) - B_a(A_a\bar{C}) \\ &= \bar{A} \times (\bar{B} \times \bar{C}) - \bar{B} \times (\bar{A} \times \bar{C}).\end{aligned}$$

Thus the third fundamental theorem of Lie is obtained.

For a linear infinitesimal transformation, consider a linear operator

$$e^{\tau A} = I + \tau A + \cdots$$

as in Section 4.26, and write $A = (\alpha_{\mu\nu})$. Let $S(A) = \sum_{\tau} \alpha_{\tau\tau}$ be the trace of $A$; then $|e^{A}| = e^{S(A)}$. For if $e^{\tau A}$ is the group of the matrix, then $|e^{\tau A}|$ is the representation of this group. Since $|e^{\tau A}| = 1 + \tau S(A) + \cdots$, then, from Lie's fundamental theorem,

$$\left(\frac{\partial}{\partial \tau}\,|e^{\tau A}|\right)_0 = \left(\frac{\partial}{\partial \tau}\,e^{\tau S(A)}\right)_0 = S(A)e^{\tau S(A)} = S(A).$$

Hence $S(A)$ is the infinitesimal transformation of the representation of the group $\{e^{\tau A}\}$ by means of a determinant. By this property we can judge whether or not the group is Abelian by looking at the infinitesimal ring of that group; if it is Abelian, then we obtain $Be^{A\tau}B^{-1} = e^{A\tau}$ from an arbitrary element $B$ of the group and an arbitrary infinitesimal transformation $A$; in other words, $e^{A\tau}$ is invariant for an arbitrary transformation of the adjoint group. It is necessary and sufficient that $A$ be invariant for an arbitrary transformation of the adjoint group; that is, $A$ must be such that $CB - BC = C \times B = 0$. This should be satisfied for any two elements of the infinitesimal ring, in order that the corresponding group may be Abelian.

Another interesting application is to the composition series. By taking $G_{i+1}$ as the largest normal divisor of $G_i$, we construct Lie's subgroups $G, G_1, G_2, \ldots, G_r = I$ with a finite number of steps. If we transfer this subgroup to the infinitesimal ring belonging to $G$, the composition series is $g, g_1, g_2, \ldots, g_r = 0$. The element $g_{i+1}$ is the largest *ideal* among $g_i$. Generally it is known that to a normal divisor in group theory there corresponds an ideal in ring theory; that is, a subset $m$ of a ring is called an *ideal*, if $a - b \in m$ follows from $a \in m$, $b \in m$; and if $ar \in m$ follows from $a \in m$ with an element $r$ of the ring (Krull, 1935). Such an ideal is called a right-hand ideal. If $ra \in m$, then it is a left-hand ideal. Here $a \in m$ means that $a$ is an element contained in $m$. Thus, if the factor group $G_i/G_{i+1} = \bar{G}_i$ is Abelian, then $G$ is also Abelian. Conversely, if a group contains a composition series such that all its factor groups $\bar{G}_i$ are Abelian, then the group $G$ is said to be *integrable* (Section 4.28). The symbol $aba^{-1}b^{-1}$ formed from $a \in G$, $b \in G$ represents the *commutator* of $a$ and $b$. The expression $c = \prod_{r=1}^{k} (a_r b_r a_r^{-1} b_r^{-1})$ represents an element of the commutator group and is written $G^{(1)}$. The commutator group $G^{(1)}$ is a normal divisor of $G$ and is indeed the smallest normal divisor with Abelian factor group $G/G^{(1)}$. It can be shown that the commutator ring $g^{(1)} = g \times g$ of the infinitesimal ring $g$ of $G$ is the smallest ideal with an Abelian remainder class ring in $g$. The group $G$ is integrable if and only if $g^{(s)} = 0$ for a finite $s$, such that $g \supset g^{(1)} \supset g^{(2)} \supset \cdots \supset g^{(s)} = 0$.

The largest integrable ideal $\mathfrak{r}$ in an infinitesimal ring $g$ is called semisimple if $\mathfrak{r} = 0$. The corresponding group is called semisimple (Section 4.28). The largest integrable ideal is called the *radical*, as in the theory of hypercomplex numbers. Suppose that multiplication obeys the associative law in the hypercomplex number system $\Omega$, such that $(uv)w = u(vw)$. If a positive integer $\rho$ can be found such as $\mathfrak{a}^\rho = 0$ for an ideal $\mathfrak{a}$, then $\mathfrak{a}$ is called a *null-power ideal* in $\Omega$. A null-power ideal in $\Omega$ corresponds to an integrable ideal in $\mathfrak{g}$. Thus, the largest null-power ideal is a radical $\mathfrak{r}$. For $\mathfrak{r} = 0$ this system $\Omega$ is semisimple.

A group $G$ having no normal divisor except itself and the unit element has been called *simple* (Section 4.28). Similarly, a ring $g$, or an ideal $\mathfrak{g}$, or the system $\Omega$, is called simple if it contains no ideal other than itself and zero. A semisimple infinitesimal ring can be expressed as the sum of simple rings that vanish by pairs in multiplication.

Suppose that an integrable group $G$ is represented by a linear transformation in a vector space. It can be proved that there exists a vector $\mathbf{e}$ such that $A\mathbf{e} = \lambda\mathbf{e}$ for all $A$ of the infinitesimal ring of the representation $\mathfrak{G}$ of $G$ (Lie's theorem).

Consider the representation of a semisimple group. Suppose that the representation of an infinitesimal ring $g$ by means of a matrix $A$ is given. Addition of a quantity written as $\beta(g)$ is called *modification*. How should $\beta(g)$ be chosen in order that $\{A\}$ and $\{A + I\beta(g)\}$ shall both be representations of $g$? The necessary and sufficient condition is that the quantity

$\beta(g)$ be a representation of the ring by means of numbers. It can be shown that the representation $\mathfrak{G}$ by means of a matrix $A$ can be replaced by its representation by means of another matrix whose traces are all null. Then the theorem can be proved: that the irreducible representation $\mathfrak{G}$ of $g$ is the representation of the factor group $g/h$ and that a null matrix can be made to correspond to each element of $h$.

Let $t = \sum_i \mu_i g_i$ be a general element of $g$, and let

$$\mathfrak{f}(t, \zeta) = |I\zeta - T|.$$

Let $\mathfrak{f}^*(t, \zeta)$ be the result of dropping multiple factors from $\mathfrak{f}$ for a general value of $t$. Assume that $\mathfrak{f}^*$ contains no double root for $t = t_0$. Let $\mathfrak{r}_\alpha$ be a partial space of the root space for the root $\alpha$. We say that the element $r = r^{(\alpha)}$ of $\mathfrak{r}_\alpha$ *belongs* to the root $\alpha$, that is, $(T_0 - \alpha)^l r = 0$ for a sufficiently large $l$. It can be shown that, if $\mathfrak{r}_\alpha$ belongs to the root $\alpha$, and $\mathfrak{r}_\beta$ to the root $\beta$, then $\mathfrak{r}_\alpha \times \mathfrak{r}_\beta$ belongs to the root $\alpha + \beta$.

Suppose that $\alpha$ and $-\alpha$ are the roots. If $\Lambda = \Lambda(h)$ is a linear form in $\lambda$, then $\Lambda_\alpha(h_\alpha)$ is written $\Lambda_\alpha$. Suppose further that the representation of $g$ is given in the correspondence $h \to H$, $e_\alpha \to I_\alpha$. Then there exists an $e \neq 0$ such that $He = \Lambda e$; we say that $e$ is of *weight* $\Lambda$. The only possible weights are the roots of the characteristic equation. If $r$ is of weight $\Lambda$, then so is $Hr$, and $I_\alpha$ is of weight $\Lambda + \alpha$. If a vector of weight $\Lambda$ is the sum of vectors of other weights, that vector equals zero. If $e_\alpha^* \subset H_\alpha$, $h_\alpha^* = e_\alpha \times e_{-\alpha}$ and $\Lambda$ is a representation root, that is, $\Lambda(h_\alpha^*) = \Lambda_\alpha^*$, then $\Lambda_\alpha^*$ is a rational multiple of $e_\alpha^*$.

The ring $g$ is integrable if the equation $\phi(t) = 0$ holds for all elements of $g \times g$, where $\phi(t)$ appears in Cartan's criterion for integrability and semisimplicity of $g$. If $g$ is semisimple, then the quadratic form $\phi(t)$ is nondegenerate. Thus we obtain the important decomposition theorem that any semisimple group is the direct sum of simple groups.

Weyl (1925, 1926, 1926a) and Peter and Weyl (1927) discussed the representation. Instead of leading by linear transformations to quadratic forms, they led the representation by unitary transformations to Hermitian forms (cf., Van der Waerden, 1933; Littlewood, 1940). Cartan (1914, 1914a, 1926, 1927, 1927a, 1927b, 1929, 1938) considered the unitary geometry in the Riemannian space instead of referring to the notion of representation.

## REFERENCES

Arnold, V. I.

1963 *Usp. Mat. Nauk*, **18**, 91; trans. by U.S. Department of Commerce as *Small Denominator and Problems of Movement in Classical and Celestial Mechanics*, 1964.

Baker, H. F.
1902 *Proc. London Math. Soc.*, **35**, 347, 379; **34**, 91, 332, 335.
1904 *Ibid.*, ser. 2, **2**, 293.
Bianchi, L.
1918 *Teoria dei Gruppi Continui Finiti di Trasformazioni*, Zanichelli, Bologna.
Birkhoff, G. D.
1927 *Dynamical Systems*, Am. Math. Soc., New York.
Born, M.
1926 *Z. Physik*, **40**, 167.
Born, M., and Fock, V.
1928 *Z. Physik*, **51**, 165.
Brauer, R., and Schur, I.
1930 *Sitzber. preuss. Akad. Wiss. Berlin*, **1930**, 209.
Buhl, A.
1928 *Aperçu Moderne sur la Théorie des Groupes Finis et Continus*, Gauthier-Villars, Paris.
Burgers, J. M.
1917 *Phil. Mag.*, **33**, 514.
Campbell, J. E.
1903 *Theory of Continuous Groups*, Clarendon Press, Oxford.
Carathéodory, C.
1933 *Variationsrechnung*, Teubner, Leipzig.
Cartan, E.
1914 *J. Math. pures appl.*, ser. 6, **10**, 149.
1914a *Ann. Ecole Norm. Sup.*, ser. 3, **31**, 263.
1922 *Leçons sur les Invariants Intégraux*, Hermann, Paris.
1926 *Bull. Soc. Math. France*, **54**, 214.
1927 *Ibid.*, **55**, 114.
1927a *Ann. Ecole Norm. Sup.*, ser. 3, **44**, 345.
1927b *Annali Mat. pura appl.*, ser. 4, **4**, 209.
1929 *J. Math. pures appl.*, ser. 9, **8**, 1.
1930 *La Théorie des Groupes Finis et Continus et l'Analyse Situs*, Gauthier-Villars, Paris.
1931 *Bull. Soc. Math. France*, **59**, 88.
1933 *Sur la Structure des Groupes de Transformations Finies et Continues*, Gauthier-Villars, Paris.
1936 *La Topologie des Groupes de Lie*, Hermann, Paris.
1937 *Théorie des Groupes Finis et Continus et la Géométrie Différentielle*, Gauthier-Villars, Paris.
1938 *J. Math. pures appl.*, ser. 9, **17**, 1.
1945 *Les Systèmes Différentielles Extérieurs et leurs Applications Géométriques*, Hermann, Paris.
Cerf, G.
1929 *Transformations de Contact et Problème de Pfaff*, Gauthier-Villars, Paris.
Chevalley, C.
1946 *Theory of Lie Groups*, I, Princeton University Press, Princeton, N.J.
1951 *Théorie des Groupes de Lie*, II, Hermann, Paris.

de Donder, Th.
1901 *Rendiconti Cir. Mat. Palermo*, **15**, 65.
1902 *Ibid.*, **16**, 155.
1927 *Théorie des Invariants Intégraux*, Gauthier-Villars, Paris.

Dragoni, G. S.
1934 *Rendiconti Cir. Mat. Palermo*, **58**, 311.
1934a *Annali Mat. pura appl.*, ser. 4, **13**, 335.

Ehrenfest, P.
1917 *Phil. Mag.*, **33**, 500.

Eisenhardt, L. P.
1933 *Continuous Groups of Transformations*, Am. Math. Soc., New York.

Engel, F.
1916 *Nachr. Göttingen Ges. Wiss.*, **1916**, 270.
1917 *Ibid.*, **1917**, 189.

Engel, F. and Faber, K.
1932 *Lie'sche Theorie der partiellen Differentialgleichung erster Ordnung*, Teubner, Leipzig.

Fermi, E. and Persico, F.
1926 *Rendiconti Accad. Lincei*, ser. 6, **4**, 452.

Frobenius, G. and Schur, I.
1906 *Sitzber. preuss. Akad. Wiss. Berlin*, **1906**, 209.

Geppert, H.
1928 *Rendiconti Accad. Lincei*, **13**, 30, 191, 294.
1929 *Math. Annalen*, **102**, 193.

Goursat, H.
1908 *J. Math. pures appl.*, ser. 6, **4**, 331.
1912 *Leçons sur le Problème de Pfaff*, Gauthier-Villars, Paris.
1915 *J. Math. pures appl.*, ser. 7, **1**, 241.
1921 *Leçons sur l'Intégration des Equations Différentielles aux Dérivées Partielles du Premier Ordre*, Gauthier-Villars, Paris.
1923 *Compt. Rend.*, **147**, 1089.

Graffi, D.
1936 *Annali Mat. pura appl.*, ser. 4, **15**, 87.

Gröbner, W.
1960 *Die Lie-Reihen und ihre Anwendungen*, Deutsch. Verlag, Berlin.

Hagihara, Y.
1938 *Compt. Rend.*, **207**, 390.
1940 *Festschrift Elis Strömgren*, 58, Munksgaard, Copenhagen.
1944 *Proc. Imp. Acad. Japan*, **20**, 501.

Hille, E.
1948 *Functional Analysis and Semi-Groups*, Am. Math. Soc., New York.

Hoheisel, G.
1928 *Partielle Differentialgleichung*, Gruyter, Leipzig.

Jacobi, C. G. J.
1842 *Astron. Nachr.*, **20**, 81. Reprinted in *Jacobi'sche Gesammette Werke*, IV, 295, Reimer, Berlin, 1886.
1844 *J. reine angew. Math.*, **26**, 115. Reprinted in *Jacobi'sche Gesammette Werke*, IV, 295, Reimer, Berlin, 1886.

Kasuga, T.
1961 *Proc. Japan Acad.*, **37**, 366, 372, 377.
Knapp, H.
1964 *Monatshefte Math.*, **68**, 33.
Kneser, H.
1924 *Math. Annalen*, **91**, 155.
Koopman, B. O.
1931 *Proc. Natl. Acad. Sci. U.S.*, **17**, 315.
Kowalewski, G.
1931 *Einführung in die Theorie der Continuierlichen Gruppen*, Akad. Verlagsgesell., Leipzig.
Krull, W.
1935 *Idealtheorie*, Springer, Berlin.
Kulsrud, R. M.
1957 *Phys. Rev.*, **106**, 205.
Lenard, A.
1959 *Ann. Phys. (N.Y.)*, **6**, 261.
Levi-Civita, T.
1928 *Abhandl. Math. Sem. Hamburg*, **5**, 323.
1928a *Atti Congresso Internaz. Mat. Bologna*, new series, **3**.
1934 *J. Math. Phys. M.I.T.*, **13**, 18.
Lie, S.
1874 *Math. Annalen*, **8**, 215.
1888 *Theorie der Transformationsgruppen*, I, Teubner, Leipzig.
1890 *Ibid.*, II.
1891 *Vorlesungen über Differentialgleichungen mit bekannten infinitesimalen Transformationen*, Teubner, Leipzig.
1893 *Theorie der Transformationsgruppen*, III.
1893a *Kontinuierliche Gruppen*, Teubner, Leipzig.
1896 *Geometrie der Berührungstransformation*, Teubner, Leipzig.
1927 *Gesammelte Abhandlung*, VI, 248, Teubner, Leipzig.
1929 *Ibid.*, VI, 1.
Littlewood, D. E.
1940 *The Theory of Group Characters and Matrix Representations of Groups*, Clarendon Press, Oxford.
Mattioli, G.
1932 *Annali Mat. pura appl.*, ser. 4, **10**, 283.
Mineur, H.
1937 *J. Ecole Polytech., Paris*, ser. 3, cahier 143.
Murnaghan, F. D.
1938 *The Theory of Group Representations*, Johns Hopkins Press, Baltimore, Md.
Peter, F., and Weyl, H.
1927 *Math. Annalen*, **97**, 737.
Pfeiffer, G.
1933 *Acta Math.*, **61**, 203, 239.
Poincaré, H.
1889 *Acta Math.*, **13**, 1. Reprinted in *Oeuvres de Henri Poincaré*, VII, 286, Gauthier-Villars, Paris, 1952.

1897 *Bull. Astron.*, **4**, 53. Reprinted in *Acta Math.*, **21**, 83 (1897), and in *Oeuvres*, VII, 500.

1899 *Méthodes Nouvelles de la Mécanique Céleste*, III, Gauthier-Villars, Paris.

1905 *Leçons de la Mécanique Céleste*, I, Chap. 2, Gauthier-Villars, Paris.

Pontrjagin, L.

1939 *Topological Groups*, Princeton University Press, Princeton, N.J.

Potron, M.

1936 *Les Groupes de Lie*, Gauthier-Villars, Paris.

Saltykov, N.

1934 *Etudes sur l'évolution des méthodes modernes d'intégration des équations différentielles partielles du premier ordre*, *Mém. Acad. Roy. Belgique*, ser. 2, **13**, fasc. 6.

1935 *Méthodes modernes d'intégration des équations aux dérivées partielles du premier ordre à une fonction inconnue*, Gauthier-Villars, Paris.

1950 *Bull. Astron.*, ser. 2, **15**, 293.

Schmeidler, F.

1963 *Astron. Nachr.*, **287**, 1.

Schouten, J. A., and van der Kulk, W.

1949 *Pfaff's Problem and its Generalizations*, Clarendon Press, Oxford.

Schreier, O.

1926 *Abhandl. Math. Sem. Hamburg*, **4**, 15.

1927 *Ibid.*, **5**, 161, 233.

Schur, I.

1891 *Math. Annalen*, **38**, 263.

1905 *Sitzber. preuss. Akad. Wiss.*, **1905**, 406.

1906 *Ibid.*, **1906**, 164.

1924 *Ibid.*, **1924**, 189, 297, 346.

Sommerfeld, A.

1922 *Atombau und Spektrallinien*, 3rd ed., Vieweg, Brunswick.

Speiser, A.

1923 *Theorie der Gruppen endlicher Ordnung*, Springer, Berlin.

Stone, M. H.

1932 *Linear Transformations in Hilbert Space and their Applications to Analysis*, Am. Math. Soc., New York.

Van der Waerden, B. L.

1933 *Math. Z.*, **36**, 780; **37**, 446.

1935 *Gruppen von linearen Transformation*, Springer, Berlin.

1955 *Der Modernen Algebra*, I, 4th ed., Springer, Berlin.

1959 *Ibid.*, II, 4th ed., Springer, Berlin.

van Kampen, E. R., and Wintner, A.

1938 *Trans. Am. Math. Soc.*, **44**, 168.

Vivanti, L.

1918 *Teoria dei Gruppi Continui Finiti di Trasformazioni*, Zanichelli, Bologna.

von Neumann, J.

1929 *Math. Z.*, **30**, 3.

1932 *Annals of Math.*, **33**, 587.

Weber, H. v.
1900 *Vorlesungen über das Pfaffsche Problem*, Teubner, Leipzig.
Weyl, H.
1923 *Raum, Zeit, Materie*, Springer, Berlin.
1923a *Mathematische Analyse des Raumproblems*, Springer, Berlin.
1924 *Sitzber. preuss. Akad. Wiss. Berlin*, **1924**, 338.
1925 *Math. Z.*, **23**, 271.
1926 *Ibid.*, **24**, 328, 377.
1926a *Acta Math.*, **48**, 255.
1931 *Gruppentheorie und Quantenmechanik*, 2nd rev. ed., S. Hirzel, Leipzig.
1939 *The Classical Groups*, Princeton University Press, Princeton, N.J.
Williamson, J.
1940 *Am. J. Math.*, **62**, 881.

# Reduction of the $n$-Body Problem

## TRANSFORMATION OF LAGRANGE

### 5.1 TRANSFORMATION OF LAGRANGE

Consider at first the three-body problem and let the positions of the three bodies be $P_1$, $P_2$, $P_3$ and their masses be $m_1$, $m_2$, $m_3$. Denote the coordinates of $P_3$ relative to $P_2$, the coordinates of $P_1$ relative to $P_3$, and the coordinates of $P_2$ relative to $P_1$, by $x_1, x_2, x_3$; $x_4, x_5, x_6$; and $x_{,7}\ x_8, x_9$, respectively. The differential equations of motion of the three bodies in these relative coordinates are

$$\frac{d^2x_{3\mu-2}}{dt^2} + \left(\sum_{\mu=1}^{3} m_\mu\right)\frac{x_{3\mu-2}}{r_\mu^3} - m_\mu\left(\sum_{\mu=1}^{3}\frac{x_{3\mu-2}}{r_\mu^3}\right) = 0,$$

$$\frac{d^2x_{3\mu-1}}{dt^2} + \left(\sum_{\mu=1}^{3} m_\mu\right)\frac{x_{3\mu-1}}{r_\mu^3} - m_\mu\left(\sum_{\mu=1}^{3}\frac{x_{3\mu-1}}{r_\mu^3}\right) = 0, \tag{1}$$

$$\frac{d^2x_{3\mu}}{dt^2} + \left(\sum_{\mu=1}^{3} m_\mu\right)\frac{x_{3\mu}}{r_\mu^3} - m_\mu\left(\sum_{\mu=1}^{3}\frac{x_{3\mu}}{r_\mu^3}\right) = 0 \qquad (\mu = 1, 2, 3).$$

$$r_\mu^2 = x_{3\mu-2}^2 + x_{3\mu-1}^2 + x_{3\mu}^2, \tag{2}$$

$$\sum_{\mu=1}^{3} x_{3\mu-2} = \sum_{\mu=1}^{3} x_{3\mu-1} = \sum_{\mu=1}^{3} x_{3\mu} = 0. \tag{3}$$

If the motions of $P_2$ and $P_3$ relative to $P_1$ are known, then the motion of $P_1$ relative to $P_2$ can be obtained, since the equations (1) have been written in an explicit form.

Let the constants of the angular momentum be $c_1, c_2, c_3$; then

$$\begin{aligned}\sum_{\mu=1}^{3} \frac{1}{m_\mu}\left(x_{3\mu-1}\frac{dx_{3\mu}}{dt} - x_{3\mu}\frac{dx_{3\mu-1}}{dt}\right) &= c_1,\\ \sum_{\mu=1}^{3} \frac{1}{m_\mu}\left(x_{3\mu}\frac{dx_{3\mu-2}}{dt} - x_{3\mu-2}\frac{dx_{3\mu}}{dt}\right) &= c_2,\\ \sum_{\mu=1}^{3} \frac{1}{m_\mu}\left(x_{3\mu-2}\frac{dx_{3\mu-1}}{dt} - x_{3\mu-1}\frac{dx_{3\mu-2}}{dt}\right) &= c_3.\end{aligned} \tag{4}$$

Write

$$u_\mu^2 = \left(\frac{dx_{3\mu-2}}{dt}\right)^2 + \left(\frac{dx_{3\mu-1}}{dt}\right)^2 + \left(\frac{dx_{3\mu}}{dt}\right)^2 \qquad (\mu = 1, 2, 3); \tag{5}$$

then, if $h$ is an arbitrary constant, the energy integral is

$$\sum_{\mu=1}^{3} \frac{u_\mu^2}{m_\mu} - 2\left(\sum_{\mu=1}^{3} m_\mu\right)\left(\sum_{\mu=1}^{3} \frac{1}{m_\mu r_\mu}\right) = h. \tag{6}$$

Referring to (3) we see that the solution requires six unknown functions and their derivatives. These functions are determined by six differential equations of order two. However, since four integrals from (5) and (6) are known, eight unknown quantities are required; that is, the order of the differential equations of the three-body problem is reduced to eight from $3 \times 6 = 18$.

Lagrange tried to determine the sides of the triangle formed by the three bodies and then the position in space of the plane of the triangle (Lagrange, 1873; Tisserand, 1889; Radau, 1868, 1886; Bohlin, 1907). Let $\lambda, \mu, \nu$ be a cyclic permutation of 1, 2, 3, and put

$$\frac{1}{r_\mu^3} - \frac{1}{r_\nu^3} = q_\lambda \qquad (\lambda, \mu, \nu = 1, 2, 3); \tag{7}$$

then

$$q_1 + q_2 + q_3 = 0, \qquad \frac{q_1}{r_1^3} + \frac{q_2}{r_2^3} + \frac{q_3}{r_3^3} = 0. \tag{8}$$

Furthermore, by writing

$$-p_\lambda = x_{3\mu-2}x_{3\nu-2} + x_{3\mu-1}x_{3\nu-1} + x_{3\mu}x_{3\nu} \tag{9}$$

we obtain

$$p_\mu + p_\nu = r_\lambda^2 \tag{10}$$

$$p_\lambda = \tfrac{1}{2}(r_\mu^2 + r_\nu^2 - r_\lambda^2), \tag{11}$$

from (2) and (3). Differentiate (2) twice with respect to $t$, then, because of (5),

$$\frac{1}{2}\frac{d^2 r_1^2}{dt^2} = x_1 \frac{d^2 x_1}{dt^2} + x_2 \frac{d^2 x_2}{dt^2} + x_3 \frac{d^2 x_3}{dt^2} + u_1^2.$$

Substituting (1) in the right-hand side and referring to (9) and (10), we obtain

$$\frac{1}{2}\frac{d^2 r_1^2}{dt^2} + \frac{m_1 + m_2 + m_3}{r_1} - u_1^2 = m_1\left(\frac{p_2 + p_3}{r_1^3} - \frac{p_2}{r_3^3} - \frac{p_3}{r_2^3}\right),$$

which, by (7), reduces to

$$\frac{1}{2}\frac{d^2 r_\lambda^2}{dt^2} + \frac{m_\lambda + m_\mu + m_\nu}{r_\lambda} + m_\lambda(p_\mu q_\mu - p_\nu q_\nu) - u_\lambda^2 = 0$$

$$(\lambda, \mu, \nu = 1, 2, 3). \tag{12}$$

From (12) we can determine $u_1^2$, $u_2^2$, $u_3^2$ as functions of $r_1$, $r_2$, $r_3$; this gives

$$\sum_{\lambda=1}^{3} \frac{u_\lambda^2}{m_\lambda} = \sum_{\lambda=1}^{3} \frac{1}{2m_\lambda}\frac{d^2 r_\lambda^2}{dt^2} + \left(\sum_{\lambda=1}^{3} m_\lambda\right)\left(\sum_{\lambda=1}^{3} \frac{1}{m_\lambda r_\lambda}\right),$$

or, by (6),

$$\frac{1}{2}\frac{d^2}{dt^2}\left(\frac{r_1^2}{m_1} + \frac{r_2^2}{m_2} + \frac{r_3^2}{m_3}\right) - (m_1 + m_2 + m_3)\left(\frac{1}{m_1 r_1} + \frac{1}{m_2 r_2} + \frac{1}{m_3 r_3}\right) = h. \tag{13}$$

This is one of the fundamental equations of Lagrange. It plays an important role in Sundman's theory of the regularization of the three-body problem, as will be shown in a later chapter. We can write, as can be easily verified,

$$\left(x_{3\mu-2}\frac{dx_{3\nu-2}}{dt} + x_{3\mu-1}\frac{dx_{3\nu-1}}{dt} + x_{3\mu}\frac{dx_{3\nu}}{dt}\right) - \left(x_{3\nu-2}\frac{dx_{3\mu-2}}{dt} + x_{3\nu-1}\frac{dx_{3\mu-1}}{dt} + x_{3\nu}\frac{dx_{3\mu}}{dt}\right) = \rho$$

$$(\mu, \nu = 1, 2, 3), \tag{14}$$

and, by differentiating (9),

$$x_{3\mu-2}\frac{dx_{3\nu-2}}{dt} + x_{3\mu-1}\frac{dx_{3\nu-1}}{dt} + x_{3\mu}\frac{dx_{3\nu}}{dt} = \frac{1}{2}\left(-\frac{dp_\lambda}{dt} + \rho\right)$$

$$(\lambda, \mu, \nu = 1, 2, 3). \quad (15)$$

Denote the cosines of the angles between the vectors

$$\mathfrak{U}\left(-\frac{x_4}{r_2}, -\frac{x_5}{r_2}, -\frac{x_6}{r_2}\right), \qquad \mathfrak{B}\left(-\frac{1}{u_2}\frac{dx_4}{dt}, -\frac{1}{u_2}\frac{dx_5}{dt}, -\frac{1}{u_2}\frac{dx_6}{dt}\right),$$

$$\mathfrak{U}'\left(\frac{x_7}{r_3}, \frac{x_8}{r_3}, \frac{x_9}{r_3}\right), \qquad \mathfrak{B}'\left(\frac{1}{u_3}\frac{dx_7}{dt}, \frac{1}{u_3}\frac{dx_8}{dt}, \frac{1}{u_3}\frac{dx_9}{dt}\right),$$

by

$$\alpha' = \frac{\mathfrak{U}\cdot\mathfrak{U}'}{|\mathfrak{U}|\,|\mathfrak{U}'|}, \qquad \beta' = \frac{\mathfrak{U}\cdot\mathfrak{B}'}{|\mathfrak{U}|\,|\mathfrak{B}'|}, \qquad \gamma' = \frac{\mathfrak{U}\cdot\mathfrak{B}}{|\mathfrak{U}|\,|\mathfrak{B}|},$$

$$\alpha'' = \frac{\mathfrak{B}\cdot\mathfrak{B}'}{|\mathfrak{B}|\,|\mathfrak{B}'|}, \qquad \beta'' = \frac{\mathfrak{U}'\cdot\mathfrak{B}}{|\mathfrak{U}'|\,|\mathfrak{B}|}, \qquad \gamma'' = \frac{\mathfrak{U}'\cdot\mathfrak{B}'}{|\mathfrak{U}'|\,|\mathfrak{B}'|}. \quad (16)$$

Making the substitution

$$2v_\lambda = u_\mu^2 + u_\nu^2 - u_\lambda^2, \quad (17)$$

we obtain, after some computation,

$$\alpha' = \frac{p_1}{\sqrt{(p_1 + p_2)(p_1 + p_3)}}, \qquad \beta' = \frac{dp_1/dt - \rho}{2\sqrt{(p_1 + p_3)(v_1 + v_2)}},$$

$$\gamma' = \frac{dp_1/dt + dp_3/dt}{2\sqrt{(p_1 + p_3)(v_1 + v_3)}}, \quad (18)$$

$$\alpha'' = \frac{v_1}{\sqrt{(v_1 + v_2)(v_1 + v_3)}}, \qquad \beta'' = \frac{dp_1/dt + \rho}{2\sqrt{(p_1 + p_2)(v_1 + v_3)}},$$

$$\gamma'' = \frac{dp_1/dt + dp_2/dt}{2\sqrt{(p_1 + p_2)(v_1 + v_2)}}.$$

These six cosines are not independent. Because—$\gamma'$, $\beta''$, $\gamma''$, $\beta'$, $\alpha'$, $\alpha''$ being the cosines of the four sides and the diagonals of the spherical quadrangle formed by the ends of the four vectors $\mathfrak{U}$, $\mathfrak{U}'$, $\mathfrak{B}$, $\mathfrak{B}'$ on the spherical surface of unit radius—we have the relation (Tisserand, 1889)

$$\begin{aligned}1 &- (\alpha'^2 + \beta'^2 + \gamma'^2 + \alpha''^2 + \beta''^2 + \gamma''^2)\\ &+ 2(\alpha'\beta''\gamma' + \alpha'\beta'\gamma'' + \alpha''\beta'\gamma' + \alpha'\beta'\gamma' + \alpha''\beta''\gamma'') + \alpha'^2\alpha''^2 + \beta'^2\beta''^2\\ &+ \gamma'^2\gamma''^2 - 2\alpha'\alpha''\beta'\beta'' - 2\beta'\beta''\gamma'\gamma'' - 2\gamma'\gamma''\alpha'\alpha'' = 0. \quad (19)\end{aligned}$$

Put

$$4\Sigma_\lambda = \rho^2(p_\mu + p_\nu) + 2\rho\left(p_\mu \frac{dp_\nu}{dt} - p_\nu \frac{dp_\mu}{dt}\right)$$

$$+ p_\mu\left(\frac{dp_\nu}{dt}\right)^2 + p_\nu\left(\frac{dp_\mu}{dt}\right)^2 + p_\lambda\left(\frac{dp_\mu}{dt} + \frac{dp_\nu}{dt}\right)^2; \tag{20}$$

then

$$(v_2v_3 + v_3v_1 + v_1v_2)(p_2p_3 + p_3p_1 + p_1p_2) - (v_1\Sigma_1 + v_2\Sigma_2 + v_3\Sigma_3)$$

$$+ \frac{1}{16}\left(\rho^2 + \frac{dp_2}{dt}\frac{dp_3}{dt} + \frac{dp_3}{dt}\frac{dp_1}{dt} + \frac{dp_1}{dt}\frac{dp_2}{dt}\right)^2 = 0. \tag{21}$$

This is the second fundamental equation of Lagrange. $\Sigma$ of (20) is a quadratic form in $\rho$ containing terms of $\rho$ and $\rho^2$ but no term of $\rho^3$. $u$ can be expressed in terms of $r$, $\dot{r}$, and $\ddot{r}$ from (7), (11), (12); and $v$ can be expressed in terms of $r$, $\dot{r}$, and $\ddot{r}$ from (17). Hence, (21) expresses the auxiliary function $\rho$ in terms of $r_1, r_2, r_3, \dot{r}_1, \dot{r}_2, \dot{r}_3, \ddot{r}_1, \ddot{r}_2, \ddot{r}_3$.

Differentiate (14) with respect to $t$, eliminate the derivatives of $x$ with respect to $t$ by means of (1), and refer to the definition of $p$ and $q$ in (7) and (9). Then we obtain

$$\frac{d\rho}{dt} + m_1p_1q_1 + m_2p_2q_2 + m_3p_3q_3 = 0. \tag{22}$$

This expresses $d\rho/dt$ in terms of $r_1, r_2, r_3$.

Next put

$$\Pi_\mu = \left(x_{3\mu-1}\frac{dx_{3\mu}}{dt} - x_{3\mu}\frac{dx_{3\mu-1}}{dt}\right)^2 + \left(x_{3\mu}\frac{dx_{3\mu-2}}{dt} - x_{3\mu-2}\frac{dx_{3\mu}}{dt}\right)^2$$

$$+ \left(x_{3\mu-2}\frac{dx_{3\mu-1}}{dt} - x_{3\mu-1}\frac{dx_{3\mu-2}}{dt}\right)^2,$$

$$\Psi_\mu = \left(x_{3\mu+2}\frac{dx_{3\mu+3}}{dt} - x_{3\mu+3}\frac{dx_{3\mu+2}}{dt}\right)\left(x_{3\mu+5}\frac{dx_{3\mu+6}}{dt} - x_{3\mu+6}\frac{dx_{3\mu+5}}{dt}\right)$$

$$+ \left(x_{3\mu+3}\frac{dx_{3\mu+1}}{dt} - x_{3\mu+1}\frac{dx_{3\mu+3}}{dt}\right)\left(x_{3\mu+6}\frac{dx_{3\mu+4}}{dt} - x_{3\mu+4}\frac{dx_{3\mu+6}}{dt}\right)$$

$$+ \left(x_{3\mu+1}\frac{dx_{3\mu+2}}{dt} - x_{3\mu+2}\frac{dx_{3\mu+1}}{dt}\right)\left(x_{3\mu+4}\frac{dx_{3\mu+5}}{dt} - x_{3\mu+5}\frac{dx_{3\mu+4}}{dt}\right), \tag{23}$$

where $3\mu + \nu$ ($\nu = 1, 2, 3, 4, 5, 6; \mu = 1, 2, 3$), denote integers congruent by mod 9; then we have

$$\frac{\Pi_1}{m_1} + \frac{\Pi_2}{m_2} + \frac{\Pi_3}{m_3} + \frac{2\Psi_1}{m_2 m_3} + \frac{2\Psi_2}{m_3 m_1} + \frac{2\Psi_3}{m_1 m_2} = k^2 = c_1^2 + c_2^2 + c_3^2. \quad (24)$$

Equations (23) can be transformed by the identities

$$\begin{aligned}(bC - cB)^2 + (cA - aC)^2 + (aB - bA)^2 \\ \equiv (a^2 + b^2 + c^2)(A^2 + B^2 + C^2) - (aA + bB + cC)^2,\end{aligned}$$

$$\begin{aligned}(bC - cB)(b'C' - c'B') + (cA - aC)(c'A' - a'C') \\ + (aB - bA)(a'B' - b'A') \\ \equiv (aa' + bb' + cc')(AA' + BB' + CC') \\ - (aA' + bB' + cC')(a'A + b'B + c'C),\end{aligned}$$

and by (2), (5), (9), and (15) to the form

$$\Pi_\mu = u_\mu^2 r_\mu^2 - r_\mu^2 \left(\frac{dr_\mu}{dt}\right)^2, \qquad \Psi_\mu = p_\mu v_\mu + \frac{1}{4}\rho^2 - \frac{1}{4}\left(\frac{dp_\mu}{dt}\right)^2. \quad (25)$$

From (25) and (24) we obtain

$$\sum_{\nu=1}^{3} \frac{1}{m_\nu^2}\left[u_\nu^2 r_\nu^2 - r_\nu^2\left(\frac{dr_\nu}{dt}\right)^2\right] + \sum_{\nu=1}^{3} \frac{2}{m_{\nu+1} m_{\nu+2}}\left[p_\nu v_\nu - \frac{1}{4}\left(\frac{dp_\nu}{dt}\right)^2\right] + \frac{m_1 + m_2 + m_3}{2 m_1 m_2 m_3}\rho^2 = k^2. \quad (26)$$

The left-hand members of (25) and (26) are expressed in terms of $r_1, r_2, r_3$ and their first- and second-order derivatives.

Thus we have obtained four fundamental equations (13), (21), (22), and (26). $p_1, p_2, p_3, u_1^2, u_2^2, u_3^2, v_1, v_2, v_3, \Sigma_1, \Sigma_2, \Sigma_3$ are expressed in terms of $r_1, r_2, r_3$ and their derivatives because of (11), (12), (17), and (20). Then (21) expresses $\rho$ in terms of these functions. Hence, the differential equations of the three-body problem reduce to the three equations (13), (22), and (26) for the three unknown functions $r_1, r_2, r_3$. Equations (13) and (26) are second-order differential equations and contain two constants $h$ and $k$; (22) is a third-order differential equation. Hence, the whole set of differential equations is of order 7, and we require 7 integrals. Adding $h$ and $k$ brings the number of integration constants up to nine. Although we can solve this set of differential equations with these constants, we still require three more constants: two representing the position of the plane of the triangle formed by the three bodies, and another representing the orientation of the triangle on that plane. Thus we require twelve integration constants in all, in order to determine the relative motion of the three bodies. There are six more integration constants corresponding to the rectilinear motion in space of the center of mass of the whole system. Thus, by this transformation of Lagrange we can reduce the order of the differential equations from 18 to 7. We can reduce the order still further, as will be explained in this chapter.

Differentiate (5) with respect to $t$ and refer to (1). Then, using (2), (10), (15), we obtain

$$\frac{du_\mu^2}{dt} = 2(m_1 + m_2 + m_3)\frac{d}{dt}\frac{1}{r_\mu} + m_\mu\left(q_{\mu+2}\frac{dp_{\mu+2}}{dt} - q_{\mu+1}\frac{dp_{\mu+1}}{dt} - q_\mu \rho\right). \tag{27}$$

Integrate (27) with respect to $t$ and substitute $u_1^2$, $u_2^2$, $u_3^2$ in (12) and (27); then

$$\frac{1}{2}\frac{d^2 r_\mu^2}{dt^2} - \frac{m_1 + m_2 + m_3}{r_\mu} - m_\mu\left[p_{\mu+2}q_{\mu+2} - p_{\mu+1}q_{\mu+1} + \int\left(q_{\mu+2}\frac{dp_{\mu+2}}{dt} - q_{\mu+1}\frac{dp_{\mu+1}}{dt} - q_\mu \rho\right)dt\right] = 0. \tag{28}$$

The third-order differential equation obtained by differentiating this equation is important to the theory underlying Lindstedt's solution of the three-body problem (Sundman, 1913; Lindstedt, 1883, 1884). Lagrange treated the particular solution of the three-body problem that we discussed in Chapter 3 on the basis of this set of seventh-order differential equations. Instead of the mutual distances of the three bodies in the problem of Lagrange, the distances of the three bodies from the center of mass were employed as the unknown functions by Charlier (1899).

## 5.2 LEVI-CIVITA'S TRANSFORMATION

Now we propose to reduce the order of the differential equations from 18 to 8 by a transformation arising from the geometrical considerations of Levi-Civita (1915; see also Charlier, 1899; Ronchi, 1917).

We take the rectangular coordinates $\xi$, $\eta$ in the invariable plane of the three bodies $P_1$, $P_2$, $P_3$ and the $\zeta$ axis perpendicular to that plane. The center of mass $O$ is taken as the coordinate origin. The instantaneous moving plane containing the three bodies at any instant passes always through the origin $O$. Denote the positions of the three bodies referred to the moving rectangular coordinates on this instantaneous plane at $t$ by $x_{2\nu-1}, x_{2\nu}$ $(\nu = 1, 2, 3)$. The $X$ axis is taken to be in the direction of the ascending node of this instantaneous plane on the invariable plane, in such a way that the positive direction is counted from the lower to the upper side of the invariable plane in positive rotation around the $O\zeta$ axis. The $Z$ axis is normal to this instantaneous moving plane. Let the angular distance of this ascending node from the $O\xi$ axis be $\psi$, and the angle between these two planes be $\theta$. Twice the area of the triangle formed by the three bodies is

$$V = \begin{vmatrix} x_1 & x_2 & 1 \\ x_3 & x_4 & 1 \\ x_5 & x_6 & 1 \end{vmatrix}.$$

Let the masses of the three bodies be $m_{2\nu-1} = m_{2\nu}$ $(\nu = 1, 2, 3)$, $\sum_{\nu=1}^{3} m_{2\nu-1} = \sum_{\nu=1}^{3} m_{2\nu} = m$, and the moments and the products of inertia around $OX$ and $OY$, respectively, be

$$I_X = \sum_{\nu=1}^{3} m_{2\nu} x_{2\nu}^2, \qquad I_Y = \sum_{\nu=1}^{3} m_{2\nu-1} x_{2\nu-1}^2,$$

$$J = \sum_{\nu=1}^{3} m_{2\nu} x_{2\nu-1} x_{2\nu}, \qquad D = \begin{vmatrix} I_X & J \\ J & I_Y \end{vmatrix};$$

then it can be proved that

$$m_2 m_4 m_6 V^2 = mD > 0.$$

The direction cosines of the normal to the invariable plane referred to the coordinate system $OXYZ$ defined by the instantaneous moving plane of the three bodies are 0, $\sin\theta$, $\cos\theta$. Hence, the total kinetic energy of the system is

$$T = \frac{1}{2} \sum_{\nu=1}^{3} m_{2\nu-1} [(\dot{x}_{2\nu-1} - \dot{\psi}\cos\theta\cdot x_{2\nu})^2 + (\dot{x}_{2\nu} + \dot{\psi}\cos\theta\cdot x_{2\nu-1})^2 + (\dot{\theta} x_{2\nu} - \dot{\psi}\sin\theta\cdot x_{2\nu-1})^2],$$

and the force function, assuming the gravitational constant to be unity, is

$$U = \frac{m_2 m_4}{r_{12}} + \frac{m_4 m_6}{r_{23}} + \frac{m_6 m_2}{r_{31}}. \tag{29}$$

The canonically conjugate variables to these variables $x_{2\nu-1}$, $x_{2\nu}$, $\theta$, $\psi$ are

$$y_{2\nu-1} = \frac{\partial T}{\partial \dot{x}_{2\nu-1}} = m_{2\nu-1}(\dot{x}_{2\nu-1} - \dot{\psi}\cos\theta\cdot x_{2\nu}),$$

$$y_{2\nu} = \frac{\partial T}{\partial \dot{x}_{2\nu}} = m_{2\nu}(\dot{x}_{2\nu} + \dot{\psi}\cos\theta\cdot x_{2\nu-1}),$$

$$\Theta = \frac{\partial T}{\partial \dot{\theta}} = I_X\dot{\theta} - J\dot{\psi}\sin\theta, \tag{30}$$

$$\Psi = \frac{\partial T}{\partial \dot{\psi}} = \sin\theta(-J\dot{\theta} + I_Y\dot{\psi}\sin\theta) + \cos\theta\cdot M,$$

$$M = \sum_{\nu=1}^{3}(x_{2\nu-1}y_{2\nu} - x_{2\nu}y_{2\nu-1}).$$

Thus,

$$T = \mathbf{T} + f, \tag{31}$$

$$\mathbf{T} = \frac{1}{2}\sum_{\nu=1}^{3}\frac{1}{m_{2\nu}}(y_{2\nu-1}^2 + y_{2\nu}^2),$$

$$f = \frac{1}{2}\sum_{\nu=1}^{3} m_{2\nu}(\dot{\theta}x_{2\nu} - \dot{\psi}\sin\theta\cdot x_{2\nu-1})^2.$$

Substituting $\Theta$, $\Psi$, we obtain the function $f$ in the form

$$f(\dot{\theta}, \dot{\psi}) = F(\Theta, \Psi)$$

and

$$H = \mathbf{H} + F, \qquad \mathbf{H} = \mathbf{T} - U.$$

Thus, the equations of motion are

$$\frac{dx_{2\nu-1}}{dt} = \frac{\partial H}{\partial y_{2\nu-1}}, \qquad \frac{dx_{2\nu}}{dt} = \frac{\partial H}{\partial y_{2\nu}}, \qquad \frac{d\theta}{dt} = \frac{\partial F}{\partial \Theta}, \qquad \frac{d\psi}{dt} = \frac{\partial F}{\partial \Psi},$$

$$\frac{dy_{2\nu-1}}{dt} = -\frac{\partial H}{\partial x_{2\nu-1}}, \qquad \frac{dy_{2\nu}}{dt} = -\frac{\partial H}{\partial x_{2\nu}}, \qquad \frac{d\Theta}{dt} = -\frac{\partial F}{\partial \theta}, \qquad \frac{d\Psi}{dt} = -\frac{\partial F}{\partial \psi}$$

$$(\nu = 1, 2, 3). \tag{32}$$

The total angular momentum $\mathbf{K}$ has, with respect to $OXYZ$, components

$$I_X\dot{\theta} - J\dot{\psi}\sin\theta, \qquad -J\dot{\theta} + I_Y\dot{\psi}\sin\theta, \qquad M,$$

but, on the other hand, the components are 0, $K\sin\theta$, $K\cos\theta$, where $|\mathbf{K}| = K$. Hence

$$I_X\dot{\theta} - J\dot{\psi}\sin\theta = 0, \quad -J\dot{\theta} + I_Y\sin\theta = K\sin\theta, \quad M = K\cos\theta. \quad (33)$$

Substituting the definition of $\Theta$, $\Psi$, we obtain the integrals of (32):

$$\Theta = 0, \qquad \Psi = M\cos\theta + K\sin^2\theta = K, \qquad M = K\cos\theta. \quad (34)$$

Consequently,

$$\frac{\partial F}{\partial\theta} = \frac{\partial F}{\partial\psi} = 0.$$

Eliminating $\theta$ and $\psi$ from (33), we obtain

$$\dot{\psi} = \frac{K}{D}I_X, \qquad \dot{\theta} = \frac{K}{D}J\sin\theta. \quad (35)$$

Since $F$, and accordingly $f$, is a homogeneous quadratic form in $\dot{\theta}$, $\dot{\psi}$ by (31), therefore

$$F = \frac{1}{2}\left(\frac{\partial f}{\partial\dot{\theta}}\,\dot{\theta} + \frac{\partial f}{\partial\dot{\psi}}\,\dot{\psi}\right) = \frac{1}{2}\,[\Theta\dot{\theta} + (\Psi - M\cos\theta)\dot{\psi}].$$

From (34) and (35) we obtain

$$F = \frac{1}{2}\frac{K - M\cos\theta}{D}\,KI_X,$$

or, substituting from the third equation of (34),

$$\frac{1}{2}\frac{I_X}{D}(K^2 - M^2) = \mathbf{F}, \quad (36)$$

which depends only on $x_{2\nu-1}$, $x_{2\nu}$, $y_{2\nu-1}$, $y_{2\nu}$. Hence, the first twelve equations of (32) depend only on these twelve variables, for now $H = \mathbf{H} + \mathbf{F}$. Thus the problem is reduced to the plane problem in which the inclination $\theta$ is kept constant.

Furthermore, we transform by

$$x'_{2\nu-1} = x_{2\nu-1} - x_5, \qquad x'_{2\nu} = x_{2\nu} - x_6, \qquad x'_5 = x_5, \qquad x'_6 = x_6,$$
$$y'_{2\nu-1} = y_{2\nu-1}, \quad y'_{2\nu} = y_{2\nu}, \quad y'_5 = y_1 + y_3 + y_5, \quad y'_6 = y_2 + y_4 + y_6$$
$$(\nu = 1, 2). \quad (37)$$

The transformation is canonical. $y'_5$, $y'_6$ are the components of the momentum of the center of mass, which we can equate to zero. Then

$$\frac{\partial H}{\partial x'_5} = \frac{\partial H}{\partial x'_6} = 0.$$

Thus $H$ depends neither on $x'_5$ nor on $x'_6$. Hence, the equations of motion are of the eighth order and of the form

$$\begin{aligned} \frac{dx'_{2\nu-1}}{dt} &= \frac{\partial H}{\partial y'_{2\nu-1}}, & \frac{dx'_{2\nu}}{dt} &= \frac{\partial H}{\partial y'_{2\nu}}, \\ \frac{dy'_{2\nu-1}}{dt} &= -\frac{\partial H}{\partial x'_{2\nu-1}}, & \frac{dy'_{2\nu}}{dt} &= -\frac{\partial H}{\partial x'_{2\nu}} \qquad (\nu = 1, 2). \end{aligned} \tag{38}$$

After we solve these equations for $x'_{2\nu-1}$, $x'_{2\nu}$, $y'_{2\nu-1}$, $y'_{2\nu}$, we compute $x_{2\nu-1}$, $x_{2\nu}$, $y_{2\nu-1}$, $y_{2\nu}$ by the inverse transformation of (37) and then compute $I_X$, $I_Y$, $J$, $D$. Then we obtain $\psi$ and $\theta$ from

$$\frac{d\psi}{dt} = \frac{K}{D} I_X, \qquad \frac{d\theta}{dt} = \frac{K}{D} J \sin\theta. \tag{39}$$

Instead of the transformation (37) we may use the mutual distances of the three bodies according to Lagrange. Let $\phi_\nu$ be the inclination to the $X$ axis of the side opposite the vertex $(x_{2\nu-1}, x_{2\nu})$ of the triangle formed by the three bodies; then

$$\cos\phi_\nu = \frac{x_{2\lambda-1} - x_{2\mu-1}}{\sqrt{(x_{2\lambda-1} - x_{2\mu-1})^2 + (x_{2\lambda} - x_{2\mu})^2}},$$

$$\sin\phi_\nu = \frac{x_{2\lambda} - x_{2\mu}}{\sqrt{(x_{2\lambda-1} - x_{2\mu-1})^2 + (x_{2\lambda} - x_{2\mu})^2}},$$

where $\lambda$, $\mu$, $\nu$ are a cyclic permutation of 1, 2, 3. Put $\theta_\nu = \phi_\lambda - \phi_\mu$; then $\sum_{\nu=1}^{3} \theta_\nu = 0$. Introduce

$$\omega = \frac{1}{3}\sum_{\nu=1}^{3} \phi_\nu, \quad r_\lambda \equiv r_{\mu,\nu} = [(x_{2\mu-1} - x_{2\nu-1})^2 + (x_{2\mu} - x_{2\nu})^2]^{1/2}, \tag{40}$$

and denote the corresponding momenta by $\Omega$, $P_\lambda$, respectively. Then we can prove that

$$y_{2\lambda-1} = P_\mu \cos\phi_\mu - P_\nu \cos\phi_\nu - \frac{\Omega}{3}\left(\frac{\sin\phi_\mu}{r_\mu} - \frac{\sin\phi_\nu}{r_\nu}\right),$$
$$y_{2\lambda} = P_\mu \sin\phi_\mu - P_\nu \sin\phi_\nu + \frac{\Omega}{3}\left(\frac{\cos\phi_\mu}{r_\mu} - \frac{\cos\phi_\nu}{r_\nu}\right), \tag{41}$$

where

$$\cos\theta_\lambda = \frac{r_\lambda^2 - r_\mu^2 - r_\nu^2}{2r_\mu r_\nu}, \qquad \sin\theta_\lambda = \frac{V}{r_\mu r_\nu}, \tag{42}$$

$$4V^2 = (r_1 + r_2 + r_3)(-r_1 + r_2 + r_3)(r_1 - r_2 + r_3)(r_1 + r_2 - r_3), \tag{43}$$

$$\phi_\lambda = \omega + \tfrac{1}{3}(\theta_\mu - \theta_\nu). \tag{44}$$

Hence

$$mx_{2\lambda-1} = m_\nu r_\mu \cos\phi_\mu - m_\mu r_\nu \cos\phi_\nu,$$
$$mx_{2\lambda} = m_\nu r_\mu \sin\phi_\mu - m_\mu r_\nu \sin\phi_\nu,$$

and

$$\frac{1}{2}\sum\frac{1}{m_\lambda}(y_{2\nu-1}^2 + y_{2\nu}^2) = \frac{1}{2}\sum\frac{1}{m_\lambda}(P_\mu^2 + P_\nu^2 - 2P_\mu P_\nu\cos\theta_\lambda)$$
$$+ \frac{\Omega}{3}\sum\frac{1}{m_\lambda}\left(\frac{P_\mu}{r_\nu} - \frac{P_\nu}{r_\mu}\right)\sin\theta_\lambda$$
$$+ \frac{\Omega^2}{18}\sum\frac{1}{m_\lambda}\left(\frac{1}{r_\mu^2} + \frac{1}{r_\nu^2} - \frac{2\cos\theta_\lambda}{r_\mu r_\nu}\right),$$

$$I_Y = \frac{m_2 m_4 m_6}{m}\sum\frac{1}{m_\lambda} r_\lambda^2 \sin^2[\omega + \tfrac{1}{3}(\theta_\nu - \theta_\mu)].$$

Then the Hamiltonian function of (32) is

$$H = \frac{1}{2}\sum\frac{1}{m_\lambda}(y_{2\lambda-1}^2 + y_{2\lambda}^2) - \sum\frac{m_\mu m_\nu}{r_\lambda}$$
$$+ \frac{1}{D}\left[\Theta^2 I_X - 2\Theta\frac{M\cos\theta - \Psi}{\sin\theta}J + \frac{(M\cos\theta - \Psi)^2}{\sin^2\theta}I_Y\right]. \tag{45}$$

Since the origin is the center of mass,

$$\frac{dr_\lambda}{dt} = \frac{\partial H}{\partial P_\lambda}, \qquad \frac{d\omega}{dt} = \frac{\partial H}{\partial \Omega},$$

$$\frac{dP_\lambda}{dt} = -\frac{\partial H}{\partial r_\lambda}, \qquad \frac{d\Omega}{dt} = -\frac{\partial H}{\partial \omega} \qquad (\lambda = 1, 2, 3). \tag{46}$$

If we express $H$ in terms of $\omega$, $r_\lambda$, $\Omega$, $P_\lambda$ by substituting in $H$ from equations (41), (42), (43), and (44), we obtain

$$\begin{aligned} H = {} & \sum \frac{1}{2m_\lambda}(P_\mu^2 + P_\nu^2 - 2P_\mu P_\nu \cos\theta_\lambda) \\ & + \frac{\Omega}{3}\sum\left(\frac{P_\mu}{r_\nu} - \frac{P_\nu}{r_\mu}\right)\frac{\sin\theta_\lambda}{m_\lambda} + \frac{\Omega^2}{18}\sum\frac{1}{m_\lambda}\left(\frac{1}{r_\mu^2} + \frac{1}{r_\nu^2} - \frac{2\cos\theta_\lambda}{r_\mu r_\nu}\right) \\ & + \frac{K^2 - \Omega^2}{D^2}\sum\frac{r_\lambda^2}{m_\lambda}\sin^2[\omega + \tfrac{1}{3}(\theta_\mu - \theta_\nu)] - \sum\frac{m_\mu m_\nu}{r_\lambda}, \end{aligned}$$

where

$$\lambda, \mu, \nu \text{ are a cyclic permutation of } 1, 2, 3$$

and

$$\theta_1, \theta_2, \theta_3, D \text{ are expressed in terms of } r_1, r_2, r_3$$

by (42) and (43). Thus the order of the differential equations (46) is reduced to eight.

Van Kampen and Wintner (1937) and Murnaghan (1936) applied this transformation of Levi-Civita to the plane motion of three bodies.

Lemaître (1952; see also Deprit and Delie, 1961, 1963; Deprit and Roels, 1962) discovered an elegant symmetric coordinate system for the reduction of the three-body problem. Write

$$\mu = m_1 + m_2 + m_3, \qquad \mu m_0^2 = m_1 m_2 m_3, \qquad \tan\alpha_i = \frac{m_i}{m_0} \quad (i = 1, 2, 3),$$

$$m_1 m_2 m_3 A^2 = (m_1 + m_2)(m_2 + m_3)(m_3 + m_1),$$

$$\chi = \tfrac{1}{3}(\chi_1 + \chi_2 + \chi_3), \qquad \chi_1 = \chi + \tfrac{1}{3}(\alpha_3 - \alpha_2),$$

$$\chi_2 = \chi + \tfrac{1}{3}(\alpha_1 - \alpha_3), \qquad \chi_3 = \chi + \tfrac{1}{3}(\alpha_2 - \alpha_1).$$

It can be shown that

$$\sum_i \sin p\alpha_i \cos p\chi_i \equiv 0, \qquad \sum_i \sin p\alpha_i \sin p\chi_i \equiv 0$$

$$(p = 1, 2, \ldots),$$

$$\sum_i m_i \cos^2 \alpha_i \cos^2 \chi_i = \frac{\mu}{A^2}, \qquad \sum_i m_i \cos^2 \alpha_i \sin^2 \chi_i \equiv \frac{\mu}{A^2},$$

$$\sum_i m_i \cos^2 \alpha_i \cos \chi_i \sin \chi_i \equiv 0.$$

In the plane of the three bodies, Lemaître introduced

$$x_i = Ar \cos \alpha_i \cos \sigma \cos \chi_i, \qquad y_i = Ar \cos \alpha_i \sin \sigma \sin \chi_i \qquad (i = 1, 2, 3),$$

so that

$$\sum_i m_i x_i = 0, \qquad \sum_i m_i y_i = 0, \qquad \sum_i m_i x_i y_i = 0;$$

that is, the $xyz$ coordinate axes coincide with the principal axes of inertia. Let the moments of inertia be

$$I_{xx} = \sum_i m_i x_i^2 = \mu r^2 \cos^2 \sigma, \qquad I_{yy} = \sum_i m_i y_i^2 = \mu r^2 \sin^2 \sigma,$$

$$J = I_{xx} + I_{yy} = \mu r^2,$$

where $r$ denotes the radius of gyration with respect to the center of mass of the three masses in the plane of the triangle. The area of the triangle $m_1 m_2 m_3$ is

$$\Delta = \frac{1}{4} \frac{J}{m_0^2} \sin 2\sigma,$$

which is a symmetric function of the three masses. Denote the angle of the vector $m_j m_k$ with the $x$ axis by $\phi_i$ $(i, j, k = 1, 2, 3)$; then

$$\tan \phi_i = -\tan \sigma \cot \chi_i \qquad (i = 1, 2, 3).$$

The angle $\chi$ gives the position of the triangle with respect to the $x$ axis in the plane of the three bodies.

Let $\rho_i$ be the distance of $m_j$ and $m_k$, and $\theta_i$ be the angle between the vectors $m_i m_k$ and $m_i m_j$. Then

$$\rho_i = \frac{1}{\sqrt{2}} Ar \sin \alpha_i \sqrt{1 - \cos 2\sigma \cos 2\chi_i},$$

$$\sin \theta_i = \frac{\sin \alpha_i \sin 2\sigma}{\sqrt{1 - \cos 2\sigma \cos 2\chi_i}\sqrt{1 - \cos 2\sigma \cos 2\chi_k}},$$

$$\cos \theta_i = -\frac{\cos \alpha_i + \cos 2\sigma \cos (\chi_i + \chi_k)}{\sqrt{1 - \cos 2\sigma \cos 2\chi_i}\sqrt{1 - \cos 2\sigma \cos 2\chi_k}}.$$

Refer the moving $x$, $y$, $z$ axes to the $X$, $Y$, $Z$ axes fixed in space by means of Euler's angles $\theta$, $\phi$, $\psi$; there the angular velocity components are

$$\omega_x = \dot{\theta} \cos \phi + \dot{\psi} \sin \theta \sin \phi, \qquad \omega_y = -\dot{\theta} \sin \phi + \dot{\psi} \sin \theta \cos \phi,$$

$$\omega_z = \dot{\phi} + \dot{\psi} \cos \theta.$$

Then,

$$T = T_r + \omega_z C_r + \tfrac{1}{2}(I_{yy}\omega_x^2 + I_{xx}\omega_y^2 + J\omega_z^2),$$

$$U = k^2\left(\frac{m_2 m_3}{\rho_1} + \frac{m_3 m_1}{\rho_2} + \frac{m_1 m_2}{\rho_3}\right)$$

$$= \frac{k^2\sqrt{2}}{Ar} \sum_i \frac{m_j m_k}{\sin \alpha_i \sqrt{1 - \cos 2\sigma \cos 2\chi_i}},$$

where

$$T_r = \tfrac{1}{2} \sum_i m_i(\dot{x}_i^2 + \dot{y}_i^2) = \tfrac{1}{2}\mu(\dot{r}^2 + r^2\dot{\sigma}^2 + r^2\dot{\chi}^2),$$

$$C_r = \sum_i m_i(x_i\dot{y}_i - \dot{x}_i y_i) = \mu r^2 \dot{\chi} \sin 2\sigma.$$

Since $L = T + U$,

$$p_\theta = \frac{\partial L}{\partial \dot{\theta}} = \frac{\partial T}{\partial \omega_x} \cos \phi - \frac{\partial T}{\partial \omega_y} \sin \phi,$$

$$p_\phi = \frac{\partial L}{\partial \dot{\phi}} = \frac{\partial T}{\partial \omega_z},$$

$$p_\psi = \frac{\partial L}{\partial \dot{\psi}} = \frac{\partial T}{\partial \omega_x} \sin \theta \sin \phi + \frac{\partial T}{\partial \omega_y} \sin \theta \cos \phi + \frac{\partial T}{\partial \omega_z} \cos \theta.$$

We compute the Routh function

$$
\begin{aligned}
R &= \dot{\theta} p_\theta + \dot{\phi} p_\phi + \dot{\psi} p_\psi - L \\
&= \tfrac{1}{2}(I_{yy}\omega_x^2 + I_{xx}\omega_y^2 + J\omega_z^2) - T_r - U \\
&= \frac{1}{2I_{yy}} [p_\theta \cos \phi + (p_\psi - p_\phi \cos \theta) \operatorname{cosec} \theta \sin \phi]^2 \\
&\quad + \frac{1}{2I_{xx}} [-p_\theta \sin \phi + (p_\psi - p_\phi \cos \theta) \operatorname{cosec} \theta \cos \phi]^2 \\
&\quad + \frac{1}{2J} (p_\phi - C_r)^2 - T_r - U.
\end{aligned}
$$

Denote the magnitude of the angular momentum vector $C$ by $K$; then we have integrals

$$p_\theta = 0, \qquad p_\phi = K \cos \theta, \qquad p_\psi = K.$$

The third equation shows that $\psi$ is an ignorable coordinate. The first integral shows that $\theta$ can be replaced by its value $\cos^{-1}(p_\phi/K)$ in $R$. Thus,

$$
\begin{aligned}
R = \frac{1 - \cos 2\sigma \cos 2\phi}{J \sin^2 2\sigma} (K^2 - p_\phi^2) + \frac{1}{2}\frac{p_\phi^2}{J} \\
-\dot{\chi} p_\phi \sin 2\sigma - \tfrac{1}{2}\mu(\dot{r}^2 + r^2\dot{\sigma}^2 + r^2\dot{\chi}^2 \cos^2 2\sigma) - U.
\end{aligned}
$$

The equations of motion are then reduced to those of the eighth order

$$\dot{\phi} = \frac{\partial R}{\partial p_\phi}, \qquad \dot{p}_\phi = -\frac{\partial R}{\partial \phi},$$

$$\frac{d}{dt}\frac{\partial R}{\partial \dot{r}} - \frac{\partial R}{\partial r} = 0, \qquad \frac{d}{dt}\frac{\partial R}{\partial \dot{\sigma}} - \frac{\partial R}{\partial \sigma} = 0,$$

$$\frac{d}{dt}\frac{\partial R}{\partial \dot{\chi}} - \frac{\partial R}{\partial \chi} = 0.$$

We have the energy integral with a constant $h$. The third equation for $r$ gives

$$\ddot{r} + \frac{\dot{r}^2}{r} - \frac{1}{\mu r}(U + 2h) = 0,$$

from which we obtain the relation due to Lagrange

$$\ddot{J} = 2U + 4h.$$

The Hamiltonian function is

$$\begin{aligned} H &= \dot{r}p_r + \dot{\sigma}p_\sigma + \dot{\chi}p_\chi + R \\ &= \frac{1}{2\mu}p_r^2 + \frac{1}{2J}\left(p_\sigma^2 + \frac{1}{\cos^2 2\sigma}p_\chi^2 - 2\frac{\sin 2\sigma}{\cos^2 2\sigma}p_\chi p_\phi\right) \\ &\quad + \frac{1}{2J\cos^2 2\sigma}p_\phi^2 + \frac{1 - \cos 2\sigma \cos 2\phi}{\sin^2 2\sigma}\frac{K^2 - p_\phi^2}{J} - U. \end{aligned}$$

Put

$$q_1 = q\cos\lambda\cos\beta, \qquad q_2 = q\sin\lambda\cos\beta, \qquad q_3 = q\sin\beta.$$

Then

$$\rho_1 = q^2(1 - \cos^2\lambda\cos^2\beta), \quad \rho_2 = q^2(1 - \sin^2\lambda\cos^2\beta), \quad \rho_3 = q^2\cos^2\beta,$$

and the canonical variables are

$$\begin{matrix} J & \lambda & \beta & \phi \\ p_J & \varpi_\lambda & \varpi_\beta & p_\phi, \end{matrix}$$

where

$$\begin{aligned} \varpi_\lambda &= p_\lambda - A_\lambda p_\phi, \qquad \varpi_\beta = p_\beta - A_\beta p_\phi, \\ A_\lambda &= \frac{2m_0^3 q_1 q_2 \sin 2\sigma}{J^2\cos^2 2\sigma}\Phi, \qquad A_\beta = \frac{2m_0^3 q q_3 \sin 2\sigma}{J^2 \cos^2 2\sigma \cos\beta}\Psi, \\ \Phi &= \frac{J}{m_0^2}(q^2 + q_3^2) - 4\left(\frac{\rho_1}{m_1} + \frac{\rho_2}{m_2}\right)q^2 q_3^2, \\ \Psi &= \frac{J}{m_0^2}(q_3^2 - q_1^2) + 4\left(\frac{\rho_1}{m_1} - \frac{\rho_2}{m_2}\right)q_1^2 q_2^2, \end{aligned}$$

with the Hamiltonian

$$\begin{aligned} H' &= \frac{1}{2}\left(\frac{\varpi_\lambda^2}{\cos^2\beta} + \varpi_\beta^2\right) + 2m_0^2 q^2 \rho_1\rho_2\rho_3\left(4p_J^2 + \frac{p_\phi^2}{J^2} - 2\frac{U}{J} - 2\frac{h}{J}\right) \\ &\quad + \frac{4m_0^2 q^2 \rho_1\rho_2\rho_3}{J^2}\frac{1 - \cos 2\sigma\cos 2\phi}{\sin^2 2\sigma}(K^2 - p_\phi^2). \end{aligned}$$

The equations of motion are now

$$\dot{r} = \frac{1}{\mu}p_r, \qquad \dot{\chi} = \frac{1}{J\cos^2 2\sigma}\varpi_\chi, \qquad \dot{\sigma} = \frac{1}{J}p_\sigma,$$

$$\dot{p}_r = \frac{1}{\mu r^3}\left(p_\sigma^2 + \frac{1}{\cos^2 2\sigma}\varpi_\chi^2 + p_\phi^2\right)$$

$$+ \frac{2(1 - \cos 2\sigma \cos 2\phi)}{\mu r^3 \sin^2 2\sigma}(K^2 - p_\phi^2) - \frac{U}{r},$$

$$\dot{p}_\sigma = -\frac{2\sin 2\sigma}{J\cos^3 2\sigma}\varpi_\chi^2 + \frac{2(K^2 - p_\phi^2)}{J\sin^3 2\sigma}$$

$$\times [2\cos 2\sigma(1 - \cos 2\sigma \cos 2\phi) - \sin^2 2\sigma \cos 2\phi]$$

$$+ 2\dot{\chi}p_\phi \cos 2\sigma - \tfrac{1}{2}k^2A^2r^2 \sin 2\sigma \sum_{i,j,k} \frac{m_j m_k}{\rho_i^3} \sin^2 \alpha_i \cos 2\chi_i,$$

$$\dot{\varpi}_\chi = -2\dot{\sigma}p_\phi \cos 2\sigma - \tfrac{1}{2}k^2A^2r^2 \cos 2\sigma \sum_{i,j,k} \frac{m_j m_k}{\rho_i^3} \sin^2 \alpha_i \sin 2\chi_i,$$

$$\varpi_\chi = p_\chi - p_\phi \sin 2\sigma,$$

where the summation is to be extended to three terms obtained by the cyclic permutation of $i, j, k$ among 1, 2, 3.

Lemaître further introduced $\rho_i = q_j^2 + q_k^2$ with cyclic permutation of $i, j, k$. The new coordinates $q_1, q_2, q_3$ are such that: (i) $q_1 = q_2 = q_3 = 0$ represents the triple simultaneous collision of the three bodies; (ii) $q_i = 0$ represents the straight line of the collinear configuration with $m_i$ situated between $m_j$ and $m_k$; (iii) the plane $q_j = q_k$ represents the isosceles configuration with $m_j$ and $m_k$ as base; and (iv) $q_1 = q_2 = q_3$ represents the equilateral configuration (see Chapter 3).

Put $q^2 = q_1^2 + q_2^2 + q_3^2$; then

$$2q^2 = \rho_1 + \rho_2 + \rho_3, \qquad 2q_i^2 = \rho_j + \rho_k - \rho_i, \qquad \Delta = qq_1 q_2 q_3,$$

$$1 + \cos\theta_i = \frac{2q_j^2 q_k^2}{\rho_j \rho_k}, \qquad 1 - \cos\theta_i = \frac{2q^2 q_i^2}{\rho_j \rho_k},$$

$$\cos\theta_j + \cos\theta_k = -2\frac{q_j^2 q_k^2}{\rho_j \rho_k}\frac{\rho_j + \rho_k}{\rho_i},$$

$$\cos\theta_j - \cos\theta_k = 2\frac{q^2 q_i^2}{\rho_j \rho_k}\frac{\rho_j - \rho_k}{\rho_i},$$

$$1 + \cos\theta_1 + \cos\theta_2 + \cos\theta_3 = -4\frac{q_1^2 q_2^2 q_3^2}{\rho_1 \rho_2 \rho_3},$$

$$\tan\frac{1}{2}\theta_i = \frac{qq_i}{\rho_j \rho_k},$$

with cyclic permutation of $i, j, k$. This set of variables was used in the regularization of the three-body problem by Lemaître (see Section 17.15 dealing with regularizing variables).

Following Murnaghan (1936), put

$$J = \mu r^2, \qquad x = \frac{q_1}{q + q_3}, \qquad y = \frac{q_2}{q + q_3},$$

or

$$q_1 = q\,\frac{2x}{1 + x^2 + y^2}, \qquad q_2 = q\,\frac{2y}{1 + x^2 + y^2}, \qquad q_3 = q\,\frac{1 - x^2 - y^2}{1 + x^2 + y^2},$$

$$\lambda = \tan^{-1}\frac{y}{x}, \qquad \beta = \sin^{-1}\frac{1 - x^2 - y^2}{1 + x^2 + y^2},$$

and

$$w = \int^t \frac{\mu r^2(\tau)[1 + x^2(\tau) + y^2(\tau)]^2\,d\tau}{16 m_0^2 q^2(\tau)\rho_1(\tau)\rho_2(\tau)\rho_3(\tau)}.$$

Then the canonical equations with the independent variable $w$ are

$$\begin{matrix} r & x & y & \phi \\ p_r & \varpi_x & \varpi_y & p_\phi, \end{matrix}$$

where

$$\varpi_x = p_x - p_\phi A_x, \qquad \varpi_y = p_y - p_\phi A_y,$$

$$A_x = \frac{2m_0^3 q^2 x \sin 2\sigma}{J^2(x^2 + y^2)(1 + x^2 + y^2)^2\cos^2 2\sigma} \times [4y^2\Phi + (1 - x^2 - y^2)(1 + x^2 + y^2)\Psi],$$

$$A_y = \frac{2m_0^3 q^2 y \sin 2\sigma}{J^2(x^2 + y^2)(1 + x^2 + y^2)^2\cos^2 2\sigma} \times [4x^2\Phi - (1 - x^2 - y^2)(1 + x^2 + y^2)\Psi],$$

with

$$H' = \frac{1}{2}(\varpi_x^2 + \varpi_y^2) + \frac{8m_0^2 q^2\rho_1\rho_2\rho_3}{J(1 + x^2 + y^2)^2}\left(\frac{1}{\mu}p_r^2 + \frac{p_\phi^2}{J} - 2(U + h)\right)$$
$$+ \frac{16 m_0^2 q^2 \rho_1\rho_2\rho_3}{(1 + x^2 + y^2)^2}\,\frac{1 - \cos 2\sigma\cos 2\phi}{J^2\sin^2 2\sigma}\,(K^2 - p_\phi^2).$$

## TRANSFORMATION OF POINCARÉ

### 5·3 TRANSFORMATION OF POINCARÉ

Take the differential equations of motion of the $n$-body problem in the form

$$\frac{dx_i}{dt} = \frac{\partial H}{\partial y_i}, \qquad \frac{dy_i}{dt} = -\frac{\partial H}{\partial x_i} \qquad (i = 1, 2, \ldots, 3n). \tag{47}$$

Following Poincaré (1897), we denote the mass of the $\mu$th body by $m_{3\mu-2} = m_{3\mu-1} = m_{3\mu}$, its position coordinates by $x_{3\mu-2}, x_{3\mu-1}, x_{3\mu}$, and its momentum components by $y_{3\mu-2}, y_{3\mu-1}, y_{3\mu}$. If $\sum_i x'_i\, dy'_i - \sum_i x_i\, dy_i$ in the transformation from $x_i, y_i$ to $x'_i, y'_i$ is a perfect differential, then the equations of motion conserve their canonical form. If $x'_i$ ( $= 1, 2, \ldots, 3n$) are functions of $x_j$ $(j = 1, 2, \ldots, 3n)$ only, and $y'_i$ $(i = 1, 2, \ldots, 3n)$ are functions of $y_j$ $(j = 1, 2, \ldots, 3n)$ only, then the condition is written

$$\sum_i y'_i x'_i = \sum_j y_j x_j. \tag{48}$$

We make further assumptions:

(i)

$x'_{3\nu-2}$ $(\nu = 1, 2, \ldots, n)$ are functions of $x_{3\mu-2}$ $(\mu = 1, 2, \ldots, n)$ only,
$y'_{3\nu-2}$ $(\nu = 1, 2, \ldots, n)$ are functions of $y_{3\mu-2}$ $(\mu = 1, 2, \ldots, n)$ only,
$x'_{3\nu-1}$ $(\nu = 1, 2, \ldots, n)$ are functions of $x_{3\mu-1}$ $(\mu = 1, 2, \ldots, n)$ only,
$y'_{3\nu-1}$ $(\nu = 1, 2, \ldots, n)$ are functions of $y_{3\mu-1}$ $(\mu = 1, 2, \ldots, n)$ only,
$x'_{3\nu}$ $(\nu = 1, 2, \ldots, n)$ are functions of $x_{3\mu}$ $(\mu = 1, 2, \ldots, n)$ only,
$y'_{3\nu}$ $(\nu = 1, 2, \ldots, n)$ are functions of $y_{3\mu}$ $(\mu = 1, 2, \ldots, n)$ only.

(ii) The linear homogeneous relations are the same:

between
$x'_{3\nu-2}$ $(\nu = 1, 2, \ldots, n)$ and $x_{3\mu-2}$ $(\mu = 1, 2, \ldots, n)$;

between
$x'_{3\nu-1}$ $(\nu = 1, 2, \ldots, n)$ and $x_{3\mu-1}$ $(\mu = 1, 2, \ldots, n)$;

between
$x'_{3\nu}$ $(\nu = 1, 2, \ldots, n)$ and $x_{3\mu}$ $(\mu = 1, 2, \ldots, n)$.

(iii) The relations are the same:

between
$y'_{3\nu-2}$ $(\nu = 1, 2, \ldots, n)$ and $y_{3\mu-2}$ $(\mu = 1, 2, \ldots. n)$;

between
$y'_{3\nu-1}$ $(\nu = 1, 2, \ldots, n)$ and $y_{3\mu-1}$ $(\mu = 1, 2, \ldots, n)$;

between
$y'_{3\nu}$ $(\nu = 1, 2, \ldots, n)$ and $y_{3\mu}$ $(\mu, = 1, 2 \ldots, n)$.

Assumption (iii) is a consequence of assumptions (i) and (ii) and equation (48). By assumption (i), equation (48) becomes

$$\sum_{\nu} y'_{3\nu-2}x'_{3\nu-2} = \sum_{\mu} y_{3\mu-2}x_{3\mu-2},$$
$$\sum_{\nu} y'_{3\nu-1}x'_{3\nu-1} = \sum_{\mu} y_{3\mu-1}x_{3\mu-1},$$
$$\sum_{\nu} y'_{3\nu}x'_{3\nu} = \sum_{\mu} y_{3\mu}x_{3\mu},$$

or, by assumption (iii),

$$\sum_{\nu} y'_{3\nu-2}x'_{3\nu-1} = \sum_{\mu} y_{3\mu-2}x_{3\mu-1},$$
$$\sum_{\nu} y'_{3\nu-1}x'_{3\nu-2} = \sum_{\mu} y_{3\mu-1}x_{3\mu-2},$$

and, in general,

$$\sum_{\nu} y'_{3\nu-\kappa}x'_{3\nu-\lambda} = \sum_{\mu} y_{3\mu-\kappa}x_{3\mu-\lambda} \qquad (\kappa, \lambda = 0, 1, 2).$$

From these we obtain

$$\begin{aligned} \sum_{\nu} (x'_{3\nu}y'_{3\nu-1} - x'_{3\nu-1}y'_{3\nu}) &= \sum_{\mu} (x_{3\mu}y_{3\mu-1} - x_{3\mu-1}y_{3\mu}), \\ \sum_{\nu} (x'_{3\nu-2}y'_{3\nu} - x'_{3\nu}y'_{3\nu-2}) &= \sum_{\mu} (x_{3\mu-2}y_{3\mu} - x_{3\mu}y_{3\mu-2}), \\ \sum_{\nu} (x'_{3\nu-1}y'_{3\nu-2} - x'_{3\nu-2}y'_{3\nu-1}) &= \sum_{\mu} (x_{3\mu-1}y_{3\mu-2} - x_{3\mu-2}y_{3\mu-1}). \end{aligned} \tag{49}$$

Hence, the form of each angular-momentum integral remains unaltered by our transformation.

As our transformation satisfies the conditions of equation (48), we write equations (47) in the form

$$\frac{dx'_i}{dt} = \frac{\partial H}{\partial y'_i}, \qquad \frac{dy'_i}{dt} = -\frac{\partial H}{\partial x'_i} \qquad (i = 1, 2, \ldots, 3n). \tag{50}$$

We further assume that

$$ky'_{3n-2} = \sum_{\mu=1}^{n} y_{3\mu-2}, \tag{51}$$

where $k$ is constant, and that $x_{3\mu-2}$ ($\mu = 1, 2, \ldots, n$), depend only on $x_{3\mu-2} - x_{3n-2}$ ($\mu = 1, 2, \ldots, n-1$), that is, on the coordinates relative to the $n$th body. The momentum $y'_{3n-2}$ is then a constant multiple of the momentum of the center of mass. We can assume that the center of mass is at rest. Hence, $H$ does not depend on $x'_{3n-2}$, as

$$\frac{dy'_{3n-2}}{dt} = -\frac{\partial H}{\partial x'_{3n-2}}.$$

Similarly, it does not depend on $x'_{3n-1}$. This happens when $x'_{3\nu-2}$ depends only on the difference $x_{3\mu-2} - x_{3n-2}$ in the relation between the old and the new variables. Actually, $T$ depends on $y'$ and not on $x'$, since it depends on $y$ and not on $x$. $U$ depends on $x'_{3\nu-2}, x'_{3\nu-1}, x'_{3\nu}$ and not on $x'_{3n-2}, x'_{3n-1}, x'_{3n}$ ,since it depends only on $x_{3\mu-2} - x_{3n-2}, x_{3\mu-1} - x_{3n-1}, x_{3\mu} - x_{3n}$. Hence, $H$ does not contain $x'_{3n-2}, x'_{3n-1}, x'_{3n}$.

We can prove that the assumption (51) is not contradictory. If $ky'_{3n-2} = \sum_\nu y_{3\nu-2}$, then $x'_{3\nu-2}$ can be shown to depend on $x_{3\nu-2} - x_{3n-2}$, because, if $x_{3\mu-2}$ are all equal to $x_{3n-2}$ in $\sum_\nu x'_{3\nu-2}y'_{3\nu-2} = \sum_\mu x_{3\mu-2}y_{3\mu-2}$, then the right-hand side is proportional to $\sum_\nu y_{3\nu-2}$, that is, to $y'_{3n-2}$. Hence, $y'_{3\nu-2}$ on the left-hand side must be zero, and $x'_{3\mu-2}$ must vanish with $x_{3\mu-2} - x_{3n-2}$. Thus $x'_{3\nu-2}$ is a linear expression of these differences.

It is seen that $H$ does not depend on $x'_{3n-2}, x'_{3n-1}, x'_{3n}$. Hence we can suppose that $y'_{3n-2} = y'_{3n-1} = y'_{3n} = 0$, and the order of the differential equations (47) is reduced by six.

## 5.4 ELIMINATION OF THE NODES

We consider fictitious bodies according to Radau in the following manner (Bertrand, 1852). The coordinates and the momenta of the first fictitious body are $x'_1, x'_2, x'_3, y'_1, y'_2, y'_3$; its instantaneous orbital plane, that is, the plane containing the origin, the first fictitious body, and the momentum vector, is perpendicular to the vector

$$\mathfrak{v}_1 = (x'_2y'_3 - x'_3y'_2, x'_3y'_1 - x'_1y'_3, x'_1y'_2 - x'_2y'_1).$$

The position coordinates and the momenta of the second fictitious body are $x'_4, x'_5, x'_6, y'_4, y'_5, y'_6$, and its instantaneous orbital plane is perpendicular to the vector $\mathfrak{v}_2 = (x'_5y'_6 - x'_6y'_5, x'_6y'_4 - x'_4y'_6, x'_4y'_5 - x'_5y'_4)$. In general, we define the vector

$$\mathfrak{v}_\mu = (x'_{3\mu-1}y'_{3\mu} - x'_{3\mu}y'_{3\mu-1}, x'_{3\mu}y'_{3\mu-2} - x'_{3\mu-2}y'_{3\mu}, x'_{3\mu-2}y'_{3\mu-1} - x'_{3\mu-1}y'_{3\mu-2})$$

$$(\mu = 1, 2, \ldots, n-1).$$

The sum of vector $\sum_{\mu=2}^{n-1} \mathfrak{v}_\mu = \mathfrak{V}_1$ and vector $\mathfrak{v}_1$, has the components

$$\sum_{\mu=1}^{n} (x_{3\mu-1}y_{3\mu} - x_{3\mu}y_{3\mu-1}), \qquad \sum_{\mu=1}^{n} (x_{3\mu}y_{3\mu-2} - x_{3\mu-2}y_{3\mu}),$$

$$\sum_{\mu=1}^{n} (x_{3\mu-2}y_{3\mu-1} - x_{3\mu-1}y_{3\mu-2}),$$

since $y'_{3n-2} = y'_{3n-1} = y'_{3n} = 0$, and this vector sum equals the total angular momentum vector $\mathfrak{V}$ of the $n$-body system. As the total angular momentum vector is perpendicular to the invariable plane of the $n$-body system, the vector $\mathfrak{v}_1 + \mathfrak{V}_1$ is perpendicular to the invariable plane. Consequently $\mathfrak{v}_1$, $\mathfrak{V}_1$, and $\mathfrak{V}$ are all in the same plane. Hence, the intersection of the two planes perpendicular respectively to the vectors $\mathfrak{v}_1$ and $\mathfrak{V}_1$ is parallel to the invariable plane. In other words, the intersection of the instantaneous orbital plane of an arbitrary one of the $n - 1$ fictitious bodies and the plane perpendicular to the vector representing the resultant angular momentum of the remaining $n - 2$ fictitious bodies is parallel to the invariable plane of the original $n$-body system. This is an extension to the $n$-body problem of the Jacobi theorem on the *elimination of the nodes* in the three-body problem.

In particular, in the three-body problem ($n = 3$) the theorem reduces to the following: the direction of the ascending node of the instantaneous orbital plane of the first fictitious body differs by 180 degrees from the direction of the ascending node of the instantaneous orbital plane of the second fictitious body. If the triangle formed by $\mathfrak{v}_1$, $\mathfrak{V}_1$, and $\mathfrak{V}$ is acute, then the components of the vectors $\mathfrak{v}_1$ and $\mathfrak{V}_1$ in the direction of $\mathfrak{V}$ are of the same direction; that is, the two fictitious bodies revolve either both in the direct sense or both in the retrograde sense. But as the components in the direction perpendicular to $\mathfrak{V}$ are of opposite direction, the inclinations of the instantaneous orbital planes of the two fictitious bodies relative to the invariable plane (which is perpendicular to $\mathfrak{V}$) should have opposite signs. If the triangle formed of $\mathfrak{v}_1$, $\mathfrak{V}_1$, and $\mathfrak{V}$ is obtuse, then the components perpendicular to $\mathfrak{V}$ of the vectors $\mathfrak{v}_1$ and $\mathfrak{V}_1$ are of opposite directions, and one of the two fictitious bodies revolves in the direct sense and the other in the retrograde sense. In either case the ascending nodes in question should differ by 180 degrees in direction, because the intersections of the two orbital planes with the invariable plane should be parallel. This is the Jacobi theorem (Jacobi, 1842, 1844) on the elimination of the nodes, and it is an integral of the problem.

### 5·5 TRANSFORMATION $\alpha$

Poincaré has specified his transformation by further assumptions and classified the cases $\alpha$, $\beta$, and $\gamma$. The transformation, which was first discovered by Cauchy, is

$$x'_{3\mu-2} = x_{3\mu-2} - x_{3n-2}, \qquad x'_{3\mu-1} = x_{3\mu-1} - x_{3n-1}, \qquad x'_{3\mu} = x_{3\mu} - x_{3n} \qquad (\mu = 1, 2, \ldots, n-1),$$

$$x'_{3n-2} = x_{3n-2}, \qquad x'_{3n-1} = x_{3n-1}, \qquad x'_{3n} = x_{3n},$$

$$y'_i = y_i \qquad (i = 1, 2, \ldots, 3n-3),$$

$$y'_{3n-2} = \sum_{\mu=1}^{n} y_{3\mu-2}, \qquad y'_{3n-1} = \sum_{\mu=1}^{n} y_{3\mu-1}, \qquad y'_{3n} = \sum_{\mu=1}^{n} y_{3\mu}. \tag{52}$$

The motion of the $\mu$th fictitious body that we are now considering is a relative motion around the $n$th body, but its momentum is that due to its absolute motion, and $y'_{3n-2}, y'_{3n-1}, y'_{3n}$ are the components of the total momentum of the whole $n$-body system.

There is another transformation by Poincaré which is not essentially different from this:

$$x'_{3\mu-2} = x_{3\mu-2} - x_{3n-2}, \qquad x'_{3\mu-1} = x_{3\mu-1} - x_{3n-1},$$

$$x'_{3\mu} = x_{3\mu} - x_{3n} \qquad (\mu = 1, 2, \ldots, n-1),$$

$$x'_{3n-2} = \frac{\sum_{\mu=1}^{n} m_{3\mu-2}x_{3\mu-2}}{\sum_{\mu=1}^{n} m_{3\mu-2}}, \qquad x'_{3n-1} = \frac{\sum_{\mu=1}^{n} m_{3\mu-1}x_{3\mu-1}}{\sum_{\mu=1}^{n} m_{3\mu-1}},$$

$$x'_{3n} = \frac{\sum_{\mu=1}^{n} m_{3\mu}x_{3\mu}}{\sum_{\mu=1}^{n} m_{3\mu}},$$

$$y'_{3\mu-2} = y_{3\mu-2} - \frac{m_{3\mu-2}y'_{3n-2}}{\sum_{\mu=1}^{n} m_{3\mu-2}}, \qquad y'_{3\mu-1} = y_{3\mu-1} - \frac{m_{3\mu-1}y'_{3n-1}}{\sum_{\mu=1}^{n} m_{3\mu-1}}, \tag{53}$$

$$y'_{3\mu} = y_{3\mu} - \frac{m_{3\mu}y'_{3n}}{\sum_{\mu=1}^{n} m_{3\mu}},$$

$$y'_{3n-2} = \sum_{\mu=1}^{n} y_{3\mu-2}, \qquad y'_{3n-1} = \sum_{\mu=1}^{n} y_{3\mu-1}, \qquad y'_{3n} = \sum_{\mu=1}^{n} y_{3\mu}$$

$$(\mu = 1, 2, \ldots, n-1).$$

As we may suppose that the center of mass is at rest, we put

$$\sum_{\mu=1}^{n} y_{3\mu-2} = 0.$$

Then $x'_{3\mu-2}$, $y'_{3\mu-2}$, $y'_{3n-2}$ have the same values as in (52). Only $x'_{3n-2}$ is different. The Hamiltonian $H$ depends on $x'_{3\mu-2} = x_{3\mu-2} - x_{3n-2}$, but does not depend on $x'_{3n-2}$.

In order to reduce the order of the differential equations (47)

$$\frac{dx_i}{dt} = \frac{\partial H}{\partial y_i}, \qquad \frac{dy_i}{dt} = -\frac{\partial H}{\partial x_i} \qquad (i = 1, 2, \ldots, 3n), \tag{54}$$

in conjunction with

$$H = \sum_{i=1}^{3n} \frac{y_i^2}{2m_i} - \sum_{\mu,\nu}^{n} \frac{m_{3\mu} m_{3\nu}}{r_{3\mu,3\nu}},$$

from $6n$ to $6n - 6$, we make the contact transformation

$$x_i = \frac{\partial W}{\partial y_i}, \qquad y'_i = \frac{\partial W}{\partial x'_i}, \qquad (i = 1, 2, \ldots, 3n),$$

$$W = \left(\sum_{i=1}^{3n-3} x'_i y_i\right) + x'_{3n-2} \sum_{\mu=1}^{n} y_{3\mu-2} + x'_{3n-1} \sum_{\mu=1}^{n} y_{3\mu-1} + x'_{3n} \sum_{\mu=1}^{n} y_{3\mu}. \tag{55}$$

Then the transformed equations are

$$\frac{dx'_i}{dt} = \frac{\partial H}{\partial y'_i}, \qquad \frac{dy'_i}{dt} = -\frac{\partial H}{\partial x'_i} \qquad (i = 1, 2, \ldots, 3n - 3),$$

$$H = \sum_{i=1}^{3n-3} \frac{y_i'^2}{2m_i} + \frac{1}{2m_{3n}} \left[\left(y'_{3n-2} - \sum_{\mu=1}^{n-1} y'_{3\mu-2}\right)^2 + \left(y'_{3n-1} - \sum_{\mu=1}^{n-1} y'_{3\mu-1}\right)^2 + \left(y'_{3n} - \sum_{\mu=1}^{n-1} y'_{3\mu}\right)^2\right] - \sum_{\mu,\nu=1}^{n} \frac{m_{3\mu} m_{3\nu}}{r_{3\mu,3\nu}},$$

where $x_i$ in $r_{3\mu,3\nu}$ are rewritten in terms of the new variables. As $H$ is independent of $x'_{3n-2}$, $x'_{3n-1}$, $x'_{3n}$, we obtain the integrals of the total angular momentum $y'_{3n-2} =$ constant, $y'_{3n-1} =$ constant, $y'_{3n} =$ constant.

We set these constants equal to zero by considering the motion relative to the center of mass; that is, we remove the primes:

$$\frac{dx_i}{dt} = \frac{\partial H}{\partial y_i}, \qquad \frac{dy_i}{dt} = -\frac{\partial H}{\partial x_i} \qquad (i = 1, 2, \ldots, 3n-3),$$

$$H = \sum_{i=1}^{3n-3} \frac{y_i^2}{2m_i} + \frac{1}{2m_{3n}}\left[\left(\sum_{\mu=1}^{n-1} y_{3\mu-2}\right)^2 + \left(\sum_{\mu=1}^{n-1} y_{3\mu-1}\right)^2 + \left(\sum_{\mu=1}^{n-1} y_{3\mu}\right)^2\right] \tag{56}$$

$$- \sum_{\mu,\nu=1}^{n} \frac{m_{3\mu}m_{3\nu}}{r_{3\mu,3\nu}},$$

$$r^2_{3\mu,3\nu} = (x_{3\mu-2} - x_{3\nu-2})^2 + (x_{3\mu-1} - x_{3\nu-1})^2 + (x_{3\mu} - x_{3\nu})^2 \qquad (\mu, \nu \neq n),$$

$$r^2_{3\mu,3n} = x^2_{3\mu-2} + x^2_{3\mu-1} + x^2_{3\mu}.$$

## 5.6 TRANSFORMATION $\beta$

Put

$$y_i' = m_i' \frac{dx_i'}{dt}, \qquad m'_{3\mu-2} = m'_{3\mu-1} = m'_{3\mu} \qquad (\mu = 1, 2, \ldots, n); \tag{57}$$

then the relation between $y_i'$ and $y_i$ is the same as the relation between $m_i'x_i'$ and $m_ix_i$, and (48) takes the form

$$\sum_i m_i' x_i'^2 = \sum_i m_i x_i^2. \tag{58}$$

This is nothing but the kinetic energy in terms of the new variables

$$T = \sum_i \frac{m_i'}{2}\left(\frac{dx_i'}{dt}\right)^2 = \sum_i \frac{1}{2m_i'} y_i'^2.$$

Hence, the expressions not only for the total momentum but also for the kinetic energy are invariant by this transformation $\beta$. Moreover, the canonical form of the differential equations is conserved.

We propose to obtain such a transformation. Consider the transformation

$$x_{3\mu-2} = \sum_{k=1}^{n} c_{\mu\kappa}x'_{3\kappa-2}, \quad x_{3\mu-1} = \sum_{\kappa=1}^{n} c_{\mu\kappa}x'_{3\kappa-1}, \quad x_{3\mu} = \sum_{\kappa=1}^{n} c_{\mu\kappa}x'_{3\kappa}. \tag{59}$$

There are $n^2$ coefficients $c_{\mu\kappa}$. Condition (48) becomes

$$\sum_\mu m'_{3\mu-2}x'^2_{3\mu-2} = \sum_\mu m_{3\mu-2}x^2_{3\mu-2}. \tag{60}$$

Substituting (59) in this expression and equating the coefficients of $x'_{3\rho-2}x'_{3\sigma-2}$ to zero, we obtain $n(n-1)$ relations

$$\sum_{\mu} m_{3\mu-2}c_{\mu\rho}c_{\mu\sigma} = 0 \qquad (\rho, \sigma = 1, 2, \ldots, n). \tag{61}$$

From the coefficients of $x'^2_{3\kappa-2}$ we obtain $n$ relations

$$\sum_{\mu} m_{3\mu-2}c^2_{\mu\kappa} = m'_{3\kappa-2} \qquad (\kappa = 1, 2, \ldots, n). \tag{62}$$

Altogether we have $n^2$ relations among the coefficients $c_{\mu\kappa}$.

Following Jacobi, take the coordinates

$$x'_{3\kappa-2}, x'_{3\kappa-1}, x'_{3\kappa} \qquad (\kappa = 1, 2, \ldots, n-1),$$

so that the $(\kappa + 1)$th body is referred to the center of masses of the first $\kappa$ bodies, that is,

$$x'_{3\kappa-2} = x_{3(\kappa+1)-2} - \left(\sum_{\mu=1}^{\kappa} m_{3\mu-2}x_{3\mu-2}\right)\Big/\left(\sum_{\mu=1}^{\kappa} m_{3\mu-2}\right)$$
$$(\kappa = 1, 2, \ldots, n-1), \tag{63}$$
$$x'_{3n-2} = -\left(\sum_{\mu=1}^{n} m_{3\mu-2}x_{3\mu-2}\right)\Big/\left(\sum_{\mu=1}^{n} m_{3\mu-2}\right).$$

To satisfy (61) and (62), put

$$M_\kappa = \sum_{\mu=1}^{\kappa} m_{3\mu-2} \qquad (\kappa = 1, 2, \ldots, n-1); \tag{64}$$

then solve (63) for $x_{3\mu-2}$:

$$x_{3(\kappa+1)-2} = x_1 + \sum_{\mu=1}^{\kappa-1} \frac{m_{3\mu-2}x'_{3\mu-2}}{M_\mu} + x'_{3\kappa-2}. \tag{65}$$

Next substitute this expression for $x_{3\mu-2}$ in the last equation of (63):

$$M_n x'_{3n-2} = M_n x_1 + \sum_{\mu=1}^{n-1} m_{3\mu-2}(M_\mu + m_{3(\mu+1)-2} + \cdots + m_{3n-2})$$
$$\times \frac{x'_{3\mu-2}}{M_\mu} + m'_{3n-2}x'_{3n-2},$$

and substitute this $x_1$ in (65). Then,

$$x_{3\kappa-2} = x'_{3n-2} - \sum_{\mu=\kappa+1}^{n} m_{3\mu-2}\frac{x'_{3\mu-2}}{M_\mu} + \frac{M_{\kappa-1}}{M_\kappa}x'_{3\kappa-2}. \tag{66}$$

Comparing this expression with (59), we obtain

$$
\begin{aligned}
c_{\kappa\mu} &= -\frac{m_{3\mu-2}}{M_\kappa} && (\mu > \kappa), \\
c_{\kappa\mu} &= 0 && (\mu < \kappa), \\
c_{\kappa\mu} &= \frac{M_{\kappa-1}}{M_\kappa} && (\mu = \kappa).
\end{aligned} \tag{67}
$$

Equation (61) is satisfied by these coefficients. Put (67) in (62); then the right-hand member becomes

$$\frac{M_{\kappa-1}}{M_\kappa} m_{3\kappa-2}$$

and we have

$$m'_{3\kappa-2} = m_{3\kappa-2}\frac{M_{\kappa-1}}{M_\kappa} \qquad (\kappa = 1, 2, \ldots, n). \tag{68}$$

Condition (48) is satisfied if we determine $y'_{3\kappa-2}$ by (61) and (63). This is an extension of Poincaré's transformation $\beta$; the $y'_{3\kappa-2}$ are called the *relative canonical coordinates of Jacobi.*

The expression for the force function $U$ is rather complicated when (66) is used. Taking the center of masses as the origin, we have

$$x'_{3n-2} = x'_{3n-1} = x'_{3n} = 0.$$

The center of mass has zero momentum; that is, it is at rest:

$$y'_{3n-2} = y'_{3n-1} = y'_{3n} = 0.$$

By this transformation, the order of differential equations (47) is reduced from $6n$ to $6n - 6$.

$$y_i = \frac{\partial W}{\partial x_i}, \qquad x'_i = \frac{\partial W}{\partial y'_i} \qquad (i = 1, 2, \ldots, 3n), \tag{69}$$

$$
\begin{aligned}
W = \sum_{\kappa=1}^{n-1} \Bigg\{ & y'_{3\kappa-2}\left[x_{3(\kappa+1)-2} - \left(\sum_{\mu=1}^{\kappa} m_{3\mu-2}x_{3\mu-2}\right)\Big/\left(\sum_{\mu=1}^{\kappa} m_{3\mu-2}\right)\right] \\
& + y'_{3\kappa-1}\left[x_{3(\kappa+1)-1} - \left(\sum_{\mu=1}^{\kappa} m_{3\mu-1}x_{3\mu-1}\right)\Big/\left(\sum_{\mu=1}^{\kappa} m_{3\mu-1}\right)\right] \\
& + y'_{3\kappa}\left[x_{3(\kappa+1)} - \left(\sum_{\mu=1}^{\kappa} m_{3\mu}x_{3\mu}\right)\Big/\left(\sum_{\mu=1}^{\kappa} m_{3\mu}\right)\right]\Bigg\} \\
& + y'_{3n-2}\left(\sum_{\mu=1}^{n} m_{3\mu-2}x_{3\mu-2}\right) + y'_{3n-1}\left(\sum_{\mu=1}^{n} m_{3\mu-1}x_{3\mu-1}\right) \\
& + y'_{3n}\left(\sum_{\mu=1}^{n} m_{3\mu}x_{3\mu}\right).
\end{aligned}
$$

After this contact transformation, put

$$y'_{3n-2} = y'_{3n-1} = y'_{3n} = x'_{3n-2} = x'_{3n-1} = x'_{3n} = 0,$$

and drop the primes. Then,

$$\frac{dx_i}{dt} = \frac{\partial H}{\partial y_i}, \qquad \frac{dy_i}{dt} = -\frac{\partial H}{\partial x_i}, \qquad (i = 1, 2, \ldots, 3n - 3), \tag{70}$$

$$H = \sum_{\kappa=1}^{n-1} \left(\frac{y_{3\kappa-2}^2}{2m_{3\kappa-2}} + \frac{y_{3\kappa-1}^2}{2m_{3\kappa-1}} + \frac{y_{3\kappa}^2}{2m_{3\kappa}}\right) - \sum_{\mu,\nu=1}^{n} \frac{m_{3\mu}m_{3\nu}}{r_{3\mu,3\nu}},$$

where $r_{3\mu,3\nu}$ is expressed in terms of new coordinates, using (59), (67), and (68), and the primes are dropped. Jacobi and Bertrand obtained the transformation for the case $n = 3$. It is extended to the general case $n$ in the present discussion.

## 5·7 TRANSFORMATION $\gamma$

Put

$$m'_{3n-2} = \sum_{\mu=1}^{n} m_{3\mu-2}, \qquad m'_i = m_i \qquad (i = 1, 2, \ldots, 3n - 3),$$

$$y'_{3n-2} = \sum_{\mu=1}^{n-1} y_{3\mu-2} = m'_{3n-2}\frac{dx'_{3n-2}}{dt},$$

$$m'_{3n-2}x'_{3n-2} = \sum_{\mu=1}^{n-1} m_{3\mu-2}x_{3\mu-2}, \tag{71}$$

$$x'_{3\mu-2} = x_{3\mu-2} - x_{3n-2} \qquad (\mu = 1, 2, \ldots, n - 1),$$

$$y'_i = m'_i\frac{dx'_i}{dt} \qquad (i = 1, 2, \ldots, 3n).$$

The momenta $y'_{3n-1}$ and $y'_{3n}$ are transformed similarly to $y'_{3n-2}$; $x'_{3n-1}$ and $x'_{3n}$ similarly to $x'_{3n-2}$; $x'_{3\mu-1}$ and $x'_{3\mu}$ similarly to $x'_{3\mu-2}$. The sum of the masses of the system is $m'_{3n-2} = m'_{3n-1} = m'_{3n}$; the coordinates of the center of mass are $x'_{3n-2}, x'_{3n-1}, x'_{3n}$. The coordinates of the $\mu$th body relative to the $n$th body are $x'_{3\mu-2}, x'_{3\mu-1}, x'_{3\mu}$. The momentum components of the $\mu$th body relative to the $n$th body are $y'_{3\mu-2}, y'_{3\mu-1}, y'_{3\mu}$. This transformation $\gamma$ does not satisfy conditions (48) or (49). Hence, the transformed equations are not canonical. Accordingly we cannot eliminate the nodes by this transformation.

We fix the center of mass, transfer the coordinate origin to the center of mass, and put $x'_{3n-2} = x'_{3n-1} = x'_{3n} = 0$, $y'_{3n-2} = y'_{3n-1} = y'_{3n} = 0$. The transformed equations, after the primes are dropped, are

$$\frac{dx_{3\mu-2}}{dt} = \frac{\partial H_\mu}{\partial y_{3\mu-2}}, \qquad \frac{dx_{3\mu-1}}{dt} = \frac{\partial H_\mu}{\partial y_{3\mu-1}}, \qquad \frac{dx_{3\mu}}{dt} = \frac{\partial H_\mu}{\partial y_{3\mu}},$$

$$\frac{dy_{3\mu-2}}{dt} = -\frac{\partial H_\mu}{\partial x_{3\mu-2}}, \qquad \frac{dy_{3\mu-1}}{dt} = -\frac{\partial H_\mu}{\partial x_{3\mu-1}}, \qquad \frac{dy_{3\mu}}{dt} = -\frac{\partial H_\mu}{\partial x_{3\mu}},$$

$$H_\mu = \frac{1}{2}\left(\frac{y_{3\mu-2}^2}{m_{3\mu-2}} + \frac{y_{3\mu-1}^2}{m_{3\mu-1}} + \frac{y_{3\mu}^2}{m_{3\mu}}\right) - \frac{m_{3\mu}m_{3n}}{r_{3\mu,3n}} - \sum_{\substack{\nu=1 \\ (\mu\neq\nu)}}^{n-1}{}' \frac{m_{3\mu}m_{3\nu}}{r_{3\mu,3\nu}} \tag{72}$$

$$+ \sum_{\substack{\nu=1 \\ (\mu\neq\nu)}}^{n-1}{}' \frac{m_{3\mu}m_{3\nu}}{r_{3\mu,3\nu}^3}(x_{3\mu-2}x_{3\nu-2} + x_{3\mu-1}x_{3\nu-1} + x_{3\mu}x_{3\nu})$$

$$(\mu = 1, 2, \ldots, n-1).$$

These are the usual equations of motion in the perturbation theory. The Hamiltonian $H$ differs only by the last term, that is,

$$H_\mu = T - U + \sum_{\substack{\nu=1 \\ (\mu\neq\nu)}}^{n-1}{}' \frac{m_{3\mu}m_{3\nu}}{r_{3\mu,3\nu}^3}(x_{3\mu-2}x_{3\nu-2} + x_{3\mu-1}x_{3\nu-1} + x_{3\mu}x_{3\nu}), \tag{73}$$

where

$$T = \frac{1}{2}\sum_{i=1}^{3n-3} \frac{1}{m_i} y_i^2, \qquad U = \sum_{\substack{\mu,\nu \\ (\mu\neq\nu)}}{}' \frac{m_{3\mu}m_{3\nu}}{r_{3\mu,3\nu}}.$$

The perturbation of the $\nu$th body due to presence of the $\mu$th body is

$$\frac{m_{3\nu}}{r_{3\mu,3\nu}} - \frac{m_{3\nu}}{r_{3\mu,3\nu}^3}(x_{3\mu-2}x_{3\nu-2} + x_{3\mu-1}x_{3\nu-1} + x_{3\mu}x_{3\nu}),$$

and is called the *disturbing function* or the *perturbative function.* The first term is called the *direct part* and the second the *indirect part.* Such a situation occurs in the case when the mass of the $n$th body is very large, compared with the masses of the remaining $n - 1$ bodies. The motion of the $\mu$th body around the $n$th body is disturbed by the perturbative action of the remaining $n - 2$ bodies. The perturbation is included in the perturbative function and is multiplied by the disturbing mass $m_{3\nu}$.

## 5.8 REDUCTION FROM $6n - 6$ TO $6n - 10$

Using the angular momentum integrals and the elimination of the nodes, we reduce the order of the differential equations (47) from $6n - 6$ to $6n - 10$ (Whittaker, 1937; Bennett, 1905).

To (56) we apply the contact transformation

$$x_i = \frac{\partial W}{\partial y_i}, \qquad y_i' = \frac{\partial W}{\partial x_i'} \qquad (i = 1, 2, \ldots, 3n - 3), \tag{74}$$

$$\begin{aligned}
W = \sum_{\nu=1}^{n-3} [ & y_{3\nu-2}(x'_{3\nu-2} \cos x'_{3n-4} - x'_{3\nu-1} \cos x'_{3n-3} \sin x'_{3n-4} \\
& \qquad + x'_{3\nu} \sin x'_{3n-3} \sin x'_{3n-4}) \\
& + y_{3\nu-1}(x'_{3\nu-2} \sin x'_{3n-4} + x'_{3\nu-1} \cos x'_{3n-3} \cos x'_{3n-4} \\
& \qquad - x'_{3\nu} \sin x'_{3n-3} \cos x'_{3n-4}) \\
& + y_{3\nu}(x'_{3\nu-1} \sin x'_{3n-3} + x'_{3\nu} \cos x'_{3n-3})] \\
& + y_{3n-8}(x'_{3n-8} \cos x'_{3n-4} - x'_{3n-7} \cos x'_{3n-3} \sin x'_{3n-4}) \\
& + y_{3n-7}(x'_{3n-8} \sin x'_{3n-4} + x'_{3n-7} \cos x'_{3n-3} \cos x'_{3n-4}) \\
& + y_{3n-6}(x'_{3n-7} \sin x'_{3n-3}) \\
& + y_{3n-5}(x'_{3n-6} \cos x'_{3n-4} - x'_{3n-5} \cos x'_{3n-3} \sin x'_{3n-4}) \\
& + y_{3n-4}(x'_{3n-6} \sin x'_{3n-4} + x'_{3n-5} \cos x'_{3n-3} \cos x'_{3n-4}) \\
& + y_{3n-3}(x'_{3n-5} \sin x'_{3n-3}).
\end{aligned}$$

This is the coordinate transformation from $OXYZ$ to $OX'Y'Z'$ with the $n$th body as the origin, the $X'Y'$ plane containing the $(n - 1)$th and the $(n - 2)$th bodies; the direction cosines are $l_1$, $m_1$, $n_1$, etc., as shown in Table 1. $OX'$ is in the direction of the intersection of the $XY$ plane and

TABLE 1

| | $OX$ | $OY$ | $OZ$ |
|---|---|---|---|
| $OX'$ | $l_1 = \cos x'_{3n-4}$, | $m_1 = \sin x'_{3n-4}$, | $n_1 = 0$, |
| $OY'$ | $l_2 = -\sin x'_{3n-4} \cos x'_{3n-3}$, | $m_2 = \cos x'_{3n-4} \cos x'_{3n-3}$, | $n_2 = \sin x'_{3n-3}$, |
| $OZ'$ | $l_3 = \sin x'_{3n-4} \sin x'_{3n-3}$, | $m_3 = -\cos x'_{3n-4} \sin x'_{3n-3}$, | $n_3 = \cos x'_{3n-3}$ |

the $X'Y'$ plane. $x'_{3\nu-2}$, $x'_{3\nu-1}$, $x'_{3\nu}$ $(\nu = 1, 2, \ldots, n - 3)$, are the coordinates of the $\nu$th body; $x'_{3n-8}$, $x'_{3n-7}$, 0 are the coordinates of the $(n - 2)$th body; and $x'_{3n-6}$, $x'_{3n-5}$, 0 are those of the $(n - 1)$th body;

all referred to the $OX'Y'Z'$ axes. $x'_{3n-4}$ is the angle between $OX$ and $OX'$, and $90° - x'_{3n-3}$ is the angle between $OY'$ and $OZ$. $y'_{3\nu-2}, y'_{3\nu-1}, y'_{3\nu}$ ($\nu = 1, 2, \ldots, n-3$), are the components of the momentum of the $\nu$th body; $y'_{3n-8}, y'_{3n-7}, 0$ are those of the $(n-2)$th body; and $y'_{3n-6}, y'_{3n-5}, 0$ are those of the $(n-1)$th body in the directions $OX'$, $OY'$, and $OZ'$, respectively. Since

$$\begin{aligned} y'_{3n-4} = \sum_{\nu=1}^{n-3} & (-y_{3\nu-2}x_{3\nu-1} + y_{3\nu-1}x_{3\nu-2}) \\ & + (-y_{3n-8}x_{3n-7} + y_{3n-7}x_{3n-8}) \\ & + (-y_{3n-6}x_{3n-5} + y_{3n-5}x_{3n-6}), \end{aligned} \tag{75}$$

therefore, $-y'_{3n-4}$ is the total angular momentum around $OZ$. Since

$$\begin{aligned} y'_{3n-3} = \sum_{\nu=1}^{n-3} & [x'_{3\nu-1}(l_3y_{3\nu-2} + m_3y_{3\nu-1} + n_3y_{3\nu}) \\ & - x'_{3\nu}(m_3y_{3\nu-2} + m_2y_{3\nu-1} + m_1y_{3\nu})] \\ & + x'_{3n-7}(l_3y_{3n-8} + m_3y_{3n-7} + n_3y_{3n-6}) \\ & + x'_{3n-5}(l_3y_{3n-5} + m_3y_{3n-4} + n_3y_{3n-3}), \end{aligned} \tag{75a}$$

therefore $y'_{3n-3}$ is the total angular momentum around $OX'$.

Denote the momentum in the direction $OZ'$ of the $(n-2)$th and the $(n-1)$th bodies by $\Gamma$ and $\Gamma'$, respectively, and drop the primes of the new $x$ and $y$ variables. Then

$$\begin{aligned} y_{3n-3} &= \sum_{\nu=1}^{n-3} \begin{vmatrix} x_{3\nu-1} & x_{3\nu} \\ y_{3\nu-1} & y_{3\nu} \end{vmatrix} + \begin{vmatrix} x_{3n-7} & 0 \\ y_{3n-7} & \Gamma \end{vmatrix} + \begin{vmatrix} x_{3n-5} & 0 \\ y_{3n-5} & \Gamma' \end{vmatrix}, \\ -y_{3n-4} &= n_3\left\{\sum_{\nu=1}^{n-3} \begin{vmatrix} x_{3\nu-2} & x_{3\nu-1} \\ y_{3\nu-2} & y_{3\nu-1} \end{vmatrix} + \begin{vmatrix} x_{3n-8} & x_{3n-7} \\ y_{3n-8} & y_{3n-7} \end{vmatrix} + \begin{vmatrix} x_{3n-6} & x_{3n-5} \\ y_{3n-6} & y_{3n-5} \end{vmatrix}\right\} \\ &\quad + n_2\left\{\sum_{\nu=1}^{n-3} \begin{vmatrix} x_{3\nu} & x_{3\nu-2} \\ y_{3\nu} & y_{3\nu-2} \end{vmatrix} + \begin{vmatrix} 0 & x_{3n-8} \\ \Gamma & y_{3n-8} \end{vmatrix} + \begin{vmatrix} 0 & x_{3n-6} \\ \Gamma' & y_{3n-6} \end{vmatrix}\right\}. \end{aligned} \tag{76}$$

Thus $\Gamma$ and $\Gamma'$ are given by the two linear equations:

$$\Gamma x_{3n-7} + \Gamma' x_{3n-5} = y_{3n-3} - \sum_{\nu=1}^{n-3} \begin{vmatrix} x_{3\nu-1} & x_{3\nu} \\ y_{3\nu-1} & y_{3\nu} \end{vmatrix},$$

$$\Gamma x_{3n-8} + \Gamma' x_{3n-6} = y_{3n-4} \operatorname{cosec} x_{3n-3} + \sum_{\nu=1}^{n-3} \begin{vmatrix} x_{3\nu} & x_{3\nu-2} \\ y_{3\nu} & y_{3\nu-2} \end{vmatrix}$$

$$+ \cot x_{3n-3} \left\{ \sum_{\nu=1}^{n-3} \begin{vmatrix} x_{3\nu-2} & x_{3\nu-1} \\ y_{3\nu-2} & y_{3\nu-1} \end{vmatrix} + \begin{vmatrix} x_{3n-8} & x_{3n-7} \\ y_{3n-8} & y_{3n-7} \end{vmatrix} \right.$$

$$\left. + \begin{vmatrix} x_{3n-6} & x_{3n-5} \\ y_{3n-6} & y_{3n-5} \end{vmatrix} \right\}.$$

These equations do not depend on $x_{3n-4}$. Thus, the Hamiltonian in the new variables is

$$H = \sum_{i=1}^{3n-9} \frac{y_i^2}{2m_i} + \frac{1}{2m_{3n-6}} (y_{3n-8}^2 + y_{3n-7}^2 + \Gamma^2)$$

$$+ \frac{1}{2m_{3n-3}} (y_{3n-6}^2 + y_{3n-5}^2 + \Gamma'^2)$$

$$+ \frac{1}{2m_{3n}} \left[ \left( \sum_{\nu=1}^{n-3} y_{3\nu-2} + y_{3n-8} + y_{3n-6} \right)^2 \right.$$

$$+ \left( \sum_{\nu=1}^{n-3} y_{3\nu-1} + y_{3n-7} + y_{3n-5} \right)^2$$

$$\left. + \left( \sum_{\nu=1}^{n-3} y_{3\nu} + \Gamma + \Gamma' \right)^2 \right] - \sum_{\mu,\nu=1}^{n}{}' \frac{m_{3\mu} m_{3\nu}}{r_{3\mu,3\nu}}, \qquad (77)$$

where $r_{3\mu,3\nu}$ is expressed in terms of $x_1, x_2, \ldots, x_{3n-5}$. As this expression for $H$ does not depend on $x_{3n-4}$ in the canonical equations

$$\frac{dx_i}{dt} = \frac{\partial H}{\partial y_i}, \qquad \frac{dy_i}{dt} = -\frac{\partial H}{\partial x_i} \qquad (i = 1, 2, \ldots, 3n - 3), \qquad (78)$$

the canonical equations have the integral

$$y_{3n-4} = K = \text{constant}. \tag{79}$$

$x_{3n-4}$ appears only in $dx_{3n-4}/dt = \partial H/\partial y_{3n-4}$. Hence, there is no need to consider $x_{3n-4}$ in our discussion. We obtain $x_{3n-4}$ by mere quadrature in the last stage. Thus we come to the differential equations of order $6n - 8$:

$$\frac{dx_i}{dt} = \frac{\partial H}{\partial y_i}, \qquad \frac{dy_i}{dt} = -\frac{\partial H}{\partial x_i} \qquad (i = 1, 2, \ldots, 3n - 5, 3n - 3), \tag{80}$$

where we have put $K$ instead of $y_{3n-4}$.

Now we choose the $OZ$ axis parallel to the resultant angular momentum of the $n$-body system, so that the $XY$ plane is now the invariable plane. There are two integrals for the angular momentum; that is, the angular momentum around $OX'$ is zero and the angular momentum around $OZ'$ is equal to $-K \cos x_{3n-3}$; that is,

$$\begin{aligned} y_{3n-3} &= 0, \\ -K \cos x_{3n-3} &= \sum_{\nu=1}^{n-3} \begin{vmatrix} x_{3\nu-2} & x_{3\nu-1} \\ y_{3\nu-2} & y_{3\nu-1} \end{vmatrix} + \begin{vmatrix} x_{3n-8} & x_{3n-7} \\ y_{3n-8} & y_{3n-7} \end{vmatrix} \\ &\quad + \begin{vmatrix} x_{3n-6} & x_{3n-5} \\ y_{3n-6} & y_{3n-5} \end{vmatrix}. \end{aligned} \tag{81}$$

Let $H'$ be the result of substituting these values of $x_{3n-3}$ and $y_{3n-3}$ in $H$ and let one of these variables be $s$. Then,

$$\frac{\partial H'}{\partial s} = \frac{\partial H}{\partial s} + \frac{\partial H}{\partial x_{3n-3}} \frac{\partial x_{3n-3}}{\partial s}.$$

But, since $y_{3n-3} = 0$ and $\partial H/\partial x_{3n-3} + dy_{3n-3}/dt = 0$, therefore

$$\frac{\partial H'}{\partial s} = \frac{\partial H}{\partial s}.$$

The reduced differential equations are now

$$\frac{dx_i}{dt} = \frac{\partial H'}{\partial y_i}, \qquad \frac{dy_i}{dt} = -\frac{\partial H'}{\partial x_i} \qquad (i = 1, 2, \ldots, 3n - 5), \tag{82}$$

$$
\begin{aligned}
H' = {} & \sum_{i=1}^{3n-9} \frac{y_i^2}{2m_i} + \frac{1}{2m_{3n-6}} (y_{3n-8}^2 + y_{3n-7}^2 + \Gamma^2) \\
& + \frac{1}{2m_{3n-3}} (y_{3n-6}^2 + y_{3n-5}^2 + \Gamma'^2) \\
& + \frac{1}{2m_{3n}} \Bigg[ \Bigg( {\sum_{\nu=1}^{n-3}}' y_{3\nu-2} + y_{3n-8} + y_{3n-6} \Bigg)^2 \\
& \qquad + \Bigg( \sum_{\nu=1}^{n-3} y_{3\nu-1}^2 + y_{3n-7}^2 + y_{3n-5}^2 \Bigg) + \Bigg( \sum_{\nu=1}^{n-3} y_{3\nu} + \Gamma + \Gamma' \Bigg)^2 \Bigg] \\
& - {\sum_{\substack{\mu,\nu=1 \\ (\mu \neq \nu)}}^{n-3}}' m_{3\mu} m_{3\nu} [(x_{3\mu-2} - x_{3\nu-2})^2 \\
& \qquad + (x_{3\mu-1} - x_{3\nu-1})^2 + (x_{3\mu} - x_{3\nu})^2]^{-1/2} \\
& - m_{3n-6} \sum_{\nu=1}^{n-3} m_{3\nu} [(x_{3\nu-2} - x_{3n-8})^2 + (x_{3\nu-1} - x_{3n-7})^2 + x_{3\nu}^2]^{-1/2} \\
& - m_{3n-3} \sum_{\nu=1}^{n-3} m_{3\nu} [(x_{3\nu-2} - x_{3n-6})^2 + (x_{3\nu-1} - x_{3n-5})^2 + x_{3\nu}^2]^{-1/2} \\
& - m_{3n} \Bigg[ \sum_{\nu=1}^{n-3} m_{3\nu} (x_{3\nu-2}^2 + x_{3\nu-1}^2 + x_{3\nu}^2)^{-1/2} \\
& \qquad + m_{3n-3} (x_{3n-8}^2 + x_{3n-7}^2)^{-1/2} \\
& \qquad + m_{3n-3} (x_{3n-6}^2 + x_{3n-5}^2)^{-1/2} \Bigg] \\
& - m_{3n-3} m_{3n-6} [(x_{3n-8} - x_{3n-6})^2 + (x_{3n-7} - x_{3n-5})^2]^{-1/2}.
\end{aligned}
$$

By (79), the expressions for $\Gamma$ and $\Gamma'$ in the new variables become

$$
\Gamma x_{3n-7} + \Gamma' x_{3n-5} = -\sum_{\nu=1}^{n-3} \begin{vmatrix} x_{3\nu-1} & x_{3\nu} \\ y_{3\nu-1} & y_{3\nu} \end{vmatrix}, \tag{83}
$$

$$
\begin{aligned}
\Gamma x_{3n-8} + \Gamma' x_{3n-6} &= K \operatorname{cosec} x_{3n-3} + \sum_{\nu=1}^{n-3} \begin{vmatrix} x_{3\nu} & x_{3\nu-2} \\ y_{3\nu} & y_{3\nu-2} \end{vmatrix}, \\
&= \sum_{\nu=1}^{n-3} \begin{vmatrix} x_{3\nu} & x_{3\nu-2} \\ y_{3\nu} & y_{3\nu-2} \end{vmatrix} \\
&\quad + \Bigg\{ K^2 - \Bigg[ \sum_{\nu=1}^{n-3} \begin{vmatrix} x_{3\nu-2} & x_{3\nu-1} \\ y_{3\nu-2} & y_{3\nu-1} \end{vmatrix} + \begin{vmatrix} x_{3n-8} & x_{3n-7} \\ y_{3n-8} & y_{3n-7} \end{vmatrix} \\
&\qquad + \begin{vmatrix} x_{3n-6} & x_{3n-5} \\ y_{3n-6} & y_{3n-5} \end{vmatrix} \Bigg]^2 \Bigg\}^{1/2}.
\end{aligned}
$$

Thus the order of the equations of motion has been reduced to $6n - 10$.

The next step is to reduce the order to $6n - 12$, by using the energy integral and the elimination of time. This is the final stage of the reduction — using all the known first integrals — as has been shown in Section 4.10.

## ELLIPTIC MOTIONS

### 5.9 MOTION IN CONIC SECTIONS

Consider an $n$-body problem where one of the $n$ bodies has a mass much greater than the mass of any one of the other bodies. We discuss two cases. First, we investigate the motion of one of the bodies of smaller mass around the body of larger mass; the effects of the remaining bodies can then be regarded as small perturbations superposed on the motion in the two-body problem. Second, we will inquire into the case where the distance between two of the $n$ bodies is very small compared with their distances from the other bodies; the motion of these two can then be treated as a two-body problem, and the effect of the other bodies is considered a small superposed perturbation. The first case corresponds to the motion of the planets in the solar system; the second case to the motion of a satellite (including the moon and artificial earth satellites) around the planet to which the satellite belongs. We start with the rigorous solution of the two-body problem with two finite masses. The solution of the two-body problem is a motion in a conic section, that is, a Keplerian motion along a plane quadratic curve. It must be admitted that Keplerian motion is often not close enough even to serve as a first approximation to the actual complicated motions observed by modern techniques; so that we usually consider a more elaborate motion, as the first approximation. Nevertheless, here we shall begin by considering Keplerian motion, as it has long been the traditional first approximation in elementary theory.

Suppose that a mass $m_1$ is at the coordinate origin. The kinetic energy and the force function for the motion of a mass $m_2$ around the mass $m_1$ expressed in polar coordinates are

$$T = \frac{m}{2}\left[\left(\frac{dr}{dt}\right)^2 + r^2\left(\frac{d\phi}{dt}\right)^2 + r^2\cos^2\phi\left(\frac{d\theta}{dt}\right)^2\right],$$

$$U = \frac{\beta}{r}, \qquad \beta = k^2 m, \qquad m = m_1 + m_2.$$

The corresponding momenta are

$$p_r = \frac{\partial T}{\partial \dot{r}} = m\frac{dr}{dt}, \qquad p_\phi = \frac{\partial T}{\partial \dot{\phi}} = mr^2\left(\frac{d\phi}{dt}\right), \qquad p_\theta = \frac{\partial T}{\partial \dot{\theta}} = mr^2\cos^2\phi\,\frac{d\theta}{dt}.$$

Hence, if we write $m\beta$ and $mh_1$ respectively for $\beta$ and $h_1$, the Hamilton-Jacobi partial differential equation becomes

$$\left(\frac{\partial W}{\partial r}\right)^2 + \frac{1}{r^2}\left(\frac{\partial W}{\partial \phi}\right)^2 + \frac{1}{r^2 \cos^2 \phi}\left(\frac{\partial W}{\partial \theta}\right)^2 = \frac{2\beta}{r} + 2h_1.$$

By separation of variables (Section 2.5) we obtain

$$r^2\left[\left(\frac{\partial W}{\partial r}\right)^2 - \frac{2\beta}{r} - 2h_1\right] = -\left(\frac{\partial W}{\partial \phi}\right)^2 - \frac{1}{\cos^2 \phi}\left(\frac{\partial W}{\partial \theta}\right)^2.$$

As the left-hand member depends only on $r$ while the right-hand member does not depend on $r$, both members must equal a constant in order to be equal. Let the constant be $-h_2^2$; then,

$$\cos^2 \phi\left[\left(\frac{\partial W}{\partial \phi}\right)^2 - h_2^2\right] = -\left(\frac{\partial W}{\partial \theta}\right)^2,$$

$$\left(\frac{\partial W}{\partial r}\right)^2 = \frac{2\beta}{r} + 2h_1 - \frac{h_2^2}{r^2}.$$

Similarly, by equating both sides of the equation in $\phi$ and $\theta$ to a constant $-h_3^2$, we obtain

$$\left(\frac{\partial W}{\partial \phi}\right)^2 = h_2^2 - \frac{h_3^2}{\cos^2 \phi}, \qquad \left(\frac{\partial W}{\partial \theta}\right)^2 = h_3^2.$$

Hence, by the method treated in Section 1.12, we obtain

$$dt = \frac{dr}{\sqrt{\dfrac{2\beta}{r} + 2h_1 - \dfrac{h_2^2}{r^2}}},$$

$$0 = h_2 \frac{d\phi}{\sqrt{h_2^2 - \dfrac{h_3^2}{\cos^2 \phi}}} - h_2 \frac{dr}{r^2\sqrt{\dfrac{2\beta}{r} + 2h_1 - \dfrac{h_2^2}{r^2}}}, \tag{84}$$

$$0 = d\theta - h_3 \frac{d\phi}{\cos^2 \phi\sqrt{h_2^2 - \dfrac{h_3^2}{\cos^2 \phi}}},$$

and

$$W = \int_{r_0}^{r} \sqrt{\frac{2\beta}{r} + 2h_1 - \frac{h_2^2}{r^2}}\, dr + \int_0^\theta h_3\, d\theta + \int_0^\phi \sqrt{h_2^2 - \frac{h_3^2}{\cos^2 \phi}}\, d\phi.$$

Put

$$\frac{1}{r} = \rho, \qquad U(\rho) = 2h_1 + 2\beta\rho - h_2^2\rho^2,$$

$$h_3 = h_2 \cos I, \qquad \sin \phi = \sin I \sin \psi; \tag{85}$$

then, with the integration constants $\gamma_1$, $\gamma_2$, $\gamma_3$, equations (84) take the form

$$\begin{aligned}\gamma_1 + t &= -\int \frac{d\rho}{\rho^2\sqrt{U(\rho)}},\\ \gamma_2 &= \psi + h_2 \int \frac{d\rho}{\sqrt{U(\rho)}},\\ \gamma_3 &= \theta - \tan^{-1}(\tan\psi\cos I).\end{aligned} \tag{84a}$$

From the first and the third equations of (84a), we obtain the law of conservation of angular momentum, or the Keplerian second law

$$r^2 \frac{d\psi}{dt} = h_2, \tag{84b}$$

where $h_2$ is twice the areal velocity. The third equation of (84a) gives

$$\tan\psi\cos I = \tan(\theta - \gamma_3). \tag{86}$$

If we take the *ecliptic* as our reference plane, and the *vernal equinox* ♈ of an epoch as the standard direction — that is, if we take a certain fixed epoch as reference — then $\theta$ is the *longitude*, $\phi$ is the *latitude*, and (85) and (86) give their relations in a right-angled spherical triangle. The motion occurs always on the plane having *inclination* $I$ to the ecliptic and intersecting it at a point of longitude $\gamma_3$. The variation of $r$ and $\psi$ represents the motion. $\psi$ is called the *argument of latitude* (Figure 1). Hence, the problem is reduced to the discussion of

$$\begin{aligned}\gamma_1 + t &= -\int \frac{d\rho}{\rho^2\sqrt{U(\rho)}},\\ \gamma_2 - \psi &= h_2 \int \frac{d\rho}{\sqrt{U(\rho)}}.\end{aligned} \tag{87}$$

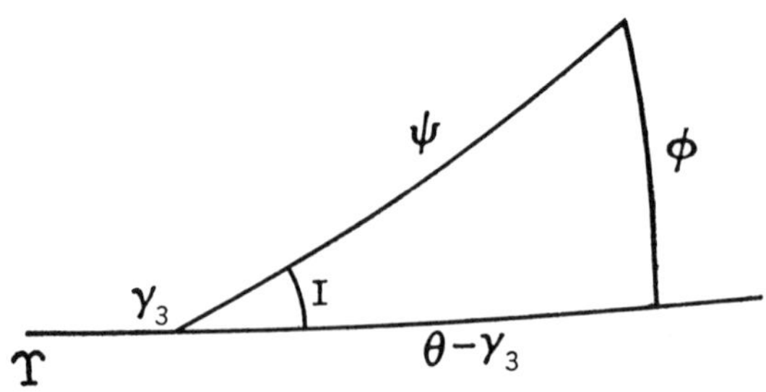


FIGURE 1

Consider

$$U(\rho) = 2h_1 + 2\beta\rho - h_2^2\rho^2. \tag{88}$$

If $U(\rho) = 0$ has no real root, then $U(\rho) > 0$ for a real motion. If $U(\rho) = 0$ has real roots $\rho_1, \rho_2$ ($\rho_1 \geqq \rho_2$), then $U(\rho) = h_2^2(\rho_1 - \rho)(\rho - \rho_2)$ and $\rho$ varies between $\rho_1$ and $\rho_2$. According as $h_1 > 0, h_1 < 0, h_1 = 0$, we have $\rho_1\rho_2 < 0$, $\rho_1\rho_2 > 0$, $\rho_1\rho_2 = 0$.

If $\rho_1\rho_2 < 0$, then one root is positive and the other negative and the motion extends to infinity, as $\rho = 1/r \geqq 0$. But, by (87), it is an asymptotic motion where $\rho$ approaches zero as $t$ approaches infinity. We have also $t = O(\rho^{-1})$ and $r = O(t)$. This is a hyperbolic motion.

If $\rho_1\rho_2 > 0$, then the two roots are real. The motion occurs in

$$r_1 \leqq r \leqq r_2,$$

where $r_1 = 1/\rho_1$, $r_2 = 1/\rho_2$. $r_1$ is the *perihelion distance* and $r_2$ is the *aphelion distance*; this is an *elliptic motion* or a *circular motion.*

If $\rho_1\rho_2 = 0$, then one root is zero. Let $\rho_2 = 0$; then, by the first equation of (87), this is an asymptotic motion where $\rho$ approaches zero. We have $t = O(\rho^{-3/2})$ or $t = O(r^{2/3})$. This is a *parabolic motion.*

Consider in particular a straight-line motion. Suppose that $h_2 = 0$. The motion occurs on the straight line $\psi = \gamma_2$. From the first equation of (87) we obtain

$$\left(\frac{dr}{dt}\right)^2 = \frac{2\beta}{r} + 2h_1.$$

If $h_1 > 0$, then $r$ must monotonically increase or decrease, as the right-hand member does not vanish for $h_1 > 0$; that is, the two bodies either recede to infinity or collide. Let $h_1 = -f < 0$. Let variable $u$ be defined by

$$\left(\frac{du}{dt}\right)^2 = \frac{\epsilon}{r}; \tag{89}$$

then

$$\left(\frac{dr}{du}\right)^2 = \frac{2f}{\epsilon}\left(\frac{\beta}{f} - r\right), \tag{90}$$

where $\epsilon$ is a positive constant to be determined later. Let $u$ increase monotonically with $t$. Put $\epsilon = f^2/(2\beta)$ and integrate (90); then

$$r = \frac{\beta}{f}(1 - u^2). \tag{91}$$

Substitution of (91) in (89) gives

$$u\sqrt{1 - u^2} + \sin^{-1} u = \frac{\sqrt{2}f^{3/2}}{\beta}(t - t_0). \tag{92}$$

Equations (91) and (92) are the solution for the straight-line case. $u < 0$ for $t < t_0$. The variable $u$ increases from zero at first, until $r$ attains its maximum, $\beta/f$. After this, $r$ decreases and then vanishes; that is, the two bodies collide in a finite time $t = t_0 + \beta\pi/(2\sqrt{2}f^{3/2})$.

Let the instant of the collision be $t = 0$ and let $r = \sqrt[3]{\beta}\, w$. Then the equation of motion becomes

$$\frac{d^2w}{dt^2} = -\frac{1}{w^2}.$$

Substitute $w = \gamma t^m$ in this equation; then $m = \frac{2}{3}$. Hence,

$$w = \sqrt[3]{\frac{9}{2}}\, t^{2/3} = \tau \tag{93}$$

is a particular solution. In order to obtain the general solution, put

$$w = \tau + \alpha_2\tau^2 + \alpha_3\tau^3 + \cdots$$

in the equation of motion and determine $\alpha_2, \alpha_3, \ldots$. This gives the general solution

$$r = \sqrt[3]{\beta}\,(\tau + \alpha_2\tau^2 - \tfrac{3}{7}\alpha_2^2\tau^2 + \tfrac{2}{6}\tfrac{3}{3}\alpha_2^3\tau^4 + \cdots). \tag{94}$$

$\alpha_2$ is one of the integration constants, such that $h_1 = 5\beta^{2/3}\alpha_2$. The other integration constant is the time of collision. This treatment is due to Burrau (1894).

An important fact known in this particular case of collision is that the mutual distance $r$ is expressed in the form of an integral power series of $t^{2/3}$, $t$ being counted backward from the instant of collision. According to Sundman (1913), a similar situation occurs in the collision of two bodies in the three-body problem, as will be discussed in Chapter 22. The differential equation remains unaltered when we change the sign of $t$. Hence, if $t$ denotes the time interval elapsed since the instant of collision, the solution can likewise be represented by the same integral power series of $t^{2/3}$. Mathematically, the solution can be *analytically continued* after the collision. Sundman shows that this is true for the three-body problem. It is also true for the isosceles-triangular solution treated in Section 3.9.

## 5.10 ELLIPTIC MOTIONS

An elliptic motion takes place for $\rho_1\rho_2 > 0$ and accordingly for $h_1 = -f < 0$. By Section 2.4, we have

$$\frac{dr}{du} = \sqrt{(r_2 - r)(r - r_1)}, \tag{95}$$

$$\frac{du}{dt} = \frac{h_2}{r}\sqrt{\rho_1\rho_2}, \tag{96}$$

or, by integrating (95),

$$\frac{1}{\rho} = r = \frac{r_1 + r_2}{2} - \frac{r_2 - r_1}{2}\cos u. \tag{97}$$

On the other hand, from (84b) we obtain

$$r^2\frac{d\psi}{dt} = h_2 = \frac{1}{\rho^2}\frac{d\psi}{dt}. \tag{98}$$

Solve (98) together with $d\rho/dt = -\rho^2 h_2\sqrt{(\rho_1 - \rho)(\rho - \rho_2)}$ and denote the integration constant by $\pi/2 + \varpi - \Omega$; then

$$\frac{\pi}{2} - \psi + \varpi - \Omega = \sin^{-1}\left[\left(\rho - \frac{\rho_1 + \rho_2}{2}\right)\Big/\left(\frac{\rho_1 - \rho_2}{2}\right)\right].$$

Put

$$\frac{1}{2}(\rho_1 + \rho_2) = \frac{1}{2}\left(\frac{1}{r_1} + \frac{1}{r_2}\right) = \frac{1}{p}, \qquad r_1 = a(1 - e), \qquad r_2 = a(1 + e),$$

then, $a\sqrt{1 - e^2} = p$ and the integral is

$$r = \frac{p}{1 + e\cos(\psi - \varpi + \Omega)}. \tag{99}$$

The constant $\Omega$ is $\gamma_3$ in the former notation. $\varpi$ is the value of $\psi + \Omega$ for $r = r_1$. Note that the counting of the angles is over two different planes, that is, $\Omega$ is counted on the ecliptic and $\psi$ on the orbital plane. $\varpi$ is called the *longitude of the perihelion*, $\Omega$ the *longitude of the ascending node*. In the case of asteroids and comets, $\omega$ is often used for $\varpi - \Omega$. Equation (99) is the equation for an ellipse and expresses the first law of Kepler; $e$ is the *eccentricity*, $a$ the *semimajor axis*, and $u$ (equation 97) the *eccentric anomaly*.

Solving (97), we obtain

$$r = a(1 - e\cos u). \tag{100}$$

Write (96) as $r\,du = \sqrt{\rho_1\rho_2}\,h_2\,dt$. Substitute this in (100); then,

$$u - e\sin u = \frac{\sqrt{\rho_1\rho_2}\,h_2}{a}(t - t_0).$$

From $h_2^2 = a(1 - e^2)\beta$, $h_1 = -\beta/2a$, we obtain $\sqrt{\rho_1\rho_2}\,h_2/a = \sqrt{\beta}/a^{3/2}$. Also $dt = \sqrt{a/\mu}\,r\,du$, by (96). The period with respect to $u$ is

$$T = \int_0^{2\pi} a(1 - e\cos u)\,du = \frac{2\pi a^{3/2}}{\sqrt{\beta}}.$$

Hence, if we put $n = 2\pi/T = \sqrt{\beta}/a^{3/2}$, we have $n^2a^3 = \beta$. This is the third law of Kepler. $n$ is called the *mean motion*. Thus

$$u - e\sin u = n(t - t_0) = M. \tag{101}$$

This is *Kepler's equation*. $M$ is the *mean anomaly* and $t_0$ *the epoch of the perihelion passage*. From (98) we have $r^2\,d\psi/dt = na^2\sqrt{1 - e^2}$. Integrate this equation by writing $\psi = v + \varpi - \Omega$ and substituting (100) for $r$; then

$$\tan\frac{v}{2} = \sqrt{\frac{1 + e}{1 - e}}\tan\frac{u}{2}. \tag{102}$$

$v$ is called the *true anomaly*. $M$, $u$, and $v$ all equal zero at the perihelion and $\pi$ at the aphelion, and have a common period $T$.

$$r\cos v = a(\cos u - e), \qquad r\sin v = a\sqrt{1 - e^2}\sin u. \tag{103}$$

$\mathscr{L} = \psi + \Omega = v + \varpi$ is called the *longitude in the orbit*. It is customary to write

$$M = nt + \epsilon - \varpi, \qquad \lambda = nt + \epsilon = M + \varpi. \tag{104}$$

$\lambda$ is the *mean longitude* or the *mean longitude in the orbit*, $\epsilon$ the *mean longitude at the epoch*, that is, for $t = 0$. These angles are counted over two different planes, just like the angle $\varpi$.

For the ecliptic we have rectangular coordinates $x_1$, $x_2$, $x_3$ with the *vernal equinox* as the $x_1$ direction, and

$$\begin{aligned} x_1 &= r(\cos\psi\cos\Omega - \sin\psi\sin\Omega\cos I),\\ x_2 &= r(\cos\psi\sin\Omega + \sin\psi\cos\Omega\cos I),\\ x_3 &= r\sin\psi\sin I. \end{aligned} \tag{105}$$

Substitute in (105) these expressions (105a) for $r\cos\psi$ and $r\sin\psi$:

$$\begin{aligned} r\cos\psi &= r\cos v\cos(\varpi - \Omega) - r\sin v\sin(\varpi - \Omega),\\ r\sin\psi &= r\cos v\sin(\varpi - \Omega) + r\sin v\cos(\varpi - \Omega); \end{aligned} \tag{105a}$$

but first substitute in (105a) the expressions for $r \cos v$ and $r \sin v$ given in (103). We thus obtain

$$\begin{aligned} x_1 &= a(\cos u - e)[\cos(\varpi - \Omega)\cos\Omega - \sin(\varpi - \Omega)\sin\Omega\cos I] \\ &\quad - a\sqrt{1-e^2}\sin u[\sin(\varpi - \Omega)\cos\Omega + \cos(\varpi - \Omega)\sin\Omega\cos I], \\ x_2 &= a(\cos u - e)[\cos(\varpi - \Omega)\sin\Omega + \sin(\varpi - \Omega)\cos\Omega\cos I] \\ &\quad + a\sqrt{1-e^2}\sin u[-\sin(\varpi - \Omega)\sin\Omega + \cos(\varpi - \Omega)\cos\Omega\cos I], \\ x_3 &= a(\cos u - e)\sin(\varpi - \Omega)\sin I + a\sqrt{1-e^2}\sin u\cos(\varpi - \Omega)\sin I; \end{aligned}$$

or, if we let

$$\begin{aligned} X &= r\cos(v - M) \\ &= a(\cos u - e)\cos M + a\sqrt{1-e^2}\sin u\sin M, \\ Y &= r\sin(v - M) \\ &= -a(\cos u - e)\sin M + a\sqrt{1-e^2}\sin u\cos M, \end{aligned} \tag{106}$$

we obtain

$$\begin{aligned} a(\cos u - e) &= X\cos M - Y\sin M, \\ a\sqrt{1-e^2}\sin u &= X\sin M + Y\cos M; \end{aligned}$$

accordingly,

$$\begin{aligned} x_1 &= X[\cos(M + \varpi - \Omega)\cos\Omega - \sin(M + \varpi - \Omega)\sin\Omega\cos I] \\ &\quad - Y[\sin(M + \varpi - \Omega)\cos\Omega + \cos(M + \varpi - \Omega)\sin\Omega\cos I], \\ x_2 &= X[\cos(M + \varpi - \Omega)\sin\Omega + \sin(M + \varpi - \Omega)\cos\Omega\cos I] \\ &\quad + Y[-\sin(M + \varpi - \Omega)\sin\Omega + \cos(M + \varpi - \Omega)\cos\Omega\cos I], \\ x_3 &= X\sin(M + \varpi - \Omega)\sin I + Y\cos(M + \varpi - \Omega)\sin I; \end{aligned} \tag{107}$$

or

$$\begin{aligned} x_1 &= X\left[\cos^2\frac{I}{2}\cos\lambda + \sin^2\frac{I}{2}\cos(\lambda - 2\Omega)\right] \\ &\quad - Y\left[\cos^2\frac{I}{2}\sin\lambda + \sin^2\frac{I}{2}\sin(\lambda - 2\Omega)\right], \\ x_2 &= X\left[\cos^2\frac{I}{2}\sin\lambda - \sin^2\frac{I}{2}\sin(\lambda - 2\Omega)\right] \\ &\quad + Y\left[\cos^2\frac{I}{2}\cos\lambda - \sin^2\frac{I}{2}\cos(\lambda - 2\Omega)\right], \\ x_3 &= X\sin I\sin(\lambda - \Omega) + Y\sin I\cos(\lambda - \Omega). \end{aligned} \tag{108}$$

The constants appearing in (84) are related to the *Keplerian constants* $a$, $e$, $I$, $\Omega$, $\varpi$, $t_0$ by

$$h_1 = -\frac{\beta}{2a}, \qquad h_2 = \sqrt{\beta a(1 - e^2)}, \qquad h_3 = \sqrt{\beta a(1 - e^2)} \cos I, \tag{109}$$

$$\gamma_1 = -t_0, \qquad \gamma_2 = \varpi - \Omega, \qquad \gamma_3 = \Omega.$$

The constants $h_1$, $h_2$, $h_3$, $\gamma_1$, $\gamma_2$, $\gamma_3$ are determined by the method of Hamilton-Jacobi and are called the *canonical constants.* $h$ and $\gamma$ are canonically conjugate.

## 5.11 CAUCHY'S FIRST THEOREM

Suppose that $S$ can be expanded in a Laurent series for all values of $z$ in the ring domain, excluding 0 and $\infty$, of a complex variable $z$ and for all values of $e$ in a domain enclosing $e = 0$:

$$S = \sum_{j=-\infty}^{+\infty} P_j z^j. \tag{110}$$

Then, by Cauchy's theorem on residues in the theory of analytic functions, we have

$$P_j = \frac{1}{2\pi\sqrt{-1}} \int^{(0+)} z^{-j-1} S\, dz \qquad (j \gtreqless 0) \tag{111}$$

for a closed integration contour surrounding the point $O$ in the positive sense. Take the variable $s$ such that

$$z = s \exp\left[-\frac{e}{2}(s - s^{-1})\right]. \tag{112}$$

The right-hand member is convergent for all values in the domain enclosing $e = 0$; indeed, it can be developed in a uniformly convergent Laurent series in the ring domain of $s$, excluding $s = 0$ and $s = \infty$. Then,

$$dz = \left[1 - \frac{e}{2}(s + s^{-1})\right] \exp\left[-\frac{e}{2}(s - s^{-1})\right] ds. \tag{113}$$

The derivative $dz/ds$ does not become infinite except for $s = 0$ or $\infty$. Hence,

$$P_j = \frac{1}{2\pi\sqrt{-1}} \int^{(0+)} s^{-j-1} U\, ds, \tag{114}$$

where

$$U = S\left[1 - \frac{e}{2}(s + s^{-1})\right] \exp\left[\frac{je}{2}(s - s^{-1})\right]. \tag{114a}$$

The integration contour for $s$ is obtained by the transformation (112) of the corresponding integration contour for $z$. But, as there is no singularity, the contour for $s$ can be deformed to the unit circle around $O$. Thus,

$$U = \sum_{j=-\infty}^{\infty} P_j s^j.$$

Since

$$P_j = \frac{1}{2\pi\sqrt{-1}} \int^{(0+)} \frac{1}{j} z^{-j} \frac{dS}{dz} dz,$$

which, by (112),

$$= \frac{1}{2\pi\sqrt{-1}\, j} \int^{(0+)} s^{-j} \exp\left[\frac{je}{2}(s - s^{-1})\right] \frac{dS}{ds} ds,$$

therefore,

$$P_j = \frac{1}{2\pi\sqrt{-1}} \int^{(0+)} s^{-j} V\, ds, \tag{115}$$

where

$$V = \frac{1}{j} \frac{dS}{ds} \exp\left[\frac{je}{2}(s - s^{-1})\right]. \tag{115a}$$

The result (114) and (115) is called *Cauchy's first theorem*. In other words, take the unit circles around $O$ as the integration contour both for $z$ and for $s$, and put

$$z = \exp(\sqrt{-1}\, M), \qquad s = \exp(\sqrt{-1}\, u);$$

then a coefficient of the Fourier series expansion of $S$ in the mean anomaly $M$ equals the corresponding coefficient of the Fourier series expansion of the function $U$ defined by (114a) in the eccentric anomaly $u$, and equals the corresponding coefficient of the Fourier series expansion of $Vs$ defined by (115a) in $u$.

As an example, take

$$S = \frac{a}{r} = (1 - e\cos u)^{-1} = \left[1 - \frac{e}{2}(s + s^{-1})\right]^{-1}.$$

Because of (114a), we have

$$U = \exp\left[\frac{je}{2}(s - s^{-1})\right]$$

and

$$P_j = P_{-j} = J_j(je)$$

with the Bessel function. Therefore

$$\frac{a}{r} = 1 + 2\sum_{j=1}^{\infty} J_j(je) \cos jM.$$

Note that the singularities

$$s_1 = \frac{e}{1 - \sqrt{1 - e^2}} \equiv \frac{1}{\omega} > 1$$

and

$$s_2 = \frac{1 - \sqrt{1 - e^2}}{e} \equiv \omega < 1,$$

obtained from

$$1 - \frac{e}{2}(s + s^{-1}) = 0,$$

occur in the transition from the complex $s$ plane to the complex $z$ plane by the condition $ds/dz = \infty$. The corresponding points in the $z$ plane are $z_1 = (1/\omega)\exp(-\sqrt{1 - e^2})$ and $z_2 = \omega \exp(\sqrt{1 - e^2})$, $z_1 > 1 > z_2$, where $dz/ds = 0$. The expansions of $z - z_1$ and $z - z_2$ in the neighborhood of these points begin with the terms $(s - s_1)^2$ and $(s - s_2)^2$, respectively. On the other hand, the expansions of $s - s_1$ and $s - s_2$ are in integral power series of $\sqrt{z - z_1}$ and $\sqrt{z - z_2}$, respectively.

Let $j$, $q$ be positive integers or zero and $p$ be a positive or negative integer or zero, and consider

$$\left(x + \frac{1}{x}\right)^j \left(x - \frac{1}{x}\right)^q = \sum_{p=-\infty}^{\infty} N_{-p,j,q} x^p. \tag{116}$$

$N_{-p,j,q}$ is called a *Cauchy number*, and we have

$$\begin{aligned} N_{-p,j,q} &= \frac{1}{2\pi\sqrt{-1}} \int^{(0+)} x^{-p-1}\left(x + \frac{1}{x}\right)^j \left(x - \frac{1}{x}\right)^q dx \\ &= \frac{1}{2\pi\sqrt{-1}} \int^{(0+)} (-1)^q x^{-p-j-q-1}(1 + x^2)^j (1 - x^2)^q \, dx \qquad (117) \\ &= \frac{1}{2\pi\sqrt{-1}} \int^{(0+)} x^{-p+j+q-1}\left(1 + \frac{1}{x^2}\right)^j \left(1 - \frac{1}{x^2}\right)^q dx. \end{aligned}$$

Insert cross cuts between $\infty$ and $-1$, between 0 and $+1$, between $+i$ and $+i\infty$, and between $-i$ and $-i\infty$, and let the integration contour surround the cross cut between 0 and $+1$. It is easily seen that $N_{-p,j,q} = 1$ for $-p + j + q = 0$, $N_{-p,j,q} = (-1)^q$ for $-p - j - q = 0$, $N_{-p,j,q} = 0$ for $-p + j + q < 0$ or $-p - j - q > 0$, and $N_{-p,j,q} = 0$ for an odd $-p + j + q > 0$ and for an odd $-p - j - q < 0$.

By formal substitution $x = 1/x'$ in (117) we obtain

$$N_{-p,j,-q} = (-1)^q N_{p,j,q}.$$

From

$$x^{-p-(j+1)-q-1}(1 + x^2)^{j+1}$$
$$\equiv x^{-p+1-j-q-1}(1 + x^2)^j + x^{-p-1-j-q-1}(1 + x^2)^j,$$

we obtain

$$N_{-p,j+1,q} = N_{-p+1,j,q} + N_{-p-1,j,q},$$
$$N_{-p,j,q+1} = N_{-p+1,j,q} - N_{-p-1,j,q}.$$

In particular, we can compute $N_{p,1,q}$ when we know $N_{p,0,q}$, because $N_{-p,1,q} = N_{-p+1,0,q} + N_{-p-1,0,q}$, and we can compute $N_{-p,j,q}$ successively by the above recurrence formula. The numerical values were given by Bourget (1863).

By formal substitution $x^2 = y$ in (117), we obtain

$$N_{-p,0,q} = \frac{(-1)^q}{2\pi\sqrt{-1}} \int^{(0+)} y^{-(p+q)/2-1}(1 - y)^q\, dy.$$

Since $q > 0$, $y = 1$ is an ordinary point, but $y = 0$ is a singular point. Consider an integration contour that surrounds $y = 0$ in the positive sense, extends from $y = 0$ to $y = +1$ along the real axis, surrounds $y = 1$ in the positive sense and returns to $y = 0$ along the real axis. If the integral taken from $y = 0$ to $y = 1$ be 1, then the integral taken from $y = 1$ to $y = 0$ is equal to

$$\exp\left[-2\pi\sqrt{-1}\left(-\frac{p+q}{2} - 1\right)\right],$$

because the contour surrounds $y = 0$ once in the positive sense (see Figure 2, Section 5.12). Hence, using Gauss's beta function $B(\xi, \eta)$ we obtain

$$N_{-p,0,q} = (-1)^q \frac{1 - \exp\{-2\pi\sqrt{-1}\,[(p+q)/2]\}}{2\pi\sqrt{-1}}$$
$$\times \int_0^1 y^{-(p+q)/2-1}(1 - y)^q\, dy$$
$$= (-1)^q(-1)^{-(p+q)/2} \frac{\sin[\pi(p+q)/2]}{\pi} B\left(-\frac{p+q}{2}, q + 1\right)$$
$$= (-1)^{(q-p)/2} \frac{1}{\Gamma[-(p+q)/2]\Gamma[1 + (p+q)/2]}$$
$$\times \frac{\Gamma[-(p+q)/2]\Gamma(q+1)}{\Gamma[-(p-q)/2 + 1]}$$
$$= (-1)^{(q-p)/2} \frac{q!}{[(q+p)/2]!\,[(q-p)/2]!}.$$

Similarly, we can prove that

$$N_{-p,j,0} = \frac{j!}{[(j+p)/2]!\,[(j-p)/2]!}.$$

Substitute $x = \exp(\sqrt{-1}\,\theta)$ in

$$N_{0,j,q} = \frac{1}{2\pi\sqrt{-1}} \int^{(0+)} x^{-1}\left(x + \frac{1}{x}\right)^j \left(x - \frac{1}{x}\right)^q dx;$$

then

$$N_{0,j,q} = \frac{1}{2\pi} \int_0^{2\pi} \cos^j \theta \sin^q \theta \, d\theta \cdot z^{j+q}(-1)^{q/2}.$$

Divide the integration path into the path from 0 to $\pi$ and the path from $\pi$ to $2\pi$, and write $\pi + \theta'$ instead of $\theta$ in the second integral. Then $j + q$ must be even. By dividing the integration path from 0 to $\pi$ into the paths from 0 to $\pi/2$ and $\pi/2$ to $\pi$, and writing $\pi/2 + \theta'$ instead of $\theta$ in the second integral, we see that $q$ must also be even. Hence,

$$\begin{aligned} N_{0,j,q} &= \frac{(-1)^{q/2}2^{j+q}}{\pi} 2 \int_0^{\pi/2} \cos^j \theta \sin^q \theta \, d\theta \\ &= \frac{(-1)^{q/2}2^{j+q}}{\pi} \frac{\Gamma[(j+1)/2]\Gamma[(q+1)/2]}{\Gamma[(j+q)/2+1]} \\ &= (-1)^{q/2} \frac{1\cdot 3\cdot 5\cdots(j-1)\cdot 1\cdot 3\cdot 5\cdots(q-1)}{[(j+q)/2]!}. \end{aligned}$$

This is a new formula due to Hagihara (1935).

For an application of Cauchy's numbers, consider

$$\frac{a^2}{r^2} = \frac{1}{2\sqrt{1-e^2}} + \sum_j c_j \cos jM.$$

We see that

$$c_j = 2 \sum_k \sum_q \frac{j^q}{q!} \left(\frac{e}{2}\right)^{k+q} N_{-j,k,q}.$$

Further, by applying this expansion to

$$\frac{dv}{dt} = \frac{a^2}{r^2}\sqrt{1-e^2},$$

we obtain

$$v - M = \sqrt{1-e^2} \sum_j c_j \frac{\sin jM}{j}.$$

This expression is called the *equation of center*. Thus the expansions in the Keplerian motion can be obtained simply by the use of Cauchy's first theorem and Cauchy's numbers.

Note that a function may have different mean values taken with respect to the mean anomaly $M$, the eccentric anomaly $u$, or the true anomaly $v$, respectively, although the three anomalies have equal periods. Kozai (1959, 1962) has given the formula (see also Tisserand, 1889):

$$\frac{1}{2\pi}\int_0^{2\pi} \cos jv\, dM = (-e)^j[1 + \sqrt{1 - e^2}]^{-j}[1 + j\sqrt{1 - e^2}]$$

$$= (-\omega)^j[1 + j\sqrt{1 - e^2}],$$

where

$$\omega = \frac{1 - \sqrt{1 - e^2}}{e} = \frac{e}{1 + \sqrt{1 - e^2}}.$$

Kozai has also given the formula

$$\gamma = \alpha + 2\sum_{j=1}^{\infty} a_j \sin j\alpha,$$

with

$$ja_j = \frac{1}{2\pi}\int_0^{2\pi} \cos j\alpha\, d\gamma \qquad (j \neq 0)$$

where $\alpha$ and $\gamma$ are any two of the three kinds of anomaly.

If $\alpha = u$ and $\gamma = M$, then, $M = u - e \sin u$,

$$\frac{1}{2\pi}\int_0^{2\pi} \cos ju\, dM = -\frac{e}{2} \qquad (j = 1),$$

$$= 0 \qquad (j > 1).$$

If $\alpha = M$ and $\gamma = v$, then the coefficient of $\sin jM$ in the equation of center can be expanded in a power series of $e$ (Tisserand, 1889).

If $\alpha = M$ and $\gamma = u$, then,

$$u = M + 2\sum_{j=1}^{\infty} \frac{1}{j} J(je) \sin jM,$$

and the mean value is

$$\frac{1}{2\pi}\int_0^{2\pi} \cos jM\, dv = J_j(je).$$

If $\alpha = u$ and $\gamma = v$, then

$$v = u + 2 \sum_{j=1}^{\infty} \frac{1}{j} \omega^j \sin ju,$$

and

$$\frac{1}{2\pi} \int_0^{2\pi} \cos ju \, dv = \omega^j.$$

If $\alpha = v$ and $\gamma = u$, then

$$u = v + 2 \sum_{j=1}^{\infty} \frac{1}{j} (-\omega)^j \sin jv,$$

and

$$\frac{1}{2\pi} \int_0^{2\pi} \cos jv \, du = (-\omega)^j.$$

Numerical tables of the coefficients in the expansions for elliptic motions have been published by Cayley (1861) and recently by Jarnagin (1965).

## 5.12 CAUCHY'S SECOND THEOREM

Put

$$t = \exp(\sqrt{-1}\, v), \qquad s = \exp(\sqrt{-1}\, u), \qquad z = \exp(\sqrt{-1}\, M) \tag{118}$$

and

$$\omega = \frac{e}{1 + \sqrt{1 - e^2}} = \frac{1 - \sqrt{1 - e^2}}{e}, \qquad \eta = \frac{1 + \sqrt{1 - e^2}}{2} = \frac{1}{1 + \omega^2},$$

$$\omega\eta = \frac{e}{2};$$

then

$$s = \frac{t + \omega}{1 + \omega t}, \qquad t = \frac{s - \omega}{1 - \omega s}.$$

This is a linear fractional transformation in the theory of modular functions. The point $s = 0$ is carried to $t = \infty$, and $s = \infty$ to $t = -1/\infty$, by this transformation. The invariant points $t = \pm 1$ are defined by

$$t = \frac{t + \omega}{1 + \omega t}.$$

We have

$$\frac{dt}{ds} = \frac{t}{s}\frac{1-\omega^2}{(1-\omega s)(1-\omega s^{-1})} = \frac{(1+\omega t)^2}{1-\omega^2} = \frac{1-\omega s+\omega\,(s-\omega)}{(1-\omega s)^2},$$

$$z = t\left(1+\frac{\omega}{t}\right)(1+\omega t)^{-1}\exp\left[-\frac{1-\omega^2}{1+\omega^2}\left(\frac{\omega t}{1+\omega t}-\frac{\omega t^{-1}}{1+\omega t^{-1}}\right)\right],$$

$$\frac{r}{a} = \frac{1}{1+\omega^2}(1-\omega s)(1-\omega s^{-1}) = \frac{(1-\omega^2)^2}{1+\omega^2}\frac{1}{(1+\omega t)(1+\omega t^{-1})},$$

$$\frac{ds}{dt} = \frac{1+\omega^2}{1-\omega^2}\frac{s}{t}\frac{r}{a},$$

$$\frac{dz}{dt} = \frac{1+\omega^2}{1-\omega^2}\frac{z}{t}\left(\frac{r}{a}\right)^2.$$

The variable $z$, regarded as a function of $s$, has singularities at $s = 0$ and $\infty$, and, as a function of $t$, has a simple pole at $t = \infty$ and an essential singularity at $t = \omega^{-1}$. Its reciprocal $1/z$, regarded as a function of $s$, has a simple pole at $t = 0$ and an essential singularity at $t = -\omega$. The radius vector $r$, regarded as a function of $s$, has no singularity, but, as a function of $t$, has simple poles at $t = -1/\omega$ and $-\omega$. The reciprocal $1/r$, regarded as a function of $s$, has simple poles at $s = \omega$ and $s = 1/\omega$, but, as a function of $t$, has no singularity. Hence, in the transition from the $z$ plane to the $t$ plane there are essential singularities at $t = -\omega$ and $t = -1/\omega$, so that we construct in the $t$ plane two cross cuts extending from $-\infty$ to $-1/\omega$ and from $-\omega$ to 0. The second cross cut is contained in the unit circle with $O$ as the center. Hence, as in Cauchy's first theorem, the coefficient $P_j$ in the expansion

$$S = \sum_{j=-\infty}^{\infty} P_j z^j$$

equals the coefficient of $t^j$ in the expansion of

$$\frac{1}{\sqrt{1-e^2}}\left(\frac{r}{a}\right)^2\left(\frac{t}{z}\right)^j S = \frac{(1-\omega^2)^3}{1+\omega^2}(1+\omega t)^{j-2}(1+\omega t^{-1})^{-j-2}$$
$$\times\exp\left[-j\frac{1-\omega^2}{1+\omega^2}\left(\frac{\omega t}{1+\omega t}\quad\frac{\omega t^{-1}}{1+\omega t^{-1}}\right)\right]S. \tag{119}$$

It also equals the coefficient of $t^j$ in the expansion of

$$\frac{t}{j}(1+\omega t)^j(1+\omega t^{-1})^{-j}\exp\left[j\frac{1-\omega^2}{1+\omega^2}\left(\frac{\omega t}{1+\omega t}-\frac{\omega t^{-1}}{1+\omega t^{-1}}\right)\right]\frac{dS}{dt}. \tag{120}$$

This is called *Cauchy's second theorem.*

In general, an expression of the form

$$f=\left(\frac{r}{a}\right)^{\rho}s^{\beta}t^{\gamma}=\left(\frac{r}{a}\right)^{\rho}[\cos(\beta u+\gamma v)+\sqrt{-1}\sin(\beta u+\gamma v)]$$

can be written

$$f=\sum_n a_n z^n=\sum_n b_n s^n=\sum_n c_n t^n,$$

where $\alpha$, $\beta$, $\gamma$ are arbitrary, with the proviso that $\beta+\gamma$ must be an integer so that $f$ will admit the period $2\pi$ with respect to one of the anomalies $M$, $u$, $v$. It can be seen (Andoyer, 1923) that

$$a_n=\sum_j b_j J_{n-1}(ne),$$

$$b_n=\eta^{\rho}K_{n-\beta-\gamma}^{\rho-\gamma,\rho+\gamma}=(-1)^{n-\beta-\gamma}\frac{(\eta')^{-\rho-1}}{\theta}K_{n-\beta-\gamma}^{-\rho-\beta+n-1,-\rho+\beta-n-1},$$

$$c_n=(-1)^{n-\beta-\gamma}(\eta')^{-\rho}K_{n-\beta-\gamma}^{-\rho-\beta,-\rho+\beta}=\theta\eta^{\rho-1}K_{n-\beta-\gamma}^{\rho-\gamma+n-1,\rho+\gamma-n-1},$$

with

$$C_p^m=\frac{p(p-1)(p-2)\cdots(p-m+1)}{1\cdot 2\cdot\cdots\cdot m},\qquad (m>0),\qquad C_p^0=1,$$

$$C_p^m=0,\qquad (m<0),$$

$$K_n^{p,q}=(-1)^n\sum_{m',m''}\omega^{m'+m''}C_p^{m'}C_q^{m''}$$

$$=(-1)^n(1-\omega^2)^{p+q+1}K_n^{n-p-1,-n-q-1},$$

where

$$e=\sin\phi,\qquad \theta=\cos\phi=\sqrt{1-e^2},$$

$$\omega=\tan(\phi/2)=e/(1+\sqrt{1-e^2}),$$

$$\eta=\cos^2(\phi/2)=\frac{1+\sqrt{1-e^2}}{2},\qquad \eta'=\frac{\cos^2(\phi/2)}{\cos^2\phi}=\frac{1+\sqrt{1-e^2}}{2(1-e^2)};$$

the last summation is to be extended to all nonnegative integers $m'$, $m''$ such that $m' - m'' = n$. If we put

$$\left[1 - \frac{e}{2}(s + s^{-1})\right]^p = \sum_n H_n^p s^n,$$

we obtain

$$b_n = H_n^p, \qquad c_n = (-1)^n \theta^{2p} H_n^{-p}, \qquad H_n^p = \eta K_n^p,$$

$$H_n^p = H_{-n}^p = (-1)^n \sum_m \left(\frac{e}{2}\right)^{n+2m} C_p^{n+2m} C_{n+2m}^m.$$

Put $(\lambda, k) = \Gamma(\lambda + k)/\Gamma(\lambda)$. The series

$$F(\alpha, \beta, \gamma, x) = \sum_{j=0}^{\infty} \frac{(\alpha, j)(\beta, j)}{(\gamma, j)(1, j)} x^j \tag{121}$$

is absolutely and uniformly convergent not only within the circle $|x| < 1$, but also on the circumference of the circle $|x| = 1$ if the real part

$$\Re(\alpha + \beta - \gamma)$$

is less than zero, and defines a function of $x$ provided that we exclude negative integral values of $\gamma$. If $\alpha$ or $\beta$ is a negative integer, then there are only a finite number of terms in the series (121). The series $F(\alpha, \beta, \gamma, x)$ is called *Gauss's hypergeometric series*. A general hypergeometric series is of the form $x^\kappa(1 - x)^\lambda F(\alpha, \beta, \gamma, x)$, with real numbers $\kappa$ and $\lambda$, and satisfies *Gauss's differential equation*

$$x(1 - x)\frac{d^2y}{dx^2} + [\gamma - (\alpha + \beta + 1)x]\frac{dy}{dx} - \alpha\beta\gamma y = 0. \tag{122}$$

The solution of this differential equation is represented by linear combinations of the four functions listed below:

$$\begin{aligned} y_1 &= F(\alpha, \beta, \gamma, x) = F(\beta, \alpha, \gamma, x), \\ y_2 &= x^{1-\gamma}F(\alpha - \gamma + 1, \beta - \gamma + 1, 2 - \gamma, x), \\ y_3 &= (1 - x)^{\gamma-\alpha-\beta}F(\gamma - \beta, \gamma - \alpha, \gamma, x), \\ y_4 &= x^{1-\gamma}(1 - x)^{\gamma-\alpha-\beta}F(1 - \beta, 1 - \alpha, 2 - \gamma, x). \end{aligned} \tag{123}$$

There are 24 such contiguous functions including these four. Kummer divided them into groups of four functions according to their domain and saw that in any domain one of the corresponding functions is expressed

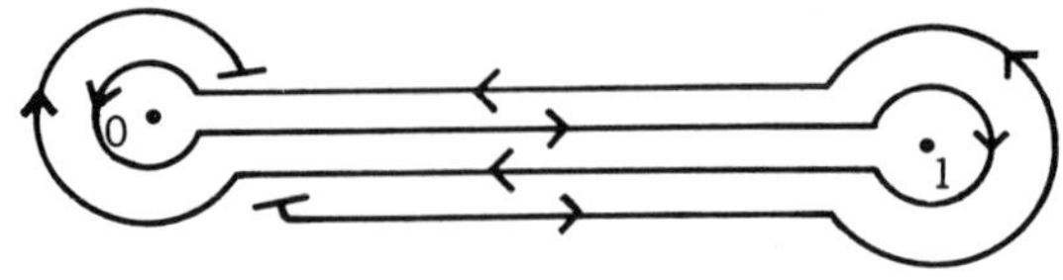


FIGURE 2

linearly by the rest of the four functions. For example, in the domain $|x| < 1$ we have $y_1 = y_3, y_2 = y_4$.

Gauss made an integral representation of the solution of the differential equation (122) in the form (Schlesinger, 1922; Klein, 1933; Poole, 1936; Goursat, 1936; Bailey, 1936)

$$f(x) = \frac{1}{2\pi\sqrt{-1}} \int_C w^{\beta-1}(1 - w)^{\gamma-\beta-1}(1 - xw)^{-\alpha}\, dw, \qquad (124)$$

where the integration contour $C$ is the one taken twice around $x = 1$ in the positive and negative senses; that is, the integration contour is such that

$$\Phi(w) = w^{\beta}(1 - w)^{\gamma-\beta}(1 - xw)^{-\alpha-1}$$

recovers its initial value when we follow along the contour $C$ (Figure 2). Put $\phi(w) = w^{\beta-1}(1 - w)^{\gamma-\beta-1}(1 - xw)^{-\alpha}$, and assume that $\Re(\beta) > 0$, $\Re(\gamma - \beta) > 0$; then,

$$f(x) = \frac{1}{2\pi\sqrt{-1}} \left\{ \int_0^1 + e^{2\pi\sqrt{-1}(\gamma-\beta)} \int_1^0 + e^{2\pi\sqrt{-1}\gamma} \int_0^1 + e^{2\pi\sqrt{-1}\beta} \int_1^0 \right\} \phi(w)\, dw.$$

We carry out the integration by noticing that $\Gamma(z)\Gamma(1 - z) = \pi/\sin \pi z$; then we obtain

$$\int_0^1 w^{\beta-1}(1 - w)^{\gamma-\beta-1}(1 - xw)^{-\alpha}\, dw = B(\beta, \gamma - \beta)F(\alpha, \beta, \gamma, x), \qquad (125)$$

where the integration contour for $B(p, q) = \int_0^1 w^{p-1}(1 - w)^{q-1}\, dw$ is $C$, just defined, and $B(p, q) = \Gamma(p)\Gamma(q)/\Gamma(p + q)$ (Nielsen, 1906; Whittaker and Watson, 1937). The expression (125) is an analytic function of $\beta$. Hence, we can drop the restriction $\Re(\beta) > 0$, by the principle of analytic continuation.

The function $f(x)$ defined by (124) remains unaltered if we make one of the transformations

$$w = 1 - w_1,$$

$$w = \frac{w_1}{1 - x + w_1 x},$$

$$w = \frac{1 - w_1}{1 - w_1 x}.$$

Thus, we obtain *Euler's transformation* of hypergeometric functions

$$\begin{aligned} F(\alpha, \beta, \gamma, x) &= (1 - x)^{-\alpha} F\left(\alpha, \gamma - \beta, \gamma, \frac{x}{x - 1}\right) \\ &= (1 - x)^{-\beta} F\left(\gamma - \alpha, \beta, \gamma, \frac{x}{x - 1}\right) \\ &= (1 - x)^{\gamma - \alpha - \beta} F(\gamma - \alpha, \gamma - \beta, \gamma, x). \end{aligned}$$

After this intermediary of hypergeometric functions we apply Cauchy's second theorem. The coefficient $X_j^{nm}$ of the expansion of

$$\left(\frac{r}{a}\right)^n \exp\left(\sqrt{-1}\, mv\right) = \sum_{j=-\infty}^{\infty} X_j^{nm} z^j$$

is called *Hansen's coefficient*. It is given by

$$X_j^{nm} = \frac{1}{2\pi\sqrt{-1}} \int_C^{(0+)} \left(\frac{r}{a}\right)^n t^m z^{-j-1}\, dz. \tag{126}$$

Writing $S = (r/a)^n t^m$ in Cauchy's second theorem, we obtain

$$X_j^{nm} = \frac{1}{2\pi\sqrt{-1}} \int_C^{(0+)} \frac{s^m}{(1 + \omega^2)^{n+1}} (1 - \omega s)^{n-m+1} \left(1 - \frac{\omega}{s}\right)^{n+m+1}$$

$$\times \exp\left[\frac{j\omega}{1 + \omega^2}\left(s - \frac{1}{s}\right)\right] s^{-j-1}\, ds, \tag{127}$$

$$X_j^{nm} = \frac{1}{2\pi\sqrt{-1}} \int_C^{(0+)} \frac{(1 - \omega^2)^{2n+3}}{(1 + \omega^2)^{2n+1}} t^m (1 + \omega t)^{-n+j-2} \left(1 + \frac{\omega}{t}\right)^{-n-j-2}$$

$$\times \exp\left[j \frac{1 - \omega^2}{1 + \omega^2}\left(\frac{\omega t}{1 + \omega t} - \frac{\omega t^{-1}}{1 + \omega t^{-1}}\right)\right] t^{-j-1}\, dt, \tag{128}$$

where the integration contour is the same as was presented in the proof of Cauchy's second theorem.

Put $j = 0$, $s = \omega\sigma$ in (127); then,

$$X_0^{nm} = \frac{1}{2\pi\sqrt{-1}} \int_C^{(0+)} \frac{(-1)^{n+m+1}}{(1+\omega^2)^{n+1}} \times \omega^m \sigma^{-n-2}(1-\sigma)^{m+n+1}(1-\omega^2\sigma)^{n-m+1}\, d\sigma.$$

We deform the integration contour so that the contour makes a single circuit around $\sigma = 0$ and a single circuit around $\sigma = 1$, both in the positive sense. We have $\sigma = 1/\omega^2 > 1$. If we suppose that $n + 2 > 0$ and $n + m + 1 > 0$, then $\sigma = 1$ is not a pole; $\sigma = 0$ is the only pole, and a function is multiplied by $\exp[-2\pi\sqrt{-1}\,(n+2)]$ after a single circuit around $\sigma = 0$. Hence,

$$\begin{aligned} X_0^{nm} &= \frac{(-1)^{n+m+1}}{2\pi\sqrt{-1}} \frac{\omega^m}{(1+\omega^2)^{n+1}} \{\exp[-2\pi\sqrt{-1}\,(n+2)] - 1\} \\ &\quad \times \int_0^1 \sigma^{-n-2}(1-\sigma)^{n+m+1}(1-\omega^2\sigma)^{n-m+1}\, d\sigma \\ &= \frac{(-1)^m \omega^m}{(1+\omega^2)^{n+1}} \frac{\Gamma(m+n+2)}{\Gamma(n+2)\Gamma(m+1)} \\ &\quad \times F(m-n-1, -n-1, m+1, \omega^2). \end{aligned}$$

There are only a finite number of terms in the expansion of the hypergeometric function if $n + m + 1 > 0$. If we suppose that $n + 2 < 0$ and $n + m + 1 > 0$, then neither $\sigma = 0$ nor $\sigma = 1$ is a pole and $X_0^{nm} = 0$. If $n + 2 < 0$ and $n + m + 1 < 0$, then $\sigma = 0$ is an ordinary point and $\sigma = 1$ is a singularity. Hence,

$$X_0^{nm} = \frac{(-1)^m \omega^m (1-\omega^2)^{2n+3}}{(1+\omega^2)^{n+1}} \frac{\Gamma(-n-1)}{\Gamma(m+1)\Gamma(-m-n-1)} \times F(n+2, m+n+2, m+1, \omega^2).$$

Substitute in (127) the expansion in Bessel functions

$$\exp\left[\frac{j\omega}{1+\omega^2}\left(s - \frac{1}{s}\right)\right] = \sum_{p=-\infty}^{\infty} J_p\left(\frac{2j\omega}{1+\omega^2}\right) s^p$$

and put $s = \omega\sigma$; then,

$$X_j^{nm} = \sum_{p=-\infty}^{\infty} J_p\left(\frac{2j\omega}{1+\omega^2}\right) X_{jp}^{nm},$$

$$X_{jp}^{nm} = \frac{(-1)^{n+m+1}}{(1+\omega^2)^{n+1}} \frac{\omega^{m+p-j}}{2\pi\sqrt{-1}} \times \int_C^{(0+)} \sigma^{p-n-j-2}(1-\sigma)^{n+m+1}(1-\omega^2\sigma)^{n-m+1}\, d\sigma.$$

If $p - n - j - 2 < 0$ and $n + m + 1 > 0$, then,

$$X_{jp}^{nm} = (-1)^{p+m-j} \frac{\omega^{m+p-j}}{(1+\omega^2)^{n+1}} \frac{\Gamma(n+m+2)}{\Gamma(-p+n+j+2)\Gamma(p+m-j+1)}$$
$$\times F(m-n-1, p-n-j-1, p+m-j+1, \omega^2),$$

and $F$ consists of a finite number of terms. If $p - n - j - 2 > 0$ and $n + m + 1 > 0$, then $X_{jp}^{nm} = 0$. If

$$p - n - j - 2 > 0 \quad \text{and} \quad n + m + 1 < 0,$$

then

$$X_{jp}^{nm} = \frac{\omega^{m+p-j}}{(1+\omega^2)^{n-1}} \frac{\Gamma(p-n-j-1)}{\Gamma(p+m-j+1)\Gamma(-n-m-1)}$$
$$\times F(m-n-1, p-n-j-1, p+m-j+1, \omega^2),$$

or, if we use (123),

$$X_{jp}^{nm} = \frac{\omega^{m+p-j}(1-\omega^2)^{2n+3}}{(1+\omega^2)^{n+1}} \frac{\Gamma(p-n-j-1)}{\Gamma(p+m-j+1)\Gamma(-n-m-1)}$$
$$\times F(p+n-j+2, -m-n-2, p+m-j+1, \omega^2),$$

which consists of a finite number of terms.

We can deduce similar formulas from (128) (Hagihara, 1935).

## 5.13 EXPANSION IN SEVERAL VARIABLES

Let $C$ and $C'$ be two simple integration contours around $z = x$ and $z' = x'$, respectively, in the complex plane of two complex variables $z$ and $z'$; and let $F(z, z')$ be a single-valued continuous analytic function inside $C$ and $C'$. Then by the residue theorem of Cauchy in the theory of analytic functions (Goursat, 1913; Osgood, 1914, 1924; Behnke and Thullen, 1934; Bochner and Martin, 1948; Forsyth, 1914; Picard, 1926; Picard and Simart, 1897), we have

$$F(x, x') = \frac{1}{(2\pi\sqrt{-1})^2} \int_C dz \int_{C'} \frac{F(z, z')\, dz'}{(z-x)(z'-x')},$$

$$\frac{\partial^{m+n} F(x, x')}{\partial x^m\, \partial x'^n} = \frac{m!\, n!}{(2\pi\sqrt{-1})^2} \int_C dz \int_{C'} \frac{F(z, z')\, dz'}{(z-x)^{m+1}(z'-x')^{n+1}}.$$

If a single-valued continuous analytic function $S(z, z')$ can be expanded in the double Laurent series

$$S(z, z') = \sum_{j=-\infty}^{\infty} \sum_{j'=-\infty}^{\infty} P_{jj'} z^j z'^{j'} \tag{129}$$

such that $0 < r < |z| < R < \infty$, $0 < r' < |z'| < R' < \infty$, then, according to this theorem, the coefficient $P_{jj'}$ can be written

$$P_{jj'} = \frac{1}{(2\pi\sqrt{-1})^2}\int_C^{(0+)}\int_{C'}^{(0+)} S(z, z')z^{-j-1}z'^{-j'-1}\,dz\,dz'. \quad (130)$$

If we transform by

$$z = s\exp\left[-\frac{e}{2}(s - s^{-1})\right], \qquad z' = s'\exp\left[-\frac{e'}{2}(s' - s'^{-1})\right],$$

then

$$P_{jj'} = \frac{1}{(2\pi\sqrt{-1})^2}\int_C^{(0+)}\int_{C'}^{(0+)} Us^{-j-1}s'^{-j'-1}\,ds\,ds', \quad (131)$$

where

$$U = S\left\{1 - \frac{e}{2}(s + s^{-1})\right\}\left\{1 - \frac{e'}{2}(s' + s'^{-1})\right\}$$
$$\times \exp\left\{\frac{je}{2}(s - s^{-1}) + \frac{j'e'}{2}(s' - s'^{-1})\right\}.$$

This is an extension of Cauchy's first theorem to two variables (Hill, 1905).

Put $|z| = 1$, $|z'| = 1$ in these integration contours and write

$$z = \exp(\sqrt{-1}\,M), \qquad z' = \exp(\sqrt{-1}\,M'),$$
$$s = \exp(\sqrt{-1}\,u), \qquad s' = \exp(\sqrt{-1}\,u');$$

then,

$$P_{jj'} = \frac{1}{(2\pi\sqrt{-1})^2}\int_0^{2\pi}\int_0^{2\pi} S\exp[-\sqrt{-1}\,(jM + j'M')]\,dM\,dM'$$
$$= \frac{1}{4\pi^2}\int_0^{2\pi}\int_0^{2\pi} U\exp[-\sqrt{-1}\,(ju + j'u')]\,du\,du', \quad (132)$$

$$U = S(1 - e\cos u)(1 - e'\cos u')\exp[\sqrt{-1}\,(je\sin u + j'e'\sin u')].$$

Hence, the coefficient of the double Fourier expansion of $S$ in the mean anomalies equals the corresponding coefficient in the double Fourier expansion of $U$ in the eccentric anomalies (Hagihara, 1936).

The relation between $S$ and $u$ can be generalized in the form

$$M = u - \alpha\phi(u),$$

replacing the Kepler equation. Put

$$\begin{aligned} F(u, u') &= \sum_{p=-\infty}^{\infty} \sum_{p'=-\infty}^{\infty} B_{pp'} \exp\left[\sqrt{-1}\,(pu + p'u')\right] \\ &= \sum_{q=-\infty}^{\infty} \sum_{q'=-\infty}^{\infty} A_{qq'} \exp\left[\sqrt{-1}\,(qM + q'M')\right]; \qquad (133) \end{aligned}$$

then,

$$\begin{aligned} A_{qq'} &= \frac{1}{4\pi^2} \int_0^{2\pi} \int_0^{2\pi} F(u, u') \exp\left[-\sqrt{-1}\,(qM + q'M')\right] dM\, dM' \\ &= \frac{1}{4\pi^2} \int_0^{2\pi} \int_0^{2\pi} F(u, u') \frac{\partial^2 \Psi}{\partial M\, \partial M'}\, dM\, dM' \\ &= \frac{1}{4\pi^2 qq'} \int_0^{2\pi} \int_0^{2\pi} \frac{\partial^2 F}{\partial u\, \partial u'} \Psi[u - e \sin u, u' - e' \sin u']\, du\, du' \\ &= \frac{1}{4\pi^2 qq'} \sum_{p=-\infty}^{\infty} \sum_{p'=-\infty}^{\infty} pp' B_{pp'} \\ &\quad \times \int_0^{2\pi} \exp\left[\sqrt{-1}\,(p - q)u + \sqrt{-1}\, qe \sin u\right] du \\ &\quad \times \int_0^{2\pi} \exp\left[\sqrt{-1}\,(p' - q')u' + \sqrt{-1}\, q'e' \sin u'\right] du', \end{aligned}$$

where

$$\Psi = -\frac{1}{qq'} \exp\left[-\sqrt{-1}\,(qM + q'M')\right].$$

Hence,

$$A_{qq'} = \sum_{p=-\infty}^{\infty} \sum_{p'=-\infty}^{\infty} \frac{pp'}{qq'} B_{pp'} J_{p-q}(qe) J_{p'-q'}(q'e'). \qquad (134)$$

In particular, if we denote the mutual distance of the two planets by $\Delta$, then $s^2 s'^2 \Delta^2 = R(s, s')$ is a sextic integral polynomial in $s$ and $s'$

(Hagihara, 1936; Poincaré, 1892, 1907; Brown and Shook, 1933; Andoyer 1923). Write

$$F = \frac{1}{\Delta} = \sum_{p=-\infty}^{\infty} \sum_{p'=-\infty}^{\infty} B_{pp'} \exp [\sqrt{-1}\,(pu + p'u')]$$

$$= \sum_{q=-\infty}^{\infty} \sum_{q'=-\infty}^{\infty} A_{qq'} \exp [\sqrt{-1}\,(qM + q'M')]; \tag{135}$$

then

$$B_{pp'} = -\frac{1}{4\pi^2} \int_C \int_{C'} \frac{ds\, ds'}{s^p s'^{p'} \sqrt{R(s, s')}}, \tag{136}$$

and

$$A_{qq'} = \frac{1}{4\pi^2 qq'} \int_0^{2\pi} \int_0^{2\pi} \frac{\partial^2(1/\Delta)}{\partial u\, \partial u'} \exp [-\sqrt{-1}\,(qM + q'M')]\, du\, du'$$

$$= -\frac{1}{4\pi^2 qq'} \int_C \int_{C'} \frac{\partial^2(1/\Delta)}{\partial s\, \partial s'} \exp [\sqrt{-1}\, qe \sin u + \sqrt{-1}\, q'e' \sin u']$$

$$\times \frac{ds}{s^q} \frac{ds'}{s'^{q'}}$$

$$= -\frac{1}{4\pi^2} \int_C \int_{C'} \frac{dM}{du} \frac{dM'}{du'} \exp [\sqrt{-1}\, qe \sin u + \sqrt{-1}\, q'e' \sin u']$$

$$\times \frac{ds\, ds'}{s^q s'^{q'} R\sqrt{(s, s')}}. \tag{137}$$

This expression has been used by Poincaré (Poincaré, 1898; Féraud, 1902).

In extending the theorems from two variables to more than two variables, put

$$\exp \left[\sum_{j=1}^{k} \frac{z_j}{2} (t^j - t^{-j})\right] = \sum_{n=-\infty}^{\infty} J_n(z_1, \ldots, z_k) t^n; \tag{138}$$

then the Laurent series converges in a ring domain excluding $t = 0$ and $\infty$. Write $t^{-1}$ in the place of $t$ and add or subtract the corresponding expressions, and put $t = \exp(-\sqrt{-1}\,\phi)$; then

$$\cos \left(\sum_{j=1}^{k} z_j \sin j\phi\right) = \sum_{n=1}^{\infty} [J_n(z_1, \ldots, z_k) + J_{-n}(z_1, \ldots, z_k)] \cos n\phi,$$

$$\sin \left(\sum_{j=1}^{k} z_j \sin j\phi\right) = \sum_{n=1}^{\infty} [J_n(z_1, \ldots, z_k) - J_{-n}(z_1, \ldots, z_k)] \sin n\phi.$$

Differentiate the first equation of (139) with respect to $\phi$ and put $\phi = 0$; then,

$$\sum_{j=1}^{k} jz_j = \sum_{n=1}^{\infty} n[J_n(z_1, \ldots, z_k) - J_{-n}(z_1, \ldots, z_k)].$$

Put $\phi = 0$ before carrying out the differentiation; then

$$1 = \sum_{n=\infty}^{\infty} J_n(z_1, \ldots, z_k).$$

Multiply (138) by the expression obtained by replacing $t$ by $t^{-1}$ in (138); then

$$1 = C_0 + \sum_{n=1}^{\infty} C_n t^n + \sum_{n=1}^{\infty} C_{-n} t^{-n}.$$

This relation is independent of $t$. Hence, $C_0 = 1$, $C_n = 0$, $C_{-n} = 0$ $(n = 1, 2, \ldots)$. The relation $C_0 = 1$ is explicitly written

$$1 = \sum_{n=-\infty}^{\infty} J_n^2(z_1, \ldots, z_k).$$

Hence

$$|J_n(z_1, \ldots, z_k)| < 1.$$

Multiply both sides of (138) by $\exp\,(-\sqrt{-1}\; 2\pi\phi)\; d\phi/2\pi$, integrate with respect to $\phi$, and take the real part; then

$$J_n(z_1, \ldots, z_k) = \frac{1}{\pi}\int_0^{\pi} \cos\left(n\phi - \sum_{j=1}^{k} z_j \sin j\phi\right) d\phi. \tag{140}$$

When there is only one $z$, this reduces to the usual integral representation of Bessel functions. Put

$$y_{k-1} = m\phi - \sum_{j=1}^{k-1} z_j \sin j\phi;$$

then (140) becomes

$$J_m(z_1, \ldots, z_k) = \frac{1}{\pi}\int_0^{\pi} \cos y_{k-1} \cos\,(z_k \sin k\phi)\; d\phi$$

$$+ \frac{1}{\pi}\int_0^{\pi} \sin y_{k-1} \sin\,(z_k \sin k\phi)\; d\phi.$$

By the usual expansion formulas in Bessel functions,

$$\frac{1}{\pi}\int_0^\pi \cos y_{k-1}\,d\phi = J_m(z_1, \ldots, z_{k-1}),$$

$$\begin{aligned}\frac{1}{\pi}\int_0^\pi & 2\cos y_{k-1}\cos 2jk\phi\,d\phi \\ &= \frac{1}{\pi}\int_0^\pi \cos\,[y_{k-1} - 2jk\phi] + \frac{1}{\pi}\int_0^\pi \cos\,(y_{k-1} + 2jk\phi)\,d\phi \\ &= J_{m-2jk}(z_1, \ldots, z_{k-1}) + J_{m+2jk}(z_1, \ldots, z_{k-1}),\end{aligned}$$

$$\begin{aligned}\frac{1}{\pi}\int_0^\pi & 2\sin y_{k-1}\sin\,(2j-1)k\phi\,d\phi \\ &= \frac{1}{\pi}\int_0^\pi \cos\,[y_{k-1} - (2j-1)k\phi]\,d\phi \\ &\quad - \frac{1}{\pi}\int_0^\pi \cos\,[y_{k-1} + (2j-1)k\phi]\,d\phi \\ &= J_{m-(2j-1)k}(z_1, \ldots, z_{k-1}) - J_{m+(2j-1)k}(z_1, \ldots, z_{k-1}).\end{aligned}$$

From these expressions we obtain

$$\begin{aligned}J_m(z_1, \ldots, z_k) &= J_0(z_k)J_m(z_1, \ldots, z_{k-1}) \\ &\quad + \sum_{j=1}^{\infty} J_{2j}(z_k)[J_{m-2jk}(z_1, \ldots, z_{k-1}) + J_{m+2jk}(z_1, \ldots, z_{k-1})] \\ &\quad + \sum_{j=1}^{\infty} J_{2j-1}(z_k)[J_{m-(2j-1)k}(z_1, \ldots, z_{k-1}) \\ &\qquad - J_{m+(2j-1)k}(z_1, \ldots, z_{k-1})].\end{aligned}$$

But, since

$$J_{2j}(z_k) = J_{-2j}(z_k), \qquad J_{2j-1}(z_k) = -J_{-(2j-1)}(z_k),$$

therefore

$$J_m(z_1, \ldots, z_k) = \sum_{q_k=-\infty}^{\infty} J_{m-kq_k}(z_1, \ldots, z_{k-1})J_{q_k}(z_k).$$

By repeated use of this formula we finally obtain

$$\begin{aligned}J_m(z_1, z_2, \ldots, z_k) \\ = \sum_{q_2=-\infty}^{\infty}\sum_{q_3=-\infty}^{\infty}\cdots\sum_{q_k=-\infty}^{\infty} J_{m-2q_2-3q_3-\cdots-kq_k}(z_1)J_{q_2}(z_2)\cdots J_{q_k}(z_k).\end{aligned} \tag{141}$$

This formula has been deduced by Jekhowsky (Appell, 1915, 1924, 1925; Pérès, 1915; Akimoff, 1915, 1916, 1917, 1924, 1927; Jekhowsky, 1916, 1916a, 1917, 1917a, 1918, 1920, 1921, 1927; Humbert, 1920).

We now expand $y = \cos pv$ with respect to $x$, such that

$$x = v - \sum_{j=1}^{k} \alpha_j \sin jv,$$

with $|\alpha_1| + |\alpha_2| + \cdots + |\alpha_k| < 1$. Since $y$ remains unaltered as $v$ increases by $2\pi/p$, the function $y$ is periodic. By the Weierstrassian approximation theorem of Section 2.6, it can be written

$$y = \tfrac{1}{2}A_0 + \sum_{j=1}^{\infty} A_j \cos jx,$$

where

$$A_j = \frac{2}{\pi}\int_0^{\pi} \cos pv \cos jx \, dx = \frac{2p}{\pi j}\int_0^{\pi} \sin pv \sin j\left(v - \sum_{i=1}^{k} \alpha_i \sin iv\right) dv$$

$$= \frac{p}{j}\,[J_{j-p}(j\alpha_1, j\alpha_2, \ldots, j\alpha_k) - J_{j+p}(j\alpha_1, \ldots, j\alpha_k)].$$

Hence,

$$\exp(\pm\sqrt{-1}\,pv) = -\frac{p\alpha_p}{2} \pm p \sum_{j=-\infty}^{\infty}{}' J_{j\mp p}(j\alpha_1, \ldots, j\alpha_k)\,\frac{1}{j}\exp(\sqrt{-1}\,jx),$$

where in the summation $j = 0$ is excluded. Put $z = \exp(\sqrt{-1}\,v)$, $s = \exp(\sqrt{-1}\,x)$; then Cauchy's theorem is generalized; that is, the coefficient $P_j$ in the expansion

$$S = \sum_{j=-\infty}^{\infty} P_j z^j$$

equals the coefficient of $s^j$ in the expansion of

$$U = S\left[1 - \frac{1}{2}\sum_{l=1}^{k} l\alpha_l(s^l + s^{-l})\right] \exp\left[\frac{j}{2}\sum_{l=1}^{k} \alpha_l(s^l - s^{-l})\right],$$

or equals the coefficient of $s^j$ in the expansion of

$$V = \frac{1}{j}\frac{dS}{ds}\, s \exp\left[\frac{j}{2}\sum_{l=1}^{k} \alpha_l(s^l - s^{-l})\right],$$

where $z$ and $s$ are related by

$$z = s \exp\left[-\tfrac{1}{2}\sum_{j=1}^{k} \alpha_j(s^j - s^{-j})\right].$$

In the case where the number of variables in the arguments of the Bessel functions is infinite, it is sufficient to have

$$|z_1| + 2|z_2| + 3|z_3| + \cdots + k|z_k| + \cdots < 1$$

in order that the series be convergent. The formulas obtained in the foregoing study are also true for $k \to \infty$. Such expansions were used by Hagihara (1931) for the expansions of quasi-periodic motions in the relativistic trajectories of Sections 2.15–2.20.

## 5.14 RADIUS OF CONVERGENCE

In Kepler's equation (101),

$$u - e \sin u = M,$$

the singularity occurs when

$$\frac{dM}{du} = 1 - e \cos u = 0 \qquad (0 < e < 1),$$

in representing the series for $u$ as power series in $M$ or in $t - t_0$.

On the complex plane for $u = z + \sqrt{-1}\, w$ we erect a rectangle of very large height between $z = m\pi$ and $z = (m + 1)\pi$ perpendicular to the real axis. Briot and Bouquet (1875) proved that there are two roots in the rectangle for an even positive or an odd negative integer $m$ and no root for an odd positive or an even negative integer $m$. For a very large $m$ the two roots whose real parts are between $2m\pi$ and $(2m + 1)\pi$ tend asymptotically (Goursat, 1913) to

$$2m\pi + \frac{\pi}{2} \pm \sqrt{-1}\left[\log\left(\frac{2}{e}\right) + \log\left(2m\pi + \frac{\pi}{2}\right)\right].$$

Moulton (1903) computed the singular points in the complex $u$-plane to be

$$z = 2m\pi, \qquad w = \pm \cosh^{-1}\left(\frac{1}{e}\right) = \pm \log \frac{1 + \sqrt{1 - e^2}}{e}$$

with an integer $m$. Using an arbitrary value $e_0$ of $e$ $(0 \leqq e_0 \leqq 1)$ as center, draw a circle through the nearest singular point. The radius of this circle is the convergence radius of the expansion at $e = e_0$.

In Lagrange's theorem of Section 2.3 put

$$t = e, \qquad \phi(z) = \sin u, \qquad z = u, \qquad a = M;$$

then a function $f$ of $u$ with Kepler's equation $u = M + e \sin u$ can be expanded in an integral power series of $e$ in the form

$$f(u) = f(M) + \frac{e}{1} f'(M) \sin M + \cdots$$
$$+ \frac{e^{n+1}}{(n+1)!} \frac{\partial^n}{\partial M^n} [f'(M) \sin^{n+1} M] + \cdots, \tag{142}$$

where

$$|e \sin u| < |u - M|.$$

Suppose that $e$ and $M$ are given, what is the maximum value of $e$ satisfying this inequality? Let

$$u = M + e \sin u \equiv M + \rho \cos \chi + \sqrt{-1}\, \rho \sin \chi,$$

where $\rho =$ constant. From this,

$$\sin u = \sin (M + \rho \cos \chi) \cosh (\rho \sin \chi)$$
$$+ \sqrt{-1} \cos (M + \rho \cos \chi) \sinh (\rho \sin \chi).$$

Hence,

$$|\sin u|^2 = \cos h^2(\rho \sin \chi) - \cos^2 (M + \rho \cos \chi).$$

The worst situation occurs when $|\sin u|$ is a maximum, that is, when

$$\sin \chi = \pm 1, \qquad \cos (M + \rho \cos \chi) = 0,$$

or $\chi = \pm\pi/2$, $M = \pm\pi/2$, or when $|\sin u| = \cosh \rho$. Hence, the required condition is

$$e < \frac{\rho}{\cosh \rho}.$$

The maximum of $\rho/\cosh \rho$ occurs when

$$\frac{d}{d\rho}\left(\frac{\rho}{\cosh \rho}\right) = 0,$$

or when $\cosh \rho = \rho \sinh \rho$, or

$$\exp (2\rho) = \frac{\rho + 1}{\rho - 1},$$

that is, $\rho = 1.9967\cdots$. Thus, if

$$e = \tan \alpha = \frac{\rho}{\cosh \rho} = \frac{1}{\sinh \rho}, \tag{143}$$

then the maximum occurs when $\alpha = 33^\circ\, 32'\, 3''.0$, $e = 0.6627434\cdots$. For any value of $e$ larger than this critical value, any expansion for an elliptic motion as an integral power series of $e$ is not convergent. This finding is due to Laplace.

Now compute the radius of convergence of an integral power series of $e$ in the complex $e$ plane, by putting $e = g + \sqrt{-1}\, h$ according to Charlier (1927). The singularities in the $M$ plane are obtained by the two equations

$$\frac{dM}{du} = 1 - e\cos u = 0, \qquad u - M - e\sin u = 0. \tag{144}$$

Since

$$\exp(\sqrt{-1}\, u) = \frac{1}{e} \mp \sqrt{\frac{1}{e^2} - 1} = \exp[\sqrt{-1}\, e\sin u + \sqrt{-1}\, M]$$
$$= \exp[\mp\sqrt{1 - e^2} + \sqrt{-1}\, M],$$

the singularities are defined by

$$\gamma = 1 \mp \sqrt{1 - e^2} = e\exp[\mp\sqrt{1 - e^2} + \sqrt{-1}\, M],$$

or by

$$\left(\frac{2}{\gamma} - 1\right)\exp(2\gamma) = \exp(2 + 2\sqrt{-1}\, M).$$

Write $\gamma = \epsilon\exp(\sqrt{-1}\,\theta)$ and take the real and imaginary parts; then the singularities are at the intersections of the two curves

$$\frac{4}{\epsilon^2} + 1 - \frac{4}{\epsilon}\cos\theta = \exp(4 - 4\epsilon\cos\theta),$$

$$\frac{2}{\epsilon}\sin(2M + \theta - 2\epsilon\sin\theta) - \sin(2M - 2\epsilon\sin\theta) = 0.$$

These equations are satisfied by

$$\epsilon\cos\theta = 1, \qquad \theta - \tan\theta = \frac{n\pi}{2} - M \qquad (n = 0, \pm 1, \pm 2, \ldots),$$

which define a set of parallel straight lines. Other solutions are obtained by the equations

$$\mu^2 = \frac{4(1 - \lambda)}{\exp[4(1 - \lambda)] - 1} - \lambda^2, \qquad (\lambda - 1)^2 = 1 + 2\mu\cot(2M - 2\mu) - \mu^2, \tag{145}$$

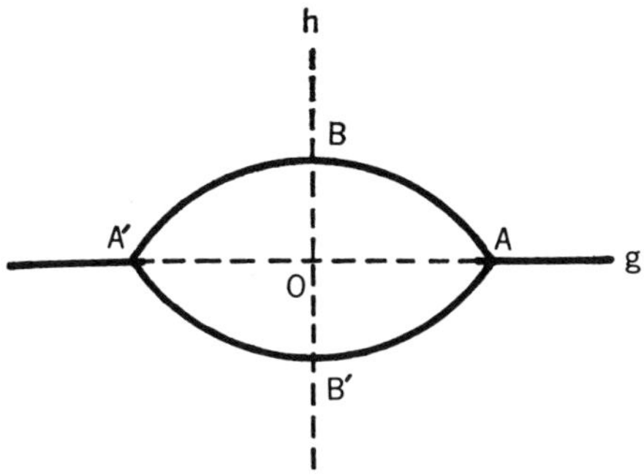


FIGURE 3

where $\lambda = \epsilon \cos \theta$, $\mu = \epsilon \sin \theta$. The solution $\lambda = 1$ corresponds to the set of straight lines just obtained. We can make the curve defined by the second equation of (145) pass through any point of the curve defined by the first equation. If we denote such values of $\lambda$, $\mu$ by $\lambda_0$, $\mu_0$, respectively, then the second equation is satisfied by

$$\tan (2M - 2\mu_0) = \frac{2\mu_0}{(\lambda_0 - 1)^2 - 1 + \mu_0}.$$

Take the real and the imaginary parts of this equation by putting

$$e = g + \sqrt{-1}\, h = \kappa \exp (\sqrt{-1}\, \tau);$$

then,

$$\begin{aligned} \kappa^2 \cos 2\tau &= -(1 - \lambda)^2 + \mu^2 + 1, \\ \kappa^2 \sin 2\tau &= 2\mu(1 - \lambda). \end{aligned} \tag{145a}$$

The solution $\lambda = 1$ corresponds to the part of the $g$ axis in the $e$ plane such that $|g| \geqq 1$. The points for $M = \pi/2$ corresponds to point $B$ in Figure 3. Put $e = \sqrt{-1}\, h$; then,

$$1 + \sqrt{1 + h^2} = h \exp (\sqrt{1 + h^2})$$

and the solution is $h = 0.6627\cdots$, which is equal to the maximum value of $e$ obtained in the foregoing discussion by means of Lagrange's theorem. All points on the singular curve of Figure 3 are singular points of the integral (144) which is expanded in the form

$$u = a_0 + a_1(e - e_0) + a_2(e - e_0)^2 + \cdots.$$

The curve is the envelope of a one-parameter family of convergence circles with variable radius centered at the point $e_0$.

Elenevskaya (1953, 1957, 1962) has written the equations (144), by separating the real and imaginary parts

$$u = x + \sqrt{-1}\, y, \qquad e = \xi + \sqrt{-1}\, \eta,$$

in the form

$$\begin{aligned} M - x &= -\xi \sin x \cosh y + \eta \cos x \sinh y, \\ y &= \xi \cos x \sinh y + \eta \sin x \cosh y, \\ 1 &= \xi \cos x \cosh y + \eta \sin x \sinh y, \\ 0 &= \xi \sin x \sinh y - \eta \cos x \cosh y. \end{aligned} \tag{146}$$

The intersections of these two curves in the $(\xi, \eta)$-plane define the singular points of the integral of the first equation of (144). The coordinates of these singular points are obtained from the last three equations of (146):

$$\xi = \frac{y \cos x}{\sinh y}, \qquad \eta = \frac{y \sin x}{\cosh y},$$

or

$$\sin^2 x = \frac{y \cosh^2 y - \sinh y \cosh y}{y}.$$

$x$ is a multi-valued function of $y$, but $\xi$ and $\eta$ are determined uniquely:

$$\begin{aligned} \xi^2 &= \frac{y}{\sinh y} (\cosh y - y \sinh y), \\ \eta^2 &= \frac{y}{\cosh y} (y \cosh y - \sinh y). \end{aligned} \tag{147}$$

It can be shown that there are no singular points other than those lying on this curve and that any point on this curve is a singular point of the integral.

Elenevskaya (1962) extended the discussion to hyperbolic motions. She derived

$$\begin{aligned} M + x &= \xi \sinh x \cos y - \eta \cosh x \sin y, \\ y &= \xi \cosh x \sin y + \eta \sinh x \cos y, \\ 1 &= \xi \cosh x \cos y - \eta \sinh x \sin y, \\ 0 &= \xi \sinh x \sin y + \eta \cosh x \cos y. \end{aligned} \tag{146a}$$

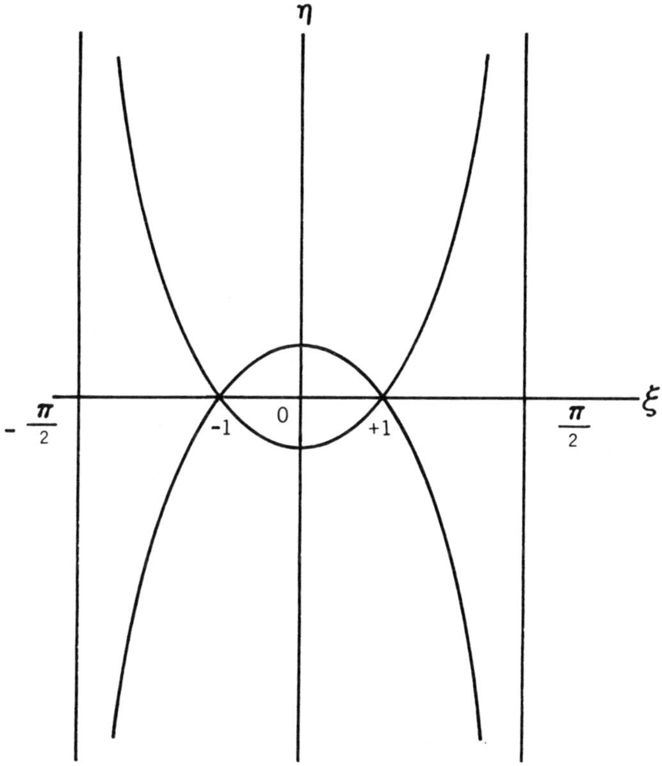


FIGURE 4

From these we obtain

$$\xi = \frac{y \cosh x}{\sin y}, \qquad \eta = \frac{y \sinh x}{\cos y},$$

$$\cosh^2 x = \frac{y \sin^2 y + \sin y \cos y}{y}.$$

With $x$ eliminated, the equation of the singular curve for a hyperbolic motion is

$$\xi^2 = \frac{y}{\sin y}(y \sin y + \cos y), \qquad \eta^2 = \frac{y}{\cos y}(\sin y - y \cos y).$$

Figure 4 shows the singular curve.

If $e$ is close to unity, then it is possible to accelerate the convergence by restricting the range of variation of the angular variable. Such an expansion is made in powers of $(1 - e)$. We have obtained the singular curve for an elliptic motion by considering (146) as a system of equations describing two curves and selecting from these the one independent of $M$. The convergence circle of the expansion of the eccentric anomaly $u$ is centered at the point $e_0$; this point is inside the singular curve. The point of contact between the convergence circle and the singular curve is the point at which the singular curve is intersected by another curve independent of $M$ and determined by (146). For a given value of $M$, the power series in $(e - e_0)$ will cease to be convergent when $e$ is outside the circle in contact with the singular curve at this point.

When $e_0 = 1$, the expansion of $u$ in powers of $1 - e$ becomes divergent for any value of $e$ except 1, at least for some values of the mean anomaly $M$. In order to determine the values of $M$ for which the divergence of the expansion does not take place, draw a circle of an arbitrary radius about a point $e_0 = 1$. This circle intersects the singular curve at two points. For convenience, consider only the part of the curve lying in the first quadrant. Each point of intersection of the circle with the singular curve is a point at which the singular curve is intersected by one of the family of curves depending on $M$ and determined by (146). We call this latter curve the *anomaly curve*. This point gives the value of $M$ for which the expansion of $u$ in powers of $1 - e$ is no longer convergent. Denote this value of $M$ by $M_{cr}$.

Consider the family of curves governed by (146) in order to determine the functional relation $M_{cr} = \phi(e)$. From (146) we obtain

$$\xi = \frac{(x - M)\cosh y}{\sin x}, \qquad \eta = \frac{(x - M)\sinh y}{\cos x}, \tag{147a}$$

so that the relation between $x$ and $y$ is

$$\sinh^2 y = \frac{\sin x \cos x - (x - M)\cos^2 x}{x - M}. \tag{148}$$

This shows that $y$ is a multi-valued function of $x$. We choose one region in which $e = \psi(u)$ is single-valued, that is, the strip of width

$$-\infty < y < +\infty, \qquad -\pi \leqq x \leqq \pi.$$

From (147a) and (148) we obtain the family of anomaly curves

$$\begin{aligned} \xi^2 &= \frac{x - M}{\sin x}[\cos x + (x - M)\sin x], \\ \eta^2 &= \frac{x - M}{\cos x}[\sin x - (x - M)\cos x]. \end{aligned} \tag{147b}$$

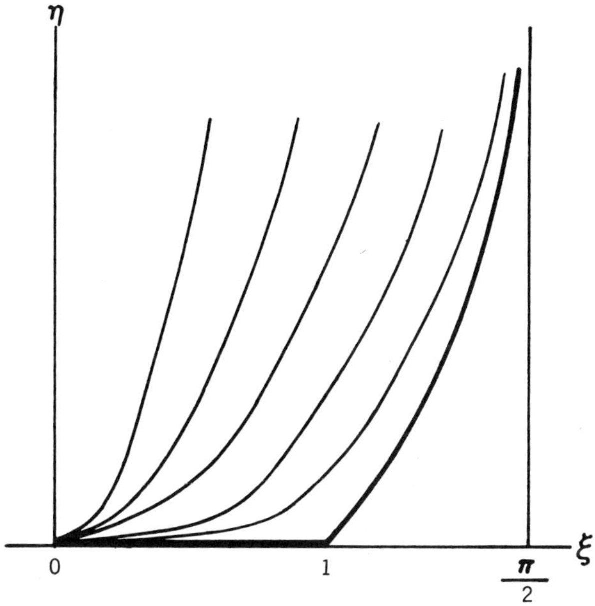


FIGURE 5

The curve of this family for $M = 0$ is identical with the singular curve of a hyperbolic motion. It shows that when $M = 0$ the expansion of $u$ in powers of $(1 - e)$ is divergent for any value of $e$. Figure 5 shows the family of anomaly curves. For $M = \pi/2$ the anomaly curve degenerates into the straight line $\xi = 0$, since $x$ in the case can only take the value $\pi/2$. Hence, for $M > \pi/2$, the expansion of $u$ will be convergent for any value of $e$.

Draw a circle of radius $R$ about the point $e_0 = 1$; at some point it touches one of the anomaly curves. Moreover, the curves of the family for which $M \leqq M_{cr}$ intersect the circle, but the curves for which $M > M_{cr}$ remain outside the circle. The intersection of the circle with the $\xi$-axis gives the real part of $e$, for which the expansion of $u$ in powers of $1 - e$ is convergent for all $M > M_{cr}$ and divergent for all $M \leqq M_{cr}$, since for a given value of $e$ these points of contact are the singular points of the equation $u - M - e \sin u = 0$.

Similarly, equations (147b) for a hyperbolic motion give

$$\begin{aligned} \xi^2 &= \frac{M + x}{\sinh x} [\cosh x - (M + x) \sinh x], \\ \eta^2 &= \frac{M + x}{\cosh x} [-\sinh x + (M + x) \cosh x]. \end{aligned} \tag{149}$$

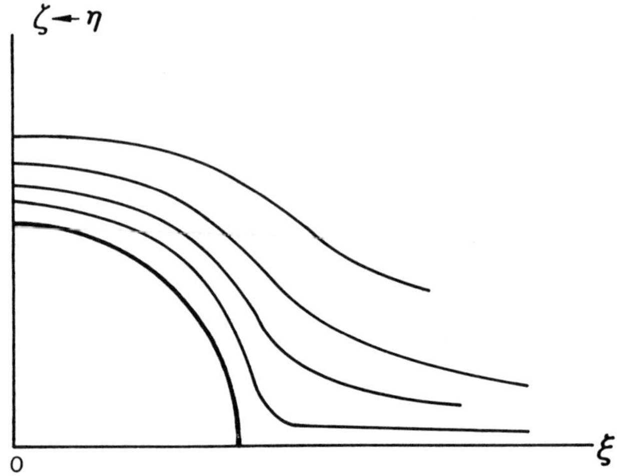


FIGURE 6

These curves are shown in Figure 6. For $M = 0$ the anomaly curves degenerate into the singular curve for an elliptic motion.

We study the general case of the convergence of the expansions of $u$ in powers of $(1 - e)$, by considering the expansion $u - M - e \sin u$ as an integral of the differential equation

$$\frac{du}{de} = \frac{\sin u}{1 - e \cos u},$$

in which $u = x + \sqrt{-1}\, y$, $e = \xi + \sqrt{-1}\, \eta$, $M = a + \sqrt{-1}\, b$. This gives as particular cases the singular and the anomaly curves for elliptic $(b = 0)$ and hyperbolic $(a = 0)$ motions. The singular curves are represented by

$$\begin{aligned} \xi^2 &= \frac{\alpha}{\sinh \alpha} (\cosh \alpha - \alpha \sinh \alpha), \\ \eta^2 &= \frac{\alpha}{\cosh \alpha} (\alpha \cosh \alpha - \sinh \alpha); \end{aligned} \tag{150}$$

and the anomaly curves by

$$\begin{aligned} \xi^2 &= \frac{\beta - c}{\sin \beta} [\cos \beta - (\beta - c) \sin \beta], \\ \eta^2 &= \frac{\beta - c}{\cos \beta} [(\beta - c) \cos \beta - \sin \beta]. \end{aligned} \tag{150a}$$

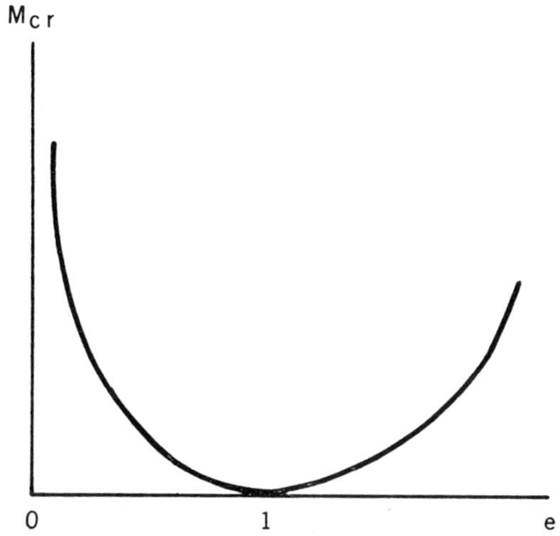


FIGURE 7

$\alpha$ and $\beta$ are parameters, real for an elliptic motion and purely imaginary for a hyperbolic motion; $c = a$ for an elliptic and $c = b$ for a hyperbolic motion. Thus, the radius of convergence of the expansions of $u$ in powers of $(1 - e)$ for all $M$ above a certain critical value is $(1 - e_{cr})$ for an elliptic and $(e_{cr} - 1)$ for a hyperbolic motion. The relation between $e$ and the critical value of $M$ is given by (150) and (150a). $M_{cr}$ as a function of $e$ is shown in Figure 7.

We have seen that, when the eccentric anomaly $u$ is expanded in a power series of $(1 - e)$, a certain region in the neighborhood of the point $(e_0 - 1)$ is cut out from the domain of the convergence of the expansion of $u$. In order to study the motion in this region, Elenevskaya has used Fourier series in the case of an elliptic motion and Laurent series in the case of a hyperbolic motion.

The coordinates of a point moving according to Kepler's laws can be represented in the form

$$\frac{f(e, u)}{(1 - e \cos u)^2},$$

where $f(e, u)$ is an integral function of $u$. If we expand the trigonometric functions in series and then substitute $z = \exp u$, we can transform this expansion for the coordinates to

$$\frac{(-1)^\alpha 2^\alpha z^\alpha f(e, u)}{e^\alpha (z - \beta)^\alpha (z - 1/\beta)^\alpha}, \qquad \beta = \tan\frac{\phi}{2}, \qquad \phi = \sin^{-1} e. \tag{151}$$

This function has no singularity apart from the poles of order $\alpha$. Hence, it can be expanded in a Laurent series within the ring domain between two circles $z = \beta$ and $z = 1/\beta$. The coefficients of such expansions are given by

$$P_n = \frac{(-1)^\alpha 2^\alpha}{e^\alpha 2\pi\sqrt{-1}} \int_C \frac{f(e, u) z^{n+\alpha-1}}{(z - \beta)^\alpha (z - 1/\beta)^\alpha} dz,$$

where $C$ is the circle of unit radius. Since

$$\beta = \tan\frac{\phi}{2} = \frac{\sin\phi_1 + \sqrt{-1}\sinh\phi_2}{\cosh\phi_2 + \cos\phi_1},$$

therefore

$$\frac{1}{\beta} = \frac{\sin\phi_1 - \sqrt{-1}\sinh\phi_2}{\cosh\phi_2 - \cos\phi_1},$$

where

$$e = \xi + \sqrt{-1}\,\eta = \sin\phi = \sin(\phi_1 + \sqrt{-1}\,\phi_2)$$
$$= \sin\phi_1\cosh\phi_2 + \sqrt{-1}\cos\phi_1\sinh\phi_2.$$

The singular points are determined by

$$z_1 = \beta = \exp(\sqrt{-1}\,x_1)\exp(-y_1) = \frac{\sin\phi_1 - \sqrt{-1}\sinh\phi_2}{\cosh\phi_2 + \cos\phi_1},$$

$$z_2 = \frac{1}{\beta} = \exp(\sqrt{-1}\,x_2)\exp(-y_2) = \frac{\sin\phi_1 + \sqrt{-1}\sinh\phi_2}{\cosh\phi_2 - \cos\phi_1}.$$

We separate the real and imaginary parts and take the ratios of corresponding parts. Then we construct the sum of the squares. We obtain the coordinates of the singular points

$$\tan x_1 = \frac{\sinh\phi_2}{\sin\phi_1}, \qquad \exp(-y_1) = \frac{\sqrt{\sin^2\phi_1 + \sinh^2\phi_2}}{\cosh\phi_2 + \cos\phi_1}$$

$$\tan x_2 = -\frac{\sinh\phi_2}{\sin\phi_1}, \qquad \exp(-y_2) = \frac{\sqrt{\sin^2\phi_1 + \sinh^2\phi_2}}{\cosh\phi_2 - \cos\phi_1}.$$

Put

$$z = \exp(-y)\exp(\sqrt{-1}\,x),$$

and transfer the coordinate origin to the point $z - 1$. Expand (151) in powers of $(1 - z)$; such expansions are convergent everywhere within the circle passing through the nearest singular point.

## 5.15 ANDOYER'S EXPANSION

Let $f(z)$ and $\phi(z)$ be two holomorphic functions in the domain of a point $a$, and $C$ be a circle with center $a$ and radius $r$ such that along the circumference of the circle

$$|\alpha f(z)| + |\beta\phi(z)| < r$$

with small constants $\alpha$ and $\beta$. The equation

$$F(z) = z - a - \alpha f(z) - \beta\phi(z) = 0$$

admits one and only one root $z = \zeta$ in the interior of the circle $C$. Any holomorphic function $\Pi(\zeta)$ of the root can be expanded in powers of $\alpha$ and $\beta$ by

$$\frac{\Pi(\zeta)}{F'(\zeta)} = \frac{1}{2\pi\sqrt{-1}}\int_C \frac{\Pi(z)\,dz}{z - a - \alpha f(z) - \beta\phi(z)},$$

which, by Lagrange's theorem (Section 2.3), gives

$$\frac{\Pi(\zeta)}{1 - \alpha f'(\zeta) - \beta\phi'(\zeta)} = \Pi(a) + \sum_{m,n}{}' \frac{\alpha^m\beta^n}{m!\,n!}\frac{d^{m+n}}{da^{m+n}}\{\Pi(a)[f(a)]^m[\phi(a)]^n\},$$

where the combination $m = n = 0$ is excluded in the summation. Put

$$\Pi(z) = \Phi(z)[1 - \alpha f'(z) - \beta\phi'(z)];$$

then,

$$\Phi(\zeta) = \Phi(a) + \sum_{m,n}{}' \frac{\alpha^m\beta^n}{m!\,n!}\frac{d^{n+m-1}}{da^{m+n-1}}\{\Phi'(a)[f(a)]^m[\phi(a)]^n\},$$

where the combination $m = n = 0$ is excluded in the summation. This gives Goursat's generalization of the Lagrange theorem from a single variable to two variables (Goursat, 1913).

Consider with Goursat (1915) two functions $z_1(x, y)$ and $z_2(x, y)$, vanishing for $x = y = 0$. Let $w = S(x, y)$ be the integral series of $x$ and $y$ representing the root $w$ of the equation

$$x \exp(w) + y \exp(-w) = w, \tag{152}$$

such that $S(x, y)$ is convergent for $|x| < R$, $|y| < R'$ and divergent for $|x| > R$, $|y| > R'$. If there are positive numbers $R$, $R'$, $\rho$ such that

$$R \exp(\rho) + R' \exp(-\rho) \leqq \rho, \qquad R \exp(-\rho) + R' \exp(\rho) \leqq \rho, \tag{153}$$

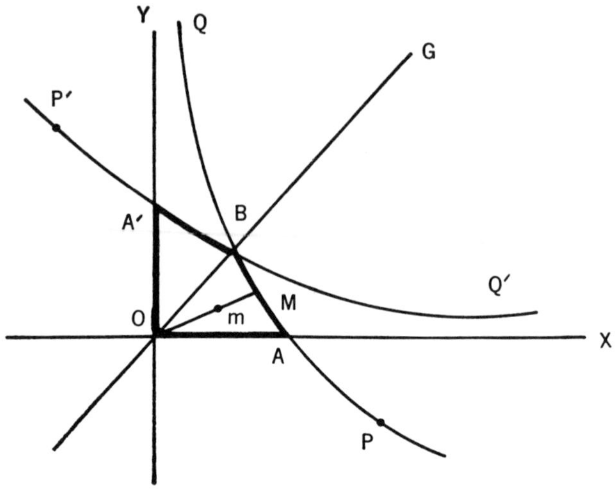


FIGURE 8

then (152) admits only one root whose modulus is smaller than $\rho$ for $|x| < R$, $|y| < R'$.

To prove this theorem of Goursat (1915), describe a circle with radius $\rho$ and with the center as origin on the complex plane of $w$. We have $|x \exp(w) + y \exp(-w)| < \rho$. Along the circumference of the circle,

$$|x \exp(w) + y \exp(-w)| < R \exp(\rho \cos \phi) + R' \exp(-\rho \cos \phi),$$

where $\phi$ is the argument of $w$. The maximum of the right-hand member is either $R \exp(\rho) + R' \exp(-\rho)$ or $R \exp(-\rho) + R' \exp(\rho)$ according as $R \gtrless R'$. If (153) are satisfied, then there is one and only one root $w_1(x, y)$ whose modulus is smaller than $\rho$. This root is a holomorphic function of $x$ and $y$ inside the circles $R$ and $R'$, drawn with $x = 0$ and $y = 0$ as centers, respectively. It can be expanded by the generalized Lagrange theorem just proved.

Let $\Pi(w)$ be a holomorphic function inside the circle $\rho$. By the residue theorem (Section 2.3), we have

$$\frac{\Pi(w_1)}{1 - x \exp(w_1) + y \exp(-w_1)} = \frac{1}{2\pi\sqrt{-1}} \int \frac{\Pi(w)\, dw}{w - x \exp(w) - y \exp(-w)},$$

the integral being taken along the circumference of the circle $\rho$.

In particular, if

$$\Pi(w) = w[1 - x \exp(w) + y \exp(-w)],$$

then

$$w_1(x, y) = \sum_{m,n} \frac{(m-n)^{m+n-1}}{m!\, n!} x^m y^n, \tag{154}$$

where the combination $m = n = 0$ is excluded from the summation. The series (154) is absolutely convergent if there exists a positive number $\rho$ such that

$$|x| \exp(\rho) + |y| \exp(-\rho) < \rho, \qquad |x| \exp(-\rho) + |y| \exp(\rho) < \rho.$$

Geometrically, consider the family of straight lines represented by

$$X \exp(\rho) + Y \exp(-\rho) = \rho \tag{155}$$

with the parameter $\rho$, $0 \leqq \rho \leqq \infty$. The envelope of these straight lines is the curve $\Gamma$ represented by

$$X = \frac{\rho + 1}{2} \exp(-\rho), \qquad Y = \frac{\rho - 1}{2} \exp(\rho).$$

The point $(X, Y)$ starts at a point $(\frac{1}{2}, -\frac{1}{2})$ and describes the infinite branch $PAQ$ which is asymptotic to the $Y$-axis (Figure 8). The abscissa of the intersection $A$ of the curve with the $X$ axis is $1/2.71828\cdots$. The intersection of this curve with the bisectrix $OG$ is the point $B$ for which the value of $\rho$ is given by $(\rho - 1) \exp(2\rho) = \rho + 1$. The envelope of the straight lines (155) is symmetric with respect to the bisectrix $OG$ of the angle $XOY$. The series (154) is absolutely convergent if the point with coordinates $(|x|, |y|)$ is in the region between the broken segment $ABA'$ and the origin because, in the region $OBA$,

$$X \exp(\rho) + Y \exp(-\rho) - \rho < 0,$$

$$X \exp(-\rho) + Y \exp(\rho) - \rho < 0,$$

since $X > Y$. The series (154) is divergent outside the region $OABA'$.

Andoyer (1923) writes

$$s_1 = \frac{e}{2} s, \qquad s_2 = \frac{e}{2} s^{-1}$$

and the function $S$ in Section 5.11 in the form

$$S = \sum_{j_1, j_2} P_{j_1 j_2} z_1^{j_1} z_2^{j_2} = \sum_{j_1, j_2} Q_{j_1 j_2} s_1^{j_1} s_2^{j_2}.$$

Put

$$z_1 = \frac{e}{2} z, \qquad z_2 = \frac{e}{2} z;$$

then, comparing with (110), that is, $S = \sum_j P_j z^j$, we obtain

$$P_j = \sum_{j_1 - j_2 = j} P_{j_1 j_2}\left(\frac{e}{2}\right)^{j_1 + j_2}, \qquad Q_j = \sum_{j_1 - j_2 - j} Q_{j_1 j_2}\left(\frac{e}{2}\right)^{j_1 + j_2}.$$

By Cauchy's first theorem the coefficient $P_{j_1 j_2}$ of the expansion of $S$ equals the coefficient $U_{j_1 j_2}$ of the expansion of

$$U(s_1, s_2) = S(1 - s_1 - s_2) \exp\,[(j_1 - j_2)(s_1 - s_2)]$$

in powers of $s_1$, $s_2$. We have

$$P_j = \sum_{j_1 - j_2 = j} U_{j_1 j_2}\left(\frac{e}{2}\right)^{j_1 + j_2},$$

or

$$P_j = \frac{1}{2\pi\sqrt{-1}} \sum_{(j_1 - j_2 = j)} \left(\frac{e}{2}\right)^{j_1 + j_2} \int^{(0+)} \int^{(0+)} S\cdot(1 - s_1 - s_2)$$
$$\times \exp\,[(j_1 - j_2)(s_1 - s_2)]\cdot s_1^{-j_1 - 1} s_2^{-j_2 - 1}\, ds_1\, ds_2.$$

Suppose that

$$f = \sum_{p_1, p_2} a_{p_1 p_2} z_1^{p_1} z_2^{p_2} = \sum_{q_1, q_2} b_{q_1 q_2} s_1^{q_1} s_2^{q_2};$$

from this, Andoyer (1923) derived

$$a_n = \sum_{(p_1 - p_2 = n)} a_{p_1 p_2}\left(\frac{e}{2}\right)^{p_1 + p_2},$$

$$a_{p_1 p_2} = \sum b_{q_1 q_2}(-1)^{p_2 - q_2} \frac{(q_1 - q_2)(p_1 - p_2)^{p_1 + p_2 - q_1 - q_2 - 1}}{(p_1 - q_1)!\,(p_2 - q_2)!},$$

where $p! = 1$ for $p = 0$ and $p! = \infty$ for a negative integer $p$. For $p_1 = p_2 = p$,

$$a_{p,p} = b_{p,p} - b_{p,p-1} - b_{p-1,p}.$$

Andoyer further computed the expansions of

$$X^{\rho,\sigma} = \left(\frac{r}{a}\right)^{\rho}\left(\frac{t}{z}\right)^{\sigma}, \qquad Y^{\rho,\sigma} = \left(\frac{r}{a}\right)^{\rho}\left(\frac{t}{s}\right)^{\sigma}.$$

Suppose that a holomorphic function is expanded in double power series of

$$z_1 = \frac{e}{2} \exp(\sqrt{-1}\, M), \qquad z_2 = \frac{e}{2} \exp(-\sqrt{-1}\, M)$$

in the form

$$\sum_{j_1, j_2} P_{j_1, j_2} z_1^{j_1} z_2^{j_2}.$$

The convergence of this series is judged by the root of

$$F = w - z_1 \exp(w) + z_2 \exp(-w) = 0, \qquad w = \sqrt{-1}\,(u - M),$$

for which

$$\frac{\partial F}{\partial w} = 1 - z_1 \exp(w) - z_2 \exp(-w) \neq 0.$$

$F(z_1, z_2)$ admits a root $w = 0$ for $z_1 = z_2 = 0$. Other roots can be expanded in integral powers of $z_1$ and $z_2$, and are real. The expansion is absolutely convergent for $|z_1| < \rho/2$, $|z_2| < \rho/2$ with a sufficiently small number $\rho$, if

$$\frac{\partial F}{\partial w} = 1 - z_1 \exp(w) - z_2 \exp(-w) \neq 0$$

for all such values of $z_1$ and $z_2$.

The critical case occurs if $F = 0$ and $\partial F/\partial w = 0$ are satisfied simultaneously. We have

$$z_1 = \tfrac{1}{2}(1 + w) \exp(-w), \qquad z_2 = \tfrac{1}{2}(1 - w) \exp(w).$$

Put $w = a + \sqrt{-1}\, b$; then,

$$|z_1|^2 = \tfrac{1}{4}[(1 + a)^2 + b^2] \exp(-2a),$$
$$|z_2|^2 = \tfrac{1}{4}[(1 - a)^2 + b^2] \exp(2a).$$

Then the foregoing condition for $z_1$ and $z_2$ is that

$$|(1 + a)^2 + b^2| \exp(-2a) < \rho^2, \qquad |(1 - a)^2 + b^2| \exp(2a) < \rho^2,$$

should admit no common root.

Consider in the complex plane of $w$ the equation

$$|(1 + a)^2 + b^2| \exp(-2a) = \rho^2,$$

or

$$b^2 = \rho^2 \exp(2a) - (1 + a)^2.$$

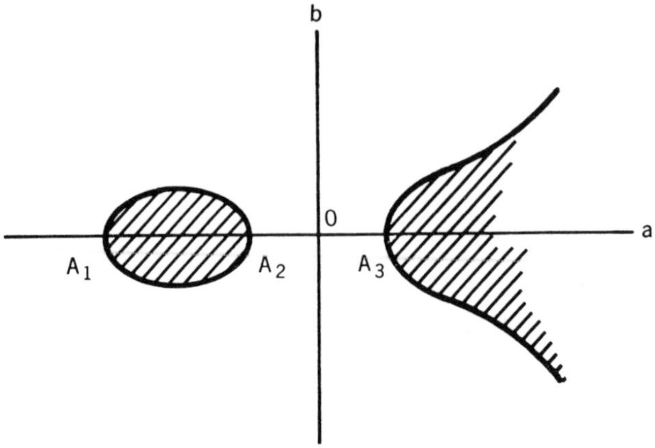


FIGURE 9

The equation $\rho^2 \exp(2a) - (1 + a)^2 = 0$, viewed as an equation in $a$, admits three real roots since $0 < \rho < 1$. One root is smaller than $-1$, the second is between $-1$ and 0, and the third is positive. The points satisfying $|(1 + a)^2 + b^2| \exp(-2a) < \rho^2$ are in the shaded area of Figure 9 on the complex plane of $w$. The area for which

$$|(1 + a)^2 + b^2| \exp(2a) < \rho^2$$

is symmetric to this area with respect to the $b$-axis. Hence the condition for convergence is $OA_1 < OA_3$. This is satisfied for small values of $\rho$, up to the value of $\rho$ such that

$$\rho^2 = (1 + a)^2 \exp(-2a) = (1 - a)^2 \exp(2a).$$

This value of $\rho$ is larger than unity for any positive value of $a$, because $a = \coth a$, $\rho = \sqrt{a^2 - 1}$. This is equal to the value $\rho$ in (143).

## VARIATION OF CONSTANTS

### 5.16 DELAUNAY'S ELEMENTS

In formula (70), Section 5.6, put

$$H' = \sum_{\kappa=1}^{n-1} \left( \frac{y_{3\kappa-2}'^2}{2m_{3\kappa-2}'} + \frac{y_{3\kappa-1}'^2}{2m_{3\kappa-1}'} + \frac{y_{3\kappa}'^2}{2m_{3\kappa}'} \right) - \sum_{\kappa=1}^{n-1} \frac{m_3 m_{3\kappa}}{(x_{3\kappa-2}'^2 + x_{3\kappa-1}'^2 + x_{3\kappa}'^2)^{1/2}}, \tag{156}$$

$$H'' = \sum_{\kappa=1}^{n-1} \frac{m_3 m_{3\kappa}}{(x_{3\kappa-2}'^2 + x_{3\kappa-1}'^2 + x_{3\kappa}'^2)^{1/2}} - \sum_{\tau,\nu=1}^{n} \frac{m_{3\tau} m_{3\nu}}{r_{3\tau,3\nu}}, \tag{157}$$

$$H = H' + H'',$$

and suppose that any mass $m_i$ other than $m_1 = m_2 = m_3$ is of magnitude $\mu \ll 1$. From (63), Section 5.6,

$$x'_{3\kappa-2} = x_{3(\kappa+1)-2} - x_1 + \frac{m_4 x_4 + \cdots + m_{3\kappa-2} x_{3\kappa-2}}{m_1 + m_4 + \cdots + m_{3\kappa-2}} - \sum_{\tau=2}^{\kappa} (m_{3\tau-2} x_{3\tau-2}) \Big/ \left(\sum_{\tau=1}^{\kappa} m_{3\tau-2}\right);$$

therefore the difference between $(x'^2_{3\kappa-2} + x'^2_{3\kappa-1} + x'^2_{3\kappa})^{1/2}$ and $r_{3,3\kappa-3}$ is of magnitude $\mu$. Hence, the difference between the first sum on the right-hand side of $H''$ and the second sum extending to $\tau = 1$, $\nu = 2, 3, \ldots, n$ is of magnitude $\mu$. A term with $\tau \neq 1$, $\nu \neq 1$ is of magnitude $\mu^2$. Hence, $H''$ has a factor $\mu$. Accordingly, $H''$ is of magnitude $\mu$ compared with $H'$, and can be regarded as the perturbation. At first we consider $H'$ only and determine the *intermediary orbit.*

We have

$$H' = \sum_{\kappa=1}^{n-1} H'_\kappa,$$

$$H'_\kappa = \frac{1}{2m'_{3\kappa}} (y'^2_{3\kappa-2} + y'^2_{3\kappa-1} + y'^2_{3\kappa}) - \frac{m_3 m_{3\kappa}}{(x'^2_{3\kappa-2} + x'^2_{3\kappa-1} + x'^2_{3\kappa})^{1/2}}. \tag{158}$$

Hence, the Hamiltonian function is separable,

$$\frac{dx_i}{dt} = \frac{\partial H'_\kappa}{\partial y_i}, \qquad \frac{dy_i}{dt} = -\frac{\partial H'_\kappa}{\partial x_i} \qquad \begin{pmatrix} i = 3\kappa - 2, 3\kappa - 1, 3\kappa; \\ \kappa = 1, 2, \ldots, n - 1 \end{pmatrix}. \tag{159}$$

Referring to the notation of Sections 5.9 and 5.10, write $\beta_\kappa = m'_{3\kappa} m_3 m_{3\kappa}$ in place of $\beta$ and $m'_{3\kappa} h_1$ in place of $h_1$; this gives us a set of canonical variables derived from (109), Section 5.10,

$$h_{3\kappa-2} = -\frac{\beta_\kappa}{2m'_{3\kappa} a}, \qquad h_{3\kappa-1} = \sqrt{\beta_\kappa a_\kappa (1 - e_\kappa^2)},$$

$$h_{3\kappa} = \sqrt{\beta_\kappa a_\kappa (1 - e_\kappa^2)} \cos I_\kappa, \tag{160}$$

$$\gamma_{3\kappa-2} = -t_\kappa, \qquad \gamma_{3\kappa-1} = \varpi_\kappa - \Omega_\kappa, \qquad \gamma_{3\kappa} = \Omega_\kappa.$$

Hence, by the principle of variation of constants stated in Sections 1.12–1.17, the differential equations for the variation of elements are

$$\frac{d\gamma_i}{dt} = \frac{\partial H''}{\partial h_i}, \qquad \frac{dh_i}{dt} = -\frac{\partial H''}{\partial \gamma_i} \qquad (i = 1, 2, \ldots, 3n - 3). \tag{161}$$

Now transform the variables in (160). Omit the suffixes $\kappa$. From the relations $n = \sqrt{\beta m'}/a^{3/2}$, $h_1 = -\beta/2a$ of Section 5.10 we obtain $n = \sqrt{m'}(-2h_1)^{3/2}/\beta$. The Hamiltonian $H''$ is a function of $x'_1, \ldots, x'_{3n-3}$. The latter variables are expressed in terms of elliptic motions, and hence are periodic functions of $M$ in the expansions described in Sections 5.9–5.15, in particular (105) of Section 5.10.

In the present notation they are multiply periodic functions of arguments $n_\kappa(1 + \gamma_{3\kappa})$. But, as $n_\kappa$ is a function of $h_{3\kappa-2}$, the time $t$ appears explicitly in front of the periodic function in the right-hand member of (161). In order to avoid this inconvenience we make the following transformation:

First, put $l = n(t + \gamma)$; then

$$\frac{dh}{dt} = -n\frac{\partial H''}{\partial t}.$$

Furthermore,

$$\frac{dl}{dt} = n\left(1 + \frac{d\gamma}{dt}\right) + (t + \gamma)\frac{\partial n}{\partial h}\frac{dh}{dt}.$$

Let $(\partial H''/\partial h)$ be the result of differentiating $H''$ with respect to $h$ by supposing that $h$ is not contained in $n$. Then

$$\frac{\partial H''}{\partial h} = \left(\frac{\partial H''}{\partial h}\right) + \frac{\partial H''}{\partial l}(t + \gamma)\frac{\partial n}{\partial h}.$$

If we put $d\gamma/dt = \partial H''/\partial h$ in the above equation for $dl/dt$, then we obtain

$$\frac{dl}{dt} = n + n\left(\frac{\partial H''}{\partial h}\right) + n\frac{\partial H''}{\partial l}(t + \gamma)\frac{\partial n}{\partial h} + (t + \gamma)\frac{\partial n}{\partial h}\frac{dh}{dt}.$$

Hence

$$\frac{dl}{dt} = n + n\left(\frac{\partial H''}{\partial h}\right).$$

Put $L = \sqrt{\beta a}$; then, $h = -\beta^2/(2m'L^2)$, $n = \beta^2/(m'L^3)$, and

$$-n\frac{\partial H''}{\partial l} = \frac{dh}{dt} = \frac{\beta^2}{L^3}\frac{dL}{dt} = n\frac{dL}{dt}.$$

Hence, $dL/dt = -\partial H''/\partial l$. Furthermore, since

$$\left(\frac{\partial H''}{\partial L}\right) = \left(\frac{\partial H''}{\partial L}\right)\frac{\partial h}{\partial L} = n\left(\frac{\partial H''}{\partial h}\right),$$

we have

$$\frac{dl}{dt} = n + \left(\frac{\partial H''}{\partial L}\right),$$

where $(\partial H''/\partial L)$ denotes the result of differentiating under the supposition that $L$ is not contained in $n$. Put

$$-\frac{\beta^2}{2m'L^2} + H'' = F;$$

then

$$\frac{dL}{dt} = \frac{\partial F}{\partial l}, \qquad \frac{dl}{dt} = -\frac{\partial F}{\partial L},$$

because

$$-\frac{\partial F}{\partial L} = n + \frac{\partial H''}{\partial L}.$$

Now write

$$F = \sum_{\kappa=1}^{n-1} \frac{\beta_\kappa^2}{3m'_{3\kappa}L_3^2} - H'' \tag{162}$$

and

$$L = \sqrt{\beta a}, \qquad G = \sqrt{\beta a(1 - e^2)}, \qquad H = \sqrt{\beta a(1 - e^2)} \cos I,$$
$$l = n(t + \gamma), \qquad g = \varpi - \Omega, \qquad h = \Omega, \tag{163}$$

for each $\kappa$; then we have the canonical equations

$$\begin{aligned} \frac{dL}{dt} &= \frac{\partial F}{\partial l}, & \frac{dG}{dt} &= \frac{\partial F}{\partial g}, & \frac{dH}{dt} &= \frac{\partial F}{\partial h}, \\ \frac{dl}{dt} &= -\frac{\partial F}{\partial L}, & \frac{dg}{dt} &= -\frac{\partial F}{\partial G}, & \frac{dh}{dt} &= -\frac{\partial F}{\partial H}. \end{aligned} \tag{164}$$

In particular, when $n = 3$,

$$m'_6 = \frac{m_3 m_6}{m_3 + m_6}, \qquad m'_9 = \frac{m_9(m_3 + m_6)}{m_3 + m_6 + m_9};$$

accordingly,

$$\beta_1 = \frac{k^2 m_3^2 m_6^2}{m_3 + m_6}, \qquad \beta_2 = \frac{k^2 m_9^2 m_3(m_3 + m_6)}{m_3 + m_6 + m_9}, \tag{165}$$

$$F = \frac{\beta_1^2}{2m'_6 L_1^2} + \frac{\beta_2^2}{2m'_9 L_2^2} + \frac{m_6 m_9}{r_{6,9}} + \frac{m_3 m_9}{r_{3,9}} - \frac{m_3 m_9}{\sqrt{x_4'^2 + x_5'^2 + x_6'^2}}. \tag{166}$$

The difference of the last two terms of $F$ is of magnitude $\mu^2$; the third term is of magnitude $\mu^2$. The first two terms are of magnitude $\mu$; we say that they are of *order* 2 and 1, respectively, with regard to the disturbing masses. $L$, $G$, $H$ are of order 1 with regard to $\mu$. These canonical variables $L, G, H, l, g, h$ are called *Delaunay's elements* (Delaunay, 1860, 1867). The variables $a, e, I, t_0, \varpi, \Omega$ are called the *Keplerian elements.* Write

$$F = F_0 + \mu F_1,$$

$$F_0 = \frac{\beta_1^2}{2m_6' L_1^2} + \frac{\beta_2^2}{2m_9' L_2^2}, \tag{167}$$

$$\mu F_1 = \frac{m_6 m_9}{r_{6,9}} + \frac{m_3 m_9}{r_{3,9}} - \frac{m_3 m_9}{\sqrt{x_4'^2 + x_5'^2 + x_6'^2}}.$$

The term $\mu F_1$ indicates the disturbing function and is multiplied by the factor $\mu$ of the order of the disturbing masses. The term $F_0$ appears only in the right-hand member of $dl/dt$ in (164). Let the solution of (164) for $\mu = 0$ be denoted by $L^{(0)}, G^{(0)}, H^{(0)}, l^{(0)}, g^{(0)}, h^{(0)}$, and write

$$n_i = -\frac{2\beta_i^2}{3m_{3i}' L_i^3} = \frac{\partial F_0}{\partial L_i}.$$

Then we obtain $l_i^{(0)} = n_i(t + \gamma_i)$, with the integration constant $\gamma_i$; the other elements are now all constants. Denote the initial values for $t = 0$ of these elements by $L^{(0)}, G^{(0)}, H^{(0)}, l^{(0)}, g^{(0)}, h^{(0)}$, and consider the quantities of magnitude $\mu$. Then

$$L_i = L_i^{(0)} + \mu \int_0^t \frac{\partial F_1}{\partial l_i}\,dt, \qquad G_t = G_i^{(0)} + \mu \int_0^t \frac{\partial F_1}{\partial g_i}\,dt,$$

$$H_i = H_i^{(0)} + \mu \int_0^t \frac{\partial F_1}{\partial h_i}\,dt,$$

$$l_i = l_i^{(0)} - \int_0^t \frac{\partial F_0}{\partial L_i}\,dt - \mu \int_0^t \frac{\partial F_1}{\partial L_i}\,dt, \qquad g_i = g_i^{(0)} - \mu \int_0^t \frac{\partial F_1}{\partial G_i}\,dt,$$

$$h_i = h_i^{(0)} - \mu \int_0^t \frac{\partial F_1}{\partial H_i}\,dt.$$

The terms with the factor $\mu$ all represent the perturbation. Let $F_0^{(0)}$ be the result of substituting the values for $\mu = 0$ in $F_0$; let $\Phi$ be the aggregate of terms of the third and higher orders with regard to $\delta L_i$; let $C_{ij}$ be a function of $L_i^{(0)}$ and $L_j^{(0)}$ with $C_{ij} = C_{ji}$. Then,

$$F_0 = F_0^{(0)} + \sum_{i=1}^{n} n_i\,\delta L_i + \frac{1}{2}\sum_{i,j} C_{ij}\,\delta L_i\,\delta L_j + \Phi.$$

Indicate the effect of the perturbation on the values for $t = 0$ by putting $\delta$ in front, and refer to the relation

$$\frac{\partial F_0}{\partial L_i} = n_i + \sum_{j=0}^{n} C_{ij}\,\delta L_j + \frac{\partial \Phi}{\partial L_i} = n_i + \mu \sum_{j=1}^{n} C_{ij} \int \frac{\partial F_1}{\partial l_j}\,dt + \frac{\partial \Phi}{\partial L_i};$$

then

$$\delta L_i = \mu \int_0^t \frac{\partial F_1}{\partial l_i}\,dt, \qquad \delta G_i = \mu \int_0^t \frac{\partial F_1}{\partial g_i}\,dt, \qquad \delta H_i = \mu \int_0^t \frac{\partial F_1}{\partial h_i}\,dt,$$

$$\delta l_i = -\mu \sum_{j=1}^{n} C_{ij} \int_0^t dt \int_0^t \frac{\partial F_1}{\partial l_j}\,dt - \int_0^t \frac{\partial \Phi}{\partial L_i}\,dt - \mu \int_0^t \frac{\partial F_1}{\partial L_i}\,dt, \tag{164a}$$

$$\delta q_i = -\mu \int_0^t \frac{\partial F_1}{\partial G_i}\,dt, \qquad \delta h_i = -\mu \int_0^t \frac{\partial F_1}{\partial H_i}\,dt.$$

This result has been obtained by Poincaré (1905).

## 5.17 POINCARÉ'S ELEMENTS

We carry out the following transformation to Delaunay's elements:

$$\begin{aligned} \Lambda &= L, & \Gamma &= L - G, & Z &= G - H, \\ \lambda &= l + g + h, & \gamma &= -g - h, & z &= -h. \end{aligned} \tag{168}$$

This is a contact transformation, as we see from any of the methods described in Sections 1.7–1.11, or from the relation

$$\Lambda\lambda + \Gamma\gamma + Zz = Ll + Gg + Hh.$$

It is written in the Keplerian elements

$$\begin{aligned} \Lambda &= \sqrt{\beta a}, \qquad \Gamma = \sqrt{\beta a}(1 - \sqrt{1 - e^2}), \\ Z &= \sqrt{\beta a(1 - e^2)}(1 - \cos I), \\ \lambda &= M + \varpi, \qquad \gamma = -\varpi, \qquad z = -\Omega. \end{aligned} \tag{169}$$

$\lambda$ is the mean longitude in the orbit, $\Gamma$ is of magnitude $e^2$, and $Z$ is of magnitude $I^2$.

Put

$$\begin{aligned} \xi &= \sqrt{2\Gamma}\cos\gamma, & p &= \sqrt{2Z}\cos z, \\ \eta &= \sqrt{2\Gamma}\sin\gamma, & q &= \sqrt{2Z}\sin z. \end{aligned} \tag{170}$$

This is also a contact transformation because

$$\xi\, d\eta = \cos\gamma \sin\gamma\, d\Gamma + 2\Gamma\cos^2\gamma\, d\gamma = \tfrac{1}{2}d(\Gamma\sin 2\gamma) + \Gamma\, d\gamma.$$

$\xi$, $\eta$ are called the *eccentric variables*, and $p$, $q$ the *oblique variables*.

$$\begin{aligned}
\Lambda &= \sqrt{\beta a}, \qquad \xi = \sqrt{2\Lambda(1 - \sqrt{1-e^2})}\cos\varpi,\\
p &= \sqrt{2\Lambda\sqrt{1-e^2}(1-\cos I)}\cos\Omega,\\
\lambda &= M + \varpi, \qquad \eta = -\sqrt{2\Lambda(1-\sqrt{1-e^2})}\sin\varpi,\\
q &= -\sqrt{2\Lambda\sqrt{1-e^2}(1-\cos I)}\sin\Omega.
\end{aligned} \tag{171}$$

From these equations,

$$\left(\frac{\xi}{\sqrt{\Lambda}}\right)^2 + \left(\frac{\eta}{\sqrt{\Lambda}}\right)^2 = 2(1-\sqrt{1-e^2}) = e^2 + \tfrac{1}{4}e^4 + \cdots.$$

Therefore,

$$e^2 = \left(\frac{\xi}{\sqrt{\Lambda}}\right)^2 + \left(\frac{\eta}{\sqrt{\Lambda}}\right)^2 + \frac{1}{4}\left[\left(\frac{\xi}{\sqrt{\Lambda}}\right)^2 + \left(\frac{\eta}{\sqrt{\Lambda}}\right)^2\right]^2 + \cdots.$$

Since $\sqrt{2(1-\sqrt{1-e^2})} = e(1 + \frac{1}{8}e^2 + \cdots)$, and

$$\frac{\xi}{\sqrt{\Lambda}} = e\cos\varpi\left(1 + \sum_{j=1}^{\infty}\alpha_j e^{2j}\right), \qquad \frac{\eta}{\sqrt{\Lambda}} = -e\sin\varpi\left(1 + \sum_{j=1}^{\infty}\alpha_j e^{2j}\right)$$

with numerical coefficients $\alpha_j$, therefore,

$$e\cos\varpi = \frac{\xi}{\sqrt{\Lambda}}\left\{1 - \frac{1}{8}\left[\left(\frac{\xi}{\sqrt{\Lambda}}\right)^2 + \left(\frac{\eta}{\sqrt{\Lambda}}\right)^2\right] + \cdots\right\},$$

$$e\sin\varpi = -\frac{\eta}{\sqrt{\Lambda}}\left\{1 - \frac{1}{8}\left[\left(\frac{\xi}{\sqrt{\Lambda}}\right)^2 + \left(\frac{\eta}{\sqrt{\Lambda}}\right)^2\right] + \cdots\right\}.$$

Hence $e\cos\varpi$ and $e\sin\varpi$ can be expanded in integral power series of $\xi/\sqrt{\Lambda}$ and $\eta/\sqrt{\Lambda}$. Conversely, $\xi/\sqrt{\Lambda}$ and $\eta/\sqrt{\Lambda}$ can be expanded in integral power series of $e\cos\varpi$ and $e\sin\varpi$. Thus, $e\cos\varpi$, $e\sin\varpi$, $\sin(I/2)\cos\Omega$, $\sin(I/2)\sin\Omega$ can be expanded in integral power series of $\xi/\sqrt{\Lambda}$, $\eta/\sqrt{\Lambda}$, $p/\sqrt{\Lambda}$, $q/\sqrt{\Lambda}$. The converse is also true.

Poincaré (1905; see also Charlier, 1927; Picart, 1906) treated the three-body problem in this manner. The coefficients of the expansion in the Keplerian elements are available in tabular form, as will be seen in a later chapter, but the coefficients of the expansion in Delaunay's or

Poincaré's elements are scarcely ever published (Norén and Walberg, 1910). So the expansion in either of these latter elements is not of much use, except in theoretical discussion. Delaunay himself has avoided making complete reference to his canonical variables in his lunar theory (Delaunay, 1860, 1867).

By the expansions for an elliptic motion in Sections 5.9–5.15 the expressions $u - M$, $\cos(u - M)$, $\sin(u - M)$ can be expanded in integral power series of $e \cos M$, $e \sin M$. Similarly, $e^2 \cos(u + M)$, $e^2 \sin(u + M)$ can be expanded. Hence, the expressions $X$, $Y$ in (106) of Section 5.10

$$\frac{X}{a} = \cos(u - M)\,\frac{1 + \sqrt{1 - e^2}}{2} + e^2 \cos(u + M)\,\frac{1 - \sqrt{1 - e^2}}{2e^2} - e \cos M,$$

$$\frac{Y}{a} = \sin(u - M)\,\frac{1 + \sqrt{1 - e^2}}{2} - e^2 \sin(u + M)\,\frac{1 - \sqrt{1 - e^2}}{2e^2} + e \sin M,$$

can be expanded in integral power series of $e \cos M$, $e \sin M$ because

$$e^2 = e^2 \cos^2 M + e^2 \sin^2 M.$$

As

$$e \cos M = e \cos \varpi \cos \lambda - e \sin \varpi \sin \lambda,$$

$$e \sin M = e \sin \varpi \cos \lambda + e \cos \varpi \sin \lambda,$$

the expressions $e \cos M$, $e \sin M$ are Fourier series in $\lambda$ with the Fourier coefficients expressed as integral power series of $\xi/\sqrt{\Lambda}$, $\eta/\sqrt{\Lambda}$. We have similar Fourier expansions for $X$, $Y$.

The coefficients of $X$, $Y$ in (108) of Section 5.10 are of the form

$$\pm \cos^2 \frac{I}{2} \begin{matrix}\cos\\ \sin\end{matrix}\, \lambda \pm \sin^2 \frac{I}{2} \begin{matrix}\cos\\ \sin\end{matrix}\, (\lambda + 2h) \qquad \text{or} \qquad \sin I \begin{matrix}\cos\\ \sin\end{matrix}\, (\lambda + h).$$

Their coefficients are expanded in integral power series of

$$\sin \frac{I}{2} \cos \Omega, \qquad \sin \frac{I}{2} \sin \Omega,$$

and accordingly they are expanded in integral power series of $\xi/\sqrt{\Lambda}$, $\eta/\sqrt{\Lambda}$, $\rho/\sqrt{\Lambda}$, $q/\sqrt{\Lambda}$. Hence, $x_1$, $x_2$, $x_3$ are expanded in integral power series of these latter quantities and are of the form $\sum AN \cos (j\lambda + b)$, where: (i), $N$ is an integral power series of $\xi, \eta, p, q$; (ii), $A$ and $b$ depend only on $\Lambda$; and (iii), $j$ is a positive or negative integer. Thus using (170) we can write the expansion in the form

$$\sum A'\Gamma^{s_1}Z^{s_2} \cos (j\lambda + j_1\gamma + j_2 z + b'), \tag{172}$$

where $A'$, $b'$ depend only on $\Lambda$, and $j$, $j_1$, $j_2$ are positive or negative integers. The integers $2s_1$, $2s_2$ are positive or zero, such that

$$2s_i \geqq |j_i| \qquad (i = 1, 2).$$

A similar expansion can be obtained with respect to the Keplerian elements.

The perturbative term $\mu F_1$ of (167) can also be expanded in a similar form, because $x_{3\kappa-2}$, $x_{3\kappa-1}$, $x_{3\kappa}$ are expanded in integral powers of $\xi_\kappa$, $\eta_\kappa$, $p_\kappa$, $q_\kappa$. Actually, $\mu F_1$ ceases to be regular with respect to $x_{3\kappa-2}$, $x_{3\kappa-1}$, $x_{3\kappa}$ only when any one of the mutual distances of the $n$ bodies becomes zero or when $(x'^2_{3\kappa-2} + x'^2_{3\kappa-1} + x'^2_{3\kappa})^{1/2}$ becomes zero. For the moment we ignore such cases and leave their discussion to a later chapter. In such cases the word "perturbation" loses its meaning; we cannot consider the actual motion as a small deviation from the intermediary orbit (in the present case an elliptic motion), so that a completely different mode of treatment should be adopted. For convenience we limit ourselves to the case of three bodies; then the expansions are represented in the series form

$$F = F_0 + \mu F_1, \tag{173}$$

$$\mu F_1 = \sum A_{s_1 s_2 s_3 s_4} \Gamma^{s_1} Z^{s_2} \Gamma'^{s_3} Z'^{s_4} \cos [j\lambda + j'\lambda' + j_1\gamma + j_2\gamma' + j_3 z + j_4 z'],$$

where

$$A \text{ is a function of } \Lambda, \Lambda',$$

$$\Gamma' \text{ is of magnitude } e^2,$$

and

$$Z \text{ is of magnitude } I^2.$$

The representative term written here is of degree $2(s_1 + s_2 + s_3 + s_4)$ with respect to $e$ and $I$. This is called the *degree* of the term. Note that $F_0$ is a function of $\Lambda$ and $\Lambda'$ only and is not multiplied by any quantity of the order of the disturbing masses.

The quantities $\Lambda$ and $\Lambda'$ are constant multiples of $\sqrt{a}$ and $\sqrt{a'}$, respectively; $F_0$ is the sum of constant multiples of $L^{-2}$ and $L'^{-2}$. As $\mu F_1$ is of degree $-1$ with regard to the length, so is the function $F$ itself.

Hence, the coefficient $A$ of (173) is homogeneous of degree $-1$ with regard to $a$, $a'$, and homogeneous of degree $-2$ with regard to $\Lambda$, $\Lambda'$. Let

$$\mathscr{D} = j\lambda + j'\lambda' + j_1\gamma + j_2\gamma' + j_3 z + j_4 z',$$

that is, the quantity under the cosine in (173). We call $\mathscr{D}$ the *argument.* From (104) of Section 5.10, we have $\lambda = nt + \epsilon$, $\lambda' = n't + \epsilon'$. If $j = j' = 0$, then $t$ does not appear explicitly in the argument, and the corresponding term does not contain $t$ explicitly. Such a constant term is called a *secular term.* If $jn + j'n' = 0$, $j \neq 0$, $j' \neq 0$, then a similar situation occurs; this is when $n:n'$ is rational, and the corresponding term is called a *critical term.*

The function $F$, being a function of the mutual distances of the three bodies, does not depend on the sense of the longitude. Consequently, $F$ remains unaltered when we change the signs of $\lambda$, $\lambda'$, $\gamma$, $\gamma'$, $z$, $z'$ simultaneously. Hence $b'$, representing the phase of the argument, should be zero and each term should be of cosine form. The differences

$$2s_i - |j_i| \qquad (i = 1, 2, 3, 4),$$

are even positive integers including 0, where $s_i$ is a positive integral multiple of $\frac{1}{2}$. Since $F$ also remains unaltered by the reversal of the sense in which the inclination is taken, the sums $2s_3 + 2s_4$ and $j_3 + j_4$ should be even integers. Since $F$ remains unaltered by the rotation through $\delta$ in longitude of the coordinate axes around the origin, we obtain $j + j' + j_1 + j_2 + j_3 + j_4 = 0$. Finally, the degree of a term of $F_1$ is equal to or higher than the sum of the coefficients of $\lambda$, $\lambda'$ in the argument; that is,

$$\begin{aligned} 2s_1 + 2s_2 + 2s_3 + 2s_4 &\geqq |j_1| + |j_2| + |j_3| + |j_4| \\ &\geqq |j_1 + j_2 + j_3 + j_4| \geqq |j + j'|. \end{aligned}$$

Brown and Shook (1933) have called a series of this nature a *D'Alembert series.* If we use Poincaré's elements, then the argument is $j\lambda + j'\lambda'$ and the coefficient has the factor $\xi^{s_1}\eta^{s'_1}p^{s_3}q^{s'_3}$ for one of the planets, and the factor $\xi'^{s_2}\eta'^{s'_2}p'^{s_4}q'^{s'_4}$ for the other planet instead of $\Gamma$, $Z$. The degree of the term is $s_1 + s'_1 + s_2 + s'_2 + s_3 + s'_3 + s_4 + s'_4$. Thus the differential equations are

$$\begin{aligned} \frac{d\Lambda}{dt} &= \mu\frac{\partial F_1}{\partial \lambda}, & \frac{d\Gamma}{dt} &= \mu\frac{\partial F_1}{\partial \gamma}, & \frac{dZ}{dt} &= \mu\frac{\partial F_1}{\partial z}, \\ \frac{d\lambda}{dt} &= -\frac{\partial F_0}{\partial \Lambda} - \mu\frac{\partial F_1}{\partial \Lambda}, & \frac{d\gamma}{dt} &= -\mu\frac{\partial F_1}{\partial \Gamma}, & \frac{dz}{dt} &= -\mu\frac{\partial F_1}{\partial z}. \end{aligned} \tag{174}$$

A similar argument can be applied to other bodies. If the Hamiltonian function has the form (156) and (157), then $F$ — given in (162) — is common to all bodies, and the differential equations are in canonical form. The element $\lambda$, as can be seen from (174), varies rapidly compared with the other quantities, which vary slowly in the order of the disturbing mass $\mu$. Note that $F_0$ contains only $\Lambda$ and $\Lambda'$; also, $\mu F_1$ is a trigonometric series with the sum of integral multiples of $\lambda$, $\lambda'$, $\gamma$, $\gamma'$, $z$, $z'$ as argument, and each coefficient is an integral power series of $\Gamma$, $Z$, $\Gamma'$, $Z'$, multiplied by a function of $\Lambda$, $\Lambda'$. This is the fundamental form of the equations in the perturbation theory of celestial mechanics that will be the main topic of the following discussions.

Corresponding to (164a), we obtain from (174) the expressions for the perturbation for time $t > 0$:

$$\delta\Lambda = \mu \int_0^t \frac{\partial F_1}{\partial \lambda}\, dt, \qquad \delta\Gamma = \mu \int_0^t \frac{\partial F_1}{\partial \gamma}\, dt, \qquad \delta Z = \mu \int_0^t \frac{\partial F_1}{\partial z}\, dt,$$

$$\delta\lambda = -\mu \sum_{j=1}^{n} C_{ij} \int_0^t dt \int_0^t \frac{\partial F_1}{\partial \Lambda}\, dt - \int_0^t \frac{\partial \Phi}{\partial \Lambda}\, dt - \mu \int_0^t \frac{\partial F_1}{\partial \Lambda}\, dt, \tag{174a}$$

$$\delta\gamma = -\mu \int_0^t \frac{\partial F_1}{\partial \Gamma}\, dt, \qquad \delta z = -\mu \int_0^t \frac{\partial F_1}{\partial Z}\, dt.$$

These expressions are important for the discussion of the form of integrals in perturbation theory (Poincaré, 1905).

Each term in the solution of (174) or in (174a) in the form of trigonometric series corresponds to a term in the trigonometric series expansion of $\mu F_1$ with the same argument. According to the method of integration, the degree or order of a term may differ from the degree or order of the corresponding term in the expansion of the disturbing function. Each term in the trignometric series for the expansion of the perturbation is called an *inequality*. Even though the expansion of the disturbing function may be convergent, it does not necessarily follow that the resulting solution for the perturbation is convergent. Indeed, the trigonometric series obtained in this manner is not, as a rule, uniformly convergent. Here we meet the greatest difficulty in perturbation theory. The theory at present resolves itself into the question; what is the best method of expansion for practical use, in which we can best approximate the rigorous solution by computing as few terms as possible from the series representation thus obtained? The series representation to be studied for such a purpose may be different for different planetary systems, for different satellite systems, or for artificial celestial vehicles. The same situation occurs when we use the Keplerian elements. As a matter of fact, the ordinary perturbation theory uses the Keplerian elements, as will be seen in Volume II.

## 5.18 KEPLERIAN ELEMENTS

In the Keplerian elements the expansion of the disturbing function is, as is easily verified,

$$\sum \mathscr{C} \cos \mathscr{D},$$

$$\mathscr{C} = \mathscr{A} e^{q_1} e'^{q_2} (\tan I)^{q_3} (\tan I')^{q_4}, \tag{173a}$$

$$\mathscr{D} = j\lambda + j'\lambda' - j_1\varpi - j_2\varpi' - j_3\Omega - j_4\Omega',$$

where $\mathscr{A}$ has the same character as $\Lambda$ in (173), so that: (i), it is homogeneous of degree $-1$ with respect to the two semimajor axes $a$ and $a'$; (ii), $q_i - |j_i|$ $(i = 1, 2, 3, 4)$ are even positive integers including 0, and; (iii), $q_3 + q_4$ and $j_3 + j_4$ are both even, in $j + j' + j_1 + j_2 + j_3 + j_4 = 0$. These are the necessary characteristics of a *D'Alembert series.*

The Delaunay elements also have this character, as is evident from (168). The expressions $1 - G^2/L^2$, $1/2 - H/(2G)$ in terms of the Delaunay elements are, respectively, $e^2$, $4 \sin^2 (I/2)$ and take the place of $\Gamma$, $Z$ in terms of the Poincaré elements, which are, respectively, $e^2$, $\tan^2 I$ in terms of the Keplerian elements. Hence ,the expansion of the disturbing function is of the form

$$\sum A(L, L')\left(1 - \frac{G^2}{L^2}\right)^{s_1}\left(1 - \frac{G'^2}{L'^2}\right)^{s_2}\left(\frac{1}{2} - \frac{H}{2G}\right)^{s_3}\left(\frac{1}{2} - \frac{H'}{2G'}\right)^{s_4}$$
$$\times \cos [jl + j'l' + (j - j_1)g + (j' - j_2)g'$$
$$+ (j - j_1 - j_3)h + (j' - j_2 - j_4)h']. \tag{173b}$$

In his lunar theory Delaunay used the noncanonical elements

$$a \quad \text{for} \quad L^2/\beta,$$

$$e \quad \text{for} \quad \left(1 - \frac{G^2}{L^2}\right)^{1/2},$$

$$\gamma \quad \text{for} \quad \left(\frac{1}{2} - \frac{H}{2G}\right)^{1/2}.$$

The second set of the Delaunay canonical equations (164) is replaced by

$$\frac{\partial F}{\partial L} = \frac{2}{na}\frac{\partial F}{\partial a} + \frac{1 - e^2}{na^2 e}\frac{\partial F}{\partial e},$$

$$\frac{\partial F}{\partial G} = -\frac{\sqrt{1 - e^2}}{nae}\frac{\partial F}{\partial e} + \frac{1 - 2\gamma^2}{4na^2\gamma\sqrt{1 - e^2}}\frac{\partial F}{\partial \gamma},$$

$$\frac{\partial F}{\partial H} = -\frac{1}{4na^2\gamma\sqrt{1 - e^2}}\frac{\partial F}{\partial \gamma} \qquad (n^2a^3 = \beta).$$

Denote any one of the Keplerian elements by $E_i$; then by the formulas of Section 1.14, with the Poisson brackets $(E_i, E_j)$, we obtain the equations

$$\frac{dE_i}{dt} = -\sum_j (E_i, E_j)\frac{\partial F}{\partial E_j}$$

for the variation of the Keplerian elements. Since these formulas are not canonical, we must refer as the standard to the elliptic motion of the body in question relative to the $n$th body—see (72), Section 5.7, on Poincaré's transformation $\gamma$. Then the following fundamental formulas (the most important in practical dynamical astronomy) are obtained for the variation of the Keplerian elements, similarly to (163) and (164):

$$\begin{aligned}
\frac{da}{dt} &= 2\sqrt{\frac{a}{\beta}}\frac{\partial R}{\partial \epsilon},\\
\frac{de}{dt} &= -\frac{\cos\phi}{\sqrt{\beta a}}\frac{\partial R}{\partial \varpi} - \frac{\tan\frac{1}{2}\phi\cos\phi}{\sqrt{\beta a}}\frac{\partial R}{\partial \epsilon},\\
\frac{dI}{dt} &= -\frac{1}{\cos\phi\sin I\sqrt{\beta a}}\frac{\partial R}{\partial \Omega} - \frac{\tan\frac{1}{2}I}{\cos\phi\sqrt{\beta a}}\left(\frac{\partial R}{\partial \varpi} + \frac{\partial R}{\partial \epsilon}\right),\\
\frac{d\Omega}{dt} &= \frac{1}{\cos\phi\sin I\sqrt{\beta a}}\frac{\partial R}{\partial I},\\
\frac{d\varpi}{dt} &= \frac{\cot\phi}{\sqrt{\beta a}}\frac{\partial R}{\partial e} + \frac{\tan\frac{1}{2}I}{\cos\phi\sqrt{\beta a}}\frac{\partial R}{\partial I},\\
\frac{d\epsilon}{dt} &= -2\sqrt{\frac{a}{\beta}}\frac{\partial R}{\partial a} + \frac{\tan\frac{1}{2}\phi\cos\phi}{\sqrt{\beta a}}\frac{\partial R}{\partial e} + \frac{\tan\frac{1}{2}I}{\cos\phi\sqrt{\beta a}}\frac{\partial R}{\partial I},\\
e &= \sin\phi, \qquad n^2a^3 = \beta,\\
R &= k^2\sum_\lambda m_{3\lambda}\cdot\left(\frac{1}{r_{3n,3\lambda}} - \frac{x_{3n-2}x_{3\lambda-2} + x_{3n-1}x_{3\lambda-1} + x_{3n}x_{3\lambda}}{r_{3n,3\lambda}^3}\right).
\end{aligned} \tag{175}$$

It is strongly suggested that, when $R$ is written in series form (173a), the mean motion $n$ is supposed to be present only in the arguments of the trigonometric functions, and the semimajor axis $a$ is supposed to be present only in their coefficients. Then $R$ is differentiated partially with respect to the Keplerian elements and $\epsilon'$ substituted for $\epsilon$ in $\lambda$, such that $\epsilon + nt = \epsilon' + \int n\,dt$. Here put

$$\rho \equiv \int n\,dt = \sqrt{\beta} \int a^{-3/2}\,dt;$$

then

$$\rho = -3 \int_0^t dt \int_0^t \frac{1}{a^2} \frac{\partial R}{\partial \epsilon}\,dt.$$

Ordinarily $R$ is called the *disturbing function.* The first term of $R$ is called its *direct part* and the second term its *indirect part.* It is customary in the treatment of planetary motion to take the sun as the $n$th body, and in the treatment of satellite motion to take the planet as the $n$th body.

The variation of elements in (175) is expressed by the partial derivatives of the disturbing function $R$ with regard to the Keplerian elements. It is sometimes useful to express this variation by the components of the disturbing force. Let $S$ be the component of the disturbing force in the direction of the instantaneous radius vector; let $T$ be the component, in the instantaneous orbital plane, perpendicular to the radius vector in the sense of the motion; and let $W$ be the component perpendicular to the instantaneous orbital plane, making a right-handed system with radius vector and the instantaneous velocity vector. The partial derivatives of $R$ in these directions — that is, the components of grad $R$ — are the components of the disturbing force in those directions. As these vectors are expressed by the Keplerian elements, those partial derivatives can easily be transformed to the partial derivatives with respect to the Keplerian elements. Such formulas for the variation of the elements in terms of $S$, $T$, $W$, can be directly derived even when there does not exist the potential $-R$ for the disturbing force. The result (Tisserand, 1889; Charlier, 1927; Plummer, 1918; Brown, 1896; Brown and Shook, 1933; Andoyer, 1923; Moulton, 1923; Dziobek, 1888; Smart, 1953; Danjon, 1953; Brouwer and Clemence, 1961) is:

$$\frac{da}{dt} = 2\sqrt{\frac{a^3}{\beta}}\{S \tan\phi \cos v + T \sec\phi(1 + e\cos v)\},$$

$$\frac{de}{dt} = \sqrt{\frac{a}{\beta}} \cos\phi\{S \sin v + T(\cos v + \cos u)\},$$

$$\frac{dI}{dt} = rW \cos(v + \varpi - \Omega)/(\sqrt{\beta}a \cos\phi),$$

$$\frac{d\Omega}{dt} = rW \sin(v + \varpi - \Omega)/(\sqrt{\beta}a \cos\phi \sin I), \qquad (176)$$

$$\frac{d\varpi}{dt} = \Big\{-aS\cos^2\phi\cos v + rT\sin v(2 + e\cos v)$$

$$+ rW\sin\phi\tan\frac{I}{2}\sin(v + \varpi - \Omega)\Big\}/(\sqrt{\beta a}\sin\phi\cos\phi),$$

$$\frac{d\epsilon}{dt} = -\frac{2rS}{\sqrt{\beta a}} + 2\sin^2\frac{\phi}{2}\frac{d\varpi}{dt} + 2\cos\phi\sin^2\frac{I}{2}\frac{d\Omega}{dt},$$

$$\frac{dn}{dt} = -3\{S\sin\phi\sin v + T(1 + e\cos v)\}/(a\cos\phi). \qquad (176\text{ cont.})$$

These formulas are especially important in the numerical computation of the variation of elements. When the positions of the disturbing planets are known at each instant, the numerical values of $S$, $T$, $W$ may be computed from these positions and the differential equations of motion numerically integrated. Such a method of computing the perturbation by numerical processes is generally called the *theory of special perturbation* (Bauschinger, 1928; Stracke, 1929).

### 5.19 ANGULAR MOMENTUM INTEGRALS

The angular momentum of the $n$-body system is invariant by Poincaré's transformation $\beta$, as has been seen in Section 5.6. For these fictitious planets, the angular momentum, by (160) of Section 5.16 and (84b) of Section 5.9, is equal to $h_2 = \sqrt{\beta a(1 - e^2)}$. The orbital plane is determined by the inclination $I$ and the longitude of the node $\Omega$ with reference to the ecliptic. Hence, the components of the angular momentum (which is perpendicular to the instantaneous orbital plane) in the directions of the three coordinate axes are $\sqrt{\beta a(1 - e^2)}\sin I\sin\Omega$, $-\sqrt{\beta a(1 - e^2)}\sin I\cos\Omega$, and $\sqrt{\beta a(1 - e^2)}\cos I$. The six orbital elements at any given instant are computed from the six data for the coordinates and the momenta at that instant. If the actual motion is rigorously Keplerian, then the six orbital elements thus computed are all constant. If it is not Keplerian, then the values of these orbital elements vary from time to time, and those variations are given by the equations (164), (174), or (175). Such orbital elements are called the *osculating elements*. The actual orbit touches the Keplerian orbit at the given instant. Also, the magnitude and the direction of the momenta are the same for the two orbits; so are the expressions for the coordinates and the momenta in terms of the orbital elements and $t$. The Keplerian elements for the actual orbit vary with time but those for the Keplerian orbit remain constant. Hence, if these expressions for the coordinates are partially differentiated with respect to $t$ (which appears explicitly) they give the velocity components. If they are multiplied by the mass, they give the momentum components. On this basis, the formulas for the variation of elements of Sections 5.17–5.19 are derived; the right-hand members of these formulas are supposed to be given in terms of the elements and $t$.

Thus the angular momentum just given is the same for the actual orbit and the osculating Keplerian orbit; hence, the components of the resultant angular momentum of the whole system are constant:

$$\sum_{j=1}^{n-1} \sqrt{\beta_j a_j(1 - e_j^2)} \sin I_j \sin \Omega_j = c_1,$$

$$-\sum_{j=1}^{n-1} \sqrt{\beta_j a_j(1 - e_j^2)} \sin I_j \cos \Omega_j = c_2,$$

$$\sum_{j=1}^{n-1} \sqrt{\beta_j a_j(1 - e_j^2)} \cos I_j = c_3.$$

In particular, for the three-body problem, with the osculating elements $a, e, I, \Omega, a', e', I', \Omega'$ of the two fictitious bodies, we have

$$\begin{aligned} \sqrt{\beta a(1 - e^2)} \sin I \sin \Omega + \sqrt{\beta' a'(1 - e'^2)} \sin I' \sin \Omega' &= c_1, \\ -\sqrt{\beta a(1 - e^2)} \sin I \cos \Omega - \sqrt{\beta' a'(1 - e'^2)} \sin I' \cos \Omega' &= c_2, \\ \sqrt{\beta a(1 - e^2)} \cos I + \sqrt{\beta' a'(1 - e'^2)} \cos I' &= c_3. \end{aligned} \tag{177}$$

Take the invariable plane of the three bodies as the standard plane and denote the values of $I$, $\Omega$ in this case by $I_0$, $\Omega_0$. Put $c^2 = c_1^2 + c_2^2 + c_3^2$, and substitute $I_0$, $\Omega_0$, $I_0'$, $\Omega_0'$, 0, 0, and $c$, for $I$, $\Omega$, $I'$, $\Omega'$, $c_1$, $c_2$, and $c_3$, respectively, in (177). Divide the first and second equations side by side, to obtain $\tan \Omega_0 = \tan \Omega_0'$. Reference to the third equation shows that $\Omega_0 = \Omega_0' + \pi$. This last relation expresses the *elimination of the nodes* (Jacobi).

In terms of the Delaunay elements, (177) is written

$$\begin{aligned} \sqrt{G^2 - H^2} \sin h + \sqrt{G'^2 - H'^2} \sin h' &= c_1, \\ -\sqrt{G^2 - H^2} \cos h - \sqrt{G'^2 - H'^2} \cos h' &= c_2, \\ H + H' &= c_3. \end{aligned} \tag{178}$$

If we take the invariable plane as the standard, then

$$G^2 - H^2 = G'^2 - H'^2, \qquad h = h' + \pi, \qquad H + H' = c. \tag{178a}$$

By these relations we can reduce the order of the differential equations (164).

According to the third equation of (178a),

$$\frac{dH}{dt} + \frac{dH'}{dt} = 0.$$

Hence, from (164),

$$\frac{\partial F}{\partial h} + \frac{\partial F}{\partial h'} = 0. \tag{179}$$

Write the expression (173b) for the expansion of $F$ in the form

$$F = \sum A \frac{\cos}{\sin} (il + i'l' + jg + j'g' + kh + k'h'), \tag{180}$$

and combine with (179). Then we see that $k + k' = 0$, and that $h$, $h'$ appear in $F$ only as $h - h'$. We have just seen that $h - h' = \pi$ when we take the invariable plane as the standard. In this case $F$ depends on neither $h$ nor $h'$. Hence by solving (178a) we obtain

$$\begin{aligned} &G = \Theta, \qquad G' = \Theta', \\ &H = \frac{c}{2} + \frac{1}{2c}(\Theta^2 - \Theta'^2), \qquad H' = \frac{c}{2} - \frac{1}{2c}(\Theta^2 - \Theta'^2). \end{aligned} \tag{181}$$

Substituting (164) in $dh/dt = dh'/dt$ obtained from (178a), we obtain

$$\frac{\partial F}{\partial H} = \frac{\partial F}{\partial H'}.$$

From (181), we obtain

$$\frac{\partial F}{\partial \Theta} = \frac{\partial F}{\partial G}, \qquad \frac{\partial F}{\partial \Theta'} = \frac{\partial F}{\partial G'}.$$

Hence, the differential equations (164) of the three-body problem are reduced to

$$\begin{aligned} \frac{dL}{dt} &= \frac{\partial F}{\partial l}, & \frac{d\Theta}{dt} &= \frac{\partial F}{\partial g}, \\ \frac{dl}{dt} &= -\frac{\partial F}{\partial L}, & \frac{dg}{dt} &= -\frac{\partial F}{\partial \Theta}, \\ \frac{dL'}{dt} &= \frac{\partial F}{\partial l'}, & \frac{d\Theta'}{dt} &= \frac{\partial F}{\partial g'}, \\ \frac{dl'}{dt} &= -\frac{\partial F}{\partial L'}, & \frac{dg'}{dt} &= -\frac{\partial F}{\partial \Theta'}, \end{aligned} \tag{182}$$

which is of the eighth order and has four degrees of freedom. We obtain the common direction of the ascending nodes from $dh/dt = -\partial F/\partial H$ by substituting in (181) after solving (182).

In the plane three-body problem, $I = 0$. Put

$$G = H = \Pi, \qquad G' = H' = \Pi';$$

then the integral of the angular momentum reduces to one equation

$$\Pi + \Pi' = c, \tag{183}$$

and $\partial F/\partial g + \partial F/\partial g' = 0$. Hence, $F$ depends on $g$, $g'$ only through $g - g'$. Write

$$K = \Pi, \qquad \kappa = g - g'; \tag{184}$$

then

$$\frac{dK}{dt} = \frac{d\Pi}{dt} = \frac{\partial F}{\partial g} = \frac{\partial F}{\partial \kappa},$$

$$\frac{d\kappa}{dt} = -\frac{\partial F}{\partial \Pi} + \frac{\partial F}{\partial \Pi'} = -\frac{\partial F}{\partial K},$$

and (164) is now of the form

$$\begin{aligned} \frac{dL}{dt} &= \frac{\partial F}{\partial l}, & \frac{dL'}{dt} &= \frac{\partial F}{\partial l'}, & \frac{dK}{dt} &= \frac{\partial F}{\partial \kappa}, \\ \frac{dl}{dt} &= -\frac{\partial F}{\partial L}, & \frac{dl'}{dt} &= -\frac{\partial F}{\partial L'}, & \frac{d\kappa}{dt} &= -\frac{\partial F}{\partial K}, \end{aligned} \tag{184a}$$

which is of the sixth order and has three degrees of freedom. The longitude of the perihelion is determined from $dg/dt = -\partial F/\partial \Pi$ by (184) after we solve (184a).

Since $\Theta = L\sqrt{1 - e^2}$ in (182), we have $L^2 > \Theta^2$. Similarly, $L'^2 > \Theta'^2$. But we know that

$$\Theta \cos I + \Theta' \cos I' = c, \qquad \Theta \sin I + \Theta' \sin I' = 0.$$

Hence,

$$|\Theta| + |\Theta'| \geqq |c|, \qquad |\Theta| - |\Theta'| \leqq |c|.$$

If $\Theta$ is approximately equal to $L$, then $F$ is not regular with regard to $\Theta$. Since (173b) shows that $s_1$, $s_2$ are integral multiples of $1/2$, the function $F$ can be developed in an integral power series not of $L - \Theta$ but of $\sqrt{L - \Theta}$. In such a case we have only to use the variables $\xi$, $\eta$ given by (170) of Section 5.17. The function $F$ is regular with regard to $\xi$, $\eta$.

Furthermore, if $|\Theta| + |\Theta'| = |c|$ or $|\Theta| - |\Theta'| = |c|$ holds approximately, that is, if $I$ and $I'$ are small (for example, if the relation $\Theta + \Theta' = c$ holds approximately), then $F$ can be developed in an integral power series of $p, q, \xi, \eta, p', q', \xi', \eta'$ of Section 5.17, that is, of $\sqrt{L - G}$, $\sqrt{L' - G'}$, $\sqrt{G - H}$, $\sqrt{G' - H'}$. If $I = I' = 0$, then $G = H$, $G' = H'$, and the last two of the four square roots become zero. The function $F$ is regular with regard to $G$, $G'$, $\sqrt{G - H}$, $\sqrt{G' - H'}$. But, as we have discussed in Section 5.17, $F$ remains unaltered when we change the signs of $I$ and $I'$, that is, when we increase $\Omega$ and $\Omega'$ by 180°. Hence, $F$ remains unaltered when we change the signs of $p$ $q$, $p'$, and $q'$ simultaneously, and, consequently, the signs of $\sqrt{G - H}$ and $\sqrt{G' - H'}$ simultaneously. Thus, $F$ is regular with regard to $G$, $G'$, $\sqrt{(G - H)(G' - H')}$, when $I$ and $I'$ are small. But, by (181),

$$\sqrt{(G - H)(G' - H')} = \frac{1}{2c}\sqrt{[(\Theta - c)^2 - \Theta'^2][(\Theta' - c)^2 - \Theta^2]}$$

$$= \frac{1}{2c}(\Theta + \Theta' - c)\sqrt{(\Theta - c - \Theta')(\Theta' - c - \Theta)}.$$

Hence, $G$, $G'$, $\sqrt{(G - H)(G' - H')}$ are regular with regard to $\Theta$, $\Theta'$, $\Theta + \Theta' = c$. Accordingly, $F$ is regular with regard to $\Theta$, $\Theta'$, $\Theta + \Theta' = c$. If the relation $\Theta - \Theta' = c$ holds approximately, then it can be similarly shown that $F$ is regular with regard to $\Theta$, $\Theta'$, $\Theta - \Theta' = c$.

In the plane three-body problem, $F$ has, by (173b), the form

$$F = \sum A(L^2 - G^2)^{s_1}[L'^2 - (c - G')^2]^{s_2} \cos(il + i'l' + ig + i'g').$$

If $L \approx G$ and $L' \approx G'$, that is, if $e$ or $e'$ is small, then $F$ is developed in integral powers of $\sqrt{L^2 - G^2}$, $\sqrt{L'^2 - G'^2}$ and looks as though it were not single-valued. This situation can be avoided by transforming to the variables $\xi, \eta, \xi', \eta'$ by the transformation of Section 5.17.

In the variables (169) of Section 5.17 the integrals of the angular momentum are

$$-\sqrt{(\Lambda - \Gamma)^2 - (\Lambda - \Gamma - Z)^2}\sin z - \sqrt{(\Lambda' - \Gamma')^2 - (\Lambda' - \Gamma' - Z')^2}\sin z' = c_1,$$

$$-\sqrt{(\Lambda - \Gamma)^2 - (\Lambda - \Gamma - Z)^2}\cos z - \sqrt{(\Lambda' - \Gamma')^2 - (\Lambda' - \Gamma' - Z')^2}\cos z' = c_2,$$

$$\Lambda - \Gamma - Z + \Lambda' - \Gamma' - Z' = c_3.$$

If we refer to the invariable plane as the reference plane, the integrals are

$$z_0 = z_0' + \pi, \qquad \Lambda - \Gamma - Z_0 + \Lambda' - \Gamma' - Z_0' = c,$$

$$(\Lambda - \Gamma)^2 - (\Lambda - \Gamma - Z_0)^2 = (\Lambda' - \Gamma')^2 - (\Lambda' - \Gamma' - Z_0')^2. \tag{185}$$

The expansion of the disturbing function can be discussed as in the case of the Delaunay elements. It can be shown that $F$ is regular with regard to the variables (169) although the expansion is formally of the form (173).

In the canonical variables (170) of Section 5.17, we find the expressions for the angular momentum by taking the invariable plane as the reference plane:

$$\frac{p}{q} = \frac{p'}{q'},$$

$$\left(\Lambda - \frac{\xi^2 + \eta^2}{2}\right)^2 - \left(\Lambda - \frac{\xi^2 + \eta^2 + p_0^2 + q_0^2}{2}\right)^2$$

$$= \left(\Lambda' - \frac{\xi'^2 + \eta'^2}{2}\right)^2 - \left(\Lambda' - \frac{\xi'^2 + \eta'^2 + p_0'^2 + q_0'^2}{2}\right)^2, \tag{186}$$

$$\Lambda - \frac{\xi^2 + \eta^2 + p_0^2 + q_0^2}{2} + \Lambda' - \frac{\xi'^2 + \eta'^2 + p_0'^2 + q_0'^2}{2} = c.$$

## 5.20 OTHER KINDS OF CANONICAL ELEMENTS

We have, from (84) of Section 5.9,

$$W = \int_{r_0}^{r} dr\sqrt{\frac{2\beta}{r} + 2h_1 - \frac{h_2^2}{r^2}} + h_3\theta + \int_0^{\phi} d\phi\sqrt{h_2^2 - \frac{h_3^2}{\cos^2\phi}}. \tag{187}$$

Put

$$P = \beta, \qquad L = \sqrt{\beta a}, \qquad P > 0, \tag{188}$$

$$Q = \sqrt{\frac{\beta}{a}}, \qquad U = \sqrt{\beta a}, \qquad QU > 0, \tag{189}$$

and

$$V = \omega\sqrt{\beta a(1 - e^2)}, \qquad G = (1 + \omega)\sqrt{\beta a(1 - e^2)}, \qquad H = G\cos I, \tag{190}$$

with a constant $\omega$. Take $h_1$ of (187) to be $L$ or $V$, $h_2$ to be $G$, and $h_3$ to be $H$. Then, from (188) and (190),

$$W = \int_{r_0}^{r} dr \sqrt{-\frac{P^2}{L^2} + \frac{2P}{r} - \frac{(G - V)^2}{r^2}} + H\theta + \int_0^{\phi} d\phi \sqrt{G^2 - \frac{H^2}{\cos^2 \phi}}. \tag{191}$$

Take $h_1$ to be $U$ or $V$, $h_2$ to be $G$, and $h_3$ to be $H$. Then, from (188) and (190),

$$W = \int_{r_0}^{r} dr \sqrt{-Q^2 + \frac{2QU}{r} - \frac{(G - V)^2}{r^2}} + H\theta + \int_0^{\phi} d\phi \sqrt{G^2 - \frac{H^2}{\cos^2 \phi}}. \tag{192}$$

The lower limit $r_0$ of the integrals in (191) and (192) is the value of $r$ that makes the expression under the root sign zero. As it has been supposed that $P > 0$, $QU > 0$, we have always $r > 0$. From the known relations

$$p_r = \frac{\partial W}{\partial r}, \qquad p_\theta = \frac{\partial W}{\partial \theta}, \qquad p_\phi = \frac{\partial W}{\partial \phi},$$

where we substitute

$$\sin \phi = \sin I \sin (v + g), \qquad \tan (\theta - h) = \cos I \tan (v + g), \tag{193}$$

$$r = a(1 - e \cos u) = \frac{a(1 - e^2)}{1 + e \cos v}, \qquad M = u - e \sin u = l, \tag{194}$$

we obtain

$$\begin{aligned} p_r &= \frac{\sqrt{p}\, e \sin v}{\sqrt{a(1 - e^2)}} = \sqrt{\beta a}\, \frac{e \sin u}{r} = m' \frac{dr}{dt}, \\ p_\theta &= (1 + \omega)\sqrt{\beta a(1 - e^2)} \cos I = m' r \cos \phi \frac{d\theta}{dt}, \\ p_\phi &= (1 + \omega)\sqrt{\beta a(1 - e^2)}\, \frac{\sin I \cos (v + g)}{\cos \phi} = m' r \frac{d\phi}{dt}, \end{aligned} \tag{195}$$

where $n^2a^2 = \beta$. If we equate $\omega$ to zero in these formulas, they reduce to those for ordinary Keplerian elliptic motion. Or, if we keep $t$ in the expression for $p_r$ unaltered but substitute $t/(1 + \omega)$ for $t$ in the expressions for $p_\theta$ and $p_\phi$, then the formulas reduce to those for a quasi-Keplerian motion, the difference from a pure Keplerian motion being that the period for $r$ is $2\pi/n$, while the period for $\theta$ and $\phi$ is $2\pi/[n(1 + \omega)]$; that is, the time interval between two consecutive perihelion passages is longer than the period for the angles $\theta$ and $\phi$, for $\omega > 0$. In other words, the perihelion rotates in the positive sense at the rate

$$\frac{n}{n(1+\omega)} = \frac{dv}{dt}\bigg/\frac{d(v+g)}{dt},$$

that is, $dg = \omega\, dv$ (see Section 2.19).

With these expressions for the momenta $p_r$, $p_\theta$, $p_\phi$, the Hamiltonian function $H$ takes the form

$$H = \frac{1}{2m'}\left(p_r^2 + \frac{G}{r^2}\right) - \frac{\beta}{r},$$

where an additional force function

$$\omega\beta a\left(1 + \frac{\omega}{2}\right)(1 - e^2)/r^2$$

is superposed on the ordinary force function $\beta/r$. Thus the ordinary integral $t + \gamma_1 = \partial W/\partial h$, is replaced either by $\partial W/\partial L = l$, or by $\partial W/\partial U = u$, or by $\partial W/\partial V = v$; the ordinary integrals $\gamma_2 = \partial W/\partial h_2$ and $\gamma_3 = \partial W/\partial L$ are replaced by $\partial W/\partial G = g$ and $\partial W/\partial H = h$, respectively.

We will now consider three fundamental cases:

I. Suppose that $P$ and $V$ are constant parameters; then the canonical variables $l, g, h, L, G, H$ correspond to $\omega = 0$ and, hence, $V = 0$. This is the case of Delaunay's elements.

II. Suppose that $Q$ and $V$ are constant parameters and $U$ is variable; then $u, g, h, U, G, H$ are the canonical variables. This was discovered by Levi-Civita (1913), Andoyer (1913), and de Sitter (1913); in this case $\omega = 0$, and, hence, $V = 0$.

III. Suppose that $P$ and $L$, or $Q$ and $U$, or $\beta$ and $a$ are constant; then $v, g, h, V, G, H$ are canonical variables. This was discovered by Hill (1913), Andoyer (1913), and de Sitter (1913).

In these three cases the Hamiltonian functions are, respectively:

in case I, $$H = -\frac{P^2}{2L^2} + \frac{P}{r} + \frac{V(2G - V)}{2r^2} - \frac{\beta}{r};$$

in case II, $$H = \frac{QU}{r} + \frac{V(2G - V)}{2r^2} - \frac{\beta}{r};$$

in case III, $$H = \frac{\beta}{r} + \frac{V(2G - V)}{2r^2} - \frac{\beta}{r} = \frac{V(2G - V)}{2r^2}.$$

If we put $P = \beta$, $V = 0$ in case I, then $H = -\beta^2/(2L^2)$. If we put $V = 0$ in case II, then $H = (QU - \beta)/r$. Case I (Delaunay) corresponds to the particular solution of the differential equations such that $U = \beta/Q$. In this case,

$$\frac{du}{dt} = \frac{Q}{r} = \sqrt{\frac{\beta}{a}}\frac{1}{r},$$

and the other elements are constant in the intermediary orbit. In case III, $H = V(2G - V)/2r^2$ and case I corresponds to the particular solution for $V = 0$. In this case III we have

$$\frac{dv}{dt} = \frac{G}{r^2} = \frac{\sqrt{\beta a(1 - e^2)}}{r^2}$$

and the other elements are constant. The force function can be discussed similarly, even if the term

$$\omega\beta a\left(1 + \frac{\omega}{2}\right)\frac{1 - e^2}{r^2}$$

is added to it. In cases I and II, the assumption $V = 0$ corresponds to $\omega = 0$, and the momenta $p_r$, $p_\theta$, $p_\phi$ are the same for the intermediary and disturbed orbits; hence the elements are osculating. But they are not the same for the two orbits in case III. The elliptic orbit and the disturbed orbit have different tangents at their common point. Such properties have been studied by Martin (1959–62) in connection with the two-body problem of variable mass.

In general, there are an infinite number of elliptic orbits that have common points with a given orbit, common tangents at any of those common points, and a common focus. Case I is one of those in which the attracted point starting with the given velocity is attracted by the force $-\beta/r^2$. This is called *isodynamical.* Case II is one of those in which the attracted point starts with the given velocity at the given point, and the energy $h_1 = -\beta/2a$ is the same as that for an elliptic motion. This is called *isoenergetic.* In general, if we suppose that $H = T - \beta/r = h_1$ has the value $h_1^{(0)}$ at $t = 0$, and $\beta_0$ is determined by $T - \beta_0^2/r = h_1^{(0)}$, then

$$T - \frac{\beta}{r} = h_1^{(0)} + \frac{\beta_0^2}{r} - \frac{\beta}{r}.$$

Each of these two elliptic motions is osculating but differs from the other. Case III, however, corresponds to the motion with $a = U/Q$. It is not osculating, since the value of $a$ is prefixed. This is called *isoprotometric.*

The notion of a protometer appears in the perturbation theory of Gyldén (see Chapter 10); the protometer corresponds to the semimajor axis of an elliptic orbit but is not the same, because the intermediary orbit is not elliptic.

In such a case,

$$H = \frac{\beta^2}{2a(1-e^2)}\left[2 + 2e\cos v - e^2\sin^2 v - \frac{G^2}{\beta a(1-e^2)}(1 + e\cos v)^2\right]$$

and

$$\frac{\partial H}{\partial v} = \frac{\beta}{a}\frac{e\sin v}{1-e^2}(1 + e\cos v)\left[\frac{G^2}{\beta a(1-e^2)} - 1\right].$$

The last factor, and hence $\omega$, has the magnitude of the disturbing mass. The relation $G^2 = \beta a(1 - e^2)$ should hold always, although $a$ varies. This is the concept of Gyldén on the protometer.

Now we have, respectively, in cases II and III, corresponding to (169) in case I:

$$\begin{aligned} \Lambda_1 &= U, & \Gamma_1 &= U - G, & Z_1 &= G - H, \\ \lambda_1 &= u + g + h, & \gamma_1 &= -g - h, & z_1 &= -h, \end{aligned} \tag{196}$$

and

$$\begin{aligned} \Lambda_2 &= V, & \Gamma_2 &= V - G, & Z_2 &= G - H, \\ \lambda_2 &= v + g + h, & \gamma_2 &= -g - h, & z_2 &= -h. \end{aligned} \tag{197}$$

Furthermore, if we put

$$F = V - G, \qquad \Psi = G,$$
$$f = v, \qquad \psi = v + g,$$

then this is also a canonical transformation and was discussed by Hill (1913).

Recently Soudan (1955) extended the studies of Hill, de Sitter, Levi-Civita and Andoyer on the canonical elements. Consider two absolute constants $\mu$ and $\rho$. Put

$$\mu^2 = n^2a^3, \qquad L = na^2, \qquad Q = na, \qquad U = na^2,$$
$$F = \rho na^2\sqrt{1-e^2}, \quad G = (1+\rho)na^2\sqrt{1-e^2}, \quad H = G\cos I.$$

Soudan proposed the following systems of canonical variables:

$$\text{I}'. \qquad L = na^2, \qquad G = (1+\rho)na^2\sqrt{1-e^2}, \qquad H = G\cos I,$$
$$l = n(t - t_0), \qquad g_1 = g + \frac{\rho}{1+\rho}v, \qquad h = \Omega.$$

For $\rho = 0$ this system reduces to Delaunay's.

II′.
$$U = na^2, \qquad G, \qquad H,$$
$$u, \qquad g_1, \qquad h.$$

For $\rho = 0$ this system reduces to Levi-Civita's.

III′.
$$F = \rho na^2\sqrt{1 - e^2}, \qquad G = (1 + \rho)na^2\sqrt{1 - e^2}, \qquad H = G \cos I,$$
$$v, \qquad g, \qquad h;$$

or

$$-na^2\sqrt{1 - e^2}, \qquad G, \qquad H,$$
$$u, \qquad g + v, \qquad h,$$

which is the system studied by Hill.

IV′.
$$M = na^2, \qquad G = (1 + \rho)na^2\sqrt{1 - e^2}, \qquad H = G \cos I,$$
$$l_1 = u + e \sin u, \qquad g_1 = g + \frac{\rho}{1 + \rho}\, v, \qquad h.$$

For $\rho = 0$ this system reduces to de Sitter's.

V′.
$$L = na^2, \qquad G \qquad H,$$
$$l, \qquad g_2 = g_1 + \frac{2(u - l)}{(1 + \rho)\sqrt{1 - e^2}}, \qquad h.$$

For $\rho = 0$ this system reduces to Jekhowski's (1916a).

VI′.
$$N = na^2, \qquad G, \qquad H,$$
$$\nu = \frac{3u - l}{2}, \qquad g_1, \qquad h.$$

Several other systems can be considered, for example:

$$L = Q\sqrt{a}, \qquad G = (1 + \rho)na\sqrt{a(1 - e^2)} \qquad H = G \cos I,$$
$$= (1 + \rho)Q\sqrt{a(1 - e^2)},$$
$$l, \qquad g_1, \qquad h;$$

or

$$U = a\sqrt{\mu}, \qquad G = (1 + \rho)na\sqrt{\mu(1 - e^2)}, \qquad H = G \cos I,$$
$$u, \qquad g_1, \qquad h.$$

In (196) the eccentric longitude $\lambda_1$ in the orbit and in (197) the true longitude $\lambda_2$ in the orbit are used as the variables, in contrast to the case (174) in which the mean longitude $\lambda$ in the orbit is used. Such canonical elements are not used in ordinary perturbation theory. But Le Verrier discussed the motion of the major planets according to the idea of Lagrange, basing it on the mean longitude in the orbit. In the past, theoretical investigations have been based on the mean longitude. But, owing to the difficulty even of the formal convergence of the series obtained by employing the mean longitude in the orbit, Hansen (1857) and Newcomb (1865, 1874) have used the eccentric anomaly or the eccentric longitude in the orbit for the computation of the ephemerides of the major planets. Newcomb (1891, 1895) obtained the development of the disturbing function in terms of the eccentric anomaly. Gyldén (1893) and Brown (1930; see also Brown and Brouwer, 1937; Brouwer, 1925) discussed perturbation theory by employing the true anomaly or true longitude in the orbit. Brendel (1898–1911) applied Gyldén's theory to the motion of asteroids and Brown applied it to the motions of the Trojan group of asteroids and the eighth satellite of Jupiter, followed by Hori (1957, 1957a, 1958) on the motion of the ninth satellite of Jupiter. Such theories will be discussed in detail in later chapters.

## REFERENCES

Akimoff, M.
1915 *Compt. Rend.*, **161**, 168.
1916 *Ibid.*, **163**, 26.
1917 *Ibid.*, **165**, 23.
1924 *Ibid.*, **179**, 435.
1927 *Ibid.*, **185**, 409.

Andoyer, H.
1913 *Bull. Astron.*, **30**, 425.
1923 *Cours de Mécanique Céleste*, I, Gauthier-Villars, Paris.

Appell, P.
1915 Compt. Rend., **160**, 419.
1924 *Ibid.*, **179**, 437.
1925 *Sur les Fonctions Hypergéométriques de plusieurs Variables, les Polynomes d'Hermite et autres Fonctions Sphériques dans l'Hyper-espace*, Gauthier-Villars, Paris.

Bailey, W. N.
1936 *Generalized Hypergeometric Series*, Cambridge University Press, London.

Bauschinger, J.
1928 *Die Bahnbestimmung der Himmelskörper*, 2nd ed., Englemann, Leipzig.

Behnke, H., and Thullen, P.
1934 *Theorie der Funktionen mehrer komplexen Veränderlichen*, Springer, Berlin.
Bennett, T. L.
1905 *Messenger of Math.*, **34**, 1113.
Bertrand, J.
1852 *J. Math. pures appl.*, **17**, 393.
Bochner, S., and Martin, W. T.
1948 *Several Complex Variables*, Princeton University Press, Princeton, N.J.
Bohlin, K.
1907 *Kongl. Svenska Vetenskaps Akad. Handl.*, **42**, no. 2.
Bourget, H.
1863 *Ann. Obs. Paris*, **7**, 263.
Brendel, M.
1898 *Abhandl. Königlichen Gesell. Wiss. Göttingen*, Math.-Phys. kl., new ser., **1–8** (1898–1911). Reprinted as *Theorie der Kleinen Planeten*, I–IV, Weidmannsche Buchhandlung, 1898–1911.
Briot, C., and Bouquet, J.-C.
1875 *Théorie des Fonctions Elliptiques*, 2nd ed., Gauthier-Villars, Paris.
Brouwer, D.
1925 *Monthly Notices Roy. Astron. Soc.*, **115**, 221.
Brouwer, D., and Clemence, G. M.
1961 *Methods of Celestial Mechanics*, Academic Press, New York.
Brown, E. W.
1896 *An Introductory Treatise on Lunar Theory*, Cambridge University Press, London.
1930 *Trans. Yale Univ. Obs.*, **6**, Part 4.
Brown, E. W., and Brouwer, D.
1937 *Trans. Yale Univ. Obs.*, **6**, Part 8.
Brown, E. W., and Shook, C. A.
1933 *Planetary Theory*, Cambridge University Press, London.
Burrau, C.
1894 *Astron. Nachr.*, **135**, 233; **136**, 161.
Cayley, A.
1861 *Mem. Roy. Astron. Soc.*, **29**, 257.
Charlier, C. V. L.
1899 *Öfversigt af kongl. Vetenskaps Akad. Förhandl.*, no. 4, 263.
1927 *Die Mechanik des Himmels für Vorlesung*, I, II, 2nd ed., Gruyter, Leipzig.
Danjon, A.
1953 *Astronomie Générale*, Sernac, Paris.
Delaunay, A.
1860 *Théorie du Mouvement de la Lune*, *Mém. Acad. Sci. Paris*, no. 28.
1867 *Ibid.*, no. 29.
Deprit, A., and Delie, A.
1961 *Ann. Soc. Sci. Bruxelles*, **75**, 5.
1963 *Arch. Ratl. Mech. Anal.*, **12**, 325.

Deprit, A., and Roels, J.

1962 *Bull. Cl. Sci. Acad. Belg.*, ser. 5, **48**, 432.

de Sitter, W.

1913 *Proc. Acad. Sci. Amsterdam*, **16**, 279.

Dziobek, O.

1888 *Mathematische Theorie der Planetenbewegung*, Barth, Leipzig.

Elenevskaya, N. B.

1953 *Byull. Inst. Teor. Astron.* (Akad. Nauk SSSR), no. 15, 27.

1957 *Ibid.*, no. 64, 34.

1962 *Astron. Zh.*, **39**, 938.

Féraud, A.

1902 *Ann. Obs. Bordeaux*, **10**, 1.

Forsyth, A. R.

1914 *Introductory Lectures to the Theory of Functions of Two Complex Variables*, Cambridge University Press, London.

Goursat, H.

1913 *Cours d'Analyse Mathématique*, II, 270, Gauthier-Villars, Paris.

1915 *Bull. Soc. Math. France*, **43**, 1.

1936 *Fonctions d'Euler et de Gauss*, Gauthier-Villars, Paris.

Gyldén, H.

*Traité Analytique des Orbites Absolues des huit Planètes Principales*, I, Beijers Bokförlags, Stockholm.

Hagihara, Y.

1931 *Japan J. Astron. Geophys.*, **8**, 67.

1935 *Proc. Imp. Acad. Japan*, **11**, 93.

1936 *Astron. J.*, **45**, 129.

Hansen, P. A.

1857 *Auseinandersetzung einer zweckmässigen Methoden zur Berechnung der absoluten Störungen der kleinen Planeten, Abhandl. sachs. Akad. Wiss. Leipzig.*

Hill, G. W.

1905 *Collected Mathematical Works of G. W. Hill*, I, 206, Carnegie Institution, Washington, D.C.

1913 *Astron. J.*, **27**, 171.

Hori, G.

1957 *Proc. Astron. Soc. Japan*, **9**, 51.

1957a *Proc. Japan Acad.*, **33**, 395.

1958 *Ibid.*, **34**, 263.

Humbert, P.

1920 *Proc. Roy. Soc. Edinburgh*, **41**, 73.

Jacobi, C. G. J.

1842 *Astron. Nachr.*, **20**, 81.

1844 *J. reine angew. Math.*, **26**, 115.

Jarnagin, M. P.

1965 *Astron. Papers Am. Eph. Naut. Alm.*, **18**.

Jekhowsky, B.

1916 *Compt. Rend.*, **162**, 318.

1916a *Bull. Astron.*, **33**, 65.

1917 *Bull. Soc. Math. France*, ser. 2, **41**, 58.

1917a *Compt. Rend.*, **164**, 719.
1918 *Bull. Astron.*, **35**, 139.
1920 *Compt. Rend.*, **170**, 1042.
1921 *Ibid.*, **172**, 1331.
1927 *J. des Observateurs*, **10**, nos. 7, 9.

Klein, F.
1933 *Vorlesungen über die hypergeometrische Funktion*, Springer, Berlin.

Kozai, Y.
1959 *Astron. J.*, **64**, 370.
1962 *Ibid.*, **67**, 311.

Lagrange, J. L.
1873 *Oeuvres de Lagrange*, VI, 229, Gauthier-Villars, Paris.

Lemaître, G.
1952 *Bull. Cl. Sci. Acad. Belg.*, ser. 5, **38**, 582, 1218.

Levi-Civita, T.
1913 *Annali Mat. pura appl.*, ser. 3, **20**, 153. Reprinted in *Opere Matematiche*, III, 341, Zanichelli, Bologna, 1957.
1915 *Atti R. Ist. Veneto*, ser. 8, **17**, 907. Reprinted in *Opere Matematiche*, III, 437.

Lindstedt, A.
1883 *Compt. Rend.*, **97**, 1276.
1884 *Ann. Ecole Norm. Sup.*, ser. 3, **1**, 85.

Martin, F. L.
1959–1962 *Oss. Astron. Trieste*, nos. 293, 295, 297, 299, 311, 315.

Moulton, F. R.
1903 *Astron. J.*, **23**, 93.
1923 *An Introduction to Celestial Mechanics*, Macmillan, New York.

Murnaghan, F. D.
1936 *Am. J. Math.*, **58**, 829.

Newcomb, S.
1865 *Smithsonian Contribution to Knowledge*, no. 199, Smithsonian Institution, Washington, D.C.
1874 *Ibid.*, no. 262.
1891 *Astron. Papers Am. Eph. Naut. Alm.*, **3**, no. 1.
1895 *Ibid.*, **5**, no. 1.

Nielsen, N.
1906 *Handbuch der Gammafunktion*, Teubner, Leipzig.

Norén, G., and Walberg, J. A.
1910 *Lunds Obs. Meddelanden*, no. 10.

Osgood, W. F.
1914 *Topics in the Theory of Functions of Several Complex Variables*, Am. Math. Soc., New York.
1924 *Lehrbuch der Funktionentheorie*, II, Part 1, 16, Teubner, Leipzig. Reprinted by Chelsea Publishing, Bronx, N.Y., 1965.

Pérès, J.
1915 *Compt. Rend.*, **161**, 168.

Picard, E.
1926 *Cours d'Analyse*, II, Gauthier-Villars, Paris.

Picard, E., and Simart, G.
1897 *Théorie des Fonctions Algébriques de deux Variables indépendantes*, I, Gauthier-Villars, Paris.

Picart, L.
1906 *Bull. Astron.*, **23**, 353.

Plummer, H. C.
1918 *An Introductory Treatise on Dynamical Astronomy*, Cambridge University Press, London.

Poincaré, H.
1892 *Méthodes Nouvelles de la Mécanique Céleste*, I, 280, Gauthier-Villars, Paris.
1897 *Bull. Astron.*, **14**, 53. Reprinted in *Acta Math.*, **21**, 83 (1897), and in *Oeuvres de Henri Poincaré*, VII, 500, Gauthier-Villars, Paris, 1952.
1898 *Bull. Astron.*, **15**, 449. Reprinted in *Oeuvres*, VIII, 33.
1905 *Leçons de la Mécanique Céleste*, I, Gauthier-Villars, Paris.
1907 *Ibid.*, II, Chap. 16.

Poole, E. G. C.
1936 *Theory of Linear Differential Equations*, Clarendon Press, Oxford.

Radau, R.
1868 *Ann. Ecole Norm. Sup.*, **5**, 311.
1886 *Bull. Astron.*, **3**, 113.

Ronchi, M.
1917 *Atti R. Ist. Veneto*, ser. 9, **1**, 1221.

Schlesinger, L.
1922 *Einführung in die Theorie der gewöhnlichen Differentialgleichungen auf funktionentheoretischer Grundlage*, Gruyter, Leipzig.

Smart, W. M. H.
1953 *Celestial Mechanics*, Longmans, London.

Soudan, A.
1955 *Bull. Astron.*, ser. 2, **19**, 225, 287.

Stracke, G.
1929 *Bahnbestimmung der Planeten und Kometen*, Springer, Berlin.

Sundman, K.
1913 *Acta Math.*, **36**, 105.

Tisserand, F.
1889 *Traité de la Mécanique Céleste*, I, Gauthier-Villars, Paris.

Van Kampen, E. R., and Wintner, A.
1937 *Amer. J. Math.*, **59**, 153.

Whittaker, E. T.
1937 *A Treatise on the Analytical Dynamics of Particles and Rigid Bodies*, Chap. XIII, 4th ed., Cambridge University Press, London.

Whittaker, E. T., and Watson, G. N.
1937 *Modern Analysis*, Cambridge University Press, London.

# Algebraic Integrals and Uniform Integrals

## THEOREM OF BRUNS

### 6.1 INTEGRALS INDEPENDENT OF TIME

In Section 4.10 we have shown that the order of the differential equations of the $n$-body problem can be reduced from $6n$ to $6n - 12$ by the use of Euler's classical ten integrals. Several attempts have been made to discover other integrals. Bruns (1887, 1887a) proved that there cannot be any integral that is algebraic with respect to the coordinates and momenta, other than the classical integrals. Poincaré (1889) further proved that there cannot be any integral that is uniform with respect to the Keplerian elements, other than the classical integrals.

We have shown in Sections 5.14 and 5.19 that the position coordinates and the momenta are uniform or regular, that is, single-valued and continuous, with respect to the osculating elements computed from these coordinates and momenta, except in the case of collisions of two or more bodies of the system. It will be shown in a later chapter that a similar result can be obtained on the solution itself of the equations of motion. This theorem of Poincaré has a part of its validity domain in common with the theorem of Bruns, although each has its own validity domain. The theorem of Poincaré presupposes that the right-hand members of the differential equations are developed in ascending powers of a small parameter $\mu$, described in Sections 5.16–5.20; that is, the disturbing mass must be assumed to be so small, that it is impossible to apply the result to the general $n$-body problem — for example, to the system of three bodies with finite masses of comparable magnitude. These theorems become trivial in the case of two bodies, as the problem can then be solved by means of the classical integrals alone. Painlevé (1896, 1896a, 1896b, 1897a, 1898) extended the theorem and proved that there cannot be an integral that is algebraic with respect to the momenta alone other than the classical integrals. The proof of Bruns's theorem has been improved by Poincaré (1896; see also MacMillan, 1913; Perchot and Ebert, 1898; Whittaker, 1928; Forsyth, 1900). We propose to extend the theorem from the three-body to the $n$-body case.

Consider a system of differential equations of order $6n - 6$ for the $n$-body problem. If we write $\mu_i$ instead of $m_i'$ of Section 5.6, then the equations become

$$\frac{dx_i}{dt} = \frac{\partial H}{\partial y_i}, \qquad \frac{dy_i}{dt} = -\frac{\partial H}{\partial x_i},$$

$$H = T - U, \qquad T = \sum_{i=1}^{N} \frac{y_i^2}{2\mu_i} \quad (i = 1, 2, \ldots, N;\ N = 3n - 3), \tag{1}$$

where $U$ is supposed to be expressed in terms of Jacobi's canonical variables $x_i$, and $T$ in terms of their conjugate variables $y_i$ only. Write any integral of (1) in the form

$$f(y_1, \ldots, y_N, x_1, \ldots, x_N) = a, \tag{2}$$

with an arbitrary constant $a$. Our problem is to prove that there cannot be an integral of the form (2) in which $f$ is algebraic with respect to $y_i$ and $x_i$, other than the classical integrals.

At first an integral (2) should contain $y_i$. If $f$ were independent of $y_i$, then, by differentiating (2) with respect to $t$, we should see from

$$\frac{dx_i}{dt} = \frac{\partial H}{\partial y_i} = \frac{y_i}{\mu_i}$$

that the relation

$$0 = \sum_{i=1}^{N} \frac{\partial f}{\partial x_i} \frac{y_i}{\mu_i}$$

should hold for all values of $t$, and hence that it should hold for any value of $y_i$. Thus, we should have $\partial f/\partial x_i = 0$ $(i = 1, 2, \ldots, N)$, and $f$ should also be independent of $x_i$.

Let $s$ be the sum of the $n(n - 1)/2$ mutual distances $r_{3j,3k}$ $(j, k = 1, 2, \ldots, n)$. Any $r_{3j,3k}$, and accordingly $U$, is expressed in terms of $x_1, x_2, \ldots, x_N$ and $s$. The function $f$ is algebraic with respect to $x_1, x_2, \ldots, x_N, y_1, y_2, \ldots, y_N$ but is not necessarily rational. By rationalizing (2) and rewriting it in descending powers of the constant $a$, we see that the coefficients are rational with respect to $y_i$ and $x_i$. It can be written in the irreducible form

$$a^l + a^{l-1}\psi_1(x_1, \ldots, y_N, s) + \cdots + \psi_l(x_1, \ldots, y_N, s) = 0, \tag{3}$$

where $\psi_j$ are rational functions of $x_1, \ldots, y_N$ and $s$. Differentiate (3) with respect to $t$ and refer to the Poisson brackets by the use of (1). Then

$$a^{l-1}(\psi_1, H) + a^{l-2}(\psi_2, H) + \cdots + (\psi_l, H) = 0, \tag{4}$$

where $(\psi_j, H)$ are rational with respect to $x_1, \ldots, y_N$ and $s$. Suppose that none of the Poisson brackets $(\psi_j, H)$ is equal to zero. In order that (3) and (4) may hold simultaneously, equations (3) and (4) should have in common more than one root of $a$. But we have assumed that (3) is irreducible, and this would contradict that assumption. Hence $(\psi_j, H)$ must all equal zero. Thus, by Poisson's theorem of Section 4.3, all functions $\psi_1, \psi_2, \ldots, \psi_l$ are integrals of (1). Hence, any integral $f$ is an algebraic combination of other rational functions of $x_1, \ldots, y_N$ and $s$. Thus we need only consider integrals that are rational with respect to $x_1, \ldots, y_N$ and $s$,

$$f(x_1, \ldots, x_N, y_1, \ldots, y_N, s) = a. \tag{5}$$

Equations (1) remain unaltered when $x_i, y_i, t$ are replaced, respectively, by $x_i\alpha^2, y_i\alpha^{-1}, t\alpha^3$ with a constant $\alpha$. Hence (5) is also an integral for any value of $\alpha$. But $f$ can be written as the quotient of two integral polynomials; after this transformation it should be of the form

$$f = \frac{A_0\alpha^p + A_1\alpha^{p-1} + \cdots + A_p}{B_0\alpha^q + B_1\alpha^{q-1} + \cdots + B_q},$$

where $A_0, \ldots, A_p, B_0, \ldots, B_q$ are integral polynomials of

$$x_1, \ldots, x_N, y_1, \ldots, y_N \quad \text{and} \quad s.$$

Since $df/dt$ should equal zero, we obtain from this expression for $f$—when the coefficients of all powers of $\alpha$ are equated to zero—

$$0 = B_0 \frac{dA_0}{dt} - A_0 \frac{dB_0}{dt},$$

$$0 = B_1 \frac{dA_0}{dt} + B_0 \frac{dA_1}{dt} - A_1 \frac{dB_0}{dt} - A_0 \frac{dB_1}{dt},$$

$$\vdots$$

$$0 = B_q \frac{dA_p}{dt} - A_p \frac{dB_q}{dt}.$$

As these $(q + p + 1)$ equations are equivalent to

$$\frac{1}{A_0}\frac{dA_0}{dt} = \frac{1}{A_1}\frac{dA_1}{dt} = \cdots = \frac{1}{A_p}\frac{dA_p}{dt} = \frac{1}{B_0}\frac{dB_0}{dt} = \cdots = \frac{1}{B_q}\frac{dB_q}{dt},$$

the quotients $A_1/A_0, A_2/A_0, \ldots, A_p/A_0, B_0/A_0, \ldots, B_q/A_0$ are all integrals. Let $G_1$, $G_2$ be integral polynomials, such that they are each multiplied by a certain power of $\alpha$ by a transformation from $x_i, y_i, t$ to $x_i\alpha^2, y_i\alpha^{-1}, t\alpha^3$. Then any integral $f$ is an algebraic combination of integrals of the form

$$\frac{G_1(x_1, \ldots, y_N, s)}{G_2(x_1, \ldots, y_N, s)} = \text{constant}. \tag{6}$$

If $G_1/G_2$ contains imaginary parts, then the real part and the imaginary part should be separate integrals. Hence we can suppose that (6) is real.

## 6.2 LEMMA I

It might occur that $G_1$ is written as the product of irreducible integral polynomials of $y_1, \ldots, y_N$. In such cases the coefficient of each term of these polynomials is a rational function of $x_1, \ldots, x_N, s$. Let $G_1$ be in the form $G_1 = \psi^\lambda\chi$. From

$$\frac{d}{dt}\left(\frac{G_1}{G_2}\right) = 0$$

we obtain

$$\frac{d\psi}{dt} = \psi\left(\frac{1}{\lambda G_2}\frac{dG_2}{dt} - \frac{1}{\lambda\chi}\frac{d\chi}{dt}\right).$$

It shows that $d\psi/dt$ and $\psi$ are integral polynomials of $y_i$ and that $\psi$ is of degree one lower than $d\psi/dt$. Also, $\psi$ has no factor in common with either $G_2$ or $\chi$. Hence, the factor of $\psi$ on the right-hand side of this equation is linear with respect to $y_i$. Denote it by $\omega$; then,

$$\frac{d\psi}{dt} = \omega\psi. \tag{7}$$

Similar relations should hold for each of the remaining irreducible factors of $G_1$. Let the factors of $G_1$ be $\psi', \psi'', \ldots$; then, $G_1 = \psi'^\mu\psi''^\nu\cdots$. Put

$$\frac{1}{\psi'}\frac{d\psi'}{dt} = \omega', \qquad \frac{1}{\psi''}\frac{d\psi''}{dt} = \omega'', \ldots;$$

then

$$\frac{1}{G_1}\frac{dG_1}{dt} = \frac{\mu}{\psi'}\frac{d\psi'}{dt} + \frac{\nu}{\psi''}\frac{d\psi''}{dt} + \cdots.$$

$$= \mu\omega' + \nu\omega'' + \cdots = \omega.$$

Since the integral polynomial $\omega$ is linear in $y_i$ and rational in $x_i$ and $s$, the polynomial $G_1$ satisfies the equation $dG_1/dt = \omega G_1$. Hence, $G_2$ should satisfy $dG_2/dt = \omega G_2$. Thus, $G_1$ and $G_2$ satisfy the same equation. Let $\phi$ represent either of the two; $\phi$ is a real integral polynomial with respect to $x_i$, $s$ and satisfies (7). In the transformation from $x_i$, $y_i$, $t$ to $x_i\alpha^2$, $y_i\alpha^{-1}$, $t\alpha^3$, $\phi$ is multiplied by a certain power of $\alpha$; therefore, the polynomial $\omega$ is multiplied by $\alpha^{-3}$ because of the relation

$$\omega = \frac{1}{\phi}\frac{d\phi}{dt} = \sum_{i=1}^{N} \frac{1}{\phi}\left(\frac{\partial\phi}{\partial x_i}\frac{y_i}{\mu_i} + \frac{\partial\phi}{\partial y_i}\frac{\partial U}{\partial x_i}\right).$$

Hence there is no term independent of $y_i$ in $\omega$, since such a term, if present, would be multiplied by an even power of $\alpha$.

Thus $\omega$ is a linear homogeneous expression in $y_i$ of the form

$$\omega = \omega_1 y_1 + \omega_2 y_2 + \cdots + \omega_N y_N,$$

where each $\omega_j$ is homogeneous of degree $-1$ in $x_1, \ldots, x_N$ and $s$. By a similar consideration, $\phi$ should be an aggregate of terms of even degrees in $y_1, \ldots, y_N$; it is written

$$\phi = \phi_0 + \phi_2 + \phi_4 + \cdots,$$

where $\phi_0$ is the aggregate of the terms of the highest degree in $y_i$, $\phi_2$ the aggregate of those of degree two lower in $y_i$ than $\phi_0$, and so on. Each $\phi_j$ is a polynomial in $(y_1, \ldots, y_N, x_1, \ldots, x_N, s)$ homogeneous in $(y_1, \ldots, y_N)$ and also in $(x_1, \ldots, x_N, s)$.

Next we shall show that, if $\phi_0$ does not contain $s$, we can obtain an integral by multiplying $\phi$ by a suitable rational function of $x_1, \ldots, x_N$.

For proof, we suppose that $\phi_0$ does not contain $s$. Equating the terms of the highest degree on both sides of

$$\frac{d\phi_0}{dt} + \frac{d\phi_2}{dt} + \cdots = (\omega_1 y_1 + \cdots + \omega_N y_N)(\phi_0 + \phi_2 + \cdots),$$

we obtain

$$\sum_{i=0}^{N} \frac{y_i}{\mu_i}\frac{\partial\phi_0}{\partial x_i} = (\omega_1 y_1 + \cdots + \omega_N y_N)\phi_0. \tag{8}$$

In general, $\phi_0$ may contain $y_1$ as a factor. Put $\phi_0 = y_1^h\phi_1'$; then

$$\sum_{i=1}^{N} \frac{y_i}{\mu_i}\frac{\partial\phi_0'}{\partial x_i} = (\omega_1 y_1 + \cdots + \omega_N y_N)\phi_0'.$$

Let the terms independent of $y_1$ in $\phi_0'$ be $\phi_0''$; then

$$\sum_{i=1}^{N} \frac{y_i}{\mu_i} \frac{\partial \phi_0''}{\partial x_i} = (\omega_2 y_2 + \cdots + \omega_N y_N)\phi_0''.$$

$\phi_0''$ may be a function of $x_i$ only. In this case denote it by $R$; then

$$\frac{1}{\mu_i} \frac{\partial R}{\partial x_i} = \omega_i R.$$

Hence,

$$\frac{\partial}{\partial x_j} (\mu_k \omega_k) = \frac{\partial}{\partial x_k} (\mu_j \omega_j) \qquad (j, k = 2, 3, \ldots, N).$$

Next, $\phi_0''$ may contain one of $y_2, \ldots, y_N$, for example, $y_2$. In this case we write $\phi_0'' = y_2^k \phi_0'''$, where $\phi_0'''$ does not contain $y_2$ as a factor. We have

$$\sum_{j=1}^{N} \frac{y_j}{\mu_j} \frac{\partial \phi_0'''}{\partial x_j} = (\omega_2 y_2 + \cdots + \omega_N y_N)\phi_0'''.$$

Let $\phi_0^{IV}$ be the terms independent of $y_2$ in $\phi_0'''$. Proceeding in a similar manner, either we finally obtain

$$\frac{\partial}{\partial x_N} (\mu_{N-1}\omega_{N-1}) = \frac{\partial}{\partial x_{N-1}} (\mu_N \omega_N),$$

or there exists an integral polynomial $\psi$ with respect to

$$x_1, \ldots, x_N, y_{N-1}, y_N,$$

which is homogeneous in $x_1, \ldots, x_N$, is homogeneous in $y_{N-1}, y_N$, is free from any factor $y_1^h y_2^k \ldots$, and satisfies

$$\frac{y_{N-1}}{\mu_{N-1}} \frac{\partial \psi}{\partial x_{N-1}} + \frac{y_N}{\mu_N} \frac{\partial \psi}{\partial x_N} = (\omega_{N-1} y_{N-1} + \omega_N y_N)\psi.$$

Put $\psi = a y_N^l + b y_{N-1}^l + c y_N^{l-1} y_{N-1} + \cdots$ and equate the coefficients of $y_N^{l+1}$, $y_{N-1}^{l+1}$ on both sides; then

$$\omega_N = \frac{1}{\mu_N a} \frac{\partial a}{\partial x_N}, \qquad \omega_{N-1} = \frac{1}{\mu_{N-1} b} \frac{\partial b}{\partial x_{N-1}}.$$

The coefficients $a, b, \ldots$, are polynomials of $x_i$. They may have a common factor $Q$, such that $a = a'Q$, $b = b'Q, \ldots$. Let

$$\psi' = a' y_N^l + b' y_{N-1}^l + c' y_N^{l-1} y_{N-1} + \cdots,$$

so that $\psi = Q\psi'$; then, writing

$$\frac{1}{\psi'}\left(\frac{y_N}{\mu_N}\frac{\partial \psi'}{\partial x_N} + \frac{y_{N-1}}{\mu_{N-1}}\frac{\partial \psi'}{\partial x_{N-1}}\right)$$

$$= \frac{1}{\psi}\left(\frac{y_N}{\mu_N}\frac{\partial \psi}{\partial x_N} + \frac{y_{N-1}}{\mu_{N-1}}\frac{\partial \psi}{\partial x_{N-1}}\right) - \frac{1}{Q}\left(\frac{y_N}{\mu_N}\frac{\partial Q}{\partial x_N} + \frac{y_{N-1}}{\mu_{N-1}}\frac{\partial Q}{\partial x_{N-1}}\right)$$

$$= \left(\omega_N - \frac{1}{Q\mu_N}\frac{\partial Q}{\partial x_N}\right)y_N + \left(\omega_{N-1} - \frac{1}{Q\mu_{N-1}}\frac{\partial Q}{\partial x_{N-1}}\right)y_{N-1}$$

$$= \omega'_N y_N + \omega'_{N-1} y_{N-1},$$

we obtain

$$\omega'_N = \frac{1}{\mu_N a'}\frac{\partial a'}{\partial x_N}, \qquad \omega'_{N-1} = \frac{1}{\mu_{N-1} b'}\frac{\partial b'}{\partial x_{N-1}},$$

and

$$\frac{y_N}{\mu_N}\frac{\partial \psi'}{\partial x_N} + \frac{y_{N-1}}{\mu_{N-1}}\frac{\partial \psi'}{\partial x_{N-1}} = (\omega'_N y_N + \omega'_{N-1} y_{N-1})\psi',$$

where the left-hand member is an integral polynomial in

$$x_1, \ldots, x_N, y_{N-1}, y_N.$$

Suppose that $a'$ contains $x_N$; then $\omega'_N$ would contain $a'$ or its power as its denominator, and $\psi'$ would contain $a'$ or its power as a factor. But this is contradictory to our assumption that $a', b', \ldots,$ have no common factor. Hence, $a'$ does not contain $x_N$, and $\omega'_N$ is zero. Similarly, $\omega'_{N-1}$ is zero. Thus

$$\omega_N = \frac{1}{Q\mu_N}\frac{\partial Q}{\partial x_N}, \qquad \omega_{N-1} = \frac{1}{Q\mu_{N-1}}\frac{\partial Q}{\partial x_{N-1}},$$

and accordingly

$$\frac{\partial}{\partial x_{N-1}}(\mu_N \omega_N) = \frac{\partial}{\partial x_N}(\mu_{N-1}\omega_{N-1}).$$

Thus, in either of the two alternative cases,

$$\frac{\partial}{\partial x_j}(\mu_k \omega_k) = \frac{\partial}{\partial x_k}(\mu_j \omega_j) \qquad (k, j = 1, 2, \ldots, N).$$

Hence we can write $\omega_i$ in the form

$$\mu_i \omega_i = \frac{1}{R}\frac{\partial R}{\partial x_i} \qquad (i = 1, 2, \ldots, N), \tag{9}$$

where $R$ is a rational function of $x_1, \ldots, x_N$. Thus

$$\omega_1 y_1 + \cdots + \omega_N y_N = \sum_{i=1}^{N} \frac{y_i}{\mu_i} \frac{1}{R} \frac{\partial R}{\partial x_i} = \sum_{i=1}^{N} \frac{1}{R} \frac{\partial R}{\partial x_i} \frac{dx_i}{dt};$$

hence,

$$\frac{1}{\phi} \frac{d\phi}{dt} = \frac{1}{R} \frac{dR}{dt}.$$

Therefore,

$$\frac{\phi}{R} = \text{constant}.$$

It has thus been proved that, if the terms $\phi_0$ of the highest degree in $G_1$ and $G_2$ do not contain $s$, then any algebraic integral in question can be obtained by multiplying an integral polynomial in $y_1, \ldots, y_N$ that is rational with regard to $x_1, \ldots, x_N$ and $s$, by a rational function of $x_1, \ldots, x_N$.

## 6.3 LEMMA II

Now we propose to prove that a function $\phi_0$ that satisfies (8) does not contain the irrationality $s$. Bruns's original proof of this part contained an error, which was corrected by Poincaré.

Assume tentatively that there exists a $\phi_0$, containing $s$, that satisfies (8). If we substitute $n(n-1)/2$ values of $s$ in $\phi_0$, then $\phi_0$ takes as many values, which are mutually independent. Let them be $\phi_0', \phi_0'', \ldots$. The corresponding value of $\omega$ must satisfy the equation of the form

$$\sum_{i=1}^{N} \frac{y_i}{\mu_i} \frac{\partial \phi_0^{(j)}}{\partial x_i} = \omega^{(j)} \phi_0^{(j)} \qquad (j = 1, 2, \ldots, n(n-1)/2).$$

Write

$$\Phi = \phi_0' \phi_0'' \cdots, \qquad \Omega = \omega' + \omega'' + \cdots;$$

then

$$\frac{1}{\Phi} \sum_{i=1}^{N} \frac{y_i}{\mu_i} \frac{\partial \Phi}{\partial x_i} = \Omega,$$

where $\Omega$ is linear homogeneous in $y_1, y_2, \ldots, y_N$ and the coefficients are rational in $x_1, x_2, \ldots, x_N$. Function $\Phi$, by definition rational in $x_1, x_2, \ldots, x_N$, does not contain $s$, and is an integral polynomial in $y_1, y_2, \ldots, y_N$. Hence, in accordance with Lemma I we can make

$\Omega = 0$ by multiplying $\Phi$ by a rational function of $x_1, x_2, \ldots, x_N$. Thus, $\Phi$ must satisfy

$$\sum_{i=1}^{N} \frac{y_i}{\mu_i} \frac{\partial \Phi}{\partial x_i} = 0.$$

As this is a partial differential equation for $N$ independent variables, $N - 1$ independent solutions are obtained according to the results in Sections 4.1 and 4.17. These solutions are

$$\left(\frac{x_i y_1}{\mu_1} - \frac{x_1 y_i}{\mu_i}\right) \qquad (i = 2, 3, \ldots, N).$$

Hence, $\Phi$ is a function of these $N - 1$ solutions and $N$ variables $y_1, y_2, \ldots, y_N$. But each factor of $\Phi$ is the value of $\phi_0$ for each root of $s$. Hence, if the two corresponding roots become equal, the corresponding two factors should coincide. Thus, $\Phi = 0$ should, as an equation for $y_1$, have at least two equal roots. Let the condition for equal roots be

$$F(x_1, x_2, \ldots, x_N) = 0; \tag{9a}$$

then $\partial\Phi/\partial y_1 = 0$. Similarly, $\partial\Phi/\partial y_2 = \cdots = \partial\Phi/\partial y_N = 0$. Since $\Phi$ is homogeneous in $y_1, y_2, \ldots, y_N$, the equation

$$y_1 \frac{\partial \Phi}{\partial y_1} + y_2 \frac{\partial \Phi}{\partial y_2} + \cdots + y_N \frac{\partial \Phi}{\partial y_N} = 0$$

is equivalent to $\Phi = 0$. Thus, $\Phi = 0$ is not independent of

$$\frac{\partial \Phi}{\partial y_1} = 0, \quad \frac{\partial \Phi}{\partial y_2} = 0, \quad \ldots, \quad \frac{\partial \Phi}{\partial y_N} = 0.$$

Suppose that we make small variations satisfying $\Phi = 0$, such that

$$\sum_{i=0}^{N} \left(\frac{\partial \Phi}{\partial x_i} \delta x_i + \frac{\partial \Phi}{\partial y_i} \delta y_i\right) = 0.$$

For any set of values satisfying $\partial\Phi/\partial y_i = 0$, this equation reduces to

$$\sum_{i=0}^{N} \frac{\partial \Phi}{\partial x_i} \delta x_i = 0,$$

and this is equivalent to

$$\sum_{i=0}^{N} \frac{\partial F}{\partial x_i} \delta x_i = 0.$$

Hence

$$\frac{\partial F/\partial x_1}{\partial \Phi/\partial x_1} = \frac{\partial F/\partial x_2}{\partial \Phi/\partial x_2} = \cdots = \frac{\partial F/\partial x_N}{\partial \Phi/\partial x_N} \tag{9b}$$

should be a consequence of $\partial\Phi/\partial y_i = 0$ $(i = 1, 2, \ldots, N)$. Since

$$\sum_{i=1}^{N} \frac{y_i}{\mu_i} \frac{\partial \Phi}{\partial x_i} = 0,$$

we have

$$\sum_{i=1}^{N} \frac{y_i}{\mu_i} \frac{\partial F}{\partial x_i} = 0$$

for values of $x_1, x_2, \ldots, x_N$ that satisfy (9b). Therefore, $F = 0$ and

$$\sum_{i=1}^{N} \frac{y_i}{\mu_i} \frac{\partial F}{\partial x_i} = 0$$

are derivable algebraically from $\partial\Phi/\partial y_i = 0$ $(i = 1, 2, \ldots, N)$. Here the actual values of $x_1, x_2, \ldots, x_N$ are not essential in this process of elimination, so we replace $x_i$ by $x_i + (y_i/\mu_i)t$ in all these equations. Then

$$F\left(x_1 + \frac{y_1}{\mu_1}t, x_2 + \frac{y_2}{\mu_2}t, \ldots, x_N + \frac{y_N}{\mu_N}t\right) = 0$$

and

$$\sum_{i=1}^{N} \frac{y_i}{\mu_i} \frac{\partial}{\partial x_i} F\left(x_1 + \frac{y_1}{\mu_1}t, x_2 + \frac{y_2}{\mu_2}t, \ldots, x_N + \frac{y_N}{\mu_N}t\right) = 0$$

are algebraic combinations of $N$ equations

$$\frac{\partial}{\partial y_i} \Phi(x_1, x_2, \ldots, x_N, y_1, y_2, \ldots, y_N) - \frac{1}{\mu_i} \frac{\partial}{\partial x_i} \Phi(x_1, x_2, \ldots, x_N, y_1, y_2, \ldots, y_N) = 0 \quad (i = 1, 2, \ldots, N).$$

But one of the algebraic combinations is

$$\Phi(x_1, x_2, \ldots, x_N, y_1, y_2, \ldots, y_N) = 0,$$

because we can derive this last equation by adding the above $N$ equations after multiplying them by $y_1, y_2, \ldots, y_N$, respectively. We shall show that this is the result of the elimination just mentioned.

Denote the result of elimination by $\Psi$. The equation

$$\sum_{i=1}^{N}\left(\frac{\partial\Psi}{\partial x_i}\,\delta x_i+\frac{\partial\Psi}{\partial y_i}\,\delta y_i\right)=0$$

must be a combination of

$$\sum_{i=1}^{N}\frac{\partial F}{\partial x_i}\left(\delta x_i+\frac{t}{\mu_i}\,\delta y_i\right)=0$$

and

$$\sum_{i=1}^{N}\frac{1}{\mu_i}\frac{\partial F}{\partial x_i}\,\delta y_i+\sum_{i=1}^{N}\sum_{j=1}^{N}\frac{y_i}{\mu_i}\frac{\partial^2 F}{\partial x_i\,\partial x_j}\left(\delta x_j+\frac{t}{\mu_j}\,\delta y_j+\frac{y_j}{\mu_j}\,\delta t\right)=0.$$

Since the last equation contains $\delta t$, it cannot enter into the combination. Hence,

$$\frac{\partial\Psi/\partial x_1}{\partial F/\partial x_1}=\frac{\partial\Psi/\partial x_2}{\partial F/\partial x_2}=\cdots=\frac{\partial\Psi/\partial x_N}{\partial F/\partial x_N},$$

$$\frac{\partial\Psi}{\partial y_i}=\frac{t}{\mu_i}\frac{\partial F}{\partial x_i}\qquad(i=1,2,\ldots,N).$$

From the fact that these equations should be identical with equations (9b) already obtained for $\Phi$ we see that $\Phi=0$ and $\Psi=0$ are equivalent. Hence, $\Phi=0$ is the result of eliminating $t$ from

$$F\left(x_1+\frac{y_1}{\mu_1}t,\,x_2+\frac{y_2}{\mu_2}t,\ldots,x_N+\frac{y_N}{\mu_N}t\right)=0,$$

and

$$\frac{\partial}{\partial t}F\left(x_1+\frac{y_1}{\mu_1}t,\,x_2+\frac{y_2}{\mu_2}t,\ldots,x_N+\frac{y_N}{\mu_N}t\right)=0.$$

Now we can write down explicitly the equation $F(x_1, x_2, \ldots, x_N)=0$, which expresses the condition for equal roots in the equation for $s$. From this we can find all possible polynomials $\Phi$, and hence all possible polynomials $\phi_0$, by factorizing $\Phi$. The values of $s$ formed from the $n(n-1)/2$ mutual distances $r_{3j,3k}$ are $\pm r_{3,6}\pm r_{3,9}\pm\cdots\pm r_{3n-3,3n}$, and there are $2^{n(n-1)/2}$ such values of $s$. Hence, two roots of $s$ become equal when we have either of the relations

$$r_{3,6}=0,\qquad r_{3,9}=0,\qquad\ldots,\qquad r_{3n-3,3n}=0;$$

$$r_{3,6}=\pm r_{3,9},\qquad r_{3,6}=\pm r_{3,12},\qquad\ldots,\qquad r_{3n-3,3n-6}=\pm r_{3n-3,3n};$$

$$r_{3,6}\pm r_{3,9}\pm r_{3,12}=0,\qquad\ldots,\qquad r_{3,6}\pm r_{3,9}\pm\cdots\pm r_{3n-3,3n}=0.$$

We limit ourselves to the case $n = 3$. Then there are eight values of $s$. Two roots become equal when we have either of the relations,

$$r_{6,9} = 0, \qquad r_{9,3} = 0, \qquad r_{3,6} = 0, \qquad r_{9,3} = \pm r_{3,6},$$

$$r_{3,6} = \pm r_{6,9}, \qquad r_{6,9} = \pm r_{9,3}, \qquad r_{6,9} \pm r_{9,3} \pm r_{3,6} = 0.$$

From $r_{6,9} = 0$ we obtain $F = x_1^2 + x_2^2 + x_3^2 = 0$. If we eliminate $t$ from

$$F\left(x_1 + \frac{y_1}{\mu_1} t, x_2 + \frac{y_2}{\mu_2} t, \ldots\right) = 0$$

and

$$\frac{\partial}{\partial t} F\left(x_1 + \frac{y_1}{\mu_1} t, x_2 + \frac{y_2}{\mu_2} t, \ldots\right) = 0,$$

then we obtain

$$(x_1^2 + x_2^2 + x_3^2)\left(\frac{y_1^2}{\mu_1^2} + \frac{y_2^2}{\mu_2^2} + \frac{y_3^2}{\mu_3^2}\right) = \left(\frac{x_1 y_1}{\mu_1} + \frac{x_2 y_2}{\mu_2} + \frac{x_3 y_3}{\mu_3}\right)^2.$$

Hence, function $\Phi$ takes the form

$$\Phi = \left(\frac{x_1 y_2}{\mu_2} - \frac{x_2 y_1}{\mu_1}\right)^2 + \left(\frac{x_2 y_3}{\mu_3} - \frac{x_3 y_2}{\mu_2}\right)^2 + \left(\frac{x_3 y_1}{\mu_1} - \frac{x_1 y_3}{\mu_3}\right)^2. \tag{9c}$$

Since this cannot be factorized into real factors, there can be no real polynomial $\phi_0$ in this case. From $r_{9,3} = 0$ and from $r_{3,6} = 0$ we obtain similar results.

From $r_{9,3} = \pm r_{3,6}$ we derive $\Phi$ in the form

$$\begin{aligned}\Phi = {} & \left\{2(x_1x_4 + x_2x_5 + x_3x_6) + \frac{m_3 - m_6}{m_3 + m_6}(x_1^2 + x_2^2 + x_3^2)\right\} \\ & \times \left\{2\left(\frac{y_1y_4}{\mu_1\mu_4} + \frac{y_2y_5}{\mu_2\mu_5} + \frac{y_3y_6}{\mu_3\mu_6}\right) + \frac{m_3 - m_6}{m_3 + m_6}\left(\frac{y_1^2}{\mu_1^2} + \frac{y_2^2}{\mu_2^2} + \frac{y_3^2}{\mu_3^2}\right)\right\} \\ & - \left\{\frac{x_1y_4}{\mu_4} + \frac{x_4y_1}{\mu_1} + \frac{x_2y_5}{\mu_5} + \frac{x_5y_2}{\mu_2} + \frac{x_3y_6}{\mu_6} + \frac{x_6y_3}{\mu_3}\right. \\ & \qquad \left. + \frac{m_3 - m_6}{m_3 + m_6}\left(\frac{x_1y_1}{\mu_1} + \frac{x_2y_2}{\mu_2} + \frac{x_3y_3}{\mu_3}\right)\right\}^2.\end{aligned} \tag{9d}$$

This cannot be factorized into real polynomials $\phi_0$ that are linear in $y_1, y_2, \ldots, y_N$. Hence, there can be no real polynomial $\phi_0$ in this case. From $r_{3,6} = \pm r_{6,9}$ and from $r_{6,9} = \pm r_{9,3}$ we obtain similar results.

Lastly, the relation $r_{6,9} \pm r_{9,3} \pm r_{3,6} = 0$ is rationalized to

$$(r_{6,9}^2 - r_{9,3}^2 + r_{3,6}^2)^2 - 4r_{3,6}^2 r_{6,9}^2 = 0.$$

If $r_{6,9} = 0$, then the case reduces to the last one (9d). $\Phi$ likewise cannot be factorized. Hence, there is no real polynomial $\phi_0$ that contains $s$.

Summarizing the result so far obtained: We have shown that any algebraic integral independent of $t$ is an algebraic function of the integrals $\phi$, each of which can be written in the form $\phi_0 + \phi_2 + \phi_4 + \cdots$, where $\phi_0$ is a homogeneous polynomial of degree $k$, say, in $y$ and a homogeneous algebraic function of degree $l$, say, in $x$, and $\phi_2$ is a homogeneous polynomial of degree $k - 2$ in $y$ and a homogeneous algebraic function of degree $l - 1$ in $x$, and so on.

## 6.4 INTEGRALS OF MOMENTUM AND ANGULAR MOMENTUM

By extending the proof in the case of three bodies to the general case of $n$ bodies we shall show that such an integral $\phi_0$ is an algebraic combination of the classical integrals.

If we replace $\phi$ by $\phi_0 + \phi_2 + \phi_4 + \cdots$ in the equation

$$\frac{d\phi}{dt} = 0,$$

that is, in

$$\sum_{i=1}^{N} \left( \frac{y_i}{\mu_i} \frac{\partial \phi}{\partial x_i} + \frac{\partial U}{\partial x_i} \frac{\partial \phi}{\partial y_i} \right) = 0$$

and, it being an identity, equate terms of equal degree, then we obtain

$$\begin{aligned} 0 &= \sum_{i=1}^{N} \frac{\partial \phi_0}{\partial x_i} \frac{y_i}{\mu_i}, \\ 0 &= \sum_{i=1}^{N} \frac{\partial \phi_2}{\partial x_i} \frac{y_i}{\mu_i} + \frac{\partial \phi_0}{\partial y_i} \frac{\partial U}{\partial x_i}, \\ &\vdots \end{aligned} \tag{10}$$

The first equation is a partial differential equation of the first order for $\phi_0$ and the solution is

$$\phi_0 = f_0(Y_2, Y_3, \ldots, Y_N, y_1, y_2, \ldots, y_N),$$

where

$$Y_i = \frac{x_i y_1}{\mu_1} - \frac{x_1 y_i}{\mu_i} \qquad (i = 2, 3, \ldots, N).$$

The expression for $\phi_2$ takes the form $\phi_2 = f_2(x_1, Y_2, \ldots, Y_N, y_1, \ldots, y_N)$ in terms of $x_1, Y_2, \ldots, Y_N, y_1, \ldots, y_N$. Refer to

$$x_i = \frac{\mu_1 Y_i}{y_1} + \frac{\mu_1 y_i x_1}{\mu_i y_1}$$

and form

$$\frac{\partial f_2}{\partial x_1};$$

then we obtain from (10)

$$\frac{y_1}{\mu_1}\frac{\partial f_2}{\partial x_1} = \sum_{i=1}^{N} \frac{y_i}{\mu_i}\frac{\partial \phi_2}{\partial x_i} = -\sum_{i=1}^{N} \frac{\partial \phi_0}{\partial y_i}\frac{\partial U}{\partial x_i}.$$

The integral of this equation is

$$f_2 = \chi(Y_2, Y_3, \ldots, Y_N, y_1, \ldots, y_N) - \int \frac{\mu_1}{y_1} \sum_{i=1}^{N} \frac{\partial \phi_0}{\partial y_i}\frac{\partial U}{\partial x_i}\, dx_1.$$

If we express

$$X = \sum_{i=1}^{N} \frac{\partial \phi_0}{\partial y_i}\frac{\partial U}{\partial x_i}$$

in terms of $x_1, Y_2, \ldots, Y_N, y_1, \ldots, y_N$, then

$$X = \sum_{i=1}^{N} \frac{\partial f_0}{\partial y_i}\frac{\partial U}{\partial x_i} + \sum_{j=2}^{N} \frac{\partial f_0}{\partial Y_j}\left(\frac{x_j}{\mu_1}\frac{\partial U}{\partial x_1} - \frac{x_1}{\mu_j}\frac{\partial U}{\partial x_j}\right).$$

As $\int X\, dx_1$ should be an algebraic function, there cannot be any logarithmic term in this integral; that is, it cannot be Abel's integral of the third kind (Section 2.20). Let $V$ be the expression of $U$ in terms of $x_1, Y_2, \ldots, Y_N, y_1, \ldots, y_N$; then

$$\frac{\partial U}{\partial x_i} = \frac{y_1}{\mu_1}\frac{\partial V}{\partial Y_i} \qquad (i > 1).$$

Since

$$\frac{\partial U}{\partial x_1} = \frac{\partial V}{\partial x_1} - \sum_{j=2}^{N} \frac{\partial V}{\partial Y_j}\frac{y_j}{\mu_j},$$

the terms of $X$ that might give rise to a logarithmic term in $\int X\, dx_1$ are

$$\sum_{j=2}^{N} \frac{\partial f_0}{\partial Y_j}\left(\frac{y_j}{\mu_j}\frac{V}{y_1} + \frac{y_j x_1}{\mu_j y_1}\sum_{k=2}^{N} \frac{\partial V}{\partial Y_k}\frac{y_k}{\mu_k} + \frac{x_1 y_1}{\mu_j \mu_1}\frac{\partial V}{\partial Y_j}\right).$$

Now $V$ is the sum of $n(n-1)/2$ terms of the form

$$(A + Bx_1 + Cx_1^2)^{-1/2}.$$

Take each of these terms separately in the above expression for $X$ and equate to zero the part of the transcendental function in $\int X\,dx_1$; then we obtain the condition

$$C\sum_{j=2}^{N}\frac{\partial f_0}{\partial Y_j}\frac{y_j}{\mu_j y_1} - \sum_{j=2}^{N}\sum_{k=2}^{N}\frac{\partial f_0}{\partial Y_j}\frac{y_j y_k}{\mu_j\mu_k y_1}\frac{1}{2}\frac{\partial B}{\partial Y_k} - \sum_{j=2}^{N}\frac{\partial f_0}{\partial Y_j}\frac{y_1}{\mu_j\mu_1}\frac{1}{2}\frac{\partial B}{\partial Y_j} = 0.$$

In general, $r_{3j,3k}^2$ is of the form

$$[(x_{3j-2} - x_{3k-2})^2 + (x_{3j-1} - x_{3k-1})^2 + (x_{3j} - x_{3k})^2]$$

and is the sum of expressions of the form $x_{3j-2}^2 + x_{3j-1}^2 + x_{3j}^2$ and $x_{3j-2}x_{3k-2} + x_{3j-1}x_{3k-1} + x_{3j}x_{3k}$. For an expression of the form $x_1^2 + x_2^2 + x_3^2$, the term in (156) and (157) of Section 5.16 can be written in the present case (after some computation) in the form

$$(A + Bx_1 + Cx_1^2)^{-1/2},$$

where

$$A = \frac{\mu_1^2}{y_1^2}(Y_2^2 + Y_3^2),$$

$$\frac{1}{2}B = \frac{\mu_1^2 Y_2 y_2}{\mu_2 y_1^2} + \frac{\mu_1^2 Y_3 y_3}{\mu_3 y_1^2},$$

$$C = 1 + \frac{\mu_1^2 y_2^2}{\mu_2^2 y_1^2} + \frac{\mu_1^2 y_3^2}{\mu_3^2 y_1^2}.$$

Hence the contribution to the above condition due to such a term is

$$\sum_{i=2}^{N}\frac{\partial f_0}{\partial Y_i}\frac{y_i}{\mu_i} - \left(\frac{\partial f_0}{\partial Y_2}\frac{\mu_1 y_2}{\mu_2^2} + \frac{\partial f_0}{\partial Y_3}\frac{\mu_1 y_3}{\mu_3^2}\right) = 0,$$

or

$$\sum_{i=4}^{N} y_i\frac{\partial f_0}{\partial Y_i} = 0. \tag{11}$$

Next, for an expression of the form $x_{3j-1}^2 + x_{3j-1}^2 + x_{3j}^2$, we have

$$A = \frac{\mu_1^2}{y_1^2}(Y_{3j-2}^2 + Y_{3j-1}^2 + Y_{3j}^2),$$

$$\frac{1}{2}B = \frac{\mu_1^2 Y_{3j-2}y_{3j-2}}{\mu_{3j-2}y_1^2} + \frac{\mu_1^2 Y_{3j-1}y_{3j-1}}{\mu_{3j-1}y_1^2} + \frac{\mu_1^2 Y_{3j}y_{3j}}{\mu_{3j}y_1^2},$$

$$C = \frac{\mu_1^2}{y_1^2}\left(\frac{y_{3j-2}^2}{\mu_{3j-2}^2} + \frac{y_{3j-1}^2}{\mu_{3j-1}^2} + \frac{y_{3j}^2}{\mu_{3j}^2}\right).$$

Hence the contribution to the above condition due to such a term is

$$y_{3j-2}\frac{\partial f_0}{\partial Y_{3j-2}} + y_{3j-1}\frac{\partial f_0}{\partial Y_{3j-1}} + y_{3j}\frac{\partial f_0}{\partial Y_{3j}} = 0 \qquad (j = 2, 3, \ldots, n-1). \tag{12}$$

Combining (12) with (11), we obtain the same $n - 2$ equations (12). By solving these equations we see that $f_0$ is a function of

$$y_1, \ldots, y_N, Y_2, Y_3, x_{3j-1}y_{3j-2} - x_{3j-2}y_{3j-1}, x_{3j}y_{3j-2} - x_{3j-2}y_{3j} \qquad (j = 2, 3, \ldots, n-1),$$

or a function of

$$y_1, \ldots, y_N, Y_2, Y_3, Y_{3j-1}y_{3j-2} - Y_{3j-2}y_{3j-1}, Y_{3j}y_{3j-2} - Y_{3j-2}y_{3j}.$$

Next, from an expression of the form

$$(x_1x_{3j-2} + x_2x_{3j-1} + x_3x_{3j}) \qquad (j = 2, 3, \ldots, n-1),$$

which is written

$$x_1\left(\frac{\mu_1 Y_{3j-2}}{y_1} + \frac{\mu_1 y_{3j-2}}{\mu_{3j-2}y_1}x_1\right) + \left(\frac{\mu_1 Y_2}{y_1} + \frac{y_2x_1}{y_1}\right)\left(\frac{\mu_1 Y_{3j-1}}{y_1} + \frac{\mu_1 y_{3j-1}x_1}{\mu_{3j-1}y_1}\right)$$
$$+ \left(\frac{\mu_1 Y_3}{y_1} + \frac{y_3x_1}{y_1}\right)\left(\frac{\mu_1 Y_{3j}}{y_1} + \frac{\mu_1 y_{3j}x_1}{\mu_{3j}y_1}\right),$$

we obtain

$$A = \frac{\mu_1^2 Y_2 Y_{3j-1}}{y_1^2} + \frac{\mu_1^2 Y_3 Y_{3j}}{y_1^2},$$

$$B = \frac{\mu_1 Y_{3j-2}}{y_1} + \frac{\mu_1 Y_{3j-1}y_2}{y_1^2} + \frac{\mu_1^2 Y_2 y_{3j-1}}{\mu_{3j-1}y_1^2} + \frac{\mu_1 Y_{3j}y_3}{y_1^2} + \frac{\mu_1^2 Y_3 y_{3j}}{\mu_{3j}y_1^2},$$

$$C = \frac{\mu_1 y_{3j-2}}{\mu_{3j-2}y_1} + \frac{\mu_1 y_2 y_{3j-1}}{\mu_{3j-1}y_1^2} + \frac{\mu_1 y_3 y_{3j}}{\mu_{3j}y_1^2}.$$

Hence the corresponding condition is

$$\frac{\partial f_0}{\partial Y_2}(y_2y_{3j-2} - y_1y_{3j-1}) + \frac{\partial f_0}{\partial Y_3}(y_3y_{3j-2} - y_1y_{3j})$$
$$+ \frac{\partial f_0}{\partial Y_{3j-2}}\left(\frac{\mu_1 y_{3j-2}^2}{\mu_{3j-2}} - y_1^2\right) + \frac{\partial f_0}{\partial Y_{3j-1}}\left(\frac{\mu_1 y_{3j-2}y_{3j-1}}{\mu_{3j-2}} - y_2y_1\right)$$
$$+ \frac{\partial f_0}{\partial Y_{3j}}\left(\frac{\mu_1 y_{3j-2}y_{3j}}{\mu_{3j-2}} - y_1y_3\right) = 0$$
$$(j = 2, 3, \ldots, n-1). \tag{13}$$

Combining (13) and (12), we obtain

$$(y_2 y_{3j-2} - y_1 y_{3j-1}) \frac{\partial f_0}{\partial Y_2} + (y_{3j-2} y_3 - y_1 y_{3j}) \frac{\partial f_0}{\partial Y_3} - y_1\left(y_1 \frac{\partial f_0}{\partial Y_{3j-2}} + y_2 \frac{\partial f_0}{\partial Y_{3j-1}} + y_3 \frac{\partial f_0}{\partial Y_{3j}}\right) = 0 \quad (j = 2, 3, \ldots, n-1). \tag{14}$$

Similarly, from an expression of the form

$$(x_{3j-2}x_{3k-2} + x_{3j-1}x_{3k-1} + x_{3j}x_{3k})$$

we obtain

$$A = \frac{\mu_1^2 Y_{3j-1} Y_{3k-1}}{y_1^2} + \frac{\mu_1^2 Y_{3j} Y_{3k}}{y_1^2},$$

$$B = \frac{\mu_1^2 y_{3k-2} Y_{3j-2}}{\mu_{3k-2} y_1^2} + \frac{\mu_1^2 y_{3j-2} Y_{3k-2}}{\mu_{3j-2} y_1^2} + \frac{\mu_1^2 y_{3k-1} Y_{3j-1}}{\mu_{3k-1} y_1^2} + \frac{\mu_1^2 y_{3j-1} Y_{3k-1}}{\mu_{3j-1} y_1^2} + \frac{\mu_1^2 y_{3k} Y_{3j}}{\mu_{3k} y_1^2} + \frac{\mu_1^2 y_{3j} Y_{3k}}{\mu_{3j} y_1^2},$$

$$C = \frac{\mu_1^2 y_{3j-2} y_{3k-2}}{\mu_{3j-2}\mu_{3k-2} y_1^2} + \frac{\mu_1^2 y_{3j-1} y_{3k-1}}{\mu_{3j-1}\mu_{3k-1} y_1^2} + \frac{\mu_1^2 y_{3j} y_{3k}}{\mu_{3j}\mu_{3k} y_1^2}.$$

Hence the corresponding condition is

$$y_{3k-2} \frac{\partial f_0}{\partial Y_{3j-2}} + y_{3k-1} \frac{\partial f_0}{\partial Y_{3j-1}} + y_{3k} \frac{\partial f_0}{\partial Y_{3j}} + y_{3j-2} \frac{\partial f_0}{\partial Y_{3k-2}} + y_{3j-1} \frac{\partial f_0}{\partial Y_{3k-1}} + y_{3j} \frac{\partial f_0}{\partial Y_{3k}} = 0 \quad (j, k = 2, 3, \ldots, n-1). \tag{15}$$

We have seen that $Y_j$ enters into these equations only in the combinations

$$(Y_{3j-2} y_{3j-1} - Y_{3j-1} y_{3j-2}), \qquad (Y_{3j-1} y_{3j} - Y_{3j} y_{3j-1}),$$

and

$$(Y_{3j} y_{3j-2} - Y_{3j-2} y_{3j}).$$

In these combinations, it is evident from (15) that we can exchange the suffixes $j$ and $k$. Hence $Y_j$ enters only in the combinations

$$\sum_{j=2}^{n} (Y_{3j-2} y_{3j-1} - Y_{3j-1} y_{3j-2}), \qquad \sum_{j=2}^{n} (Y_{3j-1} y_{3j} - Y_{3j} y_{3j-1}),$$

$$\sum_{j=2}^{n} (Y_{3j} y_{3j-2} - Y_{3j-2} y_{3j}).$$

On the other hand, the angular momentum of our dynamical system expressed in terms of $Y_i$, according to Section 4.12, is

$$\begin{aligned}
\frac{\mu_1}{y_1} H_4 &= Y_3 y_2 - Y_2 y_3 + \sum_{j=2}^{n-1} (Y_{3j} y_{3j-1} - Y_{3j-1} y_{3j}),\\
\frac{\mu_1}{y_1} H_5 &= -Y_3 y_1 + \sum_{j=2}^{n-1} (Y_{3j-2} y_{3j} - Y_{3j} y_{3j-2}),\\
\frac{\mu_1}{y_1} H_6 &= Y_2 y_1 + \sum_{j=2}^{n-1} (Y_{3j-1} y_{3j-2} - Y_{3j-2} y_{3j-1}),
\end{aligned} \tag{16}$$

and each of the integrals $H_4$ = constant, $H_5$ = constant, $H_6$ = constant satisfies partial differential equations (14). Next, put

$$H_1 = -\sum_{j=1}^{n-1} y_{3j-2}, \qquad H_2 = -\sum_{j=1}^{n-1} y_{3j-1}, \qquad H_3 = -\sum_{j=1}^{n-1} y_{3j}, \tag{17}$$

$$\begin{aligned}
\frac{\mu_1}{y_1} M_1 &= H_3\left(\sum_{j=1}^{n-1} \mu_{3j-1} Y_{3j-1}\right) - H_2\left(\sum_{j=1}^{n-1} \mu_{3j} Y_{3j}\right),\\
\frac{\mu_1}{y_1} M_2 &= H_1\left(\sum_{j=1}^{n-1} \mu_{3j} Y_{3j}\right) - H_3\left(\sum_{j=2}^{n-1} \mu_{3j-2} Y_{3j-2}\right),\\
\frac{\mu_1}{y_1} M_3 &= H_2\left(\sum_{j=2}^{n-1} \mu_{3j-2} Y_{3j-2}\right) - H_1\left(\sum_{j=1}^{n-1} \mu_{3j-1} Y_{3j-1}\right).
\end{aligned} \tag{18}$$

Then $M_1$ = constant, $M_2$ = constant, $M_3$ = constant are integrals of our partial differential equations. They are not independent but connected by the relation

$$H_1 M_1 + H_2 M_2 + H_3 M_3 \equiv 0.$$

The Jacobian of $H_4$, $H_5$, $H_6$, $M_2$, $M_3$ with regard to $Y_2$, $Y_3$, $Y_4$, $Y_5$, $Y_6$ is, apart from the factor $\mu_1/y_1$, equal to (19):

$$\begin{vmatrix}
-y_3 & y_2 & 0 & -y_6 & y_5\\
0 & -y_1 & y_6 & 0 & -y_4\\
y_1 & 0 & -y_5 & y_4 & 0\\
0 & \mu_3 H_1 & -\mu_4 H_3 & 0 & \mu_6 H_1\\
-\mu_2 H_1 & 0 & \mu_4 H_2 & -\mu_5 H_1 & 0
\end{vmatrix}$$

$$= \mu_4 H_1(\mu_6 y_1 - \mu_3 y_4)\begin{vmatrix} y_1 & y_4 & H_1\\ y_2 & y_5 & H_2\\ y_3 & y_6 & H_3\end{vmatrix} \tag{19}$$

and does not vanish. Hence, $Y_2, Y_3, Y_4, Y_5, Y_6$ can be expressed in terms of $H_4, H_5, H_6, M_2, M_3$, and the remaining $Y_7, Y_8, \ldots, Y_N$.

Suppose that $Y_2, Y_3, Y_4, Y_5, Y_6$ in an integral $\chi$ of our partial differential equations are expressed in terms of $H_4, H_5, H_6, M_2, M_3$. If we equate the latter five quantities to zero, then $\chi$ can be considered independent of the functions $Y_2, Y_3, Y_4, Y_5, Y_6$. Combining (12) with (14), we obtain

$$(y_{3j-2} - y_1)\frac{\partial\chi}{\partial Y_{3j-2}} + (y_{3j-1} - y_2)\frac{\partial\chi}{\partial Y_{3j-1}} + (y_{3j} - y_3)\frac{\partial\chi}{\partial Y_{3j}} = 0 \qquad (j = 3, 4, \ldots, n-1).$$

Put $k = 2$ in (15) and subtract it from (12); then

$$(y_{3j-2} - y_4)\frac{\partial\chi}{\partial Y_{3j-2}} + (y_{3j-1} - y_5)\frac{\partial\chi}{\partial Y_{3j-1}} + (y_{3j} - y_6)\frac{\partial\chi}{\partial Y_{3j}} = 0 \qquad (j = 3, 4, \ldots, n-1).$$

From these two equations we obtain

$$\frac{\dfrac{\partial\chi}{\partial Y_{3j-2}}}{\begin{vmatrix} y_{3j-1} - y_2 & y_{3j} - y_3 \\ y_{3j-1} - y_5 & y_{3j} - y_6 \end{vmatrix}} = \frac{\dfrac{\partial\chi}{\partial Y_{3j-1}}}{\begin{vmatrix} y_{3j} - y_3 & y_{3j-2} - y_1 \\ y_{3j} - y_6 & y_{3j-2} - y_4 \end{vmatrix}} = \frac{\dfrac{\partial\chi}{\partial Y_{3j}}}{\begin{vmatrix} y_{3j-2} - y_1 & y_{3j-1} - y_2 \\ y_{3j-2} - y_4 & y_{3j-1} - y_5 \end{vmatrix}}. \tag{20}$$

Write

$$D \equiv \begin{vmatrix} y_{3j-2} - y_1 & y_{3j-1} - y_2 & y_{3j} - y_3 \\ y_{3j-2} - y_4 & y_{3j-1} - y_5 & y_{3j} - y_6 \\ y_{3j-2} - y_{3k-2} & y_{3j-1} - y_{3k-1} & y_{3j} - y_{3k} \end{vmatrix};$$

then (20) can be transformed, with constants $c_j$ and $c_k$, to

$$(y_{3j-2} - y_{3k-2})\frac{\partial\chi}{\partial Y_{3j-2}} + (y_{3j-1} - y_{3k-1})\frac{\partial\chi}{\partial Y_{3j-1}} + (y_{3j} - y_{3k})\frac{\partial\chi}{\partial Y_{3j}} = c_j D,$$

$$(y_{3j-2} - y_{3k-2})\frac{\partial\chi}{\partial Y_{3k-2}} + (y_{3j-1} - y_{3k-1})\frac{\partial\chi}{\partial Y_{3k-1}} + (y_{3j} - y_{3k})\frac{\partial\chi}{\partial Y_{3k}} = c_k D.$$

However, from (15) and (12) we obtain

$$(y_{3j-2} - y_{3k-2})\left(\frac{\partial f_0}{\partial Y_{3j-2}} - \frac{\partial f_0}{\partial Y_{3k-2}}\right) + (y_{3j-1} - y_{3k-1})\left(\frac{\partial f_0}{\partial Y_{3j-1}} - \frac{\partial f_0}{\partial Y_{3k-1}}\right) + (y_{3j} - y_{3k})\left(\frac{\partial f_0}{\partial Y_{3j}} - \frac{\partial f_0}{\partial Y_{3k}}\right) = 0;$$

therefore,

$$(c_j - c_k)D = 0;$$

that is,

$$c_j = c_k = c.$$

Hence,

$$y_{3j-2}\frac{\partial \chi}{\partial Y_{3j-2}} + y_{3j-1}\frac{\partial \chi}{\partial Y_{3j-1}} + y_{3j}\frac{\partial \chi}{\partial Y_{3j}} = c\begin{vmatrix} y_{3j-2} & y_1 & y_4 \\ y_{3j-1} & y_2 & y_5 \\ y_{3j} & y_3 & y_6 \end{vmatrix}$$

is zero because of (11). Thus, $c = 0$. Hence, (20) becomes

$$\frac{\partial \chi}{\partial Y_{3j-2}} = \frac{\partial \chi}{\partial Y_{3j-1}} = \frac{\partial \chi}{\partial Y_{3j}} = 0 \qquad (j = 3, 4, \ldots, n-1). \tag{21}$$

Thus $\chi$ does not contain $Y_{3j-2}$, $Y_{3j-1}$, $Y_{3j}$ $(j = 3, 4, \ldots, n-1)$; that is, $f_0$ contains only $x_i$ in the combinations $H_4$, $H_5$, $H_6$, $M_2$, $M_3$. Hence, if we eliminate $x_2$, $x_3$, $x_4$, $x_5$, $x_6$ from $f_0$, then the remaining

$$x_7, x_8, \ldots, x_{3n-3}$$

also disappear from $f_0$.

When we express $Y_2$, $Y_3$, $Y_4$, $Y_5$, $Y_6$ in terms of $H_4$, $H_5$, $H_6$, $M_2$, $M_3$, the corresponding Jacobian should appear in the denominator. The Jacobian determinant of $H_4$, $H_5$, $H_6$, $M_2$, $M_3$ with regard to $x_2$, $x_3$, $x_4$, $x_5$, $x_6$ is, apart from the factor $y_1/\mu_1$, equal to (19); that is,

$$\Delta \equiv \mu_4 H_1(\mu_6 y_1 - \mu_3 y_4)\begin{vmatrix} y_1 & y_4 & H_1 \\ y_2 & y_5 & H_2 \\ y_3 & y_6 & H_3 \end{vmatrix}.$$

The expression for $f_0$ in terms of $H_4$, $H_5$, $H_6$, $M_2$, $M_3$ has a power of $\Delta$ in the denominator. Let it be

$$G(H_4, H_5, H_6, M_2, M_3, y_1, y_2, \ldots, y_N)\,\Delta^{-p}.$$

If we eliminate $x_2$, $x_3$, $x_7$, $x_8$, $x_9$ instead of $x_2$, $x_3$, $x_4$, $x_5$, $x_6$ by means of $H_4$, $H_5$, $H_6$, $M_2$, $M_3$, we obtain

$$\Delta' \equiv \mu_7 H_1 \cdot (\mu_9 y_1 - y_3 y_7) \begin{vmatrix} y_1 & y_7 & H_1 \\ y_2 & y_8 & H_2 \\ y_3 & y_9 & H_3 \end{vmatrix}.$$

As $\Delta$ and $\Delta'$ have $H_1$ in common, the remaining factor of $\Delta$ or $\Delta'$ in $f_0$ cancels with the corresponding factor of the numerator, and only a power of $H_1$ remains in the denominator of $f_0$. If we eliminate $x_1$, $x_3$, $x_4$, $x_5$, $x_6$ instead of $x_2$, $x_3$, $x_4$, $x_5$, $x_6$, and $M_3$, $M_1$ instead of $M_2$, $M_3$, then $f_0$ takes the form

$$G(H_4, H_5, H_6, M_3, M_1, y_1, y_2, \ldots, y_N) \cdot H_2^{-q}.$$

Similarly, if we eliminate $x_1$, $x_2$, $x_4$, $x_5$, $x_6$, then we obtain

$$G(H_4, H_5, H_6, M_1, M_2, y_1, y_2, \ldots, y_N) \cdot H_3^{-r}.$$

Suppose that none of the exponents $p$, $q$, $r$ of the denominator vanish. There is no linear relation among the $H_4$, $H_5$, $H_6$, $M_1$, $M_2$ except the linear relation $H_1M_1 + H_2M_2 + H_3M_3 \equiv 0$ among $M_1$, $M_2$, $M_3$; therefore the functions $H_4$, $H_5$, $H_6$, $M_1$, $M_2$, $M_3$, $y_1, y_2, \ldots, y_N$ in $f_0$ can be considered independent. If we express $y_1$, $y_2$, $y_3$ by means of $H_1$, $H_2$, $H_3$, and the remaining $y_i$, then,

$$f_0 = G(H_4, H_5, H_6, M_1, M_2, M_3, H_1, H_2, H_3, y_4, y_5, \ldots, y_N) \cdot H_1^{-m}.$$

Insofar as $x_i, y_i$ $(i = 1, 2, \ldots, N)$ are finite, $f_0$ cannot become infinite. Hence, we arrange $f_0$ in the order of descending powers of $H_1$ in the form

$$f_0 = \frac{f_{00}}{H_1^m} + \frac{f_{01}}{H_1^{m-1}} + \cdots,$$

where $f_{00}, f_{01}, \ldots$ are polynomials in the variables they contain. If $H_1 = 0$, that is, if $H_2M_2 + H_3M_3 = 0$, then $f_{00}$ is zero. Hence, $f_{00}$ has the factor $H_2M_2 + H_3M_3$, that is, $f_{00}$ is of the form

$$f_{00} = (H_2M_2 + H_3M_3) f'_{00},$$

where $f'_{00}$ is another polynomial. Hence

$$f_0 = \frac{f_{01} - M_1 f'_{00}}{H_1^{m-1}} + \frac{f_{02}}{H_1^{m-2}} + \cdots.$$

By repeating this process we can always express $f_0$ in the form of a polynomial in $H_4$, $H_5$, $H_6$, $M_1$, $M_2$, $M_3$, $y_1, y_2, \ldots, y_N$ without denominator.

## 6.5 ENERGY INTEGRAL

Now we shall prove that any integral contains $y_1, y_2, \ldots, y_N$ only in the combination

$$\sum_{i=1}^{N} \frac{y_i^2}{2\mu_i},$$

that is, in the combination $T$ of the kinetic energy.

As we have shown, the function $\phi_0$ is a polynomial of the form

$$\phi_0 = G(H_4, H_5, H_6, M_1, M_2, M_3, y_1, y_2, \ldots, y_N)$$

and satisfies

$$\frac{d\phi_0}{dt} = \sum_{i=1}^{N} \frac{\partial G}{\partial y_i}\frac{dy_i}{dt} = \sum_{i=1}^{N} \frac{\partial G}{\partial y_i}\frac{\partial U}{\partial x_i}.$$

Let $X_i$ be the function $\partial U/\partial x_i$ expressed in terms of

$$x_1, Y_2, Y_3, \ldots, Y_N, y_1, y_2, \ldots, y_N;$$

then $f_2$ is of the form

$$f_2 = \chi(Y_2, Y_3, \ldots, Y_N, y_1, y_2, \ldots, y_N) - \frac{\mu_1}{y_1}\sum_{i=1}^{N} \frac{\partial G}{\partial y_i}\int X_i\, dx_1.$$

The second term on the right-hand side can be transformed to

$$\begin{aligned}
\frac{\mu_1}{y_1}\sum_{i=1}^{N} \frac{\partial G}{\partial y_i}\int X_i\, dx_1
&= \int \frac{\mu_1}{y_1}\left\{\frac{\partial G}{\partial y_1}\left(\frac{\partial V}{\partial x_1} - \sum_{i=2}^{N} \frac{\partial V}{\partial Y_i}\frac{y_i}{\mu_i}\right) + \frac{y_1}{\mu_1}\sum_{j=2}^{N} \frac{\partial V}{\partial Y_j}\frac{\partial G}{\partial y_j}\right\} dx_1 \\
&= \frac{\mu_1}{y_1}\frac{\partial G}{\partial y_1}\int \frac{\partial V}{\partial x_1}\, dx_1 + \sum_{j=2}^{N}\int \frac{\partial V}{\partial Y_j}\left(\frac{\partial G}{\partial y_j} - \frac{\mu_1 y_j}{\mu_j y_1}\frac{\partial G}{\partial y_1}\right) dx_1 \\
&= \frac{\mu_1}{y_1}\frac{\partial G}{\partial y_1} V + \sum_{j=2}^{N}\left(\frac{\partial G}{\partial y_j} - \frac{\mu_1 y_j}{\mu_j y_1}\frac{\partial G}{\partial y_1}\right)\frac{\partial}{\partial Y_j}\int V\, dx_1 \\
&= \frac{\mu_1}{y_1}\frac{\partial G}{\partial y_1}\sum_{A} \frac{m_{3l}m_{3k}}{(A + Bx_1 + Cx_1^2)^{1/2}} \\
&\quad + \sum_{j=2}^{N}\left(\frac{\partial G}{\partial y_j} - \frac{\mu_1 y_j}{\mu_j y_1}\frac{\partial G}{\partial y_1}\right) \\
&\quad \times \sum_{A} \frac{m_{3l}m_{3k}\left(-2A\dfrac{\partial B}{\partial Y_j} - x_1 B\dfrac{\partial B}{\partial Y_j} + B\dfrac{\partial A}{\partial Y_j} + 2Cx_1\dfrac{\partial A}{\partial Y_j}\right)}{(B^2 - 4AC)(A + Bx_1 + Cx_1^2)^{1/2}},
\end{aligned}$$

where $\sum_A$ denotes the sum over $n(n-1)/2$ values of the form

$$(A + Bx_1 + Cx_1^2),$$

and $m_{3l}m_{3k}$ the product of the masses of the two bodies corresponding to the distance $r_{3l,3k}$ in the denominator of (156) and (157) of Section 5.16. As $\chi(Y_2, \ldots, Y_N, y_1, \ldots, y_N)$ does not give rise to any term containing $(A + Bx_1 + Cx_1^2)$ in the same denominator, the quantities multiplying each $(A + Bx_1 + Cx_1^2)^{1/2}$ are of the same nature as the function $\phi_2$ itself; that is, they are polynomials in $y_1, \ldots, y_N$ when expressed in terms of $x_1, \ldots, x_N, y_1, \ldots, y_N$. Hence,

$$\frac{\mu_1}{y_1}\frac{\partial G}{\partial y_1} + 2\sum_{j=2}^{N}\left(\frac{\partial G}{\partial y_j} - \frac{\mu_1 y_j}{\mu_j y_1}\frac{\partial G}{\partial y_1}\right) \times \frac{-A\dfrac{\partial B}{\partial Y_j} - \dfrac{1}{2}x_1 B\dfrac{\partial B}{\partial Y_j} + \dfrac{1}{2}B\dfrac{\partial A}{\partial Y_j} + Cx_1\dfrac{\partial A}{\partial Y_j}}{B^2 - 4AC} \tag{22}$$

is a polynomial in $y_1, \ldots, y_N$ when expressed in terms of

$$y_1, \ldots, y_N, x_1, \ldots, x_N.$$

First, take $x_{3j-2}^2 + x_{3j-1}^2 + x_{3j}^2$ for $A + Bx_1 + Cx_1^2$; then, by the definition of $Y_i$,

$$A = \frac{\mu_1^2}{y_1^2}(Y_{3j-2}^2 + Y_{3j-1}^2 + Y_{3j}^2),$$

$$\frac{1}{2}B = \left(\frac{Y_{3j-2}y_{3j-2}}{\mu_{3j-2}} + \frac{Y_{3j-1}y_{3j-1}}{\mu_{3j-1}} + \frac{Y_{3j}y_{3j}}{\mu_{3j}}\right),$$

$$C = \frac{\mu_1^2}{y_1^2}\left[\left(\frac{y_{3j-2}}{\mu_{3j-2}}\right)^2 + \left(\frac{y_{3j-1}}{\mu_{3j-1}}\right)^2 + \left(\frac{y_{3j}}{\mu_{3j}}\right)^2\right].$$

Putting these in (22), we eventually obtain

$$\begin{aligned}
&\frac{1}{y_1}\frac{\partial G}{\partial y_1} - \sum_{k=0,1,2}\left(\frac{\partial G}{\partial y_{3j-k}} - \frac{\mu_1 y_{3j-k}}{\mu_{3j-k}y_1}\frac{\partial G}{\partial y_1}\right) \\
&\times [x_{3j-k}(x_{3j-2}y_{3j-2}/\mu_{3j-2} + x_{3j-1}y_{3j-1}/\mu_{3j-1} + x_{3j}y_{3j}/\mu_{3j}) \\
&\qquad - (y_{3j-k}/\mu_{3j-k})(x_{3j-2}^2 + x_{3j-1}^2 + x_{3j}^2)] \\
&\div [(x_{3j-1}y_{3j}/\mu_{3j} - x_{3j}y_{3j-1}/\mu_{3j-1})^2 + (x_{3j}y_{3j-2}/\mu_{3j-2} - x_{3j-2}y_{3j}/\mu_{3j})^2 \\
&\qquad + (x_{3j-2}y_{3j-1}/\mu_{3j-1} - x_{3j-1}y_{3j-2}/\mu_{3j-2})^2].
\end{aligned} \tag{23}$$

Next we take $x_{3j-2}x_{3k-2} + x_{3j-1}x_{3k-1} + x_{3j}x_{3k}$ $(j = 2, 3, \ldots, n - 1)$, and obtain similar expressions. For $x_1^2 + x_2^2 + x_3^2$ we have

$$A = \frac{\mu_1^2}{y_1^2}(Y_2^2 + Y_3^2),$$

$$B = 2\left(Y_2\frac{y_2}{\mu_2} + Y_3\frac{y_3}{\mu_3}\right)\left(\frac{\mu_1}{y_1}\right)^2,$$

$$C = 1 + \left(\frac{y_2^2}{\mu_2^2} + \frac{y_3^2}{\mu_3^2}\right)\left(\frac{\mu_1}{y_1}\right)^2,$$

and (22) becomes

$$\frac{1}{y_1}\frac{\partial G}{\partial y_1} - \sum_{r=2,3}\left(\frac{\partial G}{\partial y_r} - \frac{\mu_1 y_r}{\mu_r y_1}\frac{\partial G}{\partial y_1}\right)$$
$$\times \frac{2x_r\left(\dfrac{x_1y_1}{\mu_1} + \dfrac{x_2y_2}{\mu_2} + \dfrac{x_3y_3}{\mu_3}\right) - 2\dfrac{y_r}{\mu_r}(x_1^2 + x_2^2 + x_3^2)}{\left(\dfrac{x_2y_3}{\mu_3} - \dfrac{x_3y_2}{\mu_2}\right)^2 + \left(\dfrac{x_3y_1}{\mu_1} - \dfrac{x_1y_3}{\mu_3}\right)^2 + \left(\dfrac{x_1y_2}{\mu_2} - \dfrac{x_2y_1}{\mu_1}\right)^2}. \tag{24}$$

As $G$ is a polynomial in $H_4$, $H_5$, $H_6$, the expression

$$\left(\frac{\partial G}{\partial y_i} - \frac{\mu_1 y_i}{\mu_i y_1}\frac{\partial G}{\partial y_1}\right) \qquad (i = 2, 3, \ldots, N)$$

is a polynomial in $H_4$, $H_5$, $H_6$; and $x_1$, $x_2$, $x_3$ are contained only through $H_4$, $H_5$, $H_6$. Hence, (23) and (24) either have no term containing $x_1$, $x_2$, $x_3$, or have terms containing $x_1$, $x_2$, or $x_3$. In neither case can the denominator be a factor of the numerator. Hence, in order that both (23) and (24) may hold, the condition is

$$\frac{\partial G}{\partial y_2} - \frac{\mu_1 y_2}{\mu_2 y_1}\frac{\partial G}{\partial y_1} = \frac{\partial G}{\partial y_3} - \frac{\mu_1 y_3}{\mu_3 y_1}\frac{\partial G}{\partial y_1} = 0,$$

$$\frac{\partial G}{\partial y_{3j-2}} - \frac{\mu_1 y_{3j-2}}{\mu_{3j-2} y_1}\frac{\partial G}{\partial y_1} = \frac{\partial G}{\partial y_{3j-1}} - \frac{\mu_1 y_{3j-1}}{\mu_{3j-1} y_1}\frac{\partial G}{\partial y_1}$$
$$= \frac{\partial G}{\partial y_{3j}} - \frac{\mu_1 y_{3j}}{\mu_{3j} y_1}\frac{\partial G}{\partial y_1} = 0.$$

These constitute $3n - 4$ equations; indeed, they form a complete system in the sense of Section 4.9. From what was stated in Sections 4.1 and 4.17, we see, as the number of $y_i$ is $N = 3n - 3$, that there exists one and only one independent integral. This must be

$$\sum_{i=1}^{N}\frac{y_i^2}{2\mu_i} = T = \text{constant}. \tag{25}$$

This is the total kinetic energy of the dynamical system with $N$ degrees of freedom. Hence no integral can contain $y_1, \ldots, y_N$ independently, but must contain them only in the combination $T$, and indeed the integral is a polynomial in $T$.

We are discussing differential equations (1) that have been reduced to the order $6n - 6$ from the order $6n$ by Poincaré's transformation $\beta$ of Section 5.6. The function $H_4$, $H_5$, $H_6$, $M_1$, $M_2$, $M_3$ of Section 6.4 are consequently expressed in terms of $6n - 6$ variables $y_i$, $x_i$. Thus the total momentum, total angular momentum, and total kinetic energy remain unaltered after Poincaré's transformation $\beta$ and are of the forms (17), (18), and (25), respectively. Hence these expressions represent the total momentum, total angular momentum, and total kinetic energy of the whole $n$-body system and hence are the classical integrals. The only supposition is that the $n$th fictitious body after the transformation is at rest; that is,

$$y_{3n-2} = y_{3n-1} = y_{3n} = x_{3n-2} = x_{3n-1} = x_{3n} = 0.$$

Now $\phi_0$ is homogeneous in $x_i$ and $y_i$. The integrals $H_4$, $H_5$, $H_6$, $M_1$, $M_2$, $M_3$ are linear in $x_i$. The kinetic energy $T$ is quadratic in $y_i$ and does not contain $x_i$. Hence, if $T$ is contained in $\phi_0$, then $\phi_0$ should be of the form

$$\phi_0 = h(H_4, H_5, H_6, M_1, M_2, M_3) \cdot T^m,$$

where $h$ is homogeneous in the variables it contains.

As we have shown, the function $f_2$ is of the form

$$f_2 = \chi(Y_2, \ldots, Y_N, y_1, \ldots, y_N) - \frac{\mu_1}{y_1} \frac{\partial G}{\partial y_1} U,$$

where

$$\frac{\mu_1}{y_1} \frac{\partial G}{\partial y_1} = \frac{\mu_1}{y_1} \frac{\partial G}{\partial T} \frac{\partial T}{\partial y_1} = mh(H_4, H_5, H_6, M_1, M_2, M_3) \cdot T^{m-1}.$$

Hence

$$f_2 = \chi(Y_2, \ldots, Y_N, y_1, \ldots, y_N) - mh(H_4, H_5, H_6, M_1, M_2, M_3) \cdot T^{m-1}U.$$

Thus

$$\begin{aligned} \phi &= \phi_0 + \phi_2 + \phi_4 + \cdots \\ &= h(H_4, H_5, H_6, M_1, M_2, M_3)(T^m - mT^{m-1}U) \\ &\quad + \chi(Y_2, \ldots, Y_N, y_1, \ldots, y_N) + \phi_4 + \cdots. \end{aligned}$$

Accordingly, any integral can be considered to consist of two integrals; the one is the combination $h(H_4, H_5, H_6, M_1, M_2, M_3)(T - U)^m$ of the classical integrals, and the other is $\phi' = \phi_0' + \phi_2' + \phi_4' + \cdots$, where $T - U$ is the total energy and $\phi'$ is such that

$$\phi_0' = \chi(Y_2, \ldots, Y_N, y_1, \ldots, y_N),$$

$$\phi_2' = \phi_4 - \frac{m(m-1)}{2!} h(H_4, H_5, H_6, M_1, M_2, M_3) T^{m-2} U^2,$$

$$\phi_4' = \phi_6 + \frac{m(m-1)(m-2)}{3!} h(H_4, H_5, H_6, M_1, M_2, M_3) T^{m-3} U^3,$$

$$\vdots$$

But $\phi'$ is an integral of the same form as $\phi$. The difference is that its highest-degree term $\phi_0'$ with regard to $y_i$ is of degree 2 less than the highest-degree term $\phi_0$ of $\phi$. We have shown that $\phi$ is composed of $\phi'$ and the classical integrals. Similarly, $\phi'$ is composed of the classical integrals and an integral of the same nature as $\phi'$ but of degree 2 less than $\phi'$ with regard to $y_i$. By repeating this argument we show that $\phi$ is composed of the classical integrals and an integral $\phi^{(n)}$ of the same nature as $\phi$ but of degree either 1 or 0 with regard to $y_i$. If $\phi^{(n)}$ is of degree 1, then the exponent $k$ of $T$ in

$$\phi^{(n)} = \phi_0^{(n)} = h(H_4, H_5, H_6, M_1, M_2, M_3) T^k$$

is $k = 0$. Then $\phi^{(n)}$ is composed of the classical integrals. If $\phi^{(n)}$ is of degree 0, then it is a function of $x_i$ alone. However, we proved in the beginning that such an integral never exists. Hence, $\phi$ must be a combination of the classical integrals.

Thus, we have obtained Bruns's theorem that any integral of the $n$-body problem that is algebraic with respect to the coordinates and momenta and does not contain $t$ explicitly is an algebraic combination of the classical integrals.

## 6.6 INTEGRALS CONTAINING TIME

We have proved Bruns's theorem for the integrals of the form (2) of Section 6.1 on the assumption that the integral does not contain the time $t$ explicitly. Now we shall prove that the theorem is also true for integrals that do contain $t$ explicitly. We consider the original form of the differential equations of order $6n$ for the $n$-body problem,

$$\frac{dx_i}{dt} = \frac{\partial H}{\partial y_i}, \quad \frac{dy_i}{dt} = -\frac{\partial H}{\partial x_i} \quad (i = 1, 2, \ldots, 6n), \tag{26}$$

and suppose that there is an integral with an arbitrary constant

$$f(x_1, \ldots, x_{3n}, y_1, \ldots, y_{3n}, t) = a, \tag{27}$$

in which $f$ is algebraic (though not necessarily rational) in $x_i$, $y_i$ and $t$. We rationalize $f$ in the form

$$a^m + a^{m-1}\phi_1(x_1,\ldots,x_{3n},y_1,\ldots,y_{3n},t) + \cdots + \phi_m(x_1,\ldots,x_{3n},y_1,\ldots,y_{3n},t) = 0, \quad (28)$$

where the coefficients $\phi_r$ $(r = 1, 2, \ldots, m)$ are rational in $t$ and algebraic in $x_i$, $y_i$. We may consider this equation irreducible with regard to $t$.

Differentiate (28) with respect to $t$; then,

$$a^{m-1}\frac{d\phi_1}{dt} + a^{m-2}\frac{d\phi_2}{dt} + \cdots + \frac{d\phi_m}{dt} = 0; \quad (29)$$

but $d\phi_r/dt$ $(r = 1, 2, \ldots, m)$ is rational with regard to $t$ when expressed in terms of $x_i$, $y_i$, $t$. Equation (28) would not be irreducible unless this equation were identically zero. Therefore,

$$\frac{d\phi_r}{dt} = 0 \qquad (r = 1, 2,\ldots, m).$$

Hence, $\phi_r$ itself is an integral and accordingly $f$ is rational in $t$ and is a combination of other integrals $\phi$ which are algebraic in $x_i$, $y_i$.

We decompose $\phi$ into linear factors in $t$ such that

$$\frac{P(t-\phi_1)^{m_1}(t-\phi_2)^{m_2}\cdots(t-\phi_k)^{m_k}}{(t-\psi_1)^{n_1}(t-\psi_2)^{n_2}\cdots(t-\psi_l)^{n_l}},$$

where $P, \phi_1,\ldots,\phi_k, \psi_1,\ldots,\psi_l$ are algebraic in $x_i$, $y_i$. Since this is an integral, we have

$$\frac{1}{P}\frac{\partial P}{dt} + \frac{m_1}{t-\phi_1}\left(1 - \frac{d\phi_1}{dt}\right) + \cdots + \frac{m_k}{t-\phi_k}\left(1 - \frac{d\phi_k}{dt}\right) - \frac{n_1}{t-\psi_1}\left(1 - \frac{d\psi_1}{dt}\right) - \cdots - \frac{n_l}{t-\psi_l}\left(1 - \frac{d\psi_l}{dt}\right) = 0.$$

If we replace

$$\frac{dP}{dt}, \frac{d\phi_1}{dt},\ldots,\frac{d\phi_k}{dt}, \quad \text{and} \quad \frac{d\psi_1}{dt},\ldots,\frac{d\psi_l}{dt},$$

respectively, by $(P, H)$, $(\phi_1, H),\ldots,(\phi_k, H)$, and $(\psi_1, H),\ldots,(\psi_l, H)$ using Poisson's theorem (Section 4.3), then this equation becomes an identity, that is,

$$\frac{dP}{dt} = 0, \qquad 1 - \frac{d\phi_1}{dt} = 0,\ldots, \qquad 1 - \frac{d\phi_k}{dt} = 0,$$

$$1 - \frac{d\psi_1}{dt} = 0,\ldots, \qquad 1 - \frac{d\psi_l}{dt} = 0;$$

in other words, each of the expressions

$$P, t - \phi_1, \ldots, t - \phi_k, t - \psi_1, \ldots, t - \psi_l$$

must be an integral. Hence, any algebraic integral of the $n$-body problem containing $t$ consists of an algebraic integral not containing $t$ and an integral of the form $t - \phi$ = constant, where $\phi$ is algebraic in $x_i, y_i$.

Now

$$t + \frac{\Xi_1}{\mathrm{H}_1} = \text{constant},$$

where

$$\Xi_1 = \sum_{j=1}^{n} m_{3j-2} x_{3j-2}, \qquad \mathrm{H}_1 = -\sum_{j=1}^{n} y_{3j-2}$$

is the integral representing the motion of the center of masses of the $n$ bodies. Hence, any algebraic integral containing $t$ is composed of an algebraic integral not containing $t$, an integral of the form

$$\phi + \frac{\Xi_1}{\mathrm{H}_1} = \text{constant},$$

where $\phi$ is an integral algebraic in $x_i$, $y_i$, and the classical integral $t + \Xi_1/\mathrm{H}_1$. The first two of these three integrals are algebraic integrals not containing $t$.

We can similarly discuss the classical integrals for the motion of the center of mass:

$$t + \frac{\Xi_2}{\mathrm{H}_2} = \text{constant}, \qquad \Xi_2 = \sum_{j=1}^{n} m_{3j-1} x_{3j-1}, \qquad \mathrm{H}_2 = -\sum_{j=1}^{n} y_{3j-1},$$

$$t + \frac{\Xi_3}{\mathrm{H}_3} = \text{constant}, \qquad \Xi_3 = \sum_{j=1}^{n} m_{3j} x_{3j}, \qquad \mathrm{H}_3 = -\sum_{j=1}^{n} y_{3j}.$$

Hence, any algebraic integral containing $t$ is an algebraic combination of the classical integrals. The classical integrals are algebraic in $x_i$, $y_i$, $t$. Thus it must be concluded that there does not exist any integral that is algebraic in $x_i$, $y_i$, $t$, other than the classical integrals. Thus, the theorem of Bruns has been proved also in the case of integrals containing $t$ explicitly.

## PAINLEVÉ'S EXTENSION

### 6.7 PREPARATORY LEMMAS

Painlevé (1896, 1896a, 1896b, 1897a, 1898) extended Bruns's theorem to the case in which the integral contains algebraically the momentum only, while the coordinates are left arbitrary.

The equations for the $n$-body problem constitute a system here denoted by (S):

$$\text{(S)}: \quad \frac{dx_i}{dt} = \frac{\partial H}{\partial y_i}, \quad \frac{dy_i}{dt} = -\frac{\partial H}{\partial x_i} \quad (i = 1, 2, \ldots, 3n), \tag{30}$$

where $H$ does not contain $t$ explicitly, and $U$ is the sum of the products of the masses and the reciprocals of the $n(n-1)/2$ mutual distances. The sum of the $n(n-1)/2$ mutual distances is denoted by $s$. The force function $U$ is expressed by a rational function of $x_1, x_2, \ldots, x_{3n}$ and $s$. Let $(S_0)$ denote the system of differential equations for which $U = 0$, that is, for which no force is acting:

$$(\text{S}_0): \quad \frac{dx_i}{dt} = \frac{1}{m_i} y_i, \quad \frac{dy_i}{dt} = 0 \quad (i = 1, 2, \ldots, 3n).$$

Our problem is to discuss whether the system (S) admits an integral that is algebraic in $y = (y_1, y_2, \ldots, y_{3n})$. Such integrals are assumed irreducible and of the form with an arbitrary constant $a$

$$f(y_1, y_2, \ldots, y_{3n}, x_1, x_2, \ldots, x_{3n}) = a, \tag{31}$$

in which $t$ is contained only implicitly. Here we propose to extend the results of Sections 6.1 and 6.2.

*Lemma 1:* An integral of (S) that is algebraic with respect to $y$ is an algebraic combination of the integrals of (S) that are rational in $y$.

Rationalizing (31), we obtain

$$a^m + a^{m-1}\phi_1(x_i, y_i) + \cdots + a\phi_{m-1}(x_i, y_i) + \phi_m(x_i, y_i) = 0, \tag{32}$$

where the coefficients $\phi(x_i, y_i)$ are rational in $y$. Differentiating by referring to (30), we obtain

$$a^{m-1}(\phi_1, H) + a^{m-2}(\phi_2, H) + \cdots + (\phi_m, H) = 0, \tag{33}$$

where the Poisson brackets $(\phi_j, H)$ are supposed to be nonzero and algebraic in $y$. In order that (32) and (33) may hold simultaneously they should admit one or more than one common root. Then (33) would become reducible, which is contrary to the assumption. Hence $(\phi_j, H)$ should all be equal to zero. By Poisson's theorem (Section 4.3) the functions $\phi_j$ should all be integrals of (S). Thus the integrals $\phi_j$ are rational in $y$.

The highest degree of any term in a polynomial is called the degreee of the polynomial. An integral that is rational in $y$ is called a rational integral and an integral that is a polynomial in $y$ is called a polynomial integral. Write a rational integral $f$ in the form

$$f = \frac{\Pi_2 + \Pi_1}{K_2 + K_1},$$

where $\Pi_1$ and $K_1$ are of odd degree and $\Pi_2$ and $K_2$ of even degree. The system (S) remains unaltered by the change of $t, y_1, \ldots, y_{3n}$ into $-t, -y_1, \ldots, -y_{3n}$. By this transformation $f$ is transformed to $f' = (\Pi_2 - \Pi_1)/(K_2 - K_1)$ and $f'$ is still an integral of (S). The sum $f + f'$ and the product $ff'$ contain only even-degree terms in $y$ and are both integrals of (S). Hence we can limit our discussion to real and rational integrals of (S), in which the numerator and the denominator are of the same parity, that is, either both are of even degree or both are of odd degree. The sum of odd-degree terms and the sum of even-degree terms of a rational integral are also integrals.

*Lemma 2:* Any rational integral of (S) is an algebraic combination of real rational integrals that contain only terms of the same parity.

If a function $\Theta(x_1, \ldots, x_{3n}, y_1, \ldots, y_{3n})$ becomes zero when the variables $x = (x_1, x_2, \ldots, x_{3n})$ and $y = (y_1, y_2, \ldots, y_{3n})$ in $\Theta$ are replaced by the solution of a differential equation for all values of $t$, then $\Theta = 0$ is called an *invariantive relation* according to Section 4.6.

*Lemma 3:* In order that a polynomial $P$ in $y$

$$P(y_1, y_2, \ldots, y_{3n}, x_1, x_2, \ldots, x_{3n}) \equiv \Pi(y_1, y_2, \ldots, y_{3n}, x_1, x_2, \ldots, x_{3n}, s) = 0$$

be an invariantive relation for (S), it is necessary and sufficient that

$$\frac{dP}{dt} = \sum_{i=1}^{3n} \left(\frac{\partial P}{\partial x_i}\frac{y_i}{m_i} + \frac{\partial P}{\partial y_i}\frac{\partial U}{\partial x_i}\right) = \Omega(y_1, \ldots, y_{3n}, x_1, \ldots, x_{3n})P, \quad (34)$$

where $\Omega$ is a polynomial in $y$ and should be linear, because the degree of the left-hand member is one higher than that of $P$.

To prove the condition necessary, denote by $x_1^{(0)}, \ldots, x_{3n}^{(0)}$ a set of arbitrary values of $x_1, \ldots, x_{3n}$ for which the branch of $s = \phi(x_1, \ldots, x_{3n})$ under consideration is holomorphic, together with the corresponding branches of $\partial U/\partial x_i$ (Section 2.20). Put $P \equiv \Pi[y_i, x_i, \phi(x_i)]$ and suppose that $(y_i^{(0)}, x_i^{(0)})$ is an arbitrary solution of the relation $P = 0$. The equation $\Pi = 0$ is an invariantive relation for (S); that is, the solution $(y_i(t), x_i(t))$ of (S) with the initial condition $y_i^{(0)}, x_i^{(0)}, s_0 = \phi(x_i^{(0)})$ for $t = t_0$ satisfies $P = 0$ identically. In particular, the value of $dP/dt$ for $t = t_0$, that is,

$$\sum_{i=1}^{3n} \left( \frac{\partial P}{\partial x_i^{(0)}} \frac{y_i^{(0)}}{m_i} + \frac{\partial P}{\partial y_i^{(0)}} \frac{\partial U}{\partial x_i^{(0)}} \right)$$

is equal to zero. Hence $P(y_i^{(0)}, x_i^{(0)}) = 0$ should be related to

$$\sum_{i=0}^{3n} \left[ \frac{\partial}{\partial x_i^{(0)}} P(y_i^{(0)}, x_i^{(0)}) \frac{y_i^{(0)}}{m_i} + \frac{\partial}{\partial y_i^{(0)}} P(y_i^{(0)}, x_i^{(0)}) \frac{\partial}{\partial x_i^{(0)}} U(y_i^{(0)}, x_i^{(0)}) \right] = 0. \quad (35)$$

If a polynomial $P$ in $y$ is irreducible, then the left-hand member of this equation must have a factor $P(y_i^{(0)}, x_i^{(0)})$, and we obtain (34) which is written

$$\frac{d}{dt} \log P = \Omega.$$

Suppose, on the contrary, that $P$ can be decomposed into a product $P = P_1 P_2 \cdots P_l$ of irreducible polynomials of degrees $\alpha_1, \alpha_2, \ldots, \alpha_l$, respectively, of which the factors need not be mutually independent. Then, since each polynomial must satisfy (34) we have

$$\frac{d}{dt} \log P_j = \Omega_j \qquad (j = 1, 2, \ldots, l).$$

Thus

$$\frac{d}{dt} \log P = \frac{d}{dt} \log (P_1 P_2 \cdots P_l) = \Omega_1 + \Omega_2 + \cdots + \Omega_l \equiv \Omega,$$

or

$$\sum_{i=1}^{3n} \left( \frac{\partial P}{\partial x_i} \frac{y_i}{m_i} + \frac{\partial P}{\partial y_i} \frac{\partial U}{\partial x_i} \right) = \Omega P.$$

To prove the condition sufficient, suppose that $P(y_i^{(0)}, x_i^{(0)}) = 0$ for $t = t_0$. Since the solution $(y_i(t), x_i(t))$ of (S) satisfies

$$\frac{dP}{dt} = \Omega P$$

and hence

$$\frac{d^2P}{dt^2} = \Omega \frac{dP}{dt} + P \frac{d\Omega}{dt} = \left( \Omega^2 + \frac{d\Omega}{dt} \right) P,$$

the derivatives $dP/dt$, $d^2P/dt^2, \ldots$ are all zero for $t = t_0$, and we have $P \equiv 0$ identically by substituting this solution.

From this lemma it follows that, if $P_\nu$ is the aggregate of the highest-degree terms of $P$ and if $\Omega$ is written $\Omega = \Omega_1 + \Omega_0$, where $\Omega_0$ is the part of $\Omega$ independent of $y$, then

$$\sum_{i=1}^{3n} \frac{\partial P_\nu}{\partial x_i} \frac{y_i}{m_i} = \Omega_1 P_\nu;$$

in other words, $P_\nu = 0$ is an invariantive relation for $(S_0)$. If $P$ contains only terms of the same parity, then $\Omega$ cannot contain $\Omega_0$ and is linear homogeneous with respect to $y$. If $\Omega$ is of the form

$$\Omega = \frac{\partial \lambda}{\partial x_1} \frac{y_1}{m_1} + \cdots + \frac{\partial \lambda}{\partial x_{3n}} \frac{y_{3n}}{m_{3n}},$$

where $\lambda$ is a function of $x$ only, then $Pe^{-\lambda}$ is also a polynomial integral of (S).

Finally, if $P/Q$ is a rational integral of (S), and if $P_\nu$, $Q_\mu$ are the aggregates of the highest-degree terms of $P$ and $Q$, respectively, then $P = 0$, $Q = 0$ are invariantive relations of (S) and we have

$$\frac{d}{dt} \log P = \Omega, \qquad \frac{d}{dt} \log Q = \Omega'.$$

However, $\Omega$ coincides with $\Omega'$, because

$$\frac{d}{dt} \log \left(\frac{P}{Q}\right) \equiv 0.$$

Conversely, if $P$ and $Q$ satisfy the same equation (34), then $P/Q$ is an integral of (S). In this case the equations $P_\nu = 0$, $Q_\mu = 0$ are two invariantive relations for $(S_0)$, since

$$\sum_{i=1}^{3n} \frac{\partial P_\nu}{\partial x_i} \frac{y_i}{m_i} = \Omega_1 P_\nu, \qquad \sum_{i=1}^{3n} \frac{\partial Q_\mu}{\partial x_i} \frac{y_i}{m_i} = \Omega_1 Q_\mu.$$

The fraction $P_\nu/Q_\mu$ is an integral of $(S_0)$. If the integral $P/Q$ contains only terms of the same parity, then $\Omega$ coincides with $\Omega_1$. If $P_\nu$ is a polynomial integral of $(S_0)$, then $\Omega_1 = 0$ and $P$ and $Q$ are both polynomial integrals of (S). In other words, we may state

*Lemma 4:* If $f = P/Q$ is a rational integral of (S) and the denominator and the numerator both consist of terms of the same parity, and if $P_\nu$, the highest-degree term in $P$, is an integral of $(S_0)$, then $P$ and $Q$ are both polynomial integrals of (S).

## 6.8 PROPERTIES OF (S)

Consider a rational integral $f = P/Q$ of (S), where $P$ and $Q$ are supposed to be relatively prime polynomials in $y$. If a certain number of algebraic relations among the coefficients

$$\alpha(x_1, \ldots, x_{3n}), \beta(x_1, \ldots, x_{3n}), \ldots, \lambda(x_1, \ldots, x_{3n})$$

of the polynomials $P$ and $Q$, their first order derivatives

$$\frac{\partial\alpha}{\partial x_1}, \ldots, \frac{\partial\lambda}{\partial x_{3n}},$$

and the given functions $\partial U/\partial x_i$, are compatible, then the general solution may depend either on these functions or on arbitrary constants. We propose the following

*Fundamental Theorem:* If all functions $\partial U/\partial x_i$ $(i = 1, 2, \ldots, 3n)$ are holomorphic at $x_1 = x_1^{(0)}, \ldots, x_{3n} = x_{3n}^{(0)}$, then $f = P/Q$ is either holomorphic or meromorphic at

$$y_1 = y_1^{(0)}, \ldots, y_{3n} = y_{3n}^{(0)}, \qquad x_1 = x_1^{(0)}, \ldots, x_{3n} = x_{3n}^{(0)}.$$

A function holomorphic at a point can be expanded in Taylor series convergent around that point. A meromorphic function is expressible as the quotient of two holomorphic functions at that point and accordingly admits no singularity other than poles (Section 2.20). If $f$ can be separated into several branch functions, then $f$ is said to be meromorphic when each branch function is meromorphic. When $f$ is neither holomorphic nor meromorphic at a singular point, that point is said to be a nonpolar singular point. If $f = P/Q$ is meromorphic at

$$y_1 = y_1^{(0)}, \ldots, y_{3n} = y_{3n}^{(0)}, \qquad x_1 = x_1^{(0)}, \ldots, x_{3n} = x_{3n}^{(0)},$$

then the coefficients $\alpha, \beta, \ldots, \lambda$ of $P$, $Q$ are meromorphic at

$$x_1 = x_1^{(0)}, \ldots, x_{3n} = x_{3n}^{(0)}.$$

Conversely, if $\alpha, \beta, \ldots, \lambda$ are meromorphic at $x_1 = x_1^{(0)}, \ldots, x_{3n} = x_{3n}^{(0)}$, then $f = P/Q$ is meromorphic at

$$y_1 = y_1^{(0)}, \ldots, y_{3n} = y_{3n}^{(0)}, \qquad x_1 = x_1^{(0)}, \ldots, x_{3n} = x_{3n}^{(0)}.$$

Suppose that $x_1 = a_1, \ldots, x_{3n} = a_{3n}$ is a point at which the branch function of $\partial U/\partial x_i$ is holomorphic. Assume that the branch functions corresponding to an integral $f$ were neither holomorphic nor meromorphic at that point. Then there would exist a point $(b_1, \ldots, b_{3n})$ of $(x_1, \ldots, x_{3n})$ in an arbitrary neighborhood of $(a_1, \ldots, a_{3n})$ at which the functions $\alpha, \beta, \ldots, \lambda$ would all be holomorphic. The function $f(y_1, \ldots, y_{3n}, x_1, \ldots, x_{3n})$ is either holomorphic or meromorphic at $y_1 = b'_1, \ldots, y_{3n} = b'_{3n}, x_1 = b_1, \ldots, x_{3n} = b_{3n}$, whatever the values $b'_1, \ldots, b'_{3n}$ may be. Let the solution of (S) with the initial condition $x_i = a_i, y_i = y_i^{(0)}$ $(i = 1, 2, \ldots, 3n)$ for $t = t_0$ be

$$\begin{aligned} x_i &= a_i + \frac{y_i^{(0)}}{1}(t - t_0) + \frac{1}{2!}\frac{\partial}{\partial a_i} U(a_1, \ldots, a_{3n})(t - t_0)^2 + \cdots \\ &\equiv \psi_i(t, y_i^{(0)}), \\ y_i &= y_i^{(0)} + \frac{\partial}{\partial x_i} U(a_1, \ldots, a_{3n})(t - t_0) + \cdots \equiv \chi_i(t, y_i^{(0)}) \end{aligned}$$

$$(i = 1, 2, \ldots, 3n). \quad (36)$$

With a real positive number $R$ the functions $\psi_i$ and $\chi_i$ are holomorphic with respect to $t$ and $y_i^{(0)}$ in the domain denoted by $(D_R)$:

$$(D_R): \qquad |t - t_0| < R, \quad |y_i^{(0)}| < R.$$

As $t$ and $y_i^{(0)}$ vary within $(D_R)$, the variables $x_1 = \psi_1, \ldots, x_{3n} = \psi_{3n}$ coincide with the values in the neighborhood of $(a_1, \ldots, a_{3n})$. In particular, these variables coincide, for a suitably chosen value $t_1$, $a'_i$ of $t, y_i^{(0)}$, with a value $(b_1, \ldots, b_{3n})$ for which $\alpha, \beta, \ldots, \lambda$ are holomorphic. Put $b'_i = \chi_i(t_1, a'_1, a'_2, \ldots, a'_{3n})$ $(i = 1, 2, \ldots, 3n)$; then

$$f(y_1, \ldots, y_{3n}, x_1, \ldots, x_{3n})$$

is either holomorphic or meromorphic at

$$x_i = b_i, \qquad y_i = b'_i \qquad (i = 1, 2, \ldots, 3n).$$

Let the general solution of (S) with the initial condition

$$x_i = x_i^{(0)}, \qquad y_i = y_i^{(0)} \qquad (i = 1, 2, \ldots, 3n),$$

be

$$\begin{aligned} x_i &= \Psi_i(t, y_1^{(0)}, \ldots, y_{3n}^{(0)}, x_1^{(0)}, \ldots, x_{3n}^{(0)}) \\ y_i &= \Xi_i(t, y_1^{(0)}, \ldots, y_{3n}^{(0)}, x_1^{(0)}, \ldots, x_{8n}^{(0)}) \qquad (i = 1, 2, \ldots, 3n). \end{aligned} \quad (37)$$

The functions $\Xi_i$ and $\Psi_i$ are holomorphic at $t = t_1$, $x_i^{(0)} = a_i$, $y_i^{(0)} = a'_i$, and equal to $b_i$ and $b'_i$, respectively.

Let the result of replacing $y_i$, $x_i$ by $\Xi_i$, $\Psi_i$ respectively in an integral

$$f(y_1, \ldots, y_{3n}, x_1, \ldots, x_{3n})$$

be

$$F(t, y_1^{(0)}, \ldots, y_{3n}^{(0)}, x_1^{(0)}, \ldots, x_{3n}^{(0)}).$$

This is either holomorphic or equal to the quotient of two holomorphic functions at $t = t_1$, $x_i^{(0)} = a_i$, $y_i^{(0)} = a_i'$. Since $f$ is an integral, equation (37) should be satisfied identically, that is,

$$f(y_1, \ldots, y_{3n}, x_1, \ldots, x_{3n}) \equiv f(y_1^{(0)}, \ldots, y_{3n}^{(0)}, x_1^{(0)}, \ldots, x_{3n}^{(0)}),$$

and $F$ coincides with $f(y_1^{(0)}, \ldots, y_{3n}^{(0)}, x_1^{(0)}, \ldots, x_{3n}^{(0)})$. Hence $F$ does not depend on $t$. On the other hand, $f(y_1^{(0)}, \ldots, y_{3n}^{(0)}, x_1^{(0)}, \ldots, x_{3n}^{(0)})$ is either holomorphic or meromorphic at $y_i^{(0)} = a_i'$, $x_i^{(0)} = a_i$. This contradicts the assumption. Hence the theorem is proved.

From this theorem we derive the following important

*Corollary:* An integral $f(y_1, \ldots, y_{3n}, x_1, \ldots, x_{3n})$ of (S), which is rational in $y_1, \ldots, y_{3n}$ and is the most general of its degree $\nu$, can contain only a finite number of arbitrary constants, and its nonpolar singularities coincide with the singularities of $\partial U/\partial x_i$.

The coefficient functions $\alpha, \beta, \ldots, \lambda$ cannot depend on any arbitrary functions. For, consider the differential relations connecting the functions $\alpha(x_1, \ldots, x_{3n}), \beta(x_1, \ldots, x_{3n}), \ldots$. Either the general solution of this system depends on a finite number of constants, or alternatively, for any given values of the first $j$ ($j < 3n$) variables $x_1^{(0)}, \ldots, x_j^{(0)}$, we can choose an arbitrary one of the unknown functions, for example,

$$\alpha(x_1^{(0)}, \ldots, x_j^{(0)}, x_{j+1}, \ldots, x_{3n}) \equiv \alpha_0(x_{j+1}, \ldots, x_{3n}).$$

In the latter case it is reasonable to take for $\alpha_0$ a function that admits $x_{j+1} = x_{j+1}^{(0)}, \ldots, x_{3n} = x_{3n}^{(0)}$ as a nonpolar singularity, where the values $x_1^{(0)}, \ldots, x_j^{(0)}, x_{j+1}^{(0)}, \ldots, x_{3n}^{(0)}$ do not coincide with any of the singularities of $\partial U/\partial x_i$. The integral $f$ then would admit a nonpolar singularity $x_1^{(0)}, \ldots, x_{3n}^{(0)}$ that is not a singularity of $\partial U/\partial x_i$. This is contradictory, and $\alpha, \beta, \ldots, \lambda$ can depend only on a finite number of constants.

If the integral of (S) is a polynomial $f = P$, then the singularities of $f$ should coincide with those of $\partial U/\partial x_i$. The coefficients $\alpha, \beta, \ldots, \lambda$ of $P$ satisfy linear homogeneous relations with respect to $\alpha, \beta, \ldots, \lambda$ and their derivatives. The general solution of these relations depends on only a finite number of arbitrary constants of the form $c_1 f_1 + \cdots + c_j f_j$, where $c_1, c_2, \ldots, c_j$ are constants and $f_1, \ldots, f_j$ are particular solutions.

## 6.9 PROPERTIES OF $(S_0)$

In order to study the rational integrals of (S) we first discuss the system $(S_0)$. Let a rational integral of $(S_0)$ be $f(y_1, \ldots, y_{3n}, x_1, \ldots, x_{3n})$. From the theorem just proved, the integral $f$ has only polar singularities in the finite region. The points $x_1 = \infty, \ldots, x_{3n} = \infty$ may be essential singularities of $f$. It will be proved that in reality they are not essential singularities, but $f$ should be algebraic in $x$ and even rational in $x$.

It is sufficient to prove that $x_1 = \infty$, for example, is not an essential singularity of $f$; that is, $f(y_1, \ldots, y_{3n}, 1/\xi_1, x_2, \ldots, x_{3n})$ is, as a function of $\xi_1$, either holomorphic or meromorphic for $\xi_1 = 0$. We carry out Appell's transformation, which we denote by (T):

$$\text{(T)}: \quad x_1 = \frac{1}{\xi_1}, \quad x_2 = \frac{\xi_2}{\xi_1}, \quad \ldots, \quad x_{3n} = \frac{\xi_{3n}}{\xi_1},$$

$$y_1 = -\eta_1, \quad y_2 = \eta_2\xi_1 - \eta_1\xi_2, \quad \ldots, \quad y_{3n} = \eta_{3n}\xi_1 - \eta_1\xi_{3n},$$

$$\frac{dt}{d\tau} = \frac{1}{\xi_1^2}.$$

Then $(S_0)$ becomes $(S_0')$:

$$(S_0'): \qquad \frac{d\xi_i}{d\tau} = \eta_i, \qquad \frac{d\eta_i}{d\tau} = 0 \qquad (i = 1, 2, \ldots, 3n).$$

To an integral $f$ of $(S_0)$ corresponds an integral $\phi(\eta_1, \ldots, \eta_{3n}, \xi_1, \ldots, \xi_{3n})$ of $(S_0')$. This latter integral is rational with respect to $\eta_1, \ldots, \eta_{3n}$.

By the fundamental theorem this function $\phi$ should be of the form $A/B$ around $\xi_1 = 0, \ldots, \xi_{3n} = 0$, where $A$ and $B$ are polynomials in $\eta_1, \ldots, \eta_{3n}$ whose coefficients are expanded in positive ascending powers of $\xi_1, \ldots, \xi_{3n}$. By substituting $y_1, \ldots, y_{3n}, x_1, \ldots, x_{3n}$ for

$$\eta_1, \ldots, \eta_{3n}, \xi_1, \ldots, \xi_{3n}$$

in $\phi$ by means of (T), that is, by writing

$$\xi_2 = \xi_1 x_2, \quad \ldots, \quad \xi_{3n} = \xi_1 x_{3n},$$

$$\eta_1 = -y_1, \quad \eta_2 = \frac{y_2}{\xi_1} - y_1 x_2, \quad \ldots, \quad \eta_{3n} = \frac{y_{3n}}{\xi_1} - y_1 x_{3n},$$

the expression $A/B$ takes the form $A_1/B_1$, where $A_1$ and $B_1$ are both polynomials of $y_1, \ldots, y_{3n}$ whose coefficients are expanded in positive ascending powers of $\xi_1, x_2, \ldots, x_{3n}$. The fraction $A_1/B_1$, as a function of $\xi_1$, admits $\xi_1 = 0$ as an ordinary point or as a pole, because it coincides with $f(y_1, \ldots, y_{3n}, 1/\xi_1, x_2, \ldots, x_{3n})$.

Thus it has been shown that any integral of $(S_0)$ that is rational in $y$ should be rational in $x$. Accordingly, it should be represented as the quotient $P/Q$ of two relatively prime polynomials $P$ and $Q$, in

$$y_1, \ldots, y_{3n}, x_1, \ldots, x_{3n}.$$

We prove that $P$ and $Q$ are both integrals of $(S_0)$. For this purpose we prove the following

*Lemma:* Suppose that $\Pi(y_1, \ldots, y_{3n}, x_1, \ldots, x_{3n}) = 0$ is an invariantive relation for $(S_0)$ and is an irreducible polynomial of $y_1, \ldots, y_{3n}, x_1, \ldots, x_{3n}$. Then $\Pi$ is an integral of $(S_0)$.

In fact, $(S_0)$ admits integrals

$$y_1 = y_1^{(0)}, \qquad \ldots, \qquad y_{3n} = y_{3n}^{(0)},$$

$$x_i y_1 - x_1 y_i = x_i^{(0)} y_1^{(0)} - x_1^{(0)} y_i^{(0)} \qquad (i = 2, 3, \ldots, 3n).$$

From these integrals we can obtain $x_2, \ldots, x_{3n}, y_1, \ldots, y_{3n}$ as functions of $x_1, y_1^{(0)}, \ldots, y_{3n}^{(0)}, x_1^{(0)}, \ldots, x_{3n}^{(0)}$. Substituting in $\Pi$, we obtain $\Pi$ in the form

$$\Pi_1(x_1, y_1^{(0)}, \ldots, y_{3n}^{(0)}, x_1^{(0)}, \ldots, x_{3n}^{(0)})/(y_1^{(0)})^r,$$

where $\Pi_1$ is a polynomial containing the factor

$$\Pi(y_1^{(0)}, \ldots, y_{3n}^{(0)}, x_1^{(0)}, \ldots, x_{3n}^{(0)}) \equiv \Pi_0,$$

so that

$$\Pi_1 \equiv \Pi_0 L(x_1, y_1^{(0)}, \ldots, y_{3n}^{(0)}, x_1^{(0)}, \ldots, x_{3n}^{(0)}).$$

Suppose tentatively that $L$ contains $x_1$. Let

$$y_1^{(0)}, \ldots, y_{3n}^{(0)}, x_1^{(0)}, \ldots, x_{3n}^{(0)}$$

have arbitrary values that would make neither $y_1^{(0)}$ nor $\Pi_0$ vanish. Let $\bar{x}_1$ be a value of $x_1$ that would make $L$ vanish. The solution of $(S_0)$ with the initial condition that $x_1^{(0)}, \ldots, x_{3n}^{(0)}, y_1^{(0)}, \ldots, y_{3n}^{(0)}$ have the aforesaid arbitrary values would make $\Pi$ vanish for $x_1 = \bar{x}_1$; accordingly $\Pi$ would become identically zero. Since $\Pi_0$ should vanish with this initial condition, it is contradictory to assume that $L$ contains $x_1$. Hence $L$, and accordingly $\Pi_1$, does not contain $x_1$. Thus, $L \equiv 1$, and $\Pi =$ constant is an integral of $(S_0)$.

Suppose that

$$P \equiv P_1^{l_1} \cdots P_i^{l_i}, \qquad Q \equiv Q_1^{m_j} \cdots Q_j^{m_j},$$

where $P$ and $Q$ are relatively prime polynomials of $y_1, \ldots, y_{3n}, x_1, \ldots, x_{3n}$. Then $P_1, \ldots, P_i, Q_1, \ldots, Q_j$ are irreducible polynomials in

$$y_1, \ldots, y_{3n}, x_1, \ldots, x_{3n},$$

where $l_1, \ldots, l_i, m_1, \ldots, m_j$ are positive integers. The relation $P_1 = 0$ is an invariantive relation, and, according to the property just established,

$P_1$ = constant is an integral. Similarly, $P_2, \ldots, P_i, Q_1, \ldots, Q_j$ are integrals. Hence $P$ = constant, $Q$ = constant are integrals of $(S_0)$. Thus we arrive at the following

*Theorem:* Any integral of $(S_0)$ which is rational in $y_1, \ldots, y_{3n}$ equals the quotient of two relatively prime polynomials in $y_1, \ldots, y_{3n}, x_1, \ldots, x_{3n}$, whose denominator and numerator are both integrals of $(S_0)$.

Hence, any polynomial integral of $(S_0)$ is also a polynomial in $x_1, \ldots, x_{3n}$. Otherwise, the integral would be of the form $P/Q$ and $Q$ would be a polynomial in $x_1, \ldots, x_{3n}$ and an integral of $(S_0)$. We come to the important conclusion that an integral of $(S_0)$ which is algebraic in $y_1, \ldots, y_{3n}$ is also algebraic in $x_1, \ldots, x_{3n}$.

We have seen that, if $\Pi = 0$ is an invariantive relation for $(S_0)$, where $\Pi$ is a polynomial in $y_1, \ldots, y_{3n}$, with coefficients rational in $x_1, \ldots, x_{3n}$, then $\Pi$ = constant is an integral of $(S_0)$; in other words, if

$$\Phi(y_1, \ldots, y_{3n}, x_1, \ldots, x_{3n}) = 0$$

is an invariantive relation for $(S_0)$, where $\Phi$ is a polynomial whose coefficients are rational in $x_1, \ldots, x_{3n}$, then there exists a function $\Lambda(x_1, \ldots, x_{3n})$ such that $\Lambda\Phi$ = constant is also an integral of $(S_0)$. Note that this holds not necessarily only when the coefficients in $\Phi$ are rational.

We propose to examine the properties of an invariantive relation, denoted by (E), for $(S_0)$:

$$\text{(E)}: \quad \Phi(y_1, \ldots, y_{3n}, x_1, \ldots, x_{3n}) = 0,$$

where $\Phi$ is a homogeneous polynomial in $y_1, \ldots, y_{3n}$ and is algebraic in $x_1, \ldots, x_{3n}$. Suppose that one of the coefficients of $y_1, \ldots, y_{3n}$ in the polynomial $\Phi$ is unity, and that the other coefficients are algebraic but not rational in $x_1, \ldots, x_{3n}$.

Geometrically, let $x_1, \ldots, x_{3n}$ be the rectangular coordinates of a point in a $3n$-dimensional space and let

$$\text{(D)}: \quad x_i = y_i^{(0)}t + x_i^{(0)} \qquad (i = 1, 2, \ldots, 3n)$$

be a straight line in that space. If $y_i^{(0)}$, $x_i^{(0)}$ satisfy $\Phi = 0$, then this relation is satisfied all along the straight line (D). Hence (E) represents a line complex, in which (E) defines a cone with the point $(x_1, \ldots, x_{3n})$ as vertex. Let the locus of points at which various branch functions of $\Phi$ interchange, that is, the bifurcation surface, be denoted $(\Sigma)$:

$$(\Sigma): \quad \Sigma(x_1, \ldots, x_{3n}) = 0,$$

which is the equation of an algebraic surface. This leads to the following

*Theorem:* The tangents to the surface $\Sigma$ form a part of the line complex (E).

It is sufficient to prove that if we choose arbitrarily the coordinate axes $0x_1, \ldots, 0x_{3n}$, then the tangents to $(\Sigma)$ parallel to the $(3n - 2)$ planes $x_3 = 0, \ldots, x_{3n} = 0$ form a part of the complex (E). The coordinate axes do not belong to (E) because those axes are taken arbitrarily. Let $p$ be the degree of $\Phi$ with respect to $y$. Take the coefficient of $y_1^p$ as unity. Terms in $y_1^p, \ldots, y_{3n}^p$ are all present. The origin is an ordinary point for various branch functions. Consider two branch functions $\Phi_j$ and $\Phi_l$ of $\Phi$. Then $\Phi_j - \Phi_l = 0$ defines a cone $\Gamma$ with $(x_1, \ldots, x_{3n})$ as vertex. The cone $\Gamma$ corresponding to $x_1 = 0, \ldots, x_{3n} = 0$ contains $0x_1$ but no other coordinate axes. Let $\alpha(x_1, \ldots, x_{3n})$ be the coefficient of $y_2^p$ in $\Phi$. The two values $\alpha_j$ and $\alpha_l$ of $\alpha$ corresponding to these two branch functions $\Phi_j$ and $\Phi_l$ cannot coincide identically. If they coincide they should do so at the origin; then $\Phi_j - \Phi_l$ would be zero for $y_1 = y_3 = y_4 = \cdots = y_{3n} = 0$ and would vanish all along the $0x_2$ axis.

Let $P(x_1^{(0)}, \ldots, x_{3n}^{(0)})$ be an arbitrary point of $(\Sigma)$, at which the tangent plane is not parallel to the $0x_1$ axis. Suppose that

$$x_1 = g(x_2, \ldots, x_{3n})$$

is the equation of $(\Sigma)$ near $P$, where $g$ is holomorphic in $x_2, \ldots, x_{3n}$ and equal to $x_1^{(0)}$ for $x_2 = x_2^{(0)}, \ldots, x_{3n} = x_{3n}^{(0)}$. Suppose further that $j$ branch functions $\Phi_1, \ldots, \Phi_j$ of $\Phi$ interchange in the region of $P$. Put $z = (x_1 - g)^{1/j}$; then $\Phi_1$ is expanded in ascending powers of $z$ whose coefficients are polynomials in $y_1, \ldots, y_{3n}$ and holomorphic in $x_2, \ldots, x_{3n}$ for $x_2 = x_2^{(0)}, \ldots, x_{3n} = x_{3n}^{(0)}$. If we give to $x_2, \ldots, x_{3n}$ the values $x_2^{(0)}, \ldots, x_{3n}^{(0)}$, respectively, and vary $x_1$ by making a circuit around the value $x_1^{(0)}$, then $\Phi_1$ changes to $\Phi_2$, and one value $\alpha_1$ of $\alpha$ changes to the value $\alpha_2$ of another branch. If we give to $x_3, \ldots, x_{3n}$ in $\Phi$ the values $x_3^{(0)}, \ldots, x_{3n}^{(0)}$, respectively, and put $y_3 = y_4 = \cdots = y_{3n} = 0$, then Clairaut's equation (e) is obtained:

$$\text{(e):} \qquad 0 = 1 + \alpha(x_1, x_2, x_3^{(0)}, \ldots, x_{3n}^{(0)})\left(\frac{dx_2}{dx_1}\right)^2 + \cdots,$$

where the coefficient $\alpha$ admits the curve

$$\Sigma(x_1, x_2, x_3^{(0)}, \ldots, x_{3n}^{(0)}) = 0$$

on the plane $(x_1, x_2)$ as the locus of the branch point. Hence the tangents to this curve satisfy (e); in other words, the tangents to $(\Sigma)$ parallel to the planes $x_3 = 0, \ldots, x_{3n} = 0$ form a part of the line complex (E).

As a corollary to this theorem we note the following property:

The surface $(\Sigma)$ may be decomposed into several distinct surfaces, some of which may be planes. If $(\Sigma)$ consists only of planes, then the locus of the branch points of the right-hand member of (e) should consist of straight lines, which is contradictory. Hence, $(\Sigma)$ cannot consist only of planes. Thus, all tangents to $(\Sigma)$ form a part of (E), $(\Sigma)$ contains at least a nonplanar surface $(\Sigma')$, and the tangents to $(\Sigma)$ form a complex contained in (E).

Consider the consequences of this theorem. Let $G$ be an irreducible polynomial in $x_1, \ldots, x_{3n}$ and $s$, and $s$ be an algebraic function of $x_1, \ldots, x_{3n}$ defined by $G(x_1, \ldots, x_{3n}, s) = 0$. We propose to study the condition for the existence of an invariantive relation for $(S_0)$ in the form

$$\Phi(y_1, \ldots, y_{3n}, x_2, \ldots, x_{3n}) = 0,$$

where $\Phi$ is a polynomial in $y_1, \ldots, y_{3n}$ whose coefficients are rational in $x_2, \ldots, x_{3n}, s$ but irrational in $x_2, \ldots, x_{3n}$.

Let the bifurcation surface of $s(x_1, \ldots, x_{3n})$ be

$$\Sigma(x_1, \ldots, x_{3n}) = 0.$$

If $(\Sigma)$ is composed of planes, then there is no invariantive relation of the form $\Phi = 0$. Let $\Sigma_1, \Sigma_2, \ldots$ be irreducible nonplanar surfaces contained in $(\Sigma)$. Then the tangents to $\Sigma_1, \Sigma_2, \ldots$ form the complexes

$$\psi_1(y_1, \ldots, y_{3n}, x_1, \ldots, x_{3n}) = 0, \qquad \psi_2(y_1, \ldots, y_{3n}, x_1, \ldots, x_{3n}) = 0, \ldots,$$

where $\psi$ are irreducible polynomials in $y_1, \ldots, y_{3n}, x_1, \ldots, x_{3n}$. For the existence of an invariantive relation of the form $\Phi = 0$ it is necessary and sufficient that at least one of the $\psi$ can be decomposed into a product of polynomials in $y_1, \ldots, y_{3n}$ that are rational in $x_1, \ldots, x_{3n}, s$.

Let $\phi(y_1, \ldots, y_{3n}, x_1, \ldots, x_{3n})$ be one of the decomposed factors. If this condition is satisfied, then $\phi = 0$ is an invariantive relation of the form $\Phi = 0$. Hence the condition is sufficient.

We drop from $\Phi$ those polynomials in $y_1, \ldots, y_{3n}$ that are rational in $x_1, \ldots, x_{3n}$ and are contained in $\Phi$ as factors. The complex $\Phi = 0$ contains at least one of the complexes $\psi = 0$. Let it be $\psi_1$. Since the point $(x_1, \ldots, x_{3n})$ is chosen arbitrarily, the cone $\Phi = 0$ and the cone $\psi_1 = 0$ must have a common part; in other words, the polynomials $\Phi$ and $\psi_1$ in $y_1, \ldots, y_{3n}$ must have the largest common factor $\chi(y_1, \ldots, y_{3n})$, whose coefficients are expressed rationally in terms of the coefficients of $\Phi$ and $\psi_1$, and consequently as functions of $x_1, \ldots, x_{3n}, s$. Furthermore, since it divides $\Phi$, the factor $\chi$ is not rational in $x_1, \ldots, x_{3n}$. Hence the polynomial is of the form $\psi_1 = \chi\chi_1$ and the polynomials $\chi$ and $\chi_1$ in $y_1, \ldots, y_{3n}$ are rational in $x_1, \ldots, x_{3n}, s$. Thus the condition is necessary.

If the irrationality $s$ is real and the polynomial $\Phi$ is real in (D), as the variables $x_1, \ldots, x_{3n}$ vary in the real domain (D) of the $3n$-dimensional space $x_1, \ldots, x_{3n}$ then the coefficients of $G$ are real and the coefficients of the polynomials $\psi$ are real. $\chi$ is also real, together with $\Phi$ and $\psi_1$. This gives us the following

*Theorem:* If the irrationality $s$ is real in the real domain of the space $(x_1, \ldots, x_{3n})$, then, in order that there may exist a real invariantive relation $\Phi = 0$ in (D), it is necessary and sufficient that at least one of the polynomials $\psi$ should be decomposed into a product of polynomials in $y_1, \ldots, y_{3n}$ that are rational and real in $x_1, \ldots, x_{3n}, s$.

If none of the polynomials $\psi$ can be decomposed in this manner, then an invariantive relation $\Phi = 0$, which is a polynomial in $y_1, \ldots, y_{3n}$ and real and rational in $x_1, \ldots, x_{3n}, s$, should be of the form

$$\Phi = \Lambda(x_1, \ldots, x_{3n}, s)P,$$

where $P$ represents a polynomial integral of $(S_0)$ and $\Lambda$ is rational in $x_1, \ldots, x_{3n}, s$. This result plays an important role in the $n$-body problem.

## 6.10 PROPERTIES OF THE INTEGRALS OF (S)

Let $f = P/Q$ be an integral of (S) that is rational in $y$. We have seen in Section 6.8 that $f$ can contain only a finite number of arbitrary constants. We now prove that these constants can be so chosen that they are contained in $f$ algebraically.

Put

$$y_1 = \frac{1}{u}, \qquad \frac{y_2}{y_1} = v_2, \ldots, \qquad \frac{y_{3n}}{y_1} = v_{3n};$$

then $P/Q$ becomes a rational function $f_1$ of $u, v_2, \ldots, v_{3n}$ and can be expanded in the form

$$f_1 = u^l R_0 + u^{l+1} R_1 + \cdots + u^{l+i} R_i + \cdots,$$

where $u^l R_0$ is the quotient of the highest-degree terms in $P$ and $Q$ with respect to $y$, and $l$ denotes an integer, which may be negative. By Section 6.7 it is an integral $P'/Q'$ of $(S_0)$ where $P'$ and $Q'$ are relatively prime with regard to $y$. The rational function $f_1$ contains a finite number $h$ of independent constants. We can choose an integer $j$ sufficiently large so that the first $j + 1$ terms in the expansion of $f_1$, that is,

$$F_1 = u^l R_0 + u^{l+1} R_1 + \cdots + u^{l+j} R_j,$$

can be made to depend on just $h$ independent constants.

At first we prove that these constants can be chosen so that they are contained algebraically in $F_1$. The expression $u^l R_0 \equiv P_1/Q_1$ is defined by a system of differential equations $(D_1)$ which contain algebraically the coefficients of $P_1$ and $Q_1$ together with their derivatives. On the other hand, if $P'/Q'$ is a rational integral of $(S_0)$, then $P'$ and $Q'$ contain linearly a finite number of constants $a_1, a_2, \ldots$. Substituting $P'$ and $Q'$ in $(D_1)$ for $P_1$ and $Q_1$, we obtain a certain number of algebraic relations among the constants $a_1, a_2, \ldots$. Suppose that the first $p$ constants, for example, are independent; then $P_1/Q_1$ or $u^l R_0$ contain algebraically the independent constants $a_1, a_2, \ldots, a_p$.

If the following term $u^{l+1}R_1$ of $F_1$ does not contain any new constant, then the term can be obtained algebraically as a function of

$$x_1, \ldots, x_{3n}, u, v_2, \ldots, v_{3n}$$

of $R_0$ and its derivatives, that is, as a function of

$$y_1, \ldots, y_{3n}, x_1, \ldots, x_{3n}, a_1, \ldots, a_p.$$

Suppose that $u^{l+m}R_m$ is the first term that contains constants other than $a_1, \ldots, a_p$; then the first $m$ terms of $F_1$ depend algebraically on $a_1, \ldots, a_p$. The term

$$u^{l+m}R_m \equiv \rho(y_1, \ldots, y_{3n}, x_1, \ldots, x_{3n}, a_1, \ldots, a_p, b_1, \ldots, b_q)$$

is defined by a system of differential equations $(D_m)$ which contains algebraically this function $\rho$ and its derivatives, $y$, $x$, $R_0$ and its derivatives. Hence $(D_m)$ is algebraic with respect to $\rho$, its derivatives, and the constants $a_1, \ldots, a_p$. Suppose that the numerical values $b_1^{(0)}, \ldots, b_q^{(0)}$ are given to $b_1, \ldots, b_q$, respectively. Let $f'$ be the corresponding function $f$, and $\rho' = u^{l+m}R'$ be the corresponding function $\rho$. The difference $f - f'$ is still an integral of (S). If $f - f'$ is expanded in increasing powers of $u$, then the first term of the expansion is

$$u^{l+m}(R_m - R'_m) \equiv \rho - \rho',$$

which defines an integral of $(S_0)$ rational in $y$. The difference $\sigma = \rho - \rho'$ is defined by $(D'_m)$, which is an immediate transform of $(D_m)$, algebraic with respect to $\sigma$, its derivatives, and $a_1, \ldots, a_p$. Hence it is sufficient to substitute $\sigma$ for the rational integral of $(S_0)$ in order to see that $\sigma$ contains algebraically $a_1, \ldots, a_p$ and the new constants, $a_{p+1}, \ldots, a_{p+q}$, for example.

The procedure can be followed until all terms of $F_1$ are disposed of. Let $u^{l+m+n}R_{m+n}$ be the first term of $F_1$ that algebraically contains constants other than $a_1, \ldots, a_{p+q}$. The term $u^{l+m+n}R_{m+n} \equiv \rho_1$ is defined by a differential system $(D_{m+n})$, where $\rho_1$ appears algebraically with its derivatives and $a_1, \ldots, a_{p+q}$. Let $\rho_1'$ be a particular solution of $(D_{m+n})$; then $\sigma_1 = \rho_1 - \rho_1'$ is defined by a differential system $(D'_{m+n})$ which is algebraic with respect to $\sigma_1$, its derivatives, and $a_1, \ldots, a_{p+q}$. On the other hand, since $\sigma_1$ is an integral of $(S_0)$, we see that $\sigma_1$, and hence $\rho_1$, algebraically contains $a_1, \ldots, a_{p+q}$ and the new constants suitably chosen.

Thus the $h$ independent constants $a_1, \ldots, a_h$ contained in $F_1$ appear algebraically if they are chosen in a suitable manner.

Next we prove that $f_1$ contains $a_1, \ldots, a_h$ algebraically.

Let $\bar{F}_1$ be the value of $F_1$ in which $a_1, \ldots, a_h$ are given their numerical values $\bar{a}_1, \ldots, \bar{a}_h$. Let $A(u, v_2, \ldots, v_{3n}, x_1, \ldots, x_{3n})$ be a function of $u$ holomorphic at $u = 0$. Consider all integrals $f_1$ whose expansions are of the form $f_1 = \bar{F}_1 + u^{l+j}A$. None of these functions $f_1$ can contain any constant, otherwise $f_1$ would contain at least $h + 1$ constants. Denote by $(\Delta)$ the differential system that defines $f_1$. These equations are rational with respect to the coefficients $\alpha_1, \beta_1, \ldots$ and their derivatives in the function $f_1$, which is rational in $u, v_2, \ldots, v_{3n}$. The condition that $f_1 - \bar{F}_1$ is divisible by $u^{l+j}$ is expressed by algebraic relations among $\alpha_1, \beta_1, \ldots$, in which $\bar{a}_1, \ldots, \bar{a}_h$ appear algebraically. Denote by $(\Delta')$ the whole system of conditions obtained by adding this condition to $(\Delta)$. The most general solution $\alpha_1, \beta_1, \ldots$ of $(\Delta')$ cannot contain any constant. Hence $\alpha_1, \beta_1, \ldots$ contain $\bar{a}_1, \ldots, \bar{a}_h$ only algebraically. Since this is true whatever the constants $\bar{a}_1, \ldots, \bar{a}_h$ are, the integral $f_1$ is an algebraic function of its $h$ independent constants $a_1, \ldots, a_h$.

This theorem has an important consequence. We have seen that the integral is of the form $f = P/Q$, where $P$ and $Q$ are relatively prime polynomials in $y$. Denote by $P_\nu$, $Q_\mu$ the aggregate of the highest-degree terms with regard to $y$ in $P$, $Q$, respectively. We have equated one of the coefficients of $P_\nu$, $Q_\mu$ to unity. We prove that the other coefficients of $P_\nu$, $Q_\mu$ are algebraic functions of $x$.

The polynomial $P_\nu$, $Q_\mu$ need not necessarily be relatively prime. Denote by $D$ the largest common factor of these two polynomials. Writing $P_\nu/Q_\mu$ in the irreducible form $P'/Q'$, we see that $P'/Q'$ is an integral of $(S_0)$ rational in $x$. Hence, we can write $P_\nu = DP'$, $Q_\mu = DQ'$, where $P'$, $Q'$ are two polynomials in $y$, $x$. We factorize the polynomial $D(y_1, \ldots, y_{3n})$ into irreducible polynomials $D = D_1^\lambda D_2^\mu \cdots$. If $D_1 = 0$ contains at least one of the constants $a_1, \ldots, a_h$ — say $a_1$ — then $D_1$ is algebraic with respect to $x$. Indeed, $a_1$ is contained algebraically in $D_1$, so that $D_1 = 0$ is written $d_1(y_1, \ldots, y_{3n}, x_1, \ldots, x_{3n}, a_1) = 0$, which is an irreducible polynomial in $a_1$ and $y_1, \ldots, y_{3n}$. Since $d_1 = 0$ is an invariantive relation for $(S_0)$, the ratios of the coefficients of various powers of $a_1$ in $d_1$ are also integrals of $(S_0)$. Hence, they are algebraic and rational with respect to $x$. Thus $D_1 = 0$ is algebraic with respect to $x_1, \ldots, x_{3n}$. As a result we can write

$$P_\nu = C(x_i)\Pi(y_i, x_i)\Delta(y_i, x_i),$$

$$Q_\mu = C(x_i)\mathrm{K}(y_i, x_i)\Delta(y_i, x_i).$$

where the polynomials $\Pi$ and K in $y$ are algebraic in $x$ and the polynomial $\Delta$ in $y$ contains no constant except one coefficient which is unity. This polynomial $\Delta$ is defined by a differential system which algebraically contains its coefficients $\alpha_j(x_1, \ldots, x_{3n})$, their derivatives, and $x$, but not any constant. From this evolves the theorem that the polynomials $P_\nu$, $Q_\mu$ in $y_1, \ldots, y_{3n}$ become algebraic in $x_1, \ldots, x_{3n}$ by being divided by a suitably chosen function $C(x_1, \ldots, x_{3n})$.

Next, we prove that $P_\nu$, $Q_\mu$ can be regarded as rational in $x_1, \ldots, x_{3n}, s$. For, suppose that, as the variables $x_1, \ldots, x_{3n}, s$ related by the algebraic equation $G(x_1, \ldots, x_{3n}, s) = 0$, after varying in a certain manner, return to their initial values, the polynomials $P_\nu$, $Q_\mu$ take $j$ different values $[P_\nu^{(1)}, Q_\mu^{(1)}], \ldots, [P_\nu^{(j)}, Q_\mu^{(j)}]$. Denote the $j$ branch functions of $f$ corresponding to these $j$ branches by $f_1, \ldots, f_j$. These $j$ functions are each integrals of (S) with respect to one of the branches for the corresponding values of $s$. The symmetric functions

$$F_1 = \Sigma f_1, \qquad F_2 = \Sigma f_1 f_2, \qquad \ldots, \qquad F_j = \Sigma f_1 f_2, \ldots, f_j$$

are also integrals of (S). Since the terms similar to $P_\nu$, $Q_\mu$ in these $j$ new integrals are rational in $x_1, \ldots, x_{3n}, s$, a rational integral of (S) is an algebraic combination of the rational integrals where the terms $P_\nu$, $Q_\mu$ are rational with respect to $x_1, \ldots, x_{3n}, s$.

According to Section 6.7 it is reasonable to consider those integrals that contain only terms of degrees of the same parity (both odd or both even) and are real in the same $3n$-dimensional real domain as that of $\partial U/\partial x_i$. If $\partial U/\partial x_i$ are real in this real domain (D), then all rational integrals of (S) are algebraic combinations of rational integrals which are real in (D) and contain only terms of degrees of the same parity, and their terms $P_\nu$, $Q_\mu$ are rational algebraic in $x_1, \ldots, x_{3n}, s$.

Hence, we limit ourselves to considering such integrals. Let $\Sigma = 0$ be the bifurcation surface for the irrational function $s(x_1, \ldots, x_{3n})$. Let $\Sigma_1 = 0, \Sigma_2 = 0, \ldots$ be irreducible nonplanar surfaces contained in $\Sigma = 0$. Let $\psi_1 = 0, \psi_2 = 0, \ldots$ be the equations, written as polynomials in $y_i$ and $x_i$, for the line complexes of the tangents to $\Sigma_1, \Sigma_2, \ldots$.

Suppose that none of the polynomials $\psi_1, \psi_2, \ldots$ can be decomposed into a product of polynomials in $y_1, \ldots, y_{3n}$, rational in $x_1, \ldots, x_{3n}, s$ and real within (D). Put $P_\nu = 0$ in the rational integral $f = P/Q$ we are considering; then $P_\nu = 0$ defines an invariantive relation for $(S_0)$ which is a polynomial in $y_1, \ldots, y_{3n}$, rational in $x_1, \ldots, x_{3n}, s$ and real within (D). Let $\Lambda$ be a suitably chosen function; then $\Lambda(x_1, \ldots, x_{3n})P_\nu$ is an integral of $(S_0)$ according to Section 6.9. Hence, $\Lambda P$ and $\Lambda Q$ are polynomial integrals of (S), according to Section 6.7. Thus we arrive at the

*Theorem:* The algebraic functions $\partial U/\partial x_i$, which are real in the $3n$-dimensional real domain (D), are supposed to be represented by rational expressions in terms of the irrationality $s(x_1, \ldots, x_{3n})$. Let $\Sigma_1, \Sigma_2, \ldots$ be the nonplanar bifurcation surfaces of $s$; let $\psi_1 = 0, \psi_2 = 0, \ldots$, written in the form of polynomials, be the equations for the line complexes of the tangents to $\Sigma_1, \Sigma_2, \ldots$. If none of the polynomials $\psi_1, \psi_2, \ldots$ can be decomposed into a product of polynomials in $y_1, \ldots, y_{3n}$, rational in $x_1, \ldots, x_{3n}, s$ and real in (D), then all rational integrals of (S) are algebraic combinations of the polynomial integrals of (S).

### 6.11 APPLICATION TO THE $n$-BODY PROBLEM

Put

$$r_{3j,3k} = \sqrt{(x_{3j-2} - x_{3k-2})^2 + (x_{3j-1} - x_{3k-1})^2 + (x_{3j} - x_{3k})^2},$$

$$s = \sum_{j,k} r_{3j,3k} \qquad (j, k = 1, 2, \ldots, n);$$

then the function $s$ has $2^{n(n-1)/2}$ branches. To each of these branches corresponds one and only one value of $r_{3j,3k}$. The irrationality $s$ satisfies an irreducible algebraic relation of degree $2^{n(n-1)/2}$ with regard to $s$; that is, $G = 0$. The bifurcation surfaces of the function $s$ are in a $3n$-dimensional space and defined by

$$(x_{3j-2} - x_{3k-2})^2 + (x_{3j-1} - x_{3k-1})^2 + (x_{3j} - x_{3k})^2 = 0.$$

The straight lines

$$x_{3j-2} = y_{3j-2}^{(0)}t + x_{3j-2}^{(0)}, \qquad x_{3j-1}^{(0)} = y_{3j-1}^{(0)}t + x_{3j-1}^{(0)}, \qquad x_{3j}^{(0)} = y_{3j}^{(0)}t + x_{3j}^{(0)}$$

are tangent to this surface and satisfy the relation

$$\phi_{jk} = (\xi\xi' + \eta\eta' + \zeta\zeta')^2 - (\xi^2 + \eta^2 + \zeta^2)(\xi'^2 + \eta'^2 + \zeta'^2) = 0,$$

where

$$\xi = x_{3j-2} - x_{3k-2}, \qquad \eta = x_{3j-1} - x_{3k-1}, \qquad \zeta = x_{3j} - x_{3k},$$

$$\xi' = y_{3j-2} - y_{3k-2}, \qquad \eta' = y_{3j-1} - y_{3k-1}, \qquad \zeta' = y_{3j} - y_{3k}.$$

The relation is written also in the form

$$\psi_{jk} = (\xi\eta' - \xi'\eta)^2 + (\eta\zeta' - \eta'\zeta)^2 + (\zeta\xi' - \zeta'\xi)^2 = 0,$$

where the $\psi_{jk}$ are linear in $\xi'$, $\eta'$, $\zeta'$ and cannot be decomposed into a product of linear factors in $\xi'$, $\eta'$, $\zeta'$, which are real at the same time as $\xi$, $\eta$, $\zeta$. Hence we obtain the following

*Theorem:* Any integral of the $n$-body problem which is algebraic with respect to the momentum is an algebraic combination of polynomial integrals with respect to the momentum.

Denote such an integral by

$$f = P_\nu + P_{\nu-2} + P_{\nu-4} + \cdots = \text{constant},$$

where $P_\nu, P_{\nu-2}, P_{\nu-4}, \ldots$ are homogeneous polynomials in $y_1, \ldots, y_{3n}$, of degree $\nu, \nu - 2, \nu - 4, \ldots$, respectively. It is reasonable to suppose that this integral contains only terms of degrees of the same parity in $y_1, \ldots, y_{3n}$. Since the term $P_\nu$ is an integral of $(S_0)$, it is a polynomial in $x_1, \ldots, x_{3n}$. We prove that $P_{\nu-2}$ is also algebraic in $x_1, \ldots, x_{3n}$.

Let $P'_{\nu-2}$, $P''_{\nu-2}$ be two of the branch functions of function $P_{\nu-2}$ of $x_1, \ldots, x_{3n}, s$. These branch functions interchange as the variables $x_1, \ldots, x_{3n}, s$ resume their initial values after varying in a certain manner. $P''_{\nu-2} - P'_{\nu-2}$ is a polynomial in $x_1, \ldots, x_{3n}$. For, the two branch functions of $f$—$f'$ and $f''$—correspond to the two branch functions of $P_{\nu-2}$—$P'_{\nu-2}, P''_{\nu-2}$. The branch functions $f', f''$, and $f' - f''$ are all integrals of (S). Since $P''_{\nu-2} - P'_{\nu-2}$ is an integral of $(S_0)$ and is a polynomial in $y_1, \ldots, y_{3n}$, it is also a polynomial in $x_1, \ldots, x_{3n}$.

The functions $\partial U/\partial x_i$ of (S) are of the form $\sum_{j,l} A_{jl}/r^3_{3j,3l}$, where $A_{jl}$ are linear in $x_1, \ldots, x_{3n}$. The polynomials $P_\nu$, $P_{\nu-2}$ in $y_1, \ldots, y_{3n}$ satisfy identically

$$\sum_{i=1}^{3n} \frac{\partial P_{\nu-2}}{\partial x_i}\frac{y_i}{m_i} + \sum_{i=1}^{3n} \frac{\partial P_\nu}{\partial y_i}\frac{\partial U}{\partial x_i} = 0. \tag{38}$$

If we substitute $\xi_2, \ldots, \xi_{3n}$ defined by

$$x_i y_1 - x_1 y_i = m_i \xi_i \qquad (i = 2, \ldots, 3n)$$

for $x_2, \ldots, x_{3n}$, the $P_\nu$ and $P_{\nu-2}$ become $\Pi_\nu$ and $\Pi_{\nu-2}$, respectively. Therefore the function $\Pi_\nu$ is a rational fraction in $\xi_2, \ldots, \xi_{3n}, y_1, \ldots, y_{3n}$ and independent of $x_1$, and is of the form

$$M(y_1, \ldots, y_{3n}, \xi_2, \ldots, \xi_{3n}) \cdot m_1^l y_1^{-l},$$

where $M$ is a polynomial. Also,

$$\frac{\partial P_\nu}{\partial y_i} = \frac{\partial \Pi_\nu}{\partial y_i} + \frac{\partial \Pi_\nu}{\partial \xi_i}\frac{\partial \xi_i}{\partial y_i} = \frac{\partial \Pi_\nu}{\partial y_i} - \frac{x_1}{m_1}\frac{\partial \Pi_\nu}{\partial \xi_i} \qquad (i = 2, 3, \ldots, 3n),$$

$$\frac{\partial P_\nu}{\partial y_1} = \frac{\partial \Pi_\nu}{\partial y_1} + \sum_{j=2}^{3n} \frac{\partial \Pi_\nu}{\partial x_j}\frac{\partial \xi_j}{\partial y_1} = \frac{\partial \Pi_\nu}{\partial y_1} + \frac{1}{y_1}\sum_{j=2}^{3n} \frac{\partial \Pi_\nu}{\partial \xi_j}(m_j\xi_j + x_1 y_j).$$

Hence, the second sum

$$\sum_{i=1}^{3n} \frac{\partial P}{\partial y_i}\frac{\partial U}{\partial x_i}$$

on the left-hand side of (38) is of the form

$$\sum_{j,l} (U_{jl}x_1^2 + V_{jl}x_1 + W_{jl})r_{3j,3l}^{-3},$$

with

$$r_{3j,3l}^2 = \frac{1}{y_1^2}(u_{jl}x_1^2 + v_{jl}x_1 + m_{jl}),$$

where $U$, $V$, $W$ are rational fractions in $y_1, \ldots, y_{3n}, \xi_2, \ldots, \xi_{3n}$, and $u$, $v$, $w$ are polynomials in the same variables. In particular,

$$u_{jl} = (y_{3j-2} - y_{3l-2})^2 + (y_{3j-1} - y_{3l-1})^2 + (y_{3j} - y_{3l})^2.$$

Since $\xi_i$ are integrals of $(\mathrm{S}_0)$,

$$\frac{\partial \Pi_{\nu-2}}{\partial x_1}\frac{y_1}{m_1} = \frac{\partial P_{\nu-2}}{\partial x_1}\frac{y_1}{m_1} + \cdots + \frac{\partial P_{\nu-2}}{\partial x_{3n}}\frac{y_{3n}}{m_{3n}},$$

and (38) becomes

$$\frac{\partial \Pi_{\nu-2}}{\partial x_1} = \frac{1}{y_1}\sum_{j,l} (U_{jl}x_1^2 + V_{jl}x_1 + W_{jl})\cdot r_{3j,3l}^{-3} \equiv B(x_1). \tag{39}$$

Let $\Pi'_{\nu-2}$, $\Pi''_{\nu-2}$ be the two branch functions of $\Pi_{\nu-2}$ corresponding to the same values of $y_1, \ldots, y_{3n}, x_1, \ldots, x_{3n}, s$; then $\Pi''_{\nu-2} - \Pi'_{\nu-2}$ should be rational in $y_1, \ldots, y_{3n}, \xi_2, \ldots, \xi_{3n}$.

We now prove that the Abelian integral

$$I = \int B(x_1)\, dx_1$$

formed by (39) cannot possess any period.

If this integral possesses a period $\omega(y_1, \ldots, y_{3n}, \xi_2, \ldots, \xi_{3n})$, then this period should be rational in $y_1, \ldots, y_{3n}, \xi_2, \ldots, \xi_{3n}$. If $y_i$, $\xi_i$ are given their numerical values $\bar{y}_i$, $\bar{\xi}_i$, then the branch points of the radicals

$$r_{3j,3l} = \frac{1}{y_1}\sqrt{u_{jl}x_1^2 + v_{jl}x_1 + w_{jl}},$$

are all different, unless $\bar{x}_i$, $\bar{\xi}_i$ are chosen in some particular manner. The periods of $I$ are the periods of various integrals

$$I_{jl} = y_1^2 \int \frac{U_{jl}x_1^2 + V_{jl}x_1 + W_{jl}}{(u_{jl}x_1^2 + v_{jl}x_1 + w_{jl})^{3/2}}\, dx_1.$$

Such an integral $I_{jl}$ admits $2\pi\sqrt{-1}\, y_1^2 U_{jl}/u_{jl}^{3/2}$ as the unique period. Since $U_{jl}$ and $u_{jl}$ are rational in $y_i$, $\xi_j$, and $u_{jl}$ is not a perfect square, this period can be rational with respect to $y_i$, $\xi_j$, when and only when $U_{jl}$ is identically zero. Hence, the integral $I$ cannot possess any period.

The integral $I(x_1, \xi_2, \ldots, \xi_{3n}, y_1, \ldots, y_{3n})$ is called a *primitive function* of $B(x_1)$ of (39). It is algebraic in $x_1$. It can always be so chosen that it is also algebraic in $\xi_2, \ldots, \xi_{3n}, y_1, \ldots, y_{3n}$. The difference

$$\Pi_{\nu-2} - I \equiv P_{\nu-2} - J(y_1, \ldots, y_{3n}, x_1, \ldots, x_{3n})$$

is an integral of $(S_0)$. Since this difference is algebraic in $y_1, \ldots, y_{3n}$, it is necessarily algebraic in $x_1, \ldots, x_{3n}$, according to the lemma of Section 6.9. Hence, $P_{\nu-2}$ should be algebraic in $x_1, \ldots, x_{3n}$.

If $P_{\nu-2}$ is not rational in $x_1, \ldots, x_{3n}, s$, then it can be shown by an inference similar to Section 6.10 that $f$ is an algebraic combination of analogous integrals, where $P_\nu$ and $P_{\nu-2}$ are rational in $x_1, \ldots, x_{3n}, s$. It is reasonable to suppose that $P_{\nu-2}$ is rational in $x_1, \ldots, x_{3n}, s$, and that $P_{\nu-2}$ becomes infinite only for those values of $x_i$ that make one of the quantities $r_{3j,3l}$ vanish.

Furthermore, the system (S) remains unaltered when $t$, $x_i$, $y_i$ are changed to $\alpha^3 t$, $\alpha^2 x_i$, $\alpha^{-1} y_i$, respectively. By this transformation $f$ can be decomposed into a sum of integrals, each of which remains unaltered by this transformation. From this we may state the

*Theorem:* If the $n$-body problem were to admit an integral that is algebraic with regard to the momentum, other than the classical integrals, then there would exist at least one integral that is a polynomial in $y_1, \ldots, y_{3n}$, of the form

$$P = P_\nu + P_{\nu-2} + P_{\nu-4} + \cdots,$$

where $P_j$ is a homogeneous polynomial of degree $j$ in $y_1, \ldots, y_{3n}$, the coefficients of $P_\nu$ are homogeneous polynomials all of the same degree $\lambda$ in

$x_1, \ldots, x_{3n}$, and the coefficients of $P_{\nu-2}$ are algebraic, homogeneous of degree $\lambda - 1$ in $x_1, \ldots, x_{3n}$, and rational in $x_1, \ldots, x_{3n}, s$.

The conclusion that no such integral exists is a result of Bruns's theorem (Sections 6.1–6.6). Bruns used only the first two terms $P_\nu$, $P_{\nu-2}$ of the polynomial integral $P$, supposed to be algebraic in $x_1, \ldots, x_{3n}$. By expressing, on one hand, that $U_{jl}$ are zero, and, on the other hand, that

$$\frac{\partial \Pi_{\nu-2}}{\partial x_1} = \sum_{j,l} \frac{V_{jl}x_1 + W_{jl}}{(u_{jl}x_1^2 + v_{jl}x_1 + w_{jl})^{3/2}},$$

in the present terminology, Bruns proved that $P_\nu$ should necessarily be a polynomial combination of the energy $T$ and the nine classical linear integrals.

*Painlevé's Theorem:* Whatever the values of the masses may be, the differential equations of the three-body (or $n$-body) problem do not admit, other than the classical integrals, any integral that is algebraic with regard to the momentum.

That the theorem is true also for an integral containing $t$ explicitly can be shown by the method used in the proof of Bruns's theorem.

## INTEGRALS OF THE RESTRICTED THREE-BODY PROBLEM

### 6.12 POINCARÉ'S RECURRENCE THEOREM

Suppose that two finite masses are moving in uniform circular motion around their center of mass, and consider the motion of a third body of infinitely small mass under the attraction of those two finite masses. It is assumed that the mass of the third body is so small that its motion does not affect the motion of the first two bodies. This is called the *restricted problem of three bodies.* We further assume that the third body moves in the plane of motion of the first two bodies. In this plane restricted problem of three bodies we take the $x$-axis in the direction of the two finite masses and the $y$-axis in the same plane, perpendicular to the $x$-axis. If we choose the units suitably in this rotating coordinate system, then the equations of motion of the third body are (refer to Sections 3.3–3.5)

$$\frac{d^2x}{dt^2} - 2\frac{dy}{dt} = \frac{\partial U}{\partial x}, \qquad \frac{d^2y}{dt^2} + 2\frac{dx}{dt} = \frac{\partial U}{\partial y},$$

$$U = \mu_1\left(\frac{1}{2}r^2 + \frac{1}{r}\right) + \mu\left(\frac{1}{2}r_1^2 + \frac{1}{r_1}\right), \tag{40}$$

$$r^2 = (x - \mu)^2 + y^2, \qquad r_1^2 = (x + \mu_1)^2 + y^2,$$

$$\mu + \mu_1 = 1, \qquad 0 < \mu < 1.$$

The known integral is

$$\left(\frac{dx}{dt}\right)^2 + \left(\frac{dy}{dt}\right)^2 - 2U = \text{constant}, \tag{41}$$

which is called the *Jacobi integral*. We now propose to prove that, if there exists an integral $f(x, y, dx/dt, dy/dt, t) = \text{constant}$, which is algebraic in the coordinates and the momenta and time, then it is an algebraic combination of (41). This is a special case of Bruns's theorem but it is an independent theorem and we shall prove it *de novo* after Siegel (1936).

Substitute, in (40)

$$\dot{x} = \xi, \qquad \dot{y} = \eta, \qquad \dot{\xi} = X, \qquad \dot{\eta} = Y,$$

$$2\eta + \frac{\partial U}{\partial x} = \Xi, \qquad -2\xi + \frac{\partial U}{\partial y} = \mathrm{H};$$

then the equations take the form

$$\frac{dx}{dt} = X, \qquad \frac{dy}{dt} = Y, \qquad \frac{d\xi}{dt} = \Xi, \qquad \frac{d\eta}{dt} = \mathrm{H}. \tag{40a}$$

Neither $X$, $Y$, $\Xi$, nor H explicitly contains $x$, $y$, $\xi$, or $\eta$, respectively. Therefore,

$$\frac{\partial X}{\partial x} + \frac{\partial Y}{\partial y} + \frac{\partial \Xi}{\partial \xi} + \frac{\partial \mathrm{H}}{\partial \eta} = 0. \tag{42}$$

Hence, equations (40) admit an integral invariant $\iiiint dx\, dy\, d\xi\, d\eta$, as has been shown in Section 4.18. The Jacobi integral (41) can be written

$$\xi^2 + \eta^2 - 2U = \text{constant} = C. \tag{41a}$$

Thus, in order that an integral, $f(x, y, \xi, \eta) = \text{constant}$, of (40) may exist, the equation

$$X\frac{\partial f}{\partial x} + Y\frac{\partial f}{\partial y} + \Xi\frac{\partial f}{\partial \xi} + \mathrm{H}\frac{\partial f}{\partial \eta} = 0$$

should hold identically. We shall prove that, if $f$ is algebraic in $x, y, \xi, \eta, t$, then $f$ is an algebraic function of $\xi^2 + \eta^2 - 2U$.

We can suppose that $f$ is rational in $\xi$, $\eta$, $x$, $y$, $r$, $r_1$, $t$, as shown in the proof of Bruns's theorem. Let $c$ be a function that does not contain $t$ explicitly, and $g(t) = t^m + \cdots$, $h(t) = t^n + \cdots$ be two relatively prime polynomials in $t$, of which the coefficients are rational functions of $\xi$, $\eta$, $x$, $y$, $r$, $r_1$. Write $f = cg(t)/h(t)$; then

$$\frac{\dot{c}}{c} + \frac{\dot{g}}{g} + \frac{\dot{h}}{h} = 0 \tag{43}$$

should be satisfied identically by $\xi$, $\eta$, $x$, $y$, $t$, when $\dot{\xi}$, $\dot{\eta}$ are eliminated by (40a). Since $t$ is not contained explicitly in (40a), and since $\dot{c}/c$ is independent of $t$, and $\dot{g}/g$, $\dot{h}/h$ are such that their denominators and numerators are relatively prime polynomials of $t$, equations (43) should hold term by term separately, that is, any one of $c$, $g$, $h$ should be an integral. Hence, it is sufficient to consider integrals in the form of polynomials in $t$.

If we denote by $b_0, b_1, \ldots, b_m$ rational polynomials of $\xi$, $\eta$, $x$, $y$, $r$, $r_1$, then such integrals are of the form

$$f = b_0 t^m + b_1 t^{m-1} + \cdots + b_m \qquad (m \geqq 0,\ b_0 \neq 0). \tag{44}$$

Choose a curve $C$ on the $xy$ plane (41a) such that no one of $b_0, b_1^{-1}, \ldots, b_m^{-1}$ is zero at any point on $C$. Take a general point $x = x_0$, $y = y_0$ on this plane and choose $\xi_0$, $\eta_0$ so as to satisfy (41a). Suppose that none of $b_0, b_1^{-1}, \ldots, b_m^{-1}$ is zero at $(x_0, y_0, \xi_0, \eta_0)$. We shall now show that $m = 0$.

The relation (42) holds in a four-dimensional space with rectangular coordinates $x$, $y$, $\xi$, $\eta$, and the integral of volume is invariant. In the rationality domain of $x$, $y$, $\xi$, $\eta$, let a state of motion defined by (40a) correspond uniquely to an orbit of the points in the finite portion of this four-dimensional space. Since the integral invariant of volume exists and the relation (42) is valid, the motion according to Poincaré is represented by the motion of molecules of incompressible fluid enclosed in a vessel of finite four-dimensional volume, as we have described in Section 4.18. Take a small volume $V_0$ inside the vessel; consider the motion of a particular molecule in $V_0$; let $T$ be a finite time interval. Suppose that the particular molecule that was in $V_0$ at $t = 0$ occupies a volume $V_n$ at $t = nT$, where $n$ is an integer. Since the fluid is incompressible, the volumes $V_n$ $(n = 1, 2, 3, \ldots)$ are all equal to $V_0$. There should be a finite integer $m$ such that there is at least a common part between $V_0$ and $V_m$. Hence, there should be a $V_n$ such that it has parts in common with $V_0$ an infinite number of times during the time interval from $t = 0$ to $t = +\infty$. If we take $V_0$ small enough, the orbit of the molecule comes as near as we please to the initial position in a sufficiently long interval of time. This theorem is called *Poincaré's recurrence theorem* or the *quasi-ergodic theorem* and it will be discussed in detail in a later chapter (Hagihara, 1933).

According to this theorem there should be an upper bound to the time interval during which a moving point comes again to a point where it has once passed along its orbit. For, if we form (44) for those values of $x, y, \xi, \eta$ for which the moving point comes again very close to its initial point, and also for the corresponding value of $t$, then the two resulting values of $f$ in (44) should be equal because (44) is an integral. But by our assumption, $b_0, b_1, \ldots, b_m$ have nearly the same values at these two epochs. Solving these two equations, we see the finite upper bound of the time interval of the two epochs. A similar argument holds true for $m < 0$. Hence, $m = 0$. Thus we conclude that $f$ does not contain $t$ explicitly.

We have excluded in the above the cases where $x$, $y$, $\xi$, $\eta$ become infinite. $x$ and $y$ become infinite when the orbit goes to infinity. $\Xi$ and H become infinite in the finite region when $r$ or $r_1$ becomes zero, that is, in a collision. The condition for a collision is an integral of the problem; we have, however, excluded such cases. The condition for a collision is transcendental, as we shall show in a later chapter (Painlevé, 1896, 1897; Bisconcini, 1906; Levi-Civita, 1903, 1906). In order that the orbit may extend to infinity, the energy constant $C$ in (41a) must be infinite. We can exclude this case. There must also be a condition for a periodic orbit (as will be discussed in a later chapter). Since this condition is also an integral of the problem, we can exclude this case, too. Our question is whether an integral is algebraic with regard to $t$, but an algebraic expression has no period. We also exclude orbits asymptotic to either of the straight-line equilibrium points, since special conditions are necessary for such motions.

## 6.13 LEMMA

We are looking for integrals that are rational functions of $x, y, \xi, \eta, r, r_1$. Let $f = g/h$, where $g$, $h$ are relatively prime polynomials in $\xi$, $\eta$. Since the equation

$$g \frac{dh}{dt} = h \frac{dg}{dt} \tag{45}$$

is satisfied identically when the expressions for $d\xi/dt$, $d\eta/dt$ are substituted from (40a), the equation

$$\frac{dg}{dt} = \xi \frac{\partial g}{\partial x} + \eta \frac{\partial g}{\partial y} + \Xi \frac{\partial g}{\partial \xi} + \mathrm{H} \frac{\partial g}{\partial \eta} \tag{46}$$

is a polynomial in $\xi$, $\eta$ and of degree one higher than $g$ with regard to $\xi$, $\eta$. The coefficients are rational in $x, y, r, r_1$. Hence, $dg/dt$ belongs to the rational algebraic field $\Omega(x, y, r, r_1)$. But $dg/dt$ is divisible by $g$, according to (45). Hence, we have identically

$$\frac{dg}{dt} = g(u\xi + v\eta + w). \tag{47}$$

Let $G$ be the highest degree term in $\xi$, $\eta$,

$$G \equiv a\xi^k\eta^{m-k} + \cdots + b\xi^l\eta^{m-l},$$

where $a, \ldots, b$ are not identically zero. From (46) and (47) we obtain

$$u = \frac{1}{a}\frac{\partial a}{\partial x}, \qquad v = \frac{1}{b}\frac{\partial b}{\partial y}, \tag{48}$$

$$\xi\frac{\partial G}{\partial x} + \eta\frac{\partial G}{\partial y} = \left(\frac{1}{a}\frac{\partial a}{\partial x}\xi + \frac{1}{b}\frac{\partial b}{\partial y}\eta\right)G. \tag{49}$$

If we denote polynomials in $x$, $y$ by $c_1$, $c_2$, $c_3$, $c_4$, then any quantity belonging to the algebraic field $\Omega(x, y, r, r_1)$ can be written in the form

$$c_1 + c_2 r + c_3 r_1 + c_4 r r_1.$$

As $g$, $h$ are determined by $\Omega(x, y, r, r_1)$, apart from a common factor, we can assume that $a$ has one of the forms $c_1$, $c_2 r$, $c_3 r_1$, $c_4 r r_1$, that the coefficients $a, \ldots, b$ of $G$ are all polynomials, and that $r$ and $r_1$ have no common factor in the form of a polynomial in $x$, $y$. But any irreducible polynomial belonging to the field $\Omega(x, y)$, which is included in $a$, should be included by (49) in all the coefficients in $G$. A similar statement can be made for $r$ and $r_1$; that is, the polynomial should be included in $b$ when it is included in $a$. Hence, $a$ should be a constant. As far as this argument is concerned, we can exchange $a$ and $b$ and we can conclude that $b$ is also a constant.

Because of (45), (47), and (48), $g$ must be a polynomial in $\xi$, $\eta$ with the coefficients belonging to the rational algebraic field $\Omega(x, y, r, r_1)$ and must satisfy the differential equation

$$\frac{dg}{dt} = wg, \tag{50}$$

where $w$ belongs to the rational algebraic field $\Omega(x, y, r, r_1)$.

Next we shall show that $w$ in (50) is constant. Put

$$g = g^{(m)} + g^{(m-1)} + \cdots + g^{(0)}, \tag{51}$$

where $g^{(k)}$ ($k = 0, 1, \ldots, m$), is a polynomial of degree $k$ in $\xi$, $\eta$ with the coefficients belonging to the field $\Omega(x, y, r, r_1)$. Suppose that $g^{(m)} \not\equiv 0$. From (46), (50), and (51) we obtain the following equations:

$$\xi\frac{\partial g^{(m)}}{\partial x} + \eta\frac{\partial g^{(m)}}{\partial y} = 0, \tag{52}$$

$$\xi\frac{\partial g^{(m-1)}}{\partial x} + \eta\frac{\partial g^{(m-1)}}{\partial y} + 2\eta\frac{\partial g^{(m)}}{\partial \xi} - 2\xi\frac{\partial g^{(m)}}{\partial \eta} = wg^{(m)}, \tag{53}$$

$$\xi \frac{\partial g^{(k-1)}}{\partial x} + \eta \frac{\partial g^{(k-1)}}{\partial y} + 2\eta \frac{\partial g^{(k)}}{\partial \xi} - 2\xi \frac{\partial g^{(k)}}{\partial \eta}$$

$$+ \frac{\partial U}{\partial x}\frac{\partial g^{(k+1)}}{\partial \xi} + \frac{\partial U}{\partial y}\frac{\partial g^{(k+1)}}{\partial \eta} = wg^{(k)} \qquad (k = 1, 2, \ldots, m), \quad (54)$$

$$\frac{\partial U}{\partial x}\frac{\partial g^{(1)}}{\partial \xi} + \frac{\partial U}{\partial y}\frac{\partial g^{(1)}}{\partial \eta} = wg^{(0)}. \quad (55)$$

We see from (52) that $g^{(m)}$ is a function of $\xi$, $\eta$, $x\eta - y\xi$ only. On the other hand, since $g^{(m)}$ is a polynomial in $\xi$, $\eta$ with the coefficients belonging to the field $\Omega(x, y, r, r_1)$, the polynomial $g^{(m)}$ is a polynomial in $\xi$, $\eta$, $x\eta - y\xi$.

Suppose that $w$ is a function of $x$ and has a pole of order $h \geqq 1$ at $x = x_1$, where $x_1$ differs from $\mu \pm y\sqrt{-1}$ and $-\mu_1 \pm y\sqrt{-1}$. Since $g^{(m)}$ is not factorizable, $wg^{(m)}$ has a pole of order $h$ at $x = x_1$. Expand $g^{(m-1)} = c(x - x_1)^s + \cdots$ in ascending powers of $x - x_1$; then

$$\xi \frac{\partial g^{(m-1)}}{\partial x} + \eta \frac{\partial g^{(m-1)}}{\partial y} = cs\left(\xi - \eta \frac{\partial x_1}{\partial y}\right)(x - x_1)^{s-1} + \cdots.$$

As $s - 1 = -h$, $s \neq 0$, by (53), therefore, $h > 1$, and $g^{(m-1)}$ has a pole of order $h - 1$ at $x = x_1$. Assume that $g^{(m-l)}$ has a pole of order $l(h - 1)$ at $x = x_1$; then by (54) $g^{(m-l-1)}$ has a pole of order $(l + 1)(h - 1)$ at $x = x_1$. But in that case the right-hand member of (55) has a pole of order $m(h - 1)$ and the left-hand member has a pole at most of order $(m - 1)(h - 1)$. Hence, $w$ as a function of $x$ cannot have poles anywhere but at $\infty$, $\mu \pm y\sqrt{-1}$, $-\mu_1 \pm y\sqrt{-1}$.

Next, assume that $w$ has a pole of order $h$ at $x = \infty$ and that $g^{(m)}$ is of degree $p$ with regard to $x$; then, quite similarly we see that $g^{(m-l)}$ has a pole of order $l(h + 1) + p$ at $x = \infty$, which contradicts (55). Further, if $w$ becomes infinite as $r^{-h}$ at $x = \mu \pm y\sqrt{-1}$, then, as $\partial U/\partial x$ and $\partial U/\partial y$ become infinite as $r^{-3}$ for $h > 1$, the polynomial $g^{(m-l)}$ becomes infinite as $r^{-l(h-2)}$ at $x = \mu \pm y\sqrt{-1}$ and, hence, $h > 2$. Again, this contradicts (55). Similarly, we see that $w$ cannot become infinite faster than $r_1^{-1}$ at $x = -\mu_1 \pm y\sqrt{-1}$.

We can draw a similar conclusion by considering $w$ as a function of $y$. Then $w$ takes the form

$$w = \frac{c_1}{rr_1} + \frac{c_2}{r} + \frac{c_3}{r_1} + c_4,$$

where $c_1$ is (at most) of the second degree in $x, y$; also, $c_2$, $c_3$ are (at most) of the first degree; also, $c_4$ is a constant. Since $g^{(m-1)}$ is a polynomial belonging to the field $\Omega(x, y, r, r_1)$, it should be of the form

$$g^{(m-1)} = b_1 r r_1 + b_2 r + b_3 r_1 + b_4,$$

where $b_1$, $b_2$, $b_3$, $b_4$ are polynomials in $x$, $y$. Furthermore,

$$\xi \frac{\partial}{\partial x}(b_1 r r_1) + \eta \frac{\partial}{\partial y}(b_1 r r_1) = \frac{c_1}{r r_1} g^{(m)}, \tag{56}$$

$$\xi \frac{\partial}{\partial x}(b_2 r) + \eta \frac{\partial}{\partial y}(b_2 r) = \frac{c_2}{r} g^{(m)}, \tag{57}$$

$$\xi \frac{\partial}{\partial x}(b_3 r_1) + \eta \frac{\partial}{\partial y}(b_3 r_1) = \frac{c_3}{r_1} g^{(m)}. \tag{58}$$

By introducing $\alpha = x\eta - y\xi$, $\beta = x\xi + y\eta$ in the place of $x$, $y$, we see that $g^{(m)}$ does not depend on $\beta$. From (56), we obtain

$$\frac{\partial}{\partial \beta}(b_1 r r_1) = \frac{c_1}{r r_1} \frac{1}{\xi^2 + \eta^2} g^{(m)}. \tag{59}$$

If we suppose tentatively that $b_1$ is not identically zero, then $b_1 r r_1$ becomes infinite at least as fast as $\beta^2$. Hence $\partial(b_1 r r_1)/\partial\beta$ becomes infinite at least as fast as $\beta$. But the right-hand member of (59) does not become infinite, except for the particular cases we have excluded. Hence, $b_1 = 0$, $c_1 = 0$. Similarly, because of (57) and (58), $b_2$, $b_3$ are constants. Furthermore,

$$b_2(\beta - \mu\xi) = c_2 g^{(m)}, \qquad b_3(\beta + \mu_1\xi) = c_3 g^{(m)}.$$

Since $g^{(m)}$ does not depend on $\beta$, we have $c_2 = 0$, $c_3 = 0$. Thus, we have shown that $w$ is a constant.

From Poincaré's recurrence theorem we can prove that the constant $w$ is zero. Integrating (50), we obtain

$$g = Ke^{wt}, \tag{60}$$

in which $K$ is constant along the trajectory. As a special case of Section 6.12 of Poincaré's recurrence theorem, there exists, among the motions with $g \neq 0$ in the neighborhood of the trajectory $x = x_0, y = y_0, \xi = \xi_0, \eta = \eta_0$, a motion that returns to the immediate neighborhood of the initial state of motion after the lapse of an arbitrarily long time. Suppose that $w \neq 0$; then the right-hand side of (60) would tend to the limit zero or infinity as $t$ approaches infinity, and this contradicts the theorem. Hence, $w = 0$.

## 6.14 PROOF OF SIEGEL'S THEOREM

We now state *Siegel's theorem:* An integral in the form of a polynomial in $\xi, \eta$ with the coefficients belonging to the algebraic field $\Omega(x, y, r, r_1)$ is a polynomial of only one variable $\xi^2 + \eta^2 - 2U$.

To prove this theorem, we divide an integral $g$ into parts homogeneous in $\xi, \eta$ and write

$$g = g^{(m)} + \cdots + g^{(0)}.$$

Putting $w = 0$ in (52), (53), (54), (55), we obtain

$$\xi \frac{\partial}{\partial x} g^{(m)} + \eta \frac{\partial}{\partial y} g^{(m)} = 0, \tag{61}$$

$$\xi \frac{\partial}{\partial x} g^{(m-1)} + \eta \frac{\partial}{\partial y} g^{(m-1)} + 2\eta \frac{\partial}{\partial \xi} g^{(m)} - 2\xi \frac{\partial}{\partial \eta} g^{(m)} = 0, \tag{62}$$

$$\xi \frac{\partial}{\partial x} g^{(k-1)} + \eta \frac{\partial}{\partial y} g^{(k-1)} + 2\eta \frac{\partial}{\partial \xi} g^{(k)} - 2\xi \frac{\partial}{\partial \eta} g^{(k)}$$
$$+ \frac{\partial U}{\partial x} \frac{\partial}{\partial \xi} g^{(k+1)} + \frac{\partial U}{\partial y} \frac{\partial}{\partial \eta} g^{(k+1)} = 0 \qquad (k = 1, 2, \ldots, m-1), \tag{63}$$

$$\frac{\partial U}{\partial x} \frac{\partial}{\partial \xi} g^{(1)} + \frac{\partial U}{\partial y} \frac{\partial}{\partial \eta} g^{(1)} = 0. \tag{64}$$

If we introduce new variables, $\alpha = x\eta - y\xi$, $\beta = x\xi + y\eta$, instead of $x, y$, then we see by (61), that $g^{(m)} = P(\xi, \eta, \alpha)$ is a polynomial in the three variables $\xi, \eta, \alpha$, and that (62) takes the form

$$(\xi^2 + \eta^2) \frac{\partial}{\partial \beta} g^{(m-1)} = 2\left(\xi \frac{\partial P}{\partial \eta} - \eta \frac{\partial P}{\partial \xi} + \beta \frac{\partial P}{\partial \alpha}\right);$$

hence,

$$g^{(m-1)} = \frac{2\beta}{\xi^2 + \eta^2}\left(\xi \frac{\partial P}{\partial \eta} - \eta \frac{\partial P}{\partial \xi}\right) + \frac{\beta^2}{\xi^2 + \eta^2} \frac{\partial P}{\partial \alpha} + g_1(\xi, \eta, \alpha), \tag{65}$$

where $g_1$ is independent of $\beta$. Thus $g^{(m-1)}$ is quadratic in $\beta$, considered as a function of $\xi, \eta, \alpha, \beta$.

From (63),

$$(\xi^2 + \eta^2) g^{(k-1)} = \int \left[2\left(\xi \frac{\partial g^{(k)}}{\partial \eta} - \eta \frac{\partial g^{(k)}}{\partial \xi} + \beta \frac{\partial g^{(k)}}{\partial \alpha} - \alpha \frac{\partial g^{(k)}}{\partial \beta}\right)\right.$$
$$- \frac{\partial U}{\partial x}\left(\frac{\partial g^{(k+1)}}{\partial \xi} + x \frac{\partial g^{(k+1)}}{\partial \beta} - y \frac{\partial g^{(k+1)}}{\partial \alpha}\right)$$
$$\left. - \frac{\partial U}{\partial y}\left(\frac{\partial g^{(k+1)}}{\partial \eta} + x \frac{\partial g^{(k+1)}}{\partial \alpha} + y \frac{\partial g^{(k+1)}}{\partial \beta}\right)\right] d\beta$$
$$(k = 1, 2, \ldots, m-1). \tag{66}$$

If we define $g^{(-1)} = 0$, then, by (64), this formula is true also for $k = 0$. If we define $g^{(-2)} = 0$, then this formula does not give anything new for $k = -1$, but it is true. It is sufficient to discuss the formula (66) for $k = m - 1$ and $k = m - 2$. Now, referring to

$$(\xi^2 + \eta^2)r^2 = (\alpha - \mu\eta)^2 + (\beta - \mu\xi)^2,$$
$$(\xi^2 + \eta^2)r_1^2 = (\alpha + \mu_1\eta)^2 + (\beta + \mu_1\xi)^2,$$

we integrate (66) by parts for $k = m - 1$; then we obtain

$$\begin{aligned} g^{(m-2)} = {} & \frac{1}{\alpha - \mu\eta}\left[y\frac{\partial P}{\partial \xi} - (x - \mu)\left(\frac{\partial P}{\partial \eta} + \mu\frac{\partial P}{\partial \alpha}\right)\right]\frac{\mu_1}{r} \\ & + \frac{1}{\alpha + \mu_1\eta}\left[y\frac{\partial P}{\partial \xi} - (x + \mu_1)\left(\frac{\partial P}{\partial \eta} + \mu_1\frac{\partial P}{\partial \alpha}\right)\right]\frac{\mu}{r_1} + g_2, \end{aligned} \tag{67}$$

where $g_2$ is a polynomial biquadratic function of $\xi, \eta, \alpha, \beta$. For $k = m - 2$, using (67) and (65), we obtain after some calculation

$$\begin{aligned} (\xi^2 + \eta^2)g^{(m-3)} = {} & \left(\eta\frac{\partial P}{\partial \xi} - \xi\frac{\partial P}{\partial \eta} - \mu\xi\frac{\partial P}{\partial \alpha}\right)\int\frac{\mu_1}{r}\,d\beta \\ & + \left(\eta\frac{\partial P}{\partial \xi} - \xi\frac{\partial P}{\partial \eta} + \mu_1\xi\frac{\partial P}{\partial \alpha}\right)\int\frac{\mu}{r_1}\,d\beta + g_3, \end{aligned} \tag{68}$$

where $g_3$ belongs to the algebraic field $\Omega(\xi, \eta, x, y, r, r_1)$.

In addition, put

$$\begin{aligned} \mu_1\left(\eta\frac{\partial P}{\partial \xi} - \xi\frac{\partial P}{\partial \eta} - \mu\xi\frac{\partial P}{\partial \alpha}\right) &= A, \\ \mu\left(\eta\frac{\partial P}{\partial \xi} - \xi\frac{\partial P}{\partial \eta} + \mu_1\xi\frac{\partial P}{\partial \alpha}\right) &= B; \end{aligned} \tag{69}$$

then, because of (68),

$$A\int\frac{d\beta}{r} + B\int\frac{d\beta}{r_1} \tag{70}$$

belongs to the algebraic field $\Omega(\xi, \eta, x, y, r, r_1)$. When we make a single circuit around the point

$$\beta_1 = \mu\xi + \sqrt{-1}(\alpha - \mu\eta) \neq -\mu_1\xi \pm \sqrt{-1}(\alpha + \mu_1\eta)$$

in the $\beta$ plane, then $r$ changes its sign and

$$-A\int\frac{d\beta}{r} + B\int\frac{d\beta}{r_1} \tag{71}$$

also belongs to the algebraic field $\Omega(\xi, \eta, x, y, r, r_1)$. When we make a single circuit around the point at infinity, then the two integrals increase by the amount $2\pi\sqrt{-1}$ but the expressions (70) and (71) remain unaltered. Hence $A + B = 0$, $-A + B = 0$; that is,

$$A = 0, \qquad B = 0. \tag{72}$$

We obtain from (69) and (72) the two differential equations

$$\eta \frac{\partial P}{\partial \xi} - \xi \frac{\partial P}{\partial \eta} = 0, \qquad \frac{\partial P}{\partial \alpha} = 0.$$

The second equation shows that the polynomial $P(\xi, \eta, \alpha)$ depends only on $\xi, \eta$; the first equation shows that it depends only on $\xi^2 + \eta^2$. Hence, $P(\xi, \eta, \alpha)$ as a homogeneous polynomial in $\xi, \eta$ should be of the form

$$P = \kappa(\xi^2 + \eta^2)^k,$$

with a constant $\kappa$. Since $\kappa(\xi^2 + \eta^2 - 2U)^k$ is an integral,

$$g - \kappa(\xi^2 + \eta^2 - 2U)^k$$

is an integral in the form of a polynomial in $\xi, \eta$ of lower degree than $g$ and with coefficients belonging to the finite algebraic field $\Omega(x, y, r, r_1)$.

By mathematical induction Siegel (1936) has proved that $g$ is a polynomial of only one variable $\xi^2 + \eta^2 - 2U$, and then proved the theorem in question.

## THEOREM OF POINCARÉ

### 6.15 CASE OF NONVANISHING HESSIAN

We consider the canonical equations of the $n$-body problem given in Sections 5.16–5.20:

$$\frac{dx_i}{dt} = \frac{\partial F}{\partial y_i}, \qquad \frac{dy_i}{dt} = -\frac{\partial F}{\partial x_i} \qquad (i = 1, 2, \ldots, n),$$
$$F = F_0 + \mu F_1 + \mu^2 F_2 + \cdots, \tag{73}$$

where $F_0$ does not contain $y_i$ but only $x_i$, and $F_1, F_2, \ldots$ are functions of $y_i$, $x_i$ and periodic, with period $2\pi$ with regard to each $y_i$. The canonical variables $x_i, y_i$ are the elements of the osculating elliptic motion discussed in Sections 5.16–5.17. Poincaré proved that the canonical equations (73) cannot have any uniform integral, that is, any integral which is single-valued and continuous with respect to $x_i$, $y_i$, other than $F =$ constant. This is called *the theorem of Poincaré* (1889, 1892).

We assume at first that the Hessian of $F_0$ with respect to $x_i$ is not zero.

Suppose that $\Phi$ is uniform, that is, single-valued and continuous, with respect to $x_i$, $y_i$, and $\mu$, and periodic, with period $2\pi$ in each $y_i$. The function $\Phi$ need not be uniform for all values of $x_i$, $y_i$, and $\mu$. It should be uniform for all values of $y_i$, but only for the values of $\mu$ sufficiently small and only for the values of $x_i$ in a domain (D) which can be taken as small as we please. Under this condition the function $\Phi$ can be expanded in the form

$$\Phi = \Phi_0 + \mu\Phi_1 + \mu^2\Phi_2 + \cdots, \tag{74}$$

where $\Phi_0, \Phi_1, \ldots,$ are uniform functions of $x_i, y_i$ and periodic in $y_i$. We shall prove according to Poincaré that such a function $\Phi$ cannot exist.

The necessary and sufficient condition for $\Phi$ to be an integral is, from our result of Chapter 4, that the Poisson bracket be equal to zero, that is,

$$(F, \Phi) = 0,$$

or

$$(F_0, \Phi_0) + \mu\{(F_1, \Phi_0) + (F_0, \Phi_1)\} \\ + \mu^2\{(F_2, \Phi_0) + (F_1, \Phi_1) + (F_0, \Phi_2)\} + \cdots = 0,$$

or

$$(F_0, \Phi_0) = 0, \tag{75}$$

$$(F_1, \Phi_0) + (F_0, \Phi_1) = 0, \tag{76}$$

$$\vdots$$

First we shall prove a lemma: that we can suppose that $\Phi_0$ is not a function of $F_0$.

Assume tentatively that $\Phi_0 = \psi(F_0)$. So far as the value of $x_i$ is in the domain (D), $\psi$ is, in general, uniform. Indeed, we solve

$$F_0 = F_0(x_1, x_2, \ldots, x_n)$$

in the form

$$x_1 = \theta(F_0, x_2, \ldots, x_n);$$

then $\theta$ is uniform if $\partial F_0/\partial x_1 \neq 0$ in (D). Replacing $x_1$ by this expression, we see that

$$\Phi_0(x_1, x_2, \ldots, x_n, y_1, \ldots, y_n) = \psi(F_0, x_2, \ldots, x_n, y_1, \ldots, y_n);$$

hence $\psi$ is a uniform function of $F_0, x_2, \ldots, x_n, y_1, \ldots, y_n$. Since we have assumed that $\psi$ depends only on $F_0$, the function $\Phi_0 = \psi$ is uniform as a function of $F_0$. If $\partial F_0/\partial x_1 = 0$ in (D), then we have only to take one of the $\partial F_0/\partial x_i$ which is not zero in (D) and proceed similarly.

Now, if $\Phi$ is uniform, then $\Phi - \psi(F)$ is also uniform, and $\Phi - \psi(F)$ can be expanded in integral powers of $\mu$ and is divisible by $\mu$, since $\Phi_0 - \psi(F_0) = 0$. The function $\Phi'$ defined by $\Phi - \psi(F) = \mu\Phi'$ is uniform and may be expanded in the form

$$\Phi' = \Phi'_0 + \mu\Phi'_1 + \mu^2\Phi'_2 + \cdots.$$

In general, $\Phi'_0$ is not a function of $F_0$. If $\Phi'_0$ is a function of $F_0$, then we repeat the same operation and finally we arrive at an integral that is not a function of $F_0$ for $\mu = 0$. In fact, as we have assumed that $\Phi$ is not a function of $F$, there should be at least one of the Jacobians of $\Phi$ and $F$ with respect to two of the variables $x_i, y_i$, that does not vanish. We call it $J$. We have assumed that $\Phi_0$ is a function of $F_0$. Hence, $J$ has a factor $\mu$. Let the factor be $\mu^p$; then $J'$ defined by $J = \mu J'$ has a factor $\mu^{p-1}$. By repeating this process $p$ times we arrive at a Jacobian which has no factor $\mu$ and which corresponds to an integral not reducing to a function of $F_0$ for $\mu = 0$. Hence, if there exists a uniform integral $\Phi$, which is independent of $F$ but such that $\Phi_0$ is a function of $F_0$, then we can find another integral of the same form which does not reduce to a function of $F_0$ for $\mu = 0$. Hence we can assume that $\Phi_0$ is not a function of $F_0$.

Next we shall prove a second lemma: that $\Phi_0$ contains none of the $y_i$.

Assume tentatively that $\Phi_0$ contains $y_i$. As it is periodic in $y_i$, it should be of the form

$$\Phi_0 = \sum A \exp[\sqrt{-1}(m_1 y_1 + \cdots + m_n y_n)] \equiv \sum A\zeta,$$

where $m_i$ is a positive or negative integer and $A$ a function of $x_i$. As $F_0$ does not depend on $y_i$ and as $\partial F_0/\partial y_i = 0$, we obtain

$$(F_0, \Phi_0) = \sum_{i=1}^{n} \frac{\partial F_0}{\partial x_i}\frac{\partial \Phi_0}{\partial y_i}.$$

On the other hand,

$$\frac{\partial \Phi_0}{\partial y_i} = \sum \sqrt{-1}\, m_i A\zeta.$$

Hence, (75) is written

$$\sqrt{-1}\sum A\left(m_1\frac{\partial F_0}{\partial x_1} + \cdots + m_n\frac{\partial F_0}{\partial x_n}\right)\zeta = 0.$$

Since this is an identity, we have

$$A \sum_{i=1}^{n} m_i \frac{\partial F_0}{\partial x_i} = 0.$$

Hence either

$$A = 0 \qquad \text{or} \qquad \sum_{i=1}^{n} m_i \frac{\partial F_0}{\partial x_i} = 0.$$

From the latter relation we obtain

$$\sum_{i=1}^{n} m_i \frac{\partial^2 F_0}{\partial x_i\, \partial x_k} = 0 \qquad (k = 1, 2, \ldots, n),$$

and this should occur either when $m_1 = m_2 = \cdots = m_n = 0$ or when the Hessian of $F_0$ is zero. As we have assumed that the Hessian is not zero, $A = 0$ except the terms for which $m_1 = m_2 = \cdots = m_n = 0$. Hence, $\Phi_0$ should reduce to only one term that does not depend on $y$. Hence, $\Phi_0$ does not depend on $y_i$.

After these preparations we come to the proof of the theorem. Since $F_0$ and $\Phi_0$ do not depend on $y$, we obtain from (76)

$$-\sum_{i=1}^{n} \frac{\partial \Phi_0}{\partial x_i} \frac{\partial F_1}{\partial y_i} + \sum_{i=1}^{n} \frac{\partial F_0}{\partial x_i} \frac{\partial \Phi_1}{\partial y_i} = 0.$$

By our assumption, $F_1$ and $\Phi_1$ are of the form $F_1 = \sum B\zeta$ and $\Phi_1 = \sum C\zeta$, where $B$ and $C$ are functions of $x_i$ alone and

$$\zeta = \exp\,[\sqrt{-1}(m_1 y_1 + \cdots + m_n y_n)].$$

Substitute these in (76); then,

$$-\sum B\zeta\Big(\sum_{i=1}^{n} m_i \frac{\partial \Phi_0}{\partial x_i}\Big) + \sum C\zeta\Big(\sum_{i=1}^{n} m_i \frac{\partial F_0}{\partial x_i}\Big) = 0.$$

Since this is an identity, we have for any system of integers $m_i$ the relation

$$B \sum_{i=1}^{n} m_i \frac{\partial \Phi_0}{\partial x_i} = C \sum_{i=1}^{n} m_i \frac{\partial F_0}{\partial x_i}; \tag{77}$$

indeed, the relation holds for all values of $x_i$. Now, if we choose $x_i$ so as to satisfy

$$\sum_{i=1}^{n} m_i \frac{\partial F_0}{\partial x_i} = 0, \tag{78}$$

then the right-hand member of (77) vanishes. Hence, if $x_i$ satisfies (78), we have either $B = 0$ or

$$\sum_{i=1}^{n} m_i \frac{\partial \Phi_0}{\partial x_i} = 0. \tag{79}$$

As $F_1$, accordingly $B$, is given by our problem we can find out, from (78), whether $B = 0$ is satisfied. In general, such a relation is not satisfied. Hence, (79) must hold.

Let $p_1, p_2, \ldots, p_n$ be integers; choose $x$ such that

$$\frac{\partial F_0}{p_1\, \partial x_1} = \frac{\partial F_0}{p_2\, \partial x_2} = \cdots = \frac{\partial F_0}{p_n\, \partial x_n}. \tag{80}$$

Then there should be an infinite number of sets of integers $m_1, m_2, \ldots, m_n$ such that

$$m_1 p_1 + m_2 p_2 + \cdots + m_n p_n = 0. \tag{81}$$

The relation (78) must be satisfied for an infinite number of such sets of integers. Hence,

$$\sum_{i=1}^{n} m_i \frac{\partial \Phi_0}{\partial x_i} = 0.$$

This relation, together with (78), gives

$$\frac{\partial F_0/\partial x_1}{\partial \Phi_0/\partial x_1} = \frac{\partial F_0/\partial x_2}{\partial \Phi_0/\partial x_2} = \cdots = \frac{\partial F_0/\partial x_n}{\partial \Phi_0/\partial x_n};$$

that is, the Jacobians of $F_0$ and $\Phi_0$ with respect to any two of the variables $x_i$ are all zero. This is the case when the values of $x$ satisfy (80), that is, $\partial F_0/\partial x_i$ are all commensurable. As there is an infinite number of such values of $x_i$ in any arbitrarily small domain, and as the Jacobian is continuous, it identically equals zero, that is, $\Phi_0$ would be a function of $F_0$. This contradicts our assumption. Hence, (73) has no uniform integral other than $F$ = constant.

In the proof we assumed that $B \neq 0$. We must discuss the case when one or many or an infinite number of $B$ are zero.

Corresponding to any set of integers $m_1, m_2, \ldots, m_n$ there is a coefficient $B$. A value of $B$ that corresponds to the value of $x$ satisfying (78) is called *secular*. When two sets of integers $m_1, m_2, \ldots, m_n$ and $m_1', m_2', \ldots, m_n'$ are related in the form

$$\frac{m_1}{m_1'} = \frac{m_2}{m_2'} = \cdots = \frac{m_n}{m_n'},$$

we say that the corresponding two coefficients $B$ belong to the same class. In the above proof we have assumed that none of the coefficients $B$ is zero even when it is secular. For the validity of the theorem it is sufficient to know that there should be, in each class, at least one $B$ that does not vanish even when it becomes secular.

Suppose that the value of $B$ corresponding to the set $m_1, m_2, \ldots, m_n$ is zero but that the value $B'$ of $B$ corresponding to $m_1', m_2', \ldots, m_n'$ is not zero. Choose the value of $x_i$ such that

$$\sum_{i=1}^{n} m_i \frac{\partial F_0}{\partial x_i} = 0;$$

then

$$\sum_{i=1}^{n} m_i' \frac{\partial F_0}{\partial x_i} = 0,$$

and hence

$$B \sum_{i=1}^{n} m_i \frac{\partial \Phi_0}{\partial x_i} = 0, \qquad B' \sum_{i=1}^{n} m_i' \frac{\partial \Phi_0}{\partial x_i} = 0,$$

simultaneously. We do not obtain

$$\sum_{i=1}^{n} m_i \frac{\partial \Phi_0}{\partial x_i} = 0$$

from the first equation, but we do obtain

$$\sum_{i=1}^{n} m_i' \frac{\partial \Phi_0}{\partial x_i} = 0$$

from the second equation. This leads to

$$\sum_{i=1}^{n} m_i \frac{\partial \Phi_0}{\partial x_i} = 0$$

and the proof is carried out similarly to the foregoing one.

The coefficient $B$ that becomes zero when it becomes secular is called *singular*; the coefficient $B$ that does not is called *ordinary*. Let $p_1, p_2 \ldots, p_n$ be given integers. The coefficient $B$ corresponding to the values of $m_1, m_2, \ldots, m_n$ that satisfy (81) is said to belong to the same family. Take $q$ sets of integers $m_{1i}, m_{2i}, \ldots, m_{ni}$ $(i = 1, 2, \ldots, q)$. If there is no set of $q$ integers $a_1, a_2, \ldots, a_q$ that are not zero all at once, satisfying

$$\sum_{i=1}^{q} a_i m_{ki} = 0 \qquad (k = 1, 2, \ldots, n),$$

then we say that the class of these $q$ integers is *independent.* When there are $n - 1$ independent ordinary classes, the family is said to be ordinary; otherwise, it is said to be singular. When there are $n - 2$ ordinary classes, the singularity is said to be of order 1; when there are $n - q - 1$ ordinary classes, the singularity is of order $q$. If there are any domains of $x$ that satisfy (80), then the family defined by the integers $p_1, p_2, \ldots, p_n$ is said to belong to the domain (D). We can prove that there does not exist any uniform integral other than $F$ in any domain ($\delta$) contained inside (D).

For the values of $x$ corresponding to the ordinary families the proof is quite similar. The Jacobians of $F_0$ and $\Phi_0$ with respect to two of the variables $x_i$ become zero an infinite number of times in all domains ($\delta$) contained in (D). Hence, they are identically zero.

We can show that, if there are an infinite number of singular classes of order $q$ in any region ($\delta$) contained in (D), then there are at most $q + 1$ independent uniform integrals, including $F$.

In fact, assume that there were $q + 2$ such integrals. Let them be

$$F,\ \Phi^{(1)},\ \Phi^{(2)}, \ldots,\ \Phi^{(q+1)}$$

and suppose that they become

$$F_0,\ \Phi_0^{(1)},\ \Phi_0^{(2)}, \ldots,\ \Phi_0^{(q+1)} \tag{82}$$

for $\mu = 0$. Choose the set of values of $x$ corresponding to the singular family of order $q$. There are $p = n - q - 1$ ordinary classes belonging to this family. Let the corresponding set of integers be

$$m_{1k}, m_{2k}, \ldots, m_{nk} \qquad (k = 1, 2, \ldots, p).$$

Then, for these values of $x_i$,

$$\sum_{i=1}^{n} m_{ik} \frac{\partial F_0}{\partial x_i} = \sum_{i=1}^{n} m_{ik} \frac{\partial \Phi_0^{(h)}}{\partial x_i} = 0$$

$$(k = 1, 2, \ldots, p;\ h = 1, 2, \ldots, q + 1).$$

The Jacobians of $q + 2$ functions (82) with respect to any set of $q + 2$ variables $x_i$ are zero for the values of $x_i$ in question; indeed, this occurs an infinite number of times in any domain ($\delta$). Hence, these Jacobians would be zero, and the $q + 2$ integrals in question cannot be independent.

This consideration does not interest us, according to Poincaré, as such a case does not appear in any practical dynamical system. We shall nevertheless continue the discussion of such a case from time to time in later chapters.

## 6.16 CASE OF VANISHING HESSIAN

Consider the case when the Hessian of $F_0$ with respect to $x_1, x_2, \ldots, x_n$ is zero. The function $F_0$ depends only on $x_1, x_2$ and is independent of the remaining $x_3, x_4, \ldots, x_n$. In order to distinguish $x_1, x_2$ from other $x_i$, denote the variables $x_i$ that are not contained in $F_0$ by $z_1, z_2, \ldots, z_{n-2}$, and their canonically conjugate variables by $u_1, u_2, \ldots, u_{n-2}$.

By the first lemma of Section 6.15 we can suppose that, if there exists a uniform integral $\Phi$ independent of $F$, then $\Phi_0$ is not a function of $F_0$. In the relation

$$(F_0, \Phi_0) = \frac{\partial F_0}{\partial x_1}\frac{\partial \Phi_0}{\partial y_1} + \frac{\partial F_0}{\partial x_2}\frac{\partial \Phi_0}{\partial y_2} = 0$$

we put $\zeta = \exp[\sqrt{-1}(m_1 y_1 + \cdots + m_n y_n)]$ and write $\Phi_0 = \sum A\zeta$, where $A$ is a function of $x_1, x_2, z, u$. Then, from this relation, we have an identity

$$\sqrt{-1}\sum A\zeta\left[m_1\frac{\partial F_0}{\partial x_1} + m_2\frac{\partial F_0}{\partial x_2}\right] \equiv 0.$$

Hence

$$m_1\frac{\partial F_0}{\partial x_1} + m_2\frac{\partial F_0}{\partial x_2} = 0$$

cannot hold identically, unless both $m_1$ and $m_2$ are zero. Hence, we can show in a manner similar to that of Section 6.15 that $\Phi_0$ depends neither on $y_1$ nor on $y_2$. Put

$$F_1 = \sum B\zeta, \qquad \Phi_1 = \sum C\zeta,$$

or

$$F_1 = \sum B_{m_1 m_2}\exp[\sqrt{-1}(m_1 y_1 + m_2 y_2)]$$

in (76); then

$$-\sum B\zeta\left(\sum_{i=1}^{2} m_i\frac{\partial \Phi_0}{\partial x_i}\right) + \sum C\zeta\left(\sum_{i=1}^{2} m_i\frac{\partial F_0}{\partial x_i}\right)$$
$$+ \sum \zeta\sum_{j=1}^{n-2}\left(\frac{\partial B}{\partial z_j}\frac{\partial \Phi_0}{\partial u_j} - \frac{\partial B}{\partial u_j}\frac{\partial \Phi_0}{\partial z_j}\right) = 0$$

should be an identity. Thus we can equate the coefficient of any exponential $\zeta$ to zero. Choose $x_1$, $x_2$ such that

$$m_1 \frac{\partial F_0}{\partial x_1} + m_2 \frac{\partial F_0}{\partial x_2} = 0; \tag{83}$$

then

$$-B\left(m_1 \frac{\partial \Phi_0}{\partial x_1} + m_2 \frac{\partial \Phi_0}{\partial x_2}\right) + \sum_{j=1}^{n-2} \left(\frac{\partial B}{\partial z_j}\frac{\partial \Phi_0}{\partial u_j} - \frac{\partial B}{\partial u_j}\frac{\partial \Phi_0}{\partial z_j}\right) = 0. \tag{84}$$

The coefficients $B_{m_1 m_2}$, $B_{m_2' m_1'}$, such that $m_1 m_2' - m_2 m_1' = 0$, belong to the same class, and $B_{m_1 m_2}$ belongs to the class $m_1/m_2$. $B_{00}$ belongs to all classes.

Let $p$, $q$ be integers such that $m_1 : m_2 = p : q$; put

$$\zeta = \exp\,[\sqrt{-1}(py_1 + qy_2)], \qquad D_\lambda = B_{\lambda p, \lambda q}\zeta^\lambda,$$

$$-\zeta \mathrm{H} = p\frac{\partial \Phi_0}{\partial x_1} + q\frac{\partial \Phi_0}{\partial x_2}.$$

Choose the value of $x_i$ such that

$$p\frac{\partial \Phi_0}{\partial x_1} + q\frac{\partial \Phi_0}{\partial x_2} = 0; \tag{85}$$

then for any positive or negative integer $\lambda$, including 0,

$$\mathrm{H}\frac{\partial D_\lambda}{\partial \zeta} + \sum_{j=1}^{n-2} \left(\frac{\partial D_\lambda}{\partial z_j}\frac{\partial \Phi_0}{\partial u_j} - \frac{\partial D_\lambda}{\partial u_j}\frac{\partial \Phi_0}{\partial z_j}\right) = 0. \tag{86}$$

This relation holds in the following two cases. The first case occurs when

$$\mathrm{H} = 0, \qquad \frac{\partial \Phi_0}{\partial z_j} = 0, \qquad \frac{\partial \Phi_0}{\partial u_j} = 0 \qquad (j = 1, 2, \ldots, n-2).$$

In this case,

$$\frac{\partial F_0}{\partial x_1}\frac{\partial \Phi_0}{\partial x_2} - \frac{\partial F_0}{\partial x_2}\frac{\partial \Phi_0}{\partial x_1} = 0,$$

which contradicts our assumption. The second case occurs when the Jacobian of arbitrary $2n - 3$ of the expressions $D_\lambda$ with regard to $2n - 3$ variables $\zeta$, $z_j$, $u_j$ is zero. Thus only $2n - 4$ among the $2n - 3$ expressions $D_\lambda$ are independent; in other words, if we choose the constant values of $x_1$, $x_2$ so that they satisfy (85), then

$$B^{\lambda'}_{\lambda p, \lambda q} B^{-\lambda}_{\lambda' p, \lambda' q} \tag{87}$$

depends only on $2n - 4$ variables, that is, on $z_j$ and $u_j$. If there were a uniform integral, then $2n - 5$ among these expressions (87) should be independent; that is, there should be a relation among these $2n - 4$ expressions (87).

Then what is the condition for the coexistence of the three independent uniform integrals $F =$ constant, $\Phi =$ constant, and $\Psi =$ constant? Denote their respective values for $\mu = 0$ by $F_0$, $\Phi_0$, and $\Psi_0$. As in the lemma in Section 6.15, we can consider that there is no relation among $F_0$, $\Phi_0$, and $\Psi_0$. Hence putting

$$-H'\zeta = p\frac{\partial \Psi_0}{\partial x_1} + q\frac{\partial \Psi_0}{\partial x_2},$$

we obtain

$$H'\frac{\partial D_\lambda}{\partial \zeta} + \sum_{j=1}^{n-2}\left(\frac{\partial D_\lambda}{\partial z_j}\frac{\partial \Psi_0}{\partial u_j} - \frac{\partial D_\lambda}{\partial u_j}\frac{\partial \Psi_0}{\partial z_j}\right) = 0. \tag{88}$$

Thus from (85) we can derive both of the relations (86) and (88). This circumstance can arise either when there is a relation among $F_0$, $\Phi_0$, $\Psi_0$ or when the Jacobian of arbitrary $2n - 3$ expressions among the expressions $D_\lambda$, together with all its first minor determinants, vanishes. As the first case contradicts our assumption, there should be two relations among arbitrary $2n - 3$ of $D_\lambda$, if $x_1$, $x_2$ is to satisfy (85); in other words, there are only $2n - 6$ independent expressions among (87). But the expressions (87) are formed out of the coefficients of the given function $F_1$. Hence, we can judge directly whether there are one or two relations among the expressions (87). In general, there is only one such relation. Hence, there is no uniform integral other than $F$.

When there is no relation among $2n - 4$ expressions (87), we say that they are ordinary; when there are $q$ relations, we say that they are singular of order $q - 1$. When there are values of $x_1$, $x_2$ that satisfy (85) in a domain of $x_1$, $x_2$, $z$, $u$, we say that the class $p/q$ belongs to the region ($\delta$). If there are an infinite number of ordinary classes in the region ($\delta$) of the domain (D) that include an infinite number of values of $x_1$, $x_2$, $z$, $u$, then there exists no integral (other than $F$) that is uniform with respect to $x$, $y$, $z$, $u$ and periodic with respect to $y_1$, $y_2$ for all values of $y_1$, $y_2$ and all values of $x_1$, $x_2$, $z$, $u$ in (D) so long as $\mu$ is sufficiently small. If there is an infinite number of singular classes of order $q$ in ($\delta$), then there are no more than $q + 1$ uniform independent integrals including $F$.

In the foregoing discussion we have assumed that the uniform integrals $\Phi$ in question are expanded in integral powers of $\mu$. But the theorem is true also for integrals $\Phi$ that are expanded in integral powers of $\sqrt{\mu}$. Suppose that $\Phi'$ and $\Phi''$ are integral power series of $\mu$; write

$$\Phi = \Phi' + \sqrt{\mu}\,\Phi''.$$

If $\Phi$ is an integral, then we have identically

$$(F, \Phi) = (F, \Phi') + \sqrt{\mu}\,(F, \Phi'') = 0.$$

As $(F, \Phi')$ and $(F, \Phi'')$ are both expanded in integral power series of $\mu$, we have separately $(F, \Phi') = 0$, $(F, \Phi'') = 0$. Thus, $\Phi'$ and $\Phi''$ are both integrals. Since we have proved that there exists no uniform integral that can be expanded in an integral power series of $\mu$, we see now that there exists no uniform integral that can be expanded in an integral power series of $\sqrt{\mu}$.

More generally, suppose that

$$\theta_1(\mu),\ \theta_2(\mu), \ldots,\ \theta_p(\mu)$$

are arbitrary functions of $\mu$ and that there exists no relation of the form

$$\phi_1\theta_1(\mu) + \phi_2\theta_2(\mu) + \cdots + \phi_p\theta_p(\mu) = 0 \tag{89}$$

among these functions, where the functions $\phi_1, \phi_2, \ldots, \phi_p$ are expanded in integral power series of $\mu$. Suppose that one of the functions $\phi_i$, say $\phi_1$, does not have $\mu$ as a factor. Then, solving (89), we obtain

$$\theta_1(\mu) = -\frac{\phi_2}{\phi_1}\,\theta_2(\mu) - \frac{\phi_3}{\phi_1}\,\theta_3(p) - \cdots - \frac{\phi_p}{\phi_1}\,\theta_p(\mu),$$

where $\phi_2/\phi_1, \ldots, \phi_p/\phi_1$ can be expanded in integral powers of $\mu$. Hence, we can eliminate one of the functions $\theta_1(\mu), \theta_2(\mu), \ldots, \theta_p(\mu)$ by substituting this expression for $\theta_1(\mu)$. Now we can suppose that there is no relation of the form (89) among the functions $\theta_p(\mu)$. Thus $\Phi$ is considered to be of the form

$$\begin{aligned}\Phi = {} & A_{01}\theta_1(\mu) + A_{11}\mu\theta_1(\mu) + A_{21}\mu^2\theta_1(\mu) + \cdots \\ & + A_{02}\theta_2(\mu) + A_{12}\mu\theta_2(\mu) + \cdots \\ & \vdots \\ & + A_{0p}\theta_p(\mu) + A_{1p}\mu\theta_p(\mu) + \cdots,\end{aligned} \tag{90}$$

where $A$ is a function independent of $\mu$ but a function of $x$ and $y$ only. Rewrite it in the form

$$\Phi = \Phi_1\theta_1(\mu) + \Phi_2\theta_2(\mu) + \cdots + \Phi_p\theta_p(\mu),$$

where $\Phi_1, \Phi_2, \ldots, \Phi_p$ are expanded in integral powers of $\mu$. If $\Phi$ is an integral, then

$$(F, \Phi) = \theta_1(\mu)(F, \Phi_1) + \theta_2(\mu)(F, \Phi_2) + \cdots + \theta_p(\mu)(F, \Phi_p) = 0 \tag{91}$$

should hold identically. Hence,

$$(F, \Phi_1) = (F, \Phi_2) = \cdots = (F, \Phi_p) = 0.$$

Otherwise $(F, \Phi_i)$ $(i = 1, 2, \ldots, p)$ would be expanded in integral powers of $\mu$ and (91) would acquire the form (89). This would contradict our assumption. Hence, $\Phi_1, \Phi_2, \ldots, \Phi_p$ are separately integrals of the problem. Since we have proved that there exists no uniform integral that can be expanded in an integral power series of $\mu$, we see that there exists no uniform integral that can be represented in the form (90).

This proof is valid even when $\Phi$ is expanded with respect to the base field of a denumerably infinite number of linearly independent functions $\theta_1(\mu), \theta_2(\mu), \ldots$, which can be expanded in integral powers of $\mu$.

The foregoing theorem of Poincaré deals with integrals that are uniform with respect to $y_1, \ldots, y_n, x_1, \ldots, x_n$, while Bruns's theorem deals with integrals that are algebraic with respect to $y_1, \ldots, y_n, x_1, \ldots, x_n$. Painlevé (1900) generalized the theorem of Poincaré to integrals that are uniform only with respect to $y_1, \ldots, y_n$ (Sections 6.7–6.11). When he generalized Bruns's theorem, Painlevé used the theorem that an integral which is rational in the momenta has no singularity other than poles for such values of $x_{3j-2}, x_{3j-1}, x_{3j}$ that do not satisfy

$$(x_{3j-2} - x_{3k-2})^2 + (x_{3j-1} - x_{3k-1})^2 + (x_{3j} - x_{3k})^2 = 0$$
$$(j, k = 1, 2, \ldots, n).$$

Corresponding to this is the theorem that a real uniform integral has no singularity other than those for the values satisfying $x_{3j-2} = x_{3k-2}$, $x_{3j-1} = x_{3k-1}$, $x_{3j} = x_{3k}$ $(j, k = 1, 2, \ldots, n)$ in the real domain of $x_{3j-2}, x_{3j-1}, x_{3j}$. Painlevé used this latter theorem in his generalization of the theorem of Poincaré. Then he proved that there exists no integral that is real and uniform in the momenta other than the classical integrals, and thus completed the generalization of Poincaré's theorem.

## 6.17 APPLICATION TO THE $n$-BODY PROBLEM

In the restricted problem of three bodies, Poincaré (1889, 1892) established the basis of his theory of the periodic solutions and the characteristic exponents which will be discussed in later chapters, and

## 6.18 DISCUSSION OF POINCARÉ'S THEOREM

We have given in Section 6.12 an exceptional case where there exists an algebraic integral in the restricted three-body problem. When two of the three bodies in the three-body problem collide at a finite distance at the end of a finite interval of time, the differential equations cease to be regular, as will be discussed later. Otherwise, the coordinates of the three bodies can be determined uniquely by Cauchy's theorem on the existence of the solution of a differential equation, and the solution can be expanded in integral powers of $t - t_0$ around an arbitrary initial condition for $t = t_0$ (Painlevé, 1897). There are two independent conditions for such a collision, which reduce to one in the case of the plane problem. The conditions for a collision in the two-body problem are algebraic, but the conditions for the three- and $n$-body problems are transcendental according to Painlevé (1897, 1900, 1900a, 1900b). Suppose that a collision occurs when

$$\chi(t, x_1, \ldots, x_n, y_1, \ldots, y_n) = 0 \tag{92}$$

is satisfied. Since this is an invariantive relation of the motion, it should be an algebraic combination of the known classical integrals, according to Bruns's theorem. But the conditions for a collision should be satisfied when we increase $t$ by a constant amount; they should be unaltered when we increase the coordinates and the momentum by a constant amount.

We limit ourselves to the consideration of the plane three-body problem. Let the total kinetic energy of the whole three-body system around the center of mass be $T$, the total angular momentum $W$, and the force function $U$; then $T - U$ and $W$ remain unaltered by the above change in coordinates and momentum. Hence, the condition for a collision does not depend on the coordinates and the momentum of the center of mass but should be a relation between $W$ and $T - U$. The equations of motion remain unaltered by changing $t$ to $\alpha^3 t$, $x_i$ to $\alpha^2 x_i$, and $y_i$ to $\alpha^{-1} y_i$ with a constant $\alpha$; hence, the condition for a collision also remains unaltered by this transformation — in which $T - U$ is multiplied by $\alpha^{-2}$ and $W$ is multiplied by $\alpha$. Let the condition be

$$W = \phi(T - U);$$

then,

$$\alpha W = \phi[\alpha^{-2}(T - U)] = \alpha\phi(T - U).$$

Hence, the condition should be of the form

$$\phi(T - U) = \lambda/\sqrt{T - U},$$

or

$$W^2(T - U) = \lambda^2$$

with a constant $\lambda$. Whatever the value of $\lambda$, this corresponds to a motion of hyperbolic type; that is, it is the case when the three bodies recede to infinite distances from each other as $t$ approaches $\pm\infty$. Hence, it cannot be the condition for a collision. The condition should be transcendental. This is the argument of Painlevé, (1897, 1897a). On the other hand, $W$ should be finite even at the time of a collision, as has been proved by Sundman (1907, 1909, 1913) and as we shall see in a later chapter. But, since the momentum becomes infinite at the time of a collision, this relation is contradictory. Hence, according to Bruns's theorem, the condition for a collision cannot be algebraic; it should be represented by transcendental functions at least with respect to the momentum.

It was Levi-Civita and Bisconcini who obtained the condition for a collision explicitly (Levi-Civita, 1903, 1906; Bisconcini, 1906; see also Kiveliovitch, 1926; Rosenblatt, 1928; Chazy, 1932). According to these authors there exists (as will be shown in Chapter 22) a uniform integral in the case of a collision, but it is neither for $t$ nor for the Keplerian elements — with regard to which the integral is uniform. The theorem of Poincaré deals with the Keplerian elements; hence, it expresses the uniformity of the integral in the neighborhood of those values of $x_i$, $y_i$ that represent the state of motion belonging to the same osculating elliptic elements. As has been remarked by Levi-Civita, the theorem of Poincaré does not deny the existence of uniform integrals for a *part* of the trajectory. It does not deny the existence, in the neighborhood, of a state of motion very different from an elliptic orbit (Section 14.16).

In the case of a simultaneous collision of all three bodies at a point, the total angular momentum of the system must be zero, as proved by Sundman (1909). This is an algebraic invariantive relation. The above discussion can be applied to the $n$-body problem, as we will demonstrate in Chapter 22.

Consider the state of motion for $t \to \pm\infty$. When two of the three mutual distances of the three bodies become infinite of order 1 with regard to $t$ and one of the mutual distances is bounded, the motion is said to be elliptic-hyperbolic (Chazy, 1922). When two of the three bodies move in orbits of elliptic type around the center of mass and the third body moves in an orbit of parabolic type around the center of mass, that is, when two of the mutual distances becomes infinite of order $\frac{2}{3}$ and the remaining one is bounded, we say that the motion is elliptic-parabolic. When all three of the mutual distances are bounded, we say that the motion is oscillatory. In these three cases the energy constant $h$ of the whole system is negative, and the region $h < 0$ in the 16-dimensional space is divided into four subspaces. Three of these four subspaces contain the region $h = 0$ in their interiors. Each orbit in these subspaces is elliptic-hyperbolic; two of the mutual distances become infinite of order 1 with regard to $t$ as $t \to \pm\infty$; the third mutual distance, which remains the same all the time, is bounded. The three mutual distances are either bounded or oscillatory inside the subspace bounded by three analytic surfaces.

When the three mutual distances become infinite, of order $\frac{2}{3}$, 1, 1, respectively, as $t$ increases indefinitely, the motion is said to be parabolic-hyperbolic. When all three mutual distances become infinite of order 1, the motion is said to be hyperbolic. When the three bodies collide at infinity, the order of infinity is $\frac{2}{3}$ and the motion is said to be parabolic. The configuration of the three bodies in the immediate neighborhood of a simultaneous collision is that of either the equilateral triangular solution of Lagrange or the straight-line solution of Euler, as has been remarked in Chapter 3. In the first two cases the point at infinity $t = \infty$ is a transcendental singular point, but in the case of a collision it is an algebraic singularity of order two. These are the cases when the total energy constant $h$ of the whole system is positive.

The trajectories should be inside a subspace of eight dimensions embedded in the twelve-dimensional space, but in the case of a collision they should be inside a six-dimensional subspace. Two conditions are necessary for a collision and they have been shown (Chazy, 1918, 1922, 1929, 1932, 1933) to be transcendental with regard to $y_1, y_2, y_3, y_4, y_5, y_6$. The region in which the trajectories of the three bodies lie is divided into five regions, except for a simultaneous collision of the three bodies. The exterior region corresponds to hyperbolic motions, three of the five regions to elliptic-hyperbolic motions, and the remaining fifth either to oscillatory motions for which the three mutual distances are all bounded or to motions for which two of the mutual distances are arbitrarily large while the third is bounded. There are motions with elliptic osculating orbits in the fifth region. As $t$ increases indefinitely, each of the functions $y_1, y_2, y_3, y_4, y_5, y_6$ tends to a finite limit. The limits are uniform functions of the values of the twelve variables $x_i, y_i$ $(i = 1, 2, \ldots, 6)$ for a certain epoch. Hence, the squares of the sum and the difference of the limiting values of each of these six $y_i$ are two uniform integrals in the domain of hyperbolic motions; each of those integrals represents a family of surfaces, and one of them passes through an arbitrary point in that domain. In the domain of elliptic-hyperbolic motions we denote by $y_4, y_5, y_6$ the momentum of the third body (receding to infinity) relative to the center of mass of the two bodies remaining in a bounded finite domain, and define the elliptic elements by six out of the twelve osculating elements and the hyperbolic elements by the remaining six; then these elements tend to finite limits in each of the trajectories, and each of them defines a uniform integral in the corresponding domain. Those uniform integrals that can exist simultaneously in both of these two domains define a family of surfaces, only one of which can pass through an arbitrary point in the whole domain. Among these surfaces, the surface bounding one of the five domains is a particular one. It can be shown that the integrals corresponding to this bounding surface cannot be reduced to the energy integral or the angular-momentum integrals (see Chapter 22).

As will be discussed later in detail, this theorem of Poincaré deals with uniform integrals with respect to the twelve elements belonging to the two osculating elliptic motions. Such integrals are uniform in the domain of another two-dimensional space and do not enter in this theorem of Poincaré.

The theorem of Bruns, on the other hand, differs from the theorem of Poincaré and can be applied to such integrals. One such integral is neither an algebraic combination of the classical integrals, nor an algebraic combination of $y_i$ only, as in the extension of the theorem by Painlevé. The bounding surfaces of the five domains of these trajectories are not algebraic surfaces.

Painlevé (1900, 1900a, 1900b) studied the cases in which the solution can be represented by meromorphic functions of $t$ with polar singularities. He has given algebraic differential equations that admit uniform integrals, besides the known functions, such as exponential, elliptic, Abelian, and uniform integrals of a linear differential equation including Fuchsian and automorphic functions. The critical points are called *fixed* if they do not depend on the integration constants; otherwise they are called *movable*. Painlevé, extending the works of Poincaré and Picard, examined differential equations that admit the general integrals only with fixed critical points. He considered differential equations of the first, second, and third order. In the case of the third order he found a new class of singularities, for example, those with a perfect set of discontinuities (see Chapter 22).

## 6.19 INTEGRALS WITHOUT PARAMETERS

Although the theorem of Poincaré looks as if it were more general than the theorem of Bruns, it has a narrower validity domain; it is valid only for vanishingly small values of $\mu$. However, it is wider than Bruns's in the sense that it is not limited to algebraic integrals. The theorem of Poincaré cannot be valid if we take off the restriction that the integrals should be arranged in integral powers of $\mu$ or expanded in the sequence of base functions of an algebraic field of $\mu$. It may happen to be false when the masses have particular values.

We suppose that in the system of differential equations

$$\frac{d\xi_i}{dt} = \Xi_i \qquad (i = 1, 2, \ldots, m), \tag{93}$$

the right-hand members $\Xi_1, \Xi_2, \ldots, \Xi_m$ do not contain $t$ explicitly, are expanded in integral powers of $\xi_1, \xi_2, \ldots, \xi_m$, and are not all zero for $\xi_1 = \xi_2 = \cdots = \xi_m = 0$. Then there may exist $m - 1$ independent integrals that do not contain $t$ explicitly and can be expanded in convergent integral power series of $\xi_1, \xi_2, \ldots, \xi_m$ (Cherry, 1924, 1924a, 1926, 1927, 1928; Woltjer, 1924; Fermi, 1923).

Let the values of $\Xi_1, \Xi_2, \ldots, \Xi_m$ for

$$\xi_1 = \xi_2 = \cdots = \xi_m = 0$$

be, respectively, $a_1, a_2, \ldots, a_m$ and at least one of them, for example $a_1$, be different from zero. Replacing $\xi_i$ by

$$\xi_i' = \xi_i - \frac{a_i}{a_1}\,\xi_1 \qquad (i = 2, 3, \ldots, m),$$

we obtain

$$\frac{d\xi_i'}{dt} = \Xi_i - \frac{a_i}{a_1}\,\Xi_1 = \Xi_i' \qquad (i = 2, 3, \ldots, m).$$

As

$$\Xi_i' = 0 \qquad (i = 2, 3, \ldots, m) \quad \text{for} \quad \xi_2' = \xi_3' = \cdots = \xi_m' = 0$$

and

$$\frac{\partial(\xi_1, \xi_2', \xi_3', \ldots, \xi_m')}{\partial(\xi_1, \xi_2, \xi_3, \ldots, \xi_m)} = 1,$$

(93) can be written

$$\frac{d\xi_1}{dt} = \lambda + \sum_{j=1}^{m} a_j^{(1)}\xi_j + \frac{1}{2}\,a_{11}^{(1)}\xi_1^2 + a_{12}^{(1)}\xi_1\xi_2 + \cdots,$$

$$\frac{d\xi_i}{dt} = \sum_{j=1}^{m} a_j^{(i)}\xi_j + \frac{1}{2}\,a_{11}^{(i)}\xi_1^2 + a_{12}^{(i)}\xi_1\xi_2 + \cdots \qquad (i = 2, 3, \ldots, m). \tag{93a}$$

We shall solve these equations in ascending power series of

$$\xi_1, \xi_2, \ldots, \xi_m$$

of the form

$$\xi_i = \xi_i^{(1)} + \xi_i^{(2)} + \cdots \qquad (i = 1, 2, \ldots, m). \tag{94}$$

Substituting (94) in (93a) and equating the terms of the same degree, we first obtain from

$$\frac{d\xi_1^{(1)}}{dt} = \lambda, \qquad \frac{d\xi_j^{(1)}}{dt} = 0 \qquad (j = 2, 3, \ldots, m)$$

the solution

$$\xi_1^{(1)} = \lambda(t + c_1), \qquad \xi_j^{(1)} = c_j \qquad (j = 2, 3, \ldots, m). \tag{95}$$

Substituting these in the second-degree terms

$$\frac{d\xi_i^{(2)}}{dt} = \sum_{j=1}^{m} a_j^{(i)} \xi_j^{(1)} \qquad (j = 1, 2, \ldots, m),$$

and solving

$$\frac{d\xi_i^{(2)}}{dt} = a_1^{(i)}\lambda(t + c_1) + \sum_{j=2}^{m} a_j^{(i)} c_j \qquad (i = 1, 2, \ldots, m),$$

we obtain

$$\xi_i^{(2)} = \tfrac{1}{2} a_1^{(i)}\lambda(t + c_1)^2 - \sum_{j=2}^{m} a_j^{(1)} c_j (t + c_1) \qquad (i = 1, 2, \ldots, m). \tag{96}$$

The integration constants are determined so that $\xi_i^{(2)}$ is homogeneous of the second degree in $t + c_1, c_2, \ldots, c_m$, and vanishes together with $t + c_1$. By repeating this procedure we can determine $\xi_i^{(s)}$, $(s = 2, 3, \ldots)$, so as to make them vanish together with $t + c_1$. If we substitute these in (94), we obtain the solution in the form of an integral function of $t + c_1, c_2, \ldots, c_m$ satisfying (93a).

Suppose that $\mathrm{H}_1, \mathrm{H}_2, \ldots, \mathrm{H}_m$ are the majorant functions, respectively, of $\Xi_1, \Xi_2, \ldots, \Xi_m$ in $\eta_1, \eta_2, \ldots, \eta_m$; that is, the coefficient in the expansion of $\mathrm{H}_k$ with regard to $\eta_1, \eta_2, \ldots, \eta_m$ is larger than the corresponding coefficient in the expansion of $\Xi_k$ with regard to $\xi_1, \xi_2, \ldots, \xi_m$ in the absolute values within the domain of $\xi_i$, $\eta_i$ in question. Consider

$$\begin{aligned} \frac{d\eta_1}{dt} &= \mathrm{H}_1 = \mu + \sum_{j=1}^{m} b_j^{(1)}\eta_j + \frac{1}{2} b_{11}^{(1)}\eta_1^2 + b_{12}^{(1)}\eta_1\eta_2 + \cdots, \\ \frac{d\eta_i}{dt} &= \mathrm{H}_i = \sum_{j=1}^{m} b_j^{(i)}\eta_j + \frac{1}{2} b_{11}^{(i)}\eta_1^2 + b_{12}^{(i)}\eta_1\eta_2 + \cdots \\ & \qquad (i = 2, 3, \ldots, m), \end{aligned} \tag{97}$$

and try to expand the solution, similarly to (94), in the form

$$\eta_i = \eta_i^{(1)} + \eta_i^{(2)} + \cdots \qquad (i = 1, 2, \ldots, m). \tag{98}$$

Solving these equations in the same manner as we did for (93), we obtain

$$\eta_1^{(1)} = \mu(t + d_1), \qquad y_j^{(1)} = d_j \qquad (j = 2, 3, \ldots, m), \tag{99}$$

$$\eta_i^{(2)} = \tfrac{1}{2}b_1^{(i)}\mu(t + d_1)^2 + b_2^{(i)}d_2(t + d_1) + \cdots + b_m^{(i)}d_m(t + d_1)$$
$$(i = 2, 3, \cdots, m). \tag{100}$$

Choose the integration constants $t + d_1, d_2, \ldots, d_m$ so that they are positive, as are $t + c_1, c_2, \ldots, c_m$, and satisfy

$$|t + c_1| < t + d_1, \qquad |c_2| < d_2, \ldots, |c_m| < d_m.$$

Comparing (95) with (99), we have

$$|\xi_1^{(1)}| < \eta_1^{(1)}, \ldots, |\xi_m^{(1)}| < \eta_m^{(1)}.$$

Comparing (96) with (100), we have

$$|\xi_1^{(2)}| < \eta_1^{(2)}, \ldots, |\xi_m^{(2)}| < \eta_m^{(2)}.$$

In general, $\eta_i^{(s)}$ is larger than the corresponding $\xi_i^{(s)}$ in absolute value. Hence, if $t + d_1, d_2, \ldots, d_m$ are chosen so that (98) is convergent, then the series (94) is absolutely and uniformly convergent.

Now let $M$ and $\alpha$ be sufficiently large positive constants, and the expansion of

$$\frac{M(\eta_1 + \eta_2 + \cdots + \eta_m)}{1 - \alpha(\eta_1 + \eta_2 + \cdots + \eta_m)}$$

be the functions $\mathrm{H}_1 - \mu, \mathrm{H}_2, \ldots, \mathrm{H}_m$ of (97) with $\mu > |\lambda|$; then, writing $\eta_1 + \eta_2 + \cdots + \eta_m = \zeta$, we obtain

$$\frac{d(\eta_1 - \eta_j)}{dt} = \mu \qquad (j = 2, 3, \ldots, m),$$

$$\frac{d\zeta}{dt} = \frac{\mu + \zeta(mM - \alpha\mu)}{1 - \alpha\zeta}.$$

We can choose $M$ and $\alpha$ so that they satisfy $mM - \alpha\mu = 0$ and are both sufficiently large positive constants. Then the solution of these equations are

$$\zeta - \tfrac{1}{2}\alpha\zeta^2 = \mu t + \text{constant},$$

$$\eta_1 - \eta_j = \mu t + \text{constant} \qquad (j = 2, 3, \ldots, m).$$

Determine the constants so that $\eta_1 = 0, \eta_2 = d_2, \ldots, \eta_m = d_m$ for $t + d_1 = 0$; then

$$\zeta - \tfrac{1}{2}\zeta^2 = \mu(t + d_1) + d - \tfrac{1}{2}\alpha d^2 \qquad (d = d_2 + \cdots + d_m)$$

$$\eta_1 - \eta_j = \mu(t + d_1) - d_j \qquad (j = 2, 3, \ldots, m). \tag{101}$$

Hence, the value of $\zeta$ equal to $d$ for $t + d_1 = 0$ is

$$\zeta = \frac{1}{\alpha}\,[1 - \sqrt{1 - 2\alpha\mu(t + d_1) - 2\alpha d + \alpha^2 d^2}].$$

This function can be expanded in integral powers of $t + d_1, d_2, \ldots, d_m$ which is convergent for sufficiently small values of

$$|t + d_1|, |d_2|, \ldots, |d_m|.$$

Thus $\eta_1, \eta_2, \ldots, \eta_m$ can be expanded in integral power series of the same nature by (101) and are of the same form as (98), which is absolutely and uniformly convergent. Hence, (94) is absolutely and uniformly convergent so long as $|t + c_1|, |c_2|, \ldots, |c_m|$ are sufficiently small.

Thus, we have obtained the solution of (93a) in the form (94) so that it is zero for $t + c_1 = c_2 = \cdots = c_m = 0$ and

$$\frac{\partial(\xi_1, \xi_2, \ldots, \xi_m)}{\partial(c_1, c_2, \ldots, c_m)} = \lambda.$$

Hence, $t + c_1, c_2, \ldots, c_m$ are convergent as integral power series of $\xi_1, \xi_2, \ldots, \xi_m$ insofar as the latter variables are sufficiently small. Hence

$$\begin{aligned}
\lambda(t + c_1) &= \xi_1 - \frac{\xi_1}{\lambda}\,(\tfrac{1}{2}a_1^{(1)}\xi_1 + a_2^{(1)}\xi_2 + \cdots + a_m^{(1)}\xi_m) + \cdots \\
&= \Phi_1(\xi_1, \xi_2, \ldots, \xi_m), \\
c_j &= \xi_j - \frac{\xi_1}{\lambda}\,(\tfrac{1}{2}a_1^{(j)}\xi_1 + a_2^{(j)}\xi_2 + \cdots + a_m^{(j)}\xi_m) + \cdots \\
&= \Phi_j(\xi_1, \xi_2, \ldots, \xi_m) \qquad (j = 2, 3, \ldots, m). \qquad (102)
\end{aligned}$$

Each term of the second and the higher degrees has $\xi_1$ as a factor. Since $c_1, c_2, \ldots, c_m$ are arbitrary constants, $\Phi_2, \ldots, \Phi_m$ are $m - 1$ independent integrals. Thus, we have obtained $m - 1$ independent integrals of (93).

Next we consider a set of Hamiltonian canonical equations with $m = 2n$

$$\frac{d\xi_i}{dt} = \frac{\partial F}{\partial \eta_i}, \qquad \frac{d\eta_i}{dt} = -\frac{\partial F}{\partial \xi_i} \qquad (i = 1, 2, \ldots, n), \qquad (103)$$

$$F = F_0 + F_1 + \cdots$$

If we apply Poincaré's contact transformation

$$\xi_i = \sqrt{2x_i}\cos y_i, \qquad \eta_i = \sqrt{2x_i}\sin y_i \qquad (i = 1, 2, \ldots, n),$$

then the differential equations are transformed to the same form as equations (73) in the theorem of Poincaré but without the parameter $\mu$.

In order that $F_0$ shall not contain variables other than $x_1, x_2, \ldots, x_n$ as in the form of Poincaré, $F_0$ must be a power series of

$$\xi_1^2 + \eta_1^2, \ldots, \xi_n^2 + \eta_n^2,$$

and the derivatives of $F_0$ must not be zero; furthermore, the coefficients $B_{m_1 \cdots m_n}$ in the expansion of $F_1$ must not have a small divisor $\sum_{i=1}^n m_i$ $\partial F_0/\partial x_i$. These conditions are satisfied by

$$F_0 = \lambda_1(\xi_1^2 + \eta_1^2) + \cdots + \lambda_n(\xi_n^2 + \eta_n^2) + \nu_1(\xi_1^2 + \eta_1^2)^2 + \cdots + \nu_n(\xi_n^2 + \eta_n^2)^2.$$

Indeed, if we take such an $F_0$, then there exist $2n - 2$ independent integrals as given by (102) besides $F =$ constant. Hence, there exist $2n - 2$ independent integrals without the parameter $\mu$ for the equations of Poincaré. Thus the theorem of Poincaré fails when we consider integrals that do not contain the small parameter $\mu$.

Whittaker (1916, 1937) considered the differential equations of form (103) for a dynamical system of two degrees of freedom, in which the Hamiltonian function $F$ is expanded in integral powers of $\sqrt{\xi_1}$, $\sqrt{\xi_2}$ and in a trigonometric series in $\eta_1, \eta_2$ of the form

$$\sum A\xi_1^{s_1/2}\xi_2^{s_2/2} \cos (j_1\eta_1 + j_2\eta_2),$$

with the first term $s_1\xi_1 + s_2\xi_2$, where $A$ is a constant, $s_1$, $s_2$ are positive integers, and $j_1$, $j_2$ are positive or negative integers including zero, $s_1 + s_2$ being the degree of the term. Thus $F$ is written

$$F = s_1\xi_1 + s_2\xi_2 + F_3 + F_4 + \cdots.$$

Whittaker discovered the so-called *adelphic integral* for this system of differential equations (discussed in Section 14.15). The integral takes three different forms according as: (i) $s_1/s_2$ is irrational; (ii), $s_1/s_2$ is rational and equals the ratio of two integers $m/n$, and there is no term of the form $\cos (n\eta_1 - m\eta_2)$ in $F_3$; or (iii), $s_1/s_2$ equals the ratio of two integers $m/n$ but there is a term $\cos (n\eta_1 - m\eta_2)$.

Fermi (1923, 1923a) tried to relate the theorem of Poincaré to the quasi continuity of the trajectories and to base on it the foundation of the kinetic theory of gases of Boltzmann and Maxwell. Maxwell based his theory on the assumption of the so-called continuity of path of the trajectories, the assumption that a dynamical system passes through all phases permitted by the law of energy conservation in a sufficiently long interval of time (Jeans, 1921). The modern ergodic theorem has been refined from this assumption of quasi continuity of the dynamical trajectories (Hagihara, 1933; discussed in Chapter 26).

Fermi, by adopting the above proof of the theorem of Poincaré as it stands, proved the validity of quasi ergodicity for a nondegenerate dynamical system of two degrees of freedom without perturbation. A system is nondegenerate if: (i), from the relation

$$\frac{\partial F_0}{\partial x_1}\Big/\frac{\partial F_0}{\partial x_2} = \text{constant}$$

the relation

$$\frac{\partial F_0}{\partial x_3}\Big/\frac{\partial F_0}{\partial x_4} = \text{constant},$$

for example, does not follow; (ii), if $\sum_i m_i(\partial F_0/\partial x_i)$ does not vanish on any surface $F_0 = \text{constant}$, where $m_i$ are integers, not all of them zero; and (iii), the vanishing of the coefficient $B_{m_1 \cdots m_n}$ of Section 6.15 does not follow from $\sum_i m_i(\partial F_0/\partial x_i) = 0$.

Born and Pauli (1922) expanded the perturbation in a quantum-mechanical system in the present form with the angular and linear variables of Section 2.7 as the canonically conjugate variables. They tried to express the perturbation in a multiple Fourier series of the angular variables by expanding it with a small parameter. The perturbation in this case is the imposed electric field and the magnetic field. Born and Brody (1921) based the discussion on a different form of expansion. Epstein (1922) used the expansion of Delaunay (Chapter 9). Since according to the theorem of Poincaré there exists no uniform integral, it is possible to give the quantum condition to a quantum-mechanical system with perturbations. Owing to the fact that such series expansions are not uniformly convergent (as will be shown in Chapter 15), the quantum-mechanical calculation presents a grave and fundamental difficulty. According to Born and Pauli the broadening of spectral lines of a perturbed system due to electromagnetic forces is negligible compared with the broadenings due to pressure and the Doppler effect. It is the conditionally periodic motion discussed in Chapter 2 that enables us to escape from this difficulty, but, as we have argued in Section 2.7, it is necessary to have a heavy and severe condition for that purpose (See Section 4.24; also Arnold, 1963). Whether such a condition is actually satisfied in any quantum-mechanical system, whether we should adopt such a condition as an axiom in quantum mechanics, is a question for quantum mechanicians. Here appears the important role that should be played by the theory of adiabatic invariants discussed in Section 4.23. However, we have there assumed the quasi ergodicity of the dynamical system; so the question is raised again. On the other hand, a quantum-mechanical system without perturbation is degenerate in general; so the same difficulty appears. It was presumably for this reason that quantum mechanicians in the past assumed the general invalidity of classical dynamics in quantum theory.

Arnold (1963) recently solved the problem by means of quasi-periodic functions, along with the difficulty of the degenerate cases, making the severe assumption that the series representing the functions are almost everywhere convergent — by excluding the domain of Lebesgue measure zero, as will be discussed in a later chapter.

In the light of modern developments in wave mechanics and quantum mechanics, especially the theory based on Bohr's complementarity principle, our mode of thinking has changed. However, even now it is evident that only a first approximation is treated in the computation of the perturbation; indeed, the method for the second approximation fails completely owing to the nonconvergence of the process. We are in quantum mechanics confronted with the same grave difficulty of nonconvergence that makes its first appearance in classical dynamics.

## KOWALEWSKI'S TOP MOTION

### 6.20 RIGID-BODY ROTATION

Despite the theorems of Poincaré and Bruns, Kowalewski (1888, 1890) has found an algebraic integral other than the classical integrals in the motion of a rigid body. We shall discuss this case briefly.

The motion of a rigid body suspended at its center of mass has been completely solved by Euler. Denote the three moments of inertia around the center of mass by $A$, $B$, $C$. The motion under gravity of a rigid body, such that $A = B$, suspended at a point on the principal axis of inertia corresponding to $C$, has been completely solved by Lagrange. The motion in the case $A = B = 2C$ has also been solved completely by using elliptic functions and hyperelliptic functions, no matter where the center of mass is located. This is called the *Kowalewski case*. In such cases there exists an integral algebraic in the coordinates and the momentum, which is not an algebraic combination of the classical integrals (energy and angular-momentum). Poincaré has proved that there cannot exist any uniform integral, which is not an algebraic combination of the classical integrals, for a general initial condition, if we exclude the above cases of Euler, Lagrange, and Kowalewski. Several authors have discussed the case of $A = B = 2C/n$, $n$ being an integer (Husson, 1905, 1906, 1908, 1932; Liouville, 1897; Hesse, 1890; Kötter, 1892, 1893; Staude, 1894; Nekrassoff, 1896; Kolossoff, 1902, 1903; Stäckel, 1908; Oseen, 1911; Marcolongo, 1902, 1906, 1910; Levi-Civita, 1901; Lazzarino, 1919; see also Klein and Sommerfeld, 1921; Halphen, 1888; Leimanis, 1965).

The quantum-mechanical analogues of these top motions have been studied by Casimir (1931) in general, by Kramers and Ittmann (1929, 1930) and O. Klein (1929) for Euler's case, by Reiche and Rademacher (1926, 1927) and Kronig and Rabi (1927) for Lagrange's case, and by Laporte (1933) for Kowalewski's case.

We shall first show, following Poincaré (1892), that there can exist a uniform integral in the case $A = B$.

Suppose that the product of the distance between the point of suspension and the center of mass multiplied by the gravitational force is infinitesimally small and write the equations of motion of the rigid body in the form

$$\frac{dx_i}{dt} = \frac{\partial F}{\partial y_i}, \qquad \frac{dy_i}{dt} = -\frac{\partial F}{\partial x_i} \qquad (i = 1, 2, 3),$$

$$F = F_0 + \mu F_1,$$

where $F$ is the total energy, $F_0$ the kinetic energy, $\mu F_1$ is the product of the distance from the center of mass to the horizontal plane passing through the point of suspension, multiplied by the gravitational force. If $\mu = 0$, then the center of mass coincides with the point of suspension and the motion of the rigid body is that of Poinsot. Since $\mu$ is supposed to be small, the motion is the Poinsot motion, to the first approximation. In the Poinsot motion the ellipsoid of inertia makes a rolling motion by always touching a fixed plane, without slipping. Let $P$ be the foot of the perpendicular from the point of suspension to this fixed plane, and $Q$ be the point of contact at a certain epoch. This point of contact describes a fixed curve upon the ellipsoid of inertia. This is called the polhode. Suppose that after a lapse of time $\tau$ the same point $Q$ of the polhode comes to a point $Q'$ at which the ellipsoid of inertia touches the fixed plane. Denote the angle $QPQ'$ by $\alpha$ and put $n = 2\pi/\tau$, $n' = \alpha/\tau$; $n$ and $n'$ correspond to the mean motions in the Keplerian elliptic motion. Denote the coordinates of a point of the rigid body by $x$, $y$, $z$; let the point of suspension be the coordinate origin, and take the $z$ axis as the vertical. With integration constants $\epsilon$, $\epsilon'$, write $l = nt + \epsilon$, $l' = n't + \epsilon'$. Let $\xi$, $\eta$, $\psi$ be functions of $n$, $n'$, $l$ and periodic with period $2\pi$ in $l$. With the new integration constants $\theta$, $\phi$,

$$\begin{aligned} x &= \cos\theta(\xi\cos l' - \eta\sin l') - \sin\theta\cos\phi(\xi\sin l' + \eta\cos l') \\ &\quad + \psi\sin\theta\sin\phi, \\ y &= \sin\theta(\xi\cos l' - \eta\sin l') + \cos\theta\cos\phi(\xi\sin l' + \eta\cos l') \\ &\quad - \psi\cos\theta\sin\phi, \\ z &= \sin\phi(\xi\sin l' + \eta\cos l') + \psi\cos\phi. \end{aligned}$$

Let $(x, y, z)$ be the coordinates of the center of mass; then $F_1$ is reduced to $z$, apart from a constant factor.

$$F_1 = \sum B_{m,1} \exp [\sqrt{-1}\ (ml + l')] + \sum B_{m,0} \exp [\sqrt{-1}\ ml]$$
$$+ \sum B_{m,-1} \exp [\sqrt{-1}\ (ml - l')],$$

where the coefficients $B$ depend on $n$, $n'$, $\phi$. Refer to the general discussion of Section 6.15. Choose $n$, $n'$ such that

$$p \frac{\partial F_0}{\partial x} + q \frac{\partial F_0}{\partial y} = 0;$$

then $B$ depends only on $\phi$. Hence, there must be a relation between two of the coefficients. Since $D_\lambda$ defined in Section 6.15 depends only on $\phi$ and $\zeta$, there must be one relation among arbitrary $2n - 3 = 3$ of these $D_\lambda$, where we put $D_\lambda = B_{\lambda p, \lambda q} \zeta^\lambda$. The series described in Section 6.15 are all singular of order 1; hence, there may exist a uniform integral distinct from the energy integral. Actually, there is the angular-momentum integral. In order to see whether there exists a third integral we look for the classes that are singular of order 2. In that case there must necessarily be two relations among an arbitrary three of $D_\lambda$; indeed, the condition is also sufficient. Hence, $D_\lambda$ should be a function of only one of such coefficients. We list several classes:

1. Class 1/0 which contains all coefficients of the form $B_{m,0}$. This class is singular of order 2. For, we have $B_{m,0} = C_{m,0} \cos \phi$, where $C_{m,0}$ depends only on $n$, $n'$, and, as $n$, $n'$ are supposed to be constant, $C_{m,0}$ is considered as a constant. Thus, $D_\lambda = C_{\lambda,0}(\cos \phi)\zeta^\lambda$. In order that $D_\lambda$ shall be a function of only one of the coefficients, all coefficients $C_{\lambda,0}$ except one should vanish, or $\psi$ should be reduced to the form $\exp (\sqrt{-1}\ ml)$. For satisfying $p\, \partial F_0/\partial x + q\, \partial F_0/\partial y = 0$, however, we must equate $n$ to zero. The Poinsot motion for $n = 0$ is a uniform rotation about one of the principal axes of inertia, and $\psi$ is a constant independent of $l$. Hence, the values of $C_{\lambda,0}$ other than $C_{0,0}$ are all zero for such special values of $n$, $n'$. Thus this class is singular of order 2.

2. Classes of the form $m/1$ contain only three coefficients: $B_{m,1}$, $B_{0,0}$, $B_{-m,-1}$. These classes are singular of order 2 only if $B_{m,1} = B_{-m,-1} = 0$; in other words, there should be no terms $\exp (\sqrt{-1}\ ml)$ in the expansion of $\xi + \sqrt{-1}\ \eta$, $\xi - \sqrt{-1}\ \eta$ in positive and negative powers of

$$\exp (\sqrt{-1}\ l),$$

as $\xi$, $\eta$ are supposed to be real. Generally, this does not happen unless the ellipsoid of inertia is an ellipsoid of revolution. When two axes of the ellipsoid are equal, then

$$\xi = A' \cos l + B' \sin l + C', \qquad \eta = A'' \cos l + B'' \sin l + C'',$$

where $A'$, $B'$, $C'$, $A''$, $B''$, $C''$ are constants. Hence, unless $m = 0$, 1 or $-1$, we have $B_{m,1} = -B_{-m,-1} = 0$. Hence the classes $m/1$, except $1/1$, $0/1$, $-1/1$, are singular of order 2.

3. The other classes, reducing to only one coefficient $B_{0,0}$, are singular of order 2.

Summarizing, we have seen that if the ellipsoid of inertia is an ellipsoid of revolution then the singularity is of order 2, except for the classes $1/1$, $0/1$, $-1/1$. Hence, there may exist a third uniform integral; indeed, there may exist an algebraic integral, provided that the Jacobian of these three integrals is zero for $n' = 0$ or $n' = \pm n$. In the case of Lagrange, in which the point of suspension is on the $C$ axis of the ellipsoid of inertia, the last condition is not necessary since $\xi$, $\eta$ are then reduced to constants. If the ellipsoid of inertia is not an ellipsoid of revolution, then there are an infinite number of classes that are not singular of order 2, that is, the classes $m/1$. Consider a domain $(D)$ which contains an infinite number of sets of values of $n$, $n'$, $\phi$, $\theta$ and suppose that in any such set the value of $n'$ is not an integral multiple of $n$. None of the classes $m/1$ belong to this domain. In this case there may exist a third uniform integral, provided that the Jacobian of these three integrals vanishes when $n'$ is an integral multiple of $n$. Hence we can conclude that the third integral cannot be algebraic in general.

### 6.21 SOLUTION IN ALGEBRAIC INTEGRALS

Husson (1906, 1908) has proved, independently of Poincaré, on the basis of Abel's theory of algebraic functions, that any algebraic integral with arbitrary initial conditions is an algebraic combination of the classical integrals (except in the cases of Euler, Lagrange, and Kowalewski), and that consequently there cannot exist any algebraic integral other than the classical integrals in the general case of top motions (except in the aforesaid cases). Liouville's proof (1897) of his theorem on a basis similar to the proof by Bruns was incomplete. Painlevé (1900) criticized Kowalewski's proof.

Let the origin $O$ be the point of suspension; take coordinates in the directions, fixed in the rigid body, of the principal axes of inertia with regard to the point of suspension. Let the center of mass be at a point $G$; let the unit of length be such that the numerical ratio of the external force $mg$ on the body to the distance $OG$ equals 1. Let the direction cosines of the direction of gravity in these coordinates be $\gamma, \gamma', \gamma''$. Then the equations of motion of the rigid body around $O$ are

$$A\frac{dp}{dt} = (B - C)qr + y_0\gamma'' - z_0\gamma', \qquad \frac{d\gamma}{dt} = r\gamma' - q\gamma'',$$

$$B\frac{dq}{dt} = (C - A)rp + z_0\gamma - x_0\gamma'', \qquad \frac{d\gamma'}{dt} = p\gamma'' - r\gamma, \tag{104}$$

$$C\frac{dr}{dt} = (A - B)pq + x_0\gamma' - y_0\gamma, \qquad \frac{d\gamma''}{dt} = q\gamma - p\gamma'.$$

In order to separate the variables in the manner of Sections 1.15 and 1.16, transform by

$$y_1 = ap + \sqrt{-1}\,bq, \qquad y_2 = ap - \sqrt{-1}\,bq,$$

$$z_1 = \gamma + \sqrt{-1}\,\gamma', \qquad z_2 = \gamma - \sqrt{-1}\,\gamma',$$

with

$$a = \sqrt{A(A - C)}, \qquad b = \sqrt{B(B - C)},$$

and put $-\sqrt{-1}\,t$ in place of $t$. Then

$$\frac{dy_1}{dt} = -\sqrt{\frac{(A - C)(B - C)}{AB}}\,ry_1 - \left[x_0\sqrt{\frac{B - C}{B}} + \sqrt{-1}\,y_0\sqrt{\frac{A - C}{A}}\right]\gamma''$$

$$+ \frac{z_0}{2}\left[\sqrt{\frac{B - C}{B}}\,(z_1 + z_2) + \sqrt{\frac{A - C}{A}}\,(z_1 - z_2)\right],$$

$$\frac{dy_2}{dt} = \sqrt{\frac{(A - C)(B - C)}{AB}}\,ry_2 + \left[x_0\sqrt{\frac{B - C}{B}} - \sqrt{-1}\,y_0\sqrt{\frac{A - C}{A}}\right]\gamma''$$

$$+ \frac{z_0}{2}\left[\sqrt{\frac{B - C}{B}}\,(z_1 + z_2) - \sqrt{\frac{A - C}{A}}\,(z_1 - z_2)\right],$$

$$C\frac{dr}{dt} = \frac{A - B}{4\sqrt{AB(A - C)(B - C)}}\,(y_2^2 - y_1^2) - \frac{x_0}{2}\,(z_1 - z_2)$$

$$+ \frac{\sqrt{-1}\,y_0}{2}\,(z_1 + z_2), \tag{105}$$

$$\frac{dz_1}{dt} = -rz_1 + \frac{1}{2}\left[\frac{y_1 + y_2}{\sqrt{A(A - C)}} + \frac{y_1 - y_2}{\sqrt{B(B - C)}}\right]\gamma'',$$

$$\frac{dz_2}{dt} = rz_2 - \frac{1}{2}\left[\frac{y_1 + y_2}{\sqrt{A(A - C)}} - \frac{y_1 - y_2}{\sqrt{B(B - C)}}\right]\gamma'',$$

$$\frac{d\gamma''}{dt} = \frac{1}{4\sqrt{A(A - C)}}(y_1 + y_2)(z_1 - z_2) - \frac{1}{4\sqrt{B(B - C)}}(y_1 - y_2)(z_1 + z_2). \qquad \text{(105 cont.)}$$

There exist three integrals — the energy integral, the angular-momentum integral, and the relation among the direction cosines:

$$\frac{(y_1 + y_2)^2}{4(A - C)} - \frac{(y_1 - y_2)^2}{4(B - C)} + Cr^2 - [x_0(z_1 + z_2) - \sqrt{-1}\, y_0(z_1 - z_2) + 2z_0\gamma''] = \text{constant}, \qquad (106)$$

$$\frac{A}{4\sqrt{A(A - C)}}(y_1 + y_2)(z_1 + z_2) - \frac{B}{2\sqrt{B(B - C)}}(y_1 - y_2)(z_1 - z_2) + 2Cr\gamma'' = \text{constant},$$

$$z_1 z_2 + \gamma''^2 = \text{constant}.$$

Multiply $y_1$, $z_1$, $z_2$, $\gamma''$ by a constant parameter $\lambda$. Denote the integration constants on the right-hand sides of the three equations (106) after this substitution by $h_1$, $h_2$, $h_3$, respectively; these are supposed to be independent of $\lambda$. Denote the resulting equations — after this substitution in (105) and (106) — by (105a) and (106a), respectively. Since the right-hand members of (106a) are regular for $\lambda = 0$, there exists the solution of (106a) arranged in positive integral power series of $\lambda$ insofar as $\lambda$ is sufficiently small, according to Poincaré's extension of Cauchy's existence theorem of the solution of a differential equation (discussed in a later chapter). We eliminate $t$ from these equations and try to obtain $y_1$, $y_2$, $z_1$, $z_2$, $\gamma''$ as functions of $r$. In the case treated by Lagrange, we obtain $r$ = constant, because $x_0 = y_0 = 0$, so we disregard this case. Equations (106a) give $y_2$, $z_1$, $z_2$ as functions of $y_1$, $\gamma''$, $r$, and (105a) are replaced by two differential equations of order 1 for determining $y_1(r)$, $\gamma''(r)$. The derivatives $dy_1/dr$ and $d\gamma''/dr$ are algebraic with regard to $h_1$, $h_2$, $h_3$, $\lambda$, $y_1$, $\gamma''$, $r$, and regular where $\lambda = 0$.

If there exists an algebraic integral $f(y_1, z_1, z_2, \gamma'', y_2, r)$ = constant, which does not depend on $t$ explicitly other than the classical integrals for the equations (105), then there exists an algebraic integral

$$f(\lambda y_1, \lambda z_1, \lambda z_2, \lambda\gamma'', y_2, r) = \text{constant}$$

for equations (105a) for any value of $\lambda$. Substitute for $y_2$, $z_1$, $z_2$ from (106a); then we obtain two differential equations of the first order for $y_1(r)$, $\gamma''(r)$ in the form $F(h_1, h_2, h_3, \lambda, y_1, \gamma'', r) =$ constant, where $F$ is algebraic in each of the arguments. Here $F$ must contain $y_1$ and $\gamma''$. Otherwise $r$ would become a constant; then, either the case would reduce to Lagrange's, or $F$ would not contain $y_1$, $\gamma''$, $r$ and the integral $f =$ constant would be one of the combinations of the classical integrals.

If we multiply $F$ by a power of $\lambda$ such that $\lambda = 0$ is neither a zero point nor a pole of $F$ considered as a function of $\lambda$, then the algebraic function $F$ of $\lambda$ can be expanded in positive integral ascending powers of $\lambda$ or $\lambda^{1/p}$ in the neighborhood of $\lambda = 0$. The coefficients of various powers of $\lambda$ in this expansion are algebraic in $h_1$, $h_2$, $h_3$, $y_1$, $\gamma''$, $r$. But we can suppose that $F$ contains $y_1$, $\gamma''$, $r$ besides $h_1$, $h_2$, $h_3$ at $\lambda = 0$, that is, that $F$ is of the form

$$F = F_0(h_1, h_2, h_3, y_1, \gamma'', r) + \lambda^{1/p}F_1(h_1, h_2, h_3, y_1, \gamma'', r) + \cdots = \text{constant}. \tag{107}$$

We separate the equations (105a) for $\lambda = 0$ into the following:

$$\frac{dy_1}{dt} = -\alpha_0 r y_1 + z_0(\alpha_1 z_1 + \alpha_2 z_2) - \alpha_3\gamma'', \tag{108}$$

$$\frac{dy_2}{dt} = \alpha_0 r y_2, \qquad \frac{dr}{dt} = \frac{A - B}{4ABC\alpha_0} y_2^2, \tag{109}$$

$$\begin{aligned} \frac{dz_1}{dt} &= -rz_1 + \beta_1 y_2\gamma'', \\ \frac{dz_2}{dt} &= rz_2 - \beta_2 y_2\gamma'', \qquad 2\frac{d\gamma''}{dt} = y_2(\beta_2 z_1 - \beta_1 z_2), \end{aligned} \tag{110}$$

where

$$\begin{aligned} \alpha_0 &= \sqrt{\frac{(A - C)(B - C)}{AB}}, \qquad \alpha_3 = x_0\sqrt{\frac{B - C}{B}} + \sqrt{-1}\, y_0\sqrt{\frac{A - C}{A}}, \\ 2\alpha_1 &= \sqrt{\frac{B - C}{B}} + \sqrt{\frac{A - C}{A}}, \qquad 2\alpha_2 = \sqrt{\frac{B - C}{B}} - \sqrt{\frac{A - C}{A}}, \\ 2\beta_1 &= \frac{1}{\sqrt{A(A - C)}} - \frac{1}{\sqrt{B(B - C)}}, \\ 2\beta_2 &= \frac{1}{\sqrt{A(A - C)}} + \frac{1}{\sqrt{B(B - C)}}. \end{aligned} \tag{111}$$

Let $\lambda = 0$ in (106a); then, (108), (109), (110) have algebraic integrals

$$\begin{aligned} \frac{B-A}{(A-C)(B-C)} y_2^2 + 4Cr^2 &= h_1 = 4Ca^2, \\ \frac{1}{\alpha_0}(\alpha_1 z_1 + \alpha_2 z_2) y_2 + 2Cr\gamma'' &= h_2, \\ z_1 z_2 + \gamma''^2 &= h_3, \end{aligned} \tag{112}$$

where $h_1$ is replaced by $4Ca^2$. We have the following relations among the constants

$$\begin{aligned} C\beta_1 &= \alpha_2\left(\frac{1}{\alpha_0} + 1\right), \qquad C\beta_2 = \alpha_1\left(\frac{1}{\alpha_0} - 1\right), \\ \alpha_1\beta_1 - \alpha_2\beta_2 &= \frac{1}{2}\frac{B-A}{AB}, \qquad 4\alpha_1\alpha_2 = C\frac{B-A}{AB}. \end{aligned} \tag{113}$$

Let $y_1, y_2, z_1, z_2, \gamma''$ be the solution of (108), (109), (110) expressed in terms of $r$. Deriving $y_2, z_1, z_2$ from (112) in terms of $y_1, \gamma'', r$ we obtain

$$\frac{dr}{dt} = \alpha_0(r^2 - a^2).$$

Then (108), (109), (110) are replaced by

$$\frac{dy_1}{dr} = \frac{-\alpha_0 r y_1 + z_0(\alpha_1 z_1 + \alpha_2 z_2) - \alpha_3\gamma''}{\alpha_0(r^2 - a^2)}, \tag{108a}$$

$$2\frac{d\gamma''}{dr} = \frac{y_2(\beta_2 z_1 - \beta_1 z_2)}{\alpha_0(r^2 - a^2)}. \tag{110a}$$

The right-hand members are expressed by (112) in terms of $h_1, h_2, h_3, y_1, \gamma'', r$. Substituting $\lambda = 0$ in (107), we see that these equations admit an integral of the form

$$F_0(h_1, h_2, h_3, y_1, \gamma'', r) = \text{constant},$$

where $F_0$ should contain $y_1, \gamma''$; otherwise the solution $r(t)$ of (108), (109), (110) would reduce to a constant and we would have $A = B$, Lagrange's case.

The first equation of (112) expresses $y_2$ as a function of $r$, and the second and the third equations express $z_1, z_2$ as irrational functions of $\gamma''$. To avoid this inconvenience, we introduce in place of $\gamma''$ a new variable $v$ such that the differential equations become rational:

$$v = \alpha_1 z_1 - \alpha_2 z_2 + \frac{\alpha_0}{y_2}\left(2Ca\gamma'' - h_2\frac{r}{a}\right). \tag{114}$$

The second equation of (112) is then transformed to

$$\left(v + \frac{\alpha_0 h_2}{a}\frac{r}{y_2}\right)^2 - \alpha_0^2 h_0^2 \frac{1}{y_2^2} + 4\alpha_1\alpha_2 h_3 = 4Ca\alpha_0 \frac{v}{y_2}\gamma''. \tag{115}$$

These two equations replace (108a), (110a) and form a system of equations containing $y_1(r)$, $v(r)$ rationally. From (112) we obtain an invariantive relation $v = 0$ for (108a), (110a), since

$$\frac{dv}{dt} = av.$$

To simplify (108a), let

$$h = \frac{A - B}{ABh_1}(h_2^2 - h_1 h_3), \qquad y = y_1 y_2;$$

then, from (108), (109), and (110) we obtain the differential equations

$$\begin{aligned} \frac{dv}{dr} &= \frac{a}{\alpha_0}\frac{v}{r^2 - a^2}, \\ \frac{dy}{dr} &= a_1 + a_2\frac{r}{\sqrt{r^2 - a^2}} + \left(a_3\alpha_3 + a_4 z_0 \frac{r}{\sqrt{r^2 - a^2}}\right)\left(v + \frac{h}{v}\right), \end{aligned} \tag{116}$$

where $a_1$, $a_2$, $a_3$, $a_4$ are nonzero constants. It is assumed that none of $A - B$, $B - C$, $C - A$ is zero.

Thus $v(r)$, $y(r)$ are expressed algebraically in terms of $y_1$, $\gamma''$, $r$. Hence, in order that there may exist a new algebraic integral, the expressions $v(r)$, $y(r)$, $r$ should be related by algebraic equations. After integrating (116) we obtain

$$v = k\left(\frac{r - a}{r + a}\right)^{1/(2\alpha_0)},$$

$$y = a_1 r + a_2\sqrt{r^2 - a^2} + \int \left(a_3\alpha_3 + a_4 z_0 \frac{r}{\sqrt{r^2 - a^2}}\right)\left(v + \frac{h}{v}\right) dr$$

with integration constant $k$. Therefore, the two expressions

$$w = \left(\frac{r - a}{r + a}\right)^{1/(2\alpha_0)}, \qquad z = \int \left(a_3\alpha_3 + a_4 z_0 \frac{r}{\sqrt{r^2 - a^2}}\right)\left(w + \frac{h'}{w}\right) dr,$$

and $r$ are related by algebraic equations, where $h'$ is an arbitrary number. We shall prove that this condition is satisfied only if the center of gravity coincides with the point of suspension (Euler's case).

$A$, $B$, $C$ are all different. (104) is symmetric with respect to $p$, $q$, $r$. Hence, we may suppose that the value of $C$ lies between those of $A$ and

$B$. Then $\alpha_0$ is imaginary and $w(r)$ is never algebraic, so that it should be $z(r)$ that is related to $w$ and $r$ through an algebraic equation. Moreover, this equation should be valid for all values of $h'$, and $w$ does not contain $h'$. Hence,

$$J_1 = \int \left(a_3\alpha_3 + a_4 z_0 \frac{r}{\sqrt{r^2 - a^2}}\right) w\, dr,$$

$$J_2 = \int \left(a_3\alpha_3 + a_4 z_0 \frac{r}{\sqrt{r^2 - a^2}}\right) \frac{dr}{w}$$

should be expressed algebraically in terms of $r$ and $w$.

When the representative point $r$ describes a small closed contour around $r = a$ on the plane for the values of $r$, the value of $w$ is multiplied by a constant. Hence,

$$J_1' = \int \left(a_3\alpha_3 - a_4 z_0 \frac{r}{\sqrt{r^2 - a^2}}\right) w\, dr,$$

$$J_2' = \int \left(a_3\alpha_3 - a_4 z_0 \frac{r}{\sqrt{r^2 - a^2}}\right) \frac{dr}{w}$$

should be expressed algebraically in terms of $r$ and $w$. Comparing $J_1$ and and $J_2$, we see that

$$J_3 = \alpha_3 \int w\, dr, \qquad J_4 = \alpha_3 \int \frac{dr}{w}, \tag{117}$$

$$J_5 = z_0 \int \frac{rw}{\sqrt{r^2 - a^2}}\, dr, \qquad J_6 = z_0 \int \frac{r}{w\sqrt{r^2 - a^2}}\, dr,$$

should be expressed algebraically in terms of $r$ and $w$.

## 6.22 ABEL'S LEMMAS

In order to discuss the integrals of (117) we refer to *Abel's three lemmas* (Husson, 1906, 1908).

*Abel's First Lemma:* Let $P(x)$ be a rational algebraic fraction, and $Q(x)$ be a polynomial or a rational fraction in $x$ — that is, the quotient of two polynomials or two rational polynomials in $x$ — such that $Q(x)$ has as simple poles the zero points or the poles of $P(x)$. Form

$$J(x) = \int \frac{Q}{y}\, dx$$

with

$$y(x) = [P(x)]^\alpha,$$

where $\alpha$ is a real irrational number or an imaginary number. If $J(x)$ is expressed algebraically in terms of $y$ and $x$, then the function can be represented by

$$J(x) = \frac{f(x)}{y(x)} + \text{constant},$$

where $f(x)$ is either a polynomial of $x$ of degree one higher than $Q(x)$ or a constant.

This lemma follows from the properties of algebraic functions described in Section 2.20. The integral $J(x)$ is uniform for finite values of $x$, except the zero points and the poles of $P(x)$. Let $x = a$ be a zero point of $P$ of order $m$; then

$$y(x) = (x - a)^{m\alpha} y_1(x),$$

where $y_1(x)$ is uniform at $x = a$, and $1/y_1(x)$ is also uniform at $x = a$. Then we can write, with a constant $E \leqq 1$,

$$\begin{aligned} J(x) &= \int \frac{Q(x)}{(x - a)^{m\alpha}} \frac{dx}{y_1} \\ &= \int \frac{1}{(x - a)^{m\alpha + E}} [b_0 + b_1(x - a) + b_2(x - a)^2 + \cdots] \, dx. \end{aligned}$$

The result of integration gives

$$J(x) = \frac{1}{(x - a)^{m\alpha + E - 1}} [b'_0 + b'_1(x - a) + \cdots] + C,$$

with fixed constants $b'_0, b'_1, \ldots$ and an arbitrary constant $C$. Hence, in the neighborhood of $x = a$,

$$y(x)J(x) = \phi(x) + C[P(x)]^{\alpha},$$

where $\phi(x)$ is holomorphic. If $J(x)$ is algebraic in terms of $x$ and $y$, then $J(x) - C$ is also algebraic, so that we can put $C = 0$. Let

$$y(x)J(x) = f(x, y);$$

then, $f(x, y) = \phi(x)$ in the neighborhood of $x = a$. As $x$ makes a circuit from a point $x = x_0$ near $x = a$ by describing the whole circumference of a circle with $x = a$ as center in the $(x, y)$-plane, the function $\phi(x)$ again takes the initial value, but $y(x)$ takes an infinite number of values in geometrical progression. If the algebraic function $f(x_0, y)$ depends on $y$, then it should take the same value only for a finite number of values of $y$, according to Section 2.20. Hence, for $f(x, y) = \phi(x)$ to be possible, $f$ should not depend on $y$. Thus, $y(x)J(x) = f(x)$, where $f(x)$ is an algebraic function of $x$. The function $f(x)$ ceases to be holomorphic only when $y(x)$ or $J(x)$ ceases to be holomorphic, that is, at the zero points or the poles of $P$, except in the case where $x = a$ is a zero point.

If $x = b$ is a zero point of $P(x)$ other than $x = a$, then a similar proof can be carried out. If $x = a$ is not a zero point, but is a pole, then the proof can be obtained simply by changing $m$ into $-m$. Hence the algebraic function $f(x)$ is holomorphic for any finite value of $x$.

The behavior at $x = \infty$ can be studied, as in Section 2.20, by the substitution $x = 1/z$. If $P$ and $Q$ are polynomials of degree $n$ and $n_1$, respectively, then,

$$y = z^{-n\alpha}y_2(z), \qquad Q = z^{-n_1}y_3(z),$$

where $y_2(z)$ and $y_3(z)$ are holomorphic and nonzero at $z = 0$. Hence

$$J(x) = z^{n\alpha - n_1 - 1}(b_0' + b_1'z + \cdots) + C',$$

and accordingly

$$f(x) = y(x)J(x) = z^{-n_1-1}\psi(z) + C'z^{-n\alpha}y_2(x),$$

where $\psi(z)$ is holomorphic and nonzero for $z = 0$. Since the function $z^{n\alpha}$ has $z = 0$ as a transcendental singularity, $C'$ must be zero. If $n_1 > -1$, then the point at infinity is a pole of order $n_1 + 1$ of $f(x)$. Since $f(x)$ is holomorphic for finite values of $x$, this algebraic function is a polynomial in $x$ of degree $n_1 + 1$. If $n_1 < -1$, then the point at infinity is a zero point of $f(x)$, which reduces to zero identically, and $J(x)$ cannot be expressed algebraically in terms of $y$ and $x$.

*Abel's Second Lemma:* If there exists an algebraic relation among Abel's transcendental functions $J(x)$ and $J'(x)$ and the variable $x$, then the relation is of the form $J' = kJ + \phi(x)$, where $k$ is a constant and $\phi(x)$ is an algebraic function.

Suppose that the integrals $J$ and $J'$ are related by an algebraic equation $f(J, J') = 0$ as in Section 2.20. Solving this equation for $J'$, we obtain $J' = \psi(J, x)$. Since $J$ is not an algebraic expression, the integral $J$ increases by a period $\omega$ as $x$ returns to its initial value, by describing the suitably chosen closed curve $C$ on the Riemann surface of $x$, and $J'$ increases by a period $\omega_1$ at the same time, where

$$\omega_1 = \psi(J + \omega, x) - \psi(J, x).$$

But this period $\omega_1$ does not contain $J$ if $J$ and $x$ are regarded as the independent variables, because $J$ could be solved as an algebraic expression of $x$ if $\omega_1$ contained $J$. Hence

$$\frac{\partial\psi(J + \omega, x)}{\partial J} - \frac{\partial\psi(J, x)}{\partial J} \equiv 0,$$

and $\partial\psi(J, x)/\partial J$ has the period $\omega$. Consequently we can write

$$J' = \psi(J, x) = J\alpha(x) + \beta(x),$$

where $\alpha$ and $\beta$ are algebraic expressions. Thus,

$$\frac{dJ'}{dx} = J\frac{d\alpha}{dx} + \alpha\frac{dJ}{dx} + \frac{d\beta}{dx},$$

in which $J$ is transcendental and the functions other than $J$ are algebraic. Accordingly $d\alpha/dx \equiv 0$, that is, $\alpha = k =$ constant. Similarly, from $d\beta/dx = dJ'/dx - \alpha(dJ/dx)$, we see that $d\beta/dx$ is rational in $x$, and so is $\beta(x)$.

*Corollary:* If one of the two integrals $J$ and $J'$,

$$J = \int \frac{y\,dx}{x^2 - a^2}, \qquad J' = \int \frac{dx}{y(x^2 - a^2)}, \qquad y = (x^2 - a^2)^{p/m},$$

is not algebraic, where $p$, $m$ are relatively prime positive integers — that is, if $2p/m$ is not an integer — then it is impossible to have an algebraic relation among $J$, $J'$, $x$, unless $2p/m$ is an integer, that is, one of the integrals $J$ or $J'$ is algebraic.

In fact, if $J$ and $J'$ are transcendental functions, then

$$J' = kJ + \phi(x, y)$$

by the second Lemma. As $x$ describes a closed contour $C$,

$$\omega_1 = \int_C \frac{dJ}{dx}\,dx = k\int_C \frac{dJ}{da}\,dx = k\omega.$$

As $x$ makes a circuit around $x = a$ in the region of the point $x = a$, the variable $x$ returns to its initial value, but $y$ is multiplied by

$$\exp\left(2\pi\sqrt{-1}\,\frac{p}{m}\right),$$

and $\dfrac{dJ'}{dx}$ by $\exp\left(-2\pi\sqrt{-1}\,\dfrac{p}{m}\right)$, and $\dfrac{dJ}{dx}$ by $\exp\left(2\pi\sqrt{-1}\,\dfrac{p}{m}\right)$. Hence the relation between the periods is

$$\omega_1 \exp\left(-2\pi\sqrt{-1}\,\frac{p}{m}\right) = k\omega \exp\left(2\pi\sqrt{-1}\,\frac{p}{m}\right).$$

Thus we see that $2p/m$ should be equal to an integer $n$.

*Abel's Third Lemma:* Consider the integral $K(x) = \int J(x)\rho(x, y)\,dx$ with a rational function $\rho$ of $x$ and $y$, and an Abel's transcendental function

$J(x)$, where $y$ is defined by $f(x, y) = 0$ as an implicit function of $x$. If $J$, $K$, $x$ are related by an algebraic equation, then Abel's integral $\int \rho(x, y)\, dy$ is expressed algebraically in terms of $J$ and $x$.

In fact, the algebraic relation among $J$, $K$, $x$ must contain $K$. If it did not contain $K$, then $J$ would be algebraic. Hence we obtain an equation $K = \phi(J, x)$ that is algebraic in $J$ and $x$. Suppose that, as $x$ describes a closed curve $C$, the integral $J$ increases by $\omega$ and $K(x)$ becomes $K_1(x)$.

Then

$$K_1(x) - K(x) = \phi(J + \omega, x) - \phi(J, x) = \psi(J, x).$$

But

$$\frac{dK_1}{dx} - \frac{dK}{dx} = (J + \omega)\rho - J\rho = \omega\rho(x, y).$$

Hence

$$\psi(J, x) = \omega \int \rho(x, y)\, dx.$$

Since $\psi$ is algebraic both in $J$ and $x$, the Abelian integral $\int \rho(x, y)\, dx$ can be expressed algebraically in terms of $J$ and $x$.

On the basis of Abel's lemmas, we seek the condition that integrals $J_3$, $J_4$, $J_5$, $J_6$ should be expressed algebraically by $r$ and $w$. If $\alpha_3 \neq 0$, then, according to the first lemma,

$$\int \frac{dr}{w} = \frac{br + c}{w} + \text{constant}$$

with two constants $b$ and $c$. Differentiating this expression with respect to $r$ would give an impossible relation

$$\frac{w'}{w} = \frac{a}{\alpha_0}\frac{1}{r^2 - a^2} = \frac{b - 1}{br + c}.$$

Hence

$$\alpha_3 = x_0\sqrt{\frac{B - C}{B}} + \sqrt{-1}\, y_0\sqrt{\frac{A - C}{A}} = 0.$$

Exchanging $y_1$ and $y_2$, we have

$$x_0\sqrt{\frac{B - C}{B}} - \sqrt{-1}\, y_0\sqrt{\frac{A - C}{A}} = 0.$$

Either $B = C$, $A = C$, or $x_0 = y_0 = 0$ (Lagrange's or Euler's case).

Put

$$w_1 = \left(\frac{r - a}{r + a}\right)^{1/(2\alpha_0) - 1/2}$$

in $J_6$; then, according to the first lemma,

$$\int \frac{1}{w_1} \frac{r\,dr}{r - a} = \frac{br + c}{w_1} + \text{constant}$$

for $z_0 \neq 0$. Differentiate this with respect to $r$; then

$$(r + a)[(1 - b)r + ab] = a\left(1 - \frac{1}{\alpha_0}\right)(br + c),$$

that is, $b = 1$, $c = a$, and $a = a(1 - 1/\alpha_0)$. But this last equation is valid only when $A = C$ or $B = C$ (Lagrange's case).

Hence, in order that there may exist an algebraic integral other than the classical integrals, in the case when $A$, $B$, $C$ are all different, $x_0$, $y_0$, and $z_0$ must all equal zero; that is, the center of mass coincides with the point of suspension. This is Euler's case. Thus the theorem is proved.

## 6.23 KOWALEWSKI'S SOLUTION

We have $A = B$ for the cases of Lagrange and Kowalewski. Since the center of mass is on the axis of $C$ in the case of Lagrange and is arbitrary in the case of Kowalewski, the case of Lagrange is considered as a special case of Kowalewski's. If $A = B$, then $r =$ constant. Put $A_1 = A - C$,

$$y_1 = p + \sqrt{-1}\, q \quad \text{and} \quad y_2 = p - \sqrt{-1}\, q,$$

and let $\epsilon$ be the angle that the line joining the center of mass with the point of suspension makes with the $(x, y)$ plane. Then (105) can be written

$$\begin{aligned}
A\frac{dy_1}{dt} &= -A_1 r y_1 + z_1 \sin\epsilon - \gamma'' \cos\epsilon,\\
A\frac{dy_2}{dt} &= A_1 r y_2 + \gamma'' \cos\epsilon - z_2 \sin\epsilon,\\
2C\frac{dr}{dt} &= (z_2 - z_1)\cos\epsilon,\\
\frac{dz_1}{dt} &= \gamma'' y_1 - r z_1,\\
\frac{dz_2}{dt} &= r z_2 - \gamma'' y_2,\\
2\frac{d\gamma''}{dt} &= y_2 z_1 - y_1 z_2.
\end{aligned} \tag{118}$$

These equations have the classical integrals

$$
\begin{aligned}
(Cr^2 - z_1 \cos \epsilon) + (Ay_1y_2 - z_2 \cos \epsilon - 2\gamma'' \sin \epsilon) &= \text{constant},\\
A(y_1z_2 + y_2z_1) + 2Cr\gamma'' &= \text{constant},\\
z_1z_2 + \gamma''^2 &= \text{constant}.
\end{aligned}
\tag{119}
$$

Multiplying $y_1$, $z_1$, $\gamma''$ in each of these equations by $\lambda$, we obtain six equations that we call (118a) and three that we call (119a), and denote the constants on the right-hand sides of (119a) by $h_1$, $h_2$, $h_3$, respectively. In the case of Lagrange, $r =$ constant, because the center of mass is on the $z$ axis. Excluding this case of Lagrange, take $r$ as the independent variable. Put $\lambda = 0$ and

$$\gamma'' = (r^2 - a^2)^{C/A}\Gamma''; \tag{120}$$

then from (118a) and (119a) we obtain

$$
\left.
\begin{aligned}
y_2 &= (r^2 - a^2)^{A_1/A}J,\\
J &= \text{constant} - \frac{2C}{A}\tan \epsilon \int \frac{dr}{(r^2 - a^2)^{A_1/A}},
\end{aligned}
\right\}
\tag{121}
$$

$$\frac{d\Gamma''}{dr} = \frac{2h_3 \cos \epsilon}{C}\,\frac{y_2}{(r^2 - a^2)^{(C/A)+2}} - \frac{h_2}{A}\,\frac{1}{(r^2 - a^2)^{(C/A)+1}}, \tag{122}$$

$$
\left.
\begin{aligned}
Cr^2 - z_2 \cos \epsilon = h_1 &= Ca^2,\\
A(y_1z_2 + y_2z_1) + 2Cr\gamma'' &= h_2,\\
z_1z_2 &= h_3.
\end{aligned}
\right\}
\tag{123}
$$

Equations (121) and (122) should have an integral

$$F(h_1, h_2, h_3, y_2, \gamma'', r) = \text{constant}$$

for $\lambda = 0$, which expresses $\gamma''(r)$ algebraically in terms of $r$ and $y_2$. To obtain such an integral, express the three integrals,

$$J_0 = (r^2 - a^2)^{C/A}, \qquad J_1 = \int \frac{dr}{(r^2 - a^2)^{1+(C/A)}},$$

$$J_2 = \int \frac{J\,dr}{(r^2 - a^2)^{1+2(C/A)}},$$

algebraically in terms of $r$ and $y_2$, according to (120), (121), and (123). Eliminating $y_2$ by the use of the equations for $J_0$, $J_1$, we should obtain an algebraic relation among these two equations and $r$.

Suppose that this relation contains $J_1$. In Abel's first lemma the function $f$ should be of the form $f = cx + d$ for $n_1 = 0$, that is, for the particular case $Q = 1$. Differentiating $y(x)J(x) = f(x)$, with this $f$, gives

$$\frac{P'}{P} = \frac{c-1}{c\alpha}\frac{1}{cx+d},$$

that is, $P = k(cx + d)^{(c-1)/(c\alpha)}$, where $(c - 1)/(c\alpha)$ is a positive integer, in other words, $\alpha$ must be rational. Hence, in this particular case the expression $1 + C/A$, and consequently $C/A$, should be rational. This holds whether $y_2(r)$ is a transcendental function of $r$ or not.

Let $C/A$ be an irreducible fraction $C/A = p/m$. Then $J_1$ should be expressed algebraically in terms of $r$ and $y_2$, or in terms of $r$ and $J$. By the corollary to Abel's second lemma this is possible only when $2C/A$ is an integer, regardless of whether $J$ is algebraic or transcendental. If $C/A$ is an integer, then the transcendental function $J_1$ cannot be expressed as an algebraic function of $r$ and of an algebraic function $J$. If $2C/A$ is an odd integer, then $J_1$ is an algebraic function. Hence $F$ contains $\gamma''$ only when $2C/A$ is an odd integer.

Put $2C/A = 2n - 1$ and find the condition for which $2J_2$ is expressed algebraically in terms of $r$ and $y_2$, or in terms of $r$ and $J$. First suppose that $\epsilon \neq 0$. Since $J$ is transcendental,

$$J_3 = \int \frac{dr}{(r^2 - a^2)^{1+(C/A)}} = \int \frac{dr}{(r^2 - a^2)^{2n}}$$

is expressed algebraically in terms of $r$ and $J$, or in terms of $r$ and

$$J_4 = \int \frac{dr}{(r^2 - a^2)^{1-(C/A)}} = \int (r^2 - a^2)^{n-1}\frac{dr}{\sqrt{r^2 - a^2}},$$

by Abel's third lemma.

Let $\phi$, $\phi_1$ be algebraic functions and $\alpha$, $\alpha_1$ be nonzero constants; then

$$J_3 = \phi(r) + \alpha \log\frac{r-a}{r+a}, \qquad J_4 = \phi_1(r) + \alpha_1 \log(r + \sqrt{r^2 - a^2}).$$

Hence, $r = \pm a$ are algebraic singularities of $J_4$. They are algebraic singularities of the algebraic function $J_3$ of $r$ and $J$. The points are also logarithmic singularities of $J_3$. Hence, we cannot express $J_3$ algebraically in terms of $r$ and $J$. Next, suppose that $\epsilon = 0$. This gives: $J =$ constant, $J_2 = J_3$, and $J_3 =$ an algebraic function of $r$. But this is impossible.

Hence, in neither of these cases can $F$ contain $\gamma''$. Thus $y_2(r)$ is an algebraic expression. By (121), the function $y_2$ consists of a term containing the arbitrary constant and a term containing the integral. Thus

$(r^2 - a^2)^{1-(C/A)}$ and $(2C/A)J_4 \tan \epsilon$ should be algebraic expressions. To begin with, $C/A$ should be rational. If $\epsilon = 0$, then the second expression vanishes and it is obviously algebraic. For $\epsilon \neq 0$ the integral $J_4$ should be algebraic, and $C/A$ rational. Hence $J_4$ is the integral $J$ in the second lemma, and $C/A$ should be an integer. Thus, in order that there may exist other algebraic integrals besides the classical integrals, either $C/A$ should be rational and $\epsilon = 0$, or $C/A$ should be an integer.

Husson (1906, 1908) further examined the next term in the expansion in powers of $\lambda$ and showed that there exists no algebraic integral, by excluding the case of Kowalewski $2C/A = 1$. Thus it is proved that there can be no algebraic integral other than the classical integrals with arbitrary initial conditions, except for the cases of Euler, Lagrange, and Kowalewski.

Burgatti (1906, 1910) has proved the above theorem of Goriatchoff for the nonexistence of algebraic integrals, unless transcendental, other than the classical ones, by a more elementary method than Husson's.

Kowalewski (1888; Kötter, 1893) obtained the explicit form of the solution in her case by putting $z_0 = 0$; that is, $\epsilon = 0$, and, in particular, $y_0 = 0$, $C = 1$. Then $A = 2$, and (104) are of the form

$$\begin{aligned} 2\frac{dp}{dt} &= qr, & \frac{d\gamma}{dt} &= r\gamma' - q\gamma'', \\ 2\frac{dq}{dt} &= -pr - \gamma'', & \frac{d\gamma'}{dt} &= p\gamma'' - r\gamma, \\ \frac{dr}{dt} &= \gamma', & \frac{d\gamma''}{dt} &= q\gamma - p\gamma'. \end{aligned} \tag{124}$$

Put

$$\begin{aligned} y_1 &= p + \sqrt{-1}\, q, & y_2 &= p - \sqrt{-1}\, q, \\ z_1 &= \gamma + \sqrt{-1}\, \gamma', & z_2 &= \gamma - \sqrt{-1}\, \gamma', \\ \xi_1 &= y_1^2 + z_1, & \xi_2 &= y_2^2 + z_2; \end{aligned}$$

then,

$$\frac{d}{dt}(y_1^2 + z_1) = -\sqrt{-1}\, r(y_1^2 + z_1), \quad \frac{d}{dt}(y_2^2 + z_2) = \sqrt{-1}\, r(y_2^2 + z_2),$$

that is,

$$\frac{d}{dt}\log(y_1^2 + z_1) + \frac{d}{dt}\log(y_2^2 + z_2) = 0.$$

proved the theorem that there exists no integral uniform in the Keplerian elements other than the classical integrals, in the following manner.

Consider the plane case of the three-body problem, and suppose that we reduce the differential equations of order six with the variables $L, L', K, l, l', \kappa$ in the form (184a), Section 5.19. Let the expansion of the disturbing function be

$$F_1 = \sum_{m,m'} B_{mm'} \exp\left[\sqrt{-1}\,(ml + m'l')\right];$$

then the coefficient $B_{m,m'}$ is a function of $L, L', K, \kappa$. Form the expression

$$B^{\lambda'}_{\lambda p,\lambda q} B^{-\lambda}_{\lambda' p,\lambda' q} \qquad (\lambda, \lambda' = 0, \pm 1, \pm 2, \ldots) \tag{87}$$

with two relatively prime integers, and take the values of $L, L'$ such that

$$p\frac{\partial F_0}{\partial L} + q\frac{\partial F_0}{\partial L'} = 0, \tag{85a}$$

that is, such that the mean motions are in the ratio $-p:q$. The disturbing function is, as has been seen in Sections 5.16–5.18, of degree $-1/2$ with regard to $L, L'$, and homogeneous of degree $-1$ with regard to $a, a'$. Suppose that $a' > a$. Then the disturbing function, apart from the common factor $a'^{-1}$, is a function of $a/a'$ or $L/L'$ only. Since the coefficient $B_{mm'}$ of (87) can accordingly be considered as a function of this ratio only, it is uniquely determined by (85a). In order that there may exist a uniform integral other than the energy integral, there should be, as $n = 3$ and $2n - 4 = 2$, one relation between two of the expressions (87) of Section 6.16; in other words, (87) should be a function of the secular term $B_{00}$. We know (as we shall see in Chapter 7 from the property of the expansion of the disturbing function) that there is no such relation among the coefficients. Hence, there exists no uniform integral of the form

$$\Phi(L, L', K, l, l', \kappa) = \text{constant}$$

other than the energy integral.

The above proof, however, is not yet complete. We must examine an integral of the form

$$\Phi(L, L', G, G', l, l', g, g') = \text{constant}.$$

Take the canonically conjugate variables $L, L', G, G', l, l', g, g'$ of (164) in Section 5.16 for the three-body problem; then (87) depends only on $L, L', G, G', g, g'$. Take for $L, L'$ the values satisfying (85a); then (87) depends only on $G, G', g, g'$. In order that there may exist a uniform integral other than the energy integral, there should exist, as $n = 4$ and $2n - 4 = 4$, one relation among four of the expressions (87). Since (87) is a function of $G$, $G'$, $g - g'$ only, nothing prevents there being such a relation. Thus, there is one uniform integral other than the energy integral: one of the integrals of angular momentum. In order that there may exist two uniform integrals, there should be one relation among three of the expressions (87); that is, all of the expressions (87) should be expressed in terms of only two variables. But the actual expansion of the disturbing function shows that this is not the case. Hence, there exists no uniform integral other than the energy integral and the angular-momentum integral in the plane three-body problem.

We now turn to the general three-body problem, and consider the twelve canonically conjugate variables $L, L', G, G', H, H', l, l', g, g', h, h'$ as in (164) of Section 5.16. If we choose the values of $L$, $L'$ so as to satisfy (85a), then (87) still contains eight variables $G, G', H, H', g, g', h, h'$. In order that there may exist $q$ uniform integrals other than $F$, there should be one relation among $2n - 3 - q = 3 - q$ of the expressions (87). We have proved in Section 5.19 that these expressions depend only on the five variables $G, G', g, g', h - h'$. Hence, there should be one relation among $9 - 3 = 6$ of the expressions (87). Thus, $q = 3$; nothing prevents there being three independent uniform integrals, that is, the three integrals of the angular momentum discussed in Section 5.19. There is no relation among $9 - 4 = 5$ of the expressions (87). Hence, in the general three-body problem, there exists no uniform integral other than the energy integral and the three integrals of the angular momentum.

Let us next see in detail that there is no relation among an arbitrary set of $2n - 4$ of the expressions (87).

The disturbing function is expanded in the form

$$F_1 = \sum B_{mm'} \exp\left[\sqrt{-1}\,(ml + m'l')\right],$$

where $B_{mm'}$ is a function of $L, L', G, G', g, g', H, H', h, h'$. It can be seen that the same is true for the expansion in the Keplerian elements as was described in Section 5.18. The expansion is written in the form

$$F_1 = \sum C_{mm'} \exp\{\sqrt{-1}\,[m(l + g + h) + m'(l' + g' + h')]\},$$

where

$$B_{mm'} = C_{mm'} \exp\left[\sqrt{-1}\,(mg + m'g' + mh + m'h')\right].$$

Here the mean longitude is denoted by $l + g + h = l + \varpi$. $C_{mm'}$ is a function of ten variables, $a, a', e, e', I, I', \varpi, \varpi', \Omega, \Omega'$. Since the Jacobian of $L, L', G, G', H, H', g, g', h, h'$ with regard to $a, a', e, e', I, I', \varpi, \varpi', \Omega, \Omega'$ is not zero, the result of our discussion on $B_{mm'}$ is valid also for $C_{mm'}$, and (87) is replaced by

$$C^{\lambda'}_{\lambda p,\lambda q} C^{-\lambda}_{\lambda' p,\lambda' q} \qquad (\lambda, \lambda' = 0, \pm 1, \pm 2, \ldots). \tag{87a}$$

In order that there may exist a uniform integral, there should be one relation among $2n - 4$ of the expressions (87a).

In ordinary perturbation theory the coefficient $C_{mm'}$ is expanded in powers of $e, e', I, I'$. The degree of a term in the expansion is at least $|m + m'|$ as has been shown in Sections 5.16–5.18. We arrange the terms of $C_{mm'}$ according to the degree and let it be

$$C_{mm'} = C_{mm';0} + C_{mm';1} + C_{mm';2} + \cdots + C_{mm';p} + \cdots,$$

where $C_{mm';p}$ is of degree $|m + m'| + 2p$. We call $C_{mm';0}$ the *principal part* of $C_{mm'}$. The term $C_{00;0}$ depends on $a, a'$. Suppose for the moment that $a, a'$ are constant; ignore $C_{00;0}$, since it is now a constant. We call

$$C_{00;0} + C_{00;1} + \sum C_{mm';0} \times \exp\{\sqrt{-1}\,[m(l + g + h) + m'(l' + g' + h')]\}$$

the *principal part* of the expansion of $F_1$.

From (87$a$) we form a matrix with an infinite number of elements, such that, corresponding to a positive, negative, or zero value of the index $\lambda$ of each row, the first element of the row with index $\lambda$ is $\lambda C_{\lambda p,\lambda q}$ and the other elements of the same row are the partial derivatives of $C_{\lambda p,\lambda q}$ with regard to $e, e', I, I', \varpi, \varpi', \Omega, \Omega'$. In order that there may be one relation among $2n - 4$ of (87a), all the determinants obtained by taking $2n - 3 = 9$ rows from this matrix must be zero. But, as the terms of $C_{mm'}$ are at least of degree $|m + m'|$, the first element $\lambda C_{\lambda p,\lambda q}$ of the row with index $\lambda$ begins with the term of degree $|\lambda p + \lambda q|$, while the partial derivatives of $C_{\lambda p,\lambda q}$ with regard to $e, e', I, I'$ are of degree $|\lambda p + \lambda q| - 1$, and its partial derivatives with regard to $\varpi, \varpi', \Omega, \Omega'$ are of degree $|\lambda p + \lambda q|$. The first element of the row with $\lambda = 0$ is zero. The expansion of the partial derivatives of $C_{00}$ with regard to $\varpi, \varpi', \Omega, \Omega'$ begins with a term of degree 2, and the expansion of those with regard to $e, e', I, I'$ begins with a term of degree 1. Our determinant is expanded in integral powers of $e, e', I, I'$. Denote by $\Delta$ the determinant formed of the rows with indices $\lambda_1, \lambda_2, \ldots, \lambda_{2n-3}$; then the terms of its expansion are at least of degree

$$|p + q| : (|\lambda_1| + \cdots + |\lambda_{2n-3}|) - 4.$$

Denote this degree by $\alpha$. If $\lambda_{2n-3} = 0$, then we denote

$$|p + q| \cdot (|\lambda_1| + \cdots + |\lambda_{2n-4}|) - 2$$

by $\alpha$. Since $\Delta$ is identically zero, a term with degree $\alpha$ must also be zero. Such terms are obtained by replacing each $C_{\lambda p, \lambda q}$ of $\Delta$ by its principal part $C_{\lambda p, \lambda q; 0}$. If $\lambda = 0$, then we replace it by $C_{00;1}$. The determinant $\Delta_0$ thus obtained should be identically zero. We add $C_{00;1}$ corresponding to $\lambda = 0$ to the expression

$$(C_{\lambda p, \lambda q;0})^{\lambda'}(C_{\lambda' p, \lambda' q;0})^{-\lambda} \qquad (\lambda, \lambda' = 0, \pm 1, \pm 2, \ldots). \tag{87b}$$

This is of degree 2 with regard to $e$, $e'$, $I$, $I'$. The equation $\Delta_0 = 0$ represents the fact that there is at least one relation among an arbitrary set of $2n - 4 = 8$ of (87b) thus obtained. Hence, in order that there may exist a uniform integral, there should be one relation among these $2n - 4$ expressions. While (87) is an infinite series, (87b) is rational with regard to $e$, $e'$, $I$, $I'$ and the cosines of $\varpi$, $\varpi'$, $\Omega$, $\Omega'$. The question whether there is such a relation among the coefficients of the expansion will be discussed in Section 7.30 on the expansion of the disturbing function.

In the case of the $n$-body problem we write the differential equations (164), Section 5.16, for the $n - 1$ fictitious bodies. The order of the differential equations is $6n - 6$. Hence, we take $6n - 6$ as the value of $n$ in the above discussion. If we choose $L_1, \ldots, L_{n-1}$ so as to satisfy

$$m_1 \frac{\partial F_0}{\partial x_1} + \cdots + m_{n-1} \frac{\partial F_0}{\partial x_{n-1}} = 0, \tag{83a}$$

then (87) contains $n - 3$ variables $L_j$. It also contains $n - 1$ sets of variables $G_j$, $H_j$, $g_j$, $h_j$. Thus, it is a function of

$$4(n - 1) + n - 3 = 5n - 7$$

variables altogether. How many relations are there among the $5n - 6$ variables of (87), (87a), or (87b)? In order that there may exist $q$ uniform integrals other than $F$, there should be one relation among an arbitrary set of $5n - 6 - q$ variables of (87). As we have discussed in Section 5.8, the order of the differential equations is reduced to $6n - 10$. Excluding the $n - 1$ variables $L_j$, we must express the integrals as functions of $6n - 10 - n + 1 = 5n - 9$ variables. Thus $q = 3$. These integrals are the three integrals of the angular momentum. Hence, the differential equations written in the form (164) of Section 5.16 for the $n$ bodies have no uniform integral other than those of the energy and the angular momentum.

## 6.18 DISCUSSION OF POINCARÉ'S THEOREM

We have given in Section 6.12 an exceptional case where there exists an algebraic integral in the restricted three-body problem. When two of the three bodies in the three-body problem collide at a finite distance at the end of a finite interval of time, the differential equations cease to be regular, as will be discussed later. Otherwise, the coordinates of the three bodies can be determined uniquely by Cauchy's theorem on the existence of the solution of a differential equation, and the solution can be expanded in integral powers of $t - t_0$ around an arbitrary initial condition for $t = t_0$ (Painlevé, 1897). There are two independent conditions for such a collision, which reduce to one in the case of the plane problem. The conditions for a collision in the two-body problem are algebraic, but the conditions for the three- and $n$-body problems are transcendental according to Painlevé (1897, 1900, 1900a, 1900b). Suppose that a collision occurs when

$$\chi(t, x_1, \ldots, x_n, y_1, \ldots, y_n) = 0 \tag{92}$$

is satisfied. Since this is an invariantive relation of the motion, it should be an algebraic combination of the known classical integrals, according to Bruns's theorem. But the conditions for a collision should be satisfied when we increase $t$ by a constant amount; they should be unaltered when we increase the coordinates and the momentum by a constant amount.

We limit ourselves to the consideration of the plane three-body problem. Let the total kinetic energy of the whole three-body system around the center of mass be $T$, the total angular momentum $W$, and the force function $U$; then $T - U$ and $W$ remain unaltered by the above change in coordinates and momentum. Hence, the condition for a collision does not depend on the coordinates and the momentum of the center of mass but should be a relation between $W$ and $T - U$. The equations of motion remain unaltered by changing $t$ to $\alpha^3 t$, $x_i$ to $\alpha^2 x_i$, and $y_i$ to $\alpha^{-1} y_i$ with a constant $\alpha$; hence, the condition for a collision also remains unaltered by this transformation — in which $T - U$ is multiplied by $\alpha^{-2}$ and $W$ is multiplied by $\alpha$. Let the condition be

$$W = \phi(T - U);$$

then,

$$\alpha W = \phi[\alpha^{-2}(T - U)] = \alpha\phi(T - U).$$

Hence, the condition should be of the form

$$\phi(T - U) = \lambda/\sqrt{T - U},$$

or

$$W^2(T - U) = \lambda^2$$

with a constant $\lambda$. Whatever the value of $\lambda$, this corresponds to a motion of hyperbolic type; that is, it is the case when the three bodies recede to infinite distances from each other as $t$ approaches $\pm\infty$. Hence, it cannot be the condition for a collision. The condition should be transcendental. This is the argument of Painlevé, (1897, 1897a). On the other hand, $W$ should be finite even at the time of a collision, as has been proved by Sundman (1907, 1909, 1913) and as we shall see in a later chapter. But, since the momentum becomes infinite at the time of a collision, this relation is contradictory. Hence, according to Bruns's theorem, the condition for a collision cannot be algebraic; it should be represented by transcendental functions at least with respect to the momentum.

It was Levi-Civita and Bisconcini who obtained the condition for a collision explicitly (Levi-Civita, 1903, 1906; Bisconcini, 1906; see also Kiveliovitch, 1926; Rosenblatt, 1928; Chazy, 1932). According to these authors there exists (as will be shown in Chapter 22) a uniform integral in the case of a collision, but it is neither for $t$ nor for the Keplerian elements — with regard to which the integral is uniform. The theorem of Poincaré deals with the Keplerian elements; hence, it expresses the uniformity of the integral in the neighborhood of those values of $x_i$, $y_i$ that represent the state of motion belonging to the same osculating elliptic elements. As has been remarked by Levi-Civita, the theorem of Poincaré does not deny the existence of uniform integrals for a *part* of the trajectory. It does not deny the existence, in the neighborhood, of a state of motion very different from an elliptic orbit (Section 14.16).

In the case of a simultaneous collision of all three bodies at a point, the total angular momentum of the system must be zero, as proved by Sundman (1909). This is an algebraic invariantive relation. The above discussion can be applied to the $n$-body problem, as we will demonstrate in Chapter 22.

Consider the state of motion for $t \to \pm\infty$. When two of the three mutual distances of the three bodies become infinite of order 1 with regard to $t$ and one of the mutual distances is bounded, the motion is said to be elliptic-hyperbolic (Chazy, 1922). When two of the three bodies move in orbits of elliptic type around the center of mass and the third body moves in an orbit of parabolic type around the center of mass, that is, when two of the mutual distances becomes infinite of order $\frac{2}{3}$ and the remaining one is bounded, we say that the motion is elliptic-parabolic. When all three of the mutual distances are bounded, we say that the motion is oscillatory. In these three cases the energy constant $h$ of the whole system is negative, and the region $h < 0$ in the 16-dimensional space is divided into four subspaces. Three of these four subspaces contain the region $h = 0$ in their interiors. Each orbit in these subspaces is elliptic-hyperbolic; two of the mutual distances become infinite of order 1 with regard to $t$ as $t \to \pm\infty$; the third mutual distance, which remains the same all the time, is bounded. The three mutual distances are either bounded or oscillatory inside the subspace bounded by three analytic surfaces.

When the three mutual distances become infinite, of order $\frac{2}{3}$, 1, 1, respectively, as $t$ increases indefinitely, the motion is said to be parabolic-hyperbolic. When all three mutual distances become infinite of order 1, the motion is said to be hyperbolic. When the three bodies collide at infinity, the order of infinity is $\frac{2}{3}$ and the motion is said to be parabolic. The configuration of the three bodies in the immediate neighborhood of a simultaneous collision is that of either the equilateral triangular solution of Lagrange or the straight-line solution of Euler, as has been remarked in Chapter 3. In the first two cases the point at infinity $t = \infty$ is a transcendental singular point, but in the case of a collision it is an algebraic singularity of order two. These are the cases when the total energy constant $h$ of the whole system is positive.

The trajectories should be inside a subspace of eight dimensions embedded in the twelve-dimensional space, but in the case of a collision they should be inside a six-dimensional subspace. Two conditions are necessary for a collision and they have been shown (Chazy, 1918, 1922, 1929, 1932, 1933) to be transcendental with regard to $y_1, y_2, y_3, y_4, y_5, y_6$. The region in which the trajectories of the three bodies lie is divided into five regions, except for a simultaneous collision of the three bodies. The exterior region corresponds to hyperbolic motions, three of the five regions to elliptic-hyperbolic motions, and the remaining fifth either to oscillatory motions for which the three mutual distances are all bounded or to motions for which two of the mutual distances are arbitrarily large while the third is bounded. There are motions with elliptic osculating orbits in the fifth region. As $t$ increases indefinitely, each of the functions $y_1, y_2, y_3, y_4, y_5, y_6$ tends to a finite limit. The limits are uniform functions of the values of the twelve variables $x_i$, $y_i$ ($i = 1, 2, \ldots, 6$) for a certain epoch. Hence, the squares of the sum and the difference of the limiting values of each of these six $y_i$ are two uniform integrals in the domain of hyperbolic motions; each of those integrals represents a family of surfaces, and one of them passes through an arbitrary point in that domain. In the domain of elliptic-hyperbolic motions we denote by $y_4, y_5, y_6$ the momentum of the third body (receding to infinity) relative to the center of mass of the two bodies remaining in a bounded finite domain, and define the elliptic elements by six out of the twelve osculating elements and the hyperbolic elements by the remaining six; then these elements tend to finite limits in each of the trajectories, and each of them defines a uniform integral in the corresponding domain. Those uniform integrals that can exist simultaneously in both of these two domains define a family of surfaces, only one of which can pass through an arbitrary point in the whole domain. Among these surfaces, the surface bounding one of the five domains is a particular one. It can be shown that the integrals corresponding to this bounding surface cannot be reduced to the energy integral or the angular-momentum integrals (see Chapter 22).

As will be discussed later in detail, this theorem of Poincaré deals with uniform integrals with respect to the twelve elements belonging to the two osculating elliptic motions. Such integrals are uniform in the domain of another two-dimensional space and do not enter in this theorem of Poincaré.

The theorem of Bruns, on the other hand, differs from the theorem of Poincaré and can be applied to such integrals. One such integral is neither an algebraic combination of the classical integrals, nor an algebraic combination of $y_i$ only, as in the extension of the theorem by Painlevé. The bounding surfaces of the five domains of these trajectories are not algebraic surfaces.

Painlevé (1900, 1900a, 1900b) studied the cases in which the solution can be represented by meromorphic functions of $t$ with polar singularities. He has given algebraic differential equations that admit uniform integrals, besides the known functions, such as exponential, elliptic, Abelian, and uniform integrals of a linear differential equation including Fuchsian and automorphic functions. The critical points are called *fixed* if they do not depend on the integration constants; otherwise they are called *movable*. Painlevé, extending the works of Poincaré and Picard, examined differential equations that admit the general integrals only with fixed critical points. He considered differential equations of the first, second, and third order. In the case of the third order he found a new class of singularities, for example, those with a perfect set of discontinuities (see Chapter 22).

## 6.19 INTEGRALS WITHOUT PARAMETERS

Although the theorem of Poincaré looks as if it were more general than the theorem of Bruns, it has a narrower validity domain; it is valid only for vanishingly small values of $\mu$. However, it is wider than Bruns's in the sense that it is not limited to algebraic integrals. The theorem of Poincaré cannot be valid if we take off the restriction that the integrals should be arranged in integral powers of $\mu$ or expanded in the sequence of base functions of an algebraic field of $\mu$. It may happen to be false when the masses have particular values.

We suppose that in the system of differential equations

$$\frac{d\xi_i}{dt} = \Xi_i \qquad (i = 1, 2, \ldots, m), \tag{93}$$

the right-hand members $\Xi_1, \Xi_2, \ldots, \Xi_m$ do not contain $t$ explicitly, are expanded in integral powers of $\xi_1, \xi_2, \ldots, \xi_m$, and are not all zero for $\xi_1 = \xi_2 = \cdots = \xi_m = 0$. Then there may exist $m - 1$ independent integrals that do not contain $t$ explicitly and can be expanded in convergent integral power series of $\xi_1, \xi_2, \ldots, \xi_m$ (Cherry, 1924, 1924a, 1926, 1927, 1928; Woltjer, 1924; Fermi, 1923).

Let the values of $\Xi_1, \Xi_2, \ldots, \Xi_m$ for

$$\xi_1 = \xi_2 = \cdots = \xi_m = 0$$

be, respectively, $a_1, a_2, \ldots, a_m$ and at least one of them, for example $a_1$, be different from zero. Replacing $\xi_i$ by

$$\xi_i' = \xi_i - \frac{a_i}{a_1}\xi_1 \qquad (i = 2, 3, \ldots, m),$$

we obtain

$$\frac{d\xi_i'}{dt} = \Xi_i - \frac{a_i}{a_1}\Xi_1 = \Xi_i' \qquad (i = 2, 3, \ldots, m).$$

As

$$\Xi_i' = 0 \qquad (i = 2, 3, \ldots, m) \quad \text{for} \quad \xi_2' = \xi_3' = \cdots = \xi_m' = 0$$

and

$$\frac{\partial(\xi_1, \xi_2', \xi_3', \ldots, \xi_m')}{\partial(\xi_1, \xi_2, \xi_3, \ldots, \xi_m)} = 1,$$

(93) can be written

$$\frac{d\xi_1}{dt} = \lambda + \sum_{j=1}^{m} a_j^{(1)}\xi_j + \frac{1}{2}a_{11}^{(1)}\xi_1^2 + a_{12}^{(1)}\xi_1\xi_2 + \cdots,$$

$$\frac{d\xi_i}{dt} = \sum_{j=1}^{m} a_j^{(i)}\xi_j + \frac{1}{2}a_{11}^{(i)}\xi_1^2 + a_{12}^{(i)}\xi_1\xi_2 + \cdots \qquad (i = 2, 3, \ldots, m). \tag{93a}$$

We shall solve these equations in ascending power series of

$$\xi_1, \xi_2, \ldots, \xi_m$$

of the form

$$\xi_i = \xi_i^{(1)} + \xi_i^{(2)} + \cdots \qquad (i = 1, 2, \ldots, m). \tag{94}$$

Substituting (94) in (93a) and equating the terms of the same degree, we first obtain from

$$\frac{d\xi_1^{(1)}}{dt} = \lambda, \qquad \frac{d\xi_j^{(1)}}{dt} = 0 \qquad (j = 2, 3, \ldots, m)$$

the solution

$$\xi_1^{(1)} = \lambda(t + c_1), \qquad \xi_j^{(1)} = c_j \qquad (j = 2, 3, \ldots, m). \tag{95}$$

Substituting these in the second-degree terms

$$\frac{d\xi_i^{(2)}}{dt} = \sum_{j=1}^{m} a_j^{(i)} \xi_j^{(1)} \qquad (j = 1, 2, \ldots, m),$$

and solving

$$\frac{d\xi_i^{(2)}}{dt} = a_1^{(i)}\lambda(t + c_1) + \sum_{j=2}^{m} a_j^{(i)} c_j \qquad (i = 1, 2, \ldots, m),$$

we obtain

$$\xi_i^{(2)} = \tfrac{1}{2}a_1^{(i)}\lambda(t + c_1)^2 - \sum_{j=2}^{m} a_j^{(1)} c_j(t + c_1) \qquad (i = 1, 2, \ldots, m). \tag{96}$$

The integration constants are determined so that $\xi_i^{(2)}$ is homogeneous of the second degree in $t + c_1, c_2, \ldots, c_m$, and vanishes together with $t + c_1$. By repeating this procedure we can determine $\xi_i^{(s)}$, $(s = 2, 3, \ldots)$, so as to make them vanish together with $t + c_1$. If we substitute these in (94), we obtain the solution in the form of an integral function of $t + c_1, c_2, \ldots, c_m$ satisfying (93a).

Suppose that $\mathrm{H}_1, \mathrm{H}_2, \ldots, \mathrm{H}_m$ are the majorant functions, respectively, of $\Xi_1, \Xi_2, \ldots, \Xi_m$ in $\eta_1, \eta_2, \ldots, \eta_m$; that is, the coefficient in the expansion of $\mathrm{H}_k$ with regard to $\eta_1, \eta_2, \ldots, \eta_m$ is larger than the corresponding coefficient in the expansion of $\Xi_k$ with regard to $\xi_1, \xi_2, \ldots, \xi_m$ in the absolute values within the domain of $\xi_i$, $\eta_i$ in question. Consider

$$\frac{d\eta_1}{dt} = \mathrm{H}_1 = \mu + \sum_{j=1}^{m} b_j^{(1)}\eta_j + \frac{1}{2} b_{11}^{(1)}\eta_1^2 + b_{12}^{(1)}\eta_1\eta_2 + \cdots,$$

$$\frac{d\eta_i}{dt} = \mathrm{H}_i = \sum_{j=1}^{m} b_j^{(i)}\eta_j + \frac{1}{2} b_{11}^{(i)}\eta_1^2 + b_{12}^{(i)}\eta_1\eta_2 + \cdots \tag{97}$$

$$(i = 2, 3, \ldots, m),$$

and try to expand the solution, similarly to (94), in the form

$$\eta_i = \eta_i^{(1)} + \eta_i^{(2)} + \cdots \qquad (i = 1, 2, \ldots, m). \tag{98}$$

Solving these equations in the same manner as we did for (93), we obtain

$$\eta_1^{(1)} = \mu(t + d_1), \qquad y_j^{(1)} = d_j \qquad (j = 2, 3, \ldots, m), \tag{99}$$

$$\eta_i^{(2)} = \tfrac{1}{2}b_1^{(i)}\mu(t + d_1)^2 + b_2^{(i)}d_2(t + d_1) + \cdots + b_m^{(i)}d_m(t + d_1)$$
$$(i = 2, 3, \cdots, m). \quad (100)$$

Choose the integration constants $t + d_1, d_2, \ldots, d_m$ so that they are positive, as are $t + c_1, c_2, \ldots, c_m$, and satisfy

$$|t + c_1| < t + d_1, \qquad |c_2| < d_2, \ldots, |c_m| < d_m.$$

Comparing (95) with (99), we have

$$|\xi_1^{(1)}| < \eta_1^{(1)}, \ldots, |\xi_m^{(1)}| < \eta_m^{(1)}.$$

Comparing (96) with (100), we have

$$|\xi_1^{(2)}| < \eta_1^{(2)}, \ldots, |\xi_m^{(2)}| < \eta_m^{(2)}.$$

In general, $\eta_i^{(s)}$ is larger than the corresponding $\xi_i^{(s)}$ in absolute value. Hence, if $t + d_1, d_2, \ldots, d_m$ are chosen so that (98) is convergent, then the series (94) is absolutely and uniformly convergent.

Now let $M$ and $\alpha$ be sufficiently large positive constants, and the expansion of

$$\frac{M(\eta_1 + \eta_2 + \cdots + \eta_m)}{1 - \alpha(\eta_1 + \eta_2 + \cdots + \eta_m)}$$

be the functions $\mathrm{H}_1 - \mu, \mathrm{H}_2, \ldots, \mathrm{H}_m$ of (97) with $\mu > |\lambda|$; then, writing $\eta_1 + \eta_2 + \cdots + \eta_m = \zeta$, we obtain

$$\frac{d(\eta_1 - \eta_j)}{dt} = \mu \qquad (j = 2, 3, \ldots, m),$$

$$\frac{d\zeta}{dt} = \frac{\mu + \zeta(mM - \alpha\mu)}{1 - \alpha\zeta}.$$

We can choose $M$ and $\alpha$ so that they satisfy $mM - \alpha\mu = 0$ and are both sufficiently large positive constants. Then the solution of these equations are

$$\zeta - \tfrac{1}{2}\alpha\zeta^2 = \mu t + \text{constant},$$

$$\eta_1 - \eta_j = \mu t + \text{constant} \qquad (j = 2, 3, \ldots, m).$$

Determine the constants so that $\eta_1 = 0, \eta_2 = d_2, \ldots, \eta_m = d_m$ for $t + d_1 = 0$; then

$$\zeta - \tfrac{1}{2}\zeta^2 = \mu(t + d_1) + d - \tfrac{1}{2}\alpha d^2 \qquad (d = d_2 + \cdots + d_m)$$

$$\eta_1 - \eta_j = \mu(t + d_1) - d_j \qquad (j = 2, 3, \ldots, m). \tag{101}$$

Hence, the value of $\zeta$ equal to $d$ for $t + d_1 = 0$ is

$$\zeta = \frac{1}{\alpha}[1 - \sqrt{1 - 2\alpha\mu(t + d_1) - 2\alpha d + \alpha^2 d^2}].$$

This function can be expanded in integral powers of $t + d_1, d_2, \ldots, d_m$ which is convergent for sufficiently small values of

$$|t + d_1|, |d_2|, \ldots, |d_m|.$$

Thus $\eta_1, \eta_2, \ldots, \eta_m$ can be expanded in integral power series of the same nature by (101) and are of the same form as (98), which is absolutely and uniformly convergent. Hence, (94) is absolutely and uniformly convergent so long as $|t + c_1|, |c_2|, \ldots, |c_m|$ are sufficiently small.

Thus, we have obtained the solution of (93a) in the form (94) so that it is zero for $t + c_1 = c_2 = \cdots = c_m = 0$ and

$$\frac{\partial(\xi_1, \xi_2, \ldots, \xi_m)}{\partial(c_1, c_2, \ldots, c_m)} = \lambda.$$

Hence, $t + c_1, c_2, \ldots, c_m$ are convergent as integral power series of $\xi_1, \xi_2, \ldots, \xi_m$ insofar as the latter variables are sufficiently small. Hence

$$\begin{aligned} \lambda(t + c_1) &= \xi_1 - \frac{\xi_1}{\lambda}(\tfrac{1}{2}a_1^{(1)}\xi_1 + a_2^{(1)}\xi_2 + \cdots + a_m^{(1)}\xi_m) + \cdots \\ &= \Phi_1(\xi_1, \xi_2, \ldots, \xi_m), \\ c_j &= \xi_j - \frac{\xi_1}{\lambda}(\tfrac{1}{2}a_1^{(j)}\xi_1 + a_2^{(j)}\xi_2 + \cdots + a_m^{(j)}\xi_m) + \cdots \\ &= \Phi_j(\xi_1, \xi_2, \ldots, \xi_m) \qquad (j = 2, 3, \ldots, m). \end{aligned} \tag{102}$$

Each term of the second and the higher degrees has $\xi_1$ as a factor. Since $c_1, c_2, \ldots, c_m$ are arbitrary constants, $\Phi_2, \ldots, \Phi_m$ are $m - 1$ independent integrals. Thus, we have obtained $m - 1$ independent integrals of (93).

Next we consider a set of Hamiltonian canonical equations with $m = 2n$

$$\frac{d\xi_i}{dt} = \frac{\partial F}{\partial \eta_i}, \qquad \frac{d\eta_i}{dt} = -\frac{\partial F}{\partial \xi_i} \qquad (i = 1, 2, \ldots, n), \tag{103}$$

$$F = F_0 + F_1 + \cdots$$

If we apply Poincaré's contact transformation

$$\xi_i = \sqrt{2x_i}\cos y_i, \qquad \eta_i = \sqrt{2x_i}\sin y_i \qquad (i = 1, 2, \ldots, n),$$

then the differential equations are transformed to the same form as equations (73) in the theorem of Poincaré but without the parameter $\mu$.

In order that $F_0$ shall not contain variables other than $x_1, x_2, \ldots, x_n$ as in the form of Poincaré, $F_0$ must be a power series of

$$\xi_1^2 + \eta_1^2, \ldots, \xi_n^2 + \eta_n^2,$$

and the derivatives of $F_0$ must not be zero; furthermore, the coefficients $B_{m_1 \cdots m_n}$ in the expansion of $F_1$ must not have a small divisor $\sum_{i=1}^n m_i \, \partial F_0/\partial x_i$. These conditions are satisfied by

$$F_0 = \lambda_1(\xi_1^2 + \eta_1^2) + \cdots + \lambda_n(\xi_n^2 + \eta_n^2) + \nu_1(\xi_1^2 + \eta_1^2)^2 + \cdots + \nu_n(\xi_n^2 + \eta_n^2)^2.$$

Indeed, if we take such an $F_0$, then there exist $2n - 2$ independent integrals as given by (102) besides $F =$ constant. Hence, there exist $2n - 2$ independent integrals without the parameter $\mu$ for the equations of Poincaré. Thus the theorem of Poincaré fails when we consider integrals that do not contain the small parameter $\mu$.

Whittaker (1916, 1937) considered the differential equations of form (103) for a dynamical system of two degrees of freedom, in which the Hamiltonian function $F$ is expanded in integral powers of $\sqrt{\xi_1}$, $\sqrt{\xi_2}$ and in a trigonometric series in $\eta_1, \eta_2$ of the form

$$\sum A\xi_1^{s_1/2}\xi_2^{s_2/2} \cos(j_1\eta_1 + j_2\eta_2),$$

with the first term $s_1\xi_1 + s_2\xi_2$, where $A$ is a constant, $s_1$, $s_2$ are positive integers, and $j_1$, $j_2$ are positive or negative integers including zero, $s_1 + s_2$ being the degree of the term. Thus $F$ is written

$$F = s_1\xi_1 + s_2\xi_2 + F_3 + F_4 + \cdots.$$

Whittaker discovered the so-called *adelphic integral* for this system of differential equations (discussed in Section 14.15). The integral takes three different forms according as: (i) $s_1/s_2$ is irrational; (ii), $s_1/s_2$ is rational and equals the ratio of two integers $m/n$, and there is no term of the form $\cos(n\eta_1 - m\eta_2)$ in $F_3$; or (iii), $s_1/s_2$ equals the ratio of two integers $m/n$ but there is a term $\cos(n\eta_1 - m\eta_2)$.

Fermi (1923, 1923a) tried to relate the theorem of Poincaré to the quasi continuity of the trajectories and to base on it the foundation of the kinetic theory of gases of Boltzmann and Maxwell. Maxwell based his theory on the assumption of the so-called continuity of path of the trajectories, the assumption that a dynamical system passes through all phases permitted by the law of energy conservation in a sufficiently long interval of time (Jeans, 1921). The modern ergodic theorem has been refined from this assumption of quasi continuity of the dynamical trajectories (Hagihara, 1933; discussed in Chapter 26).

Fermi, by adopting the above proof of the theorem of Poincaré as it stands, proved the validity of quasi ergodicity for a nondegenerate dynamical system of two degrees of freedom without perturbation. A system is nondegenerate if: (i), from the relation

$$\frac{\partial F_0}{\partial x_1} \bigg/ \frac{\partial F_0}{\partial x_2} = \text{constant}$$

the relation

$$\frac{\partial F_0}{\partial x_3} \bigg/ \frac{\partial F_0}{\partial x_4} = \text{constant},$$

for example, does not follow; (ii), if $\sum_i m_i(\partial F_0/\partial x_i)$ does not vanish on any surface $F_0 = \text{constant}$, where $m_i$ are integers, not all of them zero; and (iii), the vanishing of the coefficient $B_{m_1 \cdots m_n}$ of Section 6.15 does not follow from $\sum_i m_i(\partial F_0/\partial x_i) = 0$.

Born and Pauli (1922) expanded the perturbation in a quantum-mechanical system in the present form with the angular and linear variables of Section 2.7 as the canonically conjugate variables. They tried to express the perturbation in a multiple Fourier series of the angular variables by expanding it with a small parameter. The perturbation in this case is the imposed electric field and the magnetic field. Born and Brody (1921) based the discussion on a different form of expansion. Epstein (1922) used the expansion of Delaunay (Chapter 9). Since according to the theorem of Poincaré there exists no uniform integral, it is possible to give the quantum condition to a quantum-mechanical system with perturbations. Owing to the fact that such series expansions are not uniformly convergent (as will be shown in Chapter 15), the quantum-mechanical calculation presents a grave and fundamental difficulty. According to Born and Pauli the broadening of spectral lines of a perturbed system due to electromagnetic forces is negligible compared with the broadenings due to pressure and the Doppler effect. It is the conditionally periodic motion discussed in Chapter 2 that enables us to escape from this difficulty, but, as we have argued in Section 2.7, it is necessary to have a heavy and severe condition for that purpose (See Section 4.24; also Arnold, 1963). Whether such a condition is actually satisfied in any quantum-mechanical system, whether we should adopt such a condition as an axiom in quantum mechanics, is a question for quantum mechanicians. Here appears the important role that should be played by the theory of adiabatic invariants discussed in Section 4.23. However, we have there assumed the quasi ergodicity of the dynamical system; so the question is raised again. On the other hand, a quantum-mechanical system without perturbation is degenerate in general; so the same difficulty appears. It was presumably for this reason that quantum mechanicians in the past assumed the general invalidity of classical dynamics in quantum theory.

Arnold (1963) recently solved the problem by means of quasi-periodic functions, along with the difficulty of the degenerate cases, making the severe assumption that the series representing the functions are almost everywhere convergent — by excluding the domain of Lebesgue measure zero, as will be discussed in a later chapter.

In the light of modern developments in wave mechanics and quantum mechanics, especially the theory based on Bohr's complementarity principle, our mode of thinking has changed. However, even now it is evident that only a first approximation is treated in the computation of the perturbation; indeed, the method for the second approximation fails completely owing to the nonconvergence of the process. We are in quantum mechanics confronted with the same grave difficulty of non-convergence that makes its first appearance in classical dynamics.

## KOWALEWSKI'S TOP MOTION

### 6.20 RIGID-BODY ROTATION

Despite the theorems of Poincaré and Bruns, Kowalewski (1888, 1890) has found an algebraic integral other than the classical integrals in the motion of a rigid body. We shall discuss this case briefly.

The motion of a rigid body suspended at its center of mass has been completely solved by Euler. Denote the three moments of inertia around the center of mass by $A$, $B$, $C$. The motion under gravity of a rigid body, such that $A = B$, suspended at a point on the principal axis of inertia corresponding to $C$, has been completely solved by Lagrange. The motion in the case $A = B = 2C$ has also been solved completely by using elliptic functions and hyperelliptic functions, no matter where the center of mass is located. This is called the *Kowalewski case*. In such cases there exists an integral algebraic in the coordinates and the momentum, which is not an algebraic combination of the classical integrals (energy and angular-momentum). Poincaré has proved that there cannot exist any uniform integral, which is not an algebraic combination of the classical integrals, for a general initial condition, if we exclude the above cases of Euler, Lagrange, and Kowalewski. Several authors have discussed the case of $A = B = 2C/n$, $n$ being an integer (Husson, 1905, 1906, 1908, 1932; Liouville, 1897; Hesse, 1890; Kötter, 1892, 1893; Staude, 1894; Nekrassoff, 1896; Kolossoff, 1902, 1903; Stäckel, 1908; Oseen, 1911; Marcolongo, 1902, 1906, 1910; Levi-Civita, 1901; Lazzarino, 1919; see also Klein and Sommerfeld, 1921; Halphen, 1888; Leimanis, 1965).

The quantum-mechanical analogues of these top motions have been studied by Casimir (1931) in general, by Kramers and Ittmann (1929, 1930) and O. Klein (1929) for Euler's case, by Reiche and Rademacher (1926, 1927) and Kronig and Rabi (1927) for Lagrange's case, and by Laporte (1933) for Kowalewski's case.

We shall first show, following Poincaré (1892), that there can exist a uniform integral in the case $A = B$.

Suppose that the product of the distance between the point of suspension and the center of mass multiplied by the gravitational force is infinitesimally small and write the equations of motion of the rigid body in the form

$$\frac{dx_i}{dt} = \frac{\partial F}{\partial y_i}, \qquad \frac{dy_i}{dt} = -\frac{\partial F}{\partial x_i} \qquad (i = 1, 2, 3),$$

$$F = F_0 + \mu F_1,$$

where $F$ is the total energy, $F_0$ the kinetic energy, $\mu F_1$ is the product of the distance from the center of mass to the horizontal plane passing through the point of suspension, multiplied by the gravitational force. If $\mu = 0$, then the center of mass coincides with the point of suspension and the motion of the rigid body is that of Poinsot. Since $\mu$ is supposed to be small, the motion is the Poinsot motion, to the first approximation. In the Poinsot motion the ellipsoid of inertia makes a rolling motion by always touching a fixed plane, without slipping. Let $P$ be the foot of the perpendicular from the point of suspension to this fixed plane, and $Q$ be the point of contact at a certain epoch. This point of contact describes a fixed curve upon the ellipsoid of inertia. This is called the polhode. Suppose that after a lapse of time $\tau$ the same point $Q$ of the polhode comes to a point $Q'$ at which the ellipsoid of inertia touches the fixed plane. Denote the angle $QPQ'$ by $\alpha$ and put $n = 2\pi/\tau$, $n' = \alpha/\tau$; $n$ and $n'$ correspond to the mean motions in the Keplerian elliptic motion. Denote the coordinates of a point of the rigid body by $x$, $y$, $z$; let the point of suspension be the coordinate origin, and take the $z$ axis as the vertical. With integration constants $\epsilon$, $\epsilon'$, write $l = nt + \epsilon$, $l' = n't + \epsilon'$. Let $\xi$, $\eta$, $\psi$ be functions of $n$, $n'$, $l$ and periodic with period $2\pi$ in $l$. With the new integration constants $\theta$, $\phi$,

$$\begin{aligned} x &= \cos\theta(\xi\cos l' - \eta\sin l') - \sin\theta\cos\phi(\xi\sin l' + \eta\cos l') \\ &\quad + \psi\sin\theta\sin\phi, \\ y &= \sin\theta(\xi\cos l' - \eta\sin l') + \cos\theta\cos\phi(\xi\sin l' + \eta\cos l') \\ &\quad - \psi\cos\theta\sin\phi, \\ z &= \sin\phi(\xi\sin l' + \eta\cos l') + \psi\cos\phi. \end{aligned}$$

Let $(x, y, z)$ be the coordinates of the center of mass; then $F_1$ is reduced to $z$, apart from a constant factor.

$$F_1 = \sum B_{m,1} \exp [\sqrt{-1}\,(ml + l')] + \sum B_{m,0} \exp [\sqrt{-1}\, ml]$$
$$+ \sum B_{m,-1} \exp [\sqrt{-1}\,(ml - l')],$$

where the coefficients $B$ depend on $n, n', \phi$. Refer to the general discussion of Section 6.15. Choose $n, n'$ such that

$$p \frac{\partial F_0}{\partial x} + q \frac{\partial F_0}{\partial y} = 0;$$

then $B$ depends only on $\phi$. Hence, there must be a relation between two of the coefficients. Since $D_\lambda$ defined in Section 6.15 depends only on $\phi$ and $\zeta$, there must be one relation among arbitrary $2n - 3 = 3$ of these $D_\lambda$, where we put $D_\lambda = B_{\lambda p, \lambda q}\zeta^\lambda$. The series described in Section 6.15 are all singular of order 1; hence, there may exist a uniform integral distinct from the energy integral. Actually, there is the angular-momentum integral. In order to see whether there exists a third integral we look for the classes that are singular of order 2. In that case there must necessarily be two relations among an arbitrary three of $D_\lambda$; indeed, the condition is also sufficient. Hence, $D_\lambda$ should be a function of only one of such coefficients. We list several classes:

1. Class 1/0 which contains all coefficients of the form $B_{m,0}$. This class is singular of order 2. For, we have $B_{m,0} = C_{m,0} \cos \phi$, where $C_{m,0}$ depends only on $n, n'$, and, as $n, n'$ are supposed to be constant, $C_{m,0}$ is considered as a constant. Thus, $D_\lambda = C_{\lambda,0}(\cos \phi)\zeta^\lambda$. In order that $D_\lambda$ shall be a function of only one of the coefficients, all coefficients $C_{\lambda,0}$ except one should vanish, or $\psi$ should be reduced to the form $\exp(\sqrt{-1}\, ml)$. For satisfying $p\, \partial F_0/\partial x + q\, \partial F_0/\partial y = 0$, however, we must equate $n$ to zero. The Poinsot motion for $n = 0$ is a uniform rotation about one of the principal axes of inertia, and $\psi$ is a constant independent of $l$. Hence, the values of $C_{\lambda,0}$ other than $C_{0,0}$ are all zero for such special values of $n, n'$. Thus this class is singular of order 2.

2. Classes of the form $m/1$ contain only three coefficients: $B_{m,1}$, $B_{0,0}$, $B_{-m,-1}$. These classes are singular of order 2 only if $B_{m,1} = B_{-m,-1} = 0$; in other words, there should be no terms $\exp(\sqrt{-1}\, ml)$ in the expansion of $\xi + \sqrt{-1}\,\eta$, $\xi - \sqrt{-1}\,\eta$ in positive and negative powers of

$$\exp(\sqrt{-1}\, l),$$

as $\xi$, $\eta$ are supposed to be real. Generally, this does not happen unless the ellipsoid of inertia is an ellipsoid of revolution. When two axes of the ellipsoid are equal, then

$$\xi = A' \cos l + B' \sin l + C', \qquad \eta = A'' \cos l + B'' \sin l + C'',$$

where $A'$, $B'$, $C'$, $A''$, $B''$, $C''$ are constants. Hence, unless $m = 0$, 1 or $-1$, we have $B_{m,1} = -B_{-m,-1} = 0$. Hence the classes $m/1$, except $1/1$, $0/1$, $-1/1$, are singular of order 2.

3. The other classes, reducing to only one coefficient $B_{0,0}$, are singular of order 2.

Summarizing, we have seen that if the ellipsoid of inertia is an ellipsoid of revolution then the singularity is of order 2, except for the classes $1/1$, $0/1$, $-1/1$. Hence, there may exist a third uniform integral; indeed, there may exist an algebraic integral, provided that the Jacobian of these three integrals is zero for $n' = 0$ or $n' = \pm n$. In the case of Lagrange, in which the point of suspension is on the $C$ axis of the ellipsoid of inertia, the last condition is not necessary since $\xi$, $\eta$ are then reduced to constants. If the ellipsoid of inertia is not an ellipsoid of revolution, then there are an infinite number of classes that are not singular of order 2, that is, the classes $m/1$. Consider a domain $(D)$ which contains an infinite number of sets of values of $n$, $n'$, $\phi$, $\theta$ and suppose that in any such set the value of $n'$ is not an integral multiple of $n$. None of the classes $m/1$ belong to this domain. In this case there may exist a third uniform integral, provided that the Jacobian of these three integrals vanishes when $n'$ is an integral multiple of $n$. Hence we can conclude that the third integral cannot be algebraic in general.

## 6.21 SOLUTION IN ALGEBRAIC INTEGRALS

Husson (1906, 1908) has proved, independently of Poincaré, on the basis of Abel's theory of algebraic functions, that any algebraic integral with arbitrary initial conditions is an algebraic combination of the classical integrals (except in the cases of Euler, Lagrange, and Kowalewski), and that consequently there cannot exist any algebraic integral other than the classical integrals in the general case of top motions (except in the aforesaid cases). Liouville's proof (1897) of his theorem on a basis similar to the proof by Bruns was incomplete. Painlevé (1900) criticized Kowalewski's proof.

Let the origin $O$ be the point of suspension; take coordinates in the directions, fixed in the rigid body, of the principal axes of inertia with regard to the point of suspension. Let the center of mass be at a point $G$; let the unit of length be such that the numerical ratio of the external force $mg$ on the body to the distance $OG$ equals 1. Let the direction cosines of the direction of gravity in these coordinates be $\gamma, \gamma', \gamma''$. Then the equations of motion of the rigid body around $O$ are

$$\begin{aligned} A\frac{dp}{dt} &= (B-C)qr + y_0\gamma'' - z_0\gamma', & \frac{d\gamma}{dt} &= r\gamma' - q\gamma'', \\ B\frac{dq}{dt} &= (C-A)rp + z_0\gamma - x_0\gamma'', & \frac{d\gamma'}{dt} &= p\gamma'' - r\gamma, \\ C\frac{dr}{dt} &= (A-B)pq + x_0\gamma' - y_0\gamma, & \frac{d\gamma''}{dt} &= q\gamma - p\gamma'. \end{aligned} \tag{104}$$

In order to separate the variables in the manner of Sections 1.15 and 1.16, transform by

$$y_1 = ap + \sqrt{-1}\,bq, \qquad y_2 = ap - \sqrt{-1}\,bq,$$

$$z_1 = \gamma + \sqrt{-1}\,\gamma', \qquad z_2 = \gamma - \sqrt{-1}\,\gamma',$$

with

$$a = \sqrt{A(A-C)}, \qquad b = \sqrt{B(B-C)},$$

and put $-\sqrt{-1}\,t$ in place of $t$. Then

$$\begin{aligned} \frac{dy_1}{dt} &= -\sqrt{\frac{(A-C)(B-C)}{AB}}\,ry_1 - \left[x_0\sqrt{\frac{B-C}{B}} + \sqrt{-1}\,y_0\sqrt{\frac{A-C}{A}}\right]\gamma'' \\ &\quad + \frac{z_0}{2}\left[\sqrt{\frac{B-C}{B}}\,(z_1+z_2) + \sqrt{\frac{A-C}{A}}\,(z_1-z_2)\right], \\ \frac{dy_2}{dt} &= \sqrt{\frac{(A-C)(B-C)}{AB}}\,ry_2 + \left[x_0\sqrt{\frac{B-C}{B}} - \sqrt{-1}\,y_0\sqrt{\frac{A-C}{A}}\right]\gamma'' \\ &\quad + \frac{z_0}{2}\left[\sqrt{\frac{B-C}{B}}\,(z_1+z_2) - \sqrt{\frac{A-C}{A}}\,(z_1-z_2)\right], \\ C\frac{dr}{dt} &= \frac{A-B}{4\sqrt{AB(A-C)(B-C)}}\,(y_2^2 - y_1^2) - \frac{x_0}{2}\,(z_1 - z_2) \\ &\quad + \frac{\sqrt{-1}\,y_0}{2}\,(z_1+z_2), \end{aligned} \tag{105}$$

$$\frac{dz_1}{dt} = -rz_1 + \frac{1}{2}\left[\frac{y_1 + y_2}{\sqrt{A(A - C)}} + \frac{y_1 - y_2}{\sqrt{B(B - C)}}\right]\gamma'',$$

$$\frac{dz_2}{dt} = rz_2 - \frac{1}{2}\left[\frac{y_1 + y_2}{\sqrt{A(A - C)}} - \frac{y_1 - y_2}{\sqrt{B(B - C)}}\right]\gamma'',$$

$$\frac{d\gamma''}{dt} = \frac{1}{4\sqrt{A(A - C)}}(y_1 + y_2)(z_1 - z_2) - \frac{1}{4\sqrt{B(B - C)}}(y_1 - y_2)(z_1 + z_2). \tag{105 cont.}$$

There exist three integrals — the energy integral, the angular-momentum integral, and the relation among the direction cosines:

$$\frac{(y_1 + y_2)^2}{4(A - C)} - \frac{(y_1 - y_2)^2}{4(B - C)} + Cr^2 - [x_0(z_1 + z_2) - \sqrt{-1}\, y_0(z_1 - z_2) + 2z_0\gamma''] = \text{constant}, \tag{106}$$

$$\frac{A}{4\sqrt{A(A - C)}}(y_1 + y_2)(z_1 + z_2) - \frac{B}{2\sqrt{B(B - C)}}(y_1 - y_2)(z_1 - z_2) + 2Cr\gamma'' = \text{constant},$$

$$z_1 z_2 + \gamma''^2 = \text{constant}.$$

Multiply $y_1$, $z_1$, $z_2$, $\gamma''$ by a constant parameter $\lambda$. Denote the integration constants on the right-hand sides of the three equations (106) after this substitution by $h_1$, $h_2$, $h_3$, respectively; these are supposed to be independent of $\lambda$. Denote the resulting equations — after this substitution in (105) and (106) — by (105a) and (106a), respectively. Since the right-hand members of (106a) are regular for $\lambda = 0$, there exists the solution of (106a) arranged in positive integral power series of $\lambda$ insofar as $\lambda$ is sufficiently small, according to Poincaré's extension of Cauchy's existence theorem of the solution of a differential equation (discussed in a later chapter). We eliminate $t$ from these equations and try to obtain $y_1$, $y_2$, $z_1$, $z_2$, $\gamma''$ as functions of $r$. In the case treated by Lagrange, we obtain $r =$ constant, because $x_0 = y_0 = 0$, so we disregard this case. Equations (106a) give $y_2$, $z_1$, $z_2$ as functions of $y_1$, $\gamma''$, $r$, and (105a) are replaced by two differential equations of order 1 for determining $y_1(r)$, $\gamma''(r)$. The derivatives $dy_1/dr$ and $d\gamma''/dr$ are algebraic with regard to $h_1$, $h_2$, $h_3$, $\lambda$, $y_1$, $\gamma''$, $r$, and regular where $\lambda = 0$.

If there exists an algebraic integral $f(y_1, z_1, z_2, \gamma'', y_2, r) =$ constant, which does not depend on $t$ explicitly other than the classical integrals for the equations (105), then there exists an algebraic integral

$$f(\lambda y_1, \lambda z_1, \lambda z_2, \lambda\gamma'', y_2, r) = \text{constant}$$

for equations (105a) for any value of $\lambda$. Substitute for $y_2$, $z_1$, $z_2$ from (106a); then we obtain two differential equations of the first order for $y_1(r)$, $\gamma''(r)$ in the form $F(h_1, h_2, h_3, \lambda, y_1, \gamma'', r) = \text{constant}$, where $F$ is algebraic in each of the arguments. Here $F$ must contain $y_1$ and $\gamma''$. Otherwise $r$ would become a constant; then, either the case would reduce to Lagrange's, or $F$ would not contain $y_1$, $\gamma''$, $r$ and the integral $f = \text{constant}$ would be one of the combinations of the classical integrals.

If we multiply $F$ by a power of $\lambda$ such that $\lambda = 0$ is neither a zero point nor a pole of $F$ considered as a function of $\lambda$, then the algebraic function $F$ of $\lambda$ can be expanded in positive integral ascending powers of $\lambda$ or $\lambda^{1/p}$ in the neighborhood of $\lambda = 0$. The coefficients of various powers of $\lambda$ in this expansion are algebraic in $h_1, h_2, h_3, y_1, \gamma'', r$. But we can suppose that $F$ contains $y_1, \gamma'', r$ besides $h_1, h_2, h_3$ at $\lambda = 0$, that is, that $F$ is of the form

$$F = F_0(h_1, h_2, h_3, y_1, \gamma'', r) + \lambda^{1/p} F_1(h_1, h_2, h_3, y_1, \gamma'', r) + \cdots = \text{constant}. \tag{107}$$

We separate the equations (105a) for $\lambda = 0$ into the following:

$$\frac{dy_1}{dt} = -\alpha_0 r y_1 + z_0(\alpha_1 z_1 + \alpha_2 z_2) - \alpha_3 \gamma'', \tag{108}$$

$$\frac{dy_2}{dt} = \alpha_0 r y_2, \qquad \frac{dr}{dt} = \frac{A - B}{4ABC\alpha_0} y_2^2, \tag{109}$$

$$\begin{aligned} \frac{dz_1}{dt} &= -rz_1 + \beta_1 y_2 \gamma'', \\ \frac{dz_2}{dt} &= rz_2 - \beta_2 y_2 \gamma'', \qquad 2\frac{d\gamma''}{dt} = y_2(\beta_2 z_1 - \beta_1 z_2), \end{aligned} \tag{110}$$

where

$$\begin{aligned} \alpha_0 &= \sqrt{\frac{(A - C)(B - C)}{AB}}, \qquad \alpha_3 = x_0\sqrt{\frac{B - C}{B}} + \sqrt{-1}\, y_0\sqrt{\frac{A - C}{A}}, \\ 2\alpha_1 &= \sqrt{\frac{B - C}{B}} + \sqrt{\frac{A - C}{A}}, \qquad 2\alpha_2 = \sqrt{\frac{B - C}{B}} - \sqrt{\frac{A - C}{A}}, \\ 2\beta_1 &= \frac{1}{\sqrt{A(A - C)}} - \frac{1}{\sqrt{B(B - C)}}, \\ 2\beta_2 &= \frac{1}{\sqrt{A(A - C)}} + \frac{1}{\sqrt{B(B - C)}}. \end{aligned} \tag{111}$$

Let $\lambda = 0$ in (106a); then, (108), (109), (110) have algebraic integrals

$$\begin{aligned} \frac{B - A}{(A - C)(B - C)} y_2^2 + 4Cr^2 &= h_1 = 4Ca^2, \\ \frac{1}{\alpha_0} (\alpha_1 z_1 + \alpha_2 z_2) y_2 + 2Cr\gamma'' &= h_2, \\ z_1 z_2 + \gamma''^2 &= h_3, \end{aligned} \tag{112}$$

where $h_1$ is replaced by $4Ca^2$. We have the following relations among the constants

$$\begin{aligned} C\beta_1 &= \alpha_2\left(\frac{1}{\alpha_0} + 1\right), & C\beta_2 &= \alpha_1\left(\frac{1}{\alpha_0} - 1\right), \\ \alpha_1\beta_1 - \alpha_2\beta_2 &= \frac{1}{2}\frac{B - A}{AB}, & 4\alpha_1\alpha_2 &= C\frac{B - A}{AB}. \end{aligned} \tag{113}$$

Let $y_1, y_2, z_1, z_2, \gamma''$ be the solution of (108), (109), (110) expressed in terms of $r$. Deriving $y_2, z_1, z_2$ from (112) in terms of $y_1, \gamma'', r$ we obtain

$$\frac{dr}{dt} = \alpha_0(r^2 - a^2).$$

Then (108), (109), (110) are replaced by

$$\frac{dy_1}{dr} = \frac{-\alpha_0 r y_1 + z_0(\alpha_1 z_1 + \alpha_2 z_2) - \alpha_3\gamma''}{\alpha_0(r^2 - a^2)}, \tag{108a}$$

$$2\frac{d\gamma''}{dr} = \frac{y_2(\beta_2 z_1 - \beta_1 z_2)}{\alpha_0(r^2 - a^2)}. \tag{110a}$$

The right-hand members are expressed by (112) in terms of $h_1$, $h_2$, $h_3$, $y_1$, $\gamma''$, $r$. Substituting $\lambda = 0$ in (107), we see that these equations admit an integral of the form

$$F_0(h_1, h_2, h_3, y_1, \gamma'', r) = \text{constant},$$

where $F_0$ should contain $y_1$, $\gamma''$; otherwise the solution $r(t)$ of (108), (109), (110) would reduce to a constant and we would have $A = B$, Lagrange's case.

The first equation of (112) expresses $y_2$ as a function of $r$, and the second and the third equations express $z_1$, $z_2$ as irrational functions of $\gamma''$. To avoid this inconvenience, we introduce in place of $\gamma''$ a new variable $v$ such that the differential equations become rational:

$$v = \alpha_1 z_1 - \alpha_2 z_2 + \frac{\alpha_0}{y_2}\left(2Ca\gamma'' - h_2\frac{r}{a}\right). \tag{114}$$

The second equation of (112) is then transformed to

$$\left(v + \frac{\alpha_0 h_2}{a}\frac{r}{y_2}\right)^2 - \alpha_0^2 h_0^2 \frac{1}{y_2^2} + 4\alpha_1\alpha_2 h_3 = 4Ca\alpha_0 \frac{v}{y_2}\gamma''. \tag{115}$$

These two equations replace (108a), (110a) and form a system of equations containing $y_1(r)$, $v(r)$ rationally. From (112) we obtain an invariantive relation $v = 0$ for (108a), (110a), since

$$\frac{dv}{dt} = av.$$

To simplify (108a), let

$$h = \frac{A - B}{ABh_1}(h_2^2 - h_1 h_3), \qquad y = y_1 y_2;$$

then, from (108), (109), and (110) we obtain the differential equations

$$\begin{aligned} \frac{dv}{dr} &= \frac{a}{\alpha_0}\frac{v}{r^2 - a^2}, \\ \frac{dy}{dr} &= a_1 + a_2\frac{r}{\sqrt{r^2 - a^2}} + \left(a_3\alpha_3 + a_4 z_0 \frac{r}{\sqrt{r^2 - a^2}}\right)\left(v + \frac{h}{v}\right), \end{aligned} \tag{116}$$

where $a_1$, $a_2$, $a_3$, $a_4$ are nonzero constants. It is assumed that none of $A - B$, $B - C$, $C - A$ is zero.

Thus $v(r)$, $y(r)$ are expressed algebraically in terms of $y_1$, $\gamma''$, $r$. Hence, in order that there may exist a new algebraic integral, the expressions $v(r)$, $y(r)$, $r$ should be related by algebraic equations. After integrating (116) we obtain

$$v = k\left(\frac{r - a}{r + a}\right)^{1/(2\alpha_0)},$$

$$y = a_1 r + a_2\sqrt{r^2 - a^2} + \int \left(a_3\alpha_3 + a_4 z_0 \frac{r}{\sqrt{r^2 - a^2}}\right)\left(v + \frac{h}{v}\right) dr$$

with integration constant $k$. Therefore, the two expressions

$$w = \left(\frac{r - a}{r + a}\right)^{1/(2\alpha_0)}, \qquad z = \int \left(a_3\alpha_3 + a_4 z_0 \frac{r}{\sqrt{r^2 - a^2}}\right)\left(w + \frac{h'}{w}\right) dr,$$

and $r$ are related by algebraic equations, where $h'$ is an arbitrary number. We shall prove that this condition is satisfied only if the center of gravity coincides with the point of suspension (Euler's case).

$A$, $B$, $C$ are all different. (104) is symmetric with respect to $p$, $q$, $r$. Hence, we may suppose that the value of $C$ lies between those of $A$ and

$B$. Then $\alpha_0$ is imaginary and $w(r)$ is never algebraic, so that it should be $z(r)$ that is related to $w$ and $r$ through an algebraic equation. Moreover, this equation should be valid for all values of $h'$, and $w$ does not contain $h'$. Hence,

$$J_1 = \int \left(a_3\alpha_3 + a_4 z_0 \frac{r}{\sqrt{r^2 - a^2}}\right) w\, dr,$$

$$J_2 = \int \left(a_3\alpha_3 + a_4 z_0 \frac{r}{\sqrt{r^2 - a^2}}\right) \frac{dr}{w}$$

should be expressed algebraically in terms of $r$ and $w$.

When the representative point $r$ describes a small closed contour around $r = a$ on the plane for the values of $r$, the value of $w$ is multiplied by a constant. Hence,

$$J_1' = \int \left(a_3\alpha_3 - a_4 z_0 \frac{r}{\sqrt{r^2 - a^2}}\right) w\, dr,$$

$$J_2' = \int \left(a_3\alpha_3 - a_4 z_0 \frac{r}{\sqrt{r^2 - a^2}}\right) \frac{dr}{w}$$

should be expressed algebraically in terms of $r$ and $w$. Comparing $J_1$ and and $J_2$, we see that

$$J_3 = \alpha_3 \int w\, dr, \qquad J_4 = \alpha_3 \int \frac{dr}{w}, \tag{117}$$

$$J_5 = z_0 \int \frac{rw}{\sqrt{r^2 - a^2}}\, dr, \qquad J_6 = z_0 \int \frac{r}{w\sqrt{r^2 - a^2}}\, dr,$$

should be expressed algebraically in terms of $r$ and $w$.

## 6.22 ABEL'S LEMMAS

In order to discuss the integrals of (117) we refer to *Abel's three lemmas* (Husson, 1906, 1908).

*Abel's First Lemma:* Let $P(x)$ be a rational algebraic fraction, and $Q(x)$ be a polynomial or a rational fraction in $x$ — that is, the quotient of two polynomials or two rational polynomials in $x$ — such that $Q(x)$ has as simple poles the zero points or the poles of $P(x)$. Form

$$J(x) = \int \frac{Q}{y}\, dx$$

with

$$y(x) = [P(x)]^\alpha,$$

where $\alpha$ is a real irrational number or an imaginary number. If $J(x)$ is expressed algebraically in terms of $y$ and $x$, then the function can be represented by

$$J(x) = \frac{f(x)}{y(x)} + \text{constant},$$

where $f(x)$ is either a polynomial of $x$ of degree one higher than $Q(x)$ or a constant.

This lemma follows from the properties of algebraic functions described in Section 2.20. The integral $J(x)$ is uniform for finite values of $x$, except the zero points and the poles of $P(x)$. Let $x = a$ be a zero point of $P$ of order $m$; then

$$y(x) = (x - a)^{m\alpha} y_1(x),$$

where $y_1(x)$ is uniform at $x = a$, and $1/y_1(x)$ is also uniform at $x = a$. Then we can write, with a constant $E \leqq 1$,

$$\begin{aligned} J(x) &= \int \frac{Q(x)}{(x - a)^{m\alpha}} \frac{dx}{y_1} \\ &= \int \frac{1}{(x - a)^{m\alpha + E}} [b_0 + b_1(x - a) + b_2(x - a)^2 + \cdots] \, dx. \end{aligned}$$

The result of integration gives

$$J(x) = \frac{1}{(x - a)^{m\alpha + E - 1}} [b_0' + b_1'(x - a) + \cdots] + C,$$

with fixed constants $b_0', b_1', \ldots$ and an arbitrary constant $C$. Hence, in the neighborhood of $x = a$,

$$y(x)J(x) = \phi(x) + C[P(x)]^\alpha,$$

where $\phi(x)$ is holomorphic. If $J(x)$ is algebraic in terms of $x$ and $y$, then $J(x) - C$ is also algebraic, so that we can put $C = 0$. Let

$$y(x)J(x) = f(x, y);$$

then, $f(x, y) = \phi(x)$ in the neighborhood of $x = a$. As $x$ makes a circuit from a point $x = x_0$ near $x = a$ by describing the whole circumference of a circle with $x = a$ as center in the $(x, y)$-plane, the function $\phi(x)$ again takes the initial value, but $y(x)$ takes an infinite number of values in geometrical progression. If the algebraic function $f(x_0, y)$ depends on $y$, then it should take the same value only for a finite number of values of $y$, according to Section 2.20. Hence, for $f(x, y) = \phi(x)$ to be possible, $f$ should not depend on $y$. Thus, $y(x)J(x) = f(x)$, where $f(x)$ is an algebraic function of $x$. The function $f(x)$ ceases to be holomorphic only when $y(x)$ or $J(x)$ ceases to be holomorphic, that is, at the zero points or the poles of $P$, except in the case where $x = a$ is a zero point.

If $x = b$ is a zero point of $P(x)$ other than $x = a$, then a similar proof can be carried out. If $x = a$ is not a zero point, but is a pole, then the proof can be obtained simply by changing $m$ into $-m$. Hence the algebraic function $f(x)$ is holomorphic for any finite value of $x$.

The behavior at $x = \infty$ can be studied, as in Section 2.20, by the substitution $x = 1/z$. If $P$ and $Q$ are polynomials of degree $n$ and $n_1$, respectively, then,

$$y = z^{-n\alpha}y_2(z), \qquad Q = z^{-n_1}y_3(z),$$

where $y_2(z)$ and $y_3(z)$ are holomorphic and nonzero at $z = 0$. Hence

$$J(x) = z^{n\alpha - n_1 - 1}(b_0' + b_1'z + \cdots) + C',$$

and accordingly

$$f(x) = y(x)J(x) = z^{-n_1-1}\psi(z) + C'z^{-n\alpha}y_2(x),$$

where $\psi(z)$ is holomorphic and nonzero for $z = 0$. Since the function $z^{n\alpha}$ has $z = 0$ as a transcendental singularity, $C'$ must be zero. If $n_1 > -1$, then the point at infinity is a pole of order $n_1 + 1$ of $f(x)$. Since $f(x)$ is holomorphic for finite values of $x$, this algebraic function is a polynomial in $x$ of degree $n_1 + 1$. If $n_1 < -1$, then the point at infinity is a zero point of $f(x)$, which reduces to zero identically, and $J(x)$ cannot be expressed algebraically in terms of $y$ and $x$.

*Abel's Second Lemma:* If there exists an algebraic relation among Abel's transcendental functions $J(x)$ and $J'(x)$ and the variable $x$, then the relation is of the form $J' = kJ + \phi(x)$, where $k$ is a constant and $\phi(x)$ is an algebraic function.

Suppose that the integrals $J$ and $J'$ are related by an algebraic equation $f(J, J') = 0$ as in Section 2.20. Solving this equation for $J'$, we obtain $J' = \psi(J, x)$. Since $J$ is not an algebraic expression, the integral $J$ increases by a period $\omega$ as $x$ returns to its initial value, by describing the suitably chosen closed curve $C$ on the Riemann surface of $x$, and $J'$ increases by a period $\omega_1$ at the same time, where

$$\omega_1 = \psi(J + \omega, x) - \psi(J, x).$$

But this period $\omega_1$ does not contain $J$ if $J$ and $x$ are regarded as the independent variables, because $J$ could be solved as an algebraic expression of $x$ if $\omega_1$ contained $J$. Hence

$$\frac{\partial\psi(J + \omega, x)}{\partial J} - \frac{\partial\psi(J, x)}{\partial J} \equiv 0,$$

and $\partial\psi(J, x)/\partial J$ has the period $\omega$. Consequently we can write

$$J' = \psi(J, x) = J\alpha(x) + \beta(x),$$

where $\alpha$ and $\beta$ are algebraic expressions. Thus,

$$\frac{dJ'}{dx} = J\frac{d\alpha}{dx} + \alpha\frac{dJ}{dx} + \frac{d\beta}{dx},$$

in which $J$ is transcendental and the functions other than $J$ are algebraic. Accordingly $d\alpha/dx \equiv 0$, that is, $\alpha = k =$ constant. Similarly, from $d\beta/dx = dJ'/dx - \alpha(dJ/dx)$, we see that $d\beta/dx$ is rational in $x$, and so is $\beta(x)$.

*Corollary:* If one of the two integrals $J$ and $J'$,

$$J = \int \frac{y\, dx}{x^2 - a^2}, \qquad J' = \int \frac{dx}{y(x^2 - a^2)}, \qquad y = (x^2 - a^2)^{p/m},$$

is not algebraic, where $p$, $m$ are relatively prime positive integers — that is, if $2p/m$ is not an integer — then it is impossible to have an algebraic relation among $J$, $J'$, $x$, unless $2p/m$ is an integer, that is, one of the integrals $J$ or $J'$ is algebraic.

In fact, if $J$ and $J'$ are transcendental functions, then

$$J' = kJ + \phi(x, y)$$

by the second Lemma. As $x$ describes a closed contour $C$,

$$\omega_1 = \int_C \frac{dJ}{dx}\, dx = k \int_C \frac{dJ}{da}\, dx = k\omega.$$

As $x$ makes a circuit around $x = a$ in the region of the point $x = a$, the variable $x$ returns to its initial value, but $y$ is multiplied by

$$\exp\left(2\pi\sqrt{-1}\,\frac{p}{m}\right),$$

and $\dfrac{dJ'}{dx}$ by $\exp\left(-2\pi\sqrt{-1}\,\dfrac{p}{m}\right)$, and $\dfrac{dJ}{dx}$ by $\exp\left(2\pi\sqrt{-1}\,\dfrac{p}{m}\right)$. Hence the relation between the periods is

$$\omega_1 \exp\left(-2\pi\sqrt{-1}\,\frac{p}{m}\right) = k\omega \exp\left(2\pi\sqrt{-1}\,\frac{p}{m}\right).$$

Thus we see that $2p/m$ should be equal to an integer $n$.

*Abel's Third Lemma:* Consider the integral $K(x) = \int J(x)\rho(x, y)\, dx$ with a rational function $\rho$ of $x$ and $y$, and an Abel's transcendental function

$J(x)$, where $y$ is defined by $f(x, y) = 0$ as an implicit function of $x$. If $J$, $K$, $x$ are related by an algebraic equation, then Abel's integral $\int \rho(x, y)\, dy$ is expressed algebraically in terms of $J$ and $x$.

In fact, the algebraic relation among $J$, $K$, $x$ must contain $K$. If it did not contain $K$, then $J$ would be algebraic. Hence we obtain an equation $K = \phi(J, x)$ that is algebraic in $J$ and $x$. Suppose that, as $x$ describes a closed curve $C$, the integral $J$ increases by $\omega$ and $K(x)$ becomes $K_1(x)$.

Then

$$K_1(x) - K(x) = \phi(J + \omega, x) - \phi(J, x) = \psi(J, x).$$

But

$$\frac{dK_1}{dx} - \frac{dK}{dx} = (J + \omega)\rho - J\rho = \omega\rho(x, y).$$

Hence

$$\psi(J, x) = \omega \int \rho(x, y)\, dx.$$

Since $\psi$ is algebraic both in $J$ and $x$, the Abelian integral $\int \rho(x, y)\, dx$ can be expressed algebraically in terms of $J$ and $x$.

On the basis of Abel's lemmas, we seek the condition that integrals $J_3$, $J_4$, $J_5$, $J_6$ should be expressed algebraically by $r$ and $w$. If $\alpha_3 \neq 0$, then, according to the first lemma,

$$\int \frac{dr}{w} = \frac{br + c}{w} + \text{constant}$$

with two constants $b$ and $c$. Differentiating this expression with respect to $r$ would give an impossible relation

$$\frac{w'}{w} = \frac{a}{\alpha_0} \frac{1}{r^2 - a^2} = \frac{b - 1}{br + c}.$$

Hence

$$\alpha_3 = x_0\sqrt{\frac{B - C}{B}} + \sqrt{-1}\, y_0\sqrt{\frac{A - C}{A}} = 0.$$

Exchanging $y_1$ and $y_2$, we have

$$x_0\sqrt{\frac{B - C}{B}} - \sqrt{-1}\, y_0\sqrt{\frac{A - C}{A}} = 0.$$

Either $B = C$, $A = C$, or $x_0 = y_0 = 0$ (Lagrange's or Euler's case).

Put

$$w_1 = \left(\frac{r-a}{r+a}\right)^{1/(2\alpha_0)-1/2}$$

in $J_6$; then, according to the first lemma,

$$\int \frac{1}{w_1}\frac{r\,dr}{r-a} = \frac{br+c}{w_1} + \text{constant}$$

for $z_0 \neq 0$. Differentiate this with respect to $r$; then

$$(r+a)[(1-b)r+ab] = a\left(1-\frac{1}{\alpha_0}\right)(br+c),$$

that is, $b = 1$, $c = a$, and $a = a(1 - 1/\alpha_0)$. But this last equation is valid only when $A = C$ or $B = C$ (Lagrange's case).

Hence, in order that there may exist an algebraic integral other than the classical integrals, in the case when $A$, $B$, $C$ are all different, $x_0$, $y_0$, and $z_0$ must all equal zero; that is, the center of mass coincides with the point of suspension. This is Euler's case. Thus the theorem is proved.

## 6.23 KOWALEWSKI'S SOLUTION

We have $A = B$ for the cases of Lagrange and Kowalewski. Since the center of mass is on the axis of $C$ in the case of Lagrange and is arbitrary in the case of Kowalewski, the case of Lagrange is considered as a special case of Kowalewski's. If $A = B$, then $r =$ constant. Put $A_1 = A - C$,

$$y_1 = p + \sqrt{-1}\,q \quad\text{and}\quad y_2 = p - \sqrt{-1}\,q,$$

and let $\epsilon$ be the angle that the line joining the center of mass with the point of suspension makes with the $(x, y)$ plane. Then (105) can be written

$$\begin{aligned}
A\frac{dy_1}{dt} &= -A_1 r y_1 + z_1 \sin\epsilon - \gamma'' \cos\epsilon,\\
A\frac{dy_2}{dt} &= A_1 r y_2 + \gamma'' \cos\epsilon - z_2 \sin\epsilon,\\
2C\frac{dr}{dt} &= (z_2 - z_1)\cos\epsilon,\\
\frac{dz_1}{dt} &= \gamma'' y_1 - r z_1,\\
\frac{dz_2}{dt} &= r z_2 - \gamma'' y_2,\\
2\frac{d\gamma''}{dt} &= y_2 z_1 - y_1 z_2.
\end{aligned} \tag{118}$$

These equations have the classical integrals

$$\begin{aligned}(Cr^2 - z_1 \cos \epsilon) + (Ay_1y_2 - z_2 \cos \epsilon - 2\gamma'' \sin \epsilon) &= \text{constant},\\ A(y_1z_2 + y_2z_1) + 2Cr\gamma'' &= \text{constant},\\ z_1z_2 + \gamma''^2 &= \text{constant}.\end{aligned} \tag{119}$$

Multiplying $y_1$, $z_1$, $\gamma''$ in each of these equations by $\lambda$, we obtain six equations that we call (118a) and three that we call (119a), and denote the constants on the right-hand sides of (119a) by $h_1$, $h_2$, $h_3$, respectively. In the case of Lagrange, $r =$ constant, because the center of mass is on the $z$ axis. Excluding this case of Lagrange, take $r$ as the independent variable. Put $\lambda = 0$ and

$$\gamma'' = (r^2 - a^2)^{C/A}\Gamma''; \tag{120}$$

then from (118a) and (119a) we obtain

$$\left.\begin{aligned} y_2 &= (r^2 - a^2)^{A_1/A}J,\\ J &= \text{constant} - \frac{2C}{A} \tan \epsilon \int \frac{dr}{(r^2 - a^2)^{A_1/A}},\end{aligned}\right\} \tag{121}$$

$$\frac{d\Gamma''}{dr} = \frac{2h_3 \cos \epsilon}{C} \frac{y_2}{(r^2 - a^2)^{(C/A)+2}} - \frac{h_2}{A} \frac{1}{(r^2 - a^2)^{(C/A)+1}}, \tag{122}$$

$$\left.\begin{aligned} Cr^2 - z_2 \cos \epsilon = h_1 &= Ca^2,\\ A(y_1z_2 + y_2z_1) + 2Cr\gamma'' &= h_2,\\ z_1z_2 &= h_3.\end{aligned}\right\} \tag{123}$$

Equations (121) and (122) should have an integral

$$F(h_1, h_2, h_3, y_2, \gamma'', r) = \text{constant}$$

for $\lambda = 0$, which expresses $\gamma''(r)$ algebraically in terms of $r$ and $y_2$. To obtain such an integral, express the three integrals,

$$J_0 = (r^2 - a^2)^{C/A}, \qquad J_1 = \int \frac{dr}{(r^2 - a^2)^{1+(C/A)}},$$

$$J_2 = \int \frac{J\,dr}{(r^2 - a^2)^{1+2(C/A)}},$$

algebraically in terms of $r$ and $y_2$, according to (120), (121), and (123). Eliminating $y_2$ by the use of the equations for $J_0$, $J_1$, we should obtain an algebraic relation among these two equations and $r$.

Thus we have the classical three integrals

$$r^2 = 6l_1 - (y_1 + y_2)^2 + \xi_1 + \xi_2,$$
$$r\gamma'' = 2l + y_1y_2(y_1 + y_2) - y_2\xi_1 - y_1\xi_2, \tag{125}$$
$$\gamma''^2 = 1 - k^2 - y_1^2y_2^2 + y_2^2\xi_1 + y_1^2\xi_2,$$

and the fourth integral

$$\xi_1\xi_2 = k^2, \tag{126}$$

where $l$, $l_1$, $k^2$ are integration constants. In addition, let

$$a = 6l_1 - (y_1 + y_2)^2, \qquad b = 2l + y_1y_2(y_1 + y_2),$$
$$c = 1 - k^2 - y_1^2y_2^2. \tag{127}$$

Then,

$$r^2 = a + \xi_1 + \xi_2, \qquad r\gamma'' = b - y_2\xi_1 - y_1\xi_2,$$
$$\gamma''^2 = c + y_2^2\xi_1 + y_1^2\xi_2, \tag{128}$$

where $y_1$, $y_2$, $\xi_1$, $\xi_2$ are related by

$$\xi_1\xi_2 = k^2, \qquad (a + \xi_1 + \xi_2)(c + y_2^2\xi_1 + y_1^2\xi_2) = (b - y_2\xi_1 - y_1\xi_2)^2. \tag{129}$$

Put

$$R(x) = ax^2 + 2bx + c, \qquad R(x_1, x_2) = ax_1x_2 + b(x_1 + x_2) + c,$$

$$s_1 = \frac{R(y_1, y_2) + \sqrt{R(y_1)}\sqrt{R(y_2)}}{2(y_1 - y_2)^2} + \frac{1}{2}l_1, \qquad k_1 = \frac{l_1 + k}{2},$$
$$s_2 = \frac{R(y_1, y_2) - \sqrt{R(y_1)}\sqrt{R(y_2)}}{2(y_1 - y_2)^2} + \frac{1}{2}l_1, \qquad k_2 = \frac{l_1 - k}{2}; \tag{130}$$

then,

$$\xi_1 = \frac{(y_1 - y_2)^2}{R(y_2)}\{[\sqrt{(s_1 - k_1)(s_2 - k_1)} + \sqrt{(s_1 - k_2)(s_2 - k_2)}]^2 - k^2\},$$

$$\xi_2 = \frac{(y_1 - y_2)^2}{R(y_1)}\{[\sqrt{(s_1 - k_1)(s_2 - k_1)} - \sqrt{(s_1 - k_2)(s_2 - k_2)}]^2 - k^2\}.$$

Write

$$g_2 = k^2 - 1 + 3l_1^2, \qquad g_3 = l_1(k^2 - 1 - l_1^2) + l^2,$$

$$S(s) = 4s^3 - g_2 s - g_3 = 4(s - e_1)(s - e_2)(s - e_3),$$

$$e_1 + e_2 + e_3 = 0;$$

then we obtain, eventually,

$$\frac{ds_1}{\sqrt{S(s_1)}} = -\frac{dy_1}{\sqrt{R(y_1)}} + \frac{dy_2}{\sqrt{R(y_2)}},$$

$$\frac{ds_2}{\sqrt{S(s_2)}} = \frac{dy_1}{\sqrt{R(y_1)}} + \frac{dy_2}{\sqrt{R(y_2)}}, \tag{131}$$

$$\frac{ds_2}{\sqrt{W(s_2)}} = \frac{dt}{s_2 - s_1};$$

or, finally,

$$0 = \frac{ds_1}{\sqrt{W(s_1)}} + \frac{ds_2}{\sqrt{W(s_2)}},$$

$$dt = \frac{s_1\, ds_1}{\sqrt{W(s_1)}} + \frac{s_2\, ds_2}{\sqrt{W(s_2)}}, \tag{132}$$

$$W(s) = -4(s - e_1)(s - e_2)(s - e_3)(s - k_1)(s - k_2).$$

The five roots of $W(s) = 0$ are all different. The inversion of these integrals gives hyperelliptic functions (Section 6.25).

Suppose that $e_3 < k_2 < e_2 < e_1 < k_1$, and define

$$du_1 = \frac{dy_1}{\sqrt{R(y_1)}}, \qquad du_2 = \frac{dy_2}{\sqrt{R(y_2)}};$$

then, with an imaginary variable $w$ defined by $\wp(w) = -l_1$,

$$\wp'^2(w) = -4l_1^3 + g_2 l_1 - g_3 = -l_0^2,$$

$$s_1 = \wp(u_1 + u_2), \qquad s_2 = \wp(u_1 - u_2),$$

we obtain, eventually,

$$p = -\sqrt{-1}\,\frac{L_1P_1 + M_1P_2 + N_1P_3}{LP_1 + MP_2 + NP_3}, \qquad q = \frac{E}{LP_1 + MP_2 + NP_3},$$

where

$$P_i = \frac{\sigma_i(u_1 + u_2)\,\sigma_i(u_1 - u_2)}{\sigma(u_1 + u_2)\,\sigma(u_1 - u_2)} \qquad (i = 1, 2, 3),$$

$$L = (e_2 - e_3)\frac{\sigma_1(w)}{\sigma(w)}, \qquad M = (e_3 - e_1)\frac{\sigma_2(w)}{\sigma(w)},$$

$$N = (e_1 - e_2)\frac{\sigma_3(w)}{\sigma(w)},$$

$$L_1 = (e_2 - e_3)\frac{\sigma_2(w)}{\sigma(w)}\frac{\sigma_3(w)}{\sigma(w)}, \qquad M_1 = (e_3 - e_1)\frac{\sigma_3(w)}{\sigma(w)}\frac{\sigma_1(w)}{\sigma(w)},$$

$$N_1 = (e_1 - e_2)\frac{\sigma_1(w)}{\sigma(w)}\frac{\sigma_2(w)}{\sigma(w)},$$

$$E = (e_2 - e_3)(e_3 - e_1)(e_1 - e_2).$$

The remaining unknown functions are computed as follows:

$$r = \frac{2}{q}\frac{dp}{dt} = -\sqrt{-1}\,\frac{LP_{23} + MP_{13} + NP_{12}}{LP_1 + MP_2 + NP_3},$$

$$\gamma'' = -\left(2\frac{dq}{dt} + rp\right) = \frac{L_1P_{23} + M_1P_{13} + N_1P_{12}}{LP_1 + MP_2 + NP_3},$$

where

$$P_i = \sqrt{(s_1 - a_i)(s_2 - a_i)},$$

$$P_{ij} = \frac{P_iP_j}{s_1 - s_2}\left[\frac{\sqrt{R(s_1)}}{(s_1 - a_i)(s_1 - a_j)} - \frac{\sqrt{R(s_2)}}{(s_2 - a_i)(s_2 - a_j)}\right]$$

$$(i, j = 0, 1, 2, 3, 4).$$

A similar computation can be carried out for $\gamma' = dr/dt$,

$$2\gamma = 2(p^2 + q^2) + r^2 - 6l_1.$$

These functions are expressed in terms of $s_1$ and $s_2$.

The next step is to express these functions in terms of $t$. To solve (132), put

$$\begin{aligned} u_1 &= \int_{a_1}^{s_1} \frac{s\,ds}{\sqrt{W(s)}} + \int_{a_2}^{s_2} \frac{s\,ds}{\sqrt{W(s)}} = t + c, \\ u_2 &= \int_{a_1}^{s_1} \frac{ds}{\sqrt{W(s)}} + \int_{a_2}^{s_2} \frac{ds}{\sqrt{W(s)}} = c_1, \end{aligned} \qquad (133)$$

where $a_0 < a_1 < a_2 < a_3 < a_4$ are, respectively, $e_3 < k_2 < e_2 < e_1 < k_1$ of the five roots of $W(s) = 0$. Denote the integrand of the first line by $F_1(s)/\sqrt{W(s)}$ and that of the second line by $F_2(s)/\sqrt{W(s)}$, and write

$$\Omega_{ij} = \int_{a_{2j-1}}^{a_{2j}} \frac{F_i(s)\,ds}{\sqrt{W(s)}}, \qquad \sqrt{-1}\;\Omega'_{ij} = \int_{a_{2j-2}}^{a_{2j-1}} \frac{F_i(s)\,ds}{\sqrt{W(s)}} \qquad (i, j = 1, 2),$$

$$u_1 = 2\Omega_{11}v_1 + 2\Omega_{12}v_2, \qquad u_2 = 2\Omega_{21}v_1 + 2\Omega_{22}v_2,$$

or

$$v_1 = g_{11}u_1 + g_{12}u_2, \qquad v_2 = g_{21}u_1 + g_{22}u_2.$$

Take Riemann's $\vartheta$-function (Section 6.24)

$$\vartheta(v_1, v_2) = \sum_{\nu_1,\nu_2}^{-\infty\cdots+\infty} \exp\{\pi\sqrt{-1}\,[\nu_1(2v_1 + \nu_1\tau_{11} + \nu_2\tau_{12}) + \nu_2(2v_2 + \nu_1\tau_{21} + \nu_2\tau_{22})]\},$$

where

$$\tau_{ij} = 2\sqrt{-1}\,(g_{1i}\Omega'_{1j} + g_{2i}\Omega'_{2j}), \qquad \tau_{ij} = \tau_{ji}.$$

Then, with arbitrary integers $p_1, p_2$, we have

$$\vartheta(v_1 + p_1, v_2 + p_2) = \vartheta(v_1, v_2),$$

$$\vartheta(v_1 + \tau_{1i}, v_2 + \tau_{2i}) = \vartheta(v_1, v_2) \exp[-\pi\sqrt{-1}\,(2v_i + \tau_{ii})] \qquad (i = 1, 2).$$

In addition, write

$$\tau_i = n_1\tau_{i1} + n_2\tau_{i2},$$

$$\vartheta(v_1, v_2/n_1, n_2) = \vartheta(v_1 + \tau_1, v_2 + \tau_2) \times \exp\{\pi\sqrt{-1}\,[n_1(2v_1 + \tau_1) + n_2(2v_2 + \tau_2)]\}$$

and choose a set of integers $m_1^\lambda, m_2^\lambda, n_1^\lambda, n_2^\lambda$ such that

$$\vartheta(v_1, v_2)_\lambda = \vartheta(v_1 + \tfrac{1}{2}m_1^\lambda, v_2 + \tfrac{1}{2}m_2^\lambda/\tfrac{1}{2}n_1^\lambda, \tfrac{1}{2}n_2^\lambda);$$

then $P_i$ and $P_{ij}$, and accordingly $p, q, r, \gamma, \gamma', \gamma''$, are all represented by functions of the form

$$\frac{\vartheta_i(ct + c', c_1t + c'_1)}{\vartheta(ct + c', c_1t + c'_1)}.$$

Thus the top motion of Kowalewski can be solved completely by the elementary method of integration; that is, by finding the necessary number of integrals. What the theorem of Poincaré states is that the differential equations of celestial mechanics cannot be solved by such an elementary method of integration (Golubev, 1953).

### 6.24 RIEMANN $\vartheta$-FUNCTIONS

In Section 2.10 we have described quasi-periodic functions that can be expanded in trigonometric series with linear functions of the independent variable as arguments. A similar periodicity with a similar form of expansion (not in the real but in the complex domain) can be obtained for the Riemann $\vartheta$-functions (Riemann, 1892; Weierstrass, 1894; Wirtinger, 1895; Baker, 1897, 1907; Krazer, 1903; Landau, 1918; Hecke, 1928; Coble, 1929).

The series

$$\sum_{m=-\infty}^{+\infty} \exp\,(am^2 + 2bm + c)$$

is absolutely convergent for finite values of $b$ and $c$ if the real part $r$ of $a = r + \sqrt{-1}\,s$ is negative. The function represented by

$$\vartheta(u) = \sum_{m=-\infty}^{\infty} \exp\,(am^2 + 2mu)$$

is called a $\vartheta$-function of a complex variable $u$, and $a$ is called the modulus. We have

$$\vartheta(u + \pi\sqrt{-1}) = \vartheta(u), \qquad \vartheta(u) = \vartheta(-u),$$
$$\vartheta(u + a) = \exp\,(-a - 2u)\cdot\vartheta(u),$$

Repeating these formulas, we obtain

$$\vartheta(u + \kappa a + \lambda\pi\sqrt{-1}) = \exp\,(-\kappa^2 a - 2\kappa u)\cdot\vartheta(u)$$

with arbitrary integers $\kappa$ and $\lambda$. If for finite values of a single-valued continuous function of a complex variable $u$ the equations

$$G(u + \pi\sqrt{-1}) = G(u), \qquad G(u + a) = \exp\,(-a - 2u)\cdot G(u)$$

are satisfied, or if the function satisfies

$$G(u + \kappa a + \lambda\pi\sqrt{-1}) = \exp\,(-\kappa^2 a - 2\kappa u)\cdot G(u),$$

then $G(u)$ and $\vartheta(u)$ differ only by a factor independent of $u$. This function $\vartheta(u)$ is called a *Riemann $\vartheta$-function.*

When this class of function is generalized, the $p$-ple infinite series

$$\sum_{m_1,\cdots,m_p}^{-\infty\cdots+\infty} \exp\left(\sum_{\mu=1}^{p}\sum_{\nu=1}^{p} a_{\mu\nu}m_\mu m_\nu + 2\sum_{\mu=1}^{p} b_\mu m_\mu + c\right)$$

with finite values of $b_1, \ldots, b_p, c$ is absolutely convergent, if the quadratic form

$$\sum_{\mu=1}^{p}\sum_{\nu=1}^{p} r_{\mu\nu}x_\mu x_\nu$$

formed of the real parts $r_{\mu\nu}$ of $a_{\mu\nu}$ is negative-definite.

A function defined by

$$\vartheta(u_1, u_2, \ldots, u_p) = \sum_{m_1,\cdots,m_p}^{-\infty\cdots+\infty} \exp\left(\sum_{\mu=1}^{p}\sum_{\nu=1}^{p} a_{\mu\nu}m_\mu m_\nu + 2\sum_{\mu=1}^{p} m_\mu u_\mu\right) \tag{133a}$$

with the above conditions for $a_{\mu\nu}$ is called a $\vartheta$-function of complex variables $u_1, \ldots, u_p$. Such a function satisfies the set of equations (A):

$$\text{(A)}: \quad \vartheta(u_1, \ldots, u_\nu + \pi\sqrt{-1}, \ldots, u_p) = \vartheta(u_1, \ldots, u_\nu, \ldots, u_p),$$

$$\vartheta(u_1 + a_{1\nu}, \ldots, u_p + a_{p\nu}) = \vartheta(u_1, \ldots, u_p)\exp(-a_{\mu\nu} - 2u_\nu) \qquad (\nu = 1, 2, \ldots, p).$$

More generally, the function (133a) satisfies

$$\vartheta\left(u_1 + \sum_{\nu=1}^{p}\kappa_\nu a_{1\nu} + \lambda_1\pi\sqrt{-1}, \ldots, u_p + \sum_{\nu=1}^{p}\kappa_\nu a_{p\nu} + \lambda_p\pi\sqrt{-1}\right)$$

$$= \exp\left(-\sum_{\mu=1}^{p}\sum_{\nu=1}^{p} a_{\mu\nu}\kappa_\mu\kappa_\nu - 2\sum_{\mu=1}^{p}\kappa_\mu u_\mu\right)\cdot\vartheta(u_1, \ldots, u_p)$$

with integers $\kappa_1, \ldots, \kappa_p, \lambda_1, \ldots, \lambda_p$. The $2p$ sequences on the left-hand side in the foregoing $2p$ equations (A) for a $\vartheta$-function such as

$$\begin{matrix} \pi\sqrt{-1} & 0 & \cdots & 0 & a_{11} & a_{21} & \cdots & a_{p1} \\ 0 & \pi\sqrt{-1} & \cdots & 0 & a_{12} & a_{22} & \cdots & a_{p2} \\ \vdots & & & & & & & \\ 0 & 0 & \cdots & \pi\sqrt{-1} & a_{1p} & a_{2p} & \cdots & a_{pp} \end{matrix}$$

are said to give sequences for the simultaneously varying values of $u_1, \ldots, u_p$. Such sequences are called the *periodicity moduli* of a $\vartheta$-function. Conversely, a function $G(u_1, \ldots, u_p)$ satisfying

$$G\left(u_1 + \sum_{\nu=1}^{p} \kappa_\nu a_{1\nu} + \lambda_1 \pi\sqrt{-1}, \ldots, u_p + \sum_{\nu=1}^{p} \kappa_\nu a_{p\nu} + \lambda_p \pi\sqrt{-1}\right)$$

$$= \exp\left(-\sum_{\mu=1}^{p}\sum_{\nu=1}^{p} a_{\mu\nu}\kappa_\mu\kappa_\nu - 2\sum_{\mu=1}^{p} \kappa_\mu u_\mu\right)\cdot G(u_1, \ldots, u_p)$$

with arbitrary integers $\kappa_1, \kappa_2, \ldots, \kappa_p, \lambda_1, \ldots, \lambda_p$, differs from $\vartheta(u_1, \ldots, u_p)$ by a factor independent of $u$.

The periodicity moduli are related uniquely with an arbitrary sequence $c_1, \ldots, c_p$ by

$$c_1 = \sum_{\mu=1}^{p} g_\mu a_{1\mu} + h_1\pi\sqrt{-1}, \qquad \ldots, \qquad c_p = \sum_{\mu=1}^{p} g_\mu a_{p\mu} + h_p\pi\sqrt{-1},$$

with real numbers $g_1, \ldots, g_p, h_1, \ldots, h_p$. These $2p$ real numbers are called the *period characteristics* of $c_1, \ldots, c_p$.

Define

$$\vartheta\begin{bmatrix} g_1 \cdots g_p \\ h_1 \cdots h_p \end{bmatrix}(u_1, \ldots, u_p) = \sum_{m_1, \cdots, m_p}^{-\infty \cdots +\infty} \exp\left[\sum_{\mu=1}^{p}\sum_{\nu=1}^{p} a_{\mu\nu}(m_\mu + g_\mu)(m_\nu + g_\nu) + 2\sum_{\mu=1}^{p}(m_\mu + g_\mu)(m_\mu + h_\mu\pi\sqrt{-1})\right].$$

This $\vartheta$-function satisfies

$$\vartheta\begin{bmatrix} g_1 \cdots g_p \\ h_1 \cdots h_p \end{bmatrix}(u_1, \ldots, u_p)$$

$$= \vartheta\left[u_1 + \sum_{\mu=1}^{p} g_\mu a_{1\mu} + h_1\pi\sqrt{-1}, \ldots, u_p + \sum_{\mu=1}^{p} g_\mu a_{p\mu} + h_p\pi\sqrt{-1}\right]$$

$$\times \exp\left[\sum_{\mu=1}^{p}\sum_{\nu=1}^{p} a_{\mu\nu} g_\mu g_\nu + 2\sum_{\mu=1}^{p} g_\mu(u_\mu + h_\mu\pi\sqrt{-1})\right],$$

and is equal to $\vartheta(u_1, \ldots, u_p)$ if all $g$ and $h$ are zero. This $\vartheta$-function is denoted by $\vartheta[{}^{g}_{h}]((u))$ to simplify the notation.

If the period characteristics of $c_1, \ldots, c_p$ are $g_1, \ldots, g_p, h_1, \ldots, h_p$, then we denote the sequence $u_1 + c_1, \ldots, u_p + c_p$ by $(u + \{{g \atop h}\})$. This function satisfies

$$\vartheta\begin{bmatrix} g \\ h \end{bmatrix}\left(\left(u + \begin{Bmatrix} \kappa \\ \lambda \end{Bmatrix}\right)\right) = \vartheta\begin{bmatrix} g \\ h \end{bmatrix}((u)) \exp\left[-\sum_{\mu=1}^{p}\sum_{\nu=1}^{p} a_{\mu\nu}\kappa_\mu\kappa_\nu - 2\sum_{\mu=1}^{p}\kappa_\mu u_\mu + 2\sum_{\mu=1}^{p}(\lambda_\mu g_\mu - \kappa_\mu h_\mu)\pi\sqrt{-1}\right].$$

Conversely, a function $G((u))$ satisfying

$$G\left(\left(u + \begin{Bmatrix} \kappa \\ \lambda \end{Bmatrix}\right)\right) = G(((u))) \exp\left[-\sum_{\mu=1}^{p}\sum_{\nu=1}^{p} a_{\mu\nu}\kappa_\mu\kappa_\nu - 2\sum_{\mu=1}^{p}\kappa_\mu u_\mu + 2\sum_{\mu=1}^{p}(\lambda_\mu g_\mu - \kappa_\mu h_\mu)\pi\sqrt{-1}\right]$$

differs from $\vartheta[{g \atop h}]((u))$ only by a factor independent of $u$. With real numbers $g_1', \ldots, g_p', h_1', \ldots, h_p'$ this function satisfies

$$\vartheta\begin{bmatrix} g \\ h \end{bmatrix}\left(\left(u + \begin{Bmatrix} g' \\ h' \end{Bmatrix}\right)\right) = g\begin{bmatrix} g + g' \\ h + h' \end{bmatrix}((u)) \exp\left[-\sum_{\mu=1}^{p}\sum_{\nu=1}^{p} a_{\mu\nu}g_\mu' g_\upsilon' - 2\sum_{\mu=1}^{p} g_\mu'(u_\mu + \lambda_\mu\pi\sqrt{-1} + h_\mu'\pi\sqrt{-1})\right],$$

$$\vartheta\begin{bmatrix} g + \kappa \\ h + \lambda \end{bmatrix}((u)) = \vartheta\begin{bmatrix} g \\ h \end{bmatrix}((u)) \exp\left(2\sum_{\mu=1}^{p}\lambda_\mu g_\mu\pi\sqrt{-1}\right),$$

$$\vartheta\begin{bmatrix} g \\ h \end{bmatrix}(-u_1, \ldots, -u_p) = \vartheta\begin{bmatrix} -g \\ -h \end{bmatrix}(u_1, \ldots, u_p).$$

If the period characteristics $g_1, \ldots, g_p, h_1, \ldots, h_p$ are half-integers, denote them by $\frac{1}{2}\alpha_1, \ldots, \frac{1}{2}\alpha_p, \frac{1}{2}\beta_1, \ldots, \frac{1}{2}\beta_p$ with integers $\alpha_1, \ldots, \alpha_p$, $\beta_1, \ldots, \beta_p$. Let $\alpha\cdot\beta = \alpha_1\beta_1 + \cdots + \alpha_p\beta_p$, the scalar product of two vectors $\alpha = (\alpha_1, \ldots, \alpha_p)$ and $\beta = (\beta_1, \ldots, \beta_p)$, and write

$$\Theta(u; \tfrac{1}{2}\beta; \tfrac{1}{2}\alpha) = \vartheta\begin{bmatrix} \frac{1}{2}\alpha_1 \cdots \frac{1}{2}\alpha_p \\ \frac{1}{2}\beta_1 \cdots \frac{1}{2}\beta_p \end{bmatrix}(\pi\sqrt{-1}\, u_1, \ldots, \pi\sqrt{-1}\, u_p).$$

We have

$$\Theta(-u; \tfrac{1}{2}\beta; \tfrac{1}{2}\alpha) = \exp(\pi\sqrt{-1}\,\beta\alpha)\cdot\Theta(u; \tfrac{1}{2}\beta; \tfrac{1}{2}\alpha).$$

$\Theta(u; \frac{1}{2}\beta; \frac{1}{2}\alpha)$ is an odd or an even function according as whether $\beta\alpha$ is odd or even.

Put

$$m_i + \alpha_{i1}m_1' + \cdots + \alpha_{ip}m_p' \equiv \Omega_{mm'} \qquad (i = 1, 2, \ldots, p)$$

with $2p$ constants $m_1, \ldots, m_p, m_1', \ldots, m_p'$, and denote the matrix products by $am'$, $m\alpha$, $m'\beta$, $a\alpha$, for example; then, from the $2p$ formulas (A) for $\vartheta(u_1, \ldots, u_p)$ we obtain

$$\Theta(u + \Omega_{mm'}; \tfrac{1}{2}\beta; \tfrac{1}{2}\alpha)$$
$$= \exp\left[-2\pi\sqrt{-1}\, m'(u + \tfrac{1}{2}am') + \pi\sqrt{-1}\,(m\alpha - m'\beta)\right]\cdot\Theta(u; \tfrac{1}{2}\beta; \tfrac{1}{2}\alpha),$$
$$\Theta(u + \tfrac{1}{2}\Omega_{\beta'\alpha'}; \tfrac{1}{2}\beta; \tfrac{1}{2}\alpha)$$
$$= \exp\left[-\pi\sqrt{-1}\,\alpha'(u + \tfrac{1}{2}\beta + \tfrac{1}{2}\beta' + \tfrac{1}{4}a\alpha')\right]\cdot\Theta(u; \tfrac{1}{2}\beta + \tfrac{1}{2}\beta'; \tfrac{1}{2}\alpha + \tfrac{1}{2}\alpha'),$$
$$\Theta(u; \tfrac{1}{2}\beta; \tfrac{1}{2}\alpha) = \exp\left[\pi\sqrt{-1}\,\alpha(u + \tfrac{1}{2}\beta + \tfrac{1}{4}a\alpha)\right]\cdot\Theta(u + \tfrac{1}{2}\Omega_{\beta\alpha}),$$
$$\Theta(u; \tfrac{1}{2}\beta + m; \tfrac{1}{2}\alpha + m') = \exp(\pi\sqrt{-1}\, m\alpha)\cdot\Theta(u; \tfrac{1}{2}\beta; \tfrac{1}{2}\alpha).$$

Let the function $\Theta(u; 0; 0)$ be abbreviated to $\Theta(u)$.

As we have seen in Section 2.20, the normal Riemann integral of the first kind $W_r(x; c)$ possesses the periods $0, 0, \ldots, 0, 1, 0, \ldots, 0, a_{r1}, \ldots, a_{rp}$ in the $r$th row, and the imaginary part of the quadratic form

$$a_{11}m_1^2 + \cdots + a_{rr}m_r^2 + \cdots + 2a_{12}m_1m_2 + \cdots + 2a_{rs}m_rm_s + \cdots$$

is positive for all constant values of $m_1, \ldots, m_p$. Hence these $\Theta$-functions are convergent.

Let $e_1, \ldots, e_p$ be the definite constant; let $m$ be a fixed place and $x$ a varying place, both on the Riemann surface. Consider the function $\Theta(u)$ as a function of the variables $u_r = W_r(x; m) + e_r$ $(r = 1, 2, \ldots, p)$ formed of the normal Riemann integrals of the first kind $W_1(x; m), \ldots, W_p(x; m)$. This function satisfies

$$\Theta(u + k) = \Theta(u),$$
$$\Theta(u_r + a_rk') = \exp\left[-2\pi\sqrt{-1}\, k'(u + \tfrac{1}{2}a_rk')\right]\cdot\Theta(u),$$

where $k = (k_1, \ldots, k_p)$, $k' = (k_1', \ldots, k_p')$, and $a_r$ is the $r$th row of the matrix $a_{rs}$. Such a function $\Theta(u)$, consisting of the integrals of the first kind (Section 2.20), does not become infinite as a function of $x$, while it possesses just $p$ simple zero points. If a zero point is $s$-ple, then it is counted as $s$ simple zero points. Denote the places of these simple zero points by $z_1, \ldots, z_p$. The Riemann surface can be partitioned so that $\log \Theta(u)$ is uniform by drawing sets of cross cuts, where a set consists of cross cuts of the $A$ type, those of the $B$ type, and the closed curves each enclosing a zero point (Section 2.20). Let each of these cross cuts correspond to one of the zero points $z_1, \ldots, z_p$. When we complete the circuit of all sets of these cross cuts, we obtain

$$\frac{1}{2\pi\sqrt{-1}}\int \log \Theta(u)\, du_s = 0,$$

because we have then completed the whole circuit along the whole boundary of the partitioned regions. The expression

$$k_s = \sum_{r=1}^{p}\int_{B_r} [W_r(x; m) + \tfrac{1}{2}a_{rr}]\, dW_s(x; m)$$

depends only on the place $m$. We determine the places $m_1, \ldots, m_p$ so that

$$W_s(m_1; P_1) + \cdots + W_s(m_p; P_p) \equiv k_s, \qquad (s = 1, 2, \ldots, p)$$

with the periods as moduli. Then we can make the $p$ places $m_1, \ldots, m_p$, which depend only on the places of $m$, correspond to an arbitrary place $m$ such that the places of the zero points of the function

$$\Theta[W(x; m) - W(z_1; m_1) - \cdots - W(z_p; m_p)],$$

as functions of $x$, are $z_1, \ldots, z_p$.

Replacing $x$ with $z_p$, we see that

$$\Theta[W(m_p; m) - W(z_1; m_1) - \cdots - W(z_{p-1}; m_{p-1})]$$

vanishes identically at all places $z_1, \ldots, z_{p-1}$. Thus the function

$$\Theta[W(x; m) - W(z_1; m_1) - \cdots - W(z_p; m_p)]$$

vanishes if and only if either $x$ coincides with one of the places $z_1, \ldots, z_p$, or the places $z_1, \ldots, z_p$ are the zero points of the $\phi$-polynomials (Section 2.20). Denote by $r$ a set of $r_1, \ldots, r_p$ for which $\Theta(r) = 0$. If $\Theta[W(x; z) + r]$ does not vanish at any place of $x$, then there exists one and only one set of places $z_1, \ldots, z_{p-1}$ for which

$$r \equiv W(m_p; m) - W(z_1; m_1) - \cdots - W(z_{p-1}; m_{p-1}).$$

In general, if $\Theta[W(x_1; z_1) + \cdots + W(x_q; z_q) + r]$ is identically zero for all places $x_1, z_1, x_2, z_2, \ldots, x_q, z_q$, and if

$$\Theta[W(x; z) + W(x_1; z_1) + \cdots + W(x_q; z_q) + r]$$

is made not identically zero with regard to $x$ by suitable choice of $z$ and is made not identically zero with regard to $z$ by suitable choice of $x$, then we can determine the set $\zeta_1, \ldots, \zeta_{p-1}$ such that

$$r \equiv W(m_p; m) - W(\zeta_1; m_1) - \cdots - W(\zeta_{p-1}; m_{p-1}),$$

in which $q$ out of these $\zeta$ places are arbitrary. We can further find the places $\xi_1, \ldots, \xi_{p-1}$ such that

$$-r \equiv W(m_p; m) - W(\xi_1; m_1) - \cdots - W(\xi_{p-1}; m_{p-1}),$$

in which $q$ out of these $\xi$ places are arbitrary. In such a case,

$$2W(m_p; m) + W(m_1; \zeta_1) + W(m_1; \xi_1) + \cdots + W(m_{p-1}; \zeta_{p-1}) + W(m_{p-1}; \xi_{p-1}) = 0,$$

or

$$(m^2, \xi_1, \ldots, \xi_{p-1}, \zeta_1, \ldots, \zeta_{p-1}) \equiv (m_1^2, m_2^2, \ldots, m_p^2),$$

which means that the residue for a closed contour describing a double circuit around the place $m$ and a single circuit around each of these places is equal to the residue for a closed curve describing double circuits around each of the places $m_1, \ldots, m_p$. Let these $2p - 2$ places $\xi_1', \ldots, \xi_{p-1}', \zeta_1', \ldots, \zeta_{p-1}'$ be the places of the zero points $a_1, \ldots, a_{2p-2}$ of the $\phi$-polynomial (Section 2.20). Then we can determine $m_1, \ldots, m_p$ from $m$ such that

$$(m^2, a_1, \ldots, a_{2p-2}) \equiv (m_1^2, m_2^2, \ldots, m_p^2), \tag{134}$$

or, if we regard $c_1, \ldots, c_p$ as independent places, there is a set of places $m_1, \ldots, m_p$ such that

$$2[W_s(m_1; c_1) + \cdots + W_s(m_p; c_p)] \equiv 2W_s(m; c_p) + W_s(a_1; c_1) + \cdots + W_s(a_{2p-2}; c_p).$$

Denoting the right-hand member by $A_s$, we obtain

$$W_s(m_1; c_1) + \cdots + W_s(m_p; c_p) \equiv \tfrac{1}{2}A_s + \tfrac{1}{2}(h_s + g_1 a_{s1} + \cdots + g_p a_{sp}) \qquad (s = 1, 2, \ldots, p),$$

with integers $g_1, \ldots, g_p, h_1, \ldots, h_p$. If the set $m_1', \ldots, m_p'$ is the set of integers of $m_1, \ldots, m_p$ satisfying (134), then

$$W_s(m_1'; m_1) + \cdots + W_s(m_p'; m_p) = \tfrac{1}{2}\Omega_{\beta\alpha} \qquad (s = 1, 2, \ldots, p),$$

with the previous notation $\Omega_{\alpha\beta}$ (Section 2.20). If we can find, with the suitably chosen integers $\alpha_1, \ldots, \alpha_p, \beta_1, \ldots, \beta_p$, more than one set of values $n_1, \ldots, n_{p-1}$ satisfying

$$-\tfrac{1}{2}\Omega_{\beta\alpha} \equiv W(m_p; m) - W(n_1; m_1) - \cdots - W(n_{p-1}; m_{p-1}),$$

then there are an infinite number of such sets. Suppose that there are $\infty^q$ of these; then, $\beta\alpha \equiv q + 1 \pmod 2$.

If the function $\Theta[W(x; z) + r]$ does not vanish identically for the values of $r$ satisfying $\Theta(r) = 0$, whatever may be the values of $x$ and $z$, then Riemann's normal integral of the third kind can be written

$$\Pi_{\alpha\beta}(x; z) = \log\left\{\frac{\Theta[W(x; \alpha) + r]}{\Theta[W(x; \beta) + r]} \div \frac{\Theta[W(z; \alpha) + r]}{\Theta[W(z; \beta) + r]}\right\}.$$

In fact, the function

$$\exp\left[-\Pi_{\alpha\beta}(x; z)\right]\cdot\frac{\Theta[W(x; \alpha) + r]}{\Theta[W(x; \beta) + r]}\,\frac{\Theta[W(z; \beta) + r]}{\Theta[W(z; \alpha) + r]}$$

is uniform on the Riemann surface which is cut along the circuits $A$ and $B$ and is neither zero nor infinite at points other than $z$. This function takes an equal value along a complete circuit around any closed cross cut of the $A$ or $B$ type. Hence, this function is also uniform on the Riemann surface before the cross cuts are made; that is, it does not depend on $x$. The function has its value corresponding to $x = z$, that is, it equals unity.

If we make the place $x_p$ tend to the place $z_p$ in the equation

$$\Theta[W(x_1; z_1) + \cdots + W(x_q; z_q) + r] = 0,$$

we obtain

$$\sum_{i=1}^{p} \Theta_i'[W(x_1; z_1) + \cdots + W(x_{q-1}; z_{q-1}) + r]\Omega_i(z_q) = 0,$$

where $\Theta_i'(u) = \partial\Theta(u)/\partial u_i$ and $\Omega_i(x) = 2\pi\sqrt{-1}\, D_x W_i(x; a)$ with an arbitrary constant $a$. This relation is valid for any place of

$$x_1, z_1, x_2, z_2, \ldots, x_{q-1}, z_{q-1}.$$

Since $\Omega_1(z_q), \ldots, \Omega_q(z_q)$ are independent, $\Theta'(u)$ vanishes for

$$u = W(x_1; z_1) + \cdots + W(x_{q-1}; z_{q-1}) + r.$$

Similarly $\Theta''(u)$ vanishes for

$$u = W(x_1; z_1) + \cdots + W(x_{q-2}; z_{q-2}) + r.$$

Hence the function $\Theta(u)$ and its derivatives up to the order $q$ vanish for $u = r$. Conversely, it can be shown that

$$\Theta[W(x_1; z_1) + \cdots + W(x_q; z_q) + r]$$

vanishes for $u = r$, whatever the values of $x_1, z_1, \ldots, x_q, z_q$, if and only if $\Theta(u)$ and its derivatives up to the order $q$ are zero.

If from the linearly independent integrals of the first kind

$$V_1(x; m), \ldots, V_p(x; m)$$

the normal integral of Riemann is obtained by means of

$$W_i(x; m) = \lambda_{i1} V_1(x; m) + \cdots + \lambda_{ip} V_p(x; m) \qquad (r = 1, 2, \ldots, p),$$

then $\Theta(u)$ is given by

$$u_i = \lambda_{i1} U_1 + \cdots + \lambda_{ip} U_p \qquad (i = 1, 2, \ldots, p)$$

as a function of $U_1, \ldots, U_p$, which we write $\vartheta(U)$. Then the $\phi$-polynomial which has the places $n_1, \ldots, n_{p-1}$ as zeros of the second order takes the form

$$\sum_{i=1}^{p} \vartheta_i' \left(\tfrac{1}{2}\bar{\Omega}_{kk'}\right) \psi_i(x),$$

where $\psi_1(x), \ldots, \psi_p(x)$ are the $\phi$-polynomials corresponding to

$$V_1(x; m), \ldots, V_p(x; m),$$

and $\frac{1}{2}\bar{\Omega}_{kk'}$ is a set of the half-periods of these $p$ functions $V_i(x; m)$. Since the half-periods $\Omega_{kk'}$ have been defined so that

$$2\lambda_{i1}\omega_{1s} + \cdots + 2\lambda_{ip}\omega_{ps} = \begin{cases} 1 & (r = s) \\ 0 & (r \neq s), \end{cases}$$

$$2\lambda_{i1}\omega_{1s}' + \cdots + 2\lambda_{ip}\omega_{ps}' = a_{is} \qquad (r, s = 1, 2, \ldots, p),$$

we have

$$\tfrac{1}{2}\bar{\Omega}_{kk'} = k_1\omega_{i1} + \cdots + k_p\omega_{ip} + k_1'\omega_{i1}' + \cdots + k_p'\omega_{ip}'$$
$$(i = 1, 2, \ldots, p).$$

Consider in general the sum over the values from $-\infty$ to $+\infty$ of $p$ integers $n_1, \ldots, n_p$ of the function

$$\vartheta(u; h; g) \equiv \vartheta(u; h)$$
$$= \sum \exp\left[\kappa u^2 + 2ku(n + g) + b(n + g)^2 + 2\pi\sqrt{-1}\, h(n + g)\right],$$

where $\kappa$ and $b$ are symmetric matrices with $p$ rows and $p$ columns, $k$ is an arbitrary matrix with $p$ rows and $p$ columns, and the quadratic form $bm^2$ is supposed to be negative for real values of $m_1, \ldots, m_p$. The letters $g$ and $h$ denote the characteristic sequences. In the place of $b$ we put $\pi\sqrt{-1}\, a$ with a symmetric matrix $a$, and denote by $U$ the $p$ quantities $ku$; then,

$$\vartheta(u; h) = \exp(\kappa u^2) \cdot \Theta(U; h, g).$$

Let $\omega$, $\omega'$, $\eta$, $\eta'$ be matrices with $p$ rows and $p$ columns; let

$$\Omega_P = 2\omega P + 2\omega' P', \qquad H_P = 2\eta P + 2\eta' P'$$

with row matrices

$$P = (P_1, \ldots, P_p), \qquad P' = (P_1', \ldots, P_p').$$

Then the expression in the generalized form

$$\begin{aligned}\vartheta(u + \Omega_P; h) = \exp\{H_P(u + \tfrac{1}{2}\Omega_P) - \pi\sqrt{-1}\, PP' + 2\pi\sqrt{-1}\,(Pg - P'h)\}\\ \times \exp\{-2\pi\sqrt{-1}\, Pg\}\cdot\vartheta(u, P + h)\end{aligned} \tag{135}$$

is valid under the condition that

$$\eta = 2\kappa\omega, \quad \eta' = 2\kappa\omega' - 2\bar{k}, \quad k\omega = \tfrac{1}{2}\pi\sqrt{-1}, \quad k'\omega' = \tfrac{1}{2}b, \tag{136}$$

where $\bar{k}$ denotes the transposed matrix of $k$. In such a case, (135) becomes

$$\begin{aligned}\vartheta(u + \Omega_P; h) &= \exp[\lambda_P(u) - 2\pi\sqrt{-1}\, P'h]\cdot\vartheta(u; h + P),\\ \lambda_P(u) &= H_P(u + \tfrac{1}{2}\Omega_P) - \pi\sqrt{-1}\, PP'.\end{aligned} \tag{137}$$

Condition (136) expresses $\omega$, $\omega'$, $\eta$, $\eta'$ in terms of $\kappa$, $k$, $b$. The remaining condition is that

$$H_P\Omega_Q - H_Q\Omega_P = 2\pi\sqrt{-1}\,(PQ' - P'Q).$$

If the matrices $P$ and $P'$ consist of integers, then (137) becomes

$$\vartheta(u + \Omega_m; h) = \exp[\lambda_m(u) + 2\pi\sqrt{-1}\,(mg - m'h)]\cdot\vartheta(u; h),$$

where $m$, $m'$ are row matrices consisting of integers. The function $\vartheta(u; h)$ defined in this manner is an odd or an even function according to whether $4gh$ is odd or even. Put

$$h = \pi\sqrt{-1}\, a, \qquad \zeta_i(u) = \frac{\partial}{\partial u_i}\log\vartheta(u);$$

then, from

$$\vartheta(u + \Omega_m) = \exp[H_m(u + \tfrac{1}{2}\Omega_m) - \pi\sqrt{-1}\, mm']\cdot\vartheta(u),$$

we obtain

$$\zeta_i(u + \Omega_m) - \zeta_i(u) = 2(\eta_{i1}m_1 + \cdots + \eta_{ip}m_p + \eta_{i1}'m_1' + \cdots + \eta_{ip}'m_p');$$

in particular,

$$\begin{aligned}\zeta_i(u_1 + 2\omega_{1r}, \ldots, u_p + 2\omega_{pr}) &= \zeta_i(u) + 2\eta_{ir},\\ \zeta_i(u_1 + 2\omega_{1r}', \ldots, u_p + 2\omega_{pr}') &= \zeta_i(u) + 2\eta_{ir}'.\end{aligned}$$

Denote by $\mu_i(x)$ the derivative $du_i(x; a)/dx$ and by $\bar{\nu}_{ri}$ a minor of the determinant formed of $\mu_i(x_r)$. Then we obtain an important relation:

$$-\zeta_i[u(x; m) - u(x_1; m_1) - \cdots - u(x_p; m_p)] + \zeta_i[u(a; m) - u(x_1; m_1) \cdots - u(x_p; m_p)] = L_i(x; a) + \sum_{r=1}^{p} \bar{\nu}_{ri}[(x_r, x) - (x_r, a)] \frac{dx_r}{dt}. \quad (138)$$

Thus $\bar{\nu}_{ri}$ depends on the places $x_1, \ldots, x_p$ exactly as the quantity $\nu_{ri}$, which is a minor of the determinant $\|\mu_i(x_r)\|$, depends on the places $c_1, \ldots, c_p$ (Section 2.20). The right-hand member has been already derived.

From (138), differentiating with regard to $x$, denoting $-\partial\zeta_i(u)/\partial u_j$ by $\wp_{ij}(u)$, and making $a$ coincide with $x$, we obtain

$$\sum_{j=1}^{p} \mu_j(x)\wp_{ij}(U) = D_z L_i(x; a) + \sum_{s=1}^{p} \nu_{si} D_z \left[(x_s, x) \frac{dx_s}{dt}\right],$$

where $U = u(z; m) - u(z_1; m_1) - \cdots - u(z_p; m_p)$ (Baker, 1897).

## 6.25 HYPERELLIPTIC FUNCTIONS

Equation (73a) of Section 2.20 in the general case can be written in the form

$$y^2 = H(x - a_1)(x - a_2) \cdots (x - a_p)(x - c_1) \cdots (x - c_{p+1})$$

and defines the Riemann surface. There exist an even number of the Riemann $\vartheta$-functions that vanish when the variable vanishes. One of the branch places, $a$ for example, is at infinity. Let $b$ and $b'$ be any two of the branch places; then the $p$ integrals $u_1(b; b'), \ldots, u_p(b; b')$ are the $p$ simultaneous components of a half-period, so that, for example

$$u_r(b; b') = m_1\omega_{r1} + \cdots + m_p\omega_{rp} + m_1'\omega_{r1}' + \cdots + m_p'\omega_{rp}' \equiv \tfrac{1}{2}\Omega \qquad (r = 1, 2, \ldots, p)$$

with integers $m_1, \ldots, m_p, m_1', \ldots, m_p'$ independent of $r$. If, $b$ remaining any branch place, $b'$ is taken to be each of the other $2p + 1$ branch places in turn, then the $2p + 1$ half-periods $u(b; b')$ thus obtained consist of $p$ odd half-periods and $p + 1$ even half-periods. Thus, if the branch places $b'$ for which $u(b; b')$ is an odd half-period be denoted by $b_1, \ldots, b_p$, then necessarily $\vartheta[u(b; b_1)] = 0, \ldots, \vartheta[u(b; b_p)] = 0$, and we can take the places $b_1, \ldots, b_p$ for the places $m, m_1, \ldots, m_p$. In particular, if the branch place $a$ is taken for $b$ and the branch places are located as

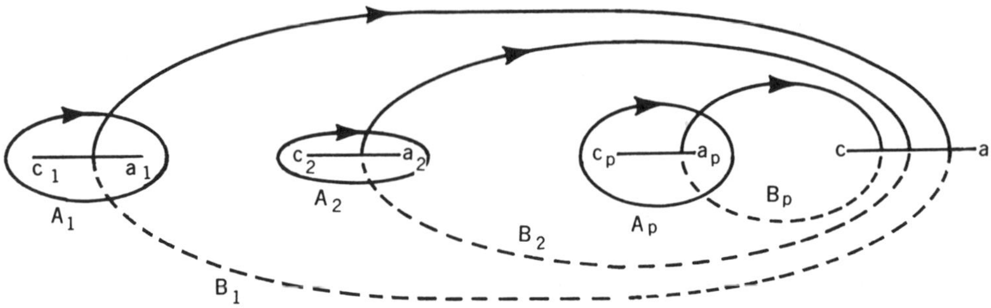


FIGURE 1

in Figure 1, then each of $u(a; a_1), \ldots, u(a; a_p)$ is an odd half-period. Hence the function $\vartheta[u(x; a) - u(x_1, a_1) - \cdots - u(x_p; a_p)]$ has the places $x_1, \ldots, x_p$ as its zero places, where it is assumed that the function $\vartheta[u(x; a)]$ does not vanish identically. This assumption is evidently justified (Baker, 1897).

Let $b, b', b'', \ldots$ be the different branch places other than $a$. From the $2p + 1$ half-periods $u(b; a)$ with respect to the branch places $b$ (not with respect to the branch places $a$), we can derive $\binom{2p+1}{2}$ half-periods of the form $u(b; a) + u(b'; a)$, and $\binom{2p+1}{3}$ half-periods of the form $u(b; a) + u(b'; a) + u(b''; a)$, and so on; in general, $\binom{2p+1}{p}$ half-periods in the form of the sum of $p$ half-periods of the type $u(b; a)$. The sum of these half-periods makes

$$\binom{2p+1}{1} + \binom{2p+1}{2} + \cdots + \binom{2p+1}{p} = 2^{2p} - 1$$

in number, the number of incongruent half-periods. The half-periods that are identically zero are not included in the sum. None of them are congruent to each other.

When, in general, the half-periods are of the form

$$\tfrac{1}{2}\Omega_{mm'} \equiv m_1\omega_{r1} + \cdots + m_p\omega_{rp} + m'_1\omega'_{r1} + \cdots + m'_p\omega'_{rp},$$

we write

$$\begin{pmatrix} k'_1 & k'_2 & \cdots & k'_p \\ k_1 & k_2 & \cdots & k_p \end{pmatrix}$$

instead of writing $m_r = k_r$, (mod 2); $m'_s = k'_s$, (mod 2); and we call that symbol the *characteristic symbol*, in which $k_s$, $k'_s$ (equal to 0 or 1) are the remainders when $m_s$, $m'_s$, respectively, are divided by 2. Each of the $2^{2p} - 1$ possible characteristic symbols, other than the one that has all its elements zero, can be obtained as the sum of not more than $p$ chosen from the $2p + 1$ so-called fundamental characteristic symbols. The sum of these $2p + 1$ fundamental characteristic symbols is equal to the symbol of which all the elements are zero in mod 2. Of these $2p + 1$ symbols, $p$ are associated with the odd half-periods $u(a; a_1), \ldots, u(a; a_p)$; the remaining $p + 1$ are associated with the even half-periods. For example, for $p = 2$,

$$\begin{pmatrix}1 & 0\\1 & 1\end{pmatrix}, \quad \begin{pmatrix}0 & 1\\0 & 1\end{pmatrix}, \quad \begin{pmatrix}0 & 0\\1 & 1\end{pmatrix}, \quad \begin{pmatrix}1 & 0\\0 & 1\end{pmatrix}, \quad \begin{pmatrix}0 & 1\\0 & 0\end{pmatrix}, \quad \begin{pmatrix}0 & 0\\0 & 0\end{pmatrix}.$$

Put

$$\tfrac{1}{2}m_s = M_s + \tfrac{1}{2}k_s, \qquad \tfrac{1}{2}m_s' = M_s' + \tfrac{1}{2}k_s',$$

where $M_s$, $M_s'$ are integers and each of $k_s$, $k_s'$ is either 1 or 0, and write

$$\lambda = [2\eta(M + \tfrac{1}{2}k) + 2\eta'(M' + \tfrac{1}{2}k')][u + \omega(M + \tfrac{1}{2}k) + \omega'(M' + \tfrac{1}{2}k')] - \pi\sqrt{-1}\,(M + \tfrac{1}{2}k)(M' + \tfrac{1}{2}k');$$

then,

$$\vartheta(u; \tfrac{1}{2}k, \tfrac{1}{2}k') = \exp[-\lambda - \pi\sqrt{-1}\,Mk']\cdot\vartheta(u + \tfrac{1}{2}\Omega_{mm'}).$$

Denote the left-hand side by $\vartheta(u/\tfrac{1}{2}\Omega_{mm'})$. If

$$\tfrac{1}{2}\Omega_{mm'} = u(b_1; a) + \cdots + u(b_s; a),$$

then we write it $\vartheta[u/u(b_1; a) + \cdots + u(b_s; a)]$. An odd half-period is represented by

$$\tfrac{1}{2}\Omega \equiv u(a_p; a) - u(n_1; a_1) - \cdots - u(n_{p-1}; a_{p-1}),$$

and an even half-period, for which $\vartheta(\tfrac{1}{2}\Omega')$ does not vanish, is represented by

$$\tfrac{1}{2}\Omega' \equiv u(b_1; a_1) + \cdots + u(b_p; a_p).$$

The zero places of $\vartheta[u(x; z)/\tfrac{1}{2}\Omega]$ are $z$ and $n_1, \ldots, n_p$. The zero places of $\vartheta[u(x; a)/\tfrac{1}{2}\Omega']$ are $b_1, \ldots, b_p$.

For $p = 2$ there is no even $\vartheta$-function that vanishes when the argument vanishes. There are six odd $\vartheta$-functions. One of them — $\vartheta_{aa_1}(u)$, for example — is of half-period $u(a; a_1) = \tfrac{1}{2}\begin{pmatrix}1 & 0\\1 & 0\end{pmatrix}$, and the corresponding half-period $u(a_2; a) - u(n_1; a_1)$ is such that $n_1 = a_2$. There are ten even $\vartheta$-functions. One of them — $\vartheta(u)$, for example — is of half-period $\tfrac{1}{2}\begin{pmatrix}0 & 0\\0 & 0\end{pmatrix}$; substituting the corresponding half-period $u(b_1; a_1) + u(b_2; a_2)$, we see that $b_1$ and $b_2$ must equal $a_1$ and $a_2$, respectively. Another function — $\vartheta_{ac}(u)$, for example — is of half-period $u(a; c) = \begin{pmatrix}1 & 1\\0 & 0\end{pmatrix}$ and the corresponding half-period is such that $b_1$ and $b_2$ must equal $c_1$ and $c_2$, respectively.

If the sum of the characteristics $q_1, \ldots, q_n$ differs from the sum of the characteristics $r_1, \ldots, r_n$ by characteristics consisting only of integers where $n \geqq 2$, then

$$f(u) = \frac{\vartheta(u; q_1)\vartheta(u; q_2)\cdots\vartheta(u; q_n)}{\vartheta(u; r_1)\vartheta(u; r_2)\cdots\vartheta(u; r_n)}$$

is a periodic function of $u$. If $u$ is given by

$$u_r = u_r(x_1; a_1) + \cdots + u_r(x_p; a_p) \qquad (r = 1, 2, \ldots, p),$$

then the function $f(u)$ is rational in $x_1, \ldots, x_p$. Accordingly $\dfrac{\vartheta^2[u/u(b; a])}{\vartheta^2(u)}$ is rational in $x_1, \ldots, x_p$. This function, regarded as a function of $x_1$, has its zero places $b$ of the second order and its infinity places $a$ of the second order. Hence, it is a product of $(b - x_1)$ with a function independent of $x_1$. Since this function is symmetric with respect to $x_1, \ldots, x_p$, it equals

$$\sqrt{(b - x_1)(b - x_2)\cdots(b - x_p)} = \frac{1}{\sqrt{K}} \frac{\vartheta[u/u(b; a)]}{\vartheta(u)}, \tag{139}$$

with a constant $K$. This function may be interpreted as single-valued of the places $x_1, \ldots, x_p$ on the Riemann surface which is dissected by the $2p$ cross cuts that are the period loops.

Put $u = u(x_1; a_1) + \cdots + u(x_p; a_p)$ and let $q$ denote in turn all possible half-integer characteristics except the identically zero one; then all the $2^{2p} - 1$ ratios $\vartheta(u; q)/\vartheta(u)$ are expressed algebraically in terms of $x_1, \ldots, x_p$. That is, if $\Delta(x_1, \ldots, x_p)$ is the product of all the differences of $x_1, \ldots, x_p$ and $\phi(x) = (x - b_1)\cdots(x - b_k)$, then the ratios can be expressed by

$$\frac{\vartheta[u/u(b_1; a) + \cdots + u(b_k; a)]\vartheta^{k-1}(u)}{\vartheta[u/u(b_1; a)]\cdots\vartheta[u/u(b_k; a)]}$$

$$= C\frac{\left\| \dfrac{x_r^\lambda y_r}{\phi(x_r)}, \dfrac{x_r^{\lambda-1} y_r}{\phi(x_r)}, \ldots, \dfrac{y_r}{\phi(x_r)}, x_r^\mu, x_r^{\mu-1}, \ldots, 1 \right\|}{\Delta(x_1, x_2, \ldots, x_p)}$$

with constant $C$, where the numerator is the determinant obtained by putting $r = 1, 2, \ldots$ for each row, and where

$$\lambda = \frac{k}{2} - 1, \qquad \mu = p - 1 - \frac{k}{2},$$

for $k$ even, and

$$\lambda = \frac{k-3}{2}, \qquad \mu = p - \frac{k+1}{2}$$

for $k$ odd. In particular, for $k = 2$ we have $\lambda = 0$, $\mu = p - 2$, giving

$$\frac{\vartheta[u/u(b_1; a) + u(b_2; a)]\vartheta(u)}{\vartheta[u/u(b_1; a)]\vartheta[u/u(b_2; a)]} = C\sum_{r=1}^{p} \frac{y_r}{(x_r - b_1)(x_r - b_2)} \frac{1}{R'(x_r)},$$

where

$$R(x) = (x - x_1)(x - x_2)\cdots(x - x_p)$$

and $C$ is a constant.

After this preparation we discuss the problem of inversion (Section 2.20) of the $p$ equations represented by

$$u_i(x_1; a_1) + \cdots + u_i(x_p; a_p) \equiv u_i \qquad (i = 1, 2, \ldots, p).$$

The solution according to (139), is given by the $2p + 1$ equations

$$\frac{\vartheta^2[u/u(b; a)]}{\vartheta^2(u)} = A(b - x_1)(b - x_2)\cdots(b - x_p).$$

From any of these $p$ equations the variables $x_1, \ldots, x_p$ can be expressed as single-valued functions of the arbitrary arguments $u_1, \ldots, u_p$. When the constant $A$ is determined we obtain

$$\frac{\vartheta^2[u/u(b; a)]}{\vartheta^2(u)} = \pm\frac{(b - x_1)(b - x_2)\cdots(b - x_p)}{[\exp(\pi\sqrt{-1}\,PP')\cdot f'(b)]^{1/2}},$$

$$f(x) = (x - a_1)\cdots(x - a_p)(x - c)(x - c_1)\cdots(x - c_p),$$

where $\exp(\pi\sqrt{-1}\,PP')$ is $+1$ or $-1$ according as $u(b; a)$ is an odd or even half-period. Put $a_1, \ldots, a_p$ in turn in the place of $b$; then it can be seen that $x_1, \ldots, x_p$ are the roots of

$$\sum_{i=1}^{p} \epsilon_i \sqrt{-\frac{Q(a_i)}{P'(a_i)}}\,\frac{\vartheta^2[u/u(a_i; a)]}{a_i - x} = \vartheta^2(u),$$

with

$$P(x) = (x - a_1)(x - a_2)\cdots(x - a_p),$$

$$Q(x) = (x - c)(x - c_1)\cdots(x - c_p),$$

since

$$\frac{(x - x_1)\cdots(x - x_p)}{P(x)} = 1 + \sum_{i=1}^{p} \frac{(a_i - x_1)\cdots(a_i - x_p)}{(x - a_i)P'(a_i)},$$

where $\epsilon_i$ stands for $+1$ or $-1$ such that

$$\frac{\vartheta^2[u/u(a_i; a)]}{\vartheta^2(u)} = \epsilon_i \frac{(a_i - x_1)\cdots(a_i - x_p)}{\sqrt{-P'(a_i)Q(a_i)}}. \tag{140}$$

This method of solution is due to Weierstrass.

Next we solve the inversion problem by a different method. Denote by $h_r$ the sum of the homogeneous products of $x_1, \ldots, x_p$, $r$ together, without repetition, and put

$$\chi_{p-1}(x; x_1, \ldots, x_p) = x^{p-i} - h_1 x^{p-i-1} + h_2 x^{p-i-2} - \cdots + (-1)^{p-i} h_{p-i}.$$

If we consider the integrals

$$\int_\mu^x \frac{dx}{y}, \quad \int_\mu^x \frac{x\,dx}{y}, \quad \ldots, \quad \int_\mu^x \frac{x^{p-1}\,dx}{y}$$

in the place of $u_1(x; \mu), \ldots, u_p(x; \mu)$, then we see that

$$\bar{\nu}_{ki} = y_k \frac{\chi_{p-i}(x_k; x_1, \ldots, x_p)}{F'(x_k)} \bigg/ \frac{dx_k}{dt},$$

$$F(x) = (x - x_1) \cdots (x - x_p).$$

We can prove from (138) that

$$\begin{aligned} -\zeta_i(u) = {} & L_i(x; \mu) + L_i(x_1; \mu) + \cdots + L_i(x_p; \mu) \\ & - \frac{y\chi_{p-i}(x; x_1, \ldots, x_p)}{2R'(x)} - \cdots - \frac{y_p\chi_{p-i}(x_p; x, x_1, \ldots, x_{p-1})}{2R'(x_p)} \\ & + \text{constant}, \end{aligned} \tag{141}$$

where

$$R(t) = (t - x)(t - x_1) \cdots (t - x_p),$$

$$u = u(x; a) + u(x_1; a_1) + \cdots + u(x_p; a_p).$$

If we make $x$ tend to the place $a$ at infinity in this formula, then

$$\begin{aligned} & -\zeta_i[u(x_1; a_1) + \cdots + u(x_p; a_p)] \\ & \quad = L_i(x_1; a_1) + \cdots + L_i(x_p; a_p) - \sum_{k=1}^{p} \frac{y_k\chi_{p-i+1}(x_k; x_1, \ldots, x_p)}{2F'(x_k)}, \end{aligned}$$

which is an odd function. This equation enables us to write $\zeta_i(u)$ for any value of $u$.

For $p = 1$ this formula reduces to the relation for the elliptic functions

$$-\zeta_1[u(x_1; a_1)] = L_1(x_1; a_1),$$

$$\begin{aligned} -\zeta_1[u(x; a) + u(x_1; a_1)] = {} & L_1(x; a_1) + L_1(x_1; a_1) \\ & - \frac{1}{2} \frac{y - y_1}{x - x_1} + \text{constant}. \end{aligned}$$

Referring to the known formulas for the elliptic functions, for which $u = u(x; a)$, giving

$$y = -\wp'(u), \qquad u(x; a_1) = \omega + \omega' - u,$$

so that

$$\zeta_1[u(x; a_1)] = -L_1(x; a_1) = -\int_{a_1}^{x} \frac{x\, dx}{y},$$

and

$$d\zeta(u) = \frac{x\, dx}{y}, \qquad \zeta_1[u(x; a_1)] = -\zeta(u),$$

we obtain the well-known addition theorem

$$\zeta(u + v) = \zeta(u) + \zeta(v) + \frac{1}{2} \frac{\wp'(u) - \wp'(v)}{\wp(u) - \wp(v)}.$$

As a particular case of (138) for $p = 2$,

$$\begin{aligned} -\zeta_p[u(x_1; a_1) + \cdots + u(x_p; a_p)] &= L_p(x_1; a_1) + \cdots + L_p(x_p; a_p) \\ &= \frac{\lambda_{2p+1}}{4} \sum_{k=1}^{p} \int_{a_k}^{x_k} \frac{x^p\, dx}{y}. \end{aligned}$$

Denote

$$-\frac{\partial}{\partial u_i} \zeta_r(u) \quad \text{by} \quad \wp_{ri}(u);$$

then,

$$\wp_{rj}(u) = \frac{1}{4} \lambda_{2p+1} \sum_{k=1}^{p} \frac{x_k^p \chi_{p-i}(x_k; x_1, \ldots, x_p)}{F'(x_k)}.$$

Finally, we can derive the formula

$$\tfrac{1}{4}\lambda_{2p+1} x_k^p = \sum_{i=1}^{p} x_k^{i-1} \wp_{pi}(u).$$

Let $\lambda_{2p+1} = 4$. Then $x_1, \ldots, x_p$ satisfying the inversion problem

$$u \equiv u(x_1; a_1) + \cdots + u(x_p; a_p)$$

are the roots of

$$F(x) = x^p - x^{p-1}\wp_{p,p}(u) - x^{p-2}\wp_{p,p-1}(u) - \cdots - \wp_{p,1}(u) = 0,$$

giving

$$h_r = (-1)^{r-1} \wp_{p,p-r+1}(u).$$

Since $F(x_k) = 0$, then, by writing

$$\frac{\partial \wp_{pq}(u)}{\partial u_r} = \wp_{pqr}(u),$$

we obtain

$$y_k = \psi(x_k),$$

$$\psi(x) = -\frac{\partial F(x)}{\partial u_p} = x^{p-1}\wp_{p,p,p}(u) + x^{p-2}\wp_{p,p,p-1}(u) + \cdots + \wp_{p,p,1}(u).$$

This formula gives the complete solution of the inversion problem.

Writing the right-hand member of (140) as

$$\frac{1}{\mu_i}(a_i - x_1)\cdots(a_i - x_p),$$

we see that

$$\mu_i \frac{\vartheta^2[u/u(a_i;\, a)]}{\vartheta^2(u)} = a_i^p - a_i^{p-1}\wp_{p,p}(u) - a_i^{p-2}\wp_{p,p-1}(u) - \cdots - \wp_{p,1}(u).$$

Denote the expression $\vartheta[u/\frac{1}{2}(\begin{smallmatrix}0 & 1\\ 1 & 1\end{smallmatrix})]$ by $\sigma(u)$, and put

$$\wp_{ij}(u) = -\frac{\partial^2}{\partial u_i\, \partial u_j}\log \sigma(u), \qquad v = u + \frac{1}{2}\,\Omega_{mm'};$$

then, for $p = 2$ we obtain

$$\frac{[\vartheta(0)]^2\mu_1\mu_2}{a_1 - a_2}\,\frac{\sigma(u+v)\sigma(u-v)}{\sigma^2(u)\sigma^2(v)}$$
$$= \wp_{11}(u) - \wp_{11}(v) + \wp_{12}(u)\wp_{22}(v) - \wp_{12}(v)\wp_{22}(u).$$

This is the generalization of the formula for the elliptic functions in the case $p = 1$, for which $\sigma(u)/u = 1$ when $u = 0$, that is, of the formula

$$\frac{\sigma(u+v)\sigma(u-v)}{\sigma^2(u)\sigma^2(v)} = \wp(v) - \wp(u).$$

There are for $p = 2$ six odd $\vartheta$-functions:

$$\vartheta_{aa_1}(u),\ \vartheta_{aa_2}(u),\ \vartheta_{a_1a_2}(u),\ \vartheta_{c_1c_2}(u),\ \vartheta_{cc_1}(u),\ \vartheta_{cc_2}(u);$$

and ten even $\vartheta$-functions:

$$\vartheta(u),\ \vartheta_{ac}(u),\ \vartheta_{ac_1}(u),\ \vartheta_{ac_2}(u),\ \vartheta_{a_1c_1}(u),\ \vartheta_{a_1c_2}(u),\ \vartheta_{a_2c_1}(u),\ \vartheta_{a_2c_2}(u),$$
$$\vartheta_{ca_2}(u),\ \vartheta_{ca_1}(u).$$

Denote the functions $\vartheta(u)$ and $\vartheta(u')$ by $\vartheta$ and $\vartheta'$, respectively; then we can derive, with a constant $C$,

$$C\frac{\vartheta[(u+u')/u(a_1;a)]}{\vartheta(u+u')} = -[\epsilon(\vartheta\vartheta_{aa_1}\vartheta'_{a_1c_1}\vartheta'_{ac_1} + \vartheta'\vartheta'_{aa_1}\vartheta_{a_1c_1}\vartheta_{ac_1})$$

$$+ \vartheta_{a_1a_2}\vartheta_{aa_2}\vartheta'_{a_2c_1}\vartheta'_{cc_2} + \vartheta'_{a_1a_2}\vartheta'_{aa_2}\vartheta_{a_2c_1}\vartheta_{cc_2}]$$

$$\div [\vartheta^2\vartheta'^2 + \vartheta^2_{aa_1}\vartheta'^2_{aa_1} + \vartheta^2_{aa_2}\vartheta'^2_{aa_2} + \vartheta^2_{a_1a_2}\vartheta'^2_{a_1a_2}].$$

This is an extension of the addition theorem for the elliptic functions

$$\operatorname{sn}(u+u') = \frac{\operatorname{sn} u \operatorname{cn} u' \operatorname{dn} u' + \operatorname{sn} u' \operatorname{cn} u \operatorname{dn} u}{1 - k^2 \operatorname{sn}^2 u \operatorname{sn}^2 u'}.$$

This second solution of the inversion problem is due to Königsberger.

Unfortunately this solution of the inversion problem is of little use in conjunction with the separation of variables in the Hamilton-Jacobi equation, except in the case of Kowalewski's top motion.

## REFERENCES


Arnold, V. I.

1963 *Usp. Mat. Nauk*, **18**, 91.

Baker, H. F.

1897 *Abel's Theorem and the Allied Theory including the Theory of the Theta Functions*, Cambridge University Press, London.

1907 *Multiply Periodic Functions*, Cambridge University Press, London.

Bisconcini, G.

1906 *Acta Math.*, **30**, 49.

Born, M., and Brody, E.

1921 *Z. Physik*, **6**, 140.

Born, M., and Pauli, W.

1922 *Z. Physik*, **10**, 137.

Bruns, H.

1887 *Acta Math.*, **11**, 25.

1887a *Berichte sächs. Gesell. Wiss. Leipzig*, **1**, 55.

Burgatti, P.

1906 *Annali Mat. pura appl.*, ser. 3, **12**, 81

1910 *Rendiconti Cir. Mat. Palermo*, **29**, 369.

Casimir, H. B. G.

1931 Dissertation, Leiden. Published as *Rotation of a Rigid Body in Quantum Mechanics*, J. B. Wolter, Groningen, 1931.

Chazy, J.

1918 *Bull. Astron.*, **35**, 321.

1922 *Ann. Ecole Norm. Sup.*, ser. 3, **39**, 29.

1929 *J. Math. pures appl.*, ser. 9, **8**, 353.

1932 *Bull. Sci. Math.*, ser. 2, **56**, 79.

1933 *Compt. Rend.*, **197**, 1391.

Cherry, T. M.
1924 *Proc. Cambridge Phil. Soc.*, **22**, 273, 287, 325, 510.
1924a *Trans. Cambridge Phil. Soc.*, **23**, 43.
1926 *Proc. London Math. Soc.*, **26**, 211.
1927 *Ibid.*, **27**, 151.
1928 *Phil. Trans. Roy. Soc. London*, ser. A, **227**, 137.

Coble, A. B.
1929 *Algebraic Geometry and Theta Functions*, Am. Math. Soc., New York.

Epstein, P. S.
1922 *Z. Physik*, **8**, 211, 305; **9**, 92.

Fermi, E.
1923 *Phys. Z.*, **24**, 261.
1923a *Nuovo Cimento*, **26**, 105.

Forsyth, A. R.
1900 *Theory of Differential Equations*, Part II, Vol. III, Chap. 18, Cambridge University Press, London.

Golubev, V. V.
1953 *Lectures on Integration of the Equations of Motion of a Rigid Body about a Fixed Point*, trans. from the Russian, published by the Israel Program for Sci. Trans., 1960.

Hagihara, Y.
1933 *J. Phys.-Math. Soc. Japan*, **7**, 253, 321.

Halphen, G.
1888 *Traité des Fonctions Elliptiques*, II, Gauthier-Villars, Paris.

Hecke, E.
1928 *Theorie der algebraischen Zahlen*, Akad. Verlagsgesell., Leipzig.

Hesse, W.
1890 *Math. Annalen*, **37**, 153.

Husson, É.
1905 *Compt. Rend.*, **141**, 100.
1906 *Ann. Fac. Sci. Toulouse*, ser. 2, **8**, 73. Also issued as the thesis, *Recherche des Intégrales algébriques dans le Mouvement d'un Solide pesant autour d'un Point fixe*, Gauthier-Villars, Paris.
1908 *Acta Math.*, **31**, 71.
1932 *Les Trajectoires de la Dynamique*, Gauthier-Villars, Paris.

Jeans, H.
1921 *The Dynamical Theory of Gases*, 101, 3rd ed., Cambridge University Press, London.

Kiveliovitch, M.
1926 *Compt. Rend.*, **182**, 116, 263.

Klein, F., and Sommerfeld, A.
1921 *Über die Theorie des Kreisels*, II, Teubner, Leipzig.

Klein, O.
1929 *Z. Physik*, **58**, 730.

Kolossoff, P. A.
1902 *Rendiconti Cir. Mat. Palermo*, **16**, 346.
1903 *Math. Annalen*, **56**, 265.

Kötter, F.
1892 *J. reine angew. Math.*, **109**, 51, 89.

1893 *Acta Math.*, **17**, 209.

Kowalewski, S.
1888 *Acta Math.*, **12**, 177; *Mem. Acad. Sci., Paris*, **31**, no. 1.
1890 *Acta Math.*, **14**, 81.

Kramers, H. A., and Ittmann, G. P.
1929 *Z. Physik*, **53**, 553; **58**, 217.
1930 *Ibid.*, **60**, 663.

Krazer, A.
1903 *Lehrbuch der Theta Funktionen*, Teubner, Leipzig.

Kronig, R. de L. and Rabi, I. I.
1927 *Phys. Rev.*, **29**, 262.

Landau, E.
1918 *Einführung in die elementare und analytische Theorie der algebraischen Zahlen und des Ideals*, Teubner, Leipzig.

Laporte, O.
1933 *Phys. Rev.*, **43**, 548.

Lazzarino, O.
1919 *Rendiconti Accad. Lincei*, ser. 5, **28**, 1° sem., 325, 341; *Ibid.*, 2° sem., 9, 259, 329, 423, 489.

Leimanis, E.
1965 *The General Problem of the Motion of Coupled Rigid Bodies About a Fixed Point*, Springer, Berlin.

Levi-Civita, T.
1901 *Rendiconti Accad. Lincei*, ser. 5, **10**, 1° sem., 338, 429, 461; reprinted in *Opere Matematiche*, II, 107 Zanichelli, Bologna, 1956.
1903 *Annali Mat. pura appl.*, ser. 3, **9**, 1; reprinted in *Opere Matematiche*, II, 279.
1906 *Acta Math.*, **30**, 305; reprinted in *Opere Matematiche*, II, 419.

Liouville, R.
1897 *Acta Math.*, **20**, 239.

MacMillan, W. D.
1913 *Bull. Amer. Math. Soc.*, **19**, 349.

Marcolongo, R.
1902 *Rendiconti Cir. Mat. Palermo*, **16**, 349.
1910 *Ibid.*, **29**, 369.
1906 *Annali Mat. pura appl.*, ser. 3, **12**, 81.

Nekrassoff, P. A.
1896 *Math. Annalen*, **47**, 445.

Oseen, C. W.
1911 *Arkiv Mat. Astron. Fysik*, **6**, no. 23.

Painlevé, P.
1896 *Compt. Rend.*, **123**, 1224.
1896a *Ibid.*, **123**, 636, 871.
1896b *Ibid.*, **124**, 136, 173, 221.
1897 *Leçons sur la Théorie Analytique des Equations Différentielles*, Hermann, Paris.
1897a *Compt. Rend.*, **125**, 1078.
1898 *Bull. Astron.*, **15**, 81.
1900 *Acta Math.*, **25**, 1.

1900a *Bull. Soc. Math. France*, **28**, 201.
1900b *Compt. Rend.*, **130**, 1699.

Perchot, J., and Ebert, W.
1898 *Compt. Rend.*, **126**, 725.

Poincaré, H.
1889 *Acta Math.*, **13**, 1. Reprinted in *Oeuvres de Henri Poincaré*, VII, 286, Gauthier-Villars, Paris, 1952.
1892 *Méthodes Nouvelles de la Mécanique Céleste*, I, Chap. 5, Gauthier-Villars, Paris.
1896 *Compt. Rend.*, **123**, 1224. Reprinted in *Oeuvres*, VII, 512.

Reiche, F. and Rademacher, H.
1926 *Z. Physik*, **39**, 444.
1927 *Ibid.*, **41**, 453.

Riemann, B.
1892 *B. Riemann's Gesammelte Mathematische Werke*, Teubner, Leipzig.

Rosenblatt, A.
1928 *Bull. Sci. Math.*, ser. 2, **52**, 195.

Siegel, C. L.
1936 *Trans. Amer. Math. Soc.*, **39**, 225.

Stäckel, P.
1908 *Math. Annalen*, **65**, 538.

Staude, O.
1894 *J. reine angew. Math.*, **113**, 318.

Sundman, K.
1907 *Acta Soc. Sci. Fennicae*, **34**, no. 7.
1909 *Ibid.*, **35**, no. 9.
1913 *Acta Math.*, **36**, 105.

Weierstrass, K.
1894 *Mathematische Werke von K. Weierstrass*, I, Mayer u. Müller, Berlin.

Whittaker, T.
1916 *Proc. Roy. Soc. Edinburgh*, **37**, 95.
1928 *Treastise on Analytical Dynamics of Particles and Rigid Bodies*, Chap. 14, 2nd ed., Cambridge University Press, London.
1937 *Ibid.*, Chap. 16, 3rd ed.

Wirtinger, W.
1895 *Untersuchungen über Theta Funktionen*, Teubner, Leipzig.

Woltjer, J.
1924 *Bull. Astron. Inst. Netherlands*, **2**, 179.

# Retrospect of Volume I

At this stage let us recapitulate the results of our discussion.

The elementary method for solving a system of differential equations of order $6n$ for the $n$-body problem lies in the discovery of as many independent integrals as the order of the differential equations. So far, the known classical integrals are: one for energy, three for angular momentum, six for the center of mass, one for the elimination of the node, and one for the elimination of time — in all, twelve integrals. The first ten are called the classical Eulerian integrals. Lie's theory on the continuous group of transformations is applied, to reduce the order of the differential equations. The concepts of contact transformation, involution, function groups, and distinguished functions are introduced for the theoretical discussion of the reduction. The order of the differential equations is reduced to $6n - 12$ by the known classical integrals. The problem is to see whether any further integrals exist, apart from some rigorous particular solutions of the problem treated in Chapter 3. The theorem of Bruns as extended by Painlevé states the nonexistence of new integrals that are algebraic with regard to the momenta. The theorem of Poincaré denies the existence of new integrals that are uniform functions with regard to the Keplerian elements. Then what would be the functional form of the new integrals? In search for such new integrals we must refer to transcendental functions, or try to solve the equations on the basis of transcendental processes.

The existence of the solution will be proved in a later volume basing it on Weierstrass's theory of analytical functions. The whole field of celestial mechanics at present is occupied by various attempts to solve the equations by an infinite process of iteration, which we treat in several chapters of a future volume. The usual method is to expand the solution in the form of an infinite multiple Fourier series proceeding in powers of the small disturbing masses, with linear functions of time as arguments. Although such a series expansion resembles the one in the theory of quasi-periodic or almost periodic functions, we know little about the condition for which the solution is a quasi-periodic function, so that the series employed in the usual perturbation theory in celestial mechanics cannot be said to represent the solution we are seeking. Can we possibly attack the problem from the function-theoretic standpoint of Poincaré and Painlevé? On the other hand, as will be demonstrated in one of the subsequent volumes, Poincaré has proved that none of the infinite processes ever proposed for solving the problem is uniformly convergent, notwithstanding all the efforts that our predecessors have concentrated on this method of solution. Is it possible to predict the future behavior of the motion on a purely qualitative basis — thus abandoning our original aim of modifying the laws of mechanics through comparing observations with theoretical computations carried out on a nonrigorous mathematical foundation? At this pessimistic stage Poincaré, followed by Birkhoff, founded his whole theory of topology and applied it to the equations of celestial mechanics, in order to study the state of motion in an infinite time interval. These will be the main topics of the future volumes.

# Index


Abelian integral, 185
Abel's lemmas, 648–653
Abel's theorem, 195
Absolute integral invariant, 58, 346
Accumulation point, 104
Action, 18
Adelphic integral, 3, 637
Adiabatic approximation, 391–401
Adiabatic invariants, 381–406
  defined, 385
Adjoint form, 371
Adjoint group, 424
Adjoint operators, 268
Algebraic function, 185
Algebraic integrals, 556–679
Almost periodic functions, 6, 149–166
  defined, 150
Andoyer's expansion, 521–526
Angular momentum integrals, 540–545
Aphelion distance, 485
Approximation theorem, 161
Argument, 535
Argument of latitude, 484
Associated integral invariants, 58
Associated Pfaffian, 44
Associated system, 367
Asymptotic motion, 121
Automorphism, 424

Baire's class, 108
Base field, 159–162
Bilinear covariants, 42–46
  defined, 43
Bohl's quasi-periodic function, 144–149
  defined, 148
Bolzano-Weierstrass theorem, 105
Borel field, 106
Borel set, 106
Bounded set, 105
Bounded variation of function, 112
Branch cuts, 186
Bruns's theorem, 556–583

Calculus of variations, 19–22
Canonical constants, 490
Canonical equations, 22–28
  conjugate variables, 23
  degenerate form, 24
Canonical parameters, 413
Canonical transformation, 51–59, 85–88
  defined, 53, 83, 86
Canonical variables, 23
Cartan's exterior multiplication, 363–381
Cauchy
  characteristic curve, 293
  first theorem, 490–496
  number, 492
  second theorm, 496–503
  sequence, 106
Celestial mechanics
  as deductive science, 1
  fundamental theory, 1–2
Central figure, 229, 240, 242–246
  properties, 246–256
Character of representation, 438
Characteristic, 293
Characteristic function of Hamilton, 26
Characteristic system, 368
Closed set, 104, 150, 165
Cogradient, 29
Commutative group, 296

Commutator, 297, 440
Compact set, 105
Complement, 105
Complete integrals, 62, 171–173, 293
Complete set, 106
Complete systems, 298, 316–317
Completeness theorem of Fourier series, 114
Composition series, 429
Condition for separability, 77–82
Conditionally periodic motion, 123–149
  defined, 135
Conjugate variable, 23
Connected set, 105
  closed, 106
Connectivity, 187
Conservative force, 15
Constraint, 15
Contact transformations, 47–48
  defined, 47
Continuous function, 107
Continuous transformation group, 297, 407
Continuum, 104, 106
Contragradient, 30
Convergence in the mean, 116, 158
Convergent set, 105
Convolution function, 155
Covariance, 29–36
Covariants, 43
Critical term, 5, 535
Cross cut, 186

D'Alembert series, 535, 537
Definite form, 90
Degenerate form, 24
Degree of term, 534
Delaunay's elements, 526–531
Dense set, 105
Denumerable numbers, 104
Derivative, 111
Derived group, 431
Derived set, 104
Differentiable function, 111
Diophantine inequalities, 162
Discontinuous function, 108
Distance, 105
Distinguished functions, 322, 333–336
Distinguished infinitesimal transformation, 425
Disturbing function, 476, 539
  argument, 4, 535
  critical term, 5, 535
  direct and indirect, 476, 539
  secular term, 535

Earth potential, 196–199
Earth satellite, 196–221
Eccentric anomaly, 487
Eccentricity, 487
Eigenvalues, 84
Elimination of nodes, 468–469, 541
Elimination of time, 303–304
Elliptic functions for solution of relativistic one-body problem, 174–178
Elliptic motions, 487–526
Energy integrals, 24, 577–581
  integration by, 205–214
Equation of center, 495
Equilateral-triangular equilibrium points, 240
Equivalence of systems, 40–42
  definition, 40
  ordinary correspondence, 39
Equivalent sets, 104
Ergodicity, 386
Euclidean space, 105
Eulerian integrals, 333
Everywhere dense set, 104
Expansion in several variables, 503–510
Exterior derivative, 365
Exterior multiplication forms, 363

Factor group, 430
Field, 125
Field of multiple periods, 124
First integrals, 291–296, 342–343
First-order perturbation, 5, 530
Fluctuation of function, 107
Force function, 15
Fourier
  coefficients, 113, 158
  exponents, 157
  series, 113–117, 158
Fuchsian type of differential equation, 260
Function groups, 320–323
  canonical forms, 323–328
Fundamental system of solutions, 260

Gauss's differential equation, 499
General integral, 63, 291, 294
Generating solution, 353
Generators of infinitesimal transformations, 413
Genus of surface, 188
Geometry of motion, 36–39

Great inequality, 7
Group, 296, 407
Group germ, 437

Hamilton-Jacobi partial differential equation, 61–98
  defined, 62
Hamilton-Jacobi theorem, 61–65
Hamiltonian matrix, 82–85
  defined, 83
Hamiltonian equations, 23
Hamiltonian function, 15–16
Hamilton's characteristic function, 26
Hamilton's principle, 15–19
Hansen's coefficient, 501–503
Heine-Borel's covering theorem, 105
Hermitian matrix, 90
  definite, 90
Hölder-Engel theorem, 431
Holonomic constraint, 15
Homomorphic group, 423
Homographic motions in many-body problems, 228–232
  plane, 237–246
  space, 256–259
  straight-line, 232–237
Homothetic configuration, 229
Hyperelliptic functions, 671–679
Hypergeometric series, 499

Ideal, 440
Imprimitive group, 420
Inclination, 484
Indicial equation, 263
Inequality, 536
Infinitesimal ring, 408, 436
Infinitesimal transformations, 291, 296–301
Integrable function, 108
Integral
  curve, 24
  function, 110, 190
Integral invariants, 19, 343–350, 359–363
  along the trajectories, 350–352
  relations, 24
Integrals
  containing time, 581–583
  independent of time, 556–559
  of momentum and angular momentum, 568–576
  of restricted three-body problem, 604–613
  without parameters, 632–639
Intermediary orbit, 527

Invariant, 29
  in the mean, 385
Invariant subgroup, 422
Invariantive relations, 305–310
Inverse matrix, 82–83
Inversion
  of Abelian integrals, 193
  in problem-solving, 135
Involution, 304–305
Irreducible function, 125
Irregular singularity of solutions, 263
Irreversible dynamical system, 19
Isodynamical element system, 548
Isoenergetic element system, 548
Isolated point, 104
Isomorphic group, 423
Isoprotometric element system, 548
Isosceles triangular solutions for three-body problem, 268–284
  kinds, 268
  nature, 276–284

Jacobi
  elimination of nodes, 468–469, 541
  integral, 245, 605
  last multiplier, 376
  lemma, 304
  relative canonical coordinates, 474
  system, 318
  theorem, 63
    hydrodynamical proof of, 65–67

Keplerian elements, 2, 69, 537–540
Kepler's equation, 488
Kernel, 437
Killing's equation, 433
Kirkwood's gaps, 7
Kowalewski's top motion, 639–661
Kronecker's theorem, 137–142

Lagrange
  bracket, 48
  equations, 17
  function, 16
  method, 68
  theorem, 117–119
  transformation, 447–453
Lebesgue's integrals, 104–112
  defined, 108
Legendre's condition, 22
Levi-Civita's theorem, 310
  transformation, 453–465
Libration, 122
Lie groups, 406–441

Lie's theorems, 312–316
  first fundamental, 411
  second fundamental, 415
Lie-Jordan's theorem, 429
Limit-periodic function, 148–149, 163
Limit point, 104, 105
Linear operations, 437
Linear operator, 297
Linearly independent, 298
Liouville's dynamical system, 348, 388
Liouville's theorem, 70–73, 310–312
Lipschitz condition, 113
Longitude of ascending node, 487
Longitude of perihelion, 487
Longitude in the orbit, 488

Many-body problem, particular solutions, 228–284
Maurer's relation, 410
Mean anomaly, 488
Mean longitude in the orbit, 488
Mean motion, 4, 488
Mean value theorem, 111, 153–155
Measurable set, 106
Measurable subset, 108
Measure function, 106
Meromorphic function, 186
Metrical set, 105
Modulus, 431
Motion
  in conic sections, 482–486
  of molecules in a cube, 142–144
Multiply periodic function, 123–128
  defined, 124
Multiplier, 84

$n$-body problem
  integrals, 328–343
  reductions, 447–482
Nonholonomic constraint, 15
Normalizing matrix, 83
Nonvanishing Hessian, 613–619
Normal
  divisor, 430
  function, 165
  matrix, 92
  sequence of functions, 116
  series, 429
Nonsingular matrix, 83
Null matrix, 83

Oblate spheroidal coordinates, 199–205
One-body problem, relativistic, 166–185
Open set, 105
Operation, 319
Order, 530
Orthogonal sequence of functions, 116
Osculating elements, 540

Painlevé's extension, 584–604
  applications to $n$-body problem, 600–604
  preparatory lemmas, 584–588
Parametrization, 336–342
  separating, 338–341
Parseval equation, 158
Parseval theorem, 116
Perfect set, 104
Perihelion distance, 485
Period, 123
Periods of integral, 187, 189
Permutability, 422
Perturbations, 68
  computation, 3–8
  first-order, 5
  second-order, 5
Perturbative function, 476
Pfaffian covariant, 43
  of degree $p$, 344
  dynamical system, 94–98
  variational principle, 95
Pfaff's criterion, 88–89
Phase space of dynamical system, 24
Plane configurations, condition for, 237–241
Plane equilibrium points, 237–256
Poincaré's elements, 531–537
Poincaré's recurrence theorem, 606
Poincaré's theorem, 613–639
  application to $n$-body problem, 624–628
  discussion, 629–632
Poincaré's transformation, 466–476
Pointwise discontinuous function, 108, 126
Poisson bracket, 49
Poisson's identity, 302
  theorems, 301–303
Polar factorization, 90–94
  defined, 93
Polar group, 321
Potential, 15
Primitive group, 420
  periods, 124
Primitivity order, 387
Principal function, 16, 52
Principle of least action, 18
Principle of least curvature, 39
Product set, 105

Quasi-elliptic trajectory, 182
Quasi-ergodic theorem, 386, 606
  of trajectory, 144
Quasi-periodic functions, 6, 127
  and almost periodic functions, 162–166
  of Bohl, 144–149
Quasi-periodic motion, 104–221

Radical, 440
Radius of convergence, 510–526
Reduction from $6n-6$ to $6n-10$, 476–482
Regional recurrence, 144
Regular singularity of solutions, 263
Relative canonical coordinates of Jacobi, 474
Relative integral invariant, 58, 346
Relativistic field of Schwarzchild, 168–171
Relativistic one-body problem, 166–185
Remainder class modulus, 431
Representation, 333, 437
Restricted problem of three bodies, 604–613
Reversible dynamical system, 19
Revolution, 120
Rheonomic constraint, 15
Riemann $\vartheta$-functions, 661–671
Riemann integral, 112
Riemann-Roch theorem, 194
Riesz-Fisher theorem, 115
Rigid-body rotation, 639–642
Ring, 125, 370
Root space, 435

Schreier's theorem, 429
Schwarz inequality, 115
Schwarzchild's relativistic field, 168–171
Scleronomic constraint, 15
Second-order perturbation, 5–6
Secular term, 5, 535
Semi-group, 407
Semiconvergent series, 5
Semimajor axis, 487
Semisimple group, 434
Separable, 106
Separating parametrization, 338–341
Separation of variables, 72–82
Set, defined, 104
Siegel's theorem, proof of, 611–613
Simply connected surface, 187
Simple group, 428
Simple periodic motion, 104–123
Singular matrix, 83
Singular integral, 294
Solving of differential equations, 2–3
  for three-body problem, 3
Space configurations, conditions for, 256–259
Special perturbation, 540
Spectrum, 84
Stäckel's theorem, 73–77
  defined, 75
Straight-line configurations, 232–237
  equilibrium point, 233
Structure constants, 298, 415
Subset, 105
Sum set, 105
Summable function, 110
Symplectic matrix, 90

Three-body problem, 2
  autologous solution, 8
  form of solution, 3
  restricted, 604
  stability of periodic solution, 9
Topological study of trajectories, 10–11
Totally discontinuous function, 108, 126
Trace, 434
Trajectory in dynamical system, 24
Transitive group, 410, 419
Translation number, 149, 150
Transposed matrix, 82
Trigonometric series, 113
True anomaly, 488
Two fixed centers, reduction to, 215–221

Ultimate world picture, amendment of, 2
Uniform convergence, 113
Uniform integrals, 613–632
Uniformization, 186
Uniformly continuous function, 107
Uniqueness theorem, 158, 159
Unitary matrix, 91

Vanishing Hessian, 620–624
Variation of constants, 4, 67–70, 526–551
  defined, 68
Variation of elements, 4
Variational equations, 307, 352–358
Variational principles, 15–42
Vitali's covering theorem, 111

Wave surface, 27
Weight, 441
Weierstrass's approximation theorem, 114
Weierstrass's auxiliary theorem, 128–132